SUBSURFACE CONTAMINATION BY IMMISCIBLE FLUIDS

PROCEEDINGS OF THE INTERNATIONAL CONFERENCE ON SUBSURFACE CONTAMINATION BY IMMISCIBLE FLUIDS / CALGARY / CANADA / 18-20 APRIL 1990

Subsurface Contamination by Immiscible Fluids

Edited by

K.U. WEYER

WDA Consultants Inc., Calgary, Canada & WKC Weyer Consultants, Düsseldorf, Germany

A.A. BALKEMA / ROTTERDAM / BROOKFIELD / 1992

International Association of Hydrogeologists
Canadian Chapter

The texts of the various papers in this volume were set individually by typists under the supervision of each of the authors concerned.

Published by
A.A. Balkema, P.O. Box 1675, 3000 BR Rotterdam, Netherlands
A.A. Balkema Publishers, Old Post Road, Brookfield, VT 05036, USA

ISBN 90 6191 175 3

Printed in the Netherlands

Subsurface Contamination by Immiscible Fluids, Weyer (ed.) © 1992 Balkema, Rotterdam. ISBN 90 6191 175 3

Table of contents

3 *Physical models*

4 *Numerical and analytical simulation*

5 *Comparison of numerical simulation to field and laboratory experiments*

8.2 *Investigation and remediation*

9 *Legislative and regulatory issues*

Subsurface Contamination by Immiscible Fluids, Weyer (ed.) © 1992 Balkema, Rotterdam. ISBN 90 6191 175 3

Preface

The Proceedings of the International Conference on Subsurface Contamination by Immiscible Fluids deal mainly with subsurface pollution by hydrocarbons. They document a substantial part of the present day knowledge with respect to the mechanisms and procedures that create, transmit and remediate subsurface pollution by immiscible fluids of inorganic and organic nature. Much progress has been made; more needs to be achieved. In my view, considerably more attention should also be directed towards thorough applications of hydrodynamical and geological methods.

The conference in Calgary was one of those rare occasions for an open and fruitful exchange of unbiased opinions and experiences. It was successful in attracting, on the one hand, many of the leading scientists from North America and Europe, and on the other hand, experienced consultants and regulators. This set the stage for presentations and discussions of unusual candidness and caliber.

As is often the case, compilation of the proceedings was not met with the same enthusiasm by some authors and reviewers, thus making substantial delays unavoidable. The most humorous case was that of a colleague who, immediately after the conference, urged me in writing to proceed with the publication as expediently as possible while retaining his papers and reviews for much more than a year. In the end several key authors' papers did not make it to the editor. This serves to show how widespread the gap between good intentions and reality can be. It also helps me to put forward my regrets for delays in getting these proceedings to print. Last but not least, I am responsible.

Much of the hard work in organizing and retrieving the reviews was taken on by Mika Madunicky of the Alberta Research Council, Edmonton. We all, and in particular I, owe her much appreciation. Torsten Geldsetzer, WDA Consultants Inc., Calgary, capably and patiently pre-read papers and reviews and organized communications with the authors. Much of the day-to-day operation of the conference itself was proficiently carried out by Barbara Dabrowski and Waltraud Weyer. The cooperation of the publisher, A.T. Balkema, Rotterdam, was very pleasant. If I ever again fall into the trap of preparing reviewed proceedings, it would have to be with his support.

I would like to point out that much of the organizational work for the conference was done outside the academic environment, by Calgary consultants. We acknowledge the support of the Groundwater Research Centre, University of Waterloo, Ontario, and of the Rawson Academy of Aquatic Science, Ottawa, Ontario.

We were successful in obtaining funds and donations from a wide variety of organizations which are listed as supporters of the conference. I would like to express IAH's and the organizing committee's sincere gratitude. Without their support, attendance fees would have been much higher.

Looking back, I feel privileged to have been involved in the conference and the preparation of the proceedings. I believe this conference has made a difference. My thanks are extended to the authors for their efforts in responding to reviewer's and editor's comments.

K.Udo Weyer
Editor

Subsurface Contamination by Immiscible Fluids, Weyer (ed.) © 1992 Balkema, Rotterdam. ISBN 90 6191 175 3

Organization

Organizing Committee:

K.U. Weyer (co-chairman, editor), WDA Consultants Inc., Calgary, Alberta, Canada
W.J. Ceroici (co-chairman), Alberta Environment, Edmonton, Alberta, Canada
J.T. Dance (treasurer), Golder Associates Ltd, Calgary, Alberta, Canada
T.L. Dabrowski (program coordinator), Piteau Engineering Ltd, Calgary, Alberta, Canada

Sponsor:

International Association of Hydrogeologists – Canadian Chapter, Markham, Ontario, Canada

Cosponsors:

Waterloo Centre for Groundwater Research, Waterloo, Ontario, Canada
Rawson Academy of Aquatic Science, Ottawa, Ontario, Canada

Donations were received from:

Acres International Ltd, Calgary, Alberta
Alberta Environment, Edmonton, Alberta
Beak Associates Consulting Ltd, Edmonton, Alberta
Chemex Labs Alberta Inc., Calgary, Alberta
Allan Dakin, Vancouver, British Columbia
EBA Engineering Consultants Ltd, Calgary, Alberta
Energy Resources Conservation Board, Calgary, Alberta
Gartner Lee Ltd, Markham, Ontario
Golder Associates Ltd, Calgary, Alberta
International Association of Hydrogeologists – Canadian Chapter, Markham, Ontario
International Association of Hydrogeologists – Halifax Conference
Jagger Hims Ltd, Newmarket, Ontario
O'Connor Associates Environmental Inc., Calgary, Alberta
Monenco Consultants Ltd, Toronto, Ontario
Piteau Engineering Ltd, Calgary, Alberta
Stanley Associates Engineering Ltd, Edmonton, Alberta

Thurber Consultants Ltd, Victoria, British Columbia
Waterloo Centre for Groundwater Research, Waterloo, Ontario
WDA Consultants Inc., Calgary, Alberta
Weyer Corp Inc., Calgary, Alberta

List of Reviewers:

Abboud, S.
Abriola, L.M.

Bachu, S.
Baechler, F.E.
Baehr, A.L.
Bedient, P.B.
Benazon, N.
Borneuf, D.

Ceroici, W.
Cherry, J.A.
Christiansen, E.A.
Cleary, M.
Croisé, J.
Cyr, T.

Dakin, R.A.
Dance, T.
Dorgarten, H.W.
Dudas, M.J.
Durgin, P.B.
Dusseault, M.

Elton, J.J.

Farquhar, G.
Falta, R.
Faust, C.
Foght, J.
Frind, E.O.

Gillham, R.
Gilliland, J.
Guarnaccia, J.

Hardisty, P.
Hiller, D.H.
Hitchon, B.
Hunt, J.R.
Huyakorn, P.

Johnson, R.L.

Kaluarachchi, J.J.
Katyal, A.K.
Kinzelbach, W.
Kohut, A.P.
Kuhnel, V.

Lappala, E.G.
Lerche, I.
Lehr, J.H.
Leech, R.E.J.
Lerner, D.H.
Lyness, L.

Maathuis, H.
MacFarlane, D.S.
Mackay, D.M.
Mendoza, C.A.
McBride, J.F.
McClymont, G.L.
McWhorter, D.B.

Nikols, D.
Novakowski, K.S.

Patrick, G.C.
Perkins, E.
Pfannkuch, H.O.
Pinder, G.
Pitts, L.V.
Pupp, C.

Raven, K.G.
Raymond, R.L.
Roeper, U.V.R.

Schwartz, F.
Setari, A.
Simmons, C.S.
Sitar, N.
Smith, B.E.
Smith, D.P.
Smith, L.
Skinner, A.C.
Staps, J.J.M.
Stein, R.
Storey, J.M.E.
Sykes, J.F.

Toussaint, B.
Trudell, M.

Udell, K.S.

van der Kamp, G.
van Everdingen, R.
van Genuchten, R.
Vonhof, A.
Vorreyer, C.

Wallick, E.
Ward, C.H.
Wehrle, K.
Werner, P.
Westerby, G.
Weth, D.
Wilson, J.,
Woeller, R.H.,

Zilliox, L.

1 Overview papers

Subsurface Contamination by Immiscible Fluids, Weyer (ed.) © 1992 Balkema, Rotterdam. ISBN 90 6191 175 3

Numerical modeling of multiphase/multicomponent flow and transport in porous media: An overview

B.H. Kueper
Waterloo Centre for Groundwater Research, University of Waterloo, Ont., Canada (Presently: Department of Civil Engineering, Queen's University, Kingston, Ont., Canada)

E.O. Frind
Waterloo Centre for Groundwater Research, University of Waterloo, Ont., Canada

ABSTRACT: This paper provides an overview of multiphase/multicomponent flow and transport modelling from a contaminant hydrogeology perspective. The local-scale balance equations governing such systems can be derived by considering a mass balance of individual components within phases. The equations are supplemented by constitutive relationships and mass transfer coefficients representing the partitioning of components between phases. The equations are typically highly non-linear, requiring in general iterative, fully-implicit numerical solution schemes that are computer intensive. Efficiency and simplification can be attained by recognizing that in certain cases the movement of non-aqueous phase liquids occurs at time scales much shorter than those associated with the evolution of contaminant plumes resulting from the partitioning of components. Experimental data for model input and validation is lacking at present, but the recent increased awareness of subsurface contamination by non-aqueous phase liquids has resulted in much activity in this area. Because large-scale effective parameters have not yet been rigorously defined for a wide variety of conditions, the application of models to simulate the field-scale behaviour of multiphase systems is at present restricted to the discretization of solution domains at the local scale, which is in general the scale at which laboratory determination of input parameters is carried out.

1 INTRODUCTION

Numerical modelling of multiphase flow originated in the oil industry during the 1950's in conjunction with the development of digital computers. Recent awareness of subsurface contamination by non-aqueous phase liquids (NAPLs) has resulted in a rapidly growing interest in multiphase/multicomponent flow and transport in the field of environmental science and engineering. This paper provides an overview of multiphase/multicomponent modelling in porous media from a contaminant hydrogeology perspective, and outlines some of the topic's more salient features.

All studies dealing with multiphase/multicomponent modelling cannot possibly be mentioned here. The exclusion of particular studies does not imply that they are not relevant to the topic, but rather that we have chosen not to address certain issues given the brevity of this paper. Treatments of the subject in related fields such as oil reservoir engineering, chemical engineering, and geothermal systems will not be addressed.

Non-aqueous phase liquids such as PCB oils, gasoline, jet fuel, creosotes, and chlorinated solvents can be classified according to whether they are more or less dense than water. This determines if they have the potential to migrate below the watertable, or if they are restricted to the vadose zone alone. Following a finite release into the subsurface, residual product in the form of disconnected ganglia is formed at the trailing edge of a migrating NAPL body. In addition, pools of NAPL may form in media of differing permeability or on the watertable. Although considered immiscible in air and water from a fluid mechanical point of view, these liquids generally have solubilities capable of bringing about significant levels of dissolved contamination. The residual ganglia and pools act as long-term sources, resulting in the evolution of vapour plumes in the vadose zone and dissolved groundwater plumes in the saturated zone. Vapours migrating in the vadose zone partition to the soil and soil moisture, and have the potential to contaminate infiltrating water. Figure 1 illustrates schematically the general contamination scenario associated with NAPLs.

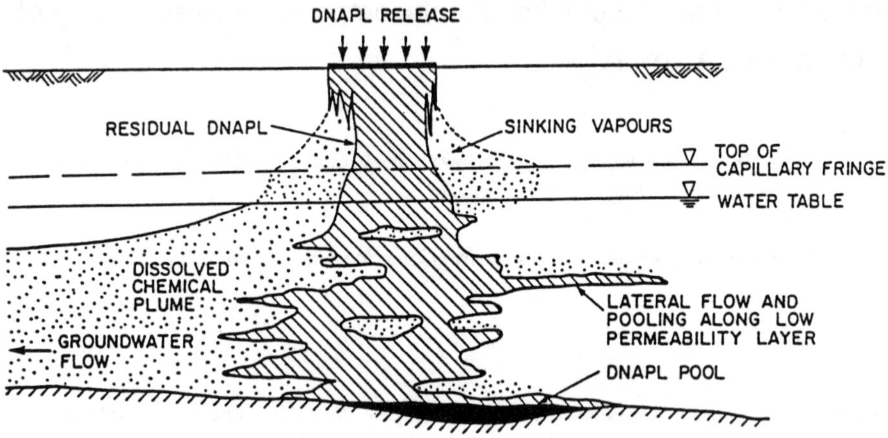

Figure 1 Subsurface NAPL Contamination Scenario. Shown is the case of a denser-than-water NAPL.

NAPL behaviour can be characterized at a variety of scales. The pore scale is usually the smallest scale at which direct observation can be made. The distribution of fluids at this scale is determined by the capillary pressures in the system and the geometry of interconnected pores and pore throats. The next scale of observation above the pore scale will be referred to as the local scale. This scale represents an averaging over 'many' pores such that some sort of average representative behaviour is achieved. The local scale is typically what is dealt with in the laboratory, and it is also a convenient scale at which to derive the differential equations governing multiphase/multicomponent flow and transport. If we average over many local scale volumes, we arrive at a representation at what we will refer to as the field scale. In practical terms, this refers to an averaging over 'many' individual lenses of porous media, each characterized by local scale properties.

In this study, attention is directed primarily to the local scale. It should be pointed out, however, that studies at the pore scale can be very fruitful in furthering our understanding of behaviour at other scales. A good example of this is the work being carried out using percolation models, which can be used to construct local-scale constitutive relationships. Larson et al. (1981) and Ferrand et al. (1990) give overviews of percolation theory including how it is used to construct capillary pressure curves, relative permeability curves, and the distribution of residual saturations. Regarding the field scale, the foremost current challenge is to rigorously define NAPL behaviour in terms of large-scale effective parameters. This is an area of intense research which should show returns in the near future.

The process of numerical modelling consists of several stages. Initially, the physical system of interest must be defined and the relevant physical characteristics and processes identified. The actual system is then replaced by a simpler conceptual model that incorporates these processes. The formation of a conceptual model usually involves the introduction of various assumptions, and it is the stage at which either a deterministic or a statistical framework is chosen. Next, a mathematical model must be formulated which expresses the conceptual model by means of partial differential equations, boundary conditions, and initial conditions. The mathematical equations are then solved in an approximate manner. This stage usually involves the development of a computer code to carry out lengthy calculations. Once the computer code is operational, it is validated to ensure that the conceptual and mathematical models correctly represent the desired physical processes. To apply the computer model to a field or laboratory setting, data must be collected and the model must often be calibrated to the specific application. The model may then be used to carry out uncertainty and sensitivity analyses. In some cases, where the system parameters and configuration are well known, the model can be used to provide predictions of anticipated future behaviour.

2 GENERAL EQUATIONS

The local-scale balance equations governing multiphase/multicomponent flow and transport in porous media can be derived by performing an appropriate averaging over the pore scale. This procedure is discussed by various authors including Marle (1982) and Hassanizadeh and Gray (1979a,b). The equations presented here are similar to those of Abriola and Pinder (1985a) and Huyakorn and Pinder (1983). We identify the water, air, NAPL, and solid phases by the subscripts W, A, N, and D, respectively. These phases are distinguished according to their fluid-mechanical behaviour, and will in general be denoted by the subscript α. The individual components which comprise these phases will be denoted by the subscript k. A typical multiphase/multicomponent system of interest to contaminant hydrogeologists may involve, for example, air, water, and gasoline phases in unsaturated porous media. The gasoline may be comprised of various chemical components such as benzene and toluene, each of which may partition to the air phase giving rise to migrating vapours, and each of which may partition to the water and soil phases, subject to biotic and abiotic transformations.

A mass balance equation can be written for component k in phase α:

$$\frac{\partial}{\partial t}(\epsilon_\alpha \rho_\alpha \omega_k^\alpha) + \frac{\partial}{\partial x_i}(\epsilon_\alpha \rho_\alpha \omega_k^\alpha \upsilon_{\alpha,i}) - \frac{\partial}{\partial x_i} J_{k,i}^\alpha = I_k^\alpha + E_k^\alpha \tag{1}$$

where ϵ_α is the fraction of bulk volume occupied by the phase, ρ_α is the average mass density of the α phase, $\omega_k{}^\alpha$ is the mass fraction of component k in the α phase, $\upsilon_{\alpha,i}$ is the velocity of the α phase in the ith direction, $J_{k,i}{}^\alpha$ is the non-advective flux of k in the α phase in the ith direction, $I_k{}^\alpha$ represents the transfer of k due to phase change and diffusion across the phase boundaries, and $E_k{}^\alpha$ the external supply of k to the α phase through biotic and abiotic transformations. It follows from (1) that the mass fractions within a given phase must sum to unity:

$$\sum_{k=1}^{n} \omega_k^\alpha = 1 \tag{2}$$

where n is the total number of components comprising phase α. It also follows that the fractional phase volumes sum to unity:

$$\sum_{\alpha=1}^{N} \epsilon_\alpha = 1 \tag{3}$$

where N is the total number of phases, in our case four. The following expresses the conservation of mass of a given component distributed amongst the phases:

$$\sum_{\alpha=1}^{N} I_k^\alpha = 0 \tag{4}$$

Equation (1) requires that the phase velocities $\upsilon_{\alpha,i}$ be defined. For single phase flow, these velocities are usually obtained from Darcy's law, which is an empirical equation relating the flux of a single fluid through a porous medium to the driving forces imposed on that fluid and a characteristic property of the medium termed its permeability. Based on laboratory experiments, workers in the oil industry (Wyckoff and Botset 1936, Leverett 1938) have suggested that Darcy's law can be extended to multiple fluid systems. Others (e.g., Allen, 1985) have shown that Darcy's law can be extended to multiphase systems by considering a momentum balance for each phase of interest. Such studies show Darcy's law to be applicable if inertial effects are negligible and there is no transfer of momentum between phases. Whitaker (1986) has examined two-phase flow through porous media and suggests that the influence of viscous drag of one phase against the other may be of relevance, particularly in cases where the two fluid viscosities are of similar magnitude and low saturations of one phase exist. It would appear that under many conditions, an extended form of Darcy's law is acceptable for use in equation (1) to define phase velocities. Darcy's law extended to multiphase flow in terms of phase velocities may be expressed as:

$$\upsilon_{\alpha,i} = -\frac{K_{ij}^\alpha}{\epsilon_\alpha}\left(\frac{\partial P_\alpha}{\partial x_j} + \rho_\alpha g \frac{\partial z}{\partial x_j}\right) \tag{5}$$

where P_α is the pressure of the α phase, g is the gravitational constant, z the vertical direction, and $K_{ij}{}^\alpha$ a second rank mobility tensor given by:

$$K_{ij}^\alpha = \frac{k_{ij} k_{r,\alpha}}{\mu_\alpha} \tag{6}$$

where μ_α is the dynamic viscosity of the α phase, k_{ij} is a second rank intrinsic permeability tensor, and $k_{r,\alpha}$ the relative permeability to the α phase accounting for the reduction in permeability due to the presence of other phases.

Equation (1) also requires that the non-advective flux vector $J_{k,i}^\alpha$ be defined. This term accounts for the molecular diffusion and mechanical dispersion of component k within phase α. It is often assumed that both of these are Fickian in nature over the scale at which averaging has been carried out to arrive at equation (1). Under this assumption the processes are additive, leading to the following representation of $J_{k,i}^\alpha$:

$$J_{k,i}^{\alpha} = -\epsilon_\alpha \tau_{\alpha,ij} D_o^{k,\alpha} \frac{\partial(\rho_\alpha \omega_k^\alpha)}{\partial x_j} - \epsilon_\alpha D_{m,ij}^{k,\alpha} \frac{\partial(\rho_\alpha \omega_k^\alpha)}{\partial x_j} \tag{7}$$

where $\tau_{\alpha,ij}$ is a tensor of phase tortuosity coefficients, $D_o^{k,\alpha}$ is the free solution molecular diffusion coefficient of k in α, and $D_{m,ij}^{k,\alpha}$ a mechanical dispersion tensor of k in α. The proper form of $D_{m,ij}^{k,\alpha}$ that should be used for a given scale of averaging and under a certain set of flow and saturation conditions is not always obvious. Various forms of the dispersion tensor for use in single phase flow are discussed by numerous authors including Bear (1972) and Gelhar (1986).

3 MULTIPHASE FLOW

In many instances, it is strictly the flow of air, water, and NAPL that is of interest. Given the low solubilities of many NAPLs and the short time scales over which NAPL may migrate in relation to the time taken for dissolved vapour and groundwater plumes to fully evolve, it is often assumed that phase flow is not affected significantly by the partitioning and transformation of components. This decoupling of phase behaviour from component behaviour greatly simplifies the required numerical formulation. In such cases, equation (1) can be summed over all components comprising a given phase to yield a mass balance equation for that particular phase:

$$\frac{\partial}{\partial t}(\epsilon_\alpha \rho_\alpha) + \frac{\partial}{\partial x_i}(\epsilon_\alpha \rho_\alpha \upsilon_{\alpha,i}) = 0 \tag{8}$$

One such equation can be written for each phase of interest. If we assume that there are no source/sink terms and that the fluid and medium are incompressible, we arrive at the following basic set of equations for the water, air, and NAPL phases:

$$\frac{\partial}{\partial x_i}\left[\frac{k_{ij}k_{r,W}}{\mu_W}\left(\frac{\partial P_W}{\partial x_j} + \rho_W g \frac{\partial z}{\partial x_j}\right)\right] = \phi \frac{\partial S_W}{\partial t} \tag{9}$$

$$\frac{\partial}{\partial x_i}\left[\frac{k_{ij}k_{r,A}}{\mu_A}\left(\frac{\partial P_A}{\partial x_j} + \rho_A g \frac{\partial z}{\partial x_j}\right)\right] = \phi \frac{\partial S_A}{\partial t} \tag{10}$$

$$\frac{\partial}{\partial x_i}\left[\frac{k_{ij}k_{r,N}}{\mu_N}\left(\frac{\partial P_N}{\partial x_j} + \rho_N g \frac{\partial z}{\partial x_j}\right)\right] = \phi \frac{\partial S_N}{\partial t} \tag{11}$$

where ϕ is the medium porosity, and S_α the phase saturation given by ϵ_α/ϕ. Note that the soil phase equation is not required in this formulation. In the general case, however, the soil phase equation serves to introduce the medium compressibility into the system. Equations (9)-(11) are subject to the constraint that phase saturations sum to unity:

$$S_W + S_A + S_N = 1 \tag{12}$$

Equations (9)-(11) are coupled through the phase pressure terms. In general, interfacial tension forces give rise to a pressure discontinuity across a curved interface separating two fluid phases. This pressure discontinuity is referred to as the capillary pressure, and is defined as the difference in pressure between the fluid on the concave side of the interface and that on the convex side of the interface. As such, capillary pressures are always positive. If averaging is carried out over the pore scale, it is found that capillary pressure is a function of fluid saturation. For an air-water-NAPL system we can define the following capillary pressures:

$$P_C^{A,W}(S_A,S_W) = P_A - P_W \tag{13}$$

$$P_C^{A,N}(S_A,S_N) = P_A - P_N \tag{14}$$

$$P_C^{N,W}(S_N,S_W) = P_N - P_W \quad (15)$$

Ideally, the relationships in (13)-(15) are determined experimentally for the media and fluids of interest. These relationships are hysteretic, showing a strong dependence on saturation path.

For the case of two-phase flow, various workers have proposed empirical relationships allowing a mathematical function to be fitted to experimental data (e.g., Brooks and Corey 1966, van Genuchten 1980, Luckner et al. 1989). Leverett (1941) presents a relationship allowing capillary pressure curves to be scaled as functions of permeability and interfacial tension. The use of interfacial tensions to scale capillary pressure curves for arbitrary fluid pairs has also been demonstrated by Schiegg (1984) and Parker et al. (1987).

Where three fluid phases are present, various authors (e.g. Leverett, 1941) have suggested using two-phase capillary pressure data to approximate three-phase capillary pressures. These studies assume that the total liquid saturation given by $S_T = S_N + S_W$ is a function of the air-NAPL capillary pressure only, and that water saturation is a function of NAPL-water capillary pressure only. These approximations assume that water is wetting with respect to NAPL, and that NAPL is wetting with respect to air on water-wet grains. Lenhard et al. (1989) use the two-phase approximation to develop hysteretic relations for three-phase systems. The use of two-phase capillary pressures to describe behaviour in three-phase systems has not yet been experimentally validated for all possible saturation paths that may occur in practice. Limited validation under specific conditions has been carried out by Lenhard and Parker (1988). The experimental work of Eckberg and Sunada (1984), on the other hand, found that air-NAPL capillary pressure did not adequately represent total liquid saturation.

As with capillary pressures, the phase relative permeabilities in equations (9)-(11) are path-dependent functions of fluid saturation. Relative permeabilities are extremely difficult to measure in practice, and as a result they are often predicted from measured capillary pressure data (e.g., Brooks and Corey, 1966). This is justified by recognizing that the distribution of pore sizes which determines largely the shape of a capillary pressure curve also determines largely the nature of the relative permeability curve. Experimental studies in three-phase systems (Leverett and Lewis 1941, Corey et al. 1956) have found that water relative permeability and air relative permeability are dependent mostly on water and air saturation, respectively. As a result, various authors (Stone 1970, Lenhard and Parker 1987) have proposed methods of predicting three-phase relative permeabilities from two-phase data. An excellent review of both two-phase and three-phase relative permeability measurements is provided by Honarpour et al. (1986).

In general, relative permeability is taken to be a scalar regardless of whether or not an isotropic intrinsic permeability is assumed. This assumption has been adopted mainly because of its inherent simplicity. Recent theoretical studies, however, show evidence that relative permeability itself has a directional dependence in anisotropic porous media, and that the degree of anisotropy is a strong function of fluid saturation. These include studies carried out at the local scale (Bear et al. 1987), as well as at the field scale (Yeh et al. 1985, Mantoglou and Gelhar 1987).

Equations (9)-(11) can be posed in terms of an alternative set of unknowns by substituting the relations given in (12) and (13)-(15). Kueper and Frind (1991a,b), for example, solve (9) and (11) for wetting-phase pressure and wetting-phase saturation to simulate the behaviour of a denser-than-water NAPL below the watertable. These variables are continuous throughout the solution domain, regardless of whether NAPL is present or not. On the other hand, if (9) and (11) are solved in their original form, it is necessary to introduce a fictitious degree of NAPL saturation to ensure an operational numerical scheme. Aziz and Settari (1979) discuss the numerical advantages of posing (9)-(11) in terms of various sets of unknowns. Letniowski (1989a) points out the advantages of keeping saturation on the time derivative in a three-phase formulation as opposed to expanding the derivative in terms of phase pressures.

In unconfined aquifers, it is often assumed that the air phase is at constant atmospheric pressure, implying that it is infinitely mobile. In such cases, equations (9) and (11) alone can be used as the basis for three-phase behaviour. This assumption is adopted in the three-phase flow models of Faust (1985), Faust et al. (1989), Forsyth (1988), Letniowski (1989a), Kuppusamy et al. (1987), and Kaluarachchi and Parker (1989).

It should be noted that the phase equations given by (9)-(11) can be formulated to account for phase transfer. In such cases, the mass transfer term represented by I_k^α in equation

(1) would be retained in performing the summation of components comprising a phase leading to equations (9)-(11).

4 MULTICOMPONENT TRANSPORT

Often, it is the fate of individual components and mass transfer between phases that is of interest. For example, one may be interested in the dissolution of a multicomponent NAPL below the watertable, or the partitioning of NAPL vapours to infiltrating water in the unsaturated zone. Models of this type are often referred to as 'compositional' models. They are based on balance equations governing the behaviour of individual components throughout the entire system. To arrive at the equation for component k, (1) is summed over all phases:

$$\sum_{\alpha=1}^{N}\left\{\frac{\partial}{\partial t}(\epsilon_\alpha \rho_\alpha \omega_k^\alpha) + \frac{\partial}{\partial x_i}(\epsilon_\alpha \rho_\alpha \omega_k^\alpha \upsilon_{\alpha,i}) - \frac{\partial}{\partial x_i} J_{k,i}^\alpha\right\} = \sum_{\alpha=1}^{N} E_k^\alpha \qquad (16)$$

One such equation can be derived for each component of interest. In addition to the fate of the individual component, (16) describes the behaviour of each phase harbouring that component.

In order to close the set of equations arising from (16), additional relationships must be specified. The capillary pressure relations in (13)-(15) are required to couple phase pressures, and relative permeabilities are required to evaluate phase velocities. In general, both capillary pressure and relative permeability relations may be functions of phase composition, as for example in the case of phase exposure to surfactants. Because of inherent experimental difficulties, however, such dependence on composition has not yet been investigated to any great degree. In addition, phase viscosities and phase densities may be functions of phase pressure and composition. Whether or not such functional dependencies need to be incorporated into the model will depend upon the specific application.

Equation (16) also requires that relationships be specified to prescribe the relative mass fractions of components among phases. Most models to date have adopted the local equilibrium assumption. This assumes that the transfer of components between phases occurs 'instantaneously' relative to the bulk phase flow rates such that equilibrium mass fractions of components are uniformly distributed within all phases at all times. The local equilibrium assumption allows partitioning coefficients of the following form to be adopted in the numerical formulation:

$$\omega_k^{\alpha_1} = Q_k^{\alpha_1,\alpha_2}\omega_k^{\alpha_2} \qquad (17)$$

where $Q_k^{\alpha 1,\alpha 2}$ is the partitioning coefficient representing the relative mass fraction of component k in the α_1 and α_2 phases. Commonly adopted equilibrium relations include Henry's law for the partitioning of NAPL components between the water and air phases, Raoult's law for partitioning between the NAPL and air phases, and linear sorption isotherms for partitioning between the water and soil phases. Raoult's law has also been found to adequately describe the equilibrium partitioning of a multicomponent NAPL to the water phase for structurally similar components (Banerjee, 1984). The use of equilibrium partitioning in a multiphase/multicomponent context is disussed further by Corapcioglu and Baehr (1987), and is incorporated into the multicomponent models of Abriola and Pinder (1985), Baehr and Corapcioglu (1987), and Baehr (1987).

Although the use of equilibrium partitioning coefficients is computationally convenient, their applicability under a wide variety of conditions has not yet been experimentally validated. Particularly, if phases are heterogeneously distributed within the volume of averaging represented by equation (1), and phase flow rates are fast relative to diffusion, then the local equilibrium assumption is likely to break down. In the case of NAPL dissolution below the watertable, for example, some studies have found that the mass fraction of NAPL components in the flowing water phase is a function of phase velocity, the geometry of NAPL residuals, and the degree of NAPL-water contact area. In such cases, a mass transfer term representing these effects must be incorporated into (16). The dissolution of NAPL to groundwater is discussed further by Feenstra (1990) and Hunt et al. (1988).

The right-hand side of (16) includes the reaction terms for processes such as biodegradation and chemical transformations. Simple first order reaction terms can be substituted, or more complicated expressions involving the solution of additional equations may be adopted. Corapcioglu and Baehr (1987) discuss the incorporation of biodegradation terms into (16).

In certain cases, the set of equations arising from (16) can be simplified by assuming that the behaviour of certain phases occurs at time scales much shorter than that of component evolution. An example of this could be the dissolution of NAPL below the watertable. Since this process can persist for decades or more, it can be represented by a numerical formulation which uses as an initial condition the static distribution of NAPL residual. The migration of the NAPL and the formation of this residual could be represented in a separate numerical formulation, completely decoupled from the dissolution process. The vapourization of NAPL residual in unsaturated porous media is a second example of where simplification can be achieved if it is assumed that NAPL migration takes place over time scales shorter than that associated with the evolution of vapour plumes. Examples of models incorporating this assumption include Falta et al. (1989) and Mendoza and Frind (1990a,b).

5 NUMERICAL SOLUTION TECHNIQUES

The differential equations governing multiphase/multicomponent flow and transport lend themselves to analytical solution in only extremely simplified cases. As a result, numerical methods such as the finite difference, finite volume, and finite element techniques are usually employed to obtain an approximate solution. The application of such methods involves discretizing the governing equations of interest, and representing the unknown variables at a discrete number of points within a prescribed solution domain. Upon specification of appropriate boundary conditions, a set of algebraic equations suitable for solution can be obtained.

The differential equations governing multiphase/multicomponent systems are generally non-linear, often requiring an implicit, iterative solution scheme. The set of discretized non-linear equations may be represented as:

$$F(X) = 0 \tag{18}$$

where **F** is a vector of non-linear algebraic equations and **X** a vector of discrete unknowns. In some systems, the non-linearity of the governing equations may be weak, allowing a simple iteration scheme or time-lagging scheme to be adopted. In cases involving the flow of fluid phases, however, the governing equations are generally highly non-linear due to saturation-dependent terms. Of the various schemes for handling sets of non-linear equations, Newton Raphson iteration is generally regarded as one of the most robust (Huyakorn and Pinder, 1983). Applied to (18), the scheme leads to the following linearized matrix equation:

$$\frac{\partial F(X)^n}{\partial X}\,\delta X = -F(X)^n \tag{19}$$

where $\frac{\partial F(X)^n}{\partial X}$ is a square matrix of Jacobian derivative terms, $F(X)^n$ a vector of residual values, n the iteration level, and δX a vector of incremental unknowns given by:

$$\delta X = X^{n+1} - X^n \tag{20}$$

The algorithm given by (19) is repeated until the residual vector has converged to within a prescribed tolerance.

In regions of sharp saturation fronts, numerical models can exhibit oscillations and possible convergence to an incorrect solution. The problem arises because numerical schemes do not represent the spatial variation of the dependent variables in a continuous manner. Rather, the variables are represented at discrete points, and it is difficult to propagate sharp fronts accurately. Upstream weighting, which refers to the weighting of phase mobilities based on the physically upstream direction, has been found to eliminate this problem. Upstream weighting is presently utilized in virtually all multiphase flow models and is discussed further by Aziz and Settari (1979).

The finite difference technique is likely the most straight-forward of available numerical methods to implement. Differential operators are approximated by a truncated Taylor series, leading to a set of approximate discretized equations. The technique is best suited to rectangular grids and was first applied to multiphase problems by workers in the oil industry (e.g., Peaceman, 1977).

Faust (1985) presents a two-dimensional finite difference model which solves for NAPL pressure and water saturation both above and below the watertable. The model has since been extended to three dimensions (Faust et al. 1989). Abriola and Pinder (1985a,b) have developed a one-dimensional finite difference model which solves for the behaviour of two organic species in a three-phase system. Forsyth (1988) presents a two-dimensional, three-phase flow model which incorporates the behaviour of one NAPL component. The model achieves cost savings by adopting an adaptive implicit method whereby the fully implicit form of the discretized equations are solved only in regions of the solution domain undergoing significant changes. Each of the above three-phase models assume an infinitely mobile gas phase, and assume that two-phase capillary pressure and relative permeability data can be used to describe three-phase behaviour. Kueper and Frind (1991a,b) present a two-dimensional, two-phase finite difference model to study the influence of fluid properties and porous media heterogeneity on NAPL migration in a spatially correlated random field.

Finite volume techniques are a variant of finite difference techniques whereby the solution domain is subdivided into a finite number of control volumes, each surrounding one grid point. The differential equation is integrated over each control volume, leading to a set of algebraic equations. The method has a distinct advantage over the traditional finite difference method in that irregular grid geometry is accomodated. Letniowski (1989a) utilizes a finite volume formulation to solve for three-phase flow with an infinitely mobile gas phase, and provides a comparison to a Galerkin finite element formulation of the same problem. Finite volume techniques are discussed further by Patankar (1980).

The finite element method involves subdividing the solution domain into individual subdomains called elements. Of the various finite element techniques available, the method of weighted residuals is commonly used in subsurface applications. Interpolation functions are chosen to represent the behaviour of the primary unknowns over each element. An approximate solution is assumed and the associated residual minimized by choosing a weighting function and integrating over the domain of interest. Depending upon the choice of weighting function, various schemes such as the Galerkin method or the Collocation method can arise. In general, the finite element method offers the attraction of irregular mesh geometry and the possibility of higher order approximations. Huyakorn and Pinder (1983) discuss finite element techniques applied to subsurface problems.

Dorgarten (1989) has developed a two-dimensional Galerkin finite element model which solves for NAPL and water movement with one chemical species subject to dissolution, biodegradation, and adsorption. Linear triangular elements are utilized, with the effects of species transport on phase flow lagged in time. Kuppusamy et al. (1987) present a two-dimensional variational finite element model to solve for three-phase flow assuming an infinitely mobile gas phase. Kaluarachchi and Parker (1989) present a two-dimensional Galerkin finite element model to simulate three-phase flow and compare the relative merits of Newton Raphson and Picard iteration. Sleep and Sykes (1989) use the Galerkin finite element method to solve for variably saturated water flow with density dependent gas flow along with water and gas phase transport by assuming negligible capillary effects. The model incorporates first-order rate constants for the partitioning of a single NAPL component between phases. Osborne and Sykes (1986) present a two-dimensional finite element model to solve for NAPL-water behaviour which incorporates linear quadrilateral isoparametric elements. Where applicable, each of the above models assume that two-phase capillary pressure and relative permeability data can be used to represent three-phase behaviour.

6 MATRIX SOLVERS

The set of algebraic equations which arises in a particular numerical scheme will in general have the following form:

$$[A]\{X\} = \{B\} \tag{21}$$

where [A] is a square, sparsely populated, non-symmetric coefficient matrix, {X} is a column vector of unknowns, and {B} a column vector of known values. Equation (21) can be solved either directly with a Gaussian elimination type of scheme, or in an iterative manner. For large sets of equations, direct solvers will be prohibitively slow and prone to machine round-off errors. Storage may also be a problem. Iterative solvers are typically more efficient at solving large sets of equations, and offer an additional advantage in that only the bands of non-zero elements of the coefficient matrix need be stored in memory.

Iterative solution methods generally begin with

an approximate guess to the solution of (21), such as an initial condition or the solution at the previous time step for time marching schemes. To arrive at a better approximation of {X}, an approximation [P] to [A] is obtained, and the following matrix equation solved:

$$[P]\{V\}^{k+1} = \{R\}^k \tag{22}$$

where $\{V\}^{k+1}$ is the incremental change in the solution vector given by:

$$\{V\}^{k+1} = \{X\}^{k+1} - \{X\}^k \tag{23}$$

and k is the iteration level. The term $\{R\}^k$ is a residual vector given by:

$$\{R\}^k = \{B\} - [A]\{X\}^k \tag{24}$$

The iterative scheme given by (22)-(24) will have converged when $\{R\}^k$ is within a prescribed tolerance. The scheme is usually accelerated by a conjugate gradient type of method and its effectiveness improved by some type of pre-conditioning. For unsymmetric matrices, Orthomin acceleration (Behie and Vinsome, 1982) has been proven effective. Letniowski (1989b) gives an excellent review of preconditioned iterative methods for solving sparse matrix equations.

7 ANALYTICAL SOLUTIONS

Analytical solutions to the equations governing multiphase/multicomponent flow and transport are usually not tractable except in extremely simplified cases. This is due to the non-linear nature of the governing equations and because it is difficult to obtain solutions that incorporate the effects of media heterogeneity. Analytical solutions are important in a numerical modelling context in that they can be used to assess the numerical error incurred by an approximate solution, and to verify whether such a solution converges to the correct result. To date, authors have chosen to either partially verify their models against the simplified analytical solutions that are available, compare their numerical results to those of others, or to simply not attempt verification.

For the particular case of incompressible two-phase flow through homogeneous porous media in the absence of gravity and mass transfer, many authors have chosen to verify their models against the well-known Buckley-Leverett solution (Buckley and Leverett, 1942). This solution assumes, however, that the total flux of the two fluids is constant and that the effects of capillarity are negligible. Recently, McWhorter and Sunada (1990) have developed a set of analytical solutions for two-phase flow in a homogeneous porous medium in the absence of gravity that account fully for the effects of both capillarity and relative permeability. Their solutions represent the cases of one-dimensional and radial unidirectional displacement of a resident wetting or non-wetting fluid, as well as counter-current displacements. The solutions allow incorporation of either the Brooks-Corey (Brooks and Corey, 1966) or van Genuchten (van Genuchten, 1980) capillary pressure - saturation functions, and therefore offer a level of sophistication beyond that of the Buckley-Leverett solution.

In the case of component transport, a numerical model can be partially verified against the variety of single-phase analytical solutions that exist to the well-known advection-dispersion equation. There exist also a variety of analytical solutions that solve for single-component diffusion through porous media. Baehr (1987), for example, presents a one-dimensional analytical solution for the diffusion of organic vapours through variably saturated porous media.

8 MODEL VALIDATION

Model validation is the process whereby it is determined whether the governing equations upon which the numerical model is based correctly represent the real-life processes that they are intended to. Despite the fact that a numerical model may correctly solve the governing equations it represents, this does not ensure that the proper choice of equations and constitutive relationships was made. Model validation should be carried out against a controlled laboratory or field experiment which was performed under conditions similar to those under which the model will be subsequently applied. Ideally, all parameters and variables required as model input should be measured independently in the validation procedure such that a unique numerical solution is obtained. Successful validation under a restricted set of conditions does not justify use of the numerical model under a different set of conditions.

At present, there is a lack of controlled laboratory and field experiments with which to fully validate multiphase/multicomponent models. Part of the reason for this is the hazardous nature of the substances involved, and the complexities of all the physical and chemical processes involved. Most multiphase models are currently being used to simulate conditions against which they have not been validated. The recent attention to subsurface contamination by NAPLs, however, has brought about an increased activity in the area of multiphase/multicomponent laboratory experimentation. It can be anticipated that more attention will be given to model validation in the near future as experimental results become available. Recent laboratory experiments involving multiple fluids are presented by Schwille (1988), Kell (1988), Kueper et al. (1989), Weth and Pielke (1989), Host-Madsen (1988), Lenhard et al. (1988), Johnson et al. (1989), and Cary et al. (1989). The literature at present does not provide the results of a well-documented field experiment involving multiple fluids.

9 FINGERING

Under certain conditions the macroscopic interface between two immiscible liquids can become unstable, leading to a phenomenon known as fingering. This will occur if there is an unstable density or viscosity ratio between the fluids of concern. The fingering process is thought to be triggered by pore-scale heterogeneity. Interfacial tension forces result in an optimum wavelength of fingering emerging from an initially planar interface. Theoretical studies of fingering (Hill 1952, Chuoke et al. 1959) allow prediction of finger spacings given knowledge of fluid and porous medium properties.

The fingering process, as defined by early treatments of the subject, will only occur in homogeneous media where the optimum wavelength of fingering can prevail. In heterogeneous porous media, where the length scale of heterogeneity is on the order of the critical wavelengths of fingering, no fingering by definition actually takes place. Rather, the advancement of a macroscopic fluid interface will be determined simply by the spatial distribution of capillary properties. In such cases, numerical models based on the equations presented in this paper should correctly predict the movement of two immiscible liquids, regardless of whether a stable or unstable ratio of fluid properties exists. This is subject to the condition that the proper scale of heterogeneity has been accounted for in the numerical formulation. If, on the other hand, such a numerical model were used to predict the advancement of a fluid interface in homogeneous media, it would not give rise to a fingered interface. This is because equation (1) is based on macroscopic averages, and therefore does not incorporate the effect of pore-scale variations needed to perturb the macroscopic interface. The fingering process is discussed further by Kueper and Frind (1988).

10 POROUS MEDIA HETEROGENEITY

In naturally occurring porous media, the parameters governing multiphase/multicomponent flow and transport exhibit spatial variability. These parameters include permeability, wettability, porosity, the relationship between capillary pressure and saturation, the relationship between relative permeability and saturation, and the various mass transfer coefficients representing the partitioning of components between phases. To carry out realistic field-scale simulations, the spatial variability of these parameters must be accounted for.

One approach to carrying out field-scale simulations is to discretize the solution domain at a fine enough scale such that a discrete representation of the heterogenous medium is attained. This approach requires large amounts of computer storage and execution time, and usually demands some sort of Monte Carlo approach since only the statistics of the local scale variation of parameters are typically known. The alternative involves using large-scale 'effective' parameters which represent an averaging over the local scale. Numerical simulations incorporating such effective parameters will provide only information regarding the average behaviour of the variables of interest, and will not provide detailed information regarding sub-scale behaviour.

It should be emphasized that the application of a numerical model in a given solution domain of interest must be consistent with the scale at which input data is represented. For example, capillary pressure - saturation curves are typically measured in the laboratory on small samples of porous media obtained from sampling an aquifer. Such curves are representative of a particular point in the aquifer, and do not represent 'effective' behaviour at a larger scale of averaging. If a numerical model is applied to the field aquifer of interest using the local-scale measured capillary pressure - saturation curve to

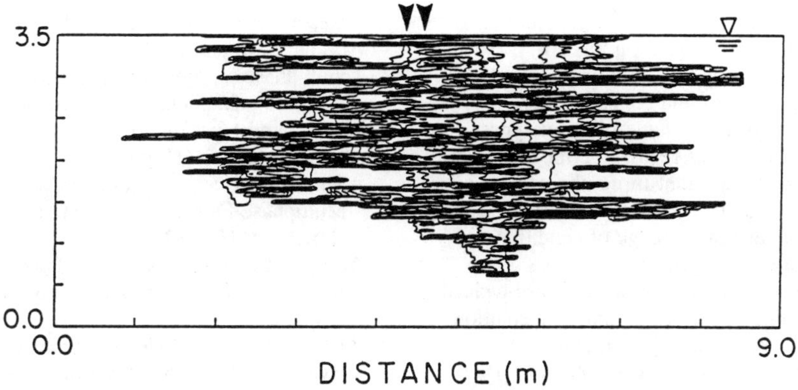

Figure 2 Tetrachloroethylene migration below the watertable in a spatially correlated random field at t = 7.40 days. The outermost contour represents 2% non-wetting phase saturation. The contour spacing is 10% saturation (after Kueper et al. 1990).

represent the entire solution domain, great errors may be introduced. This is illustrated by Kueper et al. (1990) who solve for two-phase flow below the watertable both with and without the spatial variability of local-scale capillary properties accounted for. Figure 2 illustrates one simulation from that study where the variation in permeability and capillary properties of the aquifer are represented as a spatially correlated random field.

Figure 2 illustrates the importance of the spatial variability of capillary properties in a heterogeneous porous medium on phase behaviour. When a lower permeability lens is encountered, non-wetting saturation must build up such that the capillary pressure immediately above the lens exceeds the entry pressure of the lens. This build-up of saturation is accompanied by significant lateral pressure gradients, resulting in lateral flows of non-wetting fluid. Even once the lower permeability lens is penetrated, there exists a greater relative permeability to the non-wetting fluid above the lens than within. This promotes further lateral spread of non-wetting fluid despite the fact the lens has been penetrated. It is clear that one-dimensional simulations of multiphase flow in heterogeneous porous media will be of limited value since in practice, this lateral spreading effect will result in non-wetting fluid occupying only selected portions of the porous medium. In one-dimensional simulations, the non-wetting fluid is forced to pass through all materials rather than bypassing some of the less permeable materials.

Incorporation of the small-scale variation in aquifer properties is computationally demanding. To perform large-scale average simulations, however, the appropriate large-scale effective input parameters must be defined. In the case of single-phase flow and transport, stochastic-analytic theory provides large-scale representations of various parameters including hydraulic conductivity and dispersivity (see for example Gelhar, 1986). Unfortunately, large-scale saturation-dependent parameters governing multiphase flow in heterogeneous porous media have not yet been rigorously defined except for special cases in air-water systems (e.g., Mantoglou and Gelhar, 1987). As a result, numerical models cannot in general be used at present to predict the field-scale behaviour of multiphase/multicomponent systems unless detailed, local-scale discretization is carried out. This in turn is hampered by a lack of local-scale data, particularly the statistical distribution of parameters in naturally occurring media, and by the high costs of computing associated with carrying out such simulations.

11 SUMMARY

Recent awareness of subsurface contamination by non-aqueous phase liquids has resulted in an increased activity in the area of multiphase/multicomponent flow and transport modelling. The local-scale balance equations governing such systems can be derived by performing a pore-scale averaging of the behaviour of individual components within phases. These equations are supplemented by constitutive relationships, and mass transfer coefficients representing the partitioning of components between phases. To date, most models have adopted the local-equilibrium assumption. This assumption has not, however, been experimentally validated under a wide range of conditions and scales of interest. Current models also assume that two-phase capillary pressure and relative permeability data can be used to represent three-phase behaviour. Again, this has not been experimentally validated under a wide range of conditions.

The equations governing multiphase/multicomponent systems are typically highly non-linear, requiring numerical solution schemes incorporating Newton-Raphson or a similar type of iteration. The equations are usually reduced to an algebraic set of equations through finite difference, finite volume, or finite element discretization. For large sets of equations, the use of conjugate gradient type iterative solvers is a necessity if simulations are to be performed in a cost-effective manner. Additional cost savings and simplification are attained by recognizing that NAPL movement often occurs at time scales much shorter than those of interest regarding the behaviour of dissolved contaminants.

There exists at present a lack of experimental data, particularly from controlled field experiments suitable for the validation of multiphase/multicomponent flow and transport models. Model input data in the form of saturation-dependent parameters and non-equilibrium mass transfer coefficients are also scarce. Since large-scale input parameters representing averages over the local scale have in general not yet been rigorously defined, field-scale application of multiphase models is restricted to solution domains discretized to represent the local-scale variation of porous medium and fluid properties.

ACKNOWLEDGEMENTS

This work was supported by the Waterloo Solvents-In-Groundwater Research Program through financial contributions from Ciba Geigy U.S.A., Dow Chemical U.S.A. and Canada, Eastman Kodak U.S.A., General Electric U.S.A., the province of Ontario University Research Incentive Fund, and the Natural Sciences and Engineering Research Council of Canada.

REFERENCES:

Abriola, L.M. and Pinder, G.F., 1985a. A multiphase approach to the modeling of porous media contamination by organic compounds: 1. Equation development. Wat. Resour. Res. 21(1), pp. 11-18.

Abriola, L.M. and Pinder, G.F., 1985b. A multiphase approach to the modeling of porous media contamination by organic compounds: 2: Numerical simulation. Wat. Resour. Res. 21(1), pp. 19-26.

Allen, M.B., 1985. Numerical Modelling of Multiphase Flow in Porous Media. Adv. Wat. Res. 8, pp. 162-187.

Aziz, K. and Settari, A., 1979. Petroleum Reservoir Simulation. Applied Science, London.

Baehr, A.L., 1987. Selective Transport of Hydrocarbons in the Unsaturated Zone Due to Aqueous and Vapor Phase Partitioning. Wat. Resour. Res. 23(10) pp. 1926-1938.

Baehr, A.L. and Corapcioglu, M.Y., 1987. A Compositional Multiphase Model for Groundwater Contamination by Petroleum Products 2. Numerical Solution. Wat. Resour. Res. 23(1), pp. 201-213.

Banerjee, S., 1984. Solubility of organic mixtures in water. Env. Sci. Tech. 18, pp. 587-591.

Bear, J., 1972. Dynamics of Fluids in Porous Media. Elsevier, New York.

Bear, J., Braester, C., and Menier, P.C., 1987. Effective and Relative Permeabilities of Anisotropic Porous Media. Trans. Por. Med. 2, p. 301-316.

Behie, A. and Vinsome, P.K.W., 1982. Block iterative methods for fully implicit reservoir simulation. Soc. Pet. Eng. J., Oct., pp. 658-668.

Brooks, R.H. and Corey, A.T., 1966. Properties of porous media affecting fluid flow. J. Irrig. Drain. Div., Am. Soc. Civ. Eng. 92 (IR2): 61-88.

Buckley, S.E. and Leverett, M.C., 1942. Mechanism of fluid displacement in sands. Trans. AIME, 146, pp. 107-116.

Cary, J.W., McBride, J.F. and Simmons, C.S., 1989. Observations of Water and Oil Infiltration into Soil: Some Simulation Challenges. Wat. Resour. Res. 25(1), pp. 73-80.

Chuoke, R.L., Van Meurs, P. and Van der Poel, C., 1959. The instability of slow, immiscible, viscous liquid-liquid displacements in permeable media. Petrol. Trans. AIME, 216, pp. 188-194.

Corapcioglu, M.Y. and Baehr, A.L., 1987. A Compositional Multiphase Model for Groundwater Contamination by Petroleum Products 1. Theoretical Considerations. Wat. Resour. Res. 23(1), pp. 191-200.

Corey, A.T., Rathgens, C.H., Henderson, J.H., Wyllie, M.R.J., 1956. Three-Phase Relative

Permeability. Petroleum Transactions, A.I.M.E., Vol. 207, pp. 349-351.
Dorgarten, H.W., 1989. Finite element simulation of immiscible and slightly soluble pollutants in soil and groundwater. IAHR Conference Proceedings, Contaminant Transport in Groundwater, April 4-6, Stuttgart, West Germany. pp. 389-396.
Eckberg, D.K. and Sunada D.K., 1984. Nonsteady three-phase immiscible fluid distribution in porous media. Wat. Resour. Res. 20(12), pp. 1891-1897.
Falta, R.W., Javandel, I., Pruess, K. and Witherspoon, P.A., 1989. Density-Driven Flow of Gas in the Unsaturated Zone Due to the Evaporation of Volatile Organic Compounds. Wat. Resour. Res. 25(10), pp. 2159-2169.
Faust, C.R., Guswa, J.H. and Mercer, J.W., 1989. Simulation of Three-dimensional Flow of Immiscible Fluids Within and Below the Unsaturated Zone. Wat. Resour. Res. 25(12) pp. 2449-2464.
Faust, C.R., 1985. Transport of Immiscible Fluids Within and Below the Unsaturated Zone: A Numerical Model. Wat. Resour. Res. 21(4), pp. 587-596.
Feenstra, S., 1990. The evaluation of DNAPL sources by monitoring of dissolved-phase concentrations from multi-component DNAPL. Proceedings, IAH Conference on Subsurface Contamination by Immiscible Fluids. April 18-29, Calgary, Alberta.
Ferrand, L.A., Celia, M.A., and Soll, W.E., 1990. Percolation-based models for pore-to-lab scale calculations in multifluid porous media. Chapter 16 in Dynamics of Fluids in Hierarchical Porous Media; Edited by J.H. Cushman. Academic Press.
Forsyth, P.A., 1988. Simulation of Nonaqueous Phase Groundwater Contamination. Adv. Wat. Res. 11, pp. 74-83.
Gelhar, L.W., 1986. Stochastic subsurface hydrology from theory to applications. Wat. Rcsour. Res., 22, pp. 135S-145S.
Hassanizadeh, M. and Gray, W.G., 1979a. General Conservation Equations for Multiphase Systems 1. Averaging Procedure. Adv. Wat. Res. 2, pp. 131-144.
Hassanizadeh, M. and Gray, W.G., 1979b. General Conservation Equations for Multiphase Systems 2. Mass, Momenta, Energy, and Entropy Equations. Adv. Wat. Res. 2, pp. 191-203.
Hill, L., 1952. Channelling in packed columns. Chem. Eng. Sci. 1(6), pp. 247-253.
Honarpour, M., Koederitz, L. and Harvey, A.H., 1986. Relative Permeability of Petroleum Reservoirs. CRC Press Inc.
Host-Madsen, J., 1988. Laboratory study of three-phase flow in porous media using the dual gamma attenuation technique. Proceedings Nordic Hydrological Conference, Rovaniemi, Finland, V. 2, pp. 240-262.
Hunt, J.R., Sitar, N. and Udell, K.S., 1988. Nonaqueous phase liquid transport and cleanup, 1, Analysis of Mechanisms. Wat. Resour. Res. 24(8), pp. 1247-1258.
Huyakorn, P.S. and Pinder, G.F., 1983. Computational Methods in Subsurface Flow. Academic Press.
Johnson, R.L., McGarthy, K.A., Perrott, M., and Hinman, N., 1989. Direct comparison of vapor, free product, and aqueous phase monitoring for gasoline leaks from underground storage systems. Procedings NWWA/API Petroleum Hydrocarbons and Organic Chemicals in Groundwater Conference, Housten, TX., pp. 605-615.
Kaluarachchi, J.J. and Parker, J.C., 1989. An efficient finite element method for modeling multiphase flow. Wat. Resour. Res. 25(1), pp. 43-54.
Kell, R.F., 1988. Studies on hazardous vapour transport in soil. M.A.Sc. Thesis, Dep't. Civ. Eng., Univ. of Waterloo, Waterloo, Ontario.
Kueper, B.H. and Frind, E.O., 1991b. Two-phase flow in heterogeneous porous media, 2. Model Application. Water Resources Research, 27(6), pp. 1059-1070.
Kueper, B.H. and Frind, E.O., 1991a. Two-phase flow in heterogeneous porous media, 1. Model Development. Water Resources Research, 27(6), pp. 1049-1057.
Kueper, B.H., Frind, E.O. and McWhorter, D.B.M., 1990. Application of a numerical model and laboratory parameter measurement to the movement of dense, immiscible phase liquids in a heterogeneous sand aquifer. Proceedings, IAH Conference on Subsurface Contamination by immiscible Fluids, April 18-20, Calgary, Alberta.
Kueper, B.H., Abbott, W. and Farquhar, G., 1989. Experimental Observations of Multiphase Flow in Heterogeneous Porous Media. J. Cont. Hyd. 5, pp. 83-95.
Kueper, B.H. and Frind, E.O., 1988. An Overview of Immiscible Fingering in Porous Media. J. Cont. Hyd. 2 pp.95-110.
Kuppusamy, T., Sheng, J., Parker, J.C. and Lenhard, R.J., 1987. Finite-Element Analysis of Multiphase Immiscible Flow Through Soils. Wat. Resour. Res. 23(4) pp. 625-631.
Larson, R.G., Scriven, L.E. and Davis, H.T.,

1981. Percolation theory of two phase flow in porous media. Chem. Eng. Sci., 36, pp. 57-73.
Lenhard, R.J. and Parker, J.C., 1988. Experimental Validation of the Theory of Extending Two-Phase Saturation-Pressure Relations to Three-Phase Systems for Monotonic Drainage Paths. Wat. Resour. Res. 24(3) pp. 373-380.
Lenhard, R.J. and Parker, J.C., 1987. A Model for Hysteretic Constitutive Relations Governing Multiphase Flow 2. Permeability - Saturation Relations. Wat. Resour. Res. 23(12) pp. 2197-2206.
Lenhard, R.J., Parker, J.C. and Kaluarachchi, J.J., 1989. A model for hysteretic constitutive relations governing multiphase flow 3. Refinements and numerical simulations. Wat. Resour. Res. 25(7), pp. 1727-1736.
Lenhard, R.J., Dane, J.H., Parker, J.C. and Kaluarachchi, J.J., 1988. Measurement and simulation of one-dimensional transient three-phase flow for monotonic liquid drainage. Wat. Resour. Re. 24(6), pp. 853-863.
Letniowski, F.W., 1989a. Numerical Methods for Nonaqueous Phase Liquid Groundwater Contamination in Three Dimensions. Master of Mathematics Thesis, University of Waterloo.
Letniowski, F.W., 1989b. An Overview of Preconditioned Iterative Methods for Sparse Matrix Equations. Research Report CS-89-26, Faculty of Mathematics, University of Waterloo, Waterloo, Ont.
Leverett, M.C., 1938. Flow of Oil-water Mixtures through Unconsolidated Sands. Trans. A.I.M.E. 132, pp. 149-171.
Leverett, M.C., 1941. Capillary behaviour in porous solids. Trans. Soc. Pet. Eng. AIME 142, pp. 152-169.
Leverett, M.C. and Lewis, W.B., 1941. Steady Flow of Gas-Oil-Water Mixtures Through Unconsolidated Sands. Transactions AIME, 142, p. 107.
Luckner, L., van Genuchten, M. Th. and Nielsen, D.R., 1989. A consistent set of parametric models for the two-phase flow of immiscible fluids in the subsurface. Wat. Resour. Res. 25(10) pp. 187-2193.
Mantoglou, A. and Gelhar, L.W., 1987. Stochastic Modeling of Large-Scale Transient Unsaturated Flow Systems. Wat. Resour. Res. 23(1), pp. 37-46.
Marle, C.M., 1982. On Macroscopic Equations Governing Multiphase Flow with Diffusion and Chemical Reactions in Porous Media. Int. J. Eng. Sci. 20(5) pp. 643-662.
McWhorter, D.B. and Sunada, D.K., 1990. Exact Integral Solutions for Two-Phase Flow. Wat. Resour. Res. 26(3) pp. 399-414.
Mendoza, C.A. and Frind, E.O., 1990a. Advective-Dispersive transport of dense organic vapors in the unsaturated zone, 1. Model development. Wat. Resour. Res. 26(3) pp. 378-387.
Mendoza, C.A. and Frind, E.O., 1990b. Advective-Dispersive transport of dense organic vapors in the unsaturated zone, 2. Sensitivity Analysis. Wat. Resour. Res. 26(3) pp. 388-398.
Osborne, M. and Sykes, J., 1986. Numerical modelling of immiscible organic transport at the Hyde Park Landfill. Wat. Resour. Res. 22(1), pp. 25-33.
Parker, J.C., Lenhard, R.J. and Kuppusamy, T., 1987. A Parametric Model for Constitutive Properties Governing Multiphase Flow in Porous Media. Wat. Resour. Res. 23(4) pp. 618-624.
Patankar, S.V., 1980. Numerical Heat Transfer and Fluid Flow. McGraw Hill Publishers.
Peaceman, D.W., 1977. Fundamentals of Numerical Reservoir Simulation. Elsevier, New York.
Schiegg, H.O., 1984. Considerations on water, oil and air in porous media. Wat. Sci. Tech. V. 17, pp. 467-476.
Schwille, F., 1988. Dense Chlorinated Solvents In Porous and Fractured Media - Model Experiments. Lewis Publishers.
Sleep, B.E. and Sykes, J.F., 1989. Modelling the Transport of Volatile Organics in Variably Saturated Media. Wat. Resour. Res. 25(1) pp. 81-92.
Stone, H.L., 1970. Probability Model for Estimating Three-Phase Relative Permeability. J. Pet. Tech., Feb., pp. 214-218.
van Genuchten, M.Th., 1980. A closed-form equation for predicting the hydraulic conductivity of unsaturated soils. Soil Sci. Am. J. 44(5), pp. 892-898.
Weth, D. and Pielke, M., 1989. Studies for calculating the transport of volatile chlorinated hydrocarbons by suction of subsoil air to clean contaminated soils. IAHR Conference Proceedings, Contaminant Transport in Groundwater, April 4-6, Stuttgart, West Germany. pp. 429-435.
Whitaker, S., 1986. Flow in Porous Media II: The Governing Equations for Immiscible, Two-Phase Flow. Trans. Por. Med. 1, pp. 105-125.
Wyckoff, R.D. and Botset, H.G., 1936. Physics (7), p 325.

Yeh, T.C., Gelhar, L.W., and Gutjahr, A.L., 1985. Stochastic analysis of unsaturated flow in heterogeneous soils. 2. Statistically anisotropic media with variable, Wat. Resour. Res. 21(4), pp. 457-465.

Subsurface Contamination by Immiscible Fluids, Weyer (ed.) © 1992 Balkema, Rotterdam. ISBN 90 6191 175 3

Density-driven vapor transport: Physical and numerical modeling

Richard L. Johnson, Kathleen A. McCarthy & Matthew Perrott
Oregon Graduate Institute, Beaverton, Oreg., USA

Carl A. Mendoza
University of Waterloo, Ont., Canada

ABSTRACT: Experiments were conducted in two large three-dimensional physical models to examine the density-driven advective transport of air laden with organic vapors. The experiments indicate that density-driven flow is important if the medium is coarse or if the relative vapor density is high. Observed advective velocities were less than predicted by a simple one-dimensional model (Falta et al., 1989) because of a rapid decrease in vapor concentrations resulting both from partitioning of the vapors into the pore water and gas-phase diffusion. Overall, observed vapor behavior was accurately simulated using a numerical model which accounts for phase partitioning and diffusion as well as density-driven flow (Mendoza and Frind, 1990a).

1. BACKGROUND

Vapor migration in unsaturated soil is an important pathway for transporting volatile organic contaminants in the subsurface. In the absence of external pressure gradients, this vapor transport will be dominated by either molecular diffusion or density-driven advective flow. The extent to which either of these mechanisms dominates depends upon the properties of both the compound of interest and the porous medium. A class of compounds for which vapor transport is particularly important are the volatile chlorinated hydrocarbon solvents (CHCs), such as trichloroethylene (TCE). These compounds have been, and continue to be, spilled as pure-phase liquids into the subsurface. Because of the limited solubility of the CHCs, they often move through the subsurface as a separate non-aqueous-phase liquid and a residual of the liquid becomes immobilized in the subsurface. This residual can become a long-lived source of vapor and dissolved CHCs.

Many CHCs have substantial vapor pressures, as well as molecular weights which are well above the average molecular weight of air (Table 1). As a consequence, when these compounds are present in soil pore air, the density of that air can be much greater than air without the organics (Table 1). The relative vapor density (RVD) is a measure of the density of air saturated with the compound of interest. For example, the RVD of TCE is calculated using equation 1. Numerical values of RVD for a number of environmentally important compounds are listed in Table 1.

$$RVD_{TCE}=(P_oMW+(1-P_o)M_{air})/M_{air}=1.27 \quad [1]$$

where P_o = vapor pressure in atmospheres
MW = molecular weight of the com pound of interest
M_{air} = molecular weight of clean air (e.g., 79% N_2, 21% O_2) is ~29.

Schwille [1988] examined density-driven vapor transport in a number of one-dimensional (1-D) physical model experi-ments in a large (~0.4 m diameter by 1.5 m long) column. The models were filled with a dry, medium-coarse sand (permeability (k) ~ 3 x

Table I. Physical/Chemical Parameters for the Chlorinated Hydrocarbon Solvents.

	MW	VP	K_h	RVD
TCE	131.5[1]	58	0.30	1.27
PER[2]	165.8	14	0.54	1.09
CCl_4[3]	153.8	90	0.96	1.51
DCE[4]	97	590	0.87	2.54
F22[5]	102.9	>760	>10	3.0

MW = molecular weight
VP = vapor pressure (mm Hg at 20°C)
K_h = air/water partition coefficent
[1]Data from Schwille, 1988.
[2]tetrachloroethylene
[3]carbon tetrachloride
[4]1,1-dichloroethylene
[5]dichlorofluoromethane

10^{-11} m^2), and a source of vapors was placed in a pan at the top of the column. He observed strong density effects for a variety of chlorinated solvents as well as carbon dioxide (RVD=1.52). Schwille's experiments represent optimum conditions for density-driven flow in that the soil was dry and the 1-D column did not allow the vapors to dissipate laterally. In addition, there may have been significant pressure-driven advection due to the volatilization of relatively large volumes of gas within his experimental apparatus.

Researchers at the University of Waterloo [Mendoza, 1989] conducted 1-D column experiments with pentane vapor. The column used in these experiments was 0.07 m in diameter and 0.375 m in length and was packed with dry sand (k=1.45 x 10^{-11} m^2). As in Schwille's experiments, the source (in this case liquid pentane) was placed at the top of the column. A small vent was provided at the top of the column to maintain atmospheric pressure, while the base of the column was purged with clean air. Samples were collected at 0.075 m intervals down the column. Data from these experiments show significant density-driven flow. Numerical simulations of the experiments predicted somewhat less flow than was observed. Once again, some pressure-driven advection may have been induced during the volatilization process, and may explain the observed results.

Several mathematical models have been developed to examine density-driven vapor flow. Falta et al. [1989] developed an analytical expression to estimate density-driven velocity:

$$V_p = \frac{k k_{rg} g P_o}{\phi S_g \mu_g R T} (MW - M_{air}) \qquad [2]$$

where V_p = pore velocity,
k = intrinsic permeability
k_{rg}= relative permeability of the gas phase,
g = gravitational acceleration constant
P_o = vapor pressure
ϕ = porosity
S_g = gas-phase saturation in porous medium
μ_g = viscosity of the air-vapor mix
R = ideal gas constant
T = absolute temperature

This equation does not account for the effects of diffusion, phase partitioning or boundaries on the fluid velocity. These effects can be substantial, depending upon the conditions of the release, and the authors caution that the equation gives only "order-of-magnitude" estimates. Nevertheless, it does provide a simple means of estimating the upper limit of density-driven velocity.

Two numerical models designed to simulate density-driven vapor transport in porous media have recently been reported. Falta et al. [1989] developed a numerical model to simulate density-driven flow coupled with the effects of diffusion and phase partitioning. The model is formulated in cylindrical coordinates and accounts for a variety of boundary conditions. They examined the behavior of carbon tetrachloride (CCl_4) and toluene. Both the vapor pressure and air/water partition coefficient (K_h) values for CCl_4 are ~4 times that of toluene, and the molecular weight is approximately 1.7 times greater. As a consequence, CCl_4 is expected to experience density-driven flow to a much-greater extent than toluene. Their simulations suggest that in media with $k \leq 3x10^{-11}$ m^2, toluene transport is dominated by diffusion, whereas CCl_4 transport begins to show substantial density effects at $k \geq 10^{-12}$ m^2.

Mendoza and Frind (1990a-b) have also developed a coupled numerical model which can be operated in either cylindrical or cartesian coordinates. Like the Falta, et al (1989) model, their model couples density-driven vapor transport with diffusion and phase partitioning. They examined the sensitivity of density-driven flow to a number of initial and boundary conditions, including permeability of the medium, the presence of an impervious surface boundary, and the effects of radial vs. cartesian geometry. They compared their model results to experimental data from two physical models. In both cases, agreement between the numerical and physical models was good, and both cases showed the effect of density-driven flow on vapor spreading. In the work presented here, simulations using this model have been compared to physical model experiments conducted at the Oregon Graduate Institute (OGI).

2. EXPERIMENTAL

A series of experiments were conducted in large 3-dimensional physical models at OGI to examine density-driven vapor transport. The physical models consist of rectangular steel bunkers contained within concrete bunkers. The dimensions of the inner bunker are approximately 10 m x 9 m x 4.5 m deep. The inner bunkers have been filled with either moist medium sand or moist pea gravel. A cross-sectional view of the model is shown in Figure 1. The permeability of the sand is ~5 x 10^{-11} m^2, so it is quite similar to the values used in many of the simulations in both Falta, et al (1989), and Mendoza and Frind (1990a-b). The permeability of the pea gravel is ~5 x 10^{-9} m^2. Other pertinent properties of the media are listed in Table II.

Prior to the initiation of the experiments, the water table within the model was raised to ground surface and then lowered to ~3-3.3 m below ground surface. This was done to create a zone of residual water saturation which was about 3 m thick in both of the porous media.

Table II. Physical properties of the porous media used in the experiments.

Sand
permeability=$5x10^{-11}$ m^2
total porosity=0.36
water content=0.09
fraction organic carbon content=0.0025
Pea Gravel
permeability=$5x10^{-9}$ m^2
total porosity=0.36
water content=0.05
fraction organic carbon content=0.0000

Vapor samples were collected from an array of sampling points across the aquifer (shown schematically in Figure 1). The points consisted of bundles of 3.2 mm o.d. stainless steel (SS) tubes (10 tubes per bundle), each of which terminated at a specific depth. The tip of each tube was wrapped with 100-mesh stainless-steel screen. Two different spacings were used for these experiments. For the experiments in sand, the distance between the sampling bundles was ~0.6 m. For the pea gravel experiments, the distance between the sampling bundles was ~1.4 m. Vapor samples were collected using evacuated SS cans. The cans were cleaned by heating with alternating nitrogen flushing and evacuation. Propylene (1% by volume in nitrogen) was added as an internal standard to bring the pressure of the cans to 0.2 atm (absolute). This pressure also served to insure that all vapors collected were below their saturation concentrations (including water).

Two pairs of experiments will be discussed here. The first pair involved the release of TCE into the sand and into the pea gravel. In each case, near the center of the model, one liter of liquid TCE (~1.6 kg) was poured over an excavated area ~0.3 x 0.3 m wide and ~0.5 m below ground surface over a period of ~1 minute. The excavation was then quickly filled

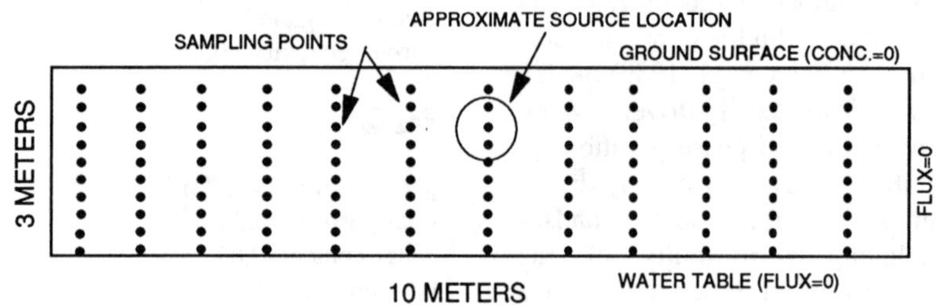

Figure 1. Schematic drawing of the 3-D physical model used for the density-driven flow experiments.

in with original material. Samples were collected at 0.33, 1, 4, 9-12, 24, and 48 hours.

In the second pair of experiments, dichlorofluoromethane vapor (Freon 22; RVD=3.0) were released into the sand. Freon 22 is a gas at 10°C, and thus can be released as pure vapor. A total of ~100 liters of Freon 22 vapor (~0.36 kg) were released over the course of ~10 minutes. To confirm that density-driven flow played an important role in the vapor distributions observed in this experiment, a subsequent experiment with dilute Freon 22 was conducted. Approximately 80 liters of a mixture of 1% Freon 22 in air (RVD=1.03. ~.003 kg) was prepared in a pressurized cylinder, and released at a similar rate to the previous experiment, and at the same location. In both experiments, vapor samples were collected at a number of locations after 0.5, 4, 9, and 22 hours.

Data from the 2-D vertical array of sampling points was used to create instantaneous concentration contour snapshots. The contour plots were produced with the Surfer graphical package (Golden Software, Inc., Golden CO) using the minimum curvature option. The normalized contours for all figures are 1, 2, 3, 4, 5, 6, 8, 10, 15, 20, 25, 30, 40, 50, 60, 70, 80, and 90% of the corresponding initial source concentrations. Temperature profiles for each of the experiments are presented along with the corresponding concentration profiles.

3. RESULTS AND DISCUSSION

3.1 Physical model experiments

Figure 2 shows a vertical cross-sections of relative vapor concentration for the release of TCE into medium sand. Samples collected at 1, 4, 12 and 48 hours after the release are shown in the figure. Despite the fact that the RVD of TCE at 10°C is ~1.15, very little density-driven flow was observed. There are three primary reasons for this behavior. First, a significant fraction of the TCE vapor partitions to the pore water as the vapor diffuses away from the source (the dimensionless air/water partition coefficient for TCE is ~0.30). This substantially reduces both

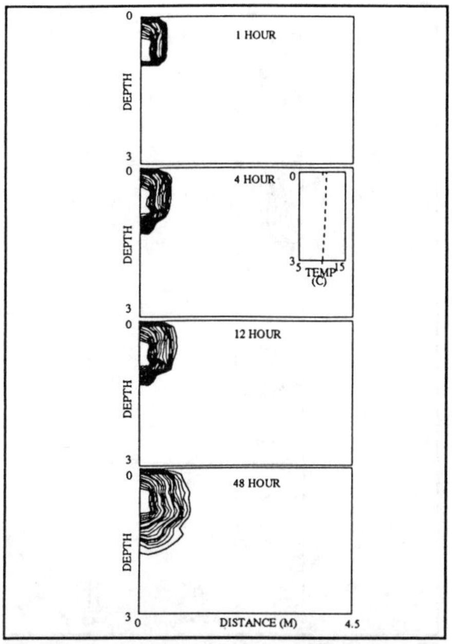

Figure 2. Vertical concentration contours at 1, 4, 12, and 48 hours after release of TCE into sand. Concentrations are normalized to the saturation vapor pressure; contour levels are listed in the text.

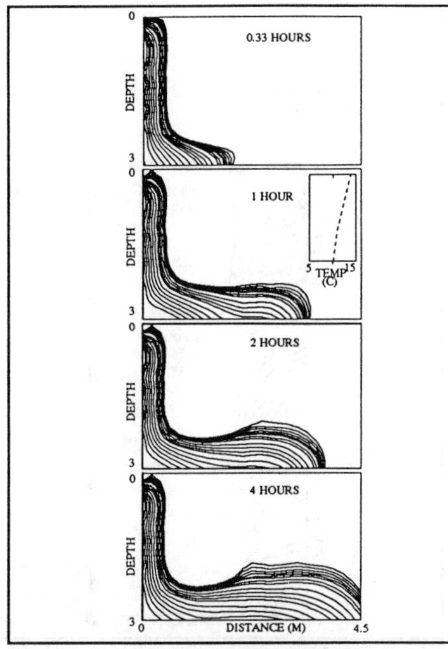

Figure 3. Vertical concentration contours at 0.33, 1, 2, and 4 hrs. after the release of TCE into pea gravel. Concentrations are normalized to the saturation vapor pressure; contours are listed in text.

the vapor concentration and the RVD. Second, diffusion radially away from the relatively small source causes concentrations to drop off quickly. Both of these phenomena limit the impact of density-driven flow. Third, given the greatly reduced TCE concentration, the permeability of the sand is too low to support any appreciable density-driven flow.

The one-liter release of TCE into the pea gravel produced a strong density effect (Figure 3). Within four hours of the release, the vapor had moved across the entire width of the model. The residual water content of the pea gravel is somewhat lower than that of the sand, thus partitioning to the aqueous phase is less in the pea gravel than in the sand. However, the primary difference in vapor transport rates between the two TCE experiments is that the permeability of the pea gravel is ~100 times greater than the sand.

As a consequence of its rapid movement, the TCE source was completely volatilized in ~8 hours, and vapor concentrations began to drop. In contrast, the TCE source in the sand experiment persisted for a number of days.

The distribution of Freon 22 within the subsurface for the 100% Freon 22 case at 0.5, 4, 9 and 22 hours are seen in Figure 4. The center of mass of the plume moved downward nearly one meter in the first four hours. During that period the maximum concentration dropped from 100% to 30%. Correspondingly, the maximum relative vapor density dropped from 3 to 1.6. Over the next 18 hours the maximum concentration dropped to 15%, and the vertical velocity decreased. By 22 hours, the zone of maximum concentration was near the bottom of the physical model, and downward migration had stopped. In contrast, the center of mass of the 1% release did not move downward during the entire sampling period (Figure 5).

3.2 Numerical modeling

The experimental data presented here were

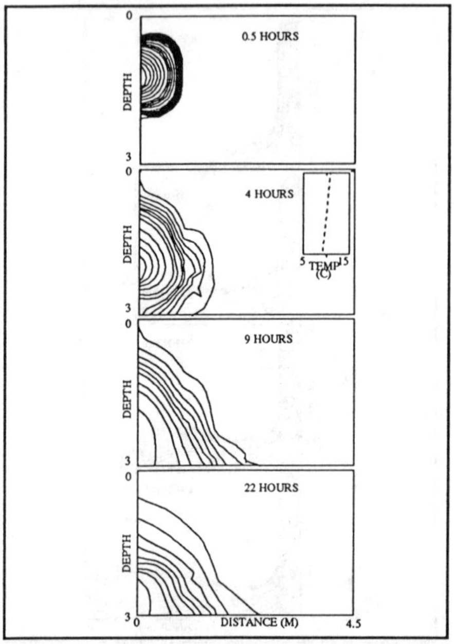

Figure 4. Vertical concentration contours at 0.5, 4, 9 and 22 hours after the release of 100% Freon 22 into pea gravel. Concentrations are normalized to 100% Freon 22; contours are listed in the text.

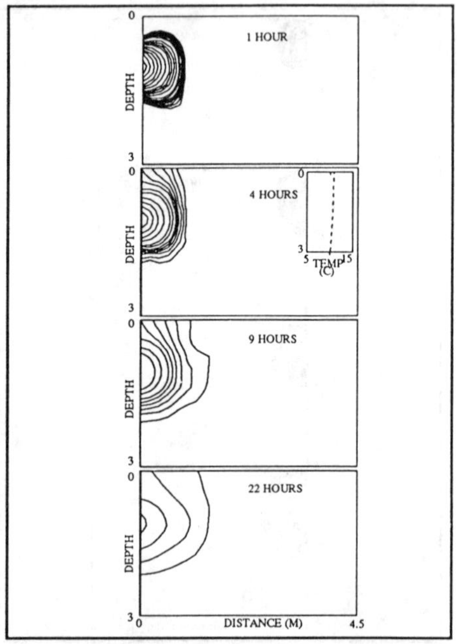

Figure 5. Vertical concentration contours at 0.5, 4, 9 and 22 hours after the release of 1% Freon 22 into sand. Concentrations are normalized to 1% Freon 22; contour levels are listed in the text.

used to evaluate the mathematical model developed by Mendoza and Frind [1990 a-b]. The model has been discussed extensively elsewhere [Mendoza, 1989], so only the aspects pertinent to the physical model experiments will be discussed here.

Because the releases were made at the center of the model and because the effects of the rectangular boundary were small, the experiments were modeled using the axis-symmetric capabilities of the model. The boundary conditions for all of the simulations were: 1) zero flux at the outer perimeter; 2) zero concentration at the ground surface; and 3) zero flux to the groundwater. The latter is not strictly true for the TCE, however mass transfer to the groundwater was quite limited during the short durations of these experiments.

No fitting of the model parameters to the experimental data was done. All of the input parameters were either determined experimentally, or obtained from the literature. This was done in order to evaluate the usefulness of the model in predicting vapor transport. Figure 6 shows the vertical concentration profiles for the simulation of the TCE experiment in the sand. As with the experimental data, all of the simulated contours represent 1, 2, 3, 4, 5, 6, 8, 10, 15, 20, 25, 30, 40, 50, 60, 70, 80, and 90% of the source values. As observed in the physical model experiments, TCE vapor movement in the sand is dominated by molecular diffusion. Simulation of this experiment by Mendoza and Frind (1990a) showed a small density effect, which did not substantially affect vapor transport. Diffusion distances were somewhat larger in the simulations, probably as the result of differences between actual parameter values and those used in the simulations. However, in the context of overall behavior, the differences are relatively minor.

Figure 7 shows concentration contours for the simulation of the TCE experiment in the pea gravel. The density-driven flow is substantially less than what was observed in the experiment. Again, this could be due to differences between actual system parameters

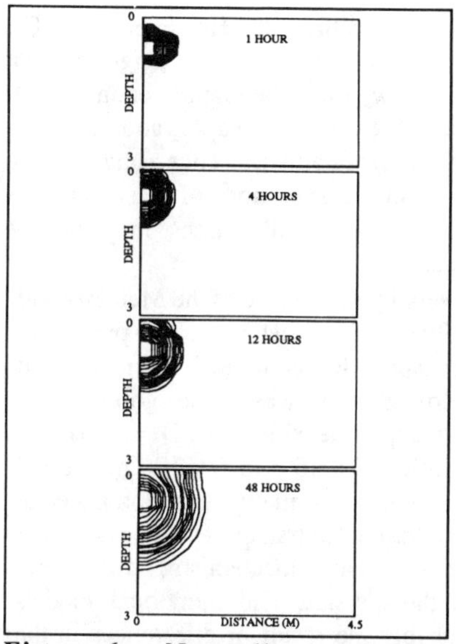

Figure 6. Normalized vertical concentration contours at 1, 4, 12, and 48 hours from the simulation of the spill of TCE into moist sand. Contour levels are listed in the text.

and the input values used in the model. However, it may also be due to non-equilibrium partitioning of TCE between the air and water phases. Laboratory experiments [Houston et al., 1989, MacPherson, 1991] suggest that phase partitioning in moist soils may take 24 hours or more to reach equilibrium. Given the very-rapid transport velocities in these experiments, it is likely that equilibrium between the phases is not achieved during the initial hours of the experiment and as a consequence, vapor densities remain higher than the modeled values. The result is that the vapor is more-rapidly transported than the simulations predict.

The simulation of the release of pure Freon 22 vapor is shown in Figure 8. As with the experimental data, strong advective flow is observed early in the experiment. Because of the instantaneous nature of the release and the rapid radial diffusion, concentrations decrease relatively quickly and little lateral spreading along the top of the capillary fringe is observed. Figure 9 shows simulation results for the 1% Freon 22 release. As expected, transport is

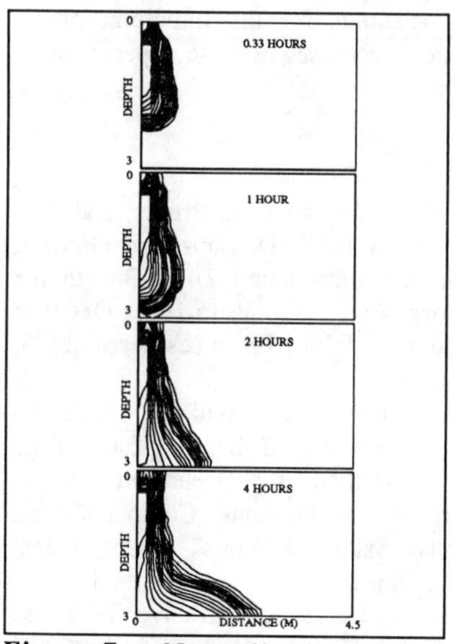

Figure 7. Normalized vertical concentration contours at 0.33, 1, 2, and 4 hours from the simulation of the spill of TCE into pea gravel. Contour levels are listed in the text.

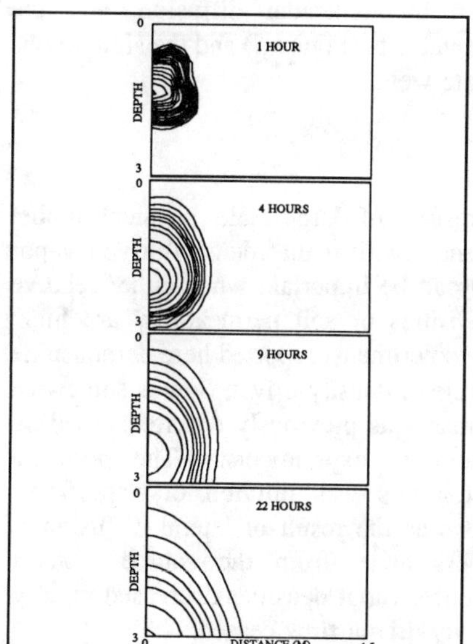

Figure 8. Normalized vertical concentration contours at 1, 4, 9, and 24 hours from the simulation of the spill of 100% Freon 22 into moist sand. Contour levels are listed in the text.

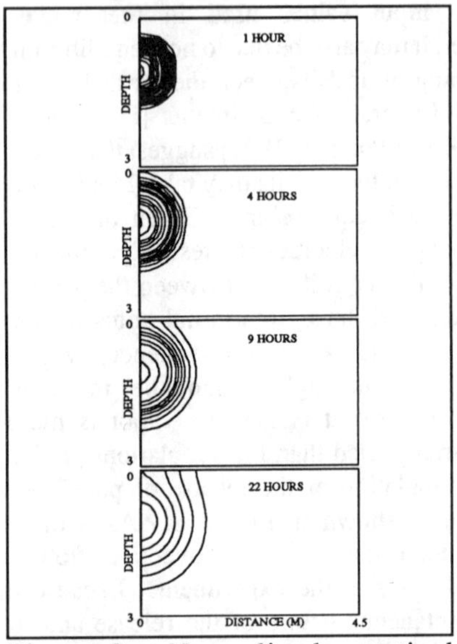

Figure 9. Normalized vertical concentration contours at 1, 4, 9, and 24 hours from the simulation of the spill of 1% Freon 22 into moist sand. Contour levels are listed in the text.

dominated by molecular diffusion, and the experimental data (Figure 5) and the simulations agree quite well.

4. CONCLUSIONS

The results of large-scale physical model experiments confirm that density-driven vapor transport can be important when either relative vapor densities or soil permeability are high The 3-D experiments reported here demonstrate that the role of density-driven flow is somewhat smaller than was previously reported based on 1-D laboratory experiments. The principal reason for this was dilution of vapor concentrations as the result of lateral diffusion of the vapors away from the source. As a consequence, vapor densities decreased rapidly, and density-driven flow ceased.

Previous laboratory experiments also used clean, dry porous media. Pore water and soil can act as significant mass reservoirs, thus reducing vapor concentrations and subsequent transport. Most numerical models assume that equilibrium among all the phases is instantaneously achieved. However, the TCE experiment conducted here suggests that advective flow may be rapid enough that equilibrium between the vapor and the pore water is not maintained. Thus, under some conditions, the assumption of instantaneous equilibrium between all phases may not be appropriate.

The overall performance of the Mendoza and Frind (1990a) numerical model in predicting vapor transport by combined diffusion and density-driven flow was quite good. The ranges of vapor densities (RVD = 1-3) and permeabilities (5×10^{-11} to 5×10^{-9} m^2) tested spanned conditions from diffusion-dominated to advection-dominated transport. When diffusion was the dominant mechanism, differences between the physical and numerical models were probably the result of differences in the partitioning parameters for the soil-vapor-water system. For the cases when both advection and diffusion were active, differences were probably the result of either non-equilibrium partitioning or incorrect permeability values. Thus, with the possible exception of non-equilibrium partitioning, it appears that the numerical model correctly accounts for the important physical/chemical processes in these experiments.

REFERENCES

Falta, R.W., I. Javandel, K. Pruess, and P.A. Witherspoon, 1989, "Density-Driven Flow of Gas in the Unsaturated Zone Due to the Evaporation of Volatile Organic Compounds", Water Resources Research, 25, 2159-2169.

Houston, Sandra L., David K. Kreamer, Randolph Marwig, 1989, "A Batch-Type Testing Method for Determination of Adsorption of Gaseous Compounds on Partially Saturated Soils", Geotechnical Testing Journal.

MacPherson, J.R., 1991, "Diffusive Gas-Phase Transport in Porous Media: Physical and Numerical Modeling", Ph.D. Dissertation, Oregon Graduate Institute, Beaverton, OR.

Mendoza, C.A. 1989, "Advective-Dispersive Transport of Heavy Organic Solvent Vapors in the Unsaturated Zone", M.S. Thesis,

University of Waterloo, Waterloo, Ontario, 104pp.
Mendoza, C.A., and E.O. Frind, 1990a, Advective-Dispersive Transport of Dense Organic Vapors in the Unsaturated Zone 1. Model Development", Water Resources Research, 26, 379-387
Mendoza, C.A., and E.O. Frind, 1990b, Advective-Dispersive Transport of Dense Organic Vapors in the Unsaturated Zone 2. Sensitivity Analysis", Water Resources Research, 26, 388-398
Schwille, F., 1988, Dense Chlorinated Solvents in Porous and Fractured Media, Translated by J.F. Pankow, Lewis Publishers, Chelsea, MI, 146pp.

Subsurface Contamination by Immiscible Fluids, Weyer (ed.) © 1992 Balkema, Rotterdam. ISBN 90 6191 175 3

Influence of hydrogeologic properties on in situ bioremediation

K.L.Duston, P.G.Allen & C.H.Ward
National Center for Ground Water Research, Department of Environmental Science and Engineering, Rice University, Houston, Tex., USA

ABSTRACT

In situ biorestoration of contaminated subsurface formations involves the stimulation of the indigenous subsurface microflora to degrade organic compounds in place. The goal is the destruction of the contaminants to produce biomass and harmless metabolic by-products. The success of this treatment is dependent on the subsurface geology of the site and the presence of metabolically appropriate microorganisms. The subsurface geology is important because the formation must be perfused with nutrients and a terminal electron acceptor to enhance microbial activity. Geologic formations which are most amenable to in situ biorestoration technology are those with low clay content and hydraulic conductivity greater than 10^{-4} cm/sec. However, even if the site is geologically ideal, biodegradation cannot occur without the presence of active contaminant-degrading microorganisms. Analysis of most subsurface materials has indicated that microorganisms are present and that they can biodegrade a variety of contaminants.

1 INTRODUCTION

Ground water is contaminated with chemicals on a daily basis from leaking underground storage tanks, improper waste disposal and agricultural practices. The available technology for aquifer remediation involves chemical, physical and biological processes. Physical methods include construction of barriers and use of hydrodynamic controls to isolate the contaminants. Chemical and biological treatments can be performed above ground or in situ. Treatment applicability depends on the hydrogeologic characteristics of the aquifer, the type and quantity of chemical contaminant present, and the presence of an active microbial population able to degrade the contaminants (Thomas et al., 1987a). Prior to the 1980's, remedial action usually involved isolation or withdrawal and treatment methods. However, often these methods did not solve the contamination problem, but instead temporarily isolated the contamination or transferred it to another part of the environment. During the 1980's, a potentially better solution to the ground water contamination problem evolved using subsurface microorganisms to degrade many of the contaminants (Lee et al., 1988). The method was first patented and used for hydrocarbon degradation by R. L. Raymond in the 1970's (Raymond, 1974; Raymond et al., 1975; Raymond et al, 1978). This potential solution, in situ biorestoration, is the process by which the indigenous subsurface microflora is stimulated to degrade the contaminants by the addition of inorganic nutrients and a terminal electron acceptor (Thomas and Ward, 1989).

Microorganisms have the potential to completely degrade, or mineralize, many of the contaminants found in ground water. In situ biorestoration may be the remediation treatment of choice, when applicable, because it offers a better solution to the contamination problem through the partial or complete destruction of the contaminants. However, the in situ biorestoration treatment method is not applicable to all contaminated subsurface formations. There are two criteria that must be met for the successful implementation of the process: 1) a subsurface material that is permeable enough to allow the transport of nutrients and a terminal electron acceptor to the microflora in the area of contamination and 2) the presence of a metabolically appropriate microflora.

2 HYDROGEOLOGY AND THE POTENTIAL FOR GROUND WATER CONTAMINATION

The hydrogeology of the subsurface formation is the primary factor governing the severity of contamination and the feasibility of in situ biorestoration. The important physical characteristics of the formation are the depth to the water table, the net recharge, the aquifer media (consolidated and unconsolidated), the soil media (rhizosphere zone), the topography, the impact of the vadose zone and the hydraulic conductivity of the aquifer (Aller et al., 1985). In Table 1, each of these physical characteristics have been weighted with respect to each other and with respect to the origin of the potential contamination (non-agricultural or agricultural sources) and rated in decreasing importance from 5 to 1. The ratings differ between the non-agricultural and agricultural sources of contamination because the soil media and topography greatly affect the application of agricultural chemicals.

Table 1. Weighted Subsurface Characteristics Which Affect Ground Water Contamination Potential (Aller et al., 1985)

CHARACTERISTIC	WEIGHT NON-AGRI-CULTURE	AGRI-CULTURE
Depth to water table	5	5
Net recharge	4	4
Aquifer media	3	3
Soil media	2	5
Topography	1	3
Impact of the vadose zone	5	4
Hydraulic conductivity of the aquifer	3	2

Shallow aquifers and those with high net recharge are obviously at greater risk of contamination. Aquifer media that are at the highest risk include unconsolidated sand and gravel, karst limestone, tubular basalt, massive permeable limestone, massive permeable sandstone, thin-bedded sandstone, limestone, and shale sequences. When the overlying soil is thin or absent or consists of sand, the aquifers are at greater risk. Less permeable soil media, such as clays and clay-rich materials, prevent or retard movement of the contaminants into the ground water. A greater potential for contamination exists when the slope of the land surface is gradual because there is less runoff and, therefore, more infiltration. The less permeable the vadose zone media, for example shale, silt, and unconsolidated clay, the lower the risk of ground water contamination. Formations with high hydraulic conductivities also have high potential for ground water contamination (Aller et al., 1985).

3 HYDROGEOLOGY AND THE POTENTIAL FOR IN SITU BIORESTORATION

The factors that are important in determining the potential for ground water contamination are also important in determining whether an aquifer is amenable to in situ biorestoration. The in situ biorestoration process is dependent upon the transport of nutrients and a terminal electron acceptor to the microorganisms in the area of contamination (Thomas et al., 1987b). The hydraulic conductivity of the formation is indicative of the rate at which a fluid moves through the formation. The perfusion of nutrients and a terminal electron acceptor through the aquifer media is more successful when the hydraulic conductivity of the media is high. The hydraulic conductivity is a function of the fluid and the type of medium and can be determined using several methods, including 1) a slug test, 2) a pump test, 3) a variable head test and 4) a tracer test. A range of values of hydraulic conductivity for various media is given in Figure 1. Subsurface formations with hydraulic conductivities of 10^{-4} cm/sec or greater are considered amenable to in situ biorestoration (Thomas et al., 1987b). In less permeable areas, the transport of the nutrients and oxygen can be limited or totally prevented. Because of the heterogeneity of the site, the formation may be permeable enough in some locations, but not in all. Therefore, the determination of the hydraulic conductivity at multiple locations and depths is desirable.

In addition to the hydraulic conductivity, the type of aquifer media is important in determining the potential for in situ biorestoration. Referring to Figure 1, several of the aquifer media have ranges of hydraulic conductivities that include the 10^{-4} cm/sec value and, therefore, appear suitable for in situ biorestoration. However, the range of some of these media may include hydraulic conductivities much lower than 10^{-4} cm/sec and may actually not be very well suited for in situ biorestoration. In addition, ground water flow through the media can vary and affect the potential for biorestoration. Flow through consolidated deposits is mainly through cracks and faults, while flow through unconsolidated deposits is through pores. Since migration of contaminants may be greatly enhanced by preferential flow paths, bioremediation of deposits with low overall permeability may have to target the highly permeable fracture zones to be successful. Several types of aquifers where in situ bioremediation has been utilized are listed (Table 2). Other important formation characteristics include the direction of ground water flow, the specific yield of the aquifer, and the depth to the water table and zone of contamination (Thomas and Ward, 1989).

Soil chemistry, particularly the interactions between the soil and nutrients, is also an important consideration. Gravel and sand deposits, which have the highest permeabilities, also have the least interaction with mineral nutrients. Clay and silt, which form less permeable deposits, tend to adsorb nutrients on the particle surfaces (Fiorenza et al., 1987).

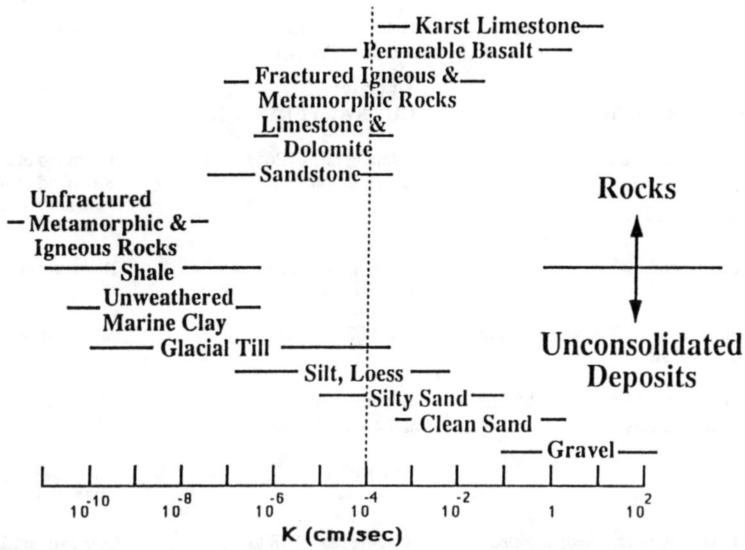

Figure 1. Range of Hydraulic Conductivity (adapted from Freeze and Cherry, 1979)

Limestone and other deposits that have high mineral contents can complex and precipitate phosphorus as orthophosphate (Brubaker and Crockett. 1986). The presence of high concentrations of aqueous reduced iron (Fe^{+2}) in the formation water can also create a precipitation problem when oxygen is introduced during biostimulation. Ferrous iron can precipitated out of solution as ferric oxide (Fe^{+3}) when an aquifer is perfused with oxygen-rich water (Hallberg and Martinell, 1976; Zienkiewicz, 1985). The oxidizing conditions created by supplying oxygen to the microflora maintain high Eh and pH values. Iron oxidizing bacteria may be involved in the oxidation reactions. Iron precipitation in the aquifer formation may result in plugging the aquifer.

4 GROUND WATER MICROBIOLOGY

The potential role of subsurface microorganisms in the degradation of ground water contaminants was addressed in a review by Dunlap and McNabb (1973), which initiated most of the research on subsurface microbiology. Prior to the publication of this review, the concept of biological activity below the rhizosphere was not widely accepted. An early study indicated that the numbers of microorganisms declined precipitously with depth (Waksman, 1916). A second study established the existence of oligotrophic, or nutrient poor, conditions below the rhizosphere (Leenheer et al., 1974), which were believed to be detrimental to the survival of microorganisms. Another review of the research performed in the area of ground water microbiology indicated that microorganisms are present in the subsurface, that they are metabolically active and that they have the ability to degrade a variety of compounds found in the subsurface (Thomas et al., 1987a).

The presence of appropriate contaminant-degrading microorganisms and their nutritional requirements must be determined at any site for which in situ biorestoration is considered. Representative samples of subsurface materials are obtained, and from them, the presence of metabolically active indigenous microorganisms can be ascertained. The method for procuring subsurface cores exclusive of surface contamination was published in 1977 (Dunlap et al., 1977) and was later modified in 1983 (Wilson et al., 1983). The method entails drilling a borehole to the desired depth with an auger and removing a soil sample using a core barrel. The core is then extruded through a sterile paring device that removes the outer layer of sediment that has come into contact with the core barrel. The pared core is therefore free of contamination introduced by the drilling process.

From the core sample, the presence and number of microorganisms can be determined by several methods. These include 1) viable counts recovered by spread plating (Balkwill and Ghiorse, 1982), 2) direct counts by epifluorescence microscopy (Ghiorse and Balkwill, 1985), 3) biochemical indicators of activity such as quantification of ATP (Webster et al., 1985), 4) fatty acid analysis of phospholipids (White et al., 1983) and 5) electron microscopy (Wilson et al., 1983).

Bacteria predominate in the subsurface, although some eukaryotic life forms have been observed (Wilson et al., 1983; Ghiorse and Balkwill, 1983; Ghiorse and Balkwill, 1985;

Table 2. Types of aquifers where in situ biorestoration has been utilized (Thomas et al., 1987a)

AQUIFER DESCRIPTION	FLOW CHARACTERISTICS	REFERENCE
high permeability dolomite	pumping rate of 265 to 378 L/min	Raymond et al., 1976 Raymond et al., 1975 Jamison et al., 1976 Jamison et al., 1975
medium to coarse sand	pumping rate of 65 to 151 L/min	Raymond et al., 1986
alluvial fan deposit of sand, gravel, and cobbles with some clay and silt	flow of 2.4 m/day	Minugh et al., 1983
poorly sorted mixture of boulders, pebbles, cobbles, sand silt and clay	hydraulic conductivity of 9.4×10^{-5}	Jhaveri and Mazzacca, 1983
clay	1.7×10^{-3}cm/sec	Jhaveri and Mazzacca, 1985
perched water table in unstratified, unsorted layer of clay, silts sands, gravels, and cobbles above a clay layer	pumping rate of 38 to to 57 L/min	Quince et al., 1985 Flathman et al., 1985
tank vault filled with pea gravel surrounded by sand and sandy clay strata	velocity in excess 100 m/yr pumping rate of 151 L/min	Westray et al., 1985 Brown et al., 1985 Brenoel and Brown, 1985
glacial outwash composed of silt, sand, and gravel	hydraulic conductivity of 8.8×10^{-4} to 1.5×10^{-3}cm/sec	Brown and Norris, 1986
coarse sands and gravel	hydraulic conductivity of 2.1 cm/sec	Nagel et al., 1982
shale and siltstone	pumping rate of 68 L/min	Brown et al., 1985 Yaniga, 1982 Yaniga et al., 1985
coarse sand with greater than 5% gravel	gradient of 0.015 to 0.02 m/m; flow of 0.61 to 0.91 m/yr	Yaniga and Mulry, 1984
glacial till composed of sand, gravel, and boulders in a silty clay matrix connected to a fractured sandstone		Yaniga et al., 1985
shallow basin containing sand and pea gravel	flow of 27 to 38 L/min	Flatham and Githens, 1985

White et al., 1983). Using a Gram staining procedure, virtually all bacteria can be classified as Gram-positive or Gram-negative dependent on their ability to retain certain dyes. Both Gram-positive and Gram-negative bacterial cell types are found in the subsurface. A size spectrum of waterborne particles (Figure 2) indicates that bacteria range in size from about 0.5 µm to about 8.0 µm. Subsurface bacteria are usually small, less than 1.0 µm in diameter. The microorganisms also tend to be few in number and oligotrophic due to the nutrient limiting conditions predominantly found in the subsurface.

Environmental factors determine whether or not microorganisms survive and/or grow in the subsurface. Temperature is one of the major factors affecting the survival and growth of bacteria. The temperature of the subsurface environment between 9 and 18 meters approximates the mean air temperature of the region (between 3 and 25°C in the United States). The temperature of the upper 10 meters of the subsurface may vary seasonally (McNabb and Dunlap, 1975). Typically, bacterial growth rates increase with increasing temperature up to about

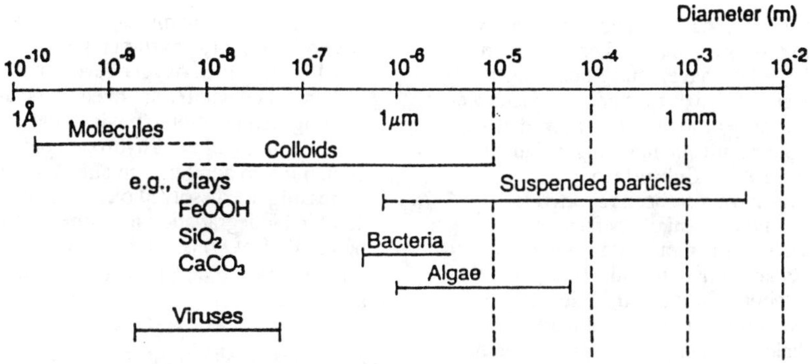

Figure 2. Size Spectrum of Waterborne Particles (McCarthy and Zachara, 1989)

40°C (Figure 3); however, viable microbial populations do exist at lower temperatures. Most bacteria tend to persist in the environment for longer periods of time at the lower temperatures (Alexander, et al., 1986), but their growth rates and subsequent biodegradation of organic contaminants may be limited.

Other environmental factors that govern the survival and growth of microorganisms include pH and adsorption to the subsurface particles and humic substances. Increasing the acidity or alkalinity of the environment can markedly affect the viability of microorganisms. Methanogenic bacteria, which may be involved in the mineralization of some aromatic hydrocarbons, are inhibited at pH values less that 6.0 (Alexander, 1977). Nitrification, the conversion of ammonia to nitrate through microbially catalyzed reactions, is also limited at pH values less that 6.0 and is inhibited at pH values less than 5.0 (Thomas et al., 1987a). In general, it is believed that the adsorption of bacteria onto particle surfaces protects the bacteria from environmental stresses and increases their survival in the subsurface (Alexander, et al., 1986).

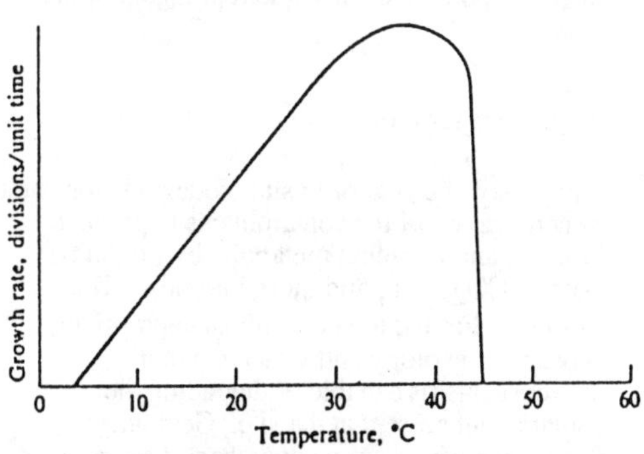

Figure 3. Bacterial Growth Rate as a Function of Temperature (Manahan, 1984)

5 METABOLIC ACTIVITY

Organic material in the subsurface that is available to microorganisms for metabolism may be in low concentrations. The majority of microbes found in the subsurface can survive in nutrient poor environments. In addition to a carbon substrate, subsurface microorganisms also need inorganic nutrients such as nitrogen and phosphorus, as well as a terminal electron acceptor. For aerobic metabolism, oxygen acts as the terminal electron acceptor. For anaerobic metabolism, nitrate, sulfate, carbon dioxide or other inorganic and organic substances may act as the electron acceptor. Presence of these limiting nutrients is necessary for continued metabolic activity and production of more biomass.

When a hydrocarbon spill occurs, the concentration of carbonaceous material available to subsurface microorganisms increases. However, the ratio of carbon to inorganic nutrients such as nitrogen, phosphorus and sulfur is too large to promote sustained bacterial growth. The concentration of oxygen, the terminal electron acceptor in aerobic metabolism, is also usually too low to support sufficient microbial growth for bioremediation. It is assumed that if metabolically appropriate microorganisms are provided with the limiting nutrients, primarily nitrogen, phosphorus and oxygen, that sustained microbial degradation of subsurface contaminants can be achieved. Since nutritional requirements for optimum growth on contaminant hydrocarbons may vary, the types and quantities of supplemental nutrients should be determined using microorganisms and solids taken from the contaminated site (Thomas et al., 1987a).

Many subsurface microorganisms are metabolically active and can utilize a variety of compounds. Simple organic compounds such as glucose, glutamate, and arginine can be degraded by subsurface microorganisms (Larson and Ventullo, 1983). Microorganisms can also be found in the subsurface that degrade a variety of compounds that are of environmental concern. Aerobic degradation of polar solvents such as acetone, isopropanol, ethanol, and methanol by subsurface microorganisms has been reported (Novak et al., 1984; Jhaveri and Mazzacca, 1983). Benzene, toluene, ethylbenzene and m-xylene were found to be partially or totally mineralized at an unleaded gasoline spill site in Granger, Indiana (Thomas et al., 1990). Subsurface microorganisms have also been reported to degrade more environmentally persistent compounds such as the methylated benzenes, chlorinated benzenes (Kuhn et al., 1985), and the chlorinated phenols (Suflita and Miller, 1985). Lipophilic compounds such as naphthalene, methylnaphthalenes, flourene, and phenanthrene are also found to be biotransformed in the subsurface (Wilson et al., 1985; Lee and Ward, 1985).

Subsurface microorganisms in uncontaminated soils may require a long period of acclimation before degrading xenobiotics. Wilson et al. (1985) reported that naphthalene, 1-methyl naphthalene, 2-methyl naphthalene, dibenzofuran and fluorene at 100 to 1000 µg/l were degraded in the subsurface soil in the contaminant plume from a creosote waste pit. Degradation of these compounds was not observed in samples of uncontaminated soil coming from the same site. On the other hand, little or no acclimation period may be required to degrade other xenobiotics. For example, benzene, toluene, and the xylenes were degraded in uncontaminated subsurface soils (Barker and Patrick, 1986). In two uncontaminated anaerobic aquifers, methanol was completely degraded after two months but tert-butanol degraded much more slowly (White et al., 1986).

6 STIMULATION BY NUTRIENT AND TERMINAL ELECTRON ACCEPTOR ADDITION

In situ biodegradation is often limited by low concentrations of the terminal electron acceptor required for metabolic activity. Aerobic biodegradation is dependent upon the concentration of oxygen in the subsurface, while anaerobic biodegradation is dependent upon the presence of other electron acceptors, for example nitrate. However, use of nitrate and oxygen together in aerobic denitrification may be possible (Major et al., 1988; Britton, 1987). In aerobic denitrification, the use of nitrate as the electron acceptor spares oxygen for use as a substrate for degradative enzymes. Emphasis has been placed on in situ biodegradation stimulated by the addition of oxygen to the subsurface since the rate and extent of aerobic hydrocarbon contaminant degradation are greater than that of anaerobic degradation.

Oxygen is often introduced into the aquifer by sparging with compressed air or pure oxygen. However, this method is limited by the low solubility of oxygen in water. The oxygen injected into a contaminated aquifer by this method is rapidly consumed by the microflora. The low dissolved oxygen levels in the contaminated aquifer are likely to be the rate-limiting step in aerobic biodegradation.

The introduction of hydrogen peroxide into a contaminated aquifer is an alternate method of increasing the dissolved oxygen concentration for aerobic biodegradation to occur. Hydrogen peroxide (H_2O_2) is soluble in water in all proportions and decomposes to produce oxygen and water:

$$H_2O_2 \rightarrow H_2O + 1/2\ O_2$$

A decrease in the number of microorganisms may indicate peroxide toxicity because hydrogen peroxide is toxic to microorganisms. Subsurface samples amended with 100 ppm exhibited decreased numbers (Texas Research Institute, 1982). Pretreatment of an aquifer with pure oxygen to acclimate the subsurface microorganisms to higher than normal oxygen concentrations, followed by a step-wise increase in peroxide concentrations may avoid peroxide toxicity problems. This procedure was performed at an aviation fuel spill in Traverse City, Michigan, but some toxicity was observed (Ward et al., 1988). Another limitation of the use of hydrogen peroxide as an alternate oxygen source is that rapid decomposition of the hydrogen peroxide can result in gas formation and plugging of the aquifer if the solubility of oxygen in water is exceeded. In general, the breakthrough of oxygen in monitoring wells downgradient from the injection point indicates that the more water soluble contaminants, such as the monoaromatics, have been removed from that area.

The addition of inorganic nutrients and oxygen into the contaminated site is usually performed using injection and production wells (Figure 4) or infiltration galleries and recovery wells (Figure 5). Nutrients are added first, followed by addition of the oxygen source. The nutrients and oxygen are not added simultaneously during start-up because that may cause microbial growth close to the injection point and subsequent plugging of the aquifer.

7 CONCLUSION

Ultimately, the goal of in situ biodegradation is the mineralization of the contaminants to produce biomass and harmless metabolic by-products, such as CO_2, CH_4 and inorganic salts. The process is limited to sites with an appropriate subsurface geology and which contain microorganisms capable of degrading the contaminant present at the site. Geologic formations which are most amenable to in situ biorestoration are those with low clay content and hydraulic conductivities greater than 10^{-4} cm/sec. Both consolidated and unconsolidated aquifer media can be treated; however, the treatment is

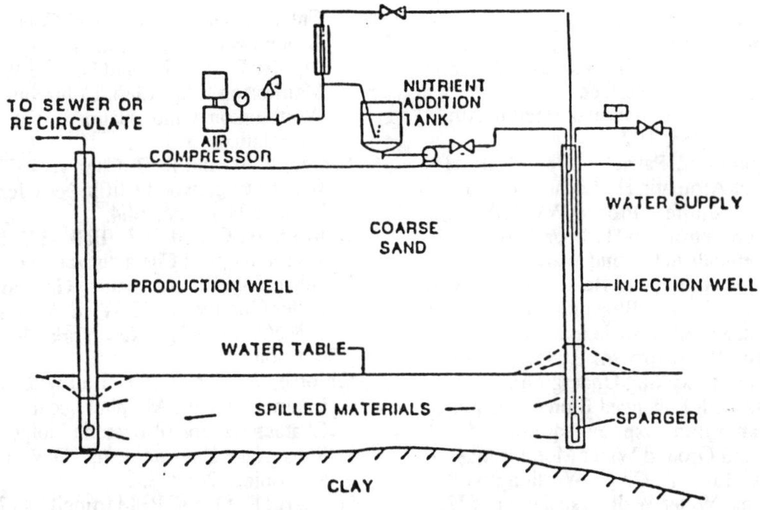

Figure 4. Typical Schematic for Aerobic Subsurface Biorestoration (Raymond et al., 1978)

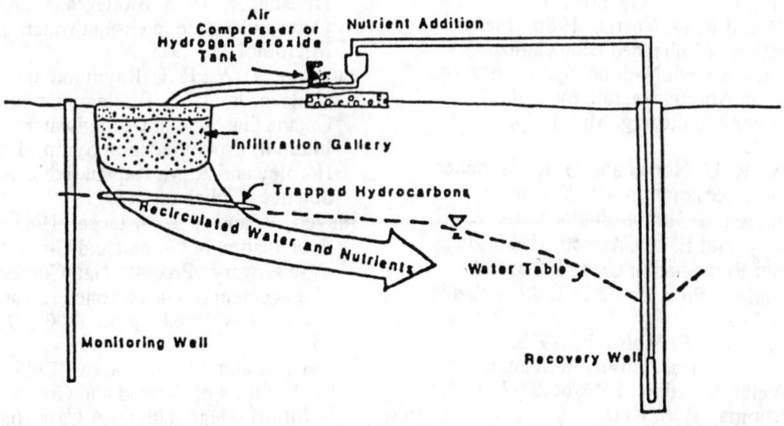

Figure 5. Use of Infiltration Gallery for Recirculation of Water and Nutrients in In Situ Biorestoration (Thomas et al., 1987a)

made more difficult when the formations are lithologically heterogeneous. The treatment is implemented through the introduction of inorganic nutrients and a terminal electron acceptor, usually oxygen, into the subsurface to stimulate the natural microflora.

ACKNOWLEDGEMENT

Although the research described in this article was supported by EPA through Assistance Agreement No. CR-812808 to Rice University, it has not been subjected to Agency review and therefore does not necessarily reflect the views of the Agency, and no official endorsement should be inferred.

The authors are members of the EPA-sponsored National Center for Ground Water Research. Research support from the Brown Family Fund, Houston, Texas, is gratefully acknowledged.

REFERENCES

Alexander, M. 1977. Soil Microbiology. New York: John Wiley and Sons. 467.

Alexander, M., D. A. Andow and J. W. Gillet. 1986. Fate and Movement of Microorganisms in the Environment. Environmental Management. 10: 4: 463-493.

Aller, L., T. Bennet, J. H. Lehr and R. Petty. 1985. DRASTIC: A Standardized System for Evaluating Ground Water Pollution Potential Using Hydrogeologic Set-Up. EPA/600/2-

85/018. Washington, D. C.: U.S.Environmental Protection Agency 163.
Balkwill, D. L. and W. C. Ghiorse. 1982. Characterization of Subsurface Microorganisms. Abst. Annual Meeting Am. Soc. Microbiol. N85: 192.
Barker, J. F. and G. C. Patrick. 1986. Natural Attenuation of Aromatic Hydrocarbons in a Shallow Sand Aquifer. Proc. NWWA/API Conference on Petroleum Hydrocarbon and Organic Chemicals in Ground Water - Prevention, Detection, and Restoration, Nov. 1984, Houston, Texas. Worthington, Ohio: National Water Well Association. 160.
Brenoel, M., and R. A. Brown. 1985. Remediation of a Leaking Underground Storage Tank with Enhanced Bioreclamation. Proc. 5th Nat. Symp. Exp. on Aquifer Restoration and Ground Water Monitoring, May 1985, Columbus, Ohio. Worthington, Ohio: National Water Well Association. 527.
Britton, L. N. 1987 Abstracts of Papers, 1987. 8th Annual Meeting of the Society of Environmental Toxicology and Chemistry, Washington, D. C., 1987. 89: 129.
Brown, R. A. and R. D. Norris. 1986. Field Demonstration of Enhanced Bioreclamation. Presented at the Sixth National Symposium and Exposition on Aquifer Restoration and Ground Water Monitoring, May 1986, Columbus, Ohio.
Brown, R. A., R. D. Norris and G. R. Brubaker. 1985 Aquifer Restoration with Enhanced Bioreclamation. Pollution Engineering. 25.
Brubaker, G. R. and E. L. Crockett. 1986. In Situ Aquifer Remediation Using Enhanced Bioreclamation. Proc. HAZMAT '86. Atlantic City, New Jersey.
Dunlap, W. J. and J. F. McNabb. 1973. Subsurface Biological Activity in Relation to Ground Water Pollution. EPA/660/2-73-014. Ada, Oklahoma: R. S. Kerr Environmental Research Laboratory. 60.
Dunlap, W. J., J. F. McNabb, M. R. Scalf and R. L. Cosby. 1977. Sampling for Organic Chemicals and Microorganisms in the Subsurface. EPA 600/2-77-176. Ada, Oklahoma: U. S. Environmental Protection Agency.
Fiorenza, S., J. M. Thomas, V. R. Gordy, P. S. Ziegenfuss, H. J. Marlow, and C. H. Ward. 1987. Approaches and Concepts for Aquifer Restoration. American Water Works Association. 1987 Annual Conference Proceedings. Part 1: Sessions 1-22.
Flathman, P. E. and G. D. Githens. 1985. In Situ Biological Treatment of Isopropanol, Acetone, and Tetrahydrofuran in the Soil/Groundwater Environment. Groundwater Treatment Technology. E. K. Nyer (ed). New York: Van Nostrand Reinhold Co. 173.
Flathman, P. E., M. J. McCloskey, J. J. Vondrick and D. M. Pimlett. 1985. In Situ Physical/Biological Treatment of Methylene Chloride (Dichloromethane) Contaminated Groundwater. Proc. 5th Nat. Symp. Exp. on Aquifer Restoration and Ground Water Monitoring, May 1985, Columbus, Ohio. Worthington, Ohio: National Water Well Association. 571.
Freeze, R. A. and R. B. Cherry. 1979. Ground Water. Englewood Cliffs, New Jersey: Prentice Hall, Inc. 604.
Ghiorse, W. C. and D. L. Balkwill. 1985. Microbiological Characterization of Subsurface Environments. Ground Water.Quality. C. H. Ward, W. Giger and P. L. McCarty, (eds.) New York: John Wiley and Sons, Inc.
Ghiorse, W. C. and D. L. Balkwill. 1983. Enumeration and Morphological Characterization of Bacteria Indigenous to Subsurface Environments. Dev. Ind. Microbiol. 24: 213.
Hallberg, R. O. and R. Martinell. 1976. Vyrodex: In Situ Purification of Ground Water. Ground Water. 14: 88.
Jamison, V. W., R. L. Raymond and J. O. Hudson, Jr. 1975 Biodegradation of High-Octane Gasoline in Groundwater. Dev. Ind. Microbiol. 16: 305
Jamison, V. W., R. L. Raymond and J. O. Hudson, Jr. 1976 Biodegradation of High-Octane Gasoline in Groundwater. Proc. 3rd Internat. Biodegradation Symp. J. M. Harpley and A. M. Kaplan (eds). Applies Science Publishers. 187.
Jhaveri, V. and A. J. Mazacca. 1983. Bio-Reclamation of Ground and Groundwater: A Case History. Proc 4th Nat. Conference on Management of Uncontrolled Hazardous Waste Sites, Washington, D. C., Oct. 1983. 242.
Jhaveri, V. and A. J. Mazacca. 1985. Bio-Reclamation of Ground and Groundwater by In-Situ Biodegradation: A Case History. Proc 6th Nat. Conference on Management of Uncontrolled Hazardous Waste Sites, Washington, D. C., Nov. 1985. 239.
Kuhn, E. P., P. J. Colberg, J. L. Schnoor, O. Warner, A. J. B. Zehnder and R. P. Schwarzenbach. 1985. Microbial Transformation of Substituted Benzenes During Infiltration of River Water to Ground Water: Laboratory Column Studies. Environ. Sci. Technol. 19: 961.
Larson, R. J. and R. M. Ventullo. 1983. Biodegradation Potential of Ground-Water Bacteria. Proc. 3rd Nat. Symp. on Aquifer Restoration and Ground Water Monitoring, May 1983, Columbus, Ohio. 402.
Lee, M. D. and C. H. Ward. 1985. Biological Methods for the Restoration of Contaminated Aquifers. J. Environ. Toxicol. Chem. 4: 743.
Lee, M. D., J. M. Thomas, R. C. Borden, P. B. Bedient, J. T. Wilson and C. H. Ward. 1988. CRC Crit. Rev. Environ. Control. 18:

29-89.
Leenheer, J. A., R. L. Malcolm, P. W. McKinley and L. A. Ecles. 1974. Occurrence of Dissolved Organic Carbon in Selected Ground-Water Samples in the United States. J. Research U. S. Geological Survey 2: 361-369.
Major, D. W., C. I. Mayfield and J. F. Barker. 1988. Biotransformation of Benzene by Denitrification in Aquifer Sand. Ground Water. 26: 8-14.
Manahan, S. E. 1984. Environmental Chemistry. Monterey, California: Brooks/Cole Publishing Company.
McCarthy, J. F. and J. M. Zachara. 1989. Subsurface Transport of Contaminants. Environ. Sci. Technol. 23: 5: 496.
McNabb J. F. and W. J. Dunlap. 1975. Subsurface Biological Activity in Relation to Ground Water Pollution. Ground Water. 13: 33.
Minugh, E. M., J. J. Patry, D. A. Keech and W. R. Leek. 1983. A Case History: Cleanup of a Subsurface Leak of Refined Product. Proc. 1983 Oil Spill Conference - Prevention, Behavior, Control, and Cleanup, San Antonio, Texas, March 1983. 397.
Nagel, G., W. Kuehn, P. Werner and H. Sontheimer. 1982. Sanitation of Ground Water by Infiltration of Ozone Treated Water. GWF-Wasser/Abwasser. 123: 399.
Novak, J. T., C. D. Goldsmith, R. E. Benoitt and J. H. O'Brien. 1984. Biodegradation of Alcohols in Subsurface Systems. Seminar Programme, Degradation, Retention and Dispersion of Pollutants in Groundwater, Copenhagen, Denmark, Sept. 1984.
Quince, J. R., R. J. Ohneck and J. J. Vondrik. 1985. Response to an Environmental Incident Affecting Ground Water. Proc. 5th Ann. Symp. Exp. on Aquifer Restoration and Ground Water Monitoring, May, 1985, Columbus, Ohio. Worthington, Ohio: National Water Well Association. 598.
Raymond, R. L. 1974. Reclamation of Hydrocarbon Contaminated Ground Water. U.S. Patent 3,846,290.
Raymond, R. L., R. A. Brown, R. D. Norris and E. T. O'Neill. 1986 Stimulation of Biooxidation Processes in Subterranean Formations. U. S. Patent 4,588,506.
Raymond, R. L., V. W. Jamison and J. O. Hudson. 1975. Beneficial Stimulation of Bacterial Activity in Ground Waters Containing Petroleum Products. API Publication No. 4427. Washington, D. C.: American Petroleum Institute.
Raymond, R. L., V. W. Jamison and J. O. Hudson. 1976. Beneficial Stimulation of Bacterial Activity in Ground Waters Containing Petroleum Products. AIChE Symposium Series. 73: 390.
Raymond, R. L., V. W. Jamison, J. O. Hudson, R. E. Mitchell and V. E. Farmer. 1978. Field Application of Subsurface Biodegradation of Gasoline in a Sand Formation. API Publication No. 4430. Washington, D. C.: American Petroleum Institute.
Suflita, J. M. and G. D. Miller. 1985. Microbial Metabolism of Chlorophenolic Compounds in Ground Water Aquifers. Environ. Toxicol. Chem. 4: 751.
Texas Research Institute. 1982. Enhancing the Microbial Degradation of Underground Gasoline By Increasing Available Oxygen. API Publication No. 4428. Washington, D.C.: American Petroleum Institute.
Thomas, J. M., V. R. Gordy, S. Fiorenza and C. H. Ward. 1990. Biodegradation of BTEX in Subsurface Materials Contaminated with Gasoline: Granger, Indiana. Wat. Sci. Tech. 22: 6: 53-62.
Thomas, J. M., M. D. Lee, P. B. Bedient, R. C. Borden, L. W. Canter and C. H. Ward. 1987a. Leaking Underground Storage Tanks: Remediation with Emphasis on In Situ Biorestoration. EPA/600/2-87/008. Ada, Oklahoma: R. S. Kerr Environmental Research Laboratory. 143 pp.
Thomas, J. M., H. J. Marlow, R. L. Raymond and C. H. Ward. 1987b. Hydrogeologic Considerations for In Situ Biorestoration. US/USSR Symposium. Fate of Pesticides and Chemicals in the Environment, University of Iowa, Iowa City, Iowa.
Thomas, J. M. and C. H. Ward. 1989. In Situ Biorestoration of Organic Contaminants in the Subsurface. Environ. Sci. Technol. 23: 760-766.
Waksman, S. A. 1916. Bacterial Numbers in Soils, at Different Depths, and in Different Seasons of the Year. Soil Sci. 1: 363-380.
Ward, C. H., J. M. Thomas, S. Fiorenza, H. S. Rifai, P. B. Bedient, J. M. Armstrong, J. T. Wilson and R. L. Raymond. 1988. A Quantitative Demonstration of the Raymond Process for In Situ Biorestoration of Contaminated Aquifers. Proceedings of the Petroleum Hydrocarbons and Organic Chemicals in Ground Water - Prevention, Detection and Restoration, November, Houston, Texas. Dublin, Ohio: NWWA/API. 723-43.
Webster, J. J., G. J. Hampton, J. T. Wilson, W. C. Ghiorse and F. R. Leach. 1985. Determination of Microbial Cell Numbers in Subsurface Samples. Ground Water. 23: 17.
Westray, M. S., R. A. Brown and R. D. Norris. 1985. Groundwater Microbiology and Pollution Control. Princeton, New Jersey: FMC Aquifer Remediation Systems. 25.
White, D. C., G. A. Smith, M. J. Gehron, J. H. Parker, R. H. Findley, R. F. Martz and H. L. Fredrickson. 1983. The Ground Water Aquifer Microbiota: Biomass, Community Structure, and Nutritional Status. Dev. Ind. Microbiol. 24: 201.
White, K. D., J. T. Novak, C. D. Goldsmith and

S. Beven. 1986. Microbial Degradation Kinetics of Alcohols in Subsurface Systems. Proc. NWWA/API Conference Petroleum Hydrocarbons and Organic Chemicals in Ground Water - Prevention, Detection, and Restoration, Nov. 1985, Houston, Texas. Worthington, Ohio: National Water Well Association.

Wilson, J. T., J. F. McNabb, D. L. Balkwill and W. C. Ghiorse. 1983. Enumeration and Characterization of Bacteria Indigenous to a Shallow Water Table Aquifer. Ground Water. 21: 134.

Wilson, J. T., J. F. McNabb, J. W. Cochran, T. H. Wang, M. B. Tomson and P. B. Bedient. 1985. Influence of Microbial Adaptation on the Fate of Organic Pollutants in Ground Water. Environ. Toxicol. Chem. 4: 721.

Yaniga, P. M. 1982. Alternatives in Decontamination for Hydrocarbon-Contaminated Aquifers. Proc. 2nd Nat. Symp. on Aquifer Restoration and Ground Water Monitoring, May 1982, Columbus, Ohio. D. M. Nielsen (ed). Worthington, Ohio: National Water Well Association. 47.

Yaniga, P. M. and J. Mulry. 1984 Accelerated Aquifer Restoration: In Situ Applied Techniques for Enhanced Free Product Recovery/Absorbed Hydrocarbon Reduction Via Bioreclamation. Proc. NWWA/API Conference on Petroleum Hydrocarbons and Organic Chemicals in Ground Water - Prevention, Detection and Restoration, Nov. 1984, Houston, Texas. Worthington, Ohio: National Water Well Association.

Yaniga, P. M., W. Smith and R. L. Raymond. 1985. Biodegradation of Organic Compounds. Enhanced Techniques for Diffusion of Oxygen in the Groundwater System. Chemical Manufacturers Association, Washington, D. C.

Zienkiewicz, A. W. 1985. Removal of Iron and Manganese from Ground Water with the Vyredox[T] Method. Proc. 2nd Internat. Conference on Ground Water Quality Research. N. N. Durham and A. E. Redelfs (eds). Stillwater, Oklahoma: University Printing Services. 74.

Subsurface Contamination by Immiscible Fluids, Weyer (ed.) © 1992 Balkema, Rotterdam. ISBN 90 6191 175 3

An overview of aquifer remediation: NAPLs, pump and treat, and bioremediation

Stephen G. Schmelling
United States Environmental Protection Agency & Robert S. Kerr Environmental Research Laboratory, Ada, Okla., USA

ABSTRACT: Pump and treat and bioremediation are two technologies that may be used for remediation of contaminated ground water. For either one to be used successfully it is important for the practitioner to understand the physical, chemical, and biological processes controlling contaminant fate and transport at the site to be remediated. Failure to do so may result in poorly designed systems that fail to perform satisfactorily. When separate, non-aqueous phase liquids are present remediation is likely to be very difficult and may require pumping for thousands of years.

1. INTRODUCTION

When contaminant liquids, such as fuels or solvents, commonly referred to as non-aqueous phase liquids (NAPLs), are released at or near the earth's surface several processes affect their rate of movement and eventual fate. One fraction of the contaminant liquid will volatilize and be transferred to the atmosphere; another fraction will be trapped in the pore spaces of the vadose zone and be held as residual saturation; and a third fraction will migrate far enough to reach ground water and become a real or potential threat to public health.

There is no single best method for cleaning up an aquifer. The choice of the method to use will depend on the contaminants to be cleaned up, the level of cleanup to be attained, and the specific characteristics of the contaminated site. Before choosing a specific remediation technology one should understand the strengths and limitations of each method, and the characteristics of the specific problem to be addressed. Two options for restoring the aquifer to its beneficial use are extraction and treatment (pump and treat) and bioremediation (McKay and Cherry 1989; Thomas and Ward 1989). Another common approach for dealing with contaminated ground water is to treat the water at the point of distribution and institute administrative controls to ensure that the water is not used without treatment. This presentation will focus on pump and treat and bioremediation.

Aquifer cleanup consists of several basic activities. One activity is the removal of any remaining free product. NAPLs that are less dense than water (LNAPLs) tend to remain near the water table and can often be found and extensively removed quite effectively. NAPLs that are more dense than water (DNAPLs) will continue to migrate through the aquifer and are much more difficult to locate and remove. In either case, the more free product that can be removed, the more effective the ground-water remediation will generally be. A second activity is the remediation of the contaminated ground water in the saturated zone. This activity is the focus of this paper. A third activity is the removal of contaminants remaining in the vadose zone. If this latter activity is not carried out, water infiltrating through the vadose zone will continue to leach the contaminants into the ground water.

An important remedial activity is to attain hydraulic control of the contaminant plume to prevent further contamination outside the source area. Hydraulic control, or limitation of the movement of the contaminant plume, must always be accomplished before any other remedial activity can succeed. This is true for both pump and treat and bioremediation.

In 1989, the EPA studied 112 pump and treat systems at Superfund sites, RCRA corrective action sites, and state-lead sites (USEPA 1989). They found 19 with sufficient operating experience to make a judgement about effectiveness of the remedial action in reaching its original goals. While almost all of the actions were successful in controlling further migration of the contaminant plume, only one claimed to have restored the aquifer to beneficial use, although several had this as their original goal. Even this claim was in doubt because of limited measurement data to back it up.

A study of bioremediation efforts (Stap 1989) showed that while it is a promising technology for aquifer cleanup, particularly for sites contaminated with petroleum hydrocarbons, the rate of success in restoring contaminated aquifers to drinking water quality is still very low.Some of the reasons for this limited rate of success for both technologies are:

1. Inadequate characterization of the site before undertaking the remedial action.

2. Failure to recognize and understand

which to base further actions.

4. Gaps in the scientific and engineering knowledge required to design and complete successful remedial actions.

Specific problems that were frequently cited in the EPA study of pump and treat systems included: (1) aquifer heterogeneity and spatial variability of aquifer parameters; (2) inadequate knowledge of the interaction of the contaminants with the geologic matrix; and (3) the presence of immiscible fluid phases.

The first part of this paper addresses each of these topics and their effect on the success or failure of aquifer remediation efforts. The paper then addresses the effect of these phenomena on pump and treat through the use of a simple example. The paper concludes with a discussion of the effect of these phenomena on in-situ bioremediation.

2. FACTORS LIMITING REMEDIATION EFFECTIVENESS

2.1 *Aquifer heterogeneity*

The geology of the subsurface is complex, and all aquifers exhibit some degree of heterogeneity and anisotropy. This complexity is a result of geologic processes that have acted over long periods of time. Some idea of the complexity of the subsurface can be gained from observations of road cuts, cliffs, and other places where the subsurface is exposed. One can see obvious layers of different types of materials such as clays, sands, and gravels. One can also see examples of rocks which have fractures running through them. An interesting aspect of this heterogeneity, and one that is important for aquifer remediation, is that it seems to persist down to very small scales, e.g. 10's of centimeters or less. Understanding this variability is important for the successful outcome of a remedial action.

Contaminants dissolved in ground water are transported principally by advective flow as the water moves through the subsurface. The rate at which water flows through the subsurface is described by Darcy's Law:

$$q = k\frac{\Delta h}{\Delta l} \qquad (1)$$

where q = specific discharge $(L^3/T)/(L^2)$; K = hydraulic conductivity (L/ T); and $(\Delta h/\Delta l)$ = hydraulic gradient (L/ L).

Variations in hydraulic conductivity between different geologic materials cover more than ten orders of magnitude. The difference in hydraulic conductivity between a sand layer and a clay layer is about four or five orders of magnitude.

A number of investigators (e.g. Molz 1990) have recently used sensitive flow meters to measure the vertical variation in hydraulic conductivity at several different sites. What they have found, even in aquifers that would have been considered to be more or less homogeneous, is that the hydraulic conductivity shows variations of a factor of ten or more over distances as small as one foot, the limit of the measuring instrument. It seems likely that if one could get better resolution in the data, this variation would continue down to much smaller intervals. Horizontal variations in hydraulic conductivity typically occur on a scale that is about ten times that found for the vertical direction.

Figure 1a shows a hypothetical vertical section through an aquifer composed of several different geologic materials. If two adjacent layers in an aquifer have hydraulic conductivities that differ by a factor of 10, then the specific discharge in the one with the larger hydraulic conductivity will be 10 times as great as that in the layer with the smaller hydraulic conductivity. This is illustrated schematically in Figure 1b. In fact, the advective flow of water through a clay layer is so small that for any reasonable value of hydraulic gradient the primary process for contaminant movement is molecular diffusion rather than advective transport. This is illustrated in Figure 1c.

For water supply purposes, a value of K averaged over the depth of the aquifer, such as measured by traditional pumping tests, is perfectly adequate. However, for purposes of aquifer remediation, the rapid variations in hydraulic conductivity can limit the efficacy of both pump and treat and bioremediation.

If the contaminants have moved into the aquifer largely through downward vertical movement, or have been in the aquifer for a long time, they will be found in both the layers with low hydraulic conductivity such as the clay, and the layers with high hydraulic conductivity such as sands. When water is pumped through the system to flush out the contaminants, much more water will flow through the regions of high hydraulic conductivity than through the regions of low hydraulic conductivity. This means that while the contaminants will be flushed from the regions of high hydraulic conductivity in a reasonably short period of time, the contaminants in the low hydraulic conductivity regions will only be removed after they have diffused out of the regions of low hydraulic conductivity into the regions of high hydraulic conductivity. Diffusion is a slow process and the contaminants may continue to bleed out for long periods of time.

The result of this phenomenon is that if one were to monitor the concentration of the contaminant in the aquifer, and plot a graph of concentration *vs.* time, one would see a rapid dropoff in concentration, followed by a slowly decreasing tail that might appear to be almost unchanging with time. If the concentration at which the graph levels off is above the maximum contaminant level (MCL) or other action level, then one might pump for years without achieving the goal of the remedial action. This rapid decrease in contaminant concentration followed by a slow or almost imperceptible decrease in concentration was observed at a number of sites included in the EPA study mentioned above.

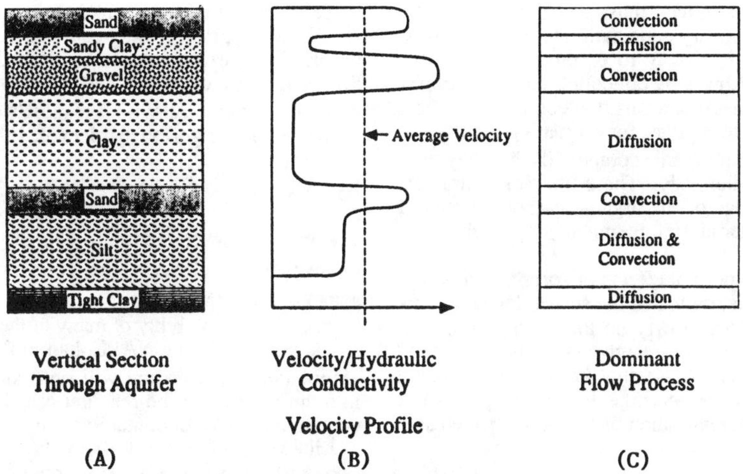

Fig. 1 Effects of geologic stratification on tailing.

The effect on biodegradation is similar. Nutrients and oxygen that are added to the system will mix with the contaminants primarily in the regions with high values of hydraulic conductivity, but will mix much less well in the regions with low values of hydraulic conductivity. Other aquifer properties that affect both pump and treat and in-situ bioremediation, although not as well studied as hydraulic conductivity, are characterized by similar variability.

A second common type of heterogeneity that can have a major effect on contaminant transport and fate, and the success of aquifer remediation, is fractured rock. In fractured rock, the water and contaminants move principally through the connected fractures while the contaminants may be found in dead-end fractures and the rock matrix as well as in the connected fractures. The effect of this on pump and treat is similar to that described above for a layered system. Characterizing the flow paths through a fractured rock formation is difficult, and the ability to predict the flow of water and contaminants in a such a geologic formation is in a primitive state Bioremediation in a fractured rock aquifer presents some very challenging problems, and there is very little experience to date in attempting to bioremediate a contaminated fractured rock aquifer.

2.2 *Sorption and partitioning*

As the contaminants in the ground water move through the geologic matrix, both metals species and organic compounds will be adsorbed on it (Weber 1990). The relationship between the concentration of the material in solution and the amount sorbed depends on the characteristics of the chemical contaminant, the chemical properties of the ground water, and the properties of the geologic matrix. In many cases this relationship is approximately linear and can be expressed as:

$$S = K_d C_a \qquad (2)$$

where S = the amount sorbed per unit mass of the solid phase (M/M); C_a = the equilibrium concentration in the solution phase (M/L^3); and K_d = the partition coefficient (L^3/M).

The value of K_d is often determined in a laboratory "batch isotherm measurement" which is a relatively straightforward procedure as long as there are no non-aqueous, separate phase liquids present. An example of the results of such a measurement is shown in Figure 2.

Sorption must be accounted for when estimating the contaminant mass to be removed. Because of sorption, the amount of material adsorbed on to the aquifer solids may be as great or greater than that in solution. When there is significant sorption,

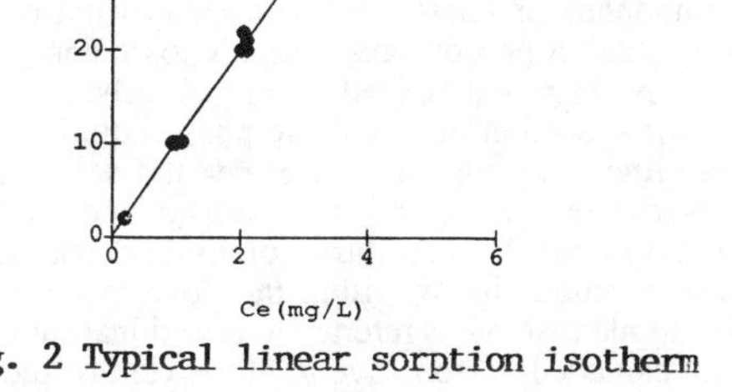

Fig. 2 Typical linear sorption isotherm

estimates of contaminant mass based only on water samples will seriously underestimate the mass to be removed. The best way to estimate the mass of sorbed contaminant is by collecting cores of aquifer material and measuring a partition coefficient. There are also indirect techniques for estimating K_D that can be used for neutral organic compounds, but they need to be applied judiciously. These indirect techniques require knowledge of the organic carbon content of the sorbing material and chemical properties of the contaminant.

One of the important effects of sorption on pump-and-treat remediation is that the sorbed contaminants tend to move more slowly on the average than the ground water or the non-reactive contaminants. If V' is the rate of advance of the sorbing contaminant front, and V is the average linear ground water velocity, then the retardation factor, R, is defined as

$$R = V/V' = 1 + (\rho/n)\,K_d \qquad (3)$$

where ρ = the bulk density of the solid, and n = the porosity of the aquifer medium.

The effect of retardation on pump and treat remediation is to extend the time required to remove the contaminants by a factor roughly equal to the retardation factor. Values of R range from one for non-reacting contaminants such as chlorides to 10 or more for hydrophobic organic compounds. Equation 2 describes an equilibrium condition. The kinetics of sorption and desorption may also be important in controlling the rate at which contaminants can be removed by pump and treat (Brusseau and Rao 1989). Sorption kinetics are not measured in batch experiments, but flow-through column experiments can provide this type of information.

The specific effects of sorption on bioremediation are the subject of current research. For instance, sorption is believed to affect the availability of contaminants for bioremediation. Finally, it should be noted that sorption to aquifer material may be strongly enhanced by small amounts (<1%) of anthropogenic organic carbon such as that which might be present at a fuel contamination site.

2.3 *Immiscible fluids*

The presence of NAPLs can have a significant effect on the outcome of a remediation project. NAPLs may be toxic in themselves, such as TCE, or they may act as a reservoir for other hydrophobic compounds. For example, gasoline consists of a large proportion of compounds of relatively low toxicity (fuel carbon) with smaller proportions of highly toxic compounds such as benzene dissolved in the fuel carbon.

Some portion of a separate phase liquid, such as gasoline, moving through either the saturated or unsaturated zone of the subsurface will be immobilized by capillary forces and trapped as discontinuous blobs within the pore spaces. This immobile material is referred to as residual saturation and can occupy from five to 40 percent of the pore space. Contaminants that are dissolved in the residual NAPL will partition between the NAPL and the water. This partitioning can be defined in terms of a dimensionless oil/water partition coefficient K_p defined in analogy with the soil/water partition coefficient K_d (e.g. Bouchard, et al. 1989).

$$K_p = C_o / C_a \qquad (4)$$

where C_o = the concentration in the non-aqueous phase (M/L^3) and C_a = the concentration in the aqueous phase (M/L^3).

Because the solubility of many of the contaminants is much higher in the NAPL than in the water (K_p > 1000), the contaminants will only slowly dissolve into the water, and the residual NAPL will act as a continuous source of contamination for a long time.

Likewise, a floating LNAPL layer or a DNAPL pool will act as a long-term source of contamination for ground water. Some of the contaminants will diffuse out of a floating NAPL layer into ground water, but the transfer of contaminants from the LNAPL to ground water appears to occur most predominantly in response to fluctuations in the water table. For this reason, it is important to remove as much free-product as possible.

Since a DNAPL is more dense than water, it will continue to move downward until it reaches an impenetrable barrier. As it does so, however, some of the DNAPL will remain trapped as residual saturation in the saturated zone. The residual saturation cannot be removed by pumping and will remain a source of contaminants for many years. Even the DNAPL that pools on the bottom of the aquifer usually cannot be effectively removed by pumping. Of all the problems encountered in implementing an aquifer remediation project, DNAPL's are among the most difficult to deal with.

NAPLS'S are not usually measured directly in water samples, although they may be, if they are present in large amounts and the monitoring well is screened in a NAPL layer or pool. Indirect evidence of the presence of NAPL's is provided by concentrations of hydrophobic contaminants that are present at or near their limit of solubility. At the present time, the best way to estimate the quantity of NAPL present in an aquifer is probably to collect cores of aquifer material and make laboratory measurements on them.

The effect of NAPL's on aquifer restoration can be disastrous, particularly if the goal of the restoration is to return the water to drinking water quality. Planning for aquifer remediation that does not take into account the presence of the NAPL's that are at the site will seriously underestimate the time required to complete the remedial effort. Because NAPL's act as continuing sources of contamination and are only very slowly removed by pumping, they can stretch out the time required to remediate an aquifer by a factor of 100 or more. Thus, a remediation which might be expected to take five years in the absence of NAPL's might take 500 or more years if NAPL's are present. NAPL's will also slow biodegradation because the NAPL's themselves are not readily biodegraded and will reduce the bioavailability of compounds dissolved in them.

3 PUMP AND TREAT SUMMARY EXAMPLE

The effects of aquifer heterogeneity, sorption, and NAPL's on a pump- and-treat cleanup can be summarized with the following example (Hall, 1988). Consider a contaminated site with the following characteristics.

Area of contaminated aquifer = 10 acres
Thickness of contaminated aquifer = 15 feet
Aquifer porosity = 0.3.

The pore volume (occupied by the water) of the aquifer is equal to (10 acres)X(43,560 ft^2/acre)X(15 ft)(0.3) = 1,960,200 ft^3. If water is pumped at a rate of 100 gal/min = (100 gal/min)X(0.134 ft^3/gal) = 13.3 ft^3/min, then the time required to flush the aquifer one time would be equal to (1,960,200 ft^3/13.3 ft^3/min) = 147,400 min = 102 days. If this aquifer was composed of fairly uniform sand and the remedial effort was attempting to remove a non-sorbing contaminant, such as chloride, most of it might be removed after flushing one pore volume of water through the aquifer. In reality, because of variability in the pore-size distribution and other heterogeneities, it would probably require flushing several pore volumes through the aquifer to reduce the contaminants to an acceptable level. This means that the cleanup time would be a year or two which is a perfectly reasonable number.

If the contaminant sorbs to the soil the time required to remove it will increase roughly in proportion to the retardation factor. If the soil has a bulk density of 1.6 g/cm^3 and a K_d of 0.75, the retardation factor, which can be calculated from the formula above, is 5.0. Thus, the one to two year cleanup time will be five to ten years which is about the upper limit of a reasonable time to conduct a pump and treat remediation.

Table 1. Results of the example calculations of remediation times.

Problem	Remediation Time
Dissolved contaminants only	1 to 2 years
Sorption on aquifer solids	5 to 10 years
Residual saturation present	>1000 years

When the contaminant is dissolved in the residual saturation of an oily phase the problem becomes much worse. If the contaminant is toluene, where the K_p is on the order of 3000, and the residual saturation occupies 30 percent of the pore space, then the time required to clean up the aquifer will increase by a factor of about 2000. Thus, the one year cleanup time will become thousands of years. This is obviously not feasible. Increasing the pumping rate to 1000 gal/min would reduce it to hundreds of years, but that is still to long to be practical. Table 1 summarizes the results of these example calculations.

4.PUMP-AND-TREAT SUMMARY

We can see from this example that while pump and treat may be a very good aquifer remediation technology in some situations, it is not effective in others. Pump and treat will be most effective when

- the aquifer is hydrologically homogeneous
- the contaminants are highly soluble
- the contaminants are non-sorbing or weakly sorbing to the aquifers solids, and
- there is no residual saturation from NAPL's.

The less well that these conditions are met, the lower the chances for a successful remediation by pump and treat.

Even when these conditions are not satisfied, pump and treat can be an effective technique for controlling the migration of the contaminant plume and for the removal of a separate phase floating on top of the water table. The first step in any remedial action where pump and treat is being considered is to conduct a thorough site characterization study to determine how well these conditions are satisfied.

5. CONSIDERATIONS FOR BIOREMEDIATION

In-situ bioremediation usually involves stimulating indigenous subsurface microorganisms to degrade organic contaminants in place by the addition of nutrients and a terminal electron acceptor. The goal of in-situ bioremediation is to convert the organic waste in to harmless end products such as CO_2, H_2O, CH_4, and N_2. The process is similar in principle to that used in surface bioreactors. But while the process is similar in principle, it is much more difficult to implement in practice.

Two reactions that lead to the mineralization of benzene are

$$C_6H_6 + 7.5O_2 \rightarrow 6CO_2 + 3H_2O \quad (5)$$

or

$$C_6H_6 + 6HNO_3 \rightarrow 6CO_2 + 6H_2O + 3N_2. \quad (6)$$

In the first example oxygen is the terminal electron

acceptor, while in the second it is nitrate. There are advantages and disadvantages to either choice.

Although shown as a single reaction, these transformations may proceed through a series of intermediate products before reaching the mineralization endpoint. There are situations when the intermediate products of biodegradation may be more hazardous than the initial contaminant. A well known example is the conversion of trichloroethene (TCE) to vinyl chloride. Finding these intermediate products in water samples from the contaminant site is good evidence that natural biodegradation is taking place.

The metabolic reactions shown above require the addition of an electron acceptor such as nitrate or oxygen and can usually be effected for simple aromatic and alkane compounds. Aerobic metabolism of chlorinated solvents such as TCE appears to require the addition of an electron donor such as methane as well as an electron acceptor. Microorganisms that use methane as an energy source can degrade the chlorinated solvent through a cometabolic process. The metabolic pathways for these cometabolic processes are the subject of current research.

For conditions where biodegradation is feasible it has some significant advantages over other remediation technologies. One advantage is that biodegradation often results in the conversion of the contaminants to non-toxic end products. A second advantage is that biodegradation is cost effective in comparison to other alternatives. Part of this cost advantage results from a third advantage, and that is that biodegradation is usually quite rapid once it starts.

Biodegradation, however, is not an appropriate technology for all contaminated ground water. There are requirements on the geology of the site and the degradability of the contaminants that must be met if bioremediation is to be successful. Even when these conditions are met, careful site characterization and system design and operation are necessary if the project is to succeed. In summary these requirements are (Thomas and Ward1989)

- A perfusable aquifer
- Adapted microorganisms
- Adequate quantities of electron acceptors
- An appropriate nutrient mix
- Appropriate conditions of pH, temperature, lack of toxicants, and so forth.

Each of these conditions is examined in more detail below.

The first essential requirement is that the aquifer have a sufficiently high hydraulic conductivity, or permeability, to allow one to perfuse the contaminated region with a solution of the necessary nutrients and electron acceptors. Typically, this requires a hydraulic conductivity of at least 10^{-4} cm/sec. Just as in the case of pump and treat, variability in hydraulic conductivity can cause an in-situ biodegradation system to operate inefficiently, and it is best to measure the value of K at several locations within the contaminated zone. In fact, the site characterization requirements for in-situ biodegradation are at least as stringent as they are for pump and treat remediation.

The second requirement for in-situ biodegradation is that there be a population of microorganisms that are capable of degrading the contaminant of interest. This is best determined by laboratory studies using samples of subsurface aquifer material from several locations at the site. These treatability studies can determine whether or not organisms in the aquifer material have the potential to degrade the contaminants, what the nutrient requirements of these organisms are, and what effect other environmental factors, such as pH and temperature, will have on the biodegradation process. Rate information from laboratory studies is not readily transferable to field conditions.

A third requirement is to provide an electron acceptor, typically oxygen. Transport of oxygen to the microorganisms is the rate-limiting factor for most in-situ bioremediation projects. For aerobic biodegradation of benzene, as illustrated above, the stoichiometric oxygen requirement is approximately 3.1 milligrams of oxygen for each milligram of benzene to be degraded. If benzene is present at a concentration near its solubility limit of 28 mg/L (from typical fuels), the amount of oxygen naturally present in ground water, ≤ 8 mg/L, will be quickly depleted, and additional oxygen must be supplied if all of the benzene is to be degraded in a reasonable period of time. This oxygen can be supplied as pure oxygen gas that is diffused into the subsurface, or as hydrogen peroxide. Hydrogen peroxide is the most common choice, although oxygen has also been used. If the contaminant concentration is sufficiently low, such as would be found away from the source area, natural biodegradation may remediate the plume without the addition of oxygen.

The fourth requirement is that the proper nutrient mix be available to the microorganisms. The appropriate mix is usually determined through laboratory treatability studies, rather than relying on theoretical ratios. It is important to consider the interactions of the oxygen and nutrients with the geologic matrix in addition to the nutritional needs of the microorganisms. For example, if oxygen is added to water with significant amounts of dissolved iron, precipitation may cause the system to fail. In addition to the proper nutrient mix, one must also have the appropriate conditions of temperature, pH, etc., and a lack of toxicants, such as heavy metals, that may inhibit biodegradation.

Finally, the electron acceptors and nutrients must be mixed with the contaminants and microorganisms. This is a difficult engineering problem since, among other reasons, the reaction is taking place in a reactor about which there is incomplete knowledge and over which there is poor control. In homogeneous sands, hydrodynamic dispersion is often not adequate to ensure good mixing, while in heterogeneous soils, preferential flowpaths will prevent effective mixing. A variety of schemes to promote mixing have been tried with varying degrees of success.

The steps required to carry out a bioremediation project can be summarized as follows:

- Physically remove any free product that is present
- Determine the geologic, hydrologic, physical, and chemical characteristics of the site.
- Demonstrate the presence of adapted or adaptable microorganisms
- Determine the requirements for electron acceptors and nutrients
- Design and build the system
- Operate the system
- Monitor the performance of the system, preferably using monitoring wells and cores from the source area

In-situ bioremediation of contaminated ground water is an emerging technology that shows a great deal of promise. Like all aquifer remediation technologies, it will be effective in some situations and not in others. In fact, many of factors that limit the effectiveness of pump and treat, such as aquifer heterogeneity and regions of low hydraulic conductivity, and the presence of separate phase fluids, also limit the effectiveness of bioremediation.

In-situ bioremediation of ground water has been most successful with contamination from petroleum hydrocarbons. Because of inadequate monitoring of the operation, it has not been possible in some cases to determine to what extent the removal of the contaminant was due to biological degradation, and to what extent it was due to physical removal as nutrients and oxygen were perfused through the system. In-situ bioremediation of chlorinated solvents is still largely in the research stage although researchers are rapidly moving up the learning curve. As with any technology, an important requirement of a successful operation is to use in-situ bioremediation in-situations where the requirements for success are present, and to choose some other technology when they are not. This, in turn, requires that one conduct a through site characterization to provide the data needed to put the decision on a sound scientific and engineering basis.

DISCLAIMER

Although the information in this document has been funded by the United States Environmental Protection Agency, it has not been subject to the Agency's peer and administrative review process. It does not necessarily reflect the views of the Agency and no official endorsement should be inferred.

ACKNOWLEDGEMENT

I would like to acknowledge the assistance of Carl Enfield for suggesting the summary example.

REFERENCES

Bouchard, D.T., C.G. Enfield, and M.D. Piwoni 1989. Transport processes involving organic chemicals. In B.L. Sawheny and K. Brown (eds.), Reactions and Movement of Organic Chemicals in Soils. Madison

Brusseau, M.L. and P.S.C. Rao 1989. Sorption nonideality during organic contaminant transport in porous media. Critical Reviews in Environmental Control. 19: 33-99.

Hall, C.W. 1988. Practical limits to pump-and-treat technology for aquifer remediatioı Groundwater Quality Protection Preconference Workshop Proceedings; Dallas. Water Pollution Control Federation

McKay, D.M. and J.A. Cherry 1989. Groundwater contamination: pump-and treat remediation. Envir. Sci. and Tech. 23: 630-636

Molz, F.J., O. Guven, and J.G. Melville 1990. A new approach and methodologies for characterizing the hydrogeologic properties of aquifers. EPA/600/2-90-02

Staps, J. J. M. 1989. International evaluation of in-situ biorestoration of contaminated soil and groundwater. In: "Report of the Third International Meeting of the NATO/CCMS Pilot Study on the Demonstration of Remedial Action Technologies for Contaminated Land and Groundwater." Montreal, Canada

Thomas, J. M., and C. H. Ward. 1989. In- situ biorestoration of organic contaminants in the subsurface. Envi. Sci. and Tech. 23: 760-766.

United States Environmental Protection Agency. 1989. Evaluation of ground-water extraction remedies; Volume 1: Summary report. EPA/540/2-89/054

Weber, W. J., Jr. 1990. Sorption processes and their effects on contaminant fate and transport in subsurface systems. Submitted to Water Research, February 1990

2 Phase behaviour of immiscible fluids

Subsurface Contamination by Immiscible Fluids, Weyer (ed.) © 1992 Balkema, Rotterdam. ISBN 90 6191 175 3

Organic liquid interactions with water-wet barrier clays

R.M.Quigley
Faculty of Engineering Science, The University of Western Ontario, London, Ont., Canada

F.Fernandez
Golder Associates Ltd, London, Ont., Canada

ABSTRACT: Hydrophobicity and surface tension effects make water-compacted clays remarkably resistant to penetration by insoluble liquid hydrocarbons. Large heads may be required to produce breakthrough of liquids like benzene, cyclohexane and toluene. After breakthrough, huge variations in **k** are produced depending on the applied gradient as the flow channels expand and contract.

Water-soluble alcohols produce dramatic increases in **k** only at concentrations exceeding 70% when the effects of viscosity are reduced and double layer contraction causes large increases in free porewater space.

The presence of alcohol, either in the soil prior to cyclohexane permeation or mixed with cyclohexane, greatly facilitates its entry and allows large immediate increases in **k** not possible with pure cyclohexane.

Effective stresses, if present on the water-wet clay, greatly inhibit the expansion of flow channels in the case of pure cyclohexane flow. In the case of ethanol or ethanol/cyclohexane flow, the presence of effective stresses permits chemically-induced consolidation that greatly restricts or eliminates the increases in **k** observed at low effective stresses.

1 INTRODUCTION

During the past decade a considerable amount of work has been published on the effect of organic liquids on the hydraulic conductivity, **k**, of clay barriers. The published studies generally indicate that certain concentrated water-soluble organic liquids can cause dramatic increases in **k** of water-compacted clays. These liquids include acetone (Brown and Anderson, 1980; Brown et al, 1983; Anderson et al, 1985; Acar et al, 1985), methanol (Fernandez and Quigley, 1985; Foreman and Daniel, 1986; Storey and Peirce, 1989; Uppot and Stephenson, 1989), ethanol (Fernandez and Quigley, 1985, 1988, 1991), ethylene glycol (Brown and Anderson, 1980) and dioxane (Michaels and Lin, 1954; Fernandez and Quigley, 1988, 1991). The increases in **k** have been largely attributed to the creation of free porewater space resulting from the collapse of double layers around clay particles in the presence of liquids less polar than water.

Studies involving water-insoluble (hydrophobic) organic liquids have shown either increases or decreases in the hydraulic conductivity of water-wet clay barriers, depending on the testing procedures and equipment used. Extensive laboratory studies carried out by Brown and his co-workers at Texas A&M University (Brown and Anderson, 1980; Anderson and Brown, 1981; Anderson et al, 1982; Brown et al, 1983) showed that insoluble solvents such as xylene and heptane apparently caused large increases in the hydraulic conductivity of water-compacted clays. However, it is not known if the testing procedures followed provided adequate control of the influent chemistry, the compaction procedures and the contact between the soil sample and the rigid-wall permeameters used. It is thus possible that the increases in **k** resulted from fracture flow, side-wall leakage or even unexpected chemical interaction discussed in this paper. Studies by other researchers have shown that forced permeation of insoluble organic permeants does not necessarily cause increases in **k** if adequate precautions are taken to prevent side-wall leakage. The permeant liquids studied include benzene and nitrobenzene (Acar et al, 1985), benzene, cyclohexane and xylene (Fernandez and Quigley, 1985), heptane

(Foreman and Daniel, 1986), and xylene (Uppot and Stephenson, 1989). It is particularly important to note that the flow of insoluble hydrocarbons through water-compacted clays is confined to channels formed by displacement of free interpedal water. At the end of testing, only 5 to 8% of the pore space is ever occupied by the organic liquid and there is no significant alteration of the soil structure (Fernandez and Quigley, 1985).

A preliminary look at the important role surfactants play in facilitating the entry of insoluble organics into and through clayey barriers was presented by Quigley and Fernandez (1989). Five percent soap added to cyclohexane allowed slow, unrestricted entry of the cyclohexane/soap mixture, whereas pure cyclohexane was held back as stationary penetration fingers until inlet heads of ~ 3.5 m were reached.

The purpose of this paper is to take a preliminary look at the equally troubling problem of association liquids. Such liquids, which include the alcohols and acetone, render insoluble organics such as cyclohexane soluble in water over certain concentration ranges.

This presentation is organized into five parts:

1) a review of the hydraulic conductivity of dry clay/organic mixtures to establish the upper-boundary **k**-values related to the fabric of flocculation;
2) the effects of pure cyclohexane permeation through compacted water-wet clay under various imposed flow rates, gradients, effective confining stresses and void ratios (the behaviour of the pure hydrophobic);
3) the effects of pure ethanol permeation through water-wet clay at low and high effective stress levels (the behaviour of the pure association liquid);
4) sequential permeation through water-wet clay by ethanol followed by cyclohexane; and
5) permeation of ethanol/cyclohexane mixtures through water-wet clay.

Where possible, the results of testing at vertical static effective stresses of both zero and 160 kPa are presented to illustrate the important role of confining stresses in self-healing wet barrier clays during exposure to organic liquids.

2 MATERIALS AND METHODS

The natural clay studied was obtained from the brown, oxidized, surface crust of a clay deposit near Sarnia, Ontario. The mineralogical composition and index properties of the Sarnia soil have been previously described (Fernandez and Quigley, 1985, 1988). Illite is the dominant clay mineral (~ 50% of the soil) followed by smectite (~ 15%), chlorite (~ 8%), carbonate (~ 6%) and quartz and feldspar (~ 20%). The near surface soil (~ 0.3 m depth) can be described as an inorganic silty clay of low to medium plasticity, consisting of 10% sand, 33% silt and 57% clay. Particles larger than 2 mm (~ 3%) were removed from the bulk soil to obtain the test soil.

A solution of 0.01 N $CaSO_4$ (in distilled water) was used as the reference permeant. The organic solvent used as the association liquid was commercial grade ethanol (95% pure). Reagent grade cyclohexane was used as the hydrophobic organic. The chemical structure and selected properties of these liquids are presented in Table 1. The behaviour of the ethanol was established in concentrated form and after mixing with both municipal solid waste (MSW) leachate and cyclohexane. The MSW leachate was obtained from the collection system of the Westminster landfill serving the City of London, Ontario, and was used after laboratory stabilization for 2 or 3 weeks.

Hydraulic conductivity testing was carried out using the fixed-ring (rigid-wall) permeameter and testing procedures described in detail by Fernandez (1989). Briefly, air dried clay samples were wetted with distilled water to obtain uniform moisture contents of about 30%, then compacted in fixed-ring permeameters using a Harvard miniature tamper. The void ratio (volume of voids/volume of solids) at this water content is about 0.82. The required static vertical effective stresses, σ_v', were applied by means of spring-loaded caps acting on the upper surface of the soil. The soil samples were then permeated with the liquid(s) of interest under conditions of either constant flow rate or constant applied head, by connecting the permeameter cells to the appropriate system. The constant flow system operates by forcing the permeant through the soil specimen at a constant flow rate, q, while the resulting hydraulic gradient, i, is simultaneously calculated from inlet pressure heads measured by pressure transducer. The constant head system, on the other hand, provides a constant head difference across the soil sample and the flow rate is measured by the displacement of a permeant/air meniscus at the influent

Table 1. Characteristics of the liquid permeants at 20°C.

LIQUID	STRUCTURAL FORMULA	DENSITY (g/mL)	DIELECTRIC CONSTANT	KINEMATIC VISCOSITY (Centistoke)
WATER	H\O/H (H–O–H)	0.99820	80.4	1.0056
ETHANOL (95%)	H – C(H_2) – C(H_2) – OH	0.81026	32	1.80559
CYCLOHEXANE	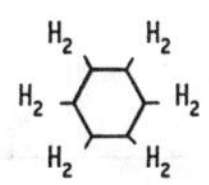	0.77855	2.0	1.246

side of the system. In both cases, the hydraulic conductivity is determined in accordance with Darcy's law; $\mathbf{k} = q/(iA)$, where A is the cross-sectional area of the soil sample.

During testing, changes in the soil volume were monitored by dial gauges and pins connected to the upper soil surface. Samples of effluent permeant were collected and chemically analyzed to assess the state of equilibrium of the clay/organic system prior to termination of testing.

3 DRY SOIL/PERMEANT MIXTURES

A series of control **k**-tests were carried out on samples of dried clay mixed in the permeameter cells with the same liquid to be used as the permeant. The results presented on Figure 1 illustrate the relationship between flow rates and hydraulic gradients for clays mixed and permeated with water, ethanol and cyclohexane. The three mixtures exhibit the same linear relationships between q and i, reported at length by Fernandez (1989) and as predicted by Darcy's law. The slope of the q versus i curves (which reflect the magnitude of the hydraulic conductivity) are very flat for water ($\mathbf{k} = 3.1 \times 10^{-9}$ cm/s), moderate for ethanol ($\mathbf{k} = 8.5 \times 10^{-8}$ cm/s) and very steep for cyclohexane ($\mathbf{k} = 3.6 \times 10^{-5}$ cm/s). Even though the three soil samples had similar values of initial void ratio ($e_o = 0.75$ to 0.78),

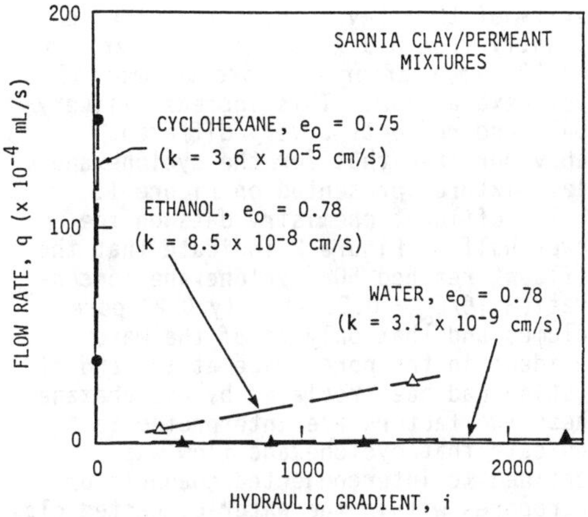

Fig.1 Flow rate versus hydraulic gradient for dry Sarnia clay mixed and permeated with water, ethanol and cyclohexane

they show a 10,000-fold variation in **k**-values reflecting the very different fabric of the soil molded in each liquid. Polar water (dielectric constant, $\varepsilon = 80$) produced a dispersed soil fabric compared to ethanol ($\varepsilon = 32$) and cyclohexane ($\varepsilon = 2$) which promoted flocculation of the soil with large interped pores (see Fernandez and Quigley (1985) and Quigley and Fernandez (1990) for detailed descriptions of this phenomenon).

4 COMPACTED WATER-WET CLAYS

If organic solvents less polar than water are capable of yielding very large values of hydraulic conductivity for permeant-mixed clays, the question arises whether these solvents will destroy the integrity of water-wet barrier clays. The following sections summarize the work carried out on water-compacted Sarnia clay samples permeated with reference water (0.01 N $CaSO_4$ solution) followed by various organic permeants. The test results are presented in separate parts for pure hydrophobic cyclohexane, concentrated hydrophilic ethanol and the ethanol/cyclohexane mixtures.

4.1 Permeation with "hydrophobic" cyclohexane

The curves presented on Figure 2 describe the results of sequential permeation of water, then cyclohexane through a water-wet sample at a constant flow rate of 1.14×10^{-4} mL/s and at a vertical static stress of 0 kPa. The hydraulic conductivity of the clay is seen to increase slightly from 0.9 x 10^{-8} cm/s (water) to 2 x 10^{-8} cm/s after ~ 2 pore volumes of cyclohexane flow. This increase is very small and reflects a very different behaviour than that for the cyclohexane clay mixtures presented on Figure 1.

The effluent chemistry data on the lower half of Figure 2 indicate that the effluent reached 50% cyclohexane concentration (C/C_o = 0.5) at only 0.27 pore volumes and that only 7% of the water resident in the pore space at the end of testing had been replaced by cyclohexane. These two factors are interpreted to indicate that cyclohexane flow was confined to interconnected channels or macropores within the water-compacted clay specimen and possibly along the cell walls. The 7% organic in the soil represents water punched out by the high gradient testing ($i_f \approx 250$).

Figure 3 taken directly from Quigley and Fernandez (1989) shows the effect of a variety of constant flow rates on the hydraulic gradient and **k** for a clay sample compacted at an initial void ratio, e_o = 0.913. At very low flow rates, the measured **k** is similar to or less than **k** for water and the gradient, i, increases rapidly with small increases in flow rate. Once a certain flow rate, q, is reached, however, a completely different flow rate vs i relationship is observed along with significant increases in **k**. As discussed at some length by Quigley and Fernandez (1989) the gradient marking the abrupt change in curvature is believed to correspond to a breakthrough gradient. On Figure 3 this apparent breakthrough gradient is about 175.

A more comprehensive look at this behaviour is illustrated on Figure 4 for three samples compacted at three different void ratios and tested at σ_v' = 0. The curve with the open circles was the one plotted on Figure 3. The three specimens were compacted wet-of-Proctor-optimum and are believed to have been fairly free of compaction-induced fractures. All three tests exhibited linear relationships between q and i for water, but for cyclohexane flow this relationship is highly non-linear. The magnitude of the apparent breakthrough gradient increases with decreasing void ratio, reflecting the inverse proportionality between capillary strength and pore diameter. It is very important to note that at low flow rates (which correspond to low inlet side pressure heads) the **k**-values are all less than water. This phenomenon may explain in part the highly variable **k**-values

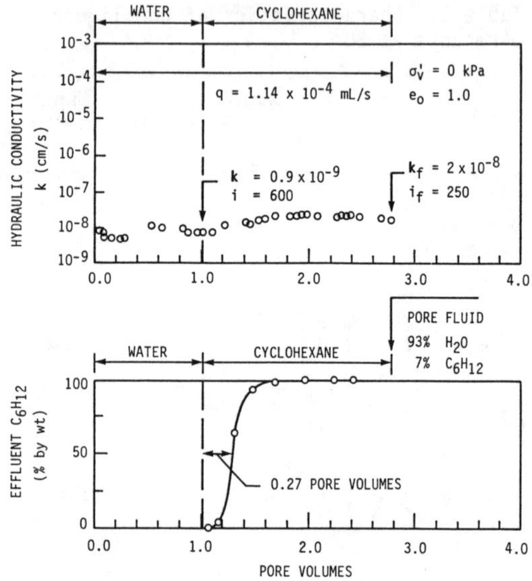

Fig.2 Hydraulic conductivity and effluent chemistry versus pore volumes for water-compacted Sarnia clay permeated with water, then cyclohexane

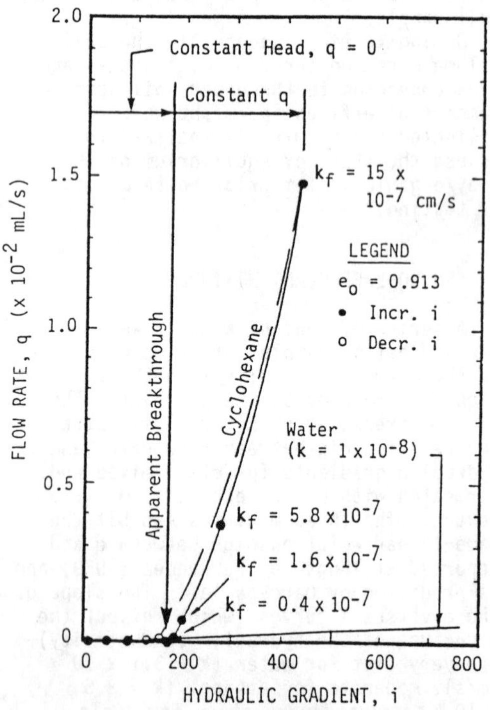

Fig.3 Flow rate versus hydraulic gradient for cyclohexane penetration into water-compacted Sarnia clay

reported in the literature for insoluble liquid organics.

This phenomenon is further illuminated on Figure 5 which shows the results for forced cyclohexane flow through three water-compacted clay samples subjected to

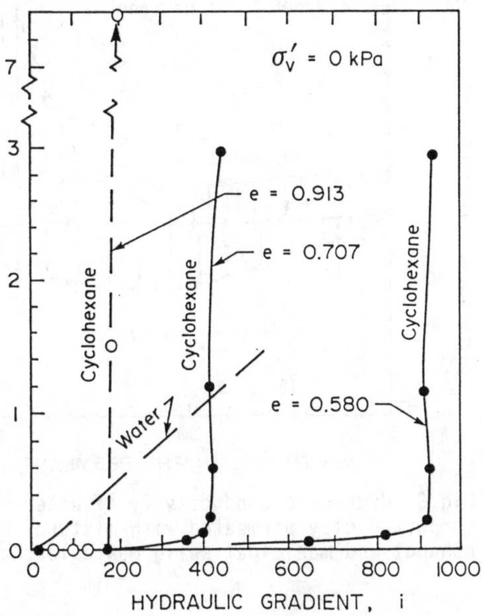

Fig.4 Flow rate versus hydraulic gradient for cyclohexane flow through water-compacted Sarnia clay at three different void ratios (σ_v' = 0)

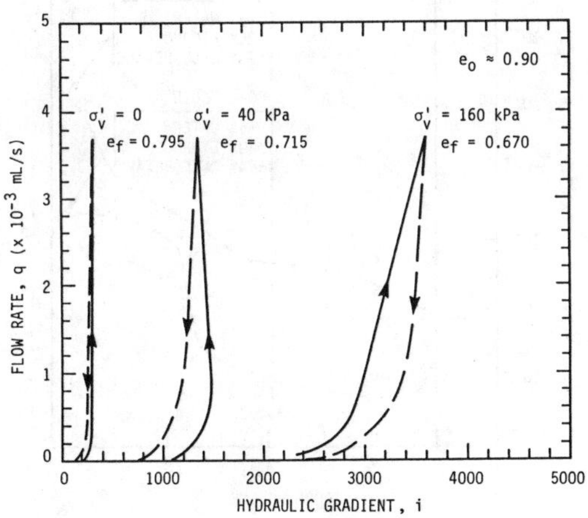

Fig.5 Flow rate versus hydraulic gradient for cyclohexane flow through water-compacted Sarnia clay at three different levels of applied effective stress ($e_o \approx$ 0.90)

initial vertical confining effective stresses, σ_{vo}', of 0, 40 and 160 kPa. All samples were originally compacted wet-of-optimum at $e_o \approx 0.90$ and decreased to values varying from 0.795 (σ_v' = 0) to 0.670 (σ_v' = 160 kPa) as a result of the static stresses and the seepage-induced drag stresses during testing. The apparent breakthrough gradient increases markedly with increasing stress level. This was probably caused in part by reductions in void ratio which resulted in increases in surface tension as described previously. Once breakthrough occurs, however, large increases in flow rate are accommodated with only minor increases in gradient.

Another striking difference between the two plots is the much higher apparent breakthrough gradient for the samples carrying static vertical effective stresses. For example, the 40 kPa sample at e_f = 0.71 yields an apparent break-through gradient of ~ 1400 (Figure 5). The unstressed sample at e_o = 0.71 yields a much smaller breakthrough gradient of only ~ 420 (Figure 4). Static effective stresses on the samples appear to greatly inhibit the expansion of the macropores required for obvious breakthrough.

4.2 Permeation with "hydrophilic" ethanol

Since ethanol is the association liquid under discussion, its own behaviour is described in this section for later reference.

A summary plot of hydraulic conductivity vs percent ethanol in a MSW leachate solution is presented on Figure 6. All samples were tested at σ_v' = 0 and σ_v' = 160 kPa. All tests started at an initial void ratio of 0.85. Reference **k**-values for water used as the initial permeant (solid circles) are shown along with the final equilibrium **k**-values for the leachate/ethanol mixtures (solid triangles). The large increases in **k** that occurred for ethanol contents greater than 70% at σ_v' = 0, are attributed to double layer contraction once ethanol entry is possible at high concentrations. Apparently the cations in the double layers prefer polar water over less polar ethanol. Below 70% ethanol, decreases in **k** were observed as a result of the increased viscosity of the ethanol/leachate mixtures. For the testing at σ_v' = 160 kPa, decreased **k**-values resulted for all ethanol mixtures. Detailed explanations of these phenomena may be found in Fernandez and Quigley, 1988 and 1991.

4.3 Permeation with ethanol/cyclohexane combinations

Association liquids such as ethanol (which are mutually soluble in both water and in non-polar liquids) seem to facilitate entry of hydrophobic liquids into the double layers of water surrounding clay minerals. The decreased dielectric constant of the ethanol/cyclohexane system seems to cause double layer contraction and opening of free pore space, leading to large increases in **k**. These phenomena are illustrated first by sequential permeation, then by permeation with ethanol/ cyclohexane mixtures.

4.3.1 Sequential permeation:

The changes in hydraulic conductivity resulting from sequential permeation of water, then ethanol, then cyclohexane, are presented on Figure 7 for molding void ratios of ~ 1.0 and ~ 0.6. Both tests were run with $\sigma_v' = 0$ kPa so that only seepage stresses were operative on the samples during permeation.

At a high void ratio of ~ 1.0, concentrated ethanol effected an increase in **k** due to double layer effects. At the lower void ratio of 0.6, a net decrease in **k** resulted after ethanol permeation apparently due to less double layer influence combined with an alcohol viscosity about 1.8 times greater than water.

Subsequent permeation with cyclohexane increased the **k** above that for ethanol by 250 to 400 times. For the "loose" sample, the final **k** was 4 orders of magnitude greater than the starting value for water. For the "denser" sample the final **k** was about 2 orders of magnitude greater than the starting value for water. These curves are very different from the one presented on Figure 2 which showed no increase resulting from cyclohexane permeation. It is suggested that the presence of ethanol in the pore spaces and double layers at a concentration of about 60 to 80% has allowed hydrophobic cyclohexane to enter the double layers also. It is suggested that this has caused significant double layer collapse and large increases in **k** even for the low void ratio sample.

4.3.2 Permeation with mixtures:

The ethanol:cyclohexane mixtures were prepared in the following ratios: 0:100, 20:80, 40:60, 60:40, 80:20 and 100:0. Two of the four mixes (20:80 and 40:60 ethanol:cyclohexane) separated into an upper, low density, cyclohexane-rich phase

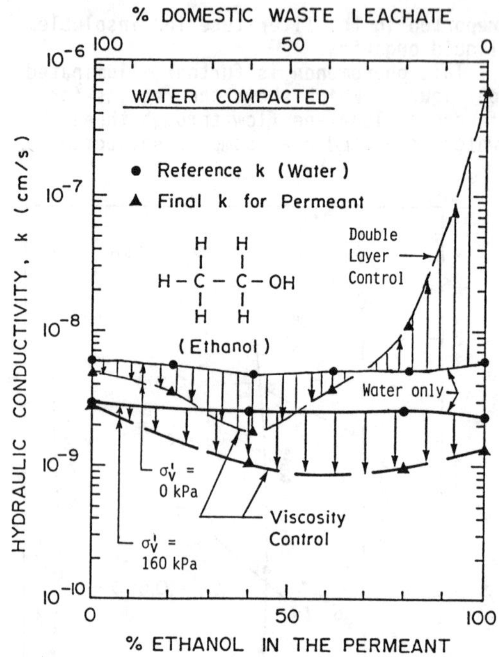

Fig.6 Hydraulic conductivity of water-compacted clay permeated with mixtures of ethanol and municipal solid waste leachate

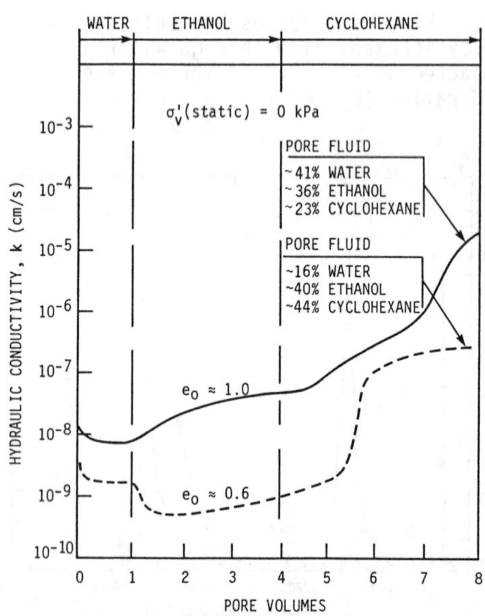

Fig.7 Hydraulic conductivity changes with sequential permeation of water, ethanol and cyclohexane

and a lower, higher density, ethanol-rich phase. In both cases the ethanol-rich phase was permeated through the sample first (top to bottom flow), followed by the cyclohexane-rich phase. These two tests, therefore, are more like the sequential permeation described previously. The phase separation greatly complicated the interpretation of the test results.

The results of testing at zero and 160 kPa are presented on Figure 8. If the 0 kPa tests on Figure 8 are compared with those on Figure 6, it is immediately obvious that the organic mixtures have generated large increases in **k** not observed for the ethanol/leachate mixtures, except at high concentrations of ethanol. In other words, association liquids appear to make clays more susceptible to damage by mixed organics than by pure NAPLs at low effective stresses and relatively low gradients.

The test results on Figure 8 for static $\sigma_v' = 160$ kPa show decreases in **k** for the single phase liquids rather than increases as observed at 0 kPa.

The test results for pure cyclohexane that appear on the left side of Figure 8 show the enormous dependence of measured **k** on applied gradient. This suggests that

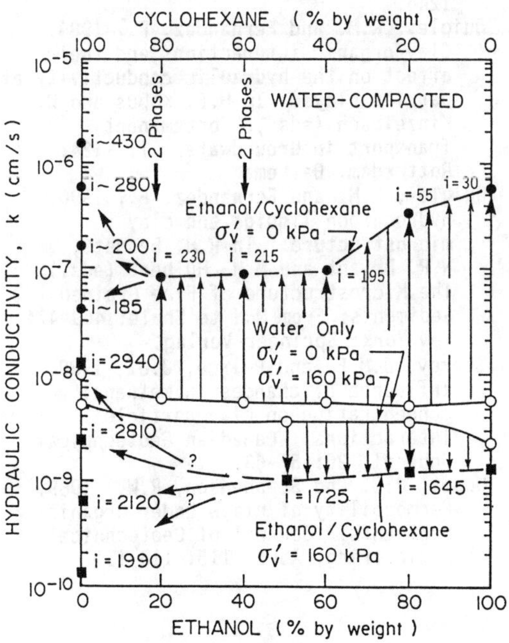

Fig.8 Hydraulic conductivity of water-compacted clay permeated with ethanol/cyclohexane mixtures at 0 and 160 kPa

all of the test results are suspect. The figure therefore demonstrates rather well the technical mess we find ourselves in with respect to reliable laboratory testing of organic liquids.

If nothing else, Figure 8 does show that the presence of static effective stresses on clay barriers greatly improves their performance with respect to permeation of mixed organics. The two mixtures produced lower **k**-values at the gradients of 1645 and 1725 generated by the flow rate used. The decreases in **k** result from chemically-induced consolidation of about 4% of the initial volume during seepage along with the increased viscosity of the mixtures (~ 60 to 80% greater than water).

5 CONCLUSIONS

The following conclusions can be made based on the laboratory study presented:

1) Dry clays mixed with water, ethanol or cyclohexane exhibit linear relationships between flow rate and hydraulic gradient and **k**-values of 3.1 x 10^{-9} cm/s for water, 8.5 x 10^{-8} cm/s for ethanol and 3.6 x 10^{-5} cm/s for cyclohexane. The larger values of hydraulic conductivity for the less polar liquids are attributed to macropores developed by flocculation of the soil particles and very thin double layers.
2) Water-compacted clay barriers, wet of Standard Proctor optimum, are very resistant to penetration by insoluble, hydrophobic, organic liquids provided they are free of contaminating surfactants and association liquids. This is due to strong aqueous surface tension effects so that large hydraulic heads or gradients are required for breakthrough. The magnitude of the apparent breakthrough gradient increases with decreasing void ratio and/or increasing effective stress on the barrier system.
3) Once breakthrough occurs, the **k**-values become very gradient dependent as the flow channels expand with increasing pressure heads.
4) Association liquids such as ethanol destroy surface tension effects facilitating the entry of hydrophobic liquids into water-compacted clays. This behaviour was illustrated for permeation with ethanol/cyclohexane mixtures and for sequential permeation

of ethanol and then cyclohexane, suggesting that little reliance should be placed on the presence of breakthrough pressures in the field.

5) Confining effective stresses acting on "soft" water-wet clay barriers prior to contact with organic liquids are very beneficial since they greatly inhibit the increases in hydraulic conductivity observed for the soluble and insoluble organics permeated at low stress levels.

6 ACKNOWLEDGEMENTS

Most of the work described above was supported by funds to the first author from the Natural Sciences and Engineering Research Council of Canada. The research involving leachate was financed by the Ontario Ministry of the Environment.

REFERENCES

Acar, Y.B., Hamidon, A., Field, S.D. and Scott, L., 1985. The effect of organic fluids on hydraulic conductivity of kaolinite. In: Hydraulic Barriers in Soil and Rock, ASTM Special Technical Publication 874: 171-187.

Anderson, D.C. and Brown, K.W., 1981. Organic leachate effects on the permeability of clay liners. Proceedings of the 7th Annual Research Symposium, Ft. Mitchell, KY, EPA-600/9-81-002b: 119-130.

Anderson, D.C., Brown, K.W. and Green, J.W., 1982. Effect of organic fluids on the permeability of clay soil liners. In: Land Disposal of Hazardous Waste. Proceedings of the 8th Annual Research Symposium, Ft. Mitchell, KY, EPA-600/9-82-002: 179-190.

Anderson, D.C., Brown, K.W. and Thomas, J.C., 1985. Conductivity of compacted clay soils to water and organic liquids. Waste Management and Research, 3: 339-349.

Brown, K.W. and Anderson, D.C., 1980. Effects of organic chemicals on clay liner permeability. Proceedings of the 6th Annual Research Symposium, Cincinnati, OH, EPA-600/9-80-010: 123-134.

Brown, K.W., Green, J.W. and Thomas, J.C., 1983. The influence of selected organic liquids on the permeability of clay liners. In: Land Disposal of Hazardous Waste. Proceedings of the 9th Annual Research Symposium, Ft. Mitchell, KY, EPA-600/9-83-018: 114-125.

Fernandez, F., 1989. The effects of waste leachates on the hydraulic conductivity of natural clays. Ph.D. Thesis, The University of Western Ontario, London, Ontario, 573 p.

Fernandez, F. and Quigley, R.M., 1985. Hydraulic conductivity of natural clays permeated with simple liquid hydrocarbons. Canadian Geotechnical Journal, 22: 205-214.

Fernandez, F. and Quigley, R.M., 1988. Viscosity and dielectric constant controls on the hydraulic conductivity of clayey soils permeated with water-soluble organics. Canadian Geotechnical Journal, 25: 582-589.

Fernandez, F. and Quigley, R.M., 1991. Controlling the destructive effects of clay-organic interactions by application of effective stresses. Canadian Geotechnical Journal, 28: 388-398.

Foreman, D.E. and Daniel, D.E., 1986. Permeation of compacted clay with organic chemicals. Journal of Geotechnical Engineering, ASCE, 112: 669-681.

Michaels, A.S. and Lin, C.S., 1954. Permeability of kaolinite. Industrial and Engineering Chemistry, 46: 1239-1246.

Quigley, R.M. and Fernandez, F., 1989. Clay/organic interactions and their effect on the hydraulic conductivity of barrier clays. In H.E. Kobus and W. Kinzelbach (eds.), Contaminant Transport in Groundwater: 117-124. Rotterdam: Balkema.

Quigley, R.M. and Fernandez, F., 1990. Hydrocarbon liquids and clay microstructure. In R.H. Bennett, W.R. Bryant and M.H. Hulbert (eds.), The Microstructure of Fine Grained Sediments: From Mud to Shale: 469-474. New York: Springer-Verlag.

Storey, J.M.E. and Peirce, J.J., 1989. Influence of changes in methanol concentration on clay particle interactions. Canadian Geotechnical Journal, 26: 57-63.

Uppot, J.O. and Stephenson, R.W., 1989. Permeability of clays under organic permeants. Journal of Geotechnical Engineering, ASCE, 115: 115-131.

Subsurface Contamination by Immiscible Fluids, Weyer (ed.) © 1992 Balkema, Rotterdam. ISBN 90 6191 175 3

The effect of tetrachloroethylene on the permeability of a fractured clay under constant stress conditions

T.A. Middleton
University of Waterloo, Ont., Canada (Currently: CH2M Hill Engineering Ltd, Waterloo, Ont., Canada)

J.A. Cherry
University of Waterloo, Ont., Canada

R.M. Quigley
University of Western Ontario, London, Ont., Canada

ABSTRACT: Water-immiscible organic fluids, such as chlorinated solvents, are believed by some researchers to cause permeability increases in clays by entering and changing the clay double layers, resulting in shrinkage and cracking of the soil. Permeability tests of specimens of naturally fractured, non-swelling clay, conducted with a flexible-wall permeameter, revealed no evidence that permeation by tetrachloroethylene (PERC) resulted in clay permeability increases. Under constant stress conditions, the permeability of clay to PERC was less than the permeability to water. All tests were performed under low effective stresses and low head drops. Theory predicts that surface tension forces will limit the penetration of immiscible-phase solvent into the clay to fractures and macropores. Even within macropores and fractures, immiscible-phase solvents (which are non-polar) should not displace water molecules (which are polar) from the charged surfaces of clays. Therefore solvents are not expected to alter the double layers of clays. This work is continuing to investigate chemical interactions between solvents and swelling clays.

1 INTRODUCTION

Over the past decade, much attention has been given to the possibility that water-immiscible organic fluids, including chlorinated solvents, may be able to increase the permeability of clay barriers through chemical interactions with the clays. Some researchers have reported that contact with water-immiscible organic liquids causes shrinkage and cracking of clay soils, which causes large increases in soil hydraulic conductivity under conditions of low confining pressures (Anderson et. al., 1985; Brown et. al., 1986; Daniel and Broderick, 1985; Foreman and Daniel, 1986). Reviews of previous work may be found in Mitchell and Madsen (1987) and Middleton (1990).

All previously reported experiments using pure-phase chlorinated solvents indicated that permeation with these fluids caused large increases in the hydraulic conductivity of compacted clay soils (Bowders, 1985; Daniel and Broderick, 1985; Evans et. al., 1985). This has important implications because chlorinated solvents (here referred to as "solvents") are more dense than water and are likely to move downward below the watertable and settle in pools on top of clay liners or natural clay deposits. Degradation of these barriers could potentially result in the subsequent migration of solvent through the barrier and the contamination of underlying groundwater resources.

There have been no previously published laboratory studies of the permeability of naturally fractured, undisturbed clays permeated by immiscible fluids such as solvents. This study used naturally fractured clays as they may be more susceptible than unfractured or compacted clays to chemically induced permeability increases caused by solvents. This is because the penetration of solvent into fractures increases the area of clay-solvent contact. Also, permeability testing of naturally fractured clay specimens will include the effects of chemical interactions between the solvent and the fracture coatings or fillings, which may result in changes to fracture apertures and therefore the permeability of the fractured clay.

This paper describes permeability tests carried out on water-saturated, naturally fractured clay samples, with sequential permeation of water and solvent of each sample. The main objective of these tests was to determine whether increases in clay permeability occur due to chemical interactions between the clay soils and the solvent. It is recognized that the measured absolute values of permeability are not directly applicable to shallow fractured clay deposits. The intent was to examine relative changes in permeability in each specimen. Tests were conducted uner low effective stress conditions. However, the precise simulation of in situ stresses was not attempted due to the difficultly of controlling the stress state of the specimens throughout the sampling and experimental periods. Throughout this paper, the term "permeability" refers to the intrinsic permeability of a porous medium.

2 MATERIALS

This study used samples taken from the shallow, fractured and weathered horizon of the Quaternary clay deposits on the property of Laidlaw Ltd. (formerly Tricil Ltd.), which operates a hazardous waste destruction and disposal facility near Sarnia, in southwestern Ontario. This site will hereafter be referred to as the "Tricil site" and the soil samples as "Tricil clay". The properties of the clay in southwestern Ontario were extensively described by McKay (1991), and Fernandez, (1989). It is a glacial till composed of 45% clay-size particles, which are primarily quartz and non-swelling clay minerals (Table 1).

The in situ hydraulic conductivity of the Tricil shallow fractured clay has been measured at approximately 10^{-5} cm/s (McKay, 1991).

Shelby tube samples of fractured clay were obtained from 0.75 m to 1.35 m below ground surface. The fracture orientations in the clay are primarily vertical, and at the depths sampled the vertical fracture spacing is approximately 1 fracture every 2 centimetres (Ruland, 1991). Clay specimens were 6.9 cm in diameter and from 3.0 to 5.0 cm in length. Each test specimen was selected to have a connected network of fractures extending along the length of the core, allowing the flow of solvent through the sample. Each sample was tested to obtain the relative permeability first to water, and then to solvent, for that particular specimen.

The chlorinated solvent used in the laboratory experiments was tetrachloroethylene (PERC). This solvent has widespread use in industry and has properties such as density, viscosity, and dielectric constant which are representative of many common chlorinated solvents (Table 2). For this reason the behaviour of PERC is expected to indicate the behaviour of other chlorinated solvents such as trichloroethylene, 1,1,1-trichloroethane, and 1,2-dichloroethane. The "water" used to permeate the specimens was de-aired tap water with a major ion content and sodium adsorption ratio similar to that of the shallow groundwater at the Tricil site.

Table 1: Description of Tricil clay soil

Depth Below Ground Surface	0.75-1.35m
Natural Moisture Content	16.5%±1.0%
Natural Void Ratio	.55 ± .05
Specific Gravity	2.78
Clay Size Fraction	45%
Silt Size Fraction	45%
Sand Size Fraction	10%

Mineralogy of clay size fraction (approximate order of decreasing abundance): quartz, illite, vermiculite, kaolinite, chlorite, calcite, minor feldspar.

Table 2. Properties of the test fluids

	Water	PERC
Density (g/cm^3)	0.998	1.62
Viscosity (cP)	1.002	0.88
Dielectric Constant	80.4	2.20
Solubility (ppm)	-	150

1. All properties at 20°C
2. Values taken from Flick (1985) and Verschueren (1983)

3 METHODS

Permeameters used for testing clay soils are of two general types: rigid-wall and flexible-wall cells. Rigid-wall permeameters are unsuitable for testing specimens of stiff, undisturbed clay such as the weathered Tricil clay till used in this study, because side-wall leakage cannot be eliminated (Daniel et al., 1985). Use of a flexible-wall, or triaxial-type, permeameter prevents the occurrence of side-wall leakage.

Most previous research using flexible-wall cells to measure the permeability of clays to organic fluids was conducted on compacted clay specimens under large hydraulic gradients, which necessitated large effective confining stresses on the effluent end of the specimen. Tests using immiscible liquids often failed to achieve penetration of these liquids into the compacted clay (Foreman and Daniel, 1986; Bowders, 1985; Uppot and Stephenson, 1989). Presumably this was because the large effective stresses on the specimens caused consolidation of the clay, and even large differences in hydraulic head across a specimen were insufficient to overcome surface tension effects and force immiscible fluids into the resulting small pore spaces. However, these fluids can enter into fractured clay under lower "entry pressures" than are required for penetration into compacted clays if the fractures are sufficiently open. Thus, permeability tests of fractured clay can be carried out under lower gradients and stress conditions, which may be more representative of those occurring in the field.

The flexible-wall permeameter (Figure 1) was similar in design to that described by Daniel et. al. (1985). The permeameter consisted of a panel board, triaxial cell, and two "bladder accumulators", manufactured by Brainard Kilman (Stone Mountain, Georgia) and modified for this

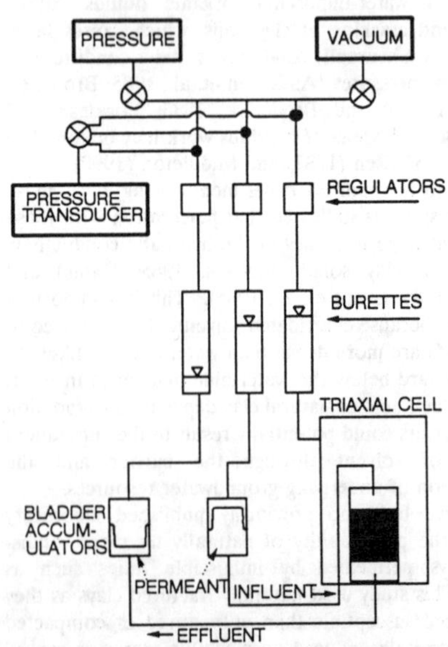

Figure 1: Schematic diagram of the permeameter used in this study

study. All components in contact with the permeant were composed of either Teflon, stainless steel or Viton.

The panel board is used to regulate and measure pressure applied to the cell chamber and the sample inlet and outlet reservoirs. Inflow and outflow volumes are measured by the movement of an air-water interface in burettes mounted on the panel board. Change in sample volume is indicated by movement of an air-water interface in an identical burette connected to the cell chamber. However, change in bulk sample volume cannot be differentiated from volume changes of the equipment itself.

The bladder accumulators are placed in-line between the panel board and sample influent and effluent ports. A concave Viton membrane separates the top and bottom chambers of the accumulator. The membrane moves with fluid flow but prevents cross-contamination of deionized water in the panel and the permeant flowing through the sample. The accumulators also prevent escape of any permeant vapours to the atmosphere.

Clay removed from the ends of each specimen, and the exposed surfaces of the prepared cores were visually examined to characterize the fracture patterns and soil fabric.

Cores of fractured clay were prepared by trimming the ends of the cores, then carefully chipping the end surfaces with a razor blade to remove the smeared layer. The core was placed in the "sample assembly" by covering each end of the core first with a water-wetted coarse filter paper, then a water-saturated stainless steel porous disc, then an end cap.

Membranes which encase the sample are made of latex, which is quickly degraded by PERC. To protect the membrane from contact with the permeant, the sample assembly was wrapped in thin (0.002 inch) Teflon film. However, Teflon film has some rigidity and could form a cylinder around the sample, allowing side-wall flow, especially under low confining pressures. To minimize problems, the Teflon film was tightly wrapped once around the sample with an overlap of about 2 cm. The sample was then encased by two latex membranes separated by a thin layer of silicon grease, and O-rings were used to seal the membrane and Teflon film to the end caps. The sample assembly was placed in the triaxial cell and the cell chamber was filled with de-aired water. De-aired water was used to increase the reliability of the measurements of change in sample volume.

Fractured clay samples were saturated using upward flow of de-aired water through the sample aided by short periods of vacuum application to the top of the outflow burette. After the saturation period, normally one day in length, downward flow of de-aired water was initiated. Back pressure saturation techniques were not used. However, generally de-aired water was permeated through the sample for over four days prior to the start of PERC flow and the specimens were likely close to complete water-saturation at that point.

Permeability measurements were obtained through "falling headwater-rising tailwater" tests. The head drop over the sample would decrease as flow through the sample progressed. The head drop was periodically recorded, and reset to the original value. Care was taken to keep the variations in head drop across the sample relatively small in order to maintain constant effective stress conditions on the specimen, and to keep a continuous flow of permeant through the sample over the testing period, typically 2 to 4 weeks.

Following permeation with water, the inlet bladder accumulator was filled with PERC. The inlet lines and upper end cap and porous plate were flushed with PERC, and flow through the sample was re-initiated. During PERC permeation, the head drop across the sample was measured in units of "equivalent head of water". Head drop measurements were not converted to gradients in order to keep consistent units throughout both the water-permeation and PERC-permeation components of each test.

The cell pressure was adjusted over the first few days of water permeation until an optimum effective stress was obtained for each specimen. The highest possible horizontal effective stress was used which would still allow a measured hydraulic conductivity of 10^{-7} to 10^{-6} cm/s. This range of hydraulic conductivity allowed hydraulic gradients of 10 or less to produce flow rates which were large enough for accurate measurement, and small enough to maintain close to a constant head drop over the sample over 12 hour periods. For comparison, the matrix permeability of the clay, determined by consolidation tests, was on the order of 10^{-8} cm/s (McKay, 1991).

The average horizontal effective stress on the tested specimens ranged from 0.6 psi to 1.2 psi (4.1 - 8.3 kPa). The highest possible effective stress was desired subject to the conditions described above, in order to aid in confining the Teflon film against the sample and prevent leaks of permeant through the overlapped edges of the Teflon sheet to the latex membrane. In fact, leaks of PERC through the overlapped edge were not uncommon, resulting in premature termination of some experiments. Upon contacting the latex membrane, the dyed PERC was immediately visible as red spots in the membrane.

During a test, the PERC was recycled between inlet and outlet reservoirs. This was to limit the amount of PERC use, and also to ensure that the properties of the PERC were not changing during a test due to interactions with the test specimen. For example, the use of recycled PERC ensured that the PERC was initially saturated with water and no additional water would dissolve into the PERC during the test. After termination of an experiment, the specimen was removed from the flexible-wall cell and broken apart to look for evidence of PERC within fractures and visible changes in the fracture patterns or clay fabric.

4 RESULTS

Initial tests were conducted in order to determine the optimum sample stress conditions described above. Figure 2 shows the response of the permeability of one sample of fractured clay to average effective stress. Water was the permeant throughout the test. Over one order of magnitude decrease in permeability resulted from increasing the average effective stress from 0.6 psi to 1.1 psi (4.1 - 7.6 kPa). Sample consolidation, and resultant decrease in hydraulic conductivity, was still occurring when this test was terminated. This and other tests indicated that the measured bulk hydraulic conductivity of a sample was very sensitive to changes in effective stress as small as 0.2 psi (1.4 kPa). It was determined that very low effective stresses, from 0.6 to 1.2 psi (4.1 - 8.3 kPa), were required to keep within the desirable hydraulic conductivity range.

During permeability testing with water and PERC, every effort was made to keep both the head drop across the sample and the total confining pressure to constant values within one test. However, in order to keep flow rates within the "workable" range defined above, and also to reach and maintain the head differential necessary for PERC entry into the specimens, it was often necessary to slightly increase the head drop across the sample upon introduction of the PERC as permeant. In these cases, the effluent (lower) fluid pressures were maintained constant and the influent (upper) fluid pressures were increased. Therefore the maximum effective stress on the sample was kept constant, the minimum effective stress on the sample decreased, and the average effective stress decreased. This probably caused swelling of the influent end of the sample resulting in some small degree of increase in hydraulic conductivity.

Figure 3b shows the results obtained from an experiment in which the head drop across the sample was systematically varied between 31 and 52 equivalent cm of water throughout the test. The cell pressure was kept constant. The total duration of the experiment was 12 days, with PERC introduced on the fifth day. It appears that the small variations in head drop did not produce significant changes in permeability compared to the scatter of permeability values caused by the imprecision of the testing apparatus.

Figure 3a shows cumulative bulk sample volume change throughout the test. Sample consolidation and/or equipment expansion of about 2 ml (1.3% total sample volume) occurred during initial water permeation. During PERC permeation, the bulk sample volume appeared to increase slightly. However, the measured increase of 0.5 ml is likely insignificant with respect to the precision of the measurements.

It is obvious that no large increases in permeability occurred after introduction of PERC into the sample. Examination of the specimen after the test revealed traces of PERC in some fractures, but no visible changes to the soil fabric or fracture pattern.

Four complete tests of water-saturated fractured clay

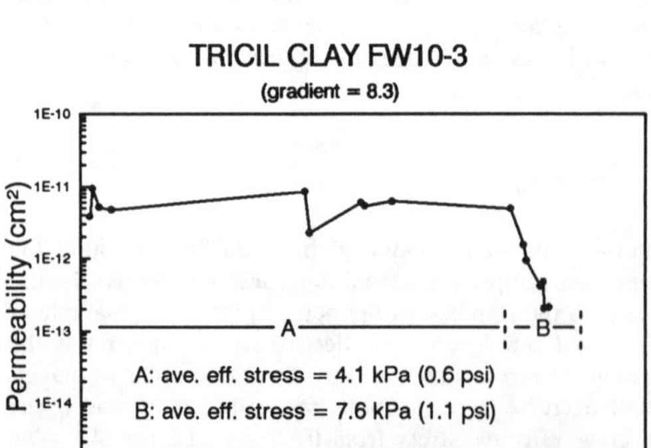

Figure 2: Permeability vs cumulative flow of water under a variable effective stress. (Note: sample was still consolidating when test was terminated)

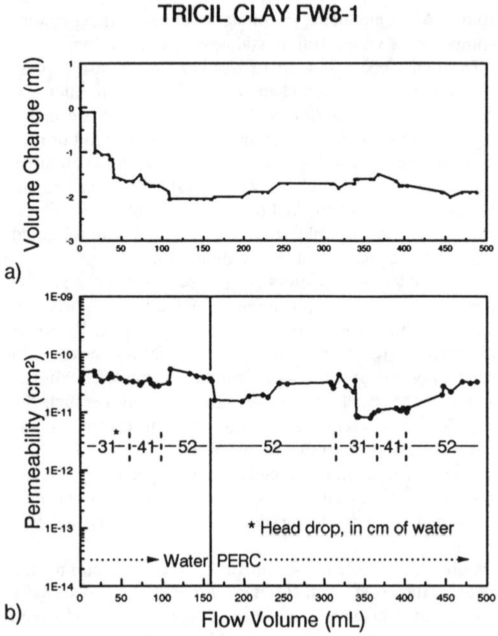

Figure 3: Cumulative bulk sample volume change (a) and permeability (b) vs cumulative flow for sequential permeation of water and PERC. Ave. Eff. Stress: 0.6 psi, Sample Length: 4.3 cm

specimens permeated with PERC were completed. The duration of each test ranged from 12 to 22 days. In every case the permeability to PERC was less than or equal to the "reference" permeability to water. Examination of the fractured clay specimens subsequent to permeability testing did not reveal visible changes in the fracture patterns or clay fabric of any sample.

In all permeability tests, the first appearance of pure PERC in the effluent lines was recorded in order to estimate the relative saturation of PERC in the specimen. However, an unknown volume of PERC travelled through the base cap and some or all of the lower porous plate before PERC was visible in the effluent. This prevented the exact determination of the amount of liquid-phase PERC in the specimen. The calculated maximum relative saturation of PERC in specimens ranged from 0.5% to 69% of the specimen void space, but the actual values are likely to be much less. In terms of the total calculated void space in a specimen, PERC breakthrough generally occurred in less than 0.2 pore volumes. Because the fracture porosity of the specimens was unknown, the amount of fracture volume occupied by PERC was also unknown.

Figure 4b presents the results of a 22 day experiment; PERC was introduced to the sample on day 11. The permeability to water and to PERC are both approximately 1×10^{-12} cm^2. This corresponds to a hydraulic conductivity to water of approximately 10^{-7} cm/s. An interruption of PERC flow during the test was apparently caused by soil particles or water bubbles clogging the effluent lines or porous plate. Flushing the effluent lines and end cap re-established PERC flow. Figure 4a shows

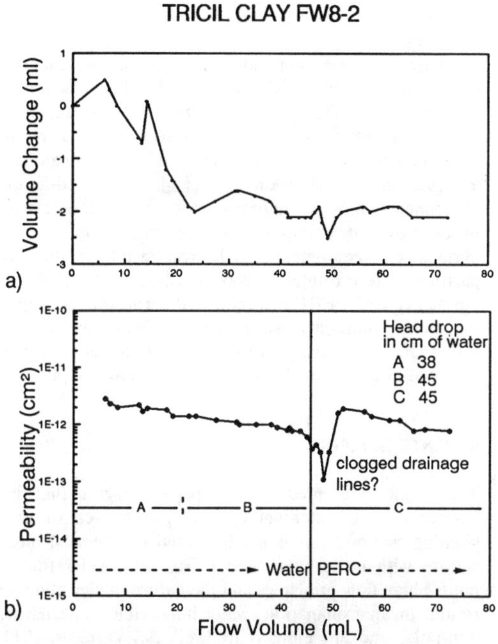

Figure 4: Cumulative bulk sample volume change (a) and permeability (b) vs cumulative flow for sequential permeation of water and PERC. Ave. Eff. Stress: 0.6 psi, Sample Length: 4.2 cm

the bulk sample volume change over the same experiment. Sample consolidation and/or equipment expansion of approximately 2 ml (1.2% total sample volume) occurred during initial permeation with water. During PERC permeation no further sample shrinkage was measured.

The results of a 12 day test (Figure 5) show that, although the head drop across the sample increased during permeation with PERC, the permeability to PERC decreased approximately one order of magnitude relative to

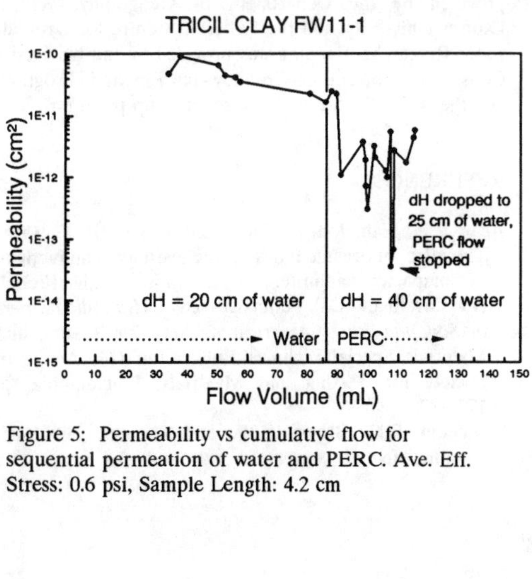

Figure 5: Permeability vs cumulative flow for sequential permeation of water and PERC. Ave. Eff. Stress: 0.6 psi, Sample Length: 4.2 cm

the permeability to water. After considerable flow of PERC, the head drop across the sample was lowered to the initial value used for water flow. The flow rate subsequently decreased over about 12 hours until flow ceased. Flow was re-initiated by increasing the head drop back to the original value of 40 cm of water. It appears that on reduction of the head drop, the continuous path of solvent in the specimen may be interrupted, perhaps by sample deformation in response to changed effective stresses. This type of behaviour was observed in some other tests as well.

5 DISCUSSION

Organic fluids may be able to cause permeability increases in clays by affecting the "double layer structure" of clay minerals. A full description of the Gouy-Chapman theory of the diffuse double layer (Van Olphen, 1977; Mitchell, 1976) is beyond the scope of this paper. Briefly, this theory suggests that intrusion of low-dielectric-constant organic fluids into clay double layers will result in shrinkage of the soil and the formation of larger pores or fractures, which would in turn result in an increased permeability. This mechanism has been used to explain permeability increases observed upon permeation of clay soils with water-miscible or immiscible organic fluids, such as methanol, xylene, and trichloroethylene (Anderson et. al., 1985; Brown et. al., 1986; Daniel and Broderick, 1985; Foreman and Daniel, 1986).

In order to examine the possible effects of immiscible liquids on clay double layers, it is necessary to consider the structure of clay minerals and the principles of multiphase flow. Clay particles carry a net negative charge (Mitchell, 1976), and as a result polar water molecules should be preferentially attracted over non-polar solvent molecules to the clay surfaces. The result is that clays are preferentially wetted by water in a clay/water/solvent system. In other words, solvent will be the non-wetting fluid and water will be the wetting fluid.

When a non-wetting fluid penetrates into a water-saturated clay, surface tension effects will restrict flow of the solvent to the largest pores in the clay, such as macropores and fractures (Bear, 1972; Schwille, 1988). Water will be displaced from the centre of pores, but a film of water will completely surround the intruded non-wetting fluid. This situation has been observed in the Athabasca tar sands where oil is the non-wetting fluid (Mossop, 1980). Therefore, within the macropores or fractures, the above conceptualization indicates that there will always be a film of water between clay surfaces and the chlorinated solvent. The liquid-phase solvent likely cannot directly contact clay surfaces or enter into the clay double layers.

The above discussion neglects the influence of the dissolved, or aqueous phase of chlorinated solvents. Low concentrations of solvent will be able to dissolve into the pore water, and will be able to penetrate throughout a clay mass by advection and diffusion in the aqueous phase. However, previous researchers have been in agreement that there is no effect on clay structure or permeability by permeation with low concentrations (e.g. less than 0.1%) of organic fluids (Daniel et. al., 1988; Acar et. al., 1985). This is expected because water containing low concentrations of organic fluids will not have a dielectric constant significantly different from pure water (Daniel et. al., 1988).

Because it is unlikely, if not impossible, that the liquid-phase solvent ever physically comes into contact with the clay minerals in a water-saturated soil, we can reasonably expect that the chemical interactions which occur between organics and clay minerals in a liquid-phase-solvent/water/clay system will be the same as those in the aqueous-phase-solvent/water/clay system. In both systems, only low concentrations of solvent actually directly contact the clay minerals. Therefore, liquid-phase solvent is not expected to cause any chemically induced changes to the permeability of water-wet clays.

The above conceptual model supports the hypothesis that water-immiscible organic fluids will not chemically increase the permeability of water-wet clays through double layer effects.

None of the tests of water-wet, fractured Tricil clay samples showed evidence of permeability increases due to chemical interactions with PERC. Under constant-stress conditions, the permeability to PERC was less than, or equal to the permeability to water. These experimental results are in agreement with the above hypothesis.

It is important to consider several experimental factors which may have prevented the direct observation of permeability increases caused by the solvent. These are: test duration, and stress conditions. Laboratory tests of relatively short duration may not show the results of time-dependent chemical reactions. However, all previously reported laboratory tests which showed large permeability increases in clays permeated by organic fluids were also conducted over similar, or shorter periods of time (Middleton, 1990). This does not exclude the possibility of time-dependent chemical interactions, but it is unlikely that the length of the testing periods can explain the difference between the results of previous researchers and the results presented here.

A constant effective stress condition was used for all tests. Conceivably, constant effective stress acting on a sample could prevent the opening of fractures or macropores during PERC permeation, and any resulting sample shrinkage and consolidation would be too small to be measured by the panel board burette. In the authors opinion, it is unlikely that the very low effective stresses used in these tests would prevent fracture formation in the case of PERC-clay chemical interactions. However, further laboratory tests are needed to eliminate this possibility.

If effective stress acting on these clay specimens did prevent solvent-induced permeability increases, can a similar behaviour be expected in the field? No laboratory test can exactly replicate field stress conditions. However, the flexible wall permeameter applies a confining stress which may be compared to in situ effective stress caused by overburden pressure. For example, a vertical effective stress of about 1 psi (8 kPa) is roughly equivalent to a depth below ground surface of 0.6 m. This assumes Tricil clay with a watertable at ground surface. Clay barriers normally occur at least several meters below ground, and it is reasonable to expect that most clay barriers will be subject to effective stresses larger than those used in these laboratory experiments. This suggests that if PERC does chemically interact with clay double layers, even very low in situ effective stresses may prevent any permeability increases.

Finally, some mention should be made of the apparent disagreement between the results of other researchers (Bowders, 1985; Daniel and Broderick, 1985; Evans et.al., 1985) and the results of these tests. Obviously, the effective stress conditions on laboratory samples can have a significant influence on permeability. The different experimental stress conditions imposed by other testing procedures (for example: high gradients) or different permeameter equipment may be responsible for the disparities in test results. Tests of compacted Tricil clay permeated with PERC in a rigid-wall permeameter suggest that permeability increases may be caused by the physical forces created by two-phase flow in clay samples. This is explored in more detail elsewhere (Middleton, 1990).

6 CONCLUSIONS

The results presented in this paper suggest that the permeability of water-wet clay soils composed of non-swelling clay minerals is not increased by chemical interactions with water-immiscible PERC. Consideration of multi-phase flow principles and the nature of clay minerals results in an explanation: water-immiscible solvents are non-polar and are unlikely to "wet" clay surfaces. This prevents any direct chemical interaction between water-immiscible solvents and water-wet clay soils. The common chlorinated solvents are likely not sufficiently soluble for the solvent in dissolved-phase to cause changes in the clay double layer, or clay permeability. Experimental work is continuing to further assess this theory and to investigate the cases of: 1)other soil types, especially swelling clays; 2)other solvents; 3)"impure" solvents such as those contaminated with other organic fluids or substances. This is necessary before this work can be confidently applied to field problems.

7 ACKNOWLEDGEMENTS

This work benefitted from the advice of R.W. Gillham, M.B. Dusseault, and B. Hughes. Laboratory facilities were provided by the Department of Geography, Wilfrid Laurier University, and the Waterloo Centre for Groundwater Research. Funding was provided by the University Consortium Solvents-in-Groundwater Research Program, and the NSERC post-graduate scholarship program.

REFERENCES

Acar, Y.B., Hamidon, A., Field, S.P. and Scott, L., 1985. The effect of organic fluids on the hydraulic conductivity of compacted kaolinite. In A. Johnson, R.K. Frobel, N.J. Cavalli and C.B. Pettersson (eds.), *Hydraulic Barriers in Soil and Rock*, American Society for Testing and Materials, Special Technical Publication 874. American Society for Testing and Materials, Philadelphia, p. 171-187.

Anderson, D.C., Brown, K.W., and Thomas, J.C. 1985. Conductivity of compacted clay soils to water and organic liquids. *Waste Management and Research*, 3: 339-349.

Bear, J. 1972. *Dynamics of Fluids in Porous Media*. American, Elsevier, New York, NY.

Bowders, J.J. 1985. The Influence of Various Concentrations of Organic Liquids on the Hydraulic Conductivity of Compacted Clay. Geotechnical Engineering Dissertation GT85-2, The University of Texas at Austin, Austin, Texas.

Brown, K.W., Thomas, J.C., Green, J.W. 1986. Field cell verification of the effects of concentrated organic solvents on the conductivity of compacted soils. *Hazardous Waste and Hazardous Materials*, 3: 1-19.

Daniel, D.E., Anderson, D.C., and Boynton, S.S. 1985. Fixed wall versus flexible wall permeameters. In A. Johnson, R.K. Frobel, N.J. Cavalli and C.B. Pettersson (eds.), *Hydraulic Barriers in Soil and Rock*, American Society for Testing and Materials, Special Technical Publication 874, American Society for Testing and Materials, Philadelphia, p. 107-126.

Daniel, D.E. and Broderick, G.P. 1985. Stabilization of compacted clay liners against attack by concentrated organic chemicals. Unpublished report to Chemical Manufacturers Association, Washington, D.C.

Daniel, D.E., Liljestrand, H.M., Broderick, G.P., and Bowders, J.J. 1988. Interaction of earthen liner materials with industrial waste leachate. *Hazardous Waste and Hazardous Materials*. 5: 93-108.

Evans, J.C., Fang, H-Y., and Kugelman, J.J. 1985. Organic fluid effects on the permeability of soil-bentonite slurry walls. Proceedings of the National Conference on Hazardous Wastes and Environmental Emergencies, May 14-16, Cincinnati, OH. p. 267-271.

Fernandez, F. 1989. The Effects of Waste Leachates on the Hydraulic Conductivity of Natural Clays. Ph.D. Thesis, Faculty of Engineering Science, University of Western Ontario, London, Ontario.

Flick, E.W. 1985. *Industrial Solvents Handbook*. Noyes Data Corporation, Park Ridge, NJ.

Foreman, D.E. and Daniel, D.E. 1986. Permeation of compacted clay with organic chemicals. *Journal of Geotechnical Engineering*, American Society of Civil Engineers, 112: 669-681.

McKay, L.D. 1991. PhD. Thesis, University of Waterloo, Waterloo, Ontario. In Preparation.

Middleton, T.A. 1990. The Effect of Chlorinated Solvents on the Permeability of Fractured Clay. M.Sc. Thesis, Department of Earth Science, University of Waterloo, Waterloo, Ontario.

Mitchell, J.K. 1976. *Fundamentals of Soil Behaviour*. John Wiley and Sons, New York, NY.

Mitchell, J.K., and Madsen, F.T. 1987. Chemical effects on clay hydraulic conductivity in R.D. Woods (ed.), Proceedings, Specialty Conference on Geotechnical Practice for Waste Disposal, 1987. American Society of Civil Engineers, Geotechnical Special Publication, No. 13, p. 87-116.

Mossop, G.D. 1980. Geology of the Athabasca Oil Sands. *Science*, 207: 145-152.

Ruland, W.W. 1991. Fracture Depths and Active Groundwater Flow in a Weathered Clayey Till in Lambton County, Ontario. *Ground Water*. In Press.

Schwille, F. 1988. *Dense Chlorinated Solvents in Porous and Fractured Media - Model Experiments*. Translated by J.F. Pankow, Lewis Publishers Inc., Chelsea, Michigan.

Uppot, J.O., and Stephenson, R.W. 1989. Permeability of clays under organic permeants. *Journal of Geotechnical Engineering*, 115: 115-131.

Van Olphen, H. 1977. *An Introduction to Clay Colloid Chemistry*. John Wiley and Sons, New York.

Verscheueren, K. 1983. *Handbook of Environmental Data on Organic Chemicals*, Second Edition. Van Nostrand Reinhold, New York, NY.

Subsurface Contamination by Immiscible Fluids, Weyer (ed.) © 1992 Balkema, Rotterdam. ISBN 90 6191 175 3

Evaluation of multi-component DNAPL sources by monitoring of dissolved-phase concentrations

Stan Feenstra
Waterloo Centre for Groundwater Research, University of Waterloo, Ont. & Applied Groundwater Research Ltd, Mississauga, Ont., Canada

ABSTRACT: Dense non-aqueous phase liquid (DNAPL) chemicals have become recognized increasingly as sources of groundwater contamination at waste disposal and chemical spill sites. For multi-component DNAPL, laboratory studies show that the concentration of chemicals dissolved in water in contact with the DNAPL can be estimated by the product of the mole fraction of the components in the DNAPL and the pure-phase solubilities of the components in water. Laboratory studies show also that the components having the highest effective solubility will be dissolved most readily. As the mass of DNAPL is depleted, preferential removal of the more soluble components may result in changes in the composition of the DNAPL, which will in turn change the dissolved concentrations in the groundwater. In many situations, it is not possible to determine the mass of DNAPL sources in the subsurface by soil sampling and analysis due to the heterogeneous distribution of the residual zones and pools. Consequently, during remedial activities such as groundwater purging, it is not generally possible to estimate the degree of mass depletion of the source. If the initial DNAPL composition is estimated, the changes in dissolved concentrations can be related to the degree of mass depletion of the DNAPL source. This may provide a useful method for estimation of DNAPL source depletion with time, particularly during groundwater purging operations.

1 INTRODUCTION

In recent years, groundwater contamination by organic chemicals from sources such as waste disposal facilities, chemical spills and leaking underground storage tanks has become a significant environmental concern. Of the organic chemical contaminants which have been detected in groundwaters, chlorinated organic solvents are the most frequently encountered. In a survey of groundwater monitoring data at 183 hazardous waste disposal sites in the United States it was found that the chlorinated solvents: trichloroethylene, dichloromethane and tetrachloroethylene and 1,1-dichloroethane were among the five most frequently identified contaminants (Plumb and Pitchford 1985).

These chlorinated solvents are dense non-aqueous phase liquids, referred to here as DNAPL chemicals. Non-aqueous phase refers to the fact that, like petroleum hydrocarbons, they are immiscible with water . Dense refers to the fact that, unlike petroleum hydrocarbons, they are more dense than water. Other organic chemicals such as halogenated benzenes, polychlorinated biphenyls (PCBs) and creosote / coal tar are also DNAPL chemicals.

The potential for groundwater contamination by DNAPL chemicals is high because of their distinctive physical and chemical properties. Their low solubility and high density enable DNAPL released into the subsurface to penetrate downward through the vadose (unsaturated) zone and the groundwater (saturated) zone as a separate non-aqueous phase. The tendency for DNAPL chemicals to sink through the saturated zone differs from that of petroleum hydrocarbons which will float on the groundwater in the saturated zone due to their lower densities. In the subsurface, small but significant quantities of chemicals can be dissolved by groundwater in contact with the DNAPL and can result in groundwater contamination.

In the past decade, DNAPL chemicals have been found to be a major and increasing cause of groundwater contamination. For example in the United States, DNAPL is present, or suspected to be present, at most of the abandoned hazardous waste disposal sites investigated for Superfund. The presence of DNAPL contamination in the subsurface is a key reason there has been so little success in the permanent clean-up of contaminated aquifers at these sites (Feenstra and Cherry 1988).

In many situations, it is not possible to determine the mass of DNAPL sources in the subsurface by soil sampling and analysis due to the heterogeneous distribution of the residual zones and pools. Consequently, during remedial activities such as groundwater purging, it is not possible to estimate the degree of mass depletion (clean-up) of the source. This paper considers how examination of dissolved-phase concentrations from multi-component DNAPL can be used to evaluate the degree of mass depletion of DNAPL sources in the subsurface.

2 SOLUBILITY AND EFFECTIVE SOLUBILITY

DNAPL chemicals can be comprised of single component liquid chemicals, mixtures of liquid chemicals or multi-component mixtures of liquids and dissolved solid-phase organics. Chlorinated solvents are liquid chemicals in their pure form and have a wide range in solubility (see Table 1).

Table 1. Solubility of common chlorinated solvents at 10°C. Data from Horvath (1982).

Chemical	Solubility in water (mg/L)
Methylene chloride	15,900
Chloroform	9,000
1,1,1-Trichloroethane	1,700
Trichloroethylene	1,060
Tetrachloroethylene	150

For mixtures of liquid chemicals, the dissolved-phase concentrations in equilibrium with the mixture can be estimated by:

$$S^e_i = X_i \ S_i$$

where S^e_i is the dissolved-phase concentration or effective solubility of compound i, X_i is the mole fraction of the compound i in the DNAPL mixture, and S_i is the pure-phase solubility of compound i.

This relationship is based on the behaviour of ideal liquids according to Raoult's Law. Laboratory experimental studies (Banerjee 1984) suggest that this relationship is a reasonable approximation for mixtures of structurally related hydrophobic organic liquids. Other laboratory studies (Leinonen and Mackay 1973) have shown that the estimation of effective solubility may be in error for complex mixtures such as petroleum hydrocarbons, but the error is unlikely to be greater than a factor of 2.

DNAPL encountered at many waste disposal sites and chemicals such as PCB Aroclors and creosote are liquids, but are comprised of many compounds which, in their pure form, are solids at environmental temperatures (see Table 2). The solubility data reported in the literature for such compounds generally relate to the solubility of the pure solid compound. This solid-phase solubility is not the appropriate value to use in the estimation of the effective solubility of such compounds when they are dissolved in a DNAPL (Shiu et al. 1988). In such a case, the liquid-phase solubility of the compound should be used.

Table 2. Melting points of several key constituents of creosote and coal tar. Data from Miller et al. (1985).

Compound	Melting Point (°C)
Naphthalene	80.2
Fluorene	117
Anthracene	217
3,4-Benzopyrene[1]	175

[1]Benzo(a)pyrene

Liquid-phase solubility values for such compounds can be estimated based on the solid-phase solubility, the entropy of fusion, and the melting point of the compound and the temperature (Miller et al. 1985), or can be measured in the laboratory. The liquid-phase solubility values can be considerably higher than solid-phase solubility values and the difference is greater for compounds having higher melting points (see Table 3).

Table 3. Comparison of solid-phase solubility values and liquid-phase solubility values for several key constituents of creosote and coal tar. Data at 25°C from Miller et al. (1985).

	Solubility (μg/L)	
Compound	Solid-Phase	Liquid-Phase
Naphthalene	30,600	102,000
Fluorene	1,840	11,600
Anthracene	731	7,090
3,4-Benzopyrene[1]	3.78	116

[1]Benzo(a)pyrene

The effective solubilities of the components of a multi-component DNAPL can be calculated readily having a complete chemical analysis of the DNAPL and suitable values of the pure-phase solubilities of the components. However, DNAPL at many waste disposal and chemical spill sites frequently contains a large proportion of organic compounds which cannot be identified and quantified readily in most analytical programs. Because of this, estimation of effective solubility requires assumptions for the molecular mass of the unidentified fraction. The estimated effective solubility values are higher for the assumption of higher molecular mass for the unidentified fraction. This is the case because for a given weight % of unidentified fraction, a higher molecular mass will result in a lower calculated mole fraction for the unidentified fraction and correspondingly higher mole fractions for the identified components. Of particular note is that the relative dissolved concentrations of the identified components are the same regardless of the assumptions made for the unidentified fraction.

A comparison of calculated effective solubility values with observed groundwater concentrations is shown in Table 4. Data for this example are derived from a Superfund site in Pennsylvania where DNAPL has been identified in numerous wells in a fractured sandstone aquifer. Two samples of DNAPL recovered from wells were analysed and show the same chemicals present although there is a considerable range in concentration. This is likely common for a waste disposal facility. Because the DNAPL is variable in composition, the dissolved concentrations in the groundwater would be expected to be variable to the same degree.

For each DNAPL sample, the effective solubility of the four principal components was calculated. The dissolved concentrations observed in groundwater close to the DNAPL source are comparable to the calculated effective solubility values. In addition, the order of abundance from the highest concentration to the lowest concentration (TCP > XYL > TOL > EB) observed in the groundwater is the same as calculated.

Table 4. Comparison of calculated effective solubility values with observed dissolved concentrations at a Superfund site in Pennsylvania.

	TCP	TOL	XYL	EB
Solubility (mg/L)	1,900	580	200	135
Molecular Mass (g/mole)	146	92	106	106
DNAPL Sample 1 (wt%)	23	4.2	17	3.8
Effective Solubility (mg/L)	400	35	43	6.5
DNAPL Sample 2 (wt%)	73	0.9	5.8	0.9
Effective Solubility (mg/L)	1,400	8.1	16	1.6
Observed in Groundwater (mg/L)				
Lower Limit	220	2.1	7.3	1.1
Upper Limit	1,200	53	74	12

For molecular mass of unidentified fraction 150 g/mole.

Key: TCP- 1,2,3-Trichloropropane
TOL- Toluene
XYL- Total xylenes
EB- Ethyl benzene

3 DISSOLUTION PROCESSES

3.1 Theory and Laboratory Studies

In the subsurface, DNAPL occurs in residual zones and in layers or pools. In the residual zones, DNAPL is present as immobile blobs and ganglia that may occupy 10% of the pore space or less. Groundwater flow occurs through the remaining pore space, although the presence of the DNAPL reduces the relative permeability of the medium to water. Dissolution of the DNAPL occurs as groundwater flows through the residual zone. Within DNAPL layers or pools, the DNAPL occupies a larger proportion of the pore space and may essentially exclude groundwater flow through these zones. Dissolution of the DNAPL in this case likely occurs only along the margins of the layers or pools.

The dissolution of DNAPL in the subsurface will depend, in fundamental terms, on the properties of the components of the DNAPL (such as effective solubility and diffusivity), the physical distribution of the DNAPL in the porous or fractured medium, and the rate of groundwater flow through and around the DNAPL.

The rate of dissolution or mass transfer of chemicals from the DNAPL to the groundwater is expressed typically as a function of a mass transfer coefficient which is related to the properties of the DNAPL components and the groundwater flow conditions, a concentration differential usually defined as the difference between the effective solubility of the component and the dissolved concentration in the groundwater in contact with the DNAPL, and the contact area between the DNAPL and the groundwater.

Mass Transfer Rate [M/T]	=	Mass Transfer Coefficient [L/T]	x	Concentration Differential [M/L^3]	x	Contact Area [L^2]

Mass Transfer Flux [M/L^2/T] (= Mass Transfer Coefficient x Concentration Differential)

The rate of mass transfer from the DNAPL to the groundwater will determine the dissolved-phase concentrations that will occur in the groundwater and the longevity of the DNAPL source.

Laboratory batch experiments indicate the dissolved chemical concentrations in water in contact with a non-aqueous phase liquid approach effective solubility values (saturated concentrations) in minutes to hours depending on the non-aqueous phase to water ratio (Shiu et al. 1988, Poulsen et al. In Preparation).

Laboratory column and tank experiments (Schwille 1988, Anderson et al. 1987, Imhoff et al. 1989) suggest that for groundwater flow rates of up to 1 m/d linear velocity, saturated dissolved concentrations should be observed in groundwater leaving zones of DNAPL contamination containing typical residual contents. Under such conditions, the dissolution of the DNAPL is not limited by the rate of mass transfer. Theoretical and laboratory studies have shown also (Pfannkuch 1984, Schwille 1988) that dissolution may be mass transfer rate limited at lower DNAPL residual contents as the mass of DNAPL is depleted. Theoretical and laboratory studies have shown also (Pfannkuch 1984, Geller and Hunt 1989) that dissolution may be mass transfer rate limited at higher groundwater velocities. However, such high groundwater velocities are likely to occur only in the immediate vicinity of pumping wells where hydraulic gradients are large.

The equilibration of groundwater with DNAPL in the residual zone can be viewed practically as being dependent on the contact time. The contact time is dependent on the groundwater velocity and the size of the DNAPL residual zone. A review of laboratory experiments on dissolution (Schwille 1988, Anderson et al. 1987, Imhoff et al. 1989) suggests that saturated dissolved concentrations should be achieved if the combination of groundwater velocity and size of the residual zone results in contact times of one half hour or longer. For groundwater flow under natural conditions and DNAPL residual zones of centimetres in size, saturated concentrations would be expected.

In situations where the DNAPL is present as layers and pools, theoretical and laboratory experiments suggest that less-than-saturated dissolved concentrations are observed (Schwille 1988, Hunt et al. 1988). Small heterogeneities in permeability or capillary properties of a porous medium will have a very significant effect on the migration and distribution of DNAPL (Kueper et al. 1989, Kueper 1989, Kueper et al. 1990). In horizontally bedded media, a large proportion of the DNAPL may be present as thin perched layers and pools.

Several models have been used to attempt to provide a description of mass transfer coefficients (Merchuk 1983). Such models include the Thin-Film Model, the Penetration Model and the Random Surface Renewal Model. These models all relate the mass transfer coefficient of a compound to its molecular diffusivity in water and an empirical parameter which is a function of the water flow conditions. These empirical parameters such as the film thickness, the contact time or surface renewal rate have no strict physical meaning and cannot be measured independently. Such models provide the basis for the description of mass transfer processes in reactors by chemical engineers but at the present time cannot be usefully applied to the dissolution of DNAPL in groundwater. However, all of these models suggest that for a given set of water flow conditions, the mass transfer coefficient should depend only on the aqueous diffusivity of the compound. Because the aqueous diffusivity of most organic chemicals of interest vary over a relatively small range (see Table 5), this means that the mass transfer coefficients for the components of a multi-component DNAPL should be roughly the same. Consequently, for a given groundwater flow regime and DNAPL residual zone, the relative rate of mass

Table 5. Aqueous diffusivity values for selected chlorinated solvents at 20°C. Benzene data from Bonolli and Witherspoon (1968) and correlation to other compounds from Wilke and Chang (1955).

Compound	Diffusivity (m^2/s)	Relative to Benzene
1,2-Dichloroethane	1.2×10^{-9}	1.14
Dichloromethane	1.3×10^{-9}	1.29
Trichloroethylene	1.1×10^{-9}	1.05
Tetrachloroethylene	9.9×10^{-10}	0.98

transfer of the components of a multi-component DNAPL, and therefore the relative dissolved concentrations (i.e. ratios), should be dependent primarily on the effective solubility of the components.

3.2 Field Observations

Saturated dissolved concentrations are not frequently observed in groundwater except in the immediate vicinity of large DNAPL sources. This has commonly been interpreted to mean that DNAPL is not present. However, evidence of other types indicate that such interpretations may be incorrect. Less-than-saturated dissolved concentrations may be a result of factors such as: dilution of the compounds of interest by other compounds in the DNAPL; the heterogeneous distribution of DNAPL as residuals and layers or pools; dilution in monitoring wells having long intake lengths; and dispersion in the aquifer.

The dilution of the compounds of interest by other compounds in the DNAPL relates to the concept of effective solubility. For example, trichloroethylene (TCE) has a pure-phase solubility of 1,100 mg/L but if the TCE released to the subsurface is a used degreasing solvent containing significant proportions of organic compounds (i.e. oils and greases) the effective solubility of TCE may be reduced substantially.

The factors of heterogeneous distribution of DNAPL and dilution in monitoring wells are related. The scale of heterogeneity of the distribution of DNAPL residual and pools will depend on the scale of heterogeneity of the geological medium. Laboratory studies (Kueper et al. 1989) and field studies (Poulsen and Kueper In Preparation) have shown that even subtle heterogeneities in the medium can control the migration and distribution of DNAPL. The scale of heterogeneity of relevance in many geological media, and consequently the scale of heterogeneity in DNAPL distribution may be measured in centimetres. This small scale contrasts markedly with the larger scale of groundwater monitoring used typically at waste disposal and chemical spill sites. Intake zones of monitoring wells are seldom shorter than a metre and in the case of fully screened wells can be tens of metres in length. Even if some zones in the medium contain groundwater having saturated concentrations as a result of DNAPL dissolution, sampling from such wells is almost certain to draw groundwater from uncontaminated and less contaminated zones at the same time, thus diluting the observed concentration.

4 MULTI-COMPONENT DNAPL

4.1 Theory and Laboratory Studies

For multi-component DNAPL, the compounds having the highest effective solubility will be removed most readily. The preferential removal of the most soluble compounds will change the composition of the DNAPL, thereby changing the effective solubility of the remaining constituent compounds. The changes in the dissolved concentrations of the constituent compounds can be estimated based on the effective solubility of the compounds with incremental depletion and change in compositions of the DNAPL source (Shiu et al. 1988). The results of this Effective Solubility Model have been shown to compare favourably with laboratory experiments on the dissolution of hydrocarbon mixtures (Shiu et al. 1988). These experiments were conducted at water flow rates low enough to yield saturated dissolved concentrations in the water.

The results of a laboratory experiment conducted by Geller and Hunt (1989) can be used to show also that the Effective Solubility Model can simulate the relative changes in dissolved concentrations in situations when water velocities are high and dissolution is limited by the rate of mass transfer. In this experiment a mixture of 50% benzene and 50% toluene was used. Although this mixture is not a DNAPL, the dissolution processes will be similar. A known volume of benzene-toluene mixture was injected into the centre of a column packed with 40-45 mesh glass beads. Water flowed through the column at a Darcy velocity of 10 m/d, equivalent to a linear velocity of approximately 25 m/d. At this high flow rate, the initial dissolved concentrations in the column effluent were approximately 55 mg/L benzene and 14 mg/L toluene, well below saturation dissolved concentrations calculated to be 900 mg/L and 250 mg/L, respectively. The low concentrations are due to mass transfer limitations resulting from the high water flow rates and the fact that the benzene-toluene residual zone did not span the complete width of the column.

As the benzene-toluene mixture dissolved, the dissolved concentrations changed as the more soluble benzene was removed preferentially. The results of the experiment are shown in Figure 1 as a plot of the toluene-benzene ratio in the water with the fraction of the source remaining in the column.

Also shown is the change in toluene-benzene ratio predicted using the Effective Solubility Model. The observed changes in ratio compare well with the predicted changes in toluene-benzene ratio.

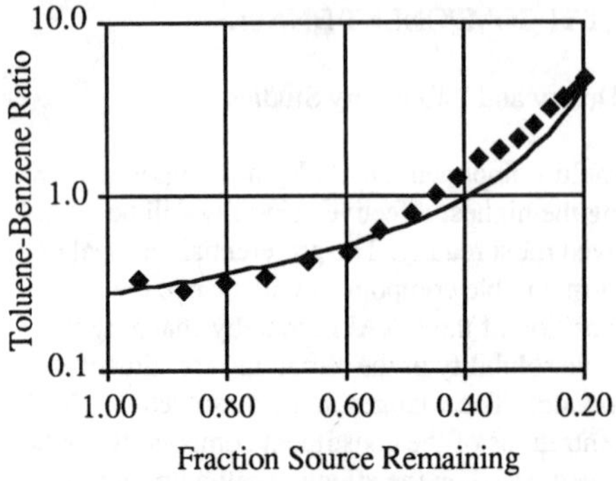

Figure 1. Comparison of benzene-toluene dissolution experiment with predicted ratios. Experimental data from Geller and Hunt (1989) shown by solid diamonds. Predicted ratios are shown by solid line.

The relationship between the relative concentrations of dissolved constituents and the fraction of the source remaining may be a useful one in the evaluation of multi-component DNAPL sources in the subsurface.

4.2 Field Observations

During the 1980's, groundwater purging operations were implemented at many waste disposal and chemical spill sites. When these purging operations were commenced, the objective of most was considered to be clean-up of the groundwater contamination. Experience has now shown us that permanent clean-up of groundwater contamination by organic chemicals to levels less than drinking water criteria or to background conditions is generally not possible. At many sites, elevated dissolved chemical concentrations continue in the groundwater even after years of purging. This is most frequently the case near the suspected source areas. A key reason for this finding is the occurrence of DNAPL in the subsurface at many sites. DNAPL in residual zones or pools represents a significant source of contamination of the groundwater which may continue for many years, decades or longer (Feenstra and Cherry 1988).

Groundwater purging operations conducted within or close to DNAPL source areas will accelerate mass depletion of the source to some degree, but the rate of depletion cannot be determined. The chemical mass removed from a purge well system can be determined from the volumes pumped and the dissolved concentrations. However, because in most situations the original mass of DNAPL residual or pools in the subsurface is unknown, there is no basis to estimate the degree of depletion of the source. The consideration of the relative concentrations of dissolved compounds derived from a DNAPL source may provide a method for the evaluation of the rate of depletion of such sources in the subsurface.

The application of this method is illustrated using the results of monitoring conducted during a groundwater purging operation at an industrial plant site in the eastern United States. The site is situated on the Atlantic Coastal Plain and is underlain by a sand and gravel aquifer approximately 15 m thick. In 1977, a dissolved plume of 1,1,1-trichloroethane (TCA) and tetrachloroethylene (PCE) was identified in the shallow aquifer. The principal source of the plume was believed to be the chemical transfer and storage area of the plant. PCE and TCA are components of the solvent products used at the plant site. In 1978, a groundwater purging system was implemented and operated until late 1984. During that time up to 14 purge wells were operated throughout the plume area and close to the suspected source area. With the exception of the wells immediately adjacent to the suspected source area, dissolved TCA and PCE concentrations in the groundwater were reduced from as high as several thousand micrograms per litres to below clean-up criteria (50 μg/L) and in some cases to below detection limits. In purge wells close to the suspected source area, dissolved concentrations were reduced significantly during purging but remained elevated (see Figure 2).

The elevated dissolved concentrations in the purge wells close to the suspected source area suggest that DNAPL remains in the subsurface. In addition,

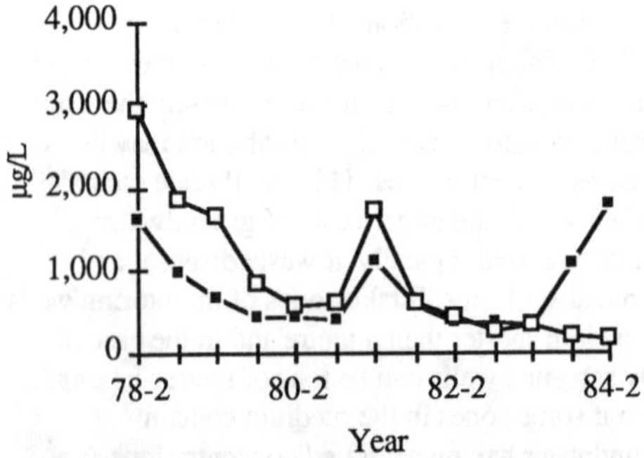

Figure 2. Dissolved concentrations in purge well GW-32 close to the suspected source during purging. TCA is shown in open squares and PCE is shown in solid squares. Data represent 6-month mean concentrations.

approximately 60% of the total chemical mass removed by the purging operations was removed from two wells close to the source. The dissolved concentrations in the purge wells close to the source area are much lower than saturation levels. This is most likely due to the heterogeneous distribution of the DNAPL source and the collection of groundwater samples from fully screened purge wells.

During the period of groundwater purging, the PCE-TCA ratio increased in the purge wells close to the suspected source area. This trend was not observed in purge wells remote from the source area. The change in PCE-TCA ratio is shown in Figure 3. TCA is more soluble than PCE (see Table 1) and the increase in PCE-TCA ratio is consistent with the preferential dissolution of TCA from the source area.

Based on the initial PCE-TCA ratio of approximately 0.5, the composition of the DNAPL is estimated to be 88% PCE and 12% TCA. With this assumption, the Effective Solubility Model can be used to predict the change in the PCE-TCA ratio as

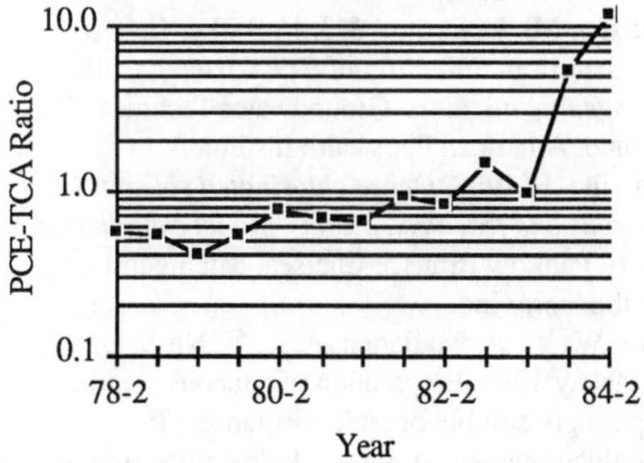

Figure 3. Change in PCE-TCA ratio in purge well GW-32 close to the suspected source area during purging operations.

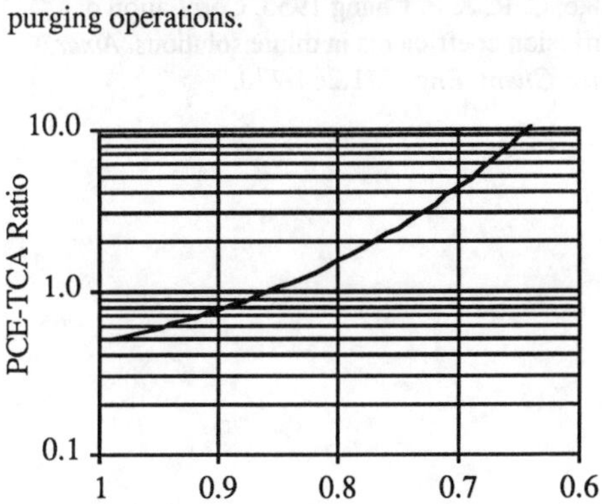

Figure 4. Predicted change in PCE-TCA ratio for dissolution of a mixture of 88% PCE and 12% TCA.

the DNAPL mass is depleted (see Figure 4).

Although the PCE-TCA ratios in the groundwater (see Figure 3) show some scatter, it is clear that there is an increase in the ratio during groundwater purging from approximately 0.5 up to approximately 5. For such a change in PCE-TCA ratio, the results of the Effective Solubility Model suggest that approximately 30% of the DNAPL source may have been removed during the groundwater purging operation. The calculations described here form the only basis for estimation of the degree of source depletion at the site because the mass of chemical released into the subsurface is not known.

5. CONCLUSIONS

For multi-component DNAPL, laboratory studies and field examples suggest that the effective solubility of the constituent compounds can be estimated by the product of the mole fraction of the chemicals in the DNAPL and the pure-phase solubilities of the chemicals in water. The chemicals having the highest effective solubility will be dissolved preferentially from the DNAPL. As the mass of DNAPL is depleted, preferential removal of these more soluble chemicals will result in changes in the composition of the DNAPL, which will in turn change the dissolved concentrations in the groundwater. For a given DNAPL composition, the changes in the ratios of the dissolved concentrations derived from the DNAPL can be related to the degree of mass depletion of the DNAPL source. The prediction of source depletion based on the changes in the ratio or relative concentration of dissolved constituents is shown to be satisfactory for dissolution of a two-component mixture in laboratory studies. The prediction of source depletion at a field site is also illustrated. The examination of changes in the relative concentrations of dissolved constituents in groundwater may provide a useful method for estimation of DNAPL source depletion with time, particularly during groundwater purging operations.

ACKNOWLEDGEMENTS

This research was supported by the University Consortium Solvents-in-Groundwater Research Program sponsored by: CIBA-GEIGY Corporation, Dow Chemical Company, Eastman Kodak Company, General Electric Corporation, the Ontario University Research Incentive Fund and the Natural Sciences and Engineering Research Council of Canada.

REFERENCES

Anderson, M. R., R. L. Johnson & J. F. Pankow 1987. The dissolution of residual dense non-aqueous phase liquid (DNAPL) from a saturated porous medium. *Proc. Petroleum Hydrocarbons and Organic Chemicals in Ground Water.* Houston, Texas, National Water Well Association.

Banerjee, S. 1984. Solubility of organic mixtures in water. *Env. Sci. & Tech..* 18:587-591.

Bonolli, L. & P. A. Witherspoon 1968. Diffusion of aromatic and cycloparaffin compounds in water at 20°C to 60°C. *J. Phys. Chem.* 72:2532- 2534.

Feenstra, S. & J. A. Cherry 1988. Subsurface contamination by dense non-aqueous phase liquid (DNAPL) chemicals. *Proc. Int. Groundwater Symp..* Halifax, Nova Scotia. International Association of Hydrogeologists.

Geller, J. T. & J. R. Hunt 1989. Non-aqueous phase organic liquids in the subsurface: Dissolution kinetics in the saturated zone. *Proc. Int. Symp. on Processes Governing the Movement and Fate of Contaminants in the Subsurface Environment.* Stanford University.

Horvath, A. L. 1982. *Halogenated Hydrocarbons: Solubility-Miscibility with Water.* New York: Marcel-Dekker, Inc.

Hunt, J. R., N. Sitar & K. S. Udell 1988. Nonaqueous phase liquid transport and cleanup 1. Analysis of mechanisms. *Wat. Resour. Res.* 24:1247-1259.

Imhoff, P. T., P. R. Jaffe & G. F. Pinder 1989. Experimental investigation of the dissolution dynamics of chlorinated hydrocarbons in porous media. *Proc. Int. Symp. on Processes Governing the Movement and Fate of Contaminants in the Subsurface Environment.* Stanford University.

Kueper, B. H. 1989. *The behaviour of dense, non-aqueous phase liquids in heterogeneous porous media.* Ph. D. Thesis, Department of Earth Sciences, University of Waterloo.

Kueper, B. H., W. Abbott & G. Farquhar 1989. Experimental observations of multiphase flow in heterogeneous porous media. *J. Cont. Hyd.* 5:83-96.

Kueper, B. H., E. O. Frind & D. B. McWhorter 1990. Application of a numerical model and laboratory parameter measurement to the movement of dense, immiscible phase liquids in a heterogeneous aquifer. *Proc. Conf. on Subsurface Contamination by Immiscible Fluids.* Calgary, Alberta. International Association of Hydrogeologists.

Leinonen, P. J. & D. Mackay 1973. The multicomponent solubility of hydrocarbons in water. *Can. J. of Chem. Eng.* 51:230-233.

Merchuk, J. C. 1983. Basic models for mass transfer. *Modular Instruction Series on Heat and Mass Transfer*, American Institute of Chemical Engineers. 33-39.

Miller, M. M., S. P. Wasik, G. L. Huang, W. Y. Shiu & D. Mackay 1985. Relationships between octanol-water partition coefficient and aqueous solubility. *Env. Sci. & Tech.* 19:522-529.

Pfannkuch, H. 1984. *Mass-exchange processes at the petroleum-water interface.* United States Geological Survey Water Resources Investigations Paper 84-4188.

Plumb, R. H. Jr. & A. M. Pitchford 1985. Volatile organic scans: Implications for ground water monitoring. *Proc. Petroleum Hydrocarbons and Organic Chemicals in Groundwater.* Houston, Texas, National Water Well Association, 207-222.

Poulsen, M. & B. H. Kueper In Preparation. *Field observations of dense, non-aqueous phase liquid behaviour in unsaturated porous media.* Waterloo Centre for Groundwater Research. University of Waterloo. Waterloo. Ontario.

Poulsen, M., L. Lemon & J. F. Barker In Preparation. *Solubility of BTEX from gasoline-oxygenate mixtures.* Groundwater Technical Task Force. American Petroleum Institute.

Schwille, F. 1988. *Dense chlorinated solvents in porous and fractured media - Model experiments.* J. F. Pankow (trans.), Chelsea, Michigan:Lewis Publishers, Inc.

Shiu, W. Y., A. Maijanen, A. L. Y. Ng & D. Mackay 1988. Preparation of aqueous solutions of sparingly soluble organic substances: II. Multicomponent systems- Hydrocarbon mixtures and petroleum products. *Env. Tox. Chem.* 7:125-137.

Wilke, C. R. & P. Chang 1955. Correlation of diffusion coefficients in dilute solutions. *Amer. Inst. Chem. Eng. J.* 1:264-270.

Subsurface Contamination by Immiscible Fluids, Weyer (ed.) © 1992 Balkema, Rotterdam. ISBN 90 6191 175 3

Some aspects of air flow through unsaturated soil by vapour extraction

J. Brauns & K. Wehrle

Institute of Soil Mechanics and Rock Mechanics, University of Karlsruhe, Germany

ABSTRACT: This paper gives a general idea of the investigations at the Institute of soil mechanics and rock mechanics (IBF) concerning the remedial actions of vapour extraction and in situ stripping. The subjects of investigations were: the air conductivity of wet soil, the flow pattern in the subsoil, the scope of vapour extraction and the general effect of in situ stripping. The investigations done so far include laboratory experiments, field experiments and numerical simulations.

1 INTRODUCTION

For some years the cleaning of contaminated soil by vapour extraction has been a widespread and rather cheap remedial action. In fact with contamination through highly volatile substances the porous structure, especially of granular soil, makes it possible to remove contaminants from the unsaturated zone using quite simple technical equipment. The success of this remedial action in principle is shown by the extraction and the recovery of a considerable amount of contaminants out of the subsoil.

Thus it can be seen that this remedial action is successful, but on the other hand there are a number of unanswered questions concerning the phase transfer processes, the transport of the contaminants, which is influenced by the technical equipment, and concerning the heterogeneous subsoil, which influences the distribution of the contaminants and the air flow in the subsoil.

Various researchers are dealing with these questions and there is some literature available on this (eg. Grathwohl 1989; Rainwater et al. 1989; Baehr et al. 1990; Johnson et al. 1990).

This paper gives a general idea of the investigations at the Institute of soil mechanics and rock mechanics (IBF) concerning the remedial actions of vapour extraction and in situ stripping. The first investigations at the IBF were done by Brauns (1987), Gottheil (1987), Engel (1988) and Mazur et al. (1988). They started out from the treatment of some contaminated sites in the valley of the river Rhine. The questions to be answered in that connection firstly concerned the air flow in the subsoil, the course of the decontamination under different operating conditions and the scope of the remedial action - whatever the scope realy means.

The goal of future activities at the IBF is to work out some basic rules for the effective and economic operation of vapour extraction. For that we are planning further field studies and laboratory experiments.

If the remediation is also to be extended to the saturated zone by in situ stripping (blowing compressed air into the contaminated groundwater), further questions concerning the principal effects should be answered.

2 AIR CONDUCTIVITY OF GRANULAR SOILS

In dry soils the air conductivity (k_a) is only dependent on the grain-size distribution, the form of the grains and the porosity (n). These parameters describe the distribution of the pore space in the soil. In wet soils on the other hand a further parameter, the water content (w) or the degree of saturation (S_r), influences the air conductivity. Figure 1 shows in two examples the relation between the relative air conductivity ($k_a/k_{a\ dry}$) and the degree of saturation (S_r) as experimental results.

The air conductivity drops with the rising degree of saturation and finally reaches zero at the critical water content (w_{crit}) or the critical degree of saturation ($S_{r\ crit}$). The pores are blocked by the pore water and the air only remains in isolated areas. In this situation, the air first has to drive out some pore water to build paths through the pore system. The dynamics of this outflow process

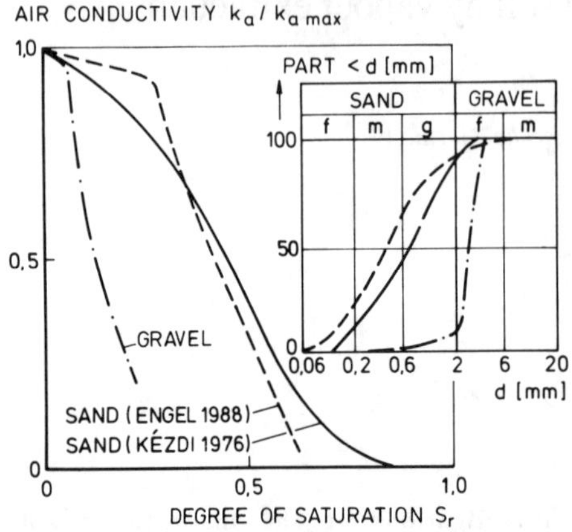

Fig.1: Relation between air conductivity (k_a) of granular soils and degree of saturation (S_r), experimental results

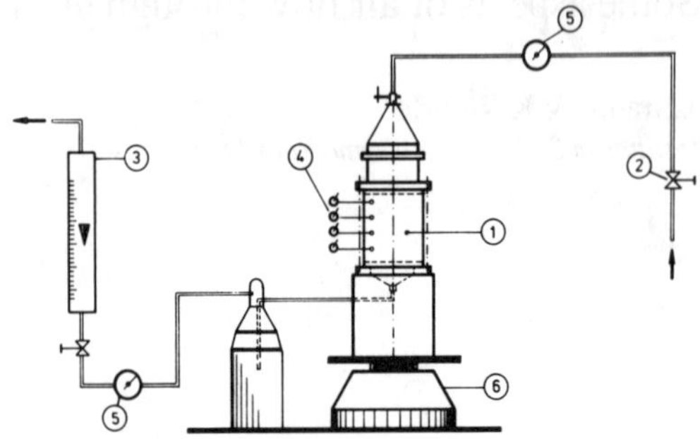

Fig.2: Equipment for measuring the air conductivity of granular soils with variable water content

1: permeability cell 4: mercury manometer
2: flow regulator 5: manometer
3: flow rate meter 6: balance

depends on the air pressure and the air pressure gradient. For measuring the air conductivity a steady air flow and a resting water phase are required. The flow rate should be low enough to leave the water motionless.

In this connection some basic questions arise concerning the vapour extraction:

- Which degree of saturation is critical and blocks the air flow through the soil and - looking at the in situ case - could the matrix potential cause such a high degree of saturation?
- And - in the case of higher in situ saturation than the critical degree of saturation - is it possible to extract pore water and to build paths through the soil by generating negative pressure and can the required conditions be achieved by using vapour extraction?

Figure 2 shows the equipment, which we use to determine the air conductivity of granular soils in relation to the water content. The flowing air causes the evaporation of pore water, therefore the water content in the soil leaves the critical point so that the soil finally becomes dry. It is not difficult to determine the varying water content during the experiment, because the total weight of the permeability cell is recorded, so that we are able to determine the evaporation of pore water and to figure out the respective water content.

As an example of the relation between air conductivity and water content, figure 3 shows the results of two experiments for different soils (sand and gravel) in a remediation area. The experiments, especially with sand, showed a very interesting phenomenon, that may also appear under field conditions and that may influence the processes during vapour extraction.

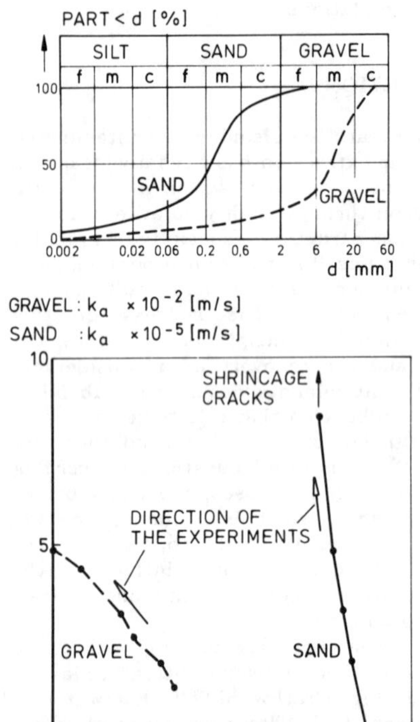

Fig.3: Results of experiments on air conductivity of two natural granular soils (Brauns et al. 1989)

The samples were first put into the permeability cell at their natural density. Then the samples were saturated and finally drained under gravity. The gravel afterwards held a water content of only w = 7% according to a degree of saturation of S_r = 48%. At the start of the experiment, the partially saturated gravel had an air conductivity of $k_a = 10^{-2}$ m/s.

On the other hand the water content of sand of w = 19.3%, held by capillary action after drainage, was higher than the critical water content of w_{crit} = 17.5%. At first the inflow of air only caused an outflow of pore water. The measurement started, when the air first broke through. The air flow was held low, so that the water had to evaporate to be removed from the sample. The air conductivity rose steadily. The interesting phenomenon mentioned above was the behaviour with a water content of about w = 15%. Shrinkage cracks appeared in the sample so that the flow rate suddenly rose rapidly.

The possibility of the appearance of this phenomenon during vapour extraction at the remediation site should be taken into special consideration. The shrinkage cracks would contradict the assumption that there is a homogeneous air flow through each layer and therefore numerical simulation would become rather difficult. Furthermore, the shrinkage could endanger buildings near the remediation area, because of soil settlements. We are going to investigate this phenomenon in further laboratory and field experiments.

3 AIR FLOW TO A VAPOUR EXTRACTION WELL

In assessing the possible effects of vapour extraction, it is, of course, interesting to see, which area of influence or which "scope" one single extraction well is attached to. This question could be answered, if the air flow in the subsoil is well known.

Investigations by various researchers (eg. Gottheil 1987, Croisé 1987, Kinzelbach and Croisé 1987) have shown, that under usual site conditions numerical solutions may be used, which start out from incompressible gas. Therefore, finite element (FE) programs developed for ground-water flow can also be applied to determine the air flow approximately.

As can be expected, the nature of the soil and the boundary conditions are very important. Looking for example at a homogeneous subsoil with constant and isotropic air conductivity where one single vapour extraction well may be installed, it can be seen, that differences in the lengths and the location of the filter cause crucially different situations of air flow in the subsoil.

Figure 4 shows - as an example - the streamlines and the potential lines of three different situations, each with an unhindered inflow of air through the surface of the soil. In all three cases, the streamlines mark areas of 10%, 20% etc. of the total volume of air flow to the extraction well. The potential lines mark the position of the lines of 90%, 80% etc. of the negative pressure, which is generated in the extraction well.

With regard to the progress of the decontamination, the determination of the scope of the action is of particular interest. In figure 4 the differences in flow situation are shown by marking the radius r_{90}, that forms the boundary of the area, 90% of the inflowing air (Q_a) passes through. This boundary does not fix the real scope of the remedial action, it only promotes the comparison of the situations shown. The scope should be defined with regard to the special problems and processes of vapour extraction. This definition should be influenced by various parameters, which are to be further investigated in combination with one another. The starting point of a new definition of the scope should be the exchange of air in the pore space. In relation to this the distribution equilibria of the contaminants, the

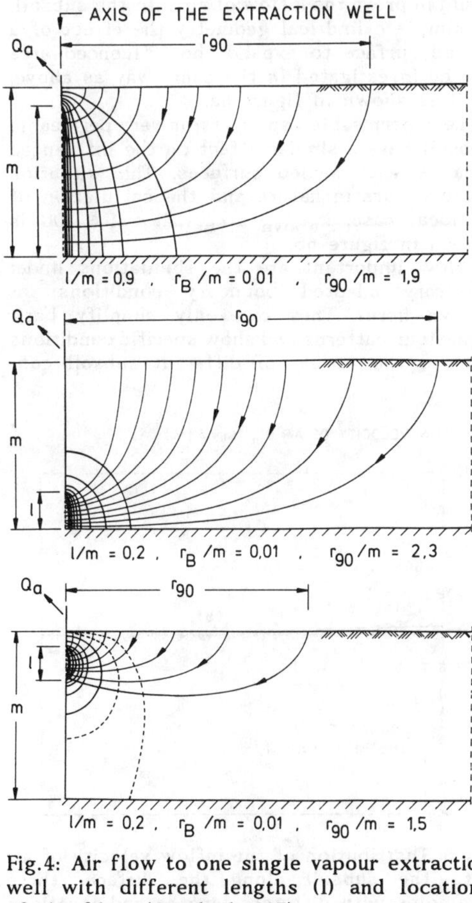

Fig.4: Air flow to one single vapour extraction well with different lengths (l) and locations of the filter (Gottheil 1987)

velocity of phase transfer and the subsoil parameters, such as water content and temperature, play a part.

For the three examples in figure 4, figure 5 shows the distribution of the velocity of air flowing into the soil to clarify the influence of the velocity on the processes.

As was to be expected in the different cases, the distribution of the velocity along the surface varies greatly. If it is taken into consideration that the flow of air on its way to the extraction well absorbs varying quantities of contaminants depending on their distribution in the subsoil and on their velocity of phase transfer and that its ability to absorb decreases, a further aspect of the complexity of the dynamics of these processes can be seen.

Conditions as shown in figures 4 and 5 suggest that investigations should be made, as to whether decontamination can be controlled or optimized by technical measures. Sealing the surface for example could expand the scope of the vapour extraction and setting up further ventilation wells around the extraction well could improve the inflow of air into the subsoil. In simple cylindrical geometry the effect of a sealed surface to expand the influenced area can be investigated in the same way as above, as it is shown in figure 6a.

Less permeable top layers on very permeable subsoil have a similar effect on the influenced area as with sealed surfaces. This situation often occurs in nature and the calculation of an ideal case, $k_{above} : k_{below} = 1 : 100$ is shown in figure 6b.

How important are the simulations under arbitrary adapted boundary conditions, as shown here? They can only simplify basic behaviour patterns and show specific conditions. Due to the variety of different subsoil con-

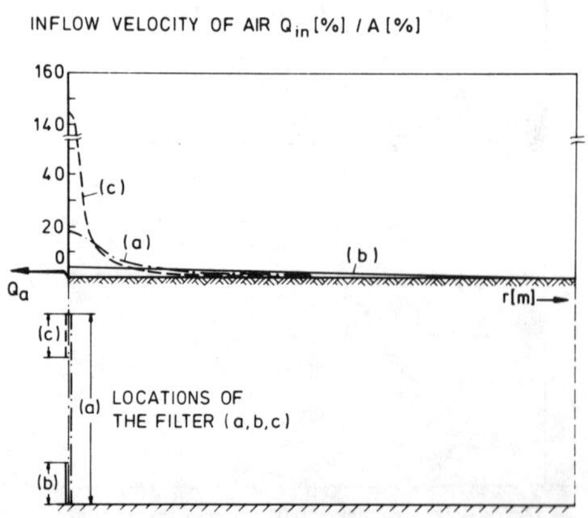

Fig.5: Distribution of the inflow velocity of air into the subsoil along the surface, three examples with different lengths and locations of the filter

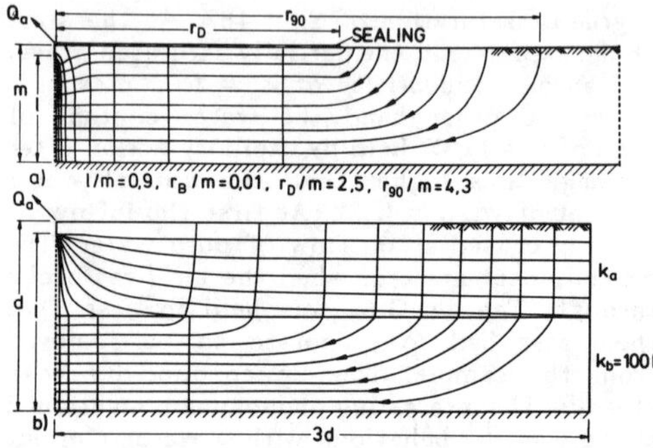

Fig.6: Two cases of airflow to an extraction well (Gottheil 1987) (a) with an artificially sealed surface (b) with layered subsoil

ditions an almost infinite number of situations can be modelled. However, the main question, which arises, is as to how far these simulations can represent the complex conditions, which occur in nature, as a rule. At the same time, the problem should also be considered, as to whether it is possible to collect sufficient data for a particular case, in order to be able to obtain a realistic model.

The following illustrates the possibilities and limitations here by means of two examples described in detail.

4 RESULTS OF INVESTIGATIONS IN SITU

4.1 Vapour extraction

Tests and measurements were carried out with an extraction well at a redeveloped site in deep permeable Rhine river sediments (see Mazur et al. 1988). Only the aspects regarding airflow will be described here in more detail, although the comprehensive results also included decontamination processes.

The unsaturated zone went down to approximately 6 m below the surface. The composition of the subsoil was investigated by means of an continuous sample boring. Figure 7 shows the result of this sample boring together with the values obtained regarding the water and air conuctivities.

Each layer was examined individually to determine the grain size distribution, from which the conductivity k could be estimated. The conductivity k_a was determined from laboratory experiments. A certain variation in both values can be seen and - on the right of the picture - it can be seen that the relation k_a/k is not constant, which could be expected solely due to the differences in the viscosity of air and water. The relation values obtained in this manner range from 24 to 107. The mean

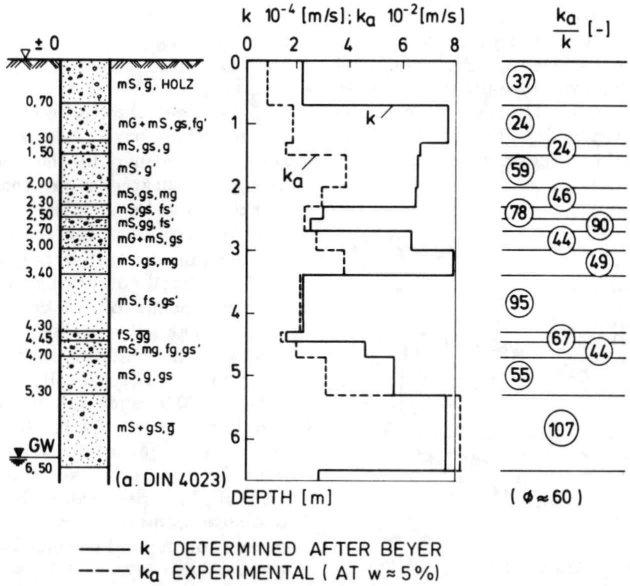

Fig.7: Soil layers and corresponding conductivity at the test site A

value of the relation figures amounts to 60.

During the tests, air was extracted from one single well. The negative pressure arising from this was measured at a particular depth, at various distances from the well.

At the same time an attempt was made to calculate the flow pattern. For this purpose the soil structure (see fig.7) was simplified as follows:

- the top layer (thickness 0.7 m) remained unchanged with an air conductivity of k_a = 0.7 * 10^{-2} m/s
- the alternating layers between the top layer and the water table, which vary by a factor less than 10, were represented by the following values:
  - $k_{a\ horizontal} = k_{ah}$ = 3.2 * 10^{-2} m/s
  - $k_{a\ vertical} = k_{av}$ = 2.6 * 10^{-2} m/s
  - corresponding anisotropic relation k_{ah} / k_{av} = 1.23

  (these figures were calculated from the layer values, taking the thicknesses of the layers into consideration).

The result of a corresponding model calculation is shown in figure 8 and is compared with the negative pressures measured.

As can be seen, the inflow area is relatively limited due to the comparatively small variations in the soil composition. The 80%-of-Q_a-line (r_{80}) is situated approximately 12m from the well, which corresponds to twice the depth of the unsaturated zone. The result of a comparative calculation for the negative pressure of Δp = -46 mbar in the well was a flow quantity of Q_a = 420 m³ /h. This calculation was 20% above the measured flow quantity of Q_a = 350 m³/h.

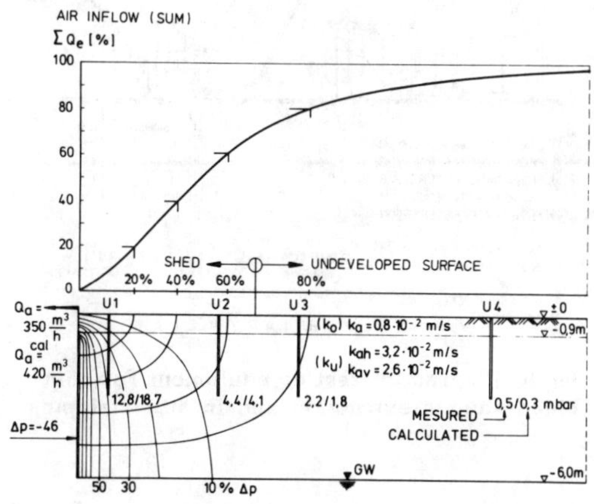

Fig.8: Structure of subsoil (simplified for calculation purposes) and flow patterns at the test site A

The negative pessures measured at the sensing probes are listed next to the calculated figures. Despite the unavoidable simplifications, the results calculated can be regarded as being quite satisfactory as a start.

No further studies regarding the dynamics of the actual decontamination have as yet been carried out. For this purpose even more detailed investigations into the behaviour of volatile substances in the various types of soils need to be carried out (see above).

The second example (see figures 9 and 10) is concerned with a site under remedial action, where extensive investigations have also been carried out. However, here we should limit ourselves mainly to the air movement processes during vapour extraction.

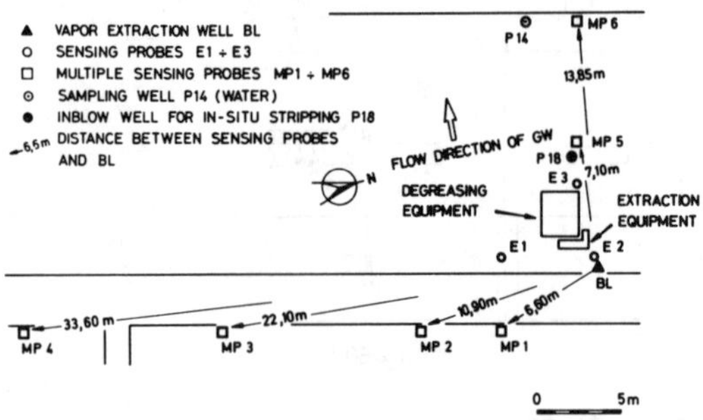

Fig. 9: Plan of the test site B

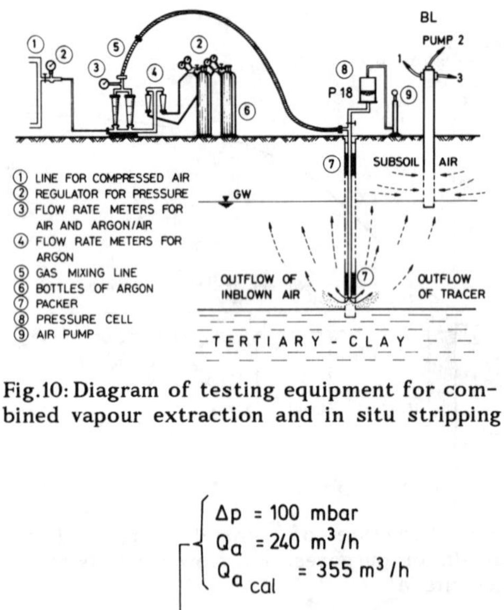

Fig.10: Diagram of testing equipment for combined vapour extraction and in situ stripping

These investigations included systematically observed tests, blowing compressed air into the saturated zone - in situ stripping (see chapt. 4.2).

This site is also to be found in the sediments of the Rhine river, although here the layers are more defined in the region of the 4.5 m deep unsaturated zone than in the example mentioned above. The subsoil conditions, sensing probe observations and results of calculations are all listed in figure 11.

The subsoil conditions were also investigated here by means of an continuous soil sample boring. The air conductivity of the granular layers was calculated from the grain size distribution. The cohesive soils at the depth of between 0.9 and 2.0 m posed a particular problem and it was assumed that these were to be found as undisturbed horizontal layers here. This material was regarded as being only very slightly permeable, however the negative pressure conditions observed did not by any means correspond to this. With an air conductivity in these layers of $k_a = 10^{-3}$ m/s the negative pressures can be modelled relatively well, as can be seen in the comparison with each of the values recorded for the sensing probes. The air flow, however, is then too large by a factor of approximately 1.5 compared to the amount in the experiment (which is even though a relatively good result).

Based on the observations made, it had to be assumed, that the cohesive top layer was on an average considerably more permeable than the undisturbed natural soil due to man-made disturbances (e g. foundations) or due to shrinkage cracks from drying. The site is in a large factory building, so access of natural moisture is prevented.

Whichever way each of the observations is interpreted, the definite influence of sealing top layers becomes clear in the examples described, as mentioned earlier. In the case

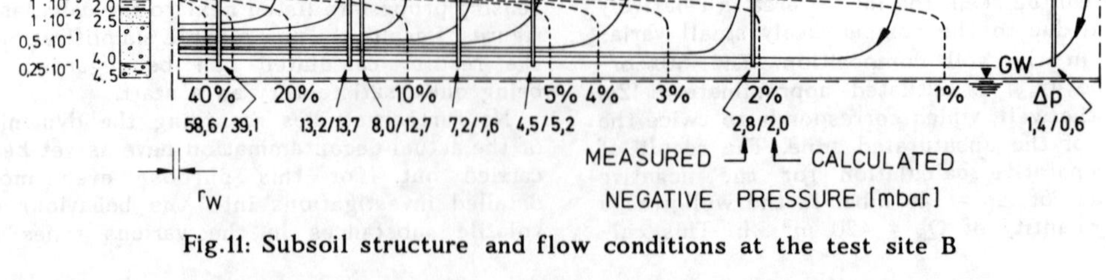

Fig.11: Subsoil structure and flow conditions at the test site B

shown in figure 11 it should be noted that the 80%-of- Q_a-line (r_{80}) was equal to six times the thickness of the unsaturated zone. If the top layer had been intact, the corresponding radius would certainly have been considerably larger.

The question as to the scope is, as already mentioned, still open as far as the main points are concerned; however, the influence of the top layers is of great importance, as can be seen. We have, therefore, looked into this problem by carrying out a parametric study of negative pressure levels and of the distribution of the air inflow depending on the ratio of the conductivity of the top layer to the soil to be decontaminated (k_{above}/k_{below}). The result of this investigation is shown in figure 12 and may be of use as a basis in future experiments.

On the one hand the negative pressure levels under the top layer are shown and, on the other hand, the cumulative curve of the airflow into the surface of the soil (in % of Q_a) along a radius is given, i.e. for 3 different ratios of k_{above}/k_{below}. The varying scope and the corresponding decrease in the negative pressure with the distance from the extraction well becomes immediately clear here.

4.2 In situ stripping

There are still many answers to be found with regard to a thorough quantitive analysis of the pressing of air into the saturated zone for the purpose of in situ stripping and its influence on the processes, which occur during vapour extraction. In this connection we have carried out several investigations on test site B, including tracer experiments to make it possible to follow the course of the air flow (see fig.9, see Böhler et al. 1989).

The experiments here have shown, that - at least under the conditions at this site - the inflow of air into the saturated zone leads to a noticeable transport of highly volatile

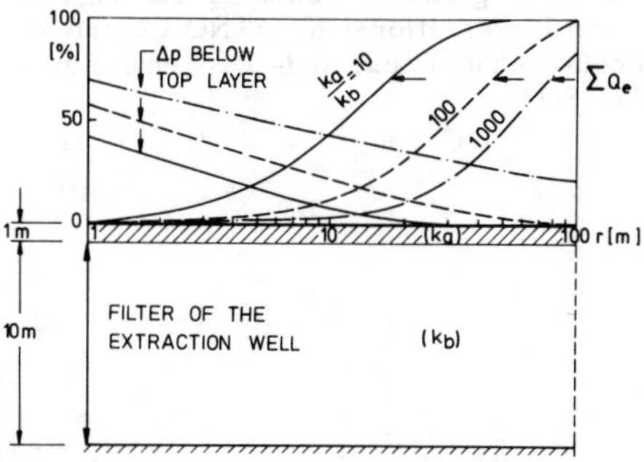

Fig.12: Negative pressure levels and distribution of inflowing air during vapour extraction from soil with top layer

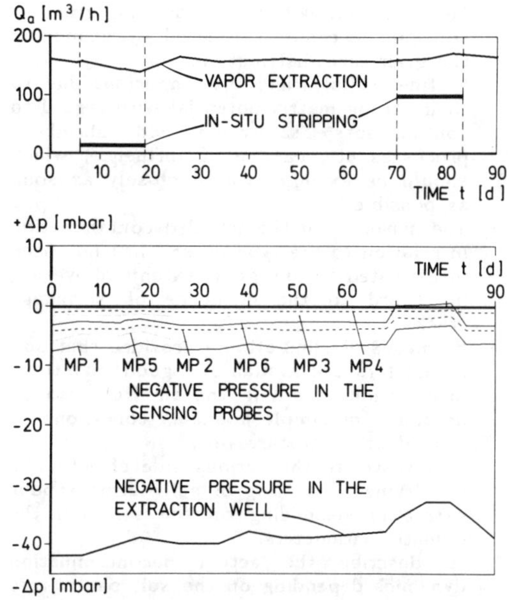

Fig.13: Lines of the air flow quantities and pressure lines of a combined vapour extraction and in situ stripping test

contaminants (CHC) out of the ground water area. However, depending on the subsoil conditions, there is a danger here that the contaminants can drift off to the side, either when particles move, due to the inflow of air, or as a result of induced ground water movements, which have been observed as a secondary effect (see Wehrle 1990).

With regard to the air pressure conditions in the unsaturated area during the combination of vapour extraction and in situ stripping, our measurements have shown, that in the region above the inflow point even over-pressure occurred, although the quantity extracted amounted to 1.6 times the inflow quantity of air (see figure 13).

From this it becomes clear, how much care has to be taken, when using this method and that, in fact, even more careful examinations are necessary, in order to be able to assess the processes with sufficient accuracy in actual cases.

5 SUMMARY AND FUTURE PROSPECTS

In the preceding sections several aspects concerning the air flow during vapour extraction from the soil have been discussed. According to the opinion of the authors, the following points can be ascertained for conditions in soil:

- The air conductivity of free draining granular soil can be easiliy estimated by examining the grain size distribution.
- In finer soils, which are important due to their strong matrix potential with regard to contaminants, saturation and shrinkage processes etc. can be of influence, which should be examined more closely as soon as possible.
- The dynamics of the actual decontamination in relation to the type of soil has not been investigated to any great extent, above all, in regard to the influence of shrinkage cracks.
- By means of simulation - provided that soil parameters are available - good quantitive information regarding the air flow can be obtained for simple boundary conditions.

The goal of our future work is:

- to investigate the various side effects like eg. drying or wetting of soil during vapour extraction depending on the soil and the climatic parameters,
- to describe the actual decontamination dynamics depending on the soil parameters and the operating conditions,
- to optimize the equipment and the operating conditions for vapour extraction and in situ stripping.

REFERENCES

Baehr, A.L., Hoag G.E.& Marley, M.C. (1990): Removing volatile contaminants from the unsaturated zone by inducing advective air-phase transport. Journal of contaminant hydrology, Vol. 4, pp. 1-26.

Böhler, U., Brauns, J.& Hötzl, H. (1989): Bodenluftabsaugung und Drucklufteinblasung zur Sanierung von CKW-Schadensfällen: Systematische Untersuchungen zu einer Sanierungsmaßnahme im Lockergestein. Mitteilungen Abteilung Erddammbau und Deponiebau am Institut für Boden- und Felsmechanik, Universität Karlsruhe, Heft 2, Karlsruhe 1989, 128 pp.

Brauns, J. (1987): Bodenmechanische Aspekte der Bodenluftabsaugung. In: Bodenluftabsaugung. Landesanstalt für Umweltschutz BW Karlsruhe (ed.), pp. 12-17.

Brauns, J., Dittmar, Ch.& Wehrle, K. (1989): Strömungsvorgänge und Bodenverhalten bei der Bodenluftabsaugung in einem flachen bedeckten Lockergesteinskörper (EVS-Tübingen). Abschlußbericht zum Forschungsvorhaben, im Auftrag der Landesanstalt für Umweltschutz BW Karlsruhe (unpubl.), 30 pp.

Croisé, J. (1987): Modelisation d'un Ecoulement gazeux en Milieu poreux non-sature. Institut für Wasserbau, Universität Stuttgart, 166 pp.

Engel, W. (1988): Meßsonde zur Erkundung der Feinschichtung von Böden auf dem Prinzip der in-situ-Bestimung der Durchlässigkeit. Diplomarbeit am Inst. f. Boden- und Felsmechanik der Universität Karlsruhe (unpubl.), 91 pp.

Gottheil, K. (1987): Numerische Berechnungen der Potentialströmung mit Hilfe der Finite-Elemente-Methode (FEM) für das Sanierungsverfahren der Bodenluftabsaugung bei Untergrundkontaminationen mit leichtflüchtigen Schadstoffen. Vertieferarbeit am Institut für Boden- und Felsmechanik der Universität Karlsruhe (unpubl.), 30 pp.

Grathwohl, P. (1989): Verteilung unpolarer organischer Verbindungen in der wasserungesättigten Bodenzone am Beispiel leichtflüchtiger aliphatischer Chlorkohlenwasserstoffe (Modellversuche). Tübinger geowissenschaftliche Arbeiten, Reihe C, Nr. 1, 102 pp.

Johnson, P.C., Stanley, C.C., Kemblowski, M.W., Byers, D.L.& Colthart, J.D. (1990): A practical approach to the design, operation, and monitoring of in situ soil venting systems. GWMR, Spring 1990, pp. 159-178.

Kézdi, A. (1976): Fragen der Bodenphysik. VDI-Verlag, Düsseldorf.

Kinzelbach, W.& Croisé, J. (1987): Zur Berechnung der Luftströmung im Untergrund bei Luftabsaugung. In: Bodenluftabsaugung. Landesanstalt für Umweltschutz BW Karlsruhe (ed.), pp. 24-31.

Mazur, W., Denzel, S., Schneider, F.& Stadler, W. (1988): Bodenluftabsaugung und In-situ-Strippen. Bericht zum Forschungsvorhaben, im Auftrag der Landesanstalt für Umweltschutz BW Karlsruhe (unpubl.), 182 pp.

Rainwater, K., Zaman, M.R., Claborn B.J.& Parker, H.W. (1989): Experimental and modelling studies of in situ volatilization: Vapor-liquid equilibrium or diffusion-controlled process? Proceedings of the petroleum hydrocarbons and organic chemicals in ground water conference, NWWA/API, pp. 457-472.

Wehrle, K. (1990): In situ cleaning of CHC-contaminated sites: Model-scale experiments using the air injection (in situ stripping) method in granular soils. Proceedings of the third international KfK/TNO Conference on contaminated soil, 10-14 Dec. 1990, Karlsruhe, FRG.

Subsurface Contamination by Immiscible Fluids, Weyer (ed.) © 1992 Balkema, Rotterdam. ISBN 90 6191 175 3

Transport of trichloroethylene vapours in the unsaturated zone: A field experiment

Bernadette M. Hughes, Robert W. Gillham & Carl A. Mendoza
Waterloo Centre for Groundwater Research, University of Waterloo, Ont., Canada

ABSTRACT: A consideration of the impact of a halogenated solvent spill on the subsurface must include the distribution of the chemical within the unsaturated zone. As a result of the volatile nature of most solvents, this distribution will be affected by the transport of vapours within the soil gas. Insight into this transport process has been provided by the results of laboratory experiments and numerical modelling, but the theory has not been tested in the field. Two experiments were carried out to investigate the transport behaviour of trichloroethylene vapours in a section of unsaturated sands and silts at a field site. Laterally extensive vapour plumes were found to form over relatively short time spans. Plume geometry is thought to have been influenced significantly by variations in temperature, soil organic matter and soil moisture. The results of these experiments were in reasonably good agreement with simulations performed using a numerical model.

1 INTRODUCTION

The widespread occurrence of halogenated solvents in groundwater, coupled with their highly toxic nature, has led to increasing concern regarding their impact on drinking water supplies. Consequently, research into the subsurface behaviour of these compounds has dealt primarily with their occurrence and migration as pure-phase or dissolved contaminants within the saturated zone (Mackay et al., 1985; Feenstra and Cherry, 1988). The fate of these chemicals within the unsaturated zone has received considerably less attention.

Most solvent releases occur at or near the ground surface and typically cause the contamination of some portion of the unsaturated zone by a residual liquid product. The high vapour pressures characteristic of most halogenated solvents provide the potential for volatilization of the liquid and subsequent transport as vapours within the soil gas. Both contamination of underlying groundwater reserves and buildup of hazardous vapour levels in nearby structures are possible consequences of such transport.

A better understanding of the vapour-phase component of contaminant transport is, therefore, essential for comprehensive site investigations and the formulation of effective spill response and remediation strategies.

Recent mathematical modelling of hypothetical solvent spills has indicated that the impact of vapour-phase transport on groundwater may indeed be significant (Mendoza and McAlary, 1990; Sleep and Sykes, 1989). The degree to which the transport theory incorporated by these models predicts actual field behaviour remains, as yet, uncertain.

Previous laboratory investigation of the controls on vapour transport of organic solvents consists mainly of batch or column studies, which allow for the isolation and quantification of one or two parameters in an otherwise constant system. Extrapolation of such research to natural systems, however, must incorporate both the combined effect of the physical and chemical properties characteristic of the site and the impact of temporal and spatial variations of these properties on the transport process.

The purpose of this paper is to provide an overview of two recently conducted experiments investigating the field-scale transport behaviour of a halogenated organic solvent vapour (Hughes, 1991). The primary objective of the research project was to provide a detailed mapping of the three-dimensional migration of a trichloroethylene (TCE) vapour plume in a natural sandy unsaturated zone. The second objective was to repeat the experiment under slightly varied conditions to provide a qualitative indication of the relative influence of the major variables affecting the transport process. The final objective was to compare the field data to a numerical model simulating vapour transport.

2 TRANSPORT THEORY

The major processes expected to affect the distribution and fate of solvent vapours in the unsaturated zone are summarized in Figure 1. Following a spill, a portion of the liquid solvent entering the unsaturated zone will become immobilized in the pore spaces as a result of capillary forces. Volatilization of this residual liquid into the adjacent soil gas will result in the formation of a vapour-phase contaminant plume. The contaminant, now remobilized, may spread vertically or laterally within the unsaturated zone. As the vapour plume comes into contact with uncontaminated soil moisture, migration will be retarded by partitioning of some of the contaminant to the soil moisture. In the case of a hydrophobic organic vapour, such as TCE, transport will also be retarded, relative to a "non-reactive" vapour, as a result of solid-phase sorption.

Vapours may ultimately be lost to the atmosphere at ground surface or to subsurface excavations. In addition, contamination of the underlying groundwater may occur due to the downward displacement of contaminated soil moisture by an infiltration event or by dissolution of vapours reaching the capillary fringe.

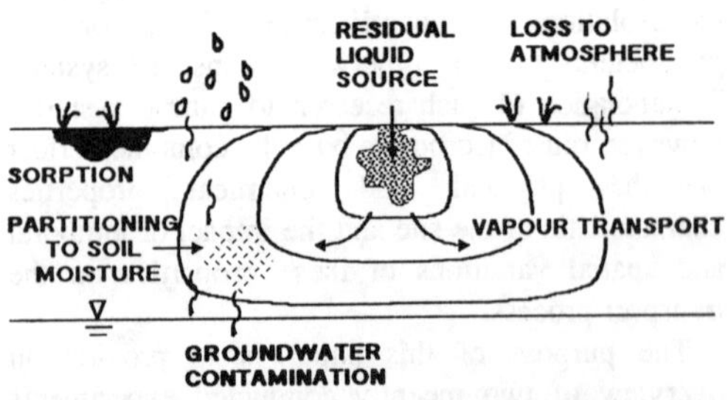

Fig. 1 Processes affecting the distribution of solvent vapours in the unsaturated zone.

2.1 *Advective Flow of Gases*

The transport of vapours within the unsaturated zone has traditionally been treated as a diffusive process. However, contamination of the soil gas by relatively high molecular weight vapours, such as halogenated solvents, will form a mixture which is denser than the surrounding uncontaminated soil gas. If the resulting density gradient is significant, it can induce advective flow of the soil gas. This will cause a "sinking" effect on the plume and increase both the rate and distance of transport relative to that predicted by diffusive transport alone. Given that this density effect is concentration-dependent and the vapour pressure of a given compound is a function of temperature, the magnitude of density-induced advection will be sensitive to subsurface temperature variations.

The application of numerical solutions to the density-dependent flow problem has been used to demonstrate, in theory, the significance of the density-induced advection of organic vapours and the sensitivity to specific physical and chemical parameters.(Mendoza and Frind, 1990b; Falta et al., 1989; Sleep and Sykes, 1989). Advective fluxes of gases may also occur in response to pressure gradients caused by barometric fluctuations, compound vapourization (Mendoza and Frind, 1990a), fluctuating water levels (Weeks et al. 1982), and subsurface depressurization due to manmade excavations (Nazaroff et al. 1987).

2.2 *Transport of Reactive Vapours*

The advection-dispersion equation for vapour contaminant transport, expressed for simplicity in the one-dimensional form, is:

$$\frac{\partial}{\partial x}(\theta_g D \frac{\partial C_g}{\partial x}) - \frac{\partial}{\partial x}(\theta_g v C_g) = \frac{\partial}{\partial t}(\theta_g R C_g) \quad (1)$$

where C_g is the concentration in the gas phase (M/L^3), v is the advective velocity of the gas mixture (L/T), θ_g is the gas-filled porosity (L^3/L^3), D is the dispersion coefficient (L^2/T), and R is the retardation factor (-). The second term describes the advective component of transport and is linked to the solution of the flow equation through the velocity term v. If density contrasts or other factors are insufficient to cause significant advective velocities, this term becomes negligible and equation

(1) reduces to the diffusion equation for reactive transport. Vapour migration will also be affected by the processes of dispersion and retardation. Expansion of the dispersion coefficient gives:

$$D = \alpha v + D_a \tau \qquad (2)$$

where α is the dispersivity of the porous medium (L), D_a is the free air diffusion coefficient for the compound (L^2/T), which varies with temperature, and τ is the tortuosity factor (-), which is a function of both the total porosity and the moisture content.

For vapour transport the retardation factor R is commonly expressed as:

$$R = 1 + \frac{\theta_w}{\theta_g}\frac{1}{H} + \frac{\rho_b}{\theta_g H} K_d \qquad (3)$$

where θ_w is the water-filled porosity (L^3/L^3), H is the Henry's constant (-), ρ_b is the bulk density of the porous medium (M/L^3), and K_d is the equilibrium solid/aqueous partitioning coefficient (L^3/M).

Use of the above formulation for the retardation factor implies that vapour sorption to solids can be described by a two-stage process: equilibrium partitioning of the contaminant between the vapour and aqueous phases followed by sorption of the aqueous-phase contaminant to the soil solids.

Equilibrium partitioning between the aqueous and vapour phases is described by Henry's law:

$$C_g = H C_{aq} \qquad (4)$$

where C_{aq} is the aqueous-phase solute concentration (M/L^3). Under ideal conditions H can also be described by:

$$H = \frac{P_v}{S} \qquad (5)$$

where P_v is the vapour pressure of the compound, and S is the compound's aqueous solubility. (In this case H has a dimensional form which depends on the units used to express P_v and S). Both vapour pressure and solubility are functions of temperature and pressure but the effects caused by pressure changes in the unsaturated zone are expected to be small. However, temperature will have a large effect on the value of H over common subsurface ranges as a result of the large increase in vapour pressure with temperature relative to smaller changes in solubility.

For the portion of the contaminant in the aqueous phase, the sorption process can be described by the solid/aqueous partitioning coefficient K_d. Use of a constant partitioning coefficient implies that the process is independent of solute concentration. This has been shown to hold true to near-saturation concentrations for several organic solvents (Chiou et al. 1979). The value of K_d is frequently calculated according to the following relationship:

$$K_d = K_{oc} f_{oc} \qquad (6)$$

where K_{oc} (L^3/M) is the organic carbon/aqueous partitioning coefficient for the solute and f_{oc} is the fractional organic carbon content of the soil. Use of equation (6) reflects the assumption that sorption of organic compounds to soils is controlled by partitioning of the solute to the organic matter in the soil. In sediments containing >0.1% organic carbon, this is generally held to be valid (Schwarzenbach and Westall, 1981; Chiou, 1981; Karickhoff et al. 1979). At the low organic carbon contents typical of most aquifer sands, however, the contribution of mineral surface adsorption may become more important and use of equation (6) has been found to underpredict sorption in these materials (Schwarzenbach and Westall, 1981; Piwoni and Banerjee, 1989; Larsen et al. 1989)

Recent research has investigated the validity of applying an aqueous partitioning approach to the prediction of vapour-phase sorption. Results indicate that at very low water contents, the sorptive capacity of the solid phase is considerably increased (Chiou and Shoup, 1985; Peterson et al., 1988; English and Loehr, 1989). The limiting water content values below which this enhanced sorptive behaviour becomes important has not yet been resolved. The work of Peterson et al. (1988) suggests that sorption from a vapour-phase source may still exceed that predicted using aqueous sorption theory at moisture contents typical of field capacity. Their research, however involved synthetic soil with a very high specific surface area (206 m^2/g). As noted by the authors, this would correspond to a much thinner coating of water on the soil grains than would be expected for a typical aquifer material at the same moisture content. Chiou and Shoup (1985) found the sorptive behaviour of benzene and chlorinated benzene vapours to a natural soil at 90% relative humidity to be very similar to the sorption of these same chemicals from an aqueous phase. For soils in humid or temperate regions with significant amounts of organic matter, the partitioning process represented by equation (6) is probably adequate.

3 EXPERIMENTAL APPROACH

The vapour transport field experiments consisted of a two-stage program conducted in the sandy unsaturated sediments at the Canadian Forces Base Borden field site, Alliston, Ontario. Both experiments involved the mapping of the spatial and temporal distribution of TCE in the soil-gas in the vicinity of an emplaced source of residual liquid product. The source was established by mixing pure TCE with native sands to approximately 4% by volume and burying the mixture in a cylindrical excavation in the middle of the site. Soil-gas samples were collected by means of a network of specially designed gas-sampling probes. The probes were concentrated along one detailed radial axis extending southwest from the center of the source. A few other samplers were installed along different axes to provide checks on symmetry of transport. TCE vapour concentrations were determined by analysis of the samples on-site with a portable gas chromatograph equipped with a photo-ionization detector (Photovac model 10S70). The layout of the samplers along the detailed profile is illustrated in Figure 2. The cross-hatched rectangle indicates the location of the emplaced source. Other labels refer to the names of the individual sampler nests.

For the first experiment, carried out in November-December 1988, the ground surface was covered with an 18 by 18 metre sheet of 0.51-mm thick PVC plastic. The purpose of placing this cover was to prevent surface loss of vapours or infiltration of precipitation to the site. This minimized the number of variables to be considered and provided the base-case for calibration of the computer model.

The second experiment was conducted in July-August, 1989. The experimental setup was

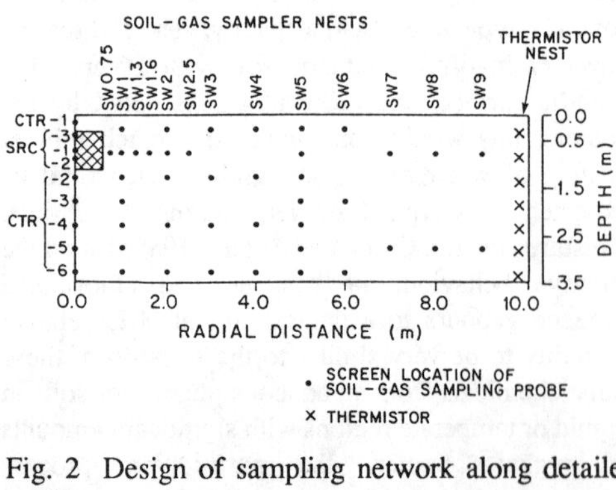

Fig. 2 Design of sampling network along detailed profile.

similar to the first experiment except that the surface cover was removed. A small cover was retained in the immediate vicinity of the source to prevent precipitation from infiltrating through the residual TCE liquid. In order to reduce the vapour contamination persisting from the first experiment, the site was remediated by vacuum extraction for 10 days prior to the start of the second experiment. Background concentrations at the beginning of the second experiment were in the range of 10 to 15 ppmv (parts per million by volume).

3.1 *Site Description*

The unsaturated zone at the site is a 3 to 4 metre section of silts and sands with annual water-table fluctuations on the order of 0.5 m to 1 m. The lower 1 to 2 metres of the section consists of bedded sands which grade upwards to more massive sands and silts. A dark-brown organic-rich layer, approximately 10 cm in thickness, occurs at a depth of about 60 cm from surface. This layer persists over the area of the site intersected by the detailed sampling profile but shallows and disappears from the section a few metres to the north of the source pit. This silty marker horizon is overlain by a very heterogeneous section of silts and sands with discontinuous lenses of more silty, organic-rich sediments, extending to surface.

This lithologic sequence is reflected in the depth variation of the organic carbon content (f_{oc}) and residual moisture content profiles measured for a suite of samples collected 2 days before the start of the second experiment (Figure 3). The occurence of the silty marker horizon at 55 cm depth is evident as a contrast in both parameters relative to the underlying sediments. At shallower depths the organic contents are relatively high but are expected to be laterally variable due to the discontinuous nature of the upper units.

A residual moisture content profile, determined from samples collected from an area of the site that had been covered by plastic for 8 months, showed only slightly lower moisture contents than those presented in Figure 3. As there was no precipitation for the first 21 days of the second experiment, a constant moisture profile at residual saturation is a reasonable assumption for this period. Hydraulic conductivity values for the sediments, as measured by falling head permeameter tests performed on two cores, range from 2.8×10^{-5} m/s to 2.4×10^{-4} m/s with an average of 1.3×10^{-4} m/s.

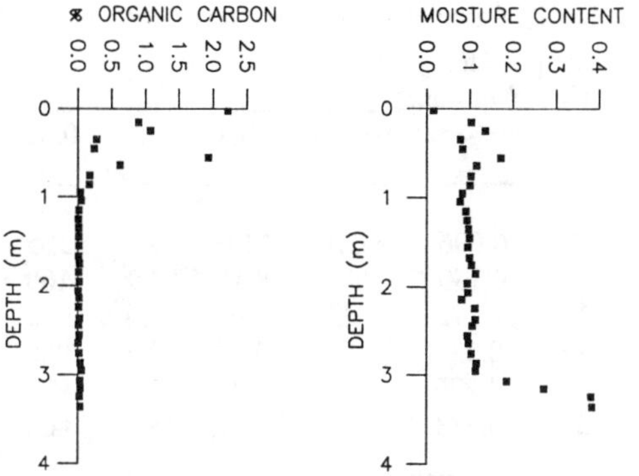

Fig. 3 Organic carbon and moisture content variation with depth.

4 RESULTS AND DISCUSSION

The TCE vapour concentration profiles for days 5 and 18 of experiments 1 and 2, are shown in Figure 4. The contours are in log of TCE concentration expressed as parts per million by volume (ppmv) and the sampling points from which concentrations were determined for each day are indicated by the solid circles on each plot. For purpose of comparison, the outer solid contour presented in each plot corresponds to a concentration of 32 ppmv TCE, slightly above the background concentrations for the second experiment. The uncontaminated background conditions for experiment 1, however, allowed for resolution of the vapour plume to lower levels as indicated by the dotted 0.1 ppmv contours.

For the plots from the second experiment, the concentration contours were not extended to ground surface. The open surface condition should result in a curvature of the concentration contours at a shallow depth to the degree that the surface approaches a zero-concentration boundary. Delineation of this vertical gradient, however, would require sampling over very small depth increments.

In both experiments the methodology appears to have provided sufficient accuracy to allow for good plume definition. Both experiments resulted in significant lateral migration within several days of source emplacement. The plume in the second experiment exhibits a more rapid rate of migration and a greater mass of contaminant near the source as reflected in the higher vapour concentrations. At first glance, these results may appear counter-intuitive given that the surface was covered in the first experiment to prevent mass loss. The explanation for this lies in the fact that there is significant seasonal temperature variation at the site as shown in Figure 5. The higher temperatures for experiment 2, relative to experiment 1, produced higher vapour concentrations within and near the source. The resulting effect is a greater amount of mass per radial distance, increased concentration gradients, and increased potential for density-induced advection. Sensitivity analysis using comparisons to computer simulations indicated that the advective component of transport was more significant for the second experiment. Diffusion was the dominant transport mechanism for the first experiment (Mendoza et al., 1992).

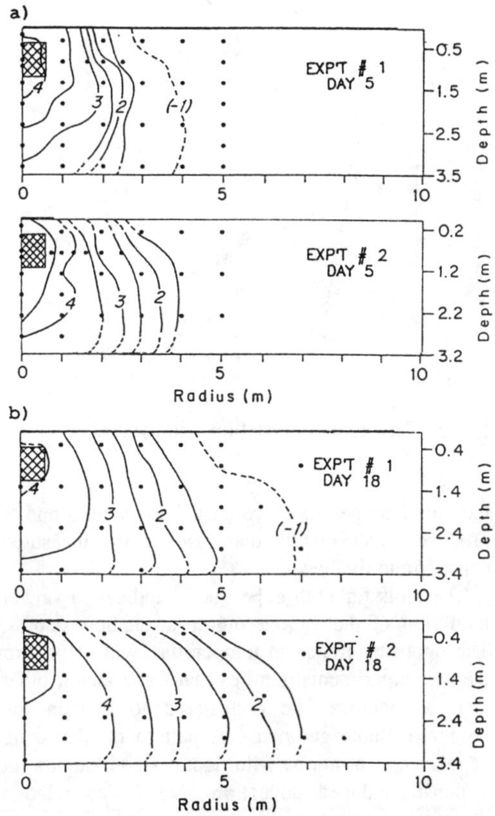

Fig. 4 TCE vapour concentration profiles for experiments 1 and 2 for a) 5 days and b) 18 days after source emplacement.

The source concentrations for experiment 1 are thought to have been higher than actually measured because of problems with vapour condensation during the extraction of samples from steel probes when surface temperatures were lower than at depth. For the purpose of modelling the experiments, source concentrations at vapour equilibrium for the

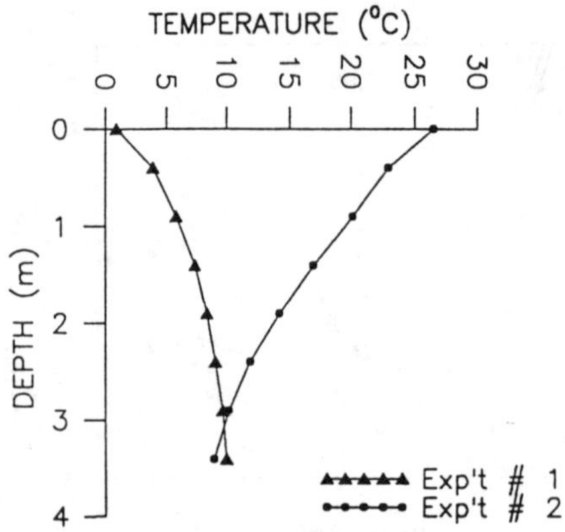

Fig. 5 Temperature variation with depth.

measured temperatures were used and were found to produce better results than use of the measured concentration values.

The plots for both experiments indicate a slower migration of the vapour plume at shallow depths. The opposing trends in temperature with depth for the two experiments requires that some factor, other than temperature, be considered to explain the observed plume geometry. A pattern of increasing vapour concentrations with depth is consistent with a density-induced advection effect. The relative density of contaminated soil gas is largely concentration dependent, however, and this density effect will become less important with distance from the source. The persistance of the vertical concentration gradients with migration distance from the source suggests that the plume geometry seen in Figure 4 is primarily a consequence of a vapour retardation process rather than density-induced advection.

As shown in Figure 3, there is an increase in the organic carbon content of the upper sediments which is over two orders of magnitude relative to the lower sands. Reference to equations (3) and (6) suggests that this will have a large affect on the amount of vapour retardation. To illustrate this, the values of the retardation factor for TCE vapour transport, calculated according to these equations, are presented for various combinations of temperature and organic carbon content over the range of values actually measured at the site (Table 1). For the purpose of these calculations, the moisture content was assumed constant. The values used in the calculation were: $\theta_g = 0.3$, $\theta_w = 0.1$, $\rho_b = 1.5$ mg/l, and $K_{oc} = 126$ ml/g (Schwille, 1988).

The upper sequence of sediments containing higher organic carbon contents is 0.6 to 1 metres

Table 1. Calculated values for retardation factor R

Temp. (°C)	Henry's* Constant (-)	f_{oc} 0.0001	0.001	0.01	0.02
		R			
0	0.096	5.13	11.0	70.1	135
5	0.130	4.05	8.41	52.0	100
10	0.172	3.30	6.60	39.6	76.2
15	0.227	2.75	5.24	30.2	58.0
20	0.296	2.34	4.25	23.4	44.7
25	0.383	2.03	3.52	18.3	34.8

*Gossett (1987)

thick in the area of the site shown in the contour plots. The values in Table 1 indicate that the amount of retardation in these sediments should be much greater than in the underlying sands, although the magnitude of the retardation will be lower in experiment 2 due to the higher values of the Henry's constant.

High moisture contents will also increase the degree of vapour retardation as shown by equation (3). In addition, increased moisture contents will slow the rate of vapour diffusion as a result of decreased values of the effective vapour diffusion coefficient. In the case of both experiments there was little temporal variation in moisture content (i.e. for the initial 21 days), although some drying of the very shallow soils was likely in experiment 2. The spatial variation in moisture content, however, is significant, particularly with respect to the silty organic-rich layer. This layer, although thin and local in its extent, may exert a significant control on transport. The high organic carbon content and moisture content result in a greater vapour retardation capacity and the high moisture content provides a greater barrier to diffusion. The apparent retardation of the TCE vapour plume above this layer is consistent with this prediction.

It should be noted that an interpretation, based on these two experiments alone, cannot be unique because the temperature profile in the case of experiment 1, and the absence of a surface cover in experiment 2 would also favour the observed trends in vertical concentration gradients. A resolution of the relative contribution of these factors would require repeating one or both of the experiments with the reverse surface-boundary condition. This was done, hypothetically, through computer simulations of the experiments and the effect of removing or adding the cover was found to have little effect on the resulting plume geometry below the level of the organic layer (Mendoza et al. 1992).

The observed concentration trends can be interpreted with recourse to transport theory using the measured parameter variations at the site. As a somewhat more quantitative test of the applicability of this theory, the experimental results were compared with simulations perfomed using a 2-dimensional, axisymmetric numerical model. The model incorporates the effects of density-induced advection and retardation. Figure 6 is an example of the match obtained for experiment 1 using model input values based on the measured site parameters and the most appropriate available literature values. Concentrations are contoured as the log of TCE concentration in ppmv. The field results were hand contoured based on values at the sampling points indicated, while the simulation results were computer contoured over the full finite-element grid.

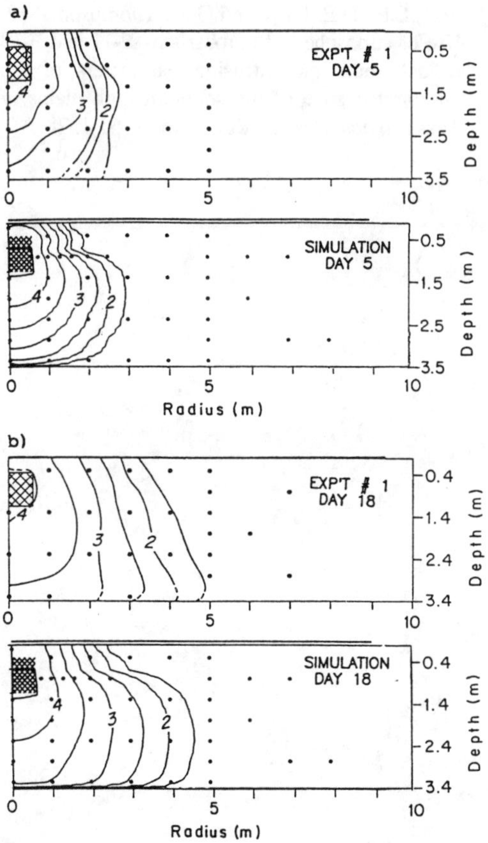

Fig. 6 Comparison of the field data to computer simulations for a) 5 days and b)18 days for experiment 1.

The match between the model simulations and the experimental data was quite reasonable, providing confidence for its use as a tool to assist in the interpretation of the experiments through sensitivity analyses (Mendoza et al. 1992).

5 CONCLUSIONS

The experimental results provide evidence that vapour transport can result in significant lateral transport of a halogenated solvent over short time periods in a relatively shallow, sandy unsaturated zone. The nature of this transport appears to be in qualitative agreement with the current transport theory and is significantly affected by the spatial and temporal variations in the transport parameters. Seasonal temperature variation has a strong effect on both the rate of vapour transport and the concentration of vapours within the unsaturated zone. Plume geometry is thought to have been influenced by several factors, in addition to temperature; particularly by moisture content and organic carbon content.

The comparison of the experimental results to simulations based on a vapour transport model were encouraging. It would appear that basic transport theory may be applied to field situations through such numerical solutions. The success of these predictions will, however, be dependent on the amount of detail available concerning the subsurface conditions at the site and the ability of the model to represent the spatial and temporal variations of the important parameters. The scale of these heterogeneities may be very local as seen in the influence of the organic-rich layer at this site.

REFERENCES

Chiou, C.T. 1981. Partition coefficient and water solubility in environmental chemistry. In J. Saxena & F.Fisher (eds.), *Hazard assessment of chemicals current developments Vol. 1.*, p. 117-153. New York: Academic Press.

Chiou, C.T., L.J. Peters & V.H. Freed 1979. Physical concept of soil-water equilibria for nonionic organic compounds. *Science* 206:831-832.

Chiou, C.T. & T.D. Shoup 1985. Soil sorption of organic vapours and effects of humidity on sorptive mechanism and capacity. *Environ. Sci. Tech.* 19:1196-1200.

English, C.W. & R.C. Loehr 1989. Sorption of

volatile organic compounds in soil. *Proceedings of the Conference on Petroleum Hydrocarbons and Organic Chemicals in Ground Water*. Houston, NWWA, p. 383-391.

Falta, R.W., I. Javandel, K. Pruess, & P.A. Witherspoon 1989. Density-driven flow of gas in the unsaturated zone due to the evaporation of volatile organic compounds. *Water Resour. Res.* 25(10):2159-2169.

Feenstra, S. & J.A. Cherry 1988. Subsurface contamination by dense non-aqueous phase liquid (DNAPL) chemicals. *Proceedings of the International Groundwater Symposium*. Halifax, Nova Scotia. International Association of Hydrogeologists, p. 61-69.

Gossett, J.M. 1987. Measurement of Henry's law constants for C_1 and C_2 chlorinated hydrocarbons. *Environ. Sci. Tech.* 21(2):202-208.

Hughes, B.M. 1991. Vapour transport of trichloroethylene in the unsaturated zone: A field experiment. MSc thesis, Dept. Earth Sciences, University of Waterloo.

Karickhoff, S.W., D.S. Brown, & T.A. Scott 1979. Sorption of hydrophobic pollutants on natural sediments. *Water. Resour. Res.* 3:241-248.

Larsen, T., P. Kjeldsen, T.H. Christensen, B. Skov & M. Refstrup 1989. Sorption of specific organics in low concentrations on aquifer materials of low organic carbon content: Laboratory experiments. In H.E. Kobus & W. Kinzelbach (eds.), *Contaminant Transport in Groundwater*. Proceedings of the International Symposium on Contaminant Transport in Groundwater. Stuttgart, p. 133-140.

Mackay, D.M., P.V. Roberts & J.A. Cherry 1985. Transport of organic contaminants in groundwater. *Environ. Sci. Tech.* 19(5):384-392.

Mendoza, C.A. & E.O. Frind 1990a. Advective-dispersive transport of dense organic vapors in the unsaturated zone 1. Model development. *Water Resour. Res.* 26(3):379-387.

Mendoza, C.A. & E.O. Frind 1990b. Advective-dispersive transport of dense organic vapors in the unsaturated zone 2. Sensitivity analysis. *Water Resour. Res.* 26(3):388-398.

Mendoza, C.A. & T.A. McAlary 1990. Modeling of ground-water contamination caused by organic solvent vapors. *Ground Water* 28(2):199-206.

Mendoza, C.A., B.M. Hughes and E.O. Frind 1992. Trichloroethylene vapours in the unsaturated zone: Numerical analysis of a field experiment. This volume.

Nazaroff, W.W., S.R. Lewis, S.M. Doyle, B.A. Moed & A.V. Nero 1987. Experiments on pollutant transport from soil in residential basements by pressure-driven airflow. *Environ. Sci. Tech.* 21(5):459-465.

Peterson, M.S., L.W. Lion & C.A. Shoemaker 1988. Influence of vapor-phase sorption and diffusion on the fate of trichloroethylene in an unsaturated aquifer system. *Environ. Sci. Tech.* 22:571-578.

Piwoni, M.D. & P. Banerjee 1989. Sorption of volatile organic solvents from aqueous solution onto subsurface solids. *J. Contam. Hydrol.* 4(2):163-179.

Schwarzenbach, R.P. & J. Westall 1981. Transport of nonpolar organic compounds from surface water to groundwater. Laboratory sorption studies. *Environ. Sci. Tech.* 15:1360-1367.

Schwille, F. 1988. *Dense chlorinated solvents in porous and fractured media - model experiments*. J.F. Pankow (trans.), Chelsea, Michigan:Lewis.

Sleep, B.E. and J.F. Sykes 1989. Modeling of transport of volatile organics in variably saturated media. *Water Resour. Res.* 25(1):81-92.

Weeks, E.P., D.E. Earp, and G.M. Thompson 1982. Use of atmospheric fluorocarbons F-11 and F-12 to determine the diffusion parameters of the unsaturated zone of the southern high plains of Texas. *Water Resour. Res.* 18(5):1365-1378.

Subsurface Contamination by Immiscible Fluids, Weyer (ed.) © 1992 Balkema, Rotterdam. ISBN 90 6191 175 3

Vapour extraction of trichloroethylene under controlled conditions at the Borden site

R. D. McClellan
Waterloo Centre for Groundwater Research, Ont., Canada (Currently: Intera Kenting, Toronto)
R. W. Gillham
Waterloo Centre for Groundwater Research, Ont., Canada

ABSTRACT: In an effort to more fully understand the micro-scale processes occurring during vapour extraction, a series of field experiments was conducted in the vadose zone at Canadian Forces Base Borden. The experiment described in this paper considered removal of a homogeneously distributed liquid phase residual of trichloroethylene in a fine-grained sand. Soil vapour concentrations within the source material and concentrations in the effluent air were monitored during vapour extraction. The relationship between the flowrate and the removal efficiency was studied as was the impact of the dissolved and sorbed phases on soil gas concentrations.

The results of this study indicate that vapour extraction can be very effective in removing residual volatile liquid contaminants from the unsaturated zone. However, persistence of the dissolved and sorbed phases of the contaminant, after the residual liquid had been removed, suggested that careful consideration of clean-up criteria is required.

1 INTRODUCTION

Following the release of a liquid volatile organic compound (VOC) at or near the ground surface, a downward migration of the liquid, induced by gravity and capillary forces, will occur. A large portion of the original spill volume may be immobilized in the unsaturated zone by capillary forces. Schwille (1988) found the minimum retention capacity of an unsaturated fine to medium grained sand was 30 L/m^3.

The presence of this residual non-aqueous phase liquid (NAPL) in the vadose zone is recognized as a potential long-term source of contamination to groundwater (Baehr *et al.*, 1989). The primary transport mechanism to the groundwater is by dissolution of the residual NAPL into infiltrating water. Another important transport mechanism is evaporation of the liquid VOC into the soil gas, and subsequent diffusive migration of the vapours through the vadose zone (Hughes *et al.*, 1992). This mechanism may lead to off-site migration of the contaminants at rates greater than those predicted from a consideration of groundwater transport alone.

Removal of the residual NAPL within the vadose zone has become an area of great interest to many regulatory agencies. Remediation of sites contaminated by VOCs has focused primarily on groundwater clean-up. Pump and treat strategies are used to arrest off-site migration of contaminated groundwater, and to slowly dissolve the residual chemical above and below the water table (Mackay and Cherry, 1989). The low solubilities of most of these compounds result in commitment to very long periods of treatment. Removal of the residual NAPL by other means would, of course, significantly reduce the time required to clean up a site. This is frequently achieved by excavation of the contaminated soils and subsequent disposal. Although this strategy addresses the source of contamination, it has the unfavourable consequence of moving the contamination to a new site. Furthermore, when excavating soils contaminated by VOCs, a significant flux of VOC vapours to the atmosphere is likely to occur. This transfer to the atmosphere of contaminant mass during excavation could exceed that allowed by regulatory agencies (Danko, 1988). If the depth of residual contamination is greater than a few metres from ground surface, excavation becomes infeasible.

A relatively new technology, vapour extraction,

has recently received considerable attention. The ability to reduce the volume of residual liquid VOCs within the vadose zone quickly and inexpensively, make it a particularly attractive method of remediation. A recent review of the literature on this topic is given by Hutzler *et al.* (1989a and b). In spite of its wide acceptance and use as a method of remediation, there have been few well-documented, controlled experiments to assess the mechanisms occurring during vapour extraction. In an effort to fill this void, a series of vapour extraction experiments was conducted at CFB Borden. The experiments were conducted in specially designed steel boxes to allow control of process variables such as contaminant distribution and flowrate. The experiment described in this paper considered the ideal conditions of one-dimensional air flow through a uniformly distributed residual fluid in a homogeneous porous medium. Specific objectives were to examine the extent to which the removal efficiency was controlled by the kinetics of mass transfer between the NAPL and the gas phase, and also to examine the rate of removal from the aqueous and solid phases, once the liquid NAPL had been removed. Trichloroethylene (TCE) was used as the non-aqueous residual liquid because of its relatively high vapour pressure and its wide-spread occurrence as a groundwater contaminant.

2 METHODOLOGY

2.1 Experimental apparatus

The experiment was conducted in a steel box measuring 1.83 m long by 1.22 m deep by 0.46 m wide which was installed in the vadose zone at CFB Borden (Figure 1). Vertical vents, screened over the entire cross-section of the steel box, were installed at each end. The seams of the box were double welded to ensure that there were no leaks. The top of the box was constructed of a 19 mm thick sheet of plywood sealed with two layers of fibreglass and resin. After filling the box with contaminated soil, the top was sealed into place using silicon sealant. One dimensional air flow was created through the box by applying a vacuum to one of the vents and allowing air to enter passively through the other.

Using a mortar mixer, 30 L of TCE was mixed with approximately 0.85 m^3 of sand from the

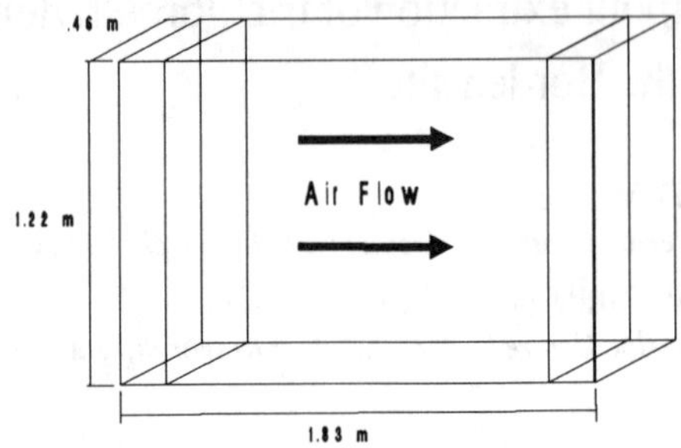

Figure 1: Schematic of vapour extraction cell.

Borden site. The mixture was packed into the cell as uniformly as possible giving a residual saturation of 3.5 % (vol_{TCE} / vol_{SOIL}). Mass losses during mixing were minimized by covering the mortar mixer with a sheet of plastic and limiting the time of mixing to 1-2 minutes.

Once the source of contamination was emplaced, 12 soil-gas sampling probes, 12 probes for measuring gas pressure, 4 thermistors, and 4 soil moisture probes were installed at various locations throughout the cell.

The soil-gas probes and pressure probes were constructed of 0.64 cm OD x 0.16 cm wall stainless steel tubing. A drive point was fitted to the bottom of each probe. Small holes were drilled through the wall of the tubing from just above the drive point to 1 cm above the drive point. Fine-mesh stainless steel screen was wrapped around the holes to prevent clogging. The screen was recessed into the wall of the probe to prevent the soil around the annulus from being disturbed. The probes were easily pushed to the desired depth in the soil. The top of each gas-sampling probe was fitted with a brass thread adapter to allow connection of a Mininert^{tm} valve. Soil gas could be sampled by inserting a syringe through the valve and withdrawing a volume of air from the probe. Pressure probes were fitted with a short length of Tygon^{tm} tubing. This was normally clamped to prevent air from being drawn through the probe. Pressure within the box was measured by connecting a probe to a U-tube water manometer.

The thermistors were installed by pushing a 0.64 cm diameter rod to the desired depth in the soil, withdrawing the rod and lowering the thermistor down the hole. Temperature was measured by connecting the thermistor to a voltage meter. Moisture content within the box was measured using Time Domain Reflectometry (TDR) (Topp *et al.*, 1982). Brass rods (0.32 cm diameter) were pushed into the soil approximately 5.6 cm apart. A cable tester was then connected to the probes to measure the dielectric constant of the soil between the brass probes. This dielectric constant was then used to estimate the average water content of the soil.

2.2 Soil-gas sampling

Soil-gas samples were withdrawn from the box through the stainless steel probes using glass syringes. Samples of the effluent gas were withdrawn from a port located at the outlet of the box, also using glass syringes. The samples were analyzed on a Photovac^{tm} 10S50 Portable Gas Chromatograph equipped with an isothermal oven and a capillary column. Concentrations were measured as mols of TCE per million mols of gas (ppm_v). The sensitivity of the chromatograph was such that the maximum concentration of an air sample which could be analyzed directly was approximately 25 ppm_v. The range in concentrations measured was 0.1 to 40,000 ppm_v, with the majority of samples exceeding 25 ppm_v. Dilution of the high-concentration samples was effected by injecting a small volume of the soil gas into a 1 L glass bottle. After equilibration, a sample of the diluted gas was analyzed.

During sampling of the soil gas probes, the vacuum was turned off to prevent air from being drawn down the probe as the syringe was inserted and withdrawn from the Mininert^{tm} valve. The probes were purged by withdrawing three internal volumes from each point before collecting the sample. This ensured that the sample being analyzed was representative of the soil gas concentration at the bottom of the probe. To check reproducibility, each diluted sample was analyzed twice. Near the end of the experiment when the concentration of gas samples was low, such that they could be injected directly to the chromatograph, two samples were collected from each probe and analyzed without dilution. If the measured difference in concentration between successive analyses was greater than 10%, the point was re-sampled.

2.3 System operation

The vapour extraction cell was allowed to equilibrate for a period of two days following placement of the contaminated soil. During this time, the monitoring probes were installed and initial readings were taken. Upon start-up of the vacuum, the effluent concentration was monitored every 10 minutes for a number of hours. Periodically, the vacuum was shut down for about 10 minutes to allow sampling of the soil gas probes. The flowrate through the box was maintained at a constant rate for 46.5 hrs. Following this period, a series of short term experiments was conducted to examine the effects of flowrate on removal efficiency. These experiments included changing the flowrate while monitoring the effluent concentration, as well as pulse tests of varying length. A pulse test consisted of turning off the vacuum for periods of from 1 hour to several days. Soil gas concentrations were monitored during the period the vacuum was turned off, and effluent concentrations were monitored upon re-starting the vacuum. The total duration of the experiment was 102 days.

3 RESULTS

3.1 TCE removal

Flowrates through the box during the experiment were varied from 18 to 71 L/min, corresponding to residence times for the air of 19 to 5 min. Upon start-up of the vacuum system, the concentration of the effluent air ranged between 35,000 to 36,000 ppm_v. The flowrate at this time was 22 L/min, giving a residence time of about 15 min. The temperature within the box ranged from 4.8 to 7.8 C while the ambient air temperature was -2 C. The corresponding range of saturation concentrations at these temperatures is 24,500 to 42,800 ppm_v, indicating that the air leaving the box was at or near saturation. This indicates that even with the relatively short residence time, the air was in equilibrium with the residual liquid TCE.

Effluent concentrations were relatively constant for the first 72 hours. During the 12 hour period shown in Figure 2, the flowrate was increased in steps from 18 to 63 L/min, resulting in a decrease in the residence time from 19 to 5.4 min. The constant concentration in the effluent (Figure 2) over this range in flow rates is further evidence that the air within the soil pores comes to equilibrium rapidly with the residual solvent. The bottom graph in Figure 2 shows the mass flowrate versus time. Clearly if the concentration remains constant, an increase in the flowrate must result in a corresponding increase in the mass removal rate.

The observations for early time (Figure 2) are substantially different from those made later in the experiment, after removal of the liquid contamination. Figure 3 shows the flowrate and effluent concentration for a 10 hour period between 380 and 390 hours. In this case, the effluent concentration is about 1.3 orders of magnitude lower than that shown in Figure 2, suggesting that the residual organic fluid has been removed. In addition, an increase in the flowrate at 382 hours caused a decrease in the effluent concentration. The decrease in concentration with increasing flowrate indicates that the gas-phase concentration was not in equilibrium with the organics remaining in the box. This is believed to be the result of the kinetics of phase transfer between the dissolved and sorbed-phase organics and the gas phase.

Further evidence of the influence of the kinetics of phase transfer, after removal of the pure phase liquid, can be seen in the results of the pulse tests shown in Figure 3. The shut down period during each pulse allowed dissolved and sorbed-phase TCE to diffuse to the air/water interface in the pores and ultimately to partition to the air phase. The vacuum was turned off for a period of one hour at hour 384. After restarting the vacuum, the effluent concentration was significantly higher than just prior to shutting down the system. This "spike" in concentration quickly decreased to levels more typical of the pre-test concentrations. This decrease indicates that the time required for the gas phase to equilibrate with the dissolved and sorbed-phase TCE across the entire length of the cell was longer than the 5 minute residence time of the moving air. A second pulse test, using a shut down period of 35 minutes (ending at hour 387), resulted in a somewhat smaller peak concentration when the vacuum was restarted. This indicates that even after 35 minutes, the soil gas was not in equilibrium with the dissolved and sorbed phase. A third pulse test, with a shut down period of one hour, at hour 388, resulted in a peak effluent concentration similar in magnitude to that seen at hour 385.

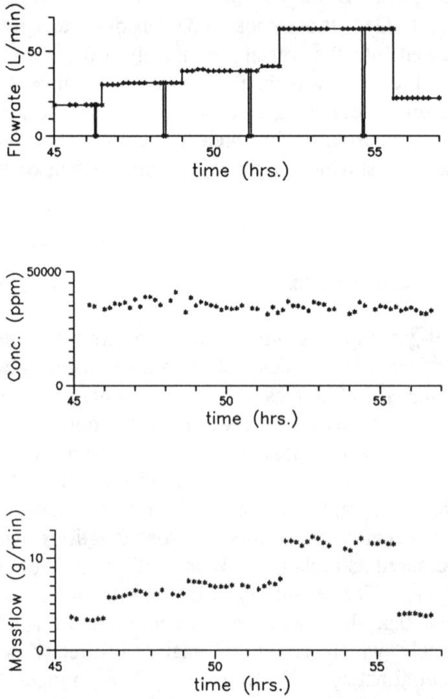

Figure 2: System flowrate, effluent concentration, and mass removal rate versus time.

The concentration of TCE in the effluent over the entire duration of the experiment is shown in Figure 4 (note the logarithmic time scale). The sharp decline in concentration corresponds to the time at which the last liquid TCE was removed from the box. The rapid decline in concentration provides further evidence of rapid equilibrium between the liquid-phase organics and the air stream, even for very low values of residual fluid. A very long tailing effect is seen in Figure 4 as the dissolved and sorbed phase TCE is slowly stripped from the soil. The persistence of the TCE in these two phases is cause for some concern. At the end of the experiment, after approximately 7,700 m^3, or 22,650 pore volumes, of air had passed through the soil, effluent concentrations were still 0.9 ppm_v. Immediately after turning the vacuum off, soil gas concentrations were less than 1.7 ppm_v. After a 19 hour equilibration period, the average soil-gas concentration had risen to 16.3 ppm_v, once again indicating that the flowing air was not at equilibrium with the dissolved and sorbed phases.

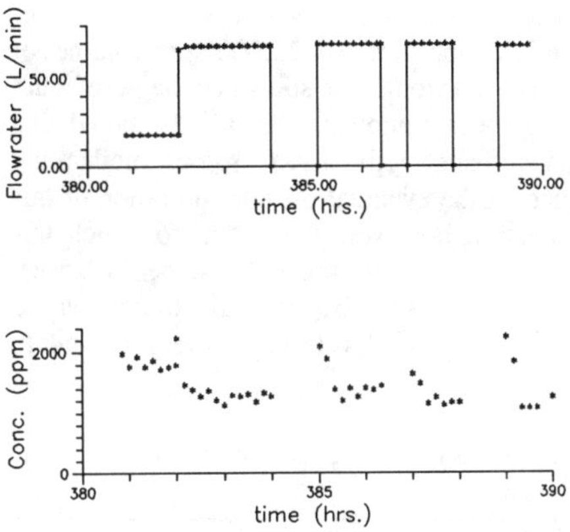

Figure 3: System flowrate and effluent concentration versus time.

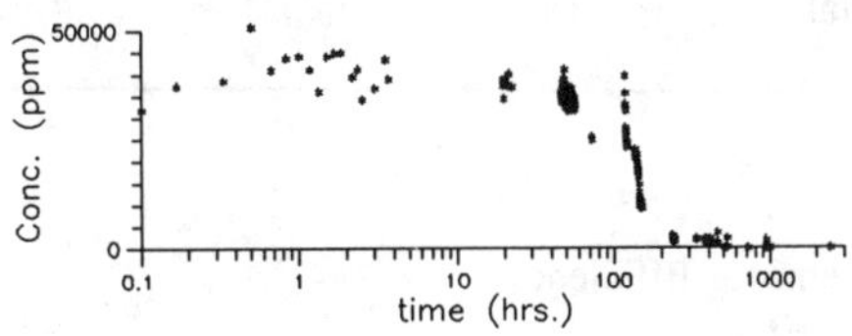

Figure 4: Effluent concentration versus time.

3.2 Mass of TCE remaining

The results of this experiment show that once the pure phase liquid has been removed, a small but environmentally significant proportion of the original mass of contaminant may still remain in the soil. The mass retained will depend on properties such as the moisture content of the soil (θ_w), the solubility of the chemical (S), the organic matter content of the soil (f_{oc}), the Henry's law partitioning coefficient for the chemical (H), and the organic carbon partitioning coefficient for the chemical (K_{oc}).

Assuming that the soil gas was at equilibrium with the dissolved and sorbed phases following the 19 hour equilibration period at the end of the experiment, an estimate of the mass of TCE remaining in the system can be made. The remaining mass will exist in the gaseous, dissolved and sorbed phases. The average soil gas concentration measured after the equilibration period was 16.3 ppm_v. The mass of TCE in the gas phase can therefore be calculated from:

$$M_g = C_g \times 10^{-6} \times \frac{P_{atm} \times V_g}{R \times T} \times M_m \quad \textbf{(1)}$$

where:
M_g = mass in gaseous phase (g)
C_g = gas phase concentration (ppm_v)
P_{atm} = atmospheric pressure (Pa)
V_g = vol. of air filled pores (m^3)
R = ideal gas constant ((Pa m^3)/(mol K))
T = temperature (°C)
M_m = molar mass (g/mol)

Knowing the concentration in the gas phase, the concentration in the dissolved phase can be calculated using Henry's Law:

$$C^* = \frac{P_p}{H} \quad \textbf{(2)}$$

where:
C^* = dissolved phase concentration (mol/m^3)
P_p = partial pressure (Pa)
H = Henry's Law constant ((Pa m^3)/mol)

Realizing that the soil gas concentration, expressed as $mols_{TCE}$ / $mols_{TOT}$, is the ratio of the vapour pressure to the total pressure, and substituting this relationship into equation 2 yields:

$$C_w = \frac{C_g \times P_{atm} \times M_m}{1 \times 10^6 \times H} \quad \textbf{(3)}$$

where:
C_w = dissolved phase concentration (mg/L)
C_g = gas phase concentration (ppm_v)
P_{atm} = atmospheric pressure (Pa)
M_m = molar mass (g/mol)

The mass contained within the dissolved phase is therefore given by:

$$M_w = C_w \times \theta_w \times V_{tot} \quad \textbf{(4)}$$

where:
M_w = mass in dissolved phase (g)
C_w = dissolved phase concentration (mg/L)
θ_w = water content
V_{tot} = volume of soil (m^3)

The mass sorbed to the solid organic carbon can be calculated according to:

$$M_s = K_d \times C_w \times \rho_b \times V_{tot} \quad \textbf{(5)}$$

where:
M_s = mass in sorbed phase (g)
K_d = solid/water partitioning coef. ((μg/g)/(μg/mL))
K_d = f_{oc} x K_{oc} (mL/g)
C_w = dissolved phase concentration (mg/L)
ρ_b = bulk density (g/cm^3)
V_{tot} = volume of soil (m^3)

Thus, for the measured average soil gas concentration of 16 ppm$_v$, and system parameters as given in Table 1, the mass remaining in the cell at the end of the experiment was calculated to be 143.2 mg. The mass remaining in each phase is given in Table 2. Although this mass appears small, in terms of drinking water limits for TCE it is significant.

Table 1. System Parameters

Parameter	Value	Unit
V_{tot}	.85	m^3
$^1\rho_b$	1.6	g/cm^3
$^1\eta$	.4	
$^1\theta_g$	.3	
$^1\theta_w$	.1	
^2C_g	16.3	ppm$_v$
V_g	.255	m^3
R	8.314	(Pa m^3)/ (mol K)
^2T	273	K
M_m	131.5	g/mol
^3H	218	(Pa m^3)/mol
$^4P_{atm}$	101300	Pa
$^3K_{oc}$	126	mL/g
$^5f_{oc}$	.0002	
^6K_d	.0252	mL/g

1 Typical value for Borden sand
2 Average measured value
3 From Schwille, 1988
4 Assumed value
5 Average measured value after Hughes et al., 1990
6 Calculated as the product $K_d = f_{oc} \times K_{oc}$

Using a drinking water limit for TCE of 5 μg/L, 143.2 mg of TCE has the potential to contaminate 28,640 L of groundwater to the drinking water limit. The calculated concentration in the soil moisture is 0.996 mg/L, or 200 times the drinking water limit. This level of contamination is significant, particularly given the fact that over 7,500 m^3 or approximately 22,000 pore volumes of air had passed through the soil since the pure liquid phase had been removed. It must be noted that dilution effects will have some moderating influence on the eventual off-site migration of this contamination however, the extent to which this will occur is site specific. In some instances, clean-up of a site by vapour extraction to environmentally safe levels may be very difficult to achieve.

Table 2. Mass of TCE remaining at the end of the experiment.

Phase	Mass (mg)
Gaseous	24.4
Dissolved	84.7
Sorbed	34.1
Total	143.2

3.3 Removal Efficiency

Removal efficiencies, expressed as grams of TCE removed per litre of air passed through the cell, can be calculated for the system operation. A comparison of the removal efficiencies for operation during pure phase removal, after pure phase removal and during pulse tests can provide an insight into the effectiveness of vapour extraction.

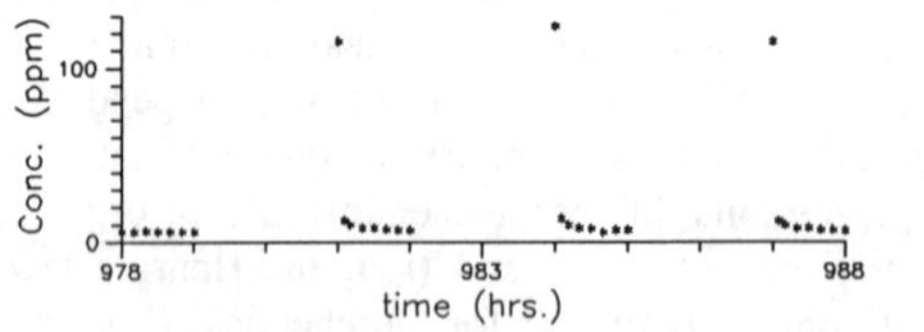

Figure 5: Pulse test effluent concentration versus time.

Removal efficiencies for the flowrates and pulse events shown in Figures 2, 3, and 5 are given in Table 3. The removal efficiency decreased by a factor of about 20 when the liquid TCE was removed from the soil (Figures 2 and 3). Once the liquid phase had been removed (Figures 3 and 5), the removal efficiency continued to decrease, but at a slower rate. It should be noted that at the end of the experiment, when there was still an environmentally significant amount of TCE in the cell, the removal efficiency was more than four orders of magnitude lower than in the case when residual organic fluid was present in the soil.

Removal efficiencies for Figure 2, suggest that there is no relation between flowrate and removal efficiency provided there is pure phase present in the system. Comparing the efficiency observed between hours 381-382 to that seen between hours 382-384 suggests that, for a system with no liquid phase solvent present, a higher efficiency can be attained by using a lower flowrate. Removal efficiencies associated with the pulse tests, (Figures 3 and 5), indicate that pulsed operation is more efficient than continuous operation after liquid phase removal is complete.

Table 3. Relative Removal Efficiencies.

Time Period (hrs.)	Mass Removed (g)	Removal Efficiency* (g/L)
Figure 2		
45.5-46.5	202.9	0.189
46.5-49.0	935.3	0.210
49.0-52.0	1346.3	0.198
52.0-55.6	2620.5	0.200
Figure 3		
381.0-382.0	11.1	0.0103
382.0-384.0	60.4	0.00741
385.0-386.4	48.1	0.00830
387.0-388.1	32.2	0.00778
Figure 5		
978.0-979.0	.134	0.0000328
981.0-982.0	.298	0.0000700
984.0-985.0	.305	0.0000713

* Removal efficiency defined as grams TCE removed per L of air passed through cell

4 CONCLUSIONS

The results of this experiment indicate that vapour extraction can be a highly effective method of removing pure-phase liquid VOC from the vadose zone. The relatively short residence times of the air (less than 15 min) was sufficient to allow equilibration between the flowing air and the liquid phase VOC. Furthermore, during the period when there was pure phase NAPL present, the relative removal efficiency was unaffected by changes in the flowrate.

A sharp decline in the concentration of the effluent marked the removal of the liquid phase. This sudden decline in effluent concentration also provides further evidence of the rapid equilibrium between the air stream and the liquid TCE. A long tailing of the effluent concentration suggests the presence of a significant mass of contaminant in the dissolved and sorbed phases. Results of flowrate and pulse tests conducted after the removal of the pure NAPL suggests that the dissolved and sorbed phases are not in equilibrium with the air stream.

Under the conditions of this experiment, when pure phase NAPL was present, relative removal efficiencies were high, and were not affected by changes in the flowrate. Following removal of the liquid-phase solvent, the relative efficiency declined sharply and the rate of removal seemed to be controlled, at least in part, by the kinetics of mass transfer from the sorbed to the dissolved to the gas phase. The removal efficiency continued to decline as the mass of TCE remaining in the cell declined. At the end of the experiment, the relative efficiency had dropped by more than 4 orders of magnitude from that observed when pure phase was present. The test results indicated that pulsed operation of the vacuum system or operation at lower flowrates could increase the efficiency during this phase of remediation.

In the case of a heterogeneous field application, system efficiency would probably not exhibit the same characteristics as observed in this study. Soil heterogeneities and non-uniform solvent distribution would likely reduce the effectiveness of the technology. This study suggests that the observed exponential decline in the effluent concentration following start-up of a commercial vapour extraction system (Applegate *et al.*, 1988, Hutzler *et al.*, 1989a, Lisiecki and Payne, 1987) is not the result of phase transfer limitations. Rather, it is likely the result of irregular air flow, caused by variations in the gas-phase permeability, and by non-uniform distribution of the NAPL phase within the geologic material.

Of particular importance, is the relatively small, but environmentally significant, mass of TCE remaining at the end of the experiment. Despite the roughly 20,000 pore volumes of air which had passed through the soil since removal of the pure liquid phase TCE, the calculated dissolved phase concentration of TCE within the soil moisture was about 200 times greater than the drinking water limit of 5 μg/L. The remaining mass of TCE in the dissolved, sorbed and gaseous phases was estimated to be 143.2 mg, or enough to contaminate 28,640 L of groundwater to the drinking water limit.

Although vapour extraction is extremely effective in removing the liquid VOC, the technology is substantially less effective in removing contaminants in the aqueous and sorbed phases. This difficulty is accentuated by the fact that the removal efficiency declines to very low values as the concentration in the dissolved and sorbed phases approaches environmentally acceptable values.

REFERENCES

Applegate, J., J.K. Gentry, and J.J. Malot. 1988. Vacuum extraction of hydrocarbons from subsurface soils at a gasoline contamination site. Volatile Organics Monitoring Remediation.

Baehr, A.L., G.E. Hoag, and M.C. Marley. 1989.Removing volatile contaminants from the unsaturated zone by inducing advective air-phase transport. J. of Cont. Hydrol. 4:1-26.

Danko, J. 1988. Soil vapor extraction at a superfund site. Proceedings, National Meeting of the Air Pollution Control Assoc., Dallas, TX.

Hughes, B.M., R.W. Gillham, and C.A. Mendoza. 1992. Transport of trichloroethylene vapours in the unsaturated zone: A field experiment. Proceedings, IAH Conference on Subsurface Contamination by Immiscible Fluids, April 18-20, Calgary, Alberta. This volume.

Hutzler, N.J., B.E. Murphy, and J.S. Gierke. 1989a. Review of soil vapor extraction system technology. Proceedings, Soil Vapor Extraction Technology for Underground Storage Tank Sites Workshop. USEPA Risk Reduction Laboratory, Edison, NJ, June 27-28.

Hutzler, N.J., B.E. Murphy, and J.S. Gierke. 1989b. State of technology review: Soil vapor extraction systems. Report to USEPA, Cincinnati, OH. Cooperative Agreement CR-814319-01-1.

Lisiecki, J.B. and F.C. Payne. 1987. Enhanced volatilization: Possibilities, practicalities, and performance. Proceedings, Engineering Foundation Conference, Mercersberg, PA, Aug. 7-12.

Mackay, D.M., and J.A. Cherry. 1989. Groundwater contamination: Pump-and-treat remediation. Environ. Sci. and Tech. 23:630-636.

Schwille, F. 1988. Dense chlorinated solvents in porous and fractured media. Chelsea, MI: Lewis.

Topp, G.C., J.L. Davis, and A.P. Annan. 1982. Electromagnetic determination of soil water content using TDR: II. Evaluation of installation and configuration of parallel transmission lines. J. Soil Sci. Soc. Am. 46:678-684.

3 Physical models

Subsurface Contamination by Immiscible Fluids, Weyer (ed.) © 1992 Balkema, Rotterdam. ISBN 90 6191 175 3

Experimental simulation of transient three-phase flow in porous media

Douglas N.Graham & Hans O.Schiegg
Simultec Ltd, Meilen/Zürich, Switzerland

John F. McBride
Pacific Northwest Laboratory, Richland, Wash., USA

ABSTRACT: Experimental data are required to verify and calibrate numerical, multiphase flow models. This paper summarizes several two-dimensional, transient, three-phase flow experiments using air, water, and a light immiscible liquid (LIL). The LIL infiltration and redistribution experiments were performed in a glass-sided flume (260 cm × 60 cm × 15 cm), filled with a uniform quartz sand (mean grain diameter of 0.33 mm). Initial water saturation distributions above the water table were either at drainage or imbibition equilibrium. After the LIL reached a near equilibrium distribution, it was redistributed by raising and lowering the water table. Saturations of the air, water and oil were measured (±2%) by using dual-source gamma-ray attenuation (cesium-137 and thulium-170 sources). Simultaneously, the pressures of the water and oil phases were measured (±0.5 mm H_2O) by using hydrophobic and hydrophilic porous cups imbedded in the sand and connected to pressure transducers. In one experiment, multiple pulses of LIL were added over several days. As expected, the first pulse was observed to infiltrate and redistribute more slowly in pristine zones than subsequent pulses into zones containing residual oil. The observed distributions of the three fluid saturations are discussed with respect to two- and three-phase capillary pressure-saturation relationships that were measured in a one-dimensional vertical column.

1 INTRODUCTION

In today's society, the production of many of the goods and services that we are accustomed to depends on a wide variety of organic chemicals immiscible with water. The solubilities of these often toxic compounds are usually low but not zero. When these chemicals are accidently released to the environment, they often provide a long-term source of contamination to groundwater. Schiegg and Schwille (1989) recommend the use of the more precise terms HILs (heavy immiscible liquids) and LILs (light immiscible liquids) in preference to the more common terms DNAPLs and LNAPLs (dense and light non-aqueous phase liquids) to class these compounds. Effective cleanup and containment at LIL spill sites will remain elusive until accurate predictive models are developed.

The infiltration and distribution of HILs and LILs in the subsurface are not easily described unless many simplifying assumptions are made. Recent attempts to more rigorously describe and predict the fate of immiscible liquids have produced increasingly more complex models (e.g., Abriola and Pinder, 1985; Kuppusamy et al., 1987; Faust et al., 1989). However, one major impediment to developing better predictive models is the lack of experimental data with which to verify numerical models and their assumptions. In an effort to resolve this dilemma, several papers have recently appeared in the literature detailing quantitative experiments of three-phase flow in porous media (e.g., Lenhard et al., 1988; Høst-Madsen, 1988; Cary et al., 1989).

European countries have been traditionally active in multiphase flow research. Much of the work, however, has not been published in English. During the late 1970s, a series of two-dimensional, transient, three-phase flow experiments were performed at the Swiss Federal Institute of Technology (ETH), in

Zürich. The results were published in two German-language reports (Schiegg, 1979 and 1980), and these experiments appear to be the most detailed published to date. The reports have been translated into English, and draft versions are available from the author. This paper is an overview of some of the work found in these reports.

2 CAPILLARY PRESSURE SATURATION

When air and water are present in a large-diameter tube, a water-air interface will exist that coincides with the elevation of the water-air table. If an immiscible liquid of intermediate density, such as an oil, is added to the tube, the water-air interface will disappear. It will be replaced by water-oil and oil-air interfaces. These two interfaces represent the water-oil and the oil-air tables, respectively. Even though the water-air interface has disappeared, the elevation of the water-air table can still be calculated from the elevations of the other two, knowing the density of the oil, based on the 'iceberg principle'.

If a capillary is placed inside the tube, the fluid levels inside the capillaries will not correspond to their respective tables. Rather, the levels in the capillary depend on several factors, including the diameter of the capillary tube, the solid and liquid properties and the elevation history of the tables. These factors manifest themselves in the capillary pressure, which equals the change in fluid pressure across an interface. The familiar capillary pressure-saturation curve describes the decrease in saturation of the wetting fluid as the distance above its table increases.

In a three-phase system, two interfaces will be present where the oil exists. Thus, in a porous medium contaminated by even a small amount of oil, the capillary pressure across the contaminated water-air meniscii will be the sum of the capillary pressures across the water-oil and the oil-air meniscii. Consequently, the height of the capillary rise of water will depend on the water-oil capillary head less the difference between the water-oil and oil-air capillary heads, the latter multiplied by the specific gravity of the oil (Schiegg, 1979: p78)

The positions of the meniscii also depend on dynamics. If the tables are moving, there will be a time lag between when the tables move and when the positions of the meniscii change, simply because of viscous flow restrictions. Schiegg (1977) showed experimentally that the capillary rise depends on the velocity of the tables. Hysteresis of the contact angle and the variable pore geometry imply that the capillary rise depends on the most recent direction of fluid table movement. Therefore, to characterize the capillary pressure-saturation relationships at equilibrium in an oil-contaminated soil, eight different capillary pressure-saturation relationships are needed: the drainage and imbibition curves for the water-oil, oil-air, and virgin and contaminated water-air curves.

One-dimensional column experiments were performed to measure these eight relationships. A 15- × 15- × 100-cm acrylic column was filled with the same, uniform, medium-sized sand as in the two-dimensional experiments. The saturation was measured every 2 cm over a 54-cm scanning range by using the gamma-ray attenuation technique discussed in the next section. The initially dry sand was wetted with water from the bottom and the oil was introduced from the top. The fluid saturations were controlled by two external reservoirs, one each for the oil and water. These reservoirs were manipulated to obtain the eight capillary pressure-saturation curves.

Figure 1 shows how these curves can be used to estimate the saturation of the three fluid phases after an oil lens has redistributed. The appropriate drainage or imbibition water-oil and oil-air curves are superimposed on their respective tables when the system is at static equilibrium. Where the oil is not present, the virgin water-air curve is used. In Figure 1, the water-air table is used as the datum for presenting the resulting fluid saturation distributions. Above and below the funicular oil region, the residual oil saturation is pendular and insular, respectively. The tube beside the saturation diagram represents the table elevations and also illustrates the thickness of oil that would accumulate in a borehole. In a transient system, this saturation estimation method can still be used, but the saturations at any point depend on the instantaneous local position of the tables. For a more detailed explanation of this procedure see Schiegg (1979, 1984) or BUNR (1985).

The saturation distribution in the column was measured several times during each experiment to obtain the eight required saturation distributions. Shown in Figure 2a is one of these saturation distributions, which was obtained by

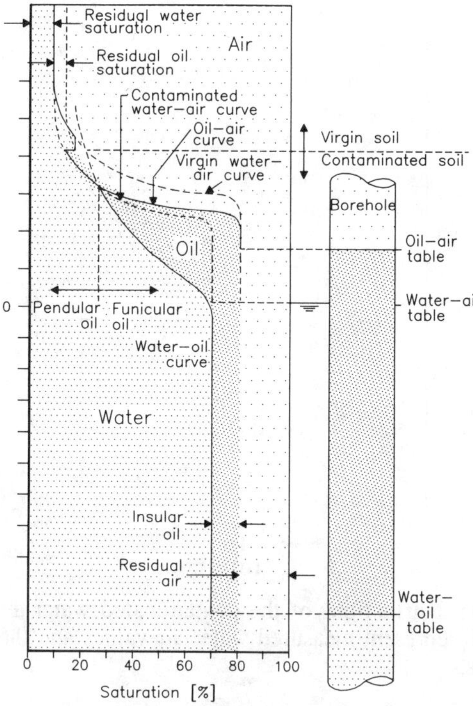

Figure 1. Theoretical construction of the three-phase saturation distribution based on the principle of tables.

first adding water to the initially dry sand until saturated (residual air saturation condition). A LIL, named SAI, was then introduced from the top of the column, displacing the water. Afterward, the water-air and the oil-air level were lowered to beneath the column, which caused air to replace the oil and water until they reached pendular saturation everywhere, except near the bottom of the column where there was a pronounced end effect.

Finally, the water-air table was raised back into the column. The resulting saturation distribution in Figure 2a shows the insular and pendular oil saturations with a funicular zone between. The additional oil, in excess of the pendular saturation, present in the funicular and insular zones was provided by the funicular oil that was trapped in the column by the end effect.

Figure 2b is the reconstruction of the saturation diagram in Figure 2a derived by superimposing the appropriate pressure-saturation curves. The two figures compare well, with the major difference occurring where funicular oil was found.

3 TWO-DIMENSIONAL EXPERIMENTAL SETUP

At the outset of the experiments, three decisions were made that dictated the design of the experimental apparatus. First of all, only LILs were to be studied. Second, the water table slope and height were to be adjustable during the experiments. Third, and most important, the saturations and pressures were to be measured simultaneously within strict tolerances (±2% saturation and ±0.5 mm H_2O pressure head).

The following description of the experimental setup and procedure is necessarily brief. A more detailed summary can be found in Schiegg and McBride (1987). The inside dimensions of the flume were 260 cm × 60 cm × 15 cm. The length and height were dictated by space restrictions. The 15-cm width was selected to reduce the measured saturation error caused by large pores located along the walls. Each side wall of the flume was constructed from a single pane of 2-cm-thick glass. Glass was chosen because no other transparent hydrophilic material could be manufactured at the ETH or found elsewhere. The tolerable outward deflection of each glass plate was 0.15 mm. The porous medium was a very uniform quartz sand, with a mean grain diameter of 0.33 mm. The water saturated hydraulic conductivity was 0.044 cm/s and the sand porosity across the flume was 0.428 ± 0.003 (as measured with the gamma-ray attenuation system).

An automated funnel and cart, travelling lengthwise, was required to achieve the desired homogeneity and isotropy in the sand. The sand was placed in 0.15-mm layers, with negligible free fall and compression. Vertically sliding end walls allowed the cart to overrun the ends to ensure a constant forward velocity during sand placement. Fine sand was placed simultaneously along the glass walls to prevent the oil from accumulating in the larger pores against the walls.

The 162 pressure probes were manufactured from plates of sintered glass, and half of them were treated with a silicon solution to render them hydrophobic. Each probe was 12 mm long, with an outside diameter of 2 mm and an inside diameter of 0.3 mm. The probes were installed in four vertical banks and were vertically separated by 4 cm. Each bank of probes was connected to one pressure transducer through a switching manifold made at the ETH.

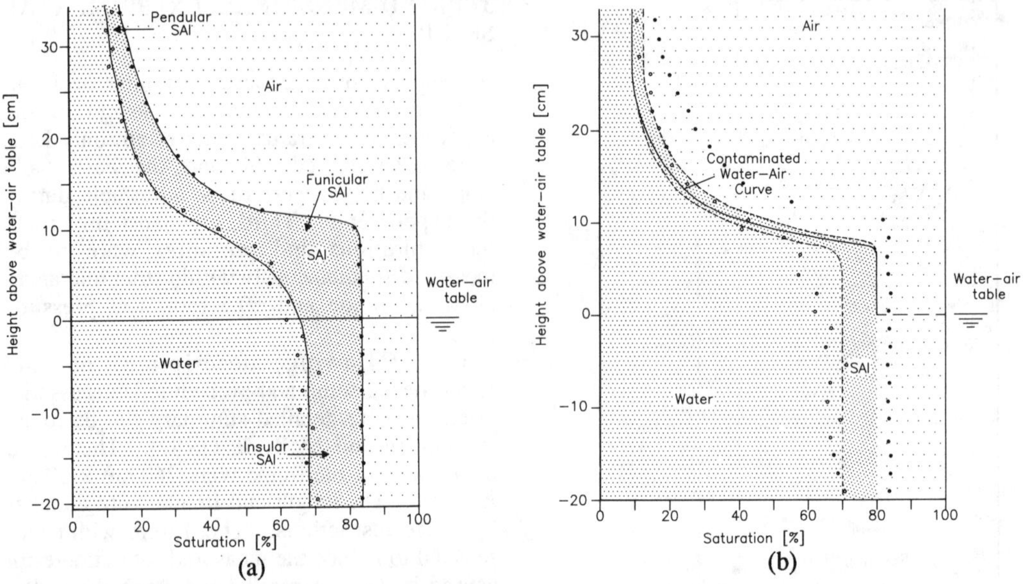

Figure 2. Measured (a) and estimated (b) saturation distributions of the contaminated water-air capillary pressure saturation curve. The open circles represent measured water-air values and the solid circles correspond to the measured oil-air values.

The three fluid saturations were measured by dual-source gamma-ray attenuation. Two sources with different attenuation characteristics for oil and water were required to differentiate the two phases. Cesium-137, a common gamma-ray source, was used as the high-energy source and thulium-170 was used as the low-energy source. Each source was collimated (0.5 cm × 2 cm), and a ten-second counting time was used. Thulium-170 was the only low-energy source that could be manufactured with a sufficiently high activity to yield 10 000 counts per second after attenuation by the glass walls and the 15-cm width filled with sand, water and LIL. Fifteen centimeters of sand seem to be near the maximum possible thickness for dual-source gamma-ray attenuation measurements when using a ten-second counting time.

The LIL was synthesized to simulate a light fuel oil and to accentuate the different attenuation characteristics between the two phases. A mercury-based fungicide (diphenyl mercury dodecenyl-succinate) dissolved in an organic solvent was diluted with an extra-light petroleum fraction until the properties approximated a light fuel oil. This LIL, named SAI, had a density of 0.85 g/cm^3, a viscosity of 0.7 cp and interfacial tensions and contact angles similiar to fuel oil.

4 EXPERIMENTAL ANALYSIS

There are nine complete experiments with data in the two reports by Schiegg (1979, 1980). Along with these, several others are discussed in more general terms. Space limitations prevent a full synopsis of all of them but the three photographs in Figure 3 illustrate the general form of the experiments. Shown in Figure 3 are two banks of pressure probes and the elevations of the water table for three of the experiments. As in all of the experiments, the oil was poured into a 2-cm wide by 2-cm deep trench, extending across the flume. The trench was located directly above one bank of pressure probes. As shown in the photographs, the form of the oil infiltration front depended on the depth to the propagation level. The lateral propagation level is located where the air saturation changes from funicular to insular. In Figure 3a, the high water table, and thus the high propagation level, caused the oil to begin flowing laterally almost immediately. In addition, the pressure of the infiltrating oil caused the oil to replace the water and thus the contaminated capillary fringe was replaced by an oil capillary fringe. When the water table was located near the bottom of the flume, a well-developed vertical infiltration front was formed (see Figure 3c) until it reached the

a)

b)

c)

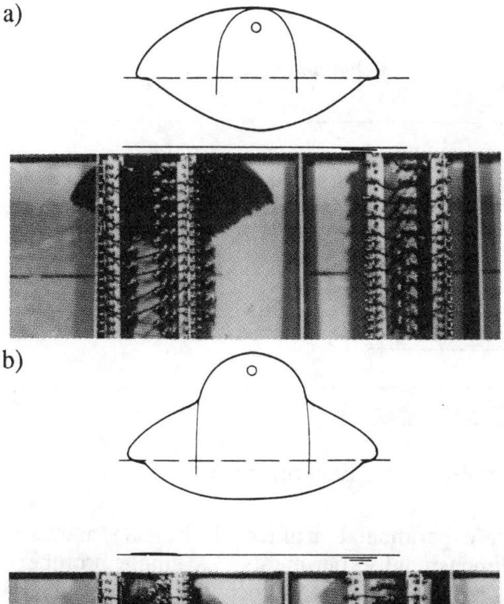

Figure 3. General forms of the first phase of vertical infiltration during the two-dimensional experiments with the horizontal propagation level: a) near the point source, b) farther from the point source, and c) farthest from the point source. The experiments were continued much beyond the time of these photographs.

propagation surface, where the oil began to flow laterally.

The various experiments included single and multiple infiltrations of SAI, with static and transient water tables that were either sloped or horizontal. One of the most detailed experiments was Experiment D from Schiegg (1979), in which one large infiltration of SAI was monitored for 79 days. Pressure data were collected at two cross sections and the saturation profiles were measured at nine cross sections.

We shall use Experiment A from Schiegg (1979) to illustrate the format of the data and the general methodology for all of the experiments. In this experiment, the water table was horizontal and 40 cm below the sand surface. The water in the sand above the water table was in the drainage state and was initially funicular throughout. SAI was poured into the trench three times at two different infiltration rates. At the start of the experiment, 2.0 liters of SAI infiltrated over 233 minutes. After one day, an additional 1.5 liters of SAI were added over a period of 114 minutes. At the beginning of Day 7, the final 1.0 liter was added in 117 minutes. Five days after the last SAI infiltration, the water table was lowered 13 cm, then raised 35 cm 22 days later.

In Experiment A, the liquid pressures and the saturations were measured only under the application site. Figure 4 consists of four selected liquid potential histograms from Experiment A. The potentials are in units of energy per unit volume ($ergs/cm^3$), where 1000 erg/cm^3 is equivalent to the potential energy in a 1-cm-high column of water or a 1.18-cm-high column of SAI. On the right side of Figure 4 is the water potential and on the left is the SAI potential. The points represent a measured potential, relative to the lowest water level. The portion of the profile containing funicular SAI or water has been shaded. At residual SAI saturation, the measured pressures are related to the capillary pressure in the isolated outer pores of the probes. Points have been used for these pressures as well. The decision on whether the saturations were funicular or not was based on observations during the experiments and analysis of photographs.

The first potential histogram was measured 90 minutes (Figure 4a) after the beginning of the experiment. The SAI was still migrating primarily vertically and the front was about 16 cm below the sand surface. The small vertical water gradient indicates that the oil was primarily displacing air. The potential gradient within the oil was approximately unity, which is expected for gravity-driven flow. The water saturation was initially funicular, but above the propagation level the SAI displaced the water to its residual saturation, thereby increasing the

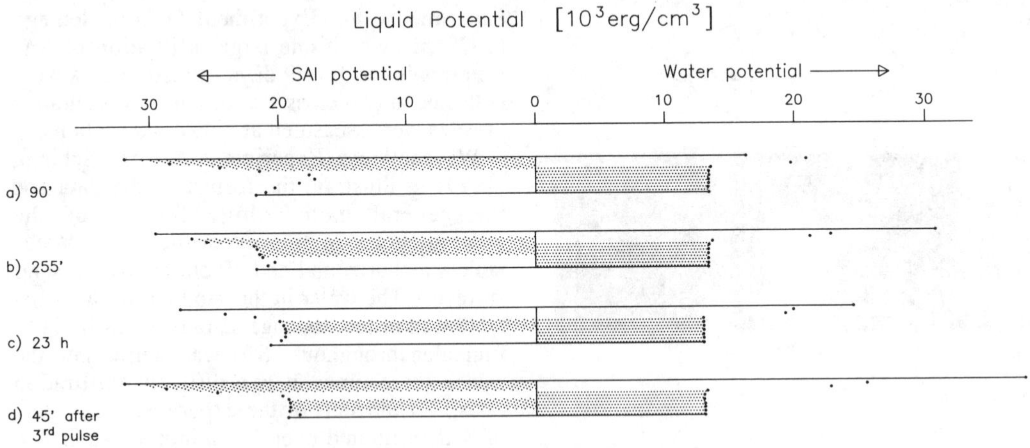

Figure 4. Selected liquid potential profiles from Experiment A.

water pressure. The SAI reached the propagation level after 120 minutes had elapsed, and lowered, by 4 cm, the virgin elevation of the transition from funicular to insular air saturation.

Twenty-two minutes after the SAI infiltration was stopped (see Figure 4b), the tail of the funicular SAI was 4 cm below the sand surface. Behind the tail, the SAI saturation was reduced to pendular saturation. Meanwhile, the SAI had dammed up on the propagation surface and had begun flowing laterally. The top of the SAI lens was located about 16 cm below the sand surface. Above the lens, the gradient in the SAI remained near unity but within the lens the vertical gradient was much smaller. The small vertical gradient within the lens implies that some vertical depression of the capillary fringe was still occurring. Within 13 hours, the vertical gradients in the SAI lens had almost completely disappeared and the potential remained nearly constant until the next SAI infiltration.

The first saturation profile was measured at the end of Day 1 and is presented in Figure 5. The vertical potential distribution corresponding to this saturation profile is given in Figure 4c. The saturation was measured every 2 cm in the vertical direction beneath the application site. In Figure 5, the SAI saturation diagram shows that the SAI saturation above the SAI lens was constant and at the residual saturation. Also note the existence of a protuberance at the 20-cm elevation in the water-oil curve. Single and multiple protuberances were observed in many of the one- and two-dimensional experiments. We think these features are not an experimental artifact of the one- and two-dimensional experiments. A single protuberance is hypothesized to occur at the elevation of the transition from funicular to insular non-wetting fluid saturation. Multiple protuberances are caused by the same process but the small density difference between water and SAI (compared to between these two liquids and air) causes the protuberances to be less pronounced. Schiegg (1978) provides a detailed explanation of this phenomenon.

Because the vertical potential gradients were zero after one day (see Figure 4c), the static capillary pressure curves measured in the one-dimensional column could be used to estimate the three-phase saturation profile. After calculating the positions of the three tables from the potential histogram in Figure 4c, the three-phase saturation profile shown in Figure 6 was constructed. The water-oil, oil-air and the contaminated water-air drainage curves were used because the initial water meniscii were in the drainage state and the oil was draining above the funicular region. The air saturation estimate in Figure 6 compares well to the measured air saturation in Figure 5, but the water saturation comparison is less favorable. The water-oil interface beneath the funicular oil is much sharper than predicted. To yield such a sharp water-oil curve, the primary drainage water-oil curve would have to be used to estimate the bottom portion of the water-oil curve, and the water-oil drainage curve would have to be used elsewhere. The presence of the protuberance in the measured water-oil saturation profile is also not predicted.

In addition to the static saturation distribution

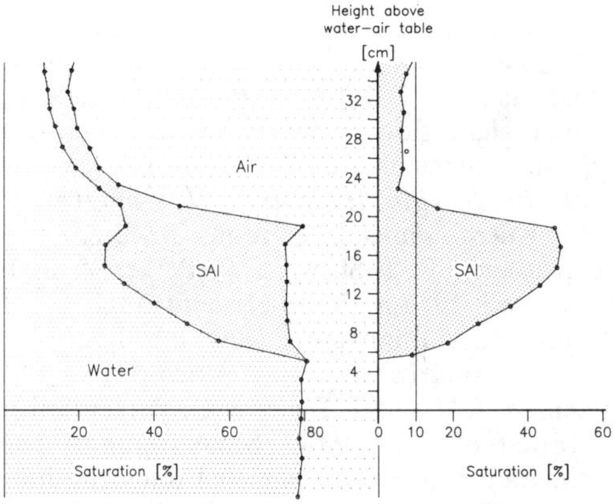

Figure 5. Measured three-phase saturation distribution in Experiment A, after one day.

the dynamic saturation distribution can also be estimated from the other potential histograms in Figure 4. The table elevations must be calculated for each potential measure-ment pair and the appropriate dynamic capillary rise-saturation curves superimposed.

At the beginning of Day 7, the final liter of SAI was poured into the trench. Although the quantity of SAI was half of the first, the rate of infiltration was the same. Figure 4d clearly shows that the residual SAI did not provide a hydraulic connection between the new SAI front and the funicular SAI below. Only when the front reached the established oil lens did a sharp increase occur in the pressure at the top of the lens. During this infiltration, the SAI reached the fourth probe after 50 minutes. This is significantly faster than during the first infiltration in a virgin water-air system when 80 minutes were required for the front to reach the same point. When the porous medium contains residual LIL, only the interconnections between the residual fractions needs to be filled; therefore, the pressure and saturation propagate more quickly, in spite of an equal infiltration rate.

5 SUMMARY AND CONCLUSIONS

During the 1970s, a unique research project to investigate the migration of LILs in the subsurface was undertaken at the Swiss Federal Institute of Technology (ETH), in Zürich. This project involved the simultaneous measurement of saturations and pressures during transient three-phase flow experiments. The results of

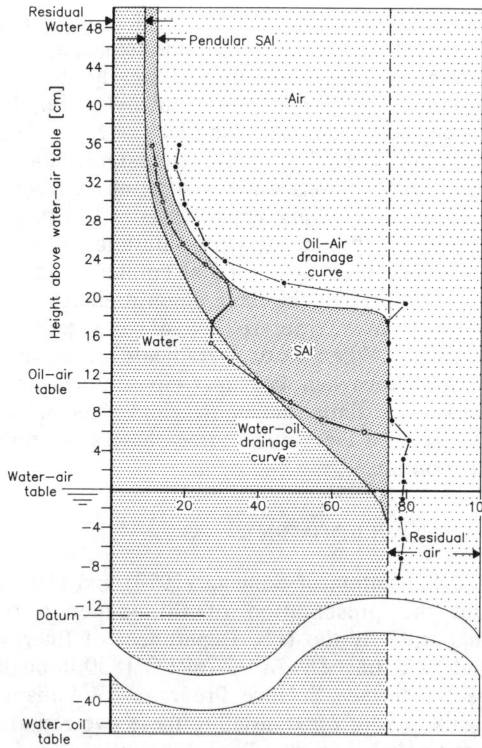

Figure 6. Saturation distribution after one day in Experiment A estimated by using the static capillary pressure curves from the column experiments.

this research effort are presented in two German-language reports (Schiegg 1979, 1980). These experiments appear to be the most detailed yet reported in the literature. The data obtained from the experiments are suitable for the preliminary validation of multiphase flow codes.

Experiment A from Schiegg (1979) is a simple experiment that illustrates the format and methodology of all of the other experiments. In this experiment, SAI was infiltrated three times. The water table was lowered and then raised to redistribute the SAI. During the initial infiltration of the SAI, the gradient behind the front was approximately unity, as was expected for gravity driven flow. Once the SAI front encountered the virgin transition elevation from insular to funicular air saturation, this transition level fell by about 4 cm, which equals the difference between the contaminated and virgin water-air capillary curves. As the SAI accumulated at the contaminated propagation level, the vertical gradients decreased and lateral flow became dominant.

Subsequent inflows of SAI clearly showed that the pendular SAI was not hydraulically connected to the underlying lens of SAI. The residual SAI above the lens significantly increased the propagation velocity of the front.

Prior to the flume experiments, a full range of static capillary pressure-saturation curves were measured in a vertical column. These curves represent the three-phase drainage and imbibition characteristics of the porous medium. When vertical gradients are absent, these curves can be superimposed on their respective static tables to yield reasonable estimates of the vertical three-phase saturation distribution.

ACKNOWLEDGMENTS

The translation of Schiegg (1979) and (1980) and the presentation of this paper were supported by the U.S. Department of Energy under Contract DE-AC06-76RLO 1830 through its Subsurface Science Program. We thank Liska Snyder for translating these two reports into English.

REFERENCES

Abriola, L. M. & G. F. Pinder 1985. A multiphase approach to the modeling of porous media contamination by organic compounds 2. Numerical simulation. *Water Resources Research.* 21:19-26

Bundesminister für Umwelt, Naturschutz, und Reaktorsicherheit (BUNR) 1985. *Beurteilung und Behandlung von Mineralölschadensfällen im Hinblick auf den Grundwasserschutz, Teil 1: Die wissenschaftlichen Grundlagen zum Verständnis des Verhaltens von Mineralöl im Untergrund* (Assessment and Treatment of Mineral Oil Spills for Groundwater Protection, Part 1: The Scientific Understanding of the Subsurface Behavior of Mineral Oil). Berlin; Umweltbundesamt

Cary, J. W., J. F. McBride & C. S. Simmons 1989. Observations of water and oil infiltration into soil: Some simulation challenges. *Water Resources Research* 1:73-80

Faust, C. R., J. H. Guswa & J. W. Mercer 1989. Simulation of three-dimensional flow of immiscible fluids within and below the unsaturated zone. *Water Resources Research* 25:2449-2464

Høst-Madsen, J. 1988. Laboratory study of three-phase flow in porous media using dual gamma attenuation technique. in *Proceedings of the Nordic Hydrological Conference*, Rovaniemi, Finland. Preprints 2:240-262

Kuppasamy, T., J. Sheng, J. C. Parker, & R. J. Lenhard 1987. Finite-element analysis of multiphase immiscible flow through soils. *Water Resources Research* 23:625-631

Lenhard, R. J., J. H. Dane, J. C. Parker & J.J. Kaluarachichi 1988. Measurement and simulation of one-dimensional transient three phase flow for monotonic liquid drainage. *Water Resources Research* 24:853-863

Schiegg, H. O. 1977. Experimental contribution to the dynamic capillary fringe. in *Proceedings of the IAHR Symposium on Hydrodynamics, Diffusion and Dispersion in Porous Media*, Pavia, Italy p.307-314

Schiegg, H. O. 1978. The hallux phenomenon and its explanation. in *Proceedings of the IAHR Symposium on Scale Effects in Porous Media*, Thessaloniki, Greece

Schiegg, H. O. 1979. *Verdrangungs-Simulation Dreier Nicht Mischbarer Fluide in Poroser Matrix* (Simulation of Three-Immiscible Fluid Displacement in Porous Media) VAW-40, Zürich: Versuchsanstalt für Wasserbau, Hydrologie, und Glaziologie: Eidgenössischen Technischen Hochschule (ETH).

Schiegg, H. O. 1980. *Grundlagen, Aufbau and Resultate von Laborexperimenten zur Erforschung von Oelausbreitungen in Grundwasssertrāgern* (Fundamentals, Setups and Results of Laboratory Experiments on Oil Propagation in Aquifers), VAW-43, Zürich: Versuchsanstalt für Wasserbau, Hydrologie und Glaziologie: Eidgenössischen Technischen Hochschule (ETH).

Schiegg, H. O. 1984. Considerations on water, oil and air in porous media. *Water Science Technolgy* 17:467-476

Schiegg, H. O. & J. F. McBride 1987. Laboratory setup to study two-dimensional multiphase flow in porous media. in *Proceedings of the Petroleum Hydrocarbons and Organic Chemicals in Ground Water Conference*, Houston, Texas p.371-385

Schiegg, H. O. & F. Schwille 1989. Hydrocarbons in porous media. in *Transport Processes in Porous Media.* NATO/ASI: in press

Subsurface Contamination by Immiscible Fluids, Weyer (ed.) © 1992 Balkema, Rotterdam. ISBN 90 6191 175 3

Pore scale behavior of spreading and non-spreading organic liquids in the vadose zone

J. L. Wilson
New Mexico Institute of Mining and Technology, N. Mex., USA

ABSTRACT: In water wet soils the propensity for an organic liquid to spread at the water–gas interface may influence behavior in the vadose zone. This propensity is measured by the 'spreading coefficient'. We investigated this hypothesis through flow visualization studies, conducted in etched glass micromodels. Positive spreading coefficients lead to an organic liquid that tends to spread out as a film along the gas–water interface. The film interconnects larger pockets of organic liquid held in pore throats and bodies, and provides capillary dominated pathways to drain or imbibe the organic liquid through the water and gas filled pore space. A negative spreading coefficient leads to a coalescence of the organic liquid into isolated pockets and tiny lenses that float at the gas–water interface. We discuss the implications of these observations to pollutant migration, aquifer remediation, and mathematical modeling.

1 INTRODUCTION

Figure 1 depicts that portion of an aquifer that makes up the vadose zone, above the capillary fringe, where a so-called immiscible organic liquid shares pore space with both gas and water. Whether an organic liquid is more (left) or less (right) dense than water, it leaves behind a trail of residual in the vadose zone as it moves downward toward the capillary fringe. This movement is more complex and diverse than in the saturated zone, especially on a pore scale, because of the presence of a non-wetting gas phase.

Soil scientists and hydrologists have studied three-phase gas–oil–water flow in column and sand tank experiments (eg., Schiegg, 1980; Schwille, 1988; Lenhard and Parker, 1989; Wilson et al., 1990; and various papers of this proceedings), but with little emphasis on pore scale phenomena. Some have simplified the system and examined two phase imbibition of organic liquids into dry soils (eg., Kia, 1988). In that case the organic liquid is assumed to be the wetting fluid, and the problem is essentially similar to that of water imbibition into a dry soil.

There are three major forces acting in both oil recovery and organic liquid behavior in groundwaters: capillary forces, viscous forces, and gravity or buoyancy forces. Capillarity is the result of the interplay of cohesive forces within each fluid phase and the adhesive forces between the solid phase and each of the fluids. The capillary force is proportional to the interfacial tension at the fluid-fluid interface and the strength of fluid wetting to the solid surface, and inversely proportional to the pore size. The vadose zone residual, which has been referred to by some researchers as the organic phase 'retention,' is held by these capillary forces. Viscous or dynamic forces within a phase are proportional to the media permeability, the fluid phase relative permeability, and the fluid phase pressure gradient. Buoyancy is a gravitational force proportional to the density difference between the fluids. The gas phase has a much lower density than either of the two other fluids, so that gravity (buoyancy) forces play a significant role in the vadose zone. Gas is almost always the non-wetting phase, so that it takes the role that the organic liquid usually takes in the saturated zone. Many organic liquids appear to be of intermediate wettability in typical aquifer materials; that is, non-wetting relative to water, but wetting relative to gas. The organic liquid phase, depending on whether it encounters gas or water, can then display either wetting or non-wetting behavior, or both

Many organic liquids also have low internal cohesion and will spread, presumably forming a film between the water phase which, because of capillarity, preferentially occupies the smallest pores, and the gas phase which preferentially fills the largest pores. This film should interconnect the pockets of organic liquid which even at residual saturation should be largely continuous. This is the usual assumption in mathematical models. One such example is described in Lenhard and Parker (1989).

The propensity to spread can be measured by the 'spreading coefficient' (Adamson, 1982; Hiemenz, 1986) :

$$S = \sigma_{aw} - (\sigma_{ow} + \sigma_{ao}) \qquad (1)$$

where S is the spreading coefficient, σ_{aw} is the gas(air)-water interfacial tension (also called the surface tension), σ_{ow} is the organic liquid–water interfacial tension, and σ_{ao} is the gas(air)–organic liquid interfacial tension. The spreading coefficient measures the difference between the work of adhesion (attraction of the

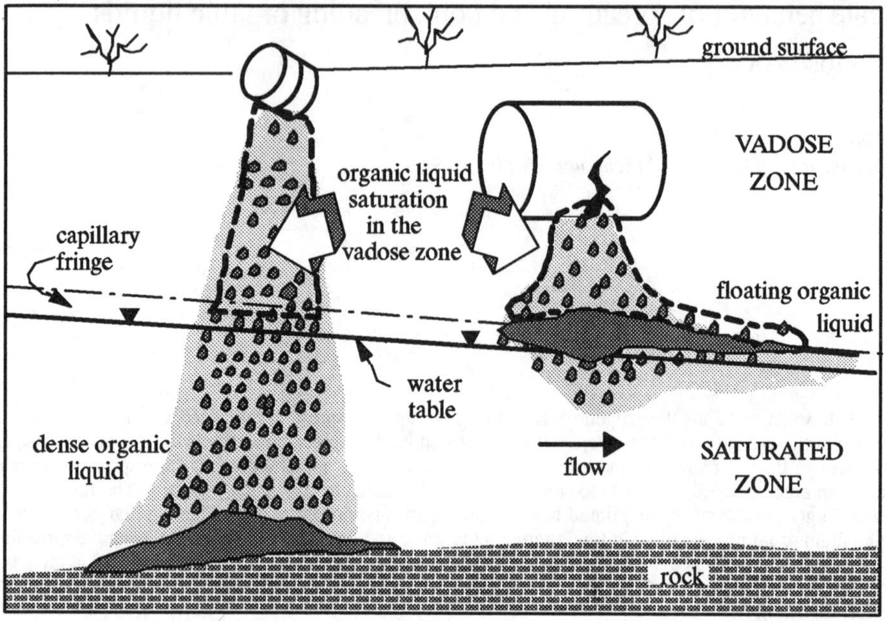

FIGURE 1. Schematic of organic liquid distribution after the initial movement through the vadose zone.

organic liquid for the water) and cohesion (attraction of organic liquid for itself). Intermediate wetting organic liquids that tend to spread as a film have a positive spreading coefficient, *S*. In this case the surface tension between the gas and water, σ_{aw}, exceeds the sum of the interfacial tensions between the organic liquid and the two other fluids, σ_{ow} & σ_{ao}. Figure 2 depicts the force balance. For spreading liquids the contact angle, θ, as measured through the organic liquid, goes to zero. Most studies of three phase flow have focused on spreading organic liquids.

Some organic liquids have more internal cohesion, resulting in a negative spreading coefficient, *S* (eg, halogenated hydrocarbons such as carbon tetrachloride and PCE). In this case the sum of the interfacial tensions between the organic liquid and the two other fluids, σ_{ow} & σ_{ao}, exceeds the surface tension between the gas and water, σ_{aw} (see Figure 2, $\theta > 0$). These organic liquids will not spread as films. On a flat water surface (the gas-water interface) non-spreading liquids tend to coalesce into lenses that float on the surface (much as depicted in Figure 2), even though many of these liquids are denser than water. When the gas-water interface is within a porous media we hypothesize that non-spreading behavior causes the organic liquid phase to become discontinuous.

We suspect that spreading and non-spreading organic liquids behave differently in the vadose zone. To investigate this issue we employed micromodel flow visualization techniques in homogeneous media to illustrate the scenario of a large slug of organic liquid percolating vertically downward into the vadose zone. Later, it's drained by air as it continues downward on its migration toward the water table. The displacement and residual saturation left behind in the micromodel 'vadose zone' was observed visually.

2 EXPERIMENTAL APPARATUS

Micromodels are physical models of a pore space network, created by etching a pattern onto two glass

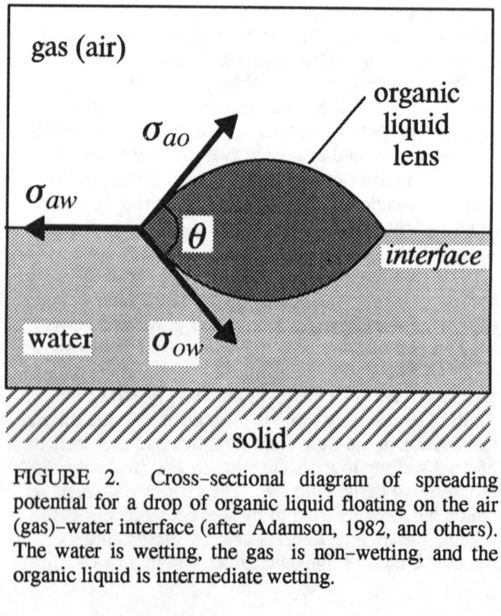

FIGURE 2. Cross-sectional diagram of spreading potential for a drop of organic liquid floating on the air (gas)-water interface (after Adamson, 1982, and others). The water is wetting, the gas is non-wetting, and the organic liquid is intermediate wetting.

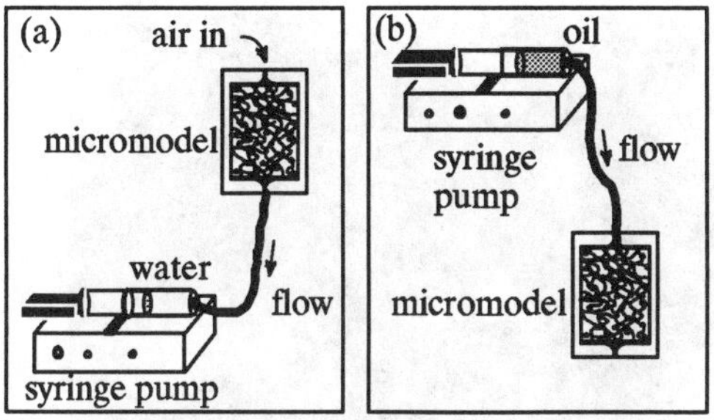

FIGURE 3. Typical micromodel set-ups.

plates which are then fused together (McKeller and Wardlaw, 1982; Conrad et al., 1989; Wilson et al., 1990; Conrad et al., 1991). The resulting pores have complex three dimensional structure, although the network is only two dimensional. The typical micro-model pore is eye-shaped in cross-section (see SEM photomicrograph in Figure 8-8 of Wilson et al., 1990). Photographs of the entire model allow examination of the bulk displacement processes, while photomicrographs taken through an optical microscope permit observation of details on a pore level.

Micromodels start with a glass mirror, stripped of its protective enamel backing to reveal a copper layer. The mirror is coated with a photosensitive resin, and a transparency of a desired pore-network pattern is placed on the coated mirror surface and exposed with ultraviolet light. The unexposed resin beneath the opaque portions of the pattern is removed with xylene. The copper beneath the pattern is removed with nitric acid, and the glass beneath the copper is etched with hydrofluoric acid (HF). A mirror-image pattern is etched on another piece of mirror, and the two etched halves are fused together in a muffle furnace to form a completed micromodel.

Each 5x8 inch micromodel was filled with water and trapped air was removed by flushing with de-aired water. Then the micromodel was positioned vertically and drained with air under a suction applied by the syringe pump or vacuum, which had been connected to the bottom fitting (Figure 3a). A syringe pump or buret was attached to the model's top fitting and organic liquid was introduced, simulating the infiltration of an organic liquid into the unsaturated zone (Figure 3b). Air and some water were displace through the bottom of the model. Once steady state conditions were reached, when fluid saturations appeared to have stabilized, a syringe pump or buret was attached to the bottom fitting and the micromodel was once again drained with air. Fluid pressures were not controlled during these experiments, although we have recently developed the techniques to so (Soll et al., 1992). The models are cleaned and reused.

Two intermediate wetting organic liquids were used in these experiments. Soltrol-130, a mixture of C_{10} to C_{13} isoparaffins, has a positive spreading coefficient and has been used in many previous experiments. The other experiments were conducted with mineral oil, a non-spreading organic liquid.

3 THE INITIAL VADOSE ZONE CONDITION

The steady-state condition achieved by draining a water saturated model with air represents the portion of the vadose zone well above the capillary fringe (see Figure 1), where water saturation is near its so-called irreducible value. The water filled some of the pore throats, an occasional by-passed pore body, and formed a film around the air-filled pores. Because the models used had fairly large pores (roughly 200 x 400 microns in cross-section) and a relatively uniform pore size distribution, not much suction was needed to achieve a fairly complete drainage. A typical view of an initial condition is shown in Figure 4.

4 SPREADING ORGANIC LIQUID

Soltrol dyed with Oil Red O was infiltrated downward into a drained micromodel at a relatively slow rate of 0.1 ml/min. A snapshot of this displacement is shown in Figure 5. Air and perhaps some water were pushed ahead of the Soltrol front. The front advanced by displacing air, in bulk, from the interconnected air-filled pore bodies and pore throats. Water that had previously been 'trapped' in pore throats was largely bypassed. During this displacement, some pockets of organic liquid were developed ahead of the essentially uniform main front of advancing organic liquid. At first glance, we were puzzled as to where this fluid had come from, but closer examination revealed that it had traveled through thin films of the organic phase formed between the air and water phase. Figure 5 illustrates several minor examples of this phenomenon. A more dramatic example is shown in Figure 10-5 of Wilson et al. (1990). Rapid advancement of organic liquids through film flow in the vadose zone is apparently a form of preferential flow. On a pore scale the front was somewhat ragged, as shown in the figure, but overall the displacement was relatively uniform and stable.

At the end of the infiltration, the Soltrol filled many pore throats and formed a film between the air and the water, as shown in Figure 6. This film penetrated somewhat into the water-filled pore throats. In short water-filled pore throats,the Soltrol from each end

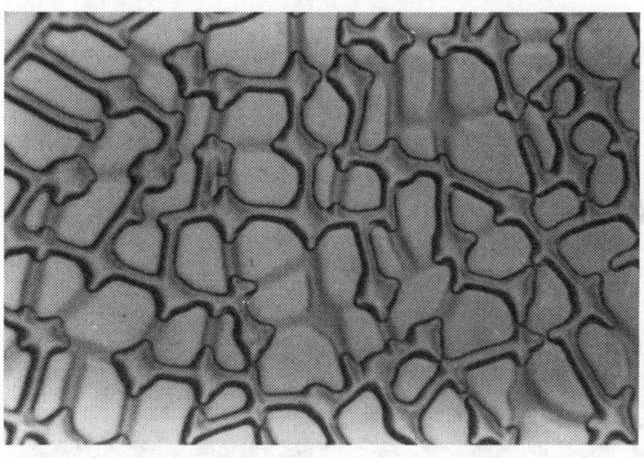

FIGURE 4. Typical view of steady state conditions after water was drained by air. The water was dyed blue (light grey), and the air was not dyed.

almost touched, and the water phase pendular ring was clearly visible (also seen later in Figure 7). Water was also present as a wetting phase film between the Soltrol and the solid surface. The wetting phase 'film' actually occupied the wedges on either side of the eye-shaped pores. The saturation and distribution of the water phase remained almost entirely unchanged when compared to the initial vadose zone condition. The non-wetting air phase was trapped in isolated bubbles or blobs, similar to those seen for the organic phase in the saturated zone (see top part of Figure 5).

In the final step of the experiment, Soltrol was drained from the model by air, representing the continued downward percolation of the organic phase toward the water table. The Soltrol was drained from the model under a small applied suction. This suction is believed to have been insufficient to completely drain the organic phase to its lowest possible saturation. The final distribution of fluids within two small sections of the model is shown in Figure 7. The amount of organic phase retained in the vadose zone was much less than residual saturation left behind in similar saturated zone experiments (Wilson et al., 1990).

In particularly dry portions of the pore space the residual is found in pore throats as pendular rings, in small pore bodies, and as a film between the air and water phases. The film helps fill the wedges of the eye-shaped pore cross-section and penetrates into the water-filled pore throats, surrounding water-filled pendular rings, as shown in the upper photo in Figure 7. This film can be quite thin. The air phase in the lower photo of Figure 7 is completely surrounded by a thin water film, sometimes barely visible in the photograph. Organic liquid was also trapped as blobs within the water phase. We found that this kind of trapping is more likely to occur when the initial water content is relatively high or when infiltrating water follows the organic liquid. Occasionally we saw a gas bubble trapped inside one of these organic liquid blobs. However, due to its intermediate wetting properties and its tendency to spread, organic liquid in the vadose zone is most commonly retained in between the water and air phases.

In contrast to organic liquid blobs trapped in the saturated zone, the residual spreading organic phase in the vadose zone remains more or less continuous, interconnected by the ubiquitous films (with the exception of any organic liquid that might have been trapped as blobs entirely within the water phase). We and others have referred to this organic liquid as 'residual' and 'trapped', but these terms must be used with care. The spreading liquid saturation is a continuous function of applied viscous and gravity forces; it can always be reduced further, albeit slowly.

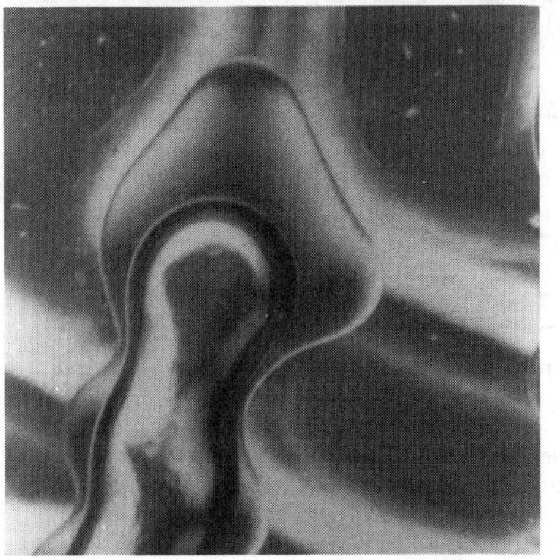

FIGURE 6. Detail of steady state conditions after the Soltrol infiltrated the model.

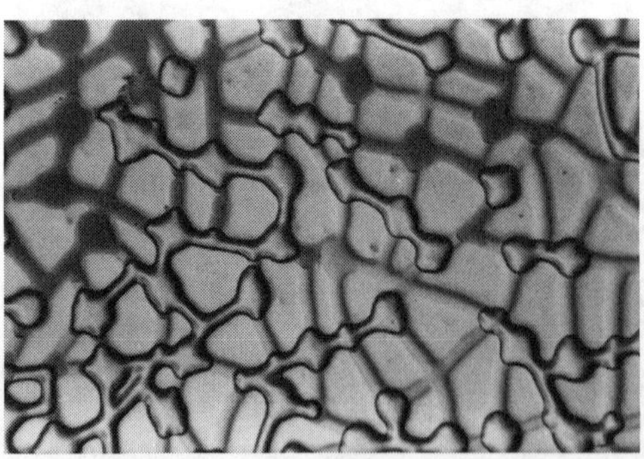

FIGURE 5. Typical view of Soltrol 'front' infiltrating the model from the top. The Soltrol was dyed red (dark grey).

The final distribution of Soltrol in this micromodel experiment, although complicated by the presence of water, was similar to drained wetting phase distributions described by Gvirtzman et al. (1987) and others. In our experiment the spreading organic phase acted much like a wetting phase both in the character of its migration and its final distribution within the pore space. This can be attributed to the relatively low initial water saturation, and because the organic phase readily spread on the water/air interface.

5 NON-SPREADING ORGANIC LIQUID

Mineral oil dyed with Sudan 4 was infiltrated downward into a drained micromodel. The infiltration front was relatively uniform and resembled the Soltrol infiltration in its movement . We suspect that there may be subtle differences and will look for them in the future using image processing techniques. Film flow did not appear to be as common as it was with Soltrol. At the end of the infiltration the final distribution of fluids was also similar, but with lesser degree of interconnection of the organic phase. Figure 8 is a photomicrograph of an entrapped air bubble, which is surrounded entirely by water, at least in the field of view. The surface of the air bubble is dimpled. Apparently these dimples are small lenses of mineral oil 'floating' on the air-water interface of the bubble. In cross-section they look like Figure 2. They were separated from the main body of oil during the infiltration, presumably by the dynamic, pulsating nature of displacement on a pore scale.

In the final step of the experiment mineral oil was drained by air. The final distribution of fluids within two small sections of the model are shown in Figure 9. The upper photo shows an air filled pore, on the left, and an oil filled pore on the right. As it leaves the photo at the top the air is surrounded by a film of oil, similar to that seen with Soltrol. As the air leaves to the left the film no longer exits. It truncates somewhere in between. In this area the air–water interface is dimpled again, with small lenses of mineral oil. These lenses are not interconnected. The various pockets of oil, such as on the right side of the figure, are also generally discontinuous and not connected. This is not always the case, however. The lower photo in Figure 9 shows an air filled pore almost completely surrounded by a thin film of mineral oil. There is some dimpling of the air–water interface in the left–top portion of the photo. The oil film stretches along both visible sides of the air phase. This stretched film does not appear to be stable. Fluctuating pressures, or simply tapping on the model, can cause it to snap, disconnecting the pockets of oil at the top and bottom of the photo.

The typical micromodel pore is eye–shaped (see Figure 8–8 of Wilson et al., 1990). The smallest dimension is from the 'floor' to the 'roof' of the pore.

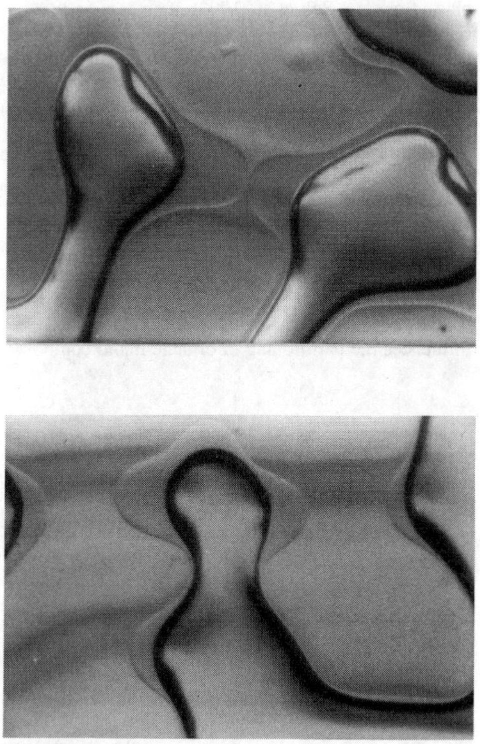

FIGURE 7. Details of steady state conditions after the Soltrol was drained by air from the model.

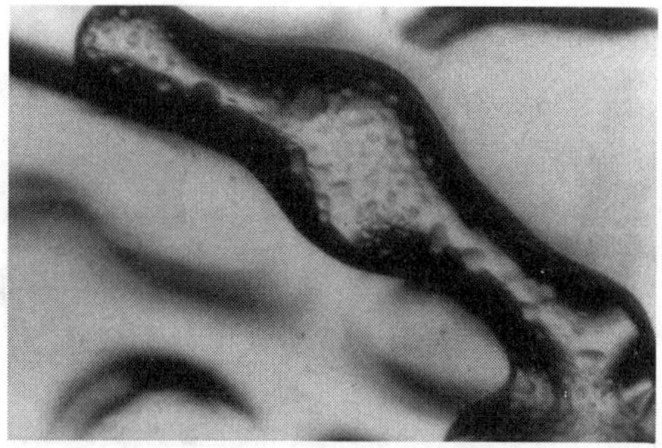

FIGURE 8. Detail of steady state conditions after the red dyed mineral oil infiltrated the model. The water was not dyed.

We infer that in the mineral oil film truncates somewhere along the floor and roof, such as in the dimpled zone of the lower photo in Figure 9. Beyond that point the mineral oil has coalesced into small lenses along the air–water interface. In the Soltrol experiment stereo microscope observations revealed that the Soltrol layer completely surrounded the air (see, eg, Figure 6).

We have performed similar experiments with PCE and obtained similar results (see Wilson et al., 1990). Figure 10 shows two discrete pockets of PCE on each side of an aqueous pendular ring. Each pocket is isolated. There are no films along the gas–water interface.

Although a non–spreading fluid may fill or penetrate into some pore throats, and may form or coat pendular rings, it does not generally spread out as a film. The organic liquid coalesces into lenses and pockets. Large portions of the pore space do not have a film of organic liquid between the air and the water.

6 DISCUSSION

Our results support the hypothesis that spreading organic liquids form a interconnected film between water and gas, while non–spreading liquids end up in disconnected pockets and small lenses floating on the gas–water interface.

We and others have quantitatively investigated the amount of residual saturation expected for spreading organic liquids under vadose zone conditions (eg, Wilson et al., 1990), but we know of none that have done so for non–spreading organics. We can speculate. Halogenated hydrocarbons such as carbon tetrachloride and PCE are non–spreading, and much denser than water. Greater density implies a greater (negative) buoyancy force, while the non–spreading nature implies a positive effective contact angle θ (see Figure 2) and therefore a lower capillary force: the result should be a lower residual saturation. The micromodel experiments also indicate that non–spreading organic liquids may have a greater tendency to by–pass portions of the pore space, leading to lower residual saturations (consider the lack of bulk mineral oil in Figure 8). The

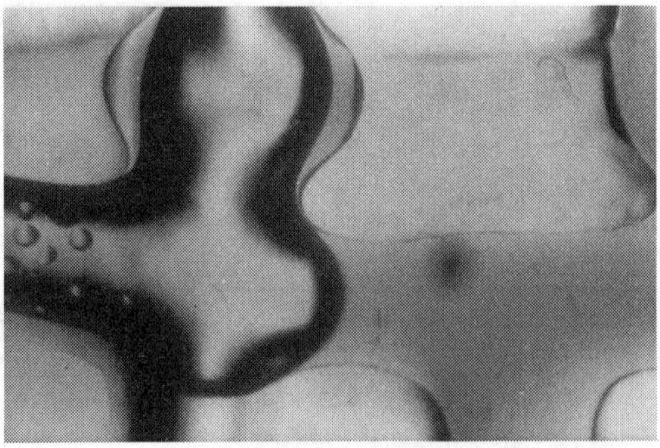

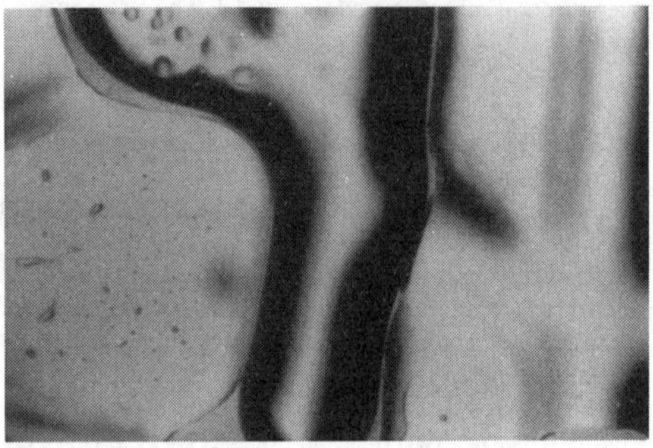

FIGURE 9. Details of steady state conditions after the mineral oil was drained by air from the model.

general absence of an interconnecting film also leads to a lower residual saturation. In summary, we hypothesize that non-spreading organic liquids have lower residual saturations in the vadose zone, than do spreading organic liquids.

Most continuum mathematical models of organic liquid migration assume that the organic liquid phase is continuous and forms a layer between the water and air phases (for one example of this model see Lenhard and Parker, 1989). Within the context of spreading organics this assumption appears valid. These models also assume that three-phase capillary pressure/ saturation relationships can be approximated by two-phase relationships. This assumption, first proposed by Leverett in the early 1940's, maintains that water saturations in a three-phase system are a function of the two-phase water-organic retention curve, while the total fluid saturations (water + organic) are functions of the organic-gas curve. Implicitly, this assumption does not allow gas-water interfaces to exist. Indeed, in our Soltrol experiments we see comparatively few water/gas interfaces. So for the majority of common organic pollutants which tend to spread between the air and water phases in the vadose zone, approximation of three-phase capillary pressure/saturation relationships using two two-phase retention curves seems to be an appropriate assumption. It appears to preserve the physics and surface chemistry of the three-phase flow system in addition to being computationally advantageous. Lenhard and Parker's (1989) model effectively assumes a zero or neutral spreading coefficient. This somewhat contradicts their other assumption that the organic liquid always forms a layer between the water and air. In any case, it is not clear whether or not this model can handle the different physics associated with a non-spreading organic liquid.

Simmons et al. (1992) have performed vadose zone sand box infiltration experiments that suggest that non-spreading organic liquids may finger due to gravity instabilities. We hypothesized that these instabilities could be due to the discrete nature of the non-spreading fluid distribution, at least at low saturations. The individual pockets and lenses of organic liquid could migrate through the pore space as a train of discrete drops, setting up the conditions necessary for gravity dominated flow along preferred paths. We saw no evidence for this hypothesis in our experiments. Indeed, within our pore space we were unable to achieve fingering. It thus appears likely that the fingering observed by Simmons et al. (1992) is due to the smaller capillary forces associated with the mineral oil they used in their experiment. The capillarity was apparently below the threshold needed to damp out the finger growth. This suggests that non-spreading fluids have a greater propensity for fingering, but do not intrinsically finger in the vadose zone.

Mathematical percolation network models of multiphase flow (eg, Soll et al., 1992) often assume that only one phase can occupy a given pore body or pore throat at any given time and neglect films and film flow. In our experiments we have commonly observed two and three phases simultaneously occupying a single pore, and we've seen that film flow can play an important role in the movement of organic liquids. The wetting and intermediate-wetting phases remain connected through the films, which maintain a continuity for drainage or imbibition that cannot be captured using a traditional percolation network approach. Future versions of these network models should explore this issue

The pore-scale distribution of the organic phase influences the partitioning of organic liquid compo-

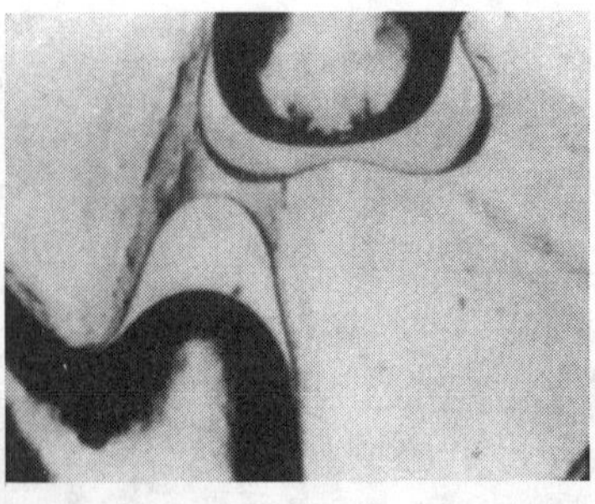

FIGURE 10. Detail of steady state conditions after PCE infiltrated the model and was drained by air. The water and PCE were not dyed.

nents. Because the organic phase is almost always less wetting than water but more wetting than air, it remains in direct contact with both the air and water allowing for both solubilization into the water phase and especially volatilization into the air phase. The formation of thin organic liquid films between the water and air increases the surface area of the organic phase, enhancing the propensity for inter-phase partitioning of organic components. Soil venting of organics in the vadose zone has become an attractive remediation strategy because volatile organics partition easily to the air phase. The thin film in Figure 7 provides a good example of the large organic-air and organic-water interfaces generated by an organic phase's tendency to spread. However, the films are not of uniform thickness, and they interconnect organic liquid filled pore throats. In a multi-component organic liquid, mass transfer rates could be limited by the diffusion of the more volatile components to the interface, from the pore throats. The potential rate of transfer would then be distributed non-uniformly along the gas-organic liquid interface, depending on the film thickness and tortuosity, and the geometry of its connection to the larger pockets of organic liquid. Non-spreading organic liquids generally do not form films, but rather coalesce into isolated pockets and small lenses that float at the water-gas interface. Thus non-spreading organics have a smaller contact area with the gas and water phases. The respective mass transfer coefficients may be significantly reduced, making soil venting and similar vadose zone remediation schemes less effective.

The distribution of the films and pockets of organic liquid can influence colonization by microorganisms, which tend to adhere to pore walls. The different environments that organisms might colonize in the vadose zone are suggested by the photomicrographs. There are largely air filled pores, with a ready access to oxygen, but with only a very thin double film of water and organic liquid. At residual saturation these films would have a very limited ability to resupply needed nutrients to microorganisms. The environment of water filled pore throats and pendular rings may be controlled by the lack of oxygen. The film of organic liquid should suppress oxygen transfer to the water phase. We speculate that transient changes in fluid saturations would improve the environmental conditions for aerobic bacteria. Biosurfactants that are a by-product of biological activity should change interfacial tensions, especially for organisms that attach at the organic liquid- water interface. These changes should reduce organic liquid saturations by drainage, and possibly cause a non-spreading organic liquid to become a spreading organic liquid. This should improve mass transfer coefficients.

7 CONCLUSIONS

The movement of organic liquid through the vadose zone is more complex than organic liquid movement in the saturated zone, especially because of the tendency for some organics to spread and others not to. The organic liquid usually advances by displacing gas, and sometimes water, from pore throats and bodies. It also advances through film flow, most easily seen in our experiments when pockets of organic liquid began to accumulate deeper in a micromodel, with no obvious connection to sources of organic liquid above. Some water is by-passed, mostly in pore throats, as is some air, mostly in pore bodies. The resulting entrapped air bubbles are essentially identical to the non-wetting phase blobs seen in two-phase saturated zone experiments.

Positive spreading coefficients lead to organic liquid films along the gas-water interface. The residual saturation of a spreading organic liquid consists of films, pendular rings, wedges surrounding aqueous pendular rings, and filled pore throats. There are a few isolated organic liquid blobs trapped in otherwise water filled pore bodies, and the occasional gas bubble trapped inside an organic liquid blob. The films connect these pockets of residual organic liquid in pore bodies or pore throats, and allow them to communicate. They are not like the isolated blobs one sees in the saturated zone.

Negative spreading coefficients lead to isolated pockets of residual organic liquid, primarily trapped as pendular rings, wedges surrounding aqueous pendular rings, and filled pore throats. These pockets are not interconnected. Some organic films are present, but appear to be less stable. In many locations the gas and water are in direct contact. Small coalesced lenses of organic liquid can be seen 'floating' on this interface.

ACKNOWLEDGEMENTS

The experiments were conducted by Bill Mason, Steve Conrad, and Paul Hoffman. The initial portion of this work was supported by the U.S. Environmental Protection Agency (Grant EPA CR-813571-01-0), and the New Mexico Water Resources Research Institute (Grant 1345648). The most recent portion was supported by the Subsurface Science Program of DOE (Grant DE-FG04-89ER60829)

REFERENCES

Adamson, A.W. 1982. *Physical Chemistry of Surfaces*, 4th ed.. Wiley, New York.

Conrad, S. H., J. L. Wilson, W. R. Mason, and W. Peplinski. 1989. Observing the transport and fate of petroleum hydrocarbons in soils and in ground water using flow visualization techniques. *Proc. Sym. on Environ. Concerns in the Petroleum Industry*, AAPG, Palm Springs, Ca, 1989.

Conrad, S.H., J.L. Wilson, W.R. Mason, and W. Peplinski. 1991. Visualization of Residual Organic Liquid Trapped in Aquifers, *Water Resour. Res.*, in press.

Gvirtzman, H., M. Magaritz, E. Klein, and A. Nader. 1987. A scanning electron microscopy study of water in soil. *Transp. In Porous Media*, 2: 83-93.

Hiemenz, P.C. 1986. *Principles of colloid and surface chemistry*, 2nd ed., Marcel Dekker, New York.

Kia, S.F. 1988. Modeling of the retention of organic contaminants in porous media of uniform spherical

particles. *Water Res.*, 22: 1301–9.
Lenhard, R. J. and J. C. Parker. 1989. A model for hysteretic constitutive relations governing multiphase flow, 3. refinements and numerical simulations. *Water Resour. Res.*, 25: 1727–36.
McKellar, M. and N.C. Wardlaw. 1982. A method of viewing water and oil distribution in native-state and restored-state reservoir core. *AAPG Bul.*, 72: 765–71.
Schiegg, H.O. 1980. Fundamentals, setups, and results of laboratory experiments on oil propagation in aquifers. VAW-Mitteilung no. 43, Versuchsanstalt für Wasserbau, Hydrologie und Glaziologie, ETH, Zurich.
Schwille, F. 1988. *Dense chlorinated solvents in porous and fractured media.* Lewis Publishers, Chelsea, Mi.
Soll, W.E., M.A. Celia, and J.L.Wilson. 1992. Quantitative comparison of computational and experimental methods for three phase immiscible transport. *Proc. Subsurface Contamination by Immiscible Fluids*, IAH, Calgary, Canada, (this proceedings).
Simmons, S., J. McBride, J. Cary and R.J. Lenhard. 1992 Organic liquid infiltration into unsaturated porous media.. *Proc. Subsurface Contamination by Immiscible Fluids*, IAH, Calgary, Canada, (this proceedings).
Wilson, J.L., S. H. Conrad, E. Hagan, W.R. Mason, W. Peplinski, and E. Hagan. 1990. Laboratory investigation of residual liquid organics. U.S. Environmental Protection Agency Rpt. CR-813571

Subsurface Contamination by Immiscible Fluids, Weyer (ed.) © 1992 Balkema, Rotterdam. ISBN 90 6191 175 3

Experiments on movement of vapour in unsaturated soil

W.B. Benson & G.J. Farquhar
Department of Civil Engineering, University of Waterloo, Ont., Canada

ABSTRACT: Forced air venting is an important technique for the removal of volatile organic liquids from soil. Field applications are common, however the design of such systems is difficult because contaminant removal efficiency and the factors which affect it are poorly understood. Such factors include air flow rate, spill size and configuration, soil properties, and boundary conditions, with efficiency expressed as rate of removal and percent of mass recovered. The research described in this paper involves laboratory scale experimentation to improve fundamental process understanding and predictability. Experiments involved vacuum gas extraction of volatile solvents, including hexane, heptane, and trichloroethylene, from sand using a range of air flow rates and various sizes and configurations of a liquid residual source. Equilibrium vapour phase concentrations were achieved in the exhaust gas only if the total air flow passed through the spill. While designs commonly assume equilibrium concentrations, this research shows that these are unlikely to occur in most field situations. A relationship for predicting exhaust vapour concentrations in the experiments was developed, based on the air phase velocity, chemical diffusion coefficient, equilibrium vapour concentration, and spill dimensions.

1 INTRODUCTION

This paper presents some recent results from a research programme studying the behaviour of hazardous immiscible liquids and their vapours in soil. This program has been in progress since 1983 in the Department of Civil Engineering at the University of Waterloo. Part of this study is currently focussed on the practice of forced air venting as a means for removing volatile contaminants from soil, and it is this work to which this paper is addressed. Presented are results of laboratory venting experiments as well as a mathematical model used for simulating the experimental results. Initial experimentation for this project was described previously by Farquhar, et al., (1988).

2 BACKGROUND

In recent years, soil venting has gained increased exposure as a remediation method for spills of volatile organic contaminants. Removal is achieved by enhanced advective vapour phase transport, induced by withdrawing or injecting air through wells screened in the unsaturated zone. Contaminant vapours are swept along the induced air flow path toward withdrawal wells, where they are removed and treated or released directly to the atmosphere. Constant recharging of the unsaturated zone with fresh air leads to increased contaminant volatilization. Several successful applications of soil venting for solvents and light hydrocarbons have been reported within the recent past, by Zimmermann and Brizgys (1986), Hoag and Cliff (1987), and Baehr, et al., (1989), to name a few. However, the predictability of the method is still very poor. Applications of soil venting as a remediation technique are performed with little advance knowledge of the extent of clean-up that is possible and the time in which it can be achieved.

Laboratory scale column experiments performed by others (Marley and Hoag (1984), Baehr, et al. (1989), Johnson, et al. (1987)) have tended to show that local liquid-vapour equilibrium exists in the porous medium, and the chemical concentrations in the exhaust vapour are at or near their saturation limits. If the chemical source is a multi-component mixture, such as gasoline, the resulting exhaust concentrations show a decline over time, due to selective volatilization of the more volatile components. In column experiments, this process has been simulated successfully (Baehr, et al, 1989) using an equilibrium-based model to predict the vapour concentration of each component in the exhaust gas, based on the vapour pressures. A feature of these column experiments was that the entire cross-section of the column was at residual saturation with the hydrocarbon source. In large scale experiments (Wootan and Voynick (1984), Rainwater, et al., (1988)) and field applications, where the contaminant source is discontinuous, concentrations in the exhaust gas are often significantly less than equilibrium saturation levels. It can be surmised that this is due to air flow through soil zones where little or no residual is present and in such cases there is no basis for predicting the chemical removal rate. Similarly, mathematical models formulated for the analysis of soil venting (Baehr, et al., (1989), Stephanatos (1988)) assume linear instantaneous equilibrium relationships for the description of interphase transfer between the liquid and aqueous phases and the gas phase.

As a consequence, this work was undertaken:

1. to examine the validity of the equilibrium saturation assumption using laboratory column experiments,
2. to devise experiments in which the chemical source would more closely simulate a contaminant spill and the resulting vapour concentrations from venting would be less than saturation levels, and
3. to develop a predictive relationship to describe this behaviour based on experimental results and relevant theoretical considerations.

3 EXPERIMENTAL WORK

Laboratory experiments were conducted to observe and to quantify the rate at which a chemical source in a porous medium would volatilize and subsequently be transported under a forced air flow. Experiments were performed in vertical columns and in a large rectangular box. The principal variables were air flow rate, chemical type, and the size and shape of the contaminant zone to be removed.

3.1 Materials and equipment

Experimentation made use of two columns of different sizes, shown in Figure 1. Air flow was introduced into the column from the bottom using positive pressure. The desired flow rate for each experiment was set initially using the inlet flow meter, and outflow was monitored over the course of the experiment since it was a function of the chemical concentration in the exhaust gas. Vapour samples were drawn from the exhaust line, immediately after the gas exited from the column, and were analysed by gas chromatography.

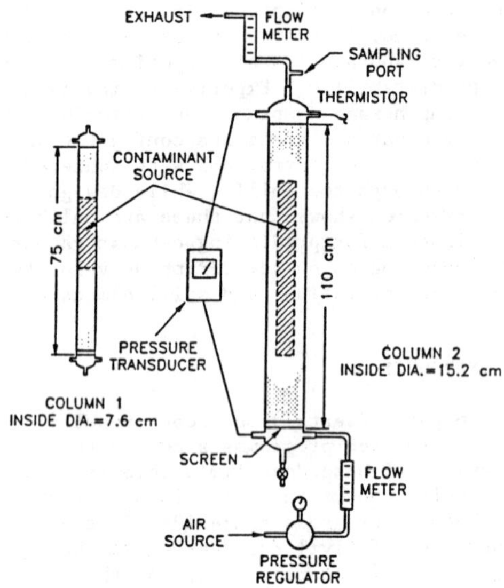

Fig. 1 Soil venting columns

Several experiments were performed in the box, illustrated in Figure 2. This box had been constructed for earlier work by Kell (1988) and was fitted with two vents to make it suitable for soil venting. Instrumentation for the box experiments was similar to that used for the columns. The design and operation of the experimental equipment is described in more detail elsewhere (Farquhar, et al., (1988), Benson, (1989).

The soil used for all experiments was a fine, uniform Ottawa sand. Soil properties are listed in Table 1.

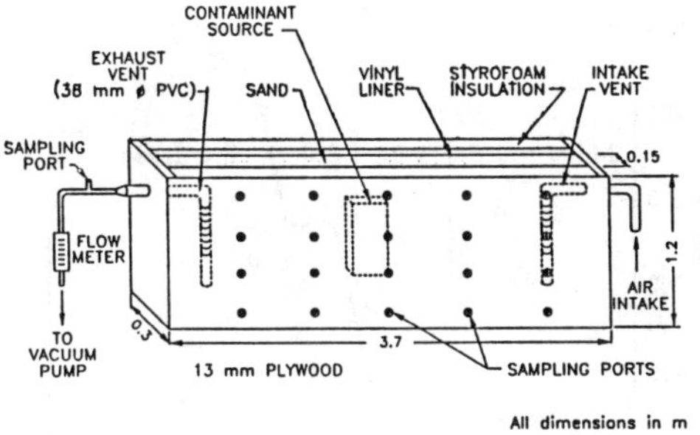

Fig. 2 Soil venting box

Table 1. Soil properties

intrinsic permeability (k)	1.45×10^{-7} cm^2
porosity	0.4
organic content	0.17% dry wt.
water content	0.035% dry wt.
bulk density	1.60 g/cm^3

Three different chemicals were used in this study, hexane, heptane, and trichloroethylene (TCE), with hexane being used predominantly. All three chemicals are considered volatile and are good candidates for removal from soil by venting. Chemical properties are summarized in Table 2.

Table 2. Chemical properties

	Hexane	Heptane	TCE
Formula	C_6H_{14}	C_7H_{16}	C_2HCl_3
Vapour Pressure (mmHg, 20°C)	124	40	60
M.W. (g/mole)	86.2	100.2	131.4
Density (g/cm^3)	0.67	0.68	1.46

3.2 Column experiments

Column 1: Chemical source occupying entire cross-section

Initial experiments were carried out in the smaller of the two columns. In these experiments, the chemical was poured into the sand-packed column from the top and allowed to migrate downward for a period of 4.5 hours. Separate experimentation showed that, after 4.5 hours, the chemical was at a residual saturation of 72 ml per litre of sand, approximately 60% of its ultimate residual saturation capacity. After 4.5 hours, the column was topped up with clean sand, capped and vented from the bottom with an air flow of 0.5 l/min (a specific discharge of 11.0 cm/min). Effluent vapour concentrations were monitored over the course of the run.

The experiment was performed using three different chemicals, hexane, heptane, and TCE, with a constant volume (90 ml) added to the column for each run. The results presented in Figure 3 show that the exhaust concentration for hexane and heptane quickly reached saturated equilibrium and remained near this level over the course of the run before declining quickly to near zero. The exhaust concentrations of the denser TCE remained at approximately 85% of the saturation concentration.

Two variations of this experiment were also performed. In the first, sand was initially mixed with hexane and packed into the column at a residual content substantially less than the estimated ultimate residual saturation capacity. In the second variation, the sand packed column was initially saturated with water and allowed to drain to its field capacity before the hexane was added. In both cases, the resulting hexane concentration in the effluent gas was essentially unchanged from that presented in Figure 3.

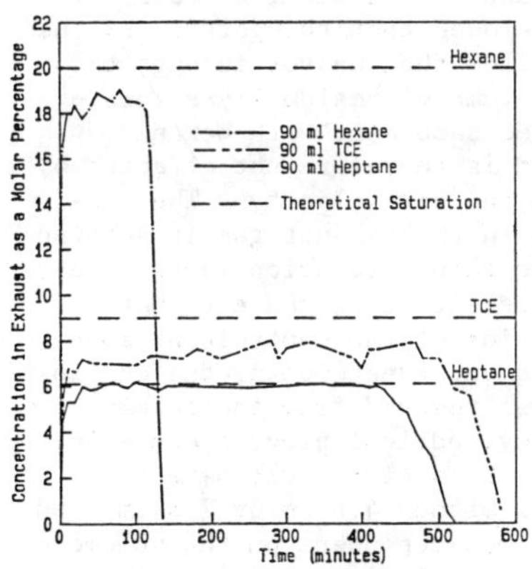

Fig. 3 Chemical concentrations in exhaust gas with entire cross-section of column at residual saturation (inflow 0.5 l/mon)

Column 2: Chemical source as a residual core in the soil

A second set of experiments was performed in the larger column with hexane exclusively as a chemical source. In order to draw conclusions regarding the effect of spill geometry on removal characteristics, it was necessary to introduce the chemical into the sand in such a way that the dimensions of the resulting contaminated zone would be known. This was accomplished by containing the hexane, as it migrated downward, within a form packed inside the column. This form could subsequently be removed to produce a cylindrical zone containing hexane at a known residual saturation.

Initial experiments were performed with the interior confining tube in place, in order to isolate the chemical removal due solely to air flowing through the zone of contamination. Air flow rates of 0.5, 1.0, 3.0, and 5.0 l/min were used. The corresponding specific discharges were 2.7, 5.5, 16.4, and 27.4 cm/min, respectively. These velocities are relatively high by field standards and would be most likely to occur in the vicinity of the discharge well. The results for a single experiment are shown in Figure 4. These results show both the exhaust hexane concentration (mg/l) and the mass removal rate of hexane (mg/min) over time, for various flow rates. It can be seen that the hexane concentration remains constant, independent of the air flow rate, and as a result, the mass removal rate is solely a function of air flow rate. It can be concluded that the portion of the total air flow that passes through the interior column of residually saturated sand emerges saturated with hexane vapours and is therefore not affected by changes in the air velocity. The concentration in the exhaust gas is substantially less than saturation levels due to the diluting effect of the air that passes outside of the contaminant zone.

In subsequent experiments, the confining form was removed from the column, leaving a cylindrical plug of contaminated sand. This plug could have a diameter of either 4.8 cm or 7.5 cm, and its height was dependent on the volume of hexane. This volume ranged from 60 ml to 120 ml which, for a 4.8 cm diameter form, would result in heights ranging from 48 cm to 96 cm. As before, the air inflow varied from 0.5 to 5.0 l/min. Results from four separate experiments are plotted in Figure 5, showing the mass removal rate of hexane versus time, for

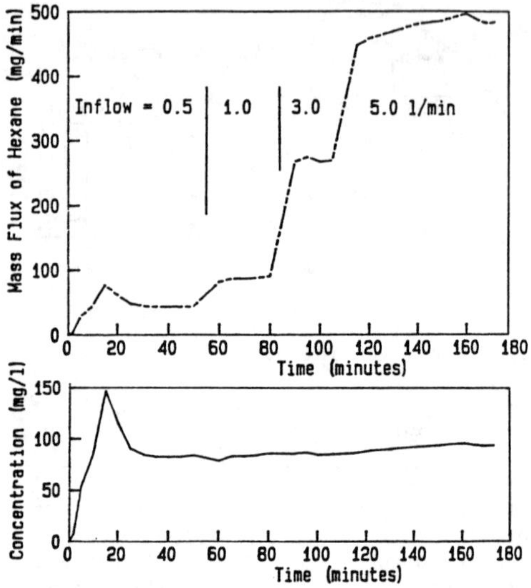

Fig. 4 Exhaust concentration and mass loss rate of hexane for 4.8 cm dia. confined plug (variable inflow, 0.5 to 5.0 l/min).

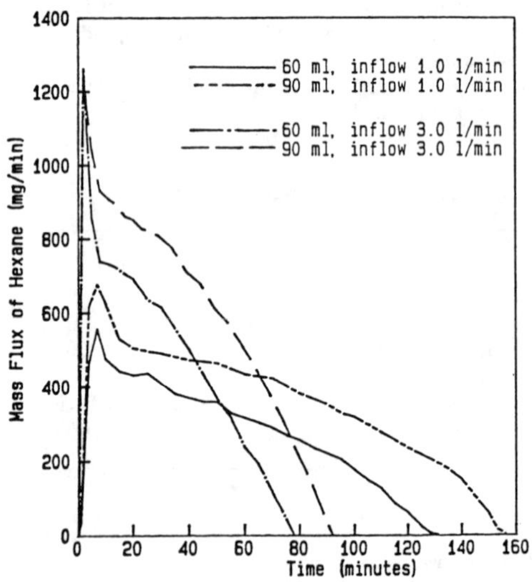

Fig. 5 Mass loss rate of hexane for 4.8 cm dia. plugs (inflows 1.0 and 3.0 l/min)

4.8 cm diameter contaminant plugs of height 48 cm and 72 cm, and air inflows of 1.0 and 3.0 l/min. The results show that the mass removal rate increases with air flow, however the actual concentration of hexane in the exhaust vapour decreases with increased air flow. Also, the concentration of hexane, and as a result the mass removal rate, increases with the axial height of the zone of contamination, for a given air flow. The actual hexane concentrations for these experiments were greater than those measured during the previous experiments, where hexane removal was due solely to air flow within the zone of contamination. In these latter experiments, there is additional removal due to outward migration of hexane into air passing along the perimeter of the contaminant plug. However, the resulting concentrations are still substantially less than the theoretical equilibrium saturation.

3.3 Box experiments

A series of experiments were performed in the large box in order to observe the phenomena studied in the column experiments on a larger scale, and to consider the effects of another dimension in the analysis. The experimental variables were air flow rate and the dimensions of the chemical residual source, in this case hexane.

As in the column experiments, the hexane was poured into a form packed within the sand so that the shape and dimensions of the resulting zone of residual contamination would be known. The form was an open-ended stainless steel box with dimensions 90 cm high by 30 cm long by 5 cm wide. It was constructed in such a way that each side could be removed independently, therefore minimizing the disturbance to the chemical source. The resulting zone of residual contamination had a length of 30 cm and a cross-sectional width of 5 cm, and the height was determined by the volume of hexane. In the experimentation, hexane volumes of 336 ml, 505 ml, and 673 ml were used which, based on a residual content of 72 ml per litre of sand, would result in heights of 31 cm, 47 cm, and 62 cm, respectively.

Air flow in the box was induced by negative pressure created with a vacuum pump. Air flow rates of 6.6, 13.2, and 26.4 l/min were used for the experimental runs, corresponding to specific discharges of 3.6, 7.2, and 14.5 cm/min, respectively. Pressure measurements in the box indicated that flow through the chemical source was essentially horizontal and uniform.

Examination of the resulting relationship between hexane mass removal rate and time showed trends similar to those observed in the column experiments.

4 MODEL DEVELOPMENT AND SIMULATIONS

After the completion of the laboratory experimentation, the second part of this study was directed towards developing a method to mathematically predict the vapour concentrations in the exhaust air. Ideally, this formulation would apply not only to the experimental work, but also would have wider applications in soil venting.

Based on experimental evidence, it was concluded that incoming air passing through the zone of residual contamination would emerge at equilibrium saturation with respect to the chemical source. It then remained to predict the additional chemical removal due to air passing along the periphery of the contaminant zone. The problem was conceptualized as a boundary layer problem, illustrated in Figure 6.

Air passing along the contaminant zone is initially free of chemical vapours, however a concentration boundary layer develops due to vapour phase diffusion perpendicular to the direction of air flow. Assuming a constant concentration C_o at the surface, the steady state mass transfer equation can be written as (Brodkey and Hershey, (1988)):

$$U_x \frac{\delta C}{\delta x} + U_y \frac{\delta C}{\delta y} = D \frac{\delta^2 C}{\delta y^2} \qquad (1)$$

U_x: vapour velocity in x-direction $(\frac{L}{T})$

U_y: vapour velocity in y-direction $(\frac{L}{T})$

C : concentration of chemical in vapour phase $(\frac{M}{L^3})$

D : diffusion coefficient $(\frac{L^2}{T})$

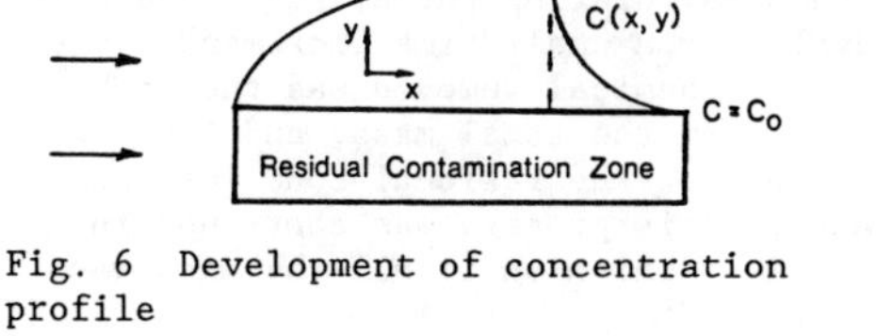

Fig. 6 Development of concentration profile

For this analysis, it was assumed that the vapour velocity through the zone of residual contamination was equal to the vapour velocity through the surrounding sand. Experimentation showed that the vapour phase conductivity in the residual zone was only 5% less than the conductivity in surrounding sand, so this assumption should have introduced little error. In this case, the velocity in the y direction was assumed to be zero, and the mass transfer equation reduced to:

$$\frac{\delta^2 C}{\delta y^2} - \frac{U_x}{D}\frac{\delta C}{\delta x} = 0 \qquad (2)$$

Boundary conditions were set as follows:

for $x > 0$: $C=C_o$ for $y=0$ and $C=0$ for $y=\infty$

for $x = 0$: $C=0$ for $y > 0$

This problem has the same form as one dimensional transient heat transfer in a semi-infinite slab, and the solution can be written as (Brodkey and Hershey, 1988; Tikhonov and Smarski, 1963):

$$C(x,y) = C_o[1 - \text{erf}\,(\frac{y}{2\sqrt{\alpha x}})] \qquad (3)$$

where α is a constant equal to D/U_x.

This formula was incorporated into a computer model to simulate the experimental conditions. It was assumed that the portion of the total air flow passing through the zone of contamination would emerge at saturation. This saturation concentration was calculated using the vapour pressure of the chemical at the appropriate temperature and pressure, and the ideal gas law. Additional concentrations were calculated using Equation 3, with the appropriate diffusion coefficient and the air phase velocity calculated from the specific discharge and the porosity of the soil. Essentially, the model worked by calculating the concentrations at an instant in time, based on the dimensions of the contaminant zone, and then calculating the average chemical removal over a small time increment. The amount of chemical removed was then subtracted from the total mass, and the new dimensions of the residual zone were calculated. This process was continued in a time-step fashion until all the mass was removed.

The results of the model simulations for three experiments performed in the box are shown in Figure 7, as well as the experimental data. These results are for contaminant zones of three different heights, vented at a constant flow rate. The model results show reasonably good agreement with the experimental results.

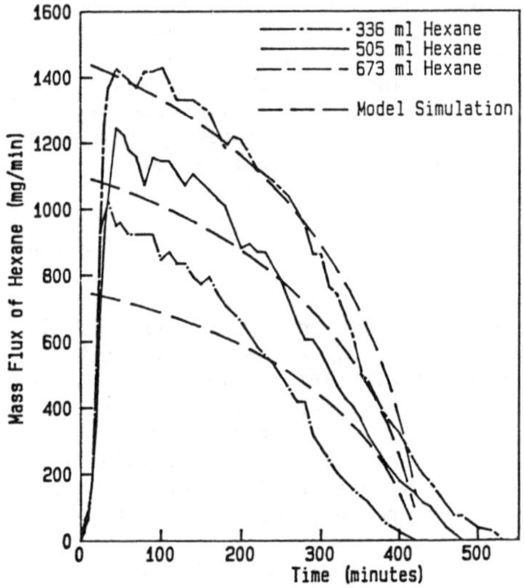

Fig. 7 Mass loss rate of hexane and model simulations for three experiments performed in the venting box (inflow 6.6 l/min)

5 DISCUSSION

The results obtained from these experiments indicate that it may be reasonable to assume equilibrium saturation concentrations in exhaust gases from soil venting, but only when all the air passes through the zone of contamination. This is unlikely to be the case in most field situations. Experimental results, and subsequent analysis, for cases where there is air flow around a simulated contaminant spill have led to a predictive relationship for mass removal that appears to be promising. This formulation relies on the equilibrium vapour saturation concentration of the contaminant, the air phase velocity, and the appropriate diffusion coefficient. To date, good simulations of laboratory experiments have been accomplished, however this analysis requires that the shape and dimensions of the contaminant zone be known, an unlikely circumstance in field applications.

REFERENCES

Baehr, A.L., G.E. Hoag, and M.C. Marley 1989. Removing volatile contaminants from the unsaturated zone by inducing advective air-phase transport. Journal of Contaminant Hydrology, 4:1:1-26.

Benson, W.B. 1989. Experiments with removal of volatile contaminants from soil by forced air venting. M.A.Sc. Thesis, Department of Civil Eng., University of Waterloo.

Farquhar, G.J., W.B. Benson, R. Kell, E. McBean, B. Stickney, and D. Farrish 1988. Immiscible liquids and vapours in soil: recent experiments on transport and control. Proceedings of the Ontario Ministry of the Environment Technology Transfer Conference, November 28-29, Toronto, Ontario:29-54.

Hoag, G.E. and B. Cliff 1987. The use of the soil venting technique for the remediation of petroleum contaminated soils. Presented at 2nd Conference on Environmental and Public Health, September 28-30, Amherst, Mass.

Kell, R.F. 1988. Studies on hazardous vapour transport. M.A.Sc. Thesis, Department of Civil Eng., University of Waterloo.

Marley, D.L. and G.E. Hoag 1984. Induced soil venting for recovery/restoration of gasoline hydrocarbons in the vadose zone. Proceedings of the NWWA/API Conference on Petroleum Hydrocarbons and Organic Chemicals in Ground Water, November 5-7, Houston, Texas:473-503.

Rainwater, K., B.J. Claborn, H.W. Parker, D. Wilkerson and M.R. Zaman 1988. Large-scale laboratory experiments for forced air volatilization of hydrocarbon liquids in soil. Proceedings of the NWWA/API Conference on Petroleum Hydrocarbons and Organic Chemicals in Ground Water, November 9- 11, Houston, Texas:501-520.

Stephanatos, B.N. 1988. Modeling transport of gasoline vapors by an advective-diffusive unsaturated zone model. Proceedings of the NWWA/API Conference on Petroleum Hydrocarbons and Organic Chemicals in Ground Water, November 9-11, Houston, Texas:591-611.

Wootan, W.L. and T. Voynick 1984. Forced venting to remove gasoline vapor from a large-scale model aquifer. American Petroleum Institute.

Zimmerman, R.E. and V.M. Brizgys 1986. Remediation of subsurface contamination from volatile organic compounds using soil gas extraction techniques. Proceedings of the 6th National Symposium and Exposition on Aquifer Restoration and Ground Water Monitoring, May 19-22, Columbus, Ohio:580-590.

Subsurface Contamination by Immiscible Fluids, Weyer (ed.) © *1992 Balkema, Rotterdam. ISBN 90 6191 175 3*

Selective retention of soluble hydrocarbon constituents: Role of the residual air compared with that of solid material

O. Razakarisoa, P. Muntzer & L. Zilliox
Institut de Mécanique des Fluides, Université Louis Pasteur, URA CNRS 854, Strasbourg, France

A. Hazemann
IUT-Département Chimie, Université Robert Schuman, Illkirch-Graffenstaden, France

ABSTRACT : Long-term contamination of aquifers by petroleum products is essentially due to the transfer and transport of dissolved hydrocarbon traces. From experimental results obtained from a dual-column model, we pointed out that liquid/gas exchange is the major factor of retention of soluble alkanes masking the effects of adsorbing materials. For the soluble monoaromatics, the main phenomenon observed is the liquid/solid exchange. Parameters such as solubility, Henry's constant and distribution coefficient must be taken into account simultaneously for studying the selective retention of constituents of complex mixtures of hydrocarbons. Transport models of soluble monoaromatics and alkanes are proposed to simulate experimental results.

1 INTRODUCTION

In the case of groundwater contamination by petroleum products, the components of the immiscible pollutant source will be selectively solubilized and transported over large distances by the water movement. The transfer and propagation of dissolved hydrocarbon traces (DHT) are the real danger. The behaviour of soluble hydrocarbons and the extent of pollution depend on hydrodynamical conditions of the infiltration area and phase partitioning between water, oil, gas and solid in the aquifer. The exchange phenomena between the pollutant and the constituents of the aquifer should be expressed by the retention of DHT (Zilliox et al 1986).

Our contribution, both experimental and theoretical, concerns the study of the retention of DHT during their transport in porous medium. The experiments which have been conducted at the laboratory with a dual-column model are intended to characterize the exchange mechanisms between liquid/ gas or liquid/ solid phases.

From the results of these investigations, we try to give some answers to questions such as :

-What should be the effect of liquid/gas or liquid/ solid exchange in contaminated aquifers, and what is the role of residual air compared with that of solid material on the exchange possibility with the water charged with DHT during the transport ?

-Is the observed exchange phenomena linked to the nature and chemical or thermodynamical properties of soluble hydrocarbons ?

-What transport model would be appropriate ?

2 MATERIALS AND METHODS

2.1 *Experimental model*

The glass column source of pollution (GCSP) (L= 75cm, ϕ=10.4cm) contained both natural quartz sand K30 and a motionless oil phase uniformly distributed in quantities below residual saturation. The second glass column (SGC) (L=70cm, ϕ= 9.3cm) was filled with the solid matrix which was not contaminated at the beginning. The solid phase in the SGC consisted of sand K30 or a mixture of soil and sand K30. The soil contained an average of 3% of organic materials, 7% of clay minerals and 26% of limestone.

The experiment consisted of getting water charged with DHT (concentration Co) coming from the GCSP to flow through the porous matrix in the SGC. The evolution of soluble hydrocarbons coming out from the SGC (concentration C) were observed to examine the retention or exchange phenomena (Fig.1).

The pore velocity varied from 0.61cm/min to 0.87cm/min according to the tests. This corresponds to a contact time of the order of 80 to 115 minutes between the water charged with DHT and the porous matrix.

2.2 *Data and studied parameters*

Continuous injections of hydrocarbon solutions made of binary mixture of alkanes at equal volumes (cyclohexane (CHX) + 2,3-dimethylbutane (2,3-DMB)), or a mixture of four aromatics (toluene (TOL) + p-xylene (p-XYL) + 1,2,3-trimethylbenzene (1,2,3-TMB) + α-methylnaphtalene (α-MNPT)) mixed with an isoparaffinic fraction were carried through the SGC. The composition in mass of the aromatic mixture was: TOL→6.5%, p-XYL→11.5%, 1,2,3-TMB→11%, α-MNPT→11%, isoparaffinic fraction →60%; that is a mixture representative of crude oil, its soluble fraction is strongly dominated by the four aromatic constituents.

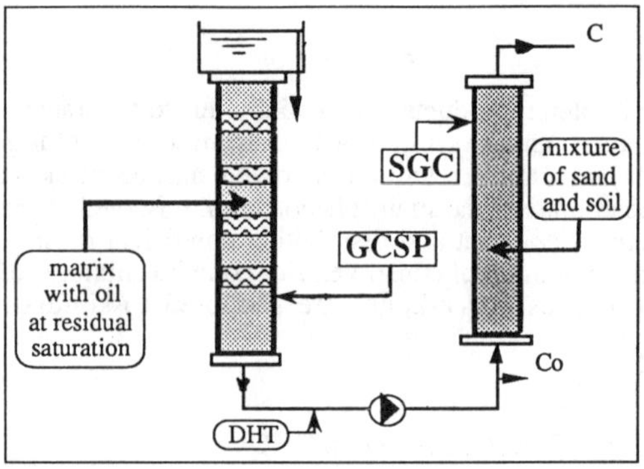

Fig.1 Scheme of the experimental installation

The DHT steming from the binary mixture of alkanes were extracted with CCl4 and analyzed by infrared spectrometry using the product itself as a standard (Ramanantsoa et al 1986). Those coming from aromatic mixture were extracted with n-pentane, by using a water/solvent ratio of 200/1 and analyzed by gas chromatography.

The variation of residual air content was obtained according to the filling method of the SGC; "dry filling" favours the presence of residual air (water introduced by imbibition); "filling by sedimentation" can reduce or partially avoid it (the SGC is filled with the porous medium under water).

It is important to note that differences were observed concerning the value of the kinematic porosity ε obtained in the same type of porous support, in each of the two filling methods. It becomes lower when the presence of residual air is favoured. Important differences were also noticed in the value of the longitudinal dispersion coefficient (D_L) which tends to increase when there is high air content in the SGC (Razakarisoa 1990). Then, those affected the values of the dispersivity (α_L) and the Brenner number (B) (Tab. 1). In other words, the residual air content played an important role on the determination of various transport parameters of the porous matrix.

Table 1. Influence of residual air on transport parameters.
DF(dry filling)
FS(filling by sedimentation)

		Q (cm^3/s)	ε	D_L (m^2/s)	α_L (cm)	B = $uL/4D_L$
Sand K30	DF	0.29	35%	3.36×10^{-7}	0.27	64
	FS	0.28	40%	1.72×10^{-7}	0.17	104
K30 + 10 % soil	DF	0.28	32.6%	12.2×10^{-7}	0.97	18
	FS	0.29	36.8%	8.46×10^{-7}	0.64	24
K30 + 20 % soil	DF	0.28	28.5%	63×10^{-7}	4.3	4
	FS	0.29	35%	35.8×10^{-7}	2.9	6

3 RESULTS AND DISCUSSIONS

3.1 *Impact of residual air*

The breakthrough curves in response to the continuous injection of alkanes show a higher retention and a more important delay when the presence of residual air is favoured in the SGC (Fig.2). This phenomenon is observed regardless of which matrix is used. When the dry filling of the SGC is adopted, no significant differences appear between the curves corresponding to tests using the matrix with 10%, 20% or 30% of soil. The role of the gaseous phase in the retention of soluble alkanes masks that of the solid material when the residual air content becomes high (Fig.3).

When the presence of residual air is minimized, fixation of hydrocarbons by the solid phase is observed and enhanced by the proportion of soil. The solid material can fully play its role when the air content is low (Fig. 4).

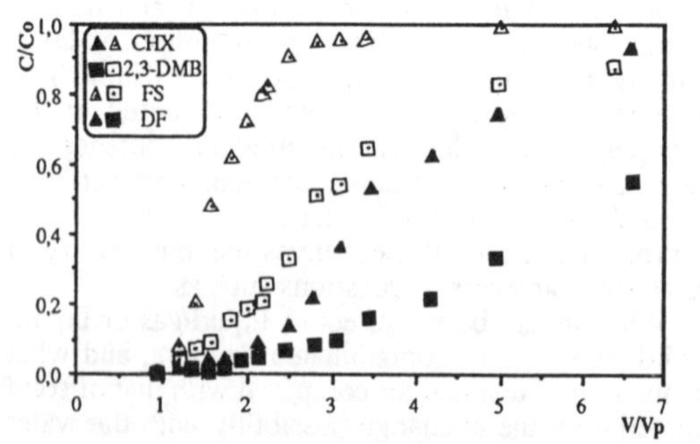

Fig.2 Curves in response to the continuous injection of alkanes when the matrix is the mixture (sand + 10% soil)

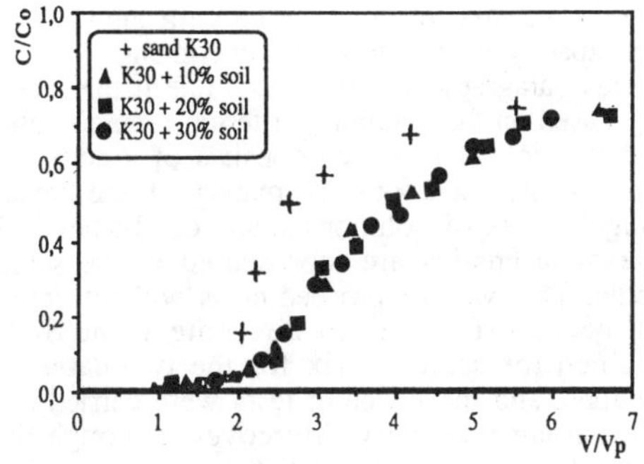

Fig.3 Breakthrough curves of alkanes when the residual air content is great. Total concentrations

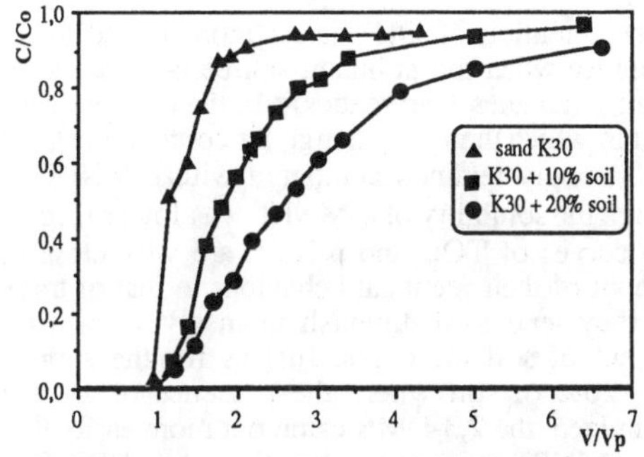

Fig.4 Breakthrough curves of alkanes when the residual air content is low. Total concentrations

For the soluble monoaromatic hydrocarbons, the evolutions of concentrations offer identical patterns of the pollutant regardless of the air content of the matrixes. In figure 5, breakthrough curves of p-XYL are shown but the same phenomenon is observed for the TOL and the 1,2,3-TMB. For a quartz sand, they practically behave as a tracer. For a porous support with 10% or 20% of soil, it appears a fixation and a delay essentially due to the retention capacity of the solid phase. The role played by the residual air is not significant in the case of monoaromatics. The main phenomenon observed is the liquid/solid exchange (Razakarisoa et al 1989, Ducreux et al 1990).

The transfer of soluble hydrocarbons in the gaseous phase varies according to their type and nature. If the exchange between the water charged with DHT and the residual air is instantaneous (Baehr 1987, Falta et al 1989), the equilibrium concentrations in each phase can be described by Henry's law

$$C_{a,k} = \frac{H_{a/w,k}}{RT^o} C_{w,k} \tag{1}$$

where $C_{a,k}$ and $C_{w,k}$ are the concentration of component k in the air and water respectively (mg/l), $H_{a/w,k}$ is Henry's constant of component k (kPa. m^3.°K^{-1}.mol^{-1}), R the constant of perfect gas (8.314 Pa.m^3.°K^{-1}. mol^{-1}) and T° the temperature (°K).

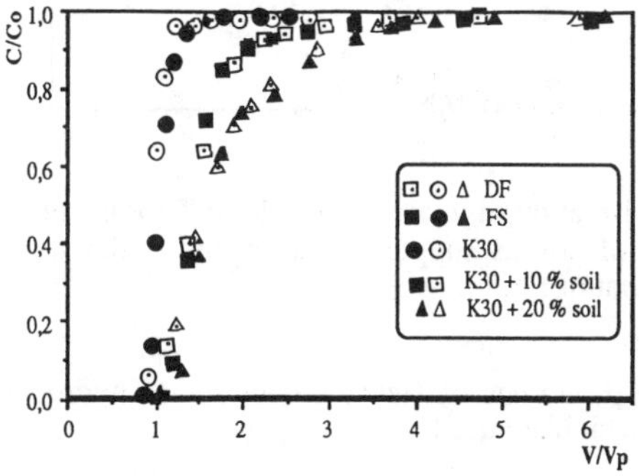

Fig.5 No impact of residual air on breakthrough curves of p-XYL

At equilibrium, $H_{a/w,k}$ can be calculated by using the following approximation (Mackay and Shiu 1981, Schwarzenbach 1986)

$$H_{a/w,k} \approx \frac{P_{v,k} M_k}{S_k} \tag{2}$$

where $P_{v,k}$, M_k, S_k are respectively the vapour pressure (kPa), the molecular weight (g/mol), and solubility (mg/l) of component k.

$H^*_{a/w,k} = H_{a/w,k}/RT^o$ is a dimensionless quantity representing the air/water partition coefficient. For the alkanes and monoaromatics, the value of $H^*_{a/w,k}$ determines if the hydrocarbon is more or less retained by the gaseous phase (in relative value). Actually, the smaller $H^*_{a/w,k}$ is, the more the transfer of the component into the air is restricted (table 2). This tendency is confirmed by our experiments for all components except for the α-MNPT. This polyaromatic hydrocarbon (PAH) has been influenced by the presence of residual air although its partition coefficient $H^*_{a/w,k}$ is very low (Fig.6). This particular behaviour is certainly due to the fact that the PAH's are organic compounds which have high hydrophobicity. The characteristic that indexes the hydrophobicity of the component is the octanol-water partition coefficient, $K_{ow,k}$ (Miller et al 1985, Schwarzenbach 1986). The table 3 indicates that the α-MNPT has the higher $K_{ow,k}$ value. Then, this implies that $H^*_{a/w,k}$ does not apply for explaining the transfer of dissolved traces of PAH's in the gaseous phase, but their hydrophobicity degree must be taken into account.

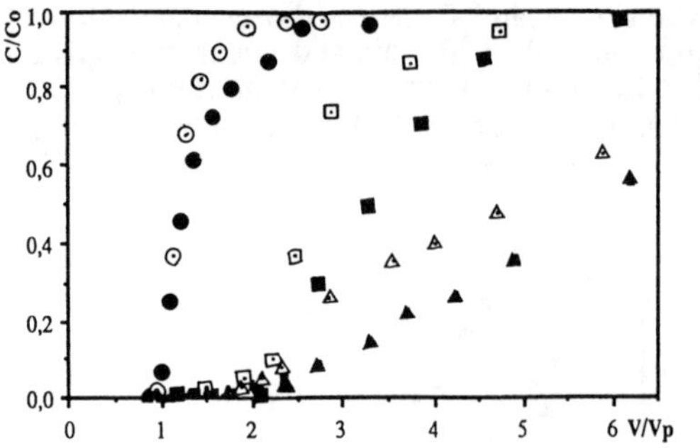

Fig.6 Breakthrough curves of α-MNPT for the three types of porous supports. The legend is the same than that of fig.5

Table 2. Values of Henry's constant of hydrocarbons which are used

	$P_{v,k}$ (kPa)	M_k (g/mol)	S_k (mg/l)	$H_{a/w,k}$ (kPa.m^3.mol^{-1})	$H^*_{a/w,k}$
2,3-DMB	32.3	86.18	20	139.2	**56.13**
CHX	12.9	84.16	56	19.4	**7.82**
TOL	3.9	92.14	515	0.70	**0.28**
p-XYL	1.2	106.17	185	0.69	**0.28**
1,2,3-TMB	0.31	120.20	75.2	0.50	**0.20**
α-MNPT	0.009	142.20	28.5	0.05	**0.02**

It should be noted that when talking about "volatilization" of pollutants from aqueous solutions, one usually has the tendency to only consider components which have high $H^*_{a/w,k}$. However, organic compounds such as pesticides and DDT with very low vapour pressures, tend to escape from the aqueous phase because of the large value of their activity coefficient corresponding to high hydrophobicity (Stumm and Morgan 1981, Mackay and Shiu 1981).

3.2 *Influence of solubility*

During their propagation through the porous matrix, the aromatic constituents are retained in a selective way. This selectivity is linked to the solubility, namely the more soluble is always the least retained (Fig.7 to 9). The retention of alkanes proves also to be selective and should be explained by the solubility criterion (Rasolofoniaina et al 1988). This phenomenon occured even in media with high adsorption capacity and whatever the air content.

In this paragraph, we try to examine if the selectivity based on the solubility criterion remains valid when the pollution source consists of a complex mixture of hydrocarbons. To manage it, the breakthrough curves of solutions made of alkanes and those of aromatics are represented in the same graphic. That way of proceeding is without drawback because for a given air content, the SGC contained the same matrix for the two cases of injections, and the different tests were carried out with the same velocity. Moreover, although the pollutants come from two different sources, that does not have consequence on their behaviour at the exit of SGC because the retention of DHT is not influenced by the value of Co (Rasolofoniaina 1985).

The retention of DHT is no more linked to the solubility when the pollutant source is made of the mixture (alkanes + aromatics) : In the case of quartz sand in which there was a high air content, the CHX was strongly retained compared with α-MNPT although the solubility of α-MNPT was lower (Fig.7). The curves of TOL and p-XYL are very close on account of their identical behaviour to that of tracer but they tend to distinguish themselves when the percent of soil increases. Just as for the support with 20% of soil where the presence of air was minimized, the 2,3-DMB came out more easily than the α-MNPT whereas solubility of 2,3-DMB is lower (Fig.8).

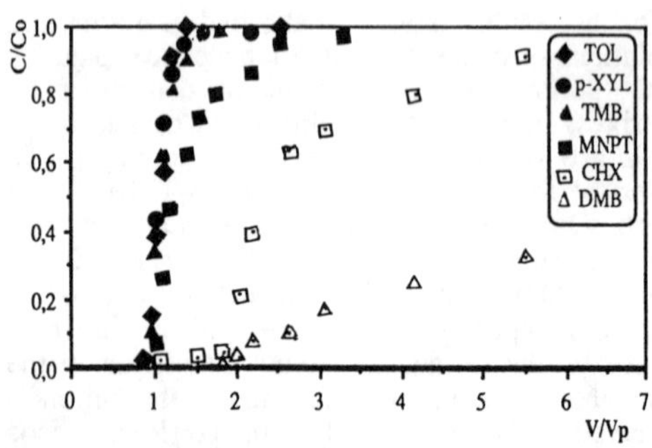

Fig.7 Curves in response to continuous injections of alkanes and aromatics through the matrix made of sand K30 when the residual air content is great

The solubility is an important parameter for studying the selective retention of components of a simple mixture of hydrocarbons belonging to a same homologous series. It is not a sufficient criterion for explaining the selective retention of various soluble components of a pollution source made of a complex mixture.

3.3 *Influence of distribution coefficient Kd,k*

From previous observations, it is shown that components which are not very soluble tend to adsorb more intensily on solid material. The partition coefficient Kd,k of each component between solid and liquid phases should probably have an influence.

Karickhoff et al (1979), Means et al (1980), Schwarzenbach and Westall (1981), Chiou et al (1982,1983), show that sorption of nonpolar organic compounds by soil (including hydrocarbons) can be normalized on the basis of both organic carbon content (fco) of the solid phase and octanol-water partition coefficient of each component k (Kow,k), regardless of the nature of the solid substrate. For a wide range of natural sorbents and organic compounds, Kd,k can be estimated from correlation proposed by Schwarzenbach and Westall (1981) for sorbents containing more than 0.1% of organic carbon :

$$\log K_{d,k} = 0.72 \log K_{ow,k} + \log f_{co} + 0.5 \qquad (3)$$

In order to illustrate the effect of Kd,k on the selective retention of various constituents during their propagation, values of Kd,k are presented in table 3. In the porous matrix with 20% of soil where the air content was low, the degree of retention and the delay of front of pollution of each constituent of the mixture (aromatics + alkanes) were in harmony with the gradation of Kd,k values (Fig.8).

Table 3. Values of octanol-water partition coefficients and distribution coefficients of various studied hydrocarbons when the porous matrix is the mixture (K30 + 20% soil) where fco ≈ 0.15%.

		S_k (mg/l)	$\log K_{ow,k}$	$K_{d,k}$ (cm³/g)
AROMATICS	TOL	515	2.65	0.39
	P-XYL	185	3.16	0.92
	1,2,3-TMB	75.2	3.55	1.75
	α-MTNPT	28.5	3.87	2.97
ALKANES	CHX	56	3.44	1.46
	2,3-DMB	20	3.81	2.70

But when the air content of the matrix became great, 1,2,3-TMB came out more easily than CHX and 2,3-DMB seemed to be more retained than α-MNPT although the partition coefficients Kd,k of these alkanes are lower. This is due to the higher $H^*_{a/w,k}$ for the CHX and the 2,3-DMB implying stronger partitioning into the gas phase (Fig.9).

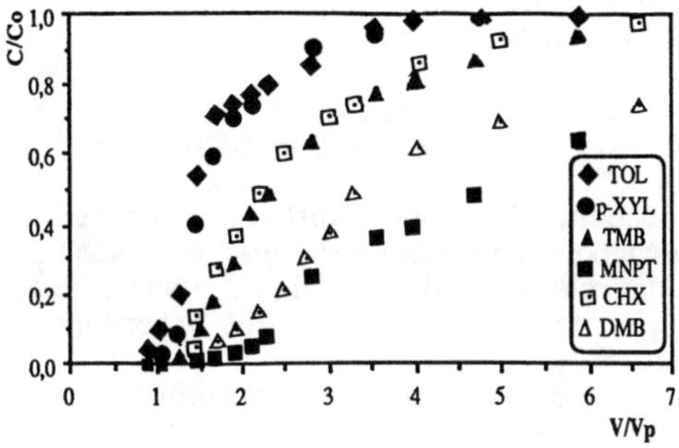

Fig.8 Curves in response to continuous injections of alkanes and aromatics through the matrix made of the mixture (sand K30 + 20% soil) when the residual air content is low

From these results, we notice that for porous support with certain adsorption capacity and exempt of air, the same sorption process as described by several authors has been verified and then, the selective retention of various components of hydrocarbon mixture should be interpreted by the knowledge of their respective coefficients Kd,k. This rule was not respected when the air content of the matrix became important because the gaseous phase played a preponderant role for the alkanes, and in that case the phenomenon should be more complex.

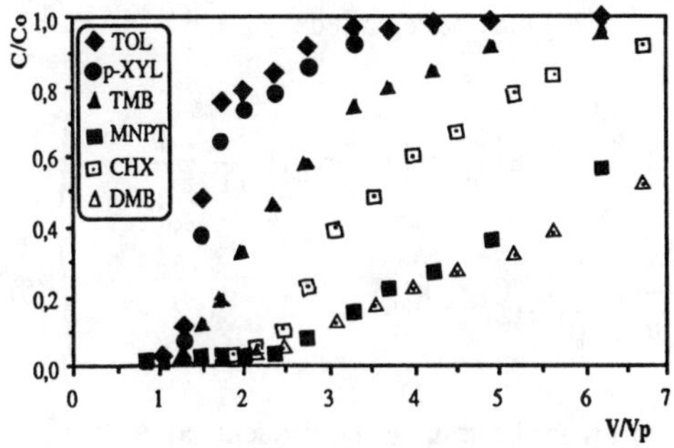

Fig.9 Curves in response to continuous injections of alkanes and aromatics through the matrix made of the mixture (sand K30 + 20% soil) when the residual air content is important

Then, neither the knowledge of Kd,k, nor $H^*_{a/w,k}$ or Sk alone can justify the real process of selective retention of constituents of a mixture of several hydrocarbon species. The simultaneous effects of these parameters must be taken into account together because they represent in a way, the physical/chemical fingerprints of the organic compound, and their simultaneous consideration is essential to assess the behaviour of each component.

4 THEORETICAL APPROACH OF OBSERVED RETENTION PHENOMENA

4.1 *Transport model of soluble monoaromatics*

On account of behaviour differences of studied hydrocarbons, concerning the impact of residual air and the role of solid material, we propose some transport models to simulate the experimental results.

For a one-dimensional and uniform flow, the equation governing the transport of DHT in saturated porous medium may be written

$$D_L\frac{\partial^2 C_w}{\partial x^2} - u.\frac{\partial C_w}{\partial x} = \frac{\partial C_w}{\partial t} + \overbrace{\frac{\gamma}{\varepsilon}.\frac{\partial C_a}{\partial t} + \frac{\rho}{\varepsilon}.\frac{\partial S}{\partial t}}^{\text{term of retention-fixation}} \quad (4)$$

where Cw and Ca are respectively the concentration of DHT in water and air (ML^{-3}), u the pore velocity (LT^{-1}), γ the air content, ρ the bulk density, ε the kinematic porosity and S the adsorbed hydrocarbon concentration (MM^{-1}). The physical/chemical reactions and mass transfer between phases are especially influenced by the term of retention-fixation which will be explained according to the nature of exchanges.

For the monoaromatics, the following transport model which is similar to that used by Cameron and Klute (1977) may be regarded :

$$\begin{cases} D_L\dfrac{\partial^2 C_w}{\partial x^2} - u\dfrac{\partial C_w}{\partial x} = \left(1+\dfrac{\rho}{\varepsilon}K_{d,k}\right)\dfrac{\partial C_w}{\partial t} + \dfrac{\rho}{\varepsilon}\dfrac{\partial S}{\partial t} \\ \dfrac{\partial S}{\partial t} = \dfrac{\varepsilon}{\rho}k_{ls}C_w - k_{sl}S \end{cases} \quad (5)$$

where kls is the transfer coefficient liquid to solid phase and ksl the transfer coefficient solid to liquid phase (T^{-1}). Two types of soil reaction sites are considered in (5). The first site comes from linear Freundlich isotherm : It assures rapid fixation and is responsable for the observed delay caused by the adsorbing effects due to the presence of organic carbon in porous matrix ($K_{d,k}$ is function of f_{co}). The second site ensures slow exchange and is represented by the first-order kinetic in (5) which governs the rate of climb of the breakthrough. The transport model (5) admit analytical solution proposed by Cameron and Klute (1977) with consideration of dimensionless transformations described in (6). The dimensionless time T is equivalent to the number of pore volumes that have passed through the SGC with length L; K1, K2 and K3 are the exchange parameters; N is a dimensionless quantity describing the mass of retained hydrocarbons.

$$\begin{cases} T=\dfrac{ut}{L};\; K_1=\dfrac{\rho}{\varepsilon}K_{d,k};\; K_2=\dfrac{L.k_{ls}}{u};\; K_3=\dfrac{L.k_{sl}}{u} \\ \mathcal{C}=\dfrac{C_w}{C_0};\quad B=\dfrac{u.L}{4.D_L};\quad X=\dfrac{x}{L};\quad N=\dfrac{\rho S}{\varepsilon C_0} \end{cases} \quad (6)$$

Values of Brenner number B and N are obtained from experiments. The boundary and initial conditions are as follows :

$$\begin{matrix} \mathcal{C}(0,T)=1 \text{ for } T>0;\quad \mathcal{C}(X,0)=0 \text{ for } X>0 \\ \mathcal{C}(X,T)=0 \text{ when } X\to\infty;\quad N(X,0)=0 \text{ for } X>0 \end{matrix} \quad (7)$$

and the analytical solution :
(Cameron and Klute 1977)

$$\begin{cases} \mathcal{C}(X,T)=\dfrac{2}{\sqrt{\pi}}\exp(2BX)\displaystyle\int_s^{\infty}\exp\left[-h^2-\dfrac{B^2X^2}{h^2}\right].J(p,q)\,dh \\ N(X,T)=K_1.\mathcal{C}(X,T)+K_2\displaystyle\int_0^T \mathcal{C}(X,u)\exp[-K_3(T-u)]\,du \end{cases} \quad (8)$$

For determination of parameters K2 and K3, a computer program was used to adjust their values until the difference between the predicted and experimentally observed breakthrough curves approached a minimum. The Brenner numbers obtained in the tests using the filling by sedimentation method gave the best simulation whatever the air content. That is an interesting observation because it supports our earlier hypothesis that the role of residual air is negligible for the monoaromatics.

Figure 10 shows an outcome of the simulation which indicates a good agreement with the experimental results : The main phenomenon observed is then the liquid/ solid exchange.

4.2 *Transport model of soluble alkanes*

Considering the important role played by residual air on the propagation of soluble alkanes, two configuration cases may be envisaged according to the air content of the matrix.

When the air content is low, liquid/gas exchange is only supposed instantaneous (air directly in contact

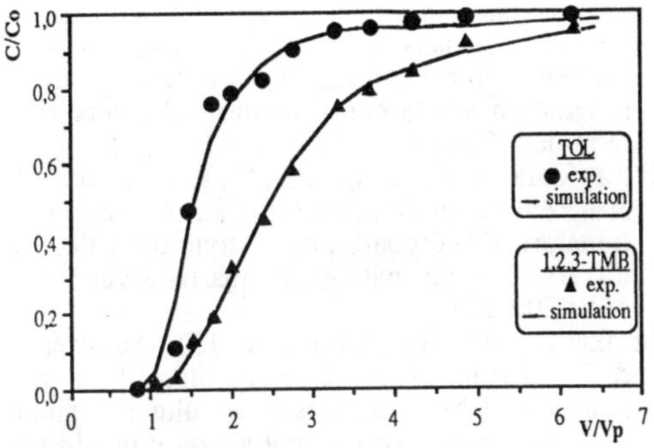

Fig.10 Simulation of the propagation of TOL and 1,2,3-TMB through the matrix (K30 + 20% soil) where the air content is high.

with the moving water charged with DHT), i.e the air eventually trapped in motionless water is negligible. On the other hand, the solid phase contributes to both rapid and slow exchange. The transport model describing this first situation is represented in (9) and will be applied in the propagation tests using the filling by sedimentation method.

$$\left\{\begin{array}{l} D_L \dfrac{\partial^2 C_w}{\partial x^2} - u\dfrac{\partial C_w}{\partial x} = \left(1+\dfrac{\rho}{\varepsilon}K_{d,k}+\dfrac{\gamma_1}{\varepsilon}H^*_{e/a,k}\right)\dfrac{\partial C_w}{\partial t}+\dfrac{\rho}{\varepsilon}\dfrac{\partial S_2}{\partial t} \\ \dfrac{\partial S_2}{\partial t} = \dfrac{\varepsilon}{\rho}k_{ls}C_w - k_{sl}S_2 \end{array}\right\} \qquad (9)$$

The subscripts 1 and 2 refer to rapid and slow exchange, respectively. γ_1 is the air content in moving water. In that case, the value of N considered is that corresponding to quantities fixed by solid phase.

When the air content is important, owing to the preponderant role of the gaseous phase, DHT of alkanes may react rapidly or slowly with the residual air, and the solid phase is supposed to contribute alone to the retardation of propagation (in our tests, the adsorbing effects of solid material are obscured). This second configuration is represented in (10) and concerns the tests where the dry filling method is adopted:

$$\left\{\begin{array}{l} D_L \dfrac{\partial^2 C_w}{\partial x^2} - u\dfrac{\partial C_w}{\partial x} = \left(1+\dfrac{\rho}{\varepsilon}K_{d,k}+\dfrac{\gamma_1}{\varepsilon}H^*_{e/a,k}\right)\dfrac{\partial C_w}{\partial t}+\dfrac{\gamma_2}{\varepsilon}\dfrac{\partial C_{a2}}{\partial t} \\ \dfrac{\partial C_{a2}}{\partial t} = \dfrac{\varepsilon}{\gamma_2}k_{lg}C_w - k_{gl}C_{a2} \end{array}\right\} \qquad (10)$$

γ_2 is the air content in motionless water, k_{lg} and k_{gl} are transfer coefficients liquid to gaseous phase and gaseous to liquid phase respectively (T^{-1}). In this second case, the value of N taken into account is that corresponding to quantities retained by residual air.

The two models (9) and (10) admit the same form of analytical solution as (8) in which values of parameters K_1, K_2, K_3 and N change physical meaning according to the models which are adopted. In figure 11, good agreement are obtained: The role played by the gaseous phase on the retention of soluble alkanes is preponderant when the air content is high.

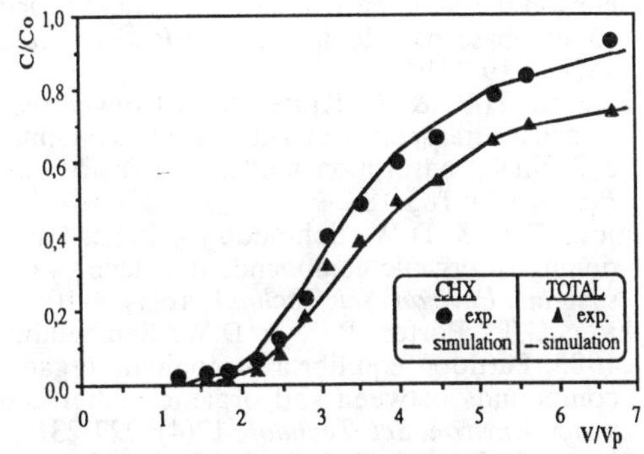

Fig.11 Simulation of the propagation of alkanes through the matrix (K30 + 20% soil) where the air content is high.

5 CONCLUSIONS

The effects of the residual air and the role of the solid material on the propagation of dissolved hydrocarbon traces were investigated. Concerning the retention of soluble alkanes, the experimental approaches with physical dual-column model have placed the impact of the gaseous phase confined within porous medium in a prominent position: When the residual air content of the matrix is important, the liquid/gas exchange may mask the effects of adsorbing materials. Then, the retardation of soluble alkanes should be more pronounced in unsaturated zone than in that saturated. The effect of adsorbing character of the matrix exists and appears clearly, only when the air content is low. On the other hand, the residual air content has no influence on the transport of soluble monoaromatics but the main mechanism observed is the liquid/solid exchange.

From experimental results, during the transport, the selective retention of various constituents of a mixture of several hydrocarbon species should be interpreted by the knowledge of their respective distribution coefficients when the air content is low. The phenomenon should be more complex when the air content is great because the gaseous phase plays a preponderant role for the alkanes. In that case, the simultaneous effects of solubility, air-water partition coefficient and distribution coefficient must be taken into account together to assess the behaviour of each component in the case of aquifer pollution.

The exchange phenomena is probably more complex than the transport model would indicate but its consideration is necessary to get a simple and usable mathematical approximation of what we think might be happening.

REFERENCES

Baehr, A.L. 1987. Selective transport of hydrocarbons in the insaturated zone due to aqueous and vapor phase partitioning. *Water Resour. Res.*, 23(10): 1926-1938.

Cameron, D.R. & A. Klute 1977. Convective-dispersive transport with a combined equilibrium and kinetic adsorption model. *Water Resour. Res.*, 13(1): 183-188.

Chiou, C.T. & D.W. Schmedding 1982. Partitioning of organic compounds in octanol-water systems. *Environ. Sci. Technol.*, 16(1): 4-10.

Chiou, C.T., Porter, P.E. & D.W. Schmedding 1983. Partition equilibria of nonionic organic compounds between soil organic matter and water. *Environ. Sci. Technol.*, 17(4): 227-231.

Ducreux, J., Bocard, C., Muntzer, P., Razakarisoa, O. & L. Zilliox 1990. Mobility of soluble and non-soluble hydrocarbons in contaminated aquifer. *Water Science and Technology*, 22(6): 27-36.

Falta, R.W., Javandel, I., Pruess, K. & P.A. Witherspoon 1989. Density-driven flow of gas in the unsaturated zone due to the evaporation of volatile organic compounds. *Water Resour. Res.*, 25(10): 2159-2169.

Karickhoff, S.W., Brown, D.S. & T.A. Scott 1979. Sorption of hydrophobic pollutants on natural sediments. *Water Res.*, 13: 241-248.

Mackay, D. & W.Y. Shiu 1981. A critical review of Henry's law constants for chemicals of environmental interest. *J. Phys. Chem. Ref. Data*, 10(4): 1175-1199.

Means, J.C., Wood, S.G., Hassett, J.J. & W. L. Banwart 1980. Sorption of polynuclear aromatic hydrocarbons by sediments and soils. *Environ. Sci. Technol.*, 14(12): 1524-1528.

Miller, M.M., Waslk, S.P., Huang, G.L., Shiu W.Y. & D. Mackay 1985. Relationships between octanol-water partition coefficient and aqueous solubility. *Environ. Sci. Technol.*, 19(6): 522-529.

Ramanantsoa, B., Muntzer, P. & L. Zilliox 1986. Dissolution sélective d'un mélange d'hydrocarbures par l'eau en milieu poreux saturé. Application à la pollution des eaux souterraines par des produits petroliers. *Sciences de l'eau*, 5(2): 149-168.

Rasolofoniaina, J.D. 1985. Transfert d'hydrocarbures solubles dans l'eau en milieu poreux saturé. Influence du milieu situé en aval du corps d'impregnation sur la contamination des eaux souterraines. *Thèse D.I.- ULP Strasbourg*, F.

Rasolofoniaina, J.D., Muntzer, P., Razakarisoa, O. & L. Zilliox 1988. Impact de l'air résiduel sur le transfert d'hydrocarbures dissous dans l'eau à travers un sable naturel de quartz. *Stygologia*, 4(3): 209-227.

Razakarisoa, O., Rasolofoniaina, J.D., Muntzer P. & L. Zilliox 1989. Selective dissolution and transport of hydrocarbons in an alluvial aquifer. Role and impact of residual air on groundwater contamination. In *"Contaminant Transport in Groundwater"*, p.405-412. Rotterdam: Balkema.

Razakarisoa, 0. 1990. Dissolution et rétention sélective d'hydrocarbures en milieu poreux saturé. Impact de l'air résiduel et rôle du matériau solide, lors de leur propagation. *Thèse de doctorat ULP Strasbourg*, F.

Schwarzenbach, R.P. & J. Westall 1981. Transport of nonpolar organic compounds from surface water to groundwater. Laboratory sorption studies. *Environ. Sci. Technol.*, 15(11): 1360-1367.

Schwarzenbach, R.P. 1986. The physical-chemical fingerprint of organic compounds. Personal communication.

Stumm, W. & J.J. Morgan 1981. *Aquatic chemistry*, New York: Wiley-Interscience.

Zilliox, L., Behra, P. & J.D. Rasolofoniaina 1986. Altération de la qualité des eaux souterraines par des rejets ou déversements accidentels de polluant. *Soc. Hyd. de France. XIXè journée de l'hydraulique, Paris.* Rapport n°III/7.

Subsurface Contamination by Immiscible Fluids, Weyer (ed.) © 1992 Balkema, Rotterdam. ISBN 90 6191 175 3

Effects of capillarity on DNAPL thickness in wells and in adjacent sands

Timothy V. Adams
ENSR Consulting and Engineering, Westmont, Ill., USA
Duane R. Hampton
Western Michigan University, Kalamazoo, Mich., USA

ABSTRACT: Physical model experiments were used to study the behavior of DNAPLs in various geologic media.

The objectives of the laboratory investigations were to compare DNAPL thickness in wells to thickness in the adjacent sands and observe and interpret dyed DNAPL migration in unsaturated and saturated sands. Two cylindrical glass columns fitted with various well screens were filled with sands, clay layers, water, and dyed DNAPLs in four experiments. Columns were packed with fine or coarse sand and clay layers. Coarse hydrophobic sand was also used.

At equilibrium, DNAPL thickness in wells exceeds that in sands. The finer the sand, the greater the difference. The thickness difference equals the DNAPL-water capillary fringe height, which varies with grain size, packing and DNAPL properties. In contrast, DNAPL thickness in the hydrophobic sand exceeds that in wells due to DNAPL capillary rise. Wells can serve as conduits for vertical DNAPL migration.

1 INTRODUCTION

Substances such as many halogenated solvents, halogenated benzenes, phthalate esters, and polychlorinated biphenyls (PCBs) comprise a group of organic chemicals which can be categorized as dense non-aqueous phase liquids (DNAPLs). Approximately one quarter of the organic chemicals on the United States Environmental Protection Agency (USEPA) List of Priority Pollutants are DNAPLs (Feenstra and Cherry 1988).

A DNAPL's relatively low solubility and high density enable it to penetrate downward through unsaturated and saturated porous media as an immiscible liquid phase. Many DNAPLs are also very volatile. DNAPL vapor may migrate away from a liquid phase source spreading contamination in the unsaturated zone (Schwille 1988). Below the water table, small quantities of DNAPL can become entrained in the groundwater flow regime by dissolution. Most DNAPLs have a relatively low solubility in water, and therefore may continue to dissolve in groundwater for extended periods of time.

An ever-increasing number of sites found contaminated with DNAPLs is forcing today's groundwater industry to understand the controlling factors that govern their behavior in the subsurface. Locating and removing pools of DNAPLs in aquifers are difficult but important steps in remediating this contamination. These steps depend upon understanding the interactions of DNAPLs in porous media and wells.

One approach to this dilemma is to physically model subsurface conditions in which DNAPLs may be present and observe their behavior when introduced under controlled conditions.

2 MATERIALS AND METHODS

Two PYREX glass cylindrical columns were fitted with various 5-cm diameter well screens. Each column was 58 cm tall and 28 cm in internal diameter. Stainless steel, PTFE (teflon), and fiber-reinforced epoxy (fiberglass) well screens were cut in half lengthwise and attached to the inside walls of the columns using a silicone vacuum grease or a bentonite powder/water paste.

The columns were filled with various geologic materials including 0.41 mm median grain size (fine) gravel pack sand, 0.77 mm median grain size (coarse) gravel pack sand, clay layers made of bentonite /sand mixtures, coarse gravel, and a hydrophobically-treated volume of the coarse gravel pack sand. A total of four experiments were conducted utilizing the two columns and injecting dyed DNAPL.

Two DNAPL mixtures were used during these experiments. Most experiments used blue-dyed perchloroethylene (PER). A red-dyed chlorobenzene and aniline mixture (CAM), comprised of equal volumes of each, was also used in experiments 1 and 2. The properties of the selected DNAPLs are given in Table 1.

Table 1. Properties of DNAPLs used.

DNAPL Type	Density (g/ml)	Absolute Viscosity (c.poise)	Solu-bility (mg/L)
PER	1.63	0.9	200
Chloro-benzene	1.11	0.8	488[a]
aniline	1.02	3.7	20,000

[a] Value measured at 25° celsius.
Note: Temperature of measurement is 20° celsius unless otherwise noted.

In the first experiment column A was fitted with PTFE and fiberglass well screens. The screens were set on a clay layer at the base of the column. The column was then packed with 10 cm of fine sand, another 1-cm thick clay layer followed by 33 cm of fine sand. Once a water table and capillary fringe were established, blue-dyed PER was injected at the surface of the sand.

The stainless steel well screen was set to the base of column B. The column was packed in the same manner as column A. The well screen was filled with granular bentonite to the top of the upper clay layer to prevent CAM migration through the well into the lower sand layer. Once a water table and capillary fringe were established CAM was injected at the sand's surface.

Both columns were monitored and photographed until migration ceased (about 3 weeks). DNAPL thickness measurements and observations were recorded and compared between the two columns.

In the second experiment, the teflon and stainless steel well screens, one per column, were set on top of a 2-cm thick clay layer with fine gravel pack sand above and below the clay. Blue-dyed PER and red-dyed CAM were injected at the sand surface in the respective columns once a water table and capillary fringe were established. DNAPL migration was photographed and monitored for one month.

At the completion of experiment 2, an extraction experiment was conducted on both columns. PER and CAM that had accumulated in each of the wells in columns A and B, respectively, was extracted using a hand pump. Product recovery in the wells was monitored over time as DNAPL from the adjacent sands reoccupied the wells.

One column was utilized for experiment 3. The teflon well screen was set to the base of the column. The column was packed with coarse gravel pack sand. No clay layers were used. The water table and capillary fringe were established by adding water to the column through the well screen. PER was injected at the surface of the sand.

Experiment 3A was conducted using the same column setup and boundary conditions that were used in experiment 3. However, the coarse sand was made hydrophobic to observe the effects of capillarity on DNAPL distribution.

The treated sand was packed in the column. Untreated coarse sand was used for the upper part (approximately 54% of the profile). Blue-dyed PER was injected at the surface of the sand once a water table and capillary fringe were established. The results of the experiment were compared with the results from experiment 3.

A coarse gravel was packed in the column for experiment 4 after the stainless steel well screen was attached to the column. Blue-dyed PER was injected at the surface of the gravel once the water level was established.

At the end of each experiment, the columns were excavated and DNAPL migration routes were observed and photographed. The columns were washed with ALCONOX detergent and water and rinsed with water and isopropyl alcohol at the end of each experiment.

3 RESULTS AND DISCUSSION

In experiment 1 several phenomena were observed. PER was trickled onto the fine sand away from the well screens. PER quickly moved vertically through the unsaturated sand. It spread horizontally along the capillary fringe until it intersected the wells. It migrated vertically downward along the well screens and quickly accumulated at the base of the wells. Eventually, most of the PER accumulated in the screens and only slowly moved out into coarser sand lenses. At equilibrium, the PER thickness in both wells was equal despite the different capillary properties of teflon and fiberglass well materials. This suggests that in wells of this diameter (5 cm) the capillarity of the well material does not cause DNAPL thickness differences in a well. PER thickness over time in both wells is shown in Figure 1a.

CAM infiltration in experiment 1 contrasted with PER infiltration. CAM accumulated for hours above the capillary fringe without penetrating it. The original CAM volume of 750 ml had to be doubled before CAM penetrated the saturated sand

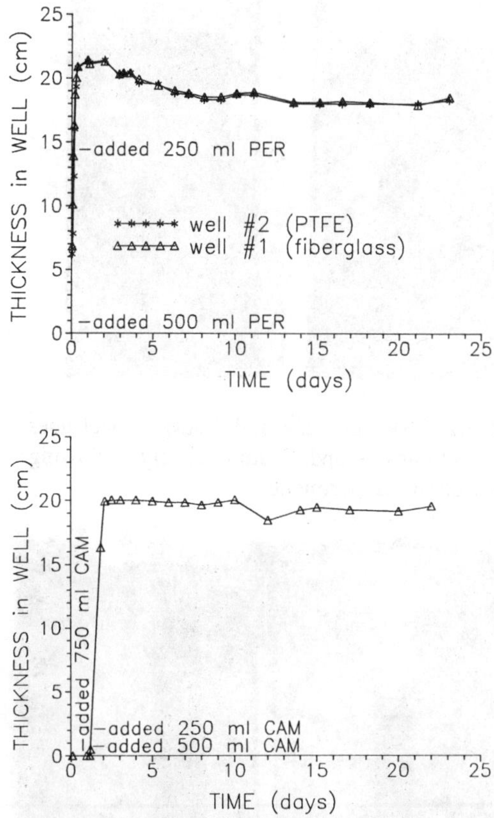

Fig.1 Experiment 1: a) PER thickness in both wells (PTFE & fiberglass) in column A b) CAM thickness in stainless steel well in column B

and entered the well (Figure 1b). The slowness of CAM penetration into the column in contrast with PER infiltration is due to its relatively lower density and higher viscosity. CAM leaked through the plugged well screen or perhaps down its side into the lower sand layer below the clay.

The same infiltration behavior in experiment 1 was also observed in experiment 2. Figure 2 shows PER starting to infiltrate along fingers while the more viscous CAM has developed a stable position above the capillary fringe (shown by the upper white line). The lower white line on the columns indicates the pre-injection water table position.

In contrast with their divergent infiltration

Fig.2 PER (on left) and CAM in backsides of columns A and B, immediately following injection; Experiment 2

Fig.3 PER (on left) and CAM thicknesses in wells and sand 48 hours into Experiment 2. Note thickness difference between PER and CAM in well and in adjacent sand

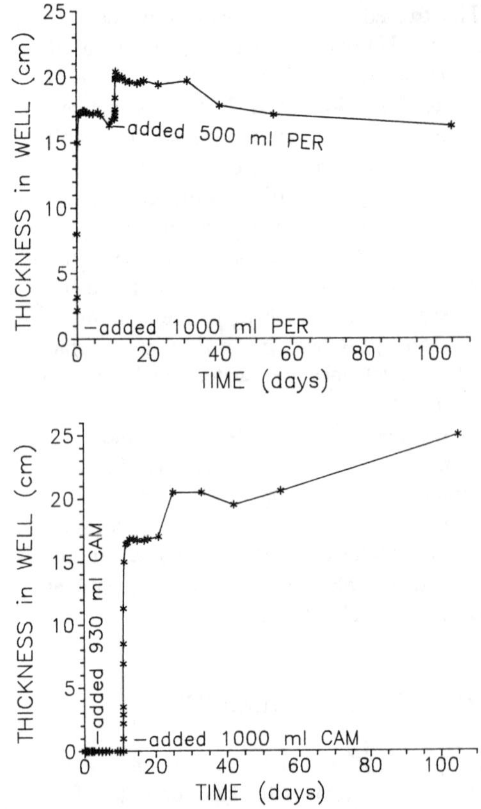

Fig.4 Experiment 2: a) PER thickness in well in column A versus time b) CAM thickness in well in column B versus time

behavior, the final DNAPL thickness in both columns in experiment 2 was similar. Figure 3 shows that the DNAPL-water interface in the teflon well screen is slightly higher than in the stainless steel screen. The interface is harder to distinguish in the latter screen since it is darker, but is just above the white line marking the original water table. The white (To) line on each column in the photo refers to the water level after DNAPLs had been added to each column, respectively. The water level is higher in column B since more DNAPL was added than in column A. The PER thickness in the sand in column A was fairly continuous around the base of the column, but was difficult to measure accurately. Very little CAM was seen in the sand outside the well, and was in irregular blobs. Hence, the CAM thickness in the sand could not be determined, but it was small. Figures 4a and b show the temporal distribution of PER and CAM thicknesses, respectively, in the wells. While PER quickly reached equilibrium, equilibration of CAM thickness in the well was slowed by its higher viscosity and smaller density contrast with water.

The well contents were extracted from columns A and B at the completion of

Fig.5 PER migration in coarse sand during Experiment 3: a) 5 minutes after injection (backside) b) after 2 hours (backside) c) 190 hours (frontside)

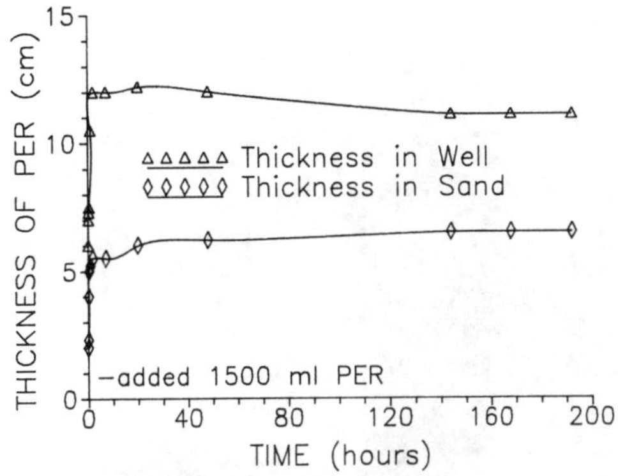

Fig.6 PER thickness in well and sand during Experiment 3

experiment 2. Water and DNAPL recoveries were measured in the respective wells. PER levels recovered more quickly than CAM. Lower viscosity enabled PER to reoccupy the well more easily than CAM.

PER migrated more easily through the coarse sand in experiment 3. In contrast with experiments 1 and 2, PER was not held above the capillary fringe (Figure 5). There was a measurable PER thickness in the sand compared to experiments 1 and 2. There was less thickness difference of PER between the well and sand. Figure 6 shows these thicknesses over the duration of experiment 3.

Conditions in experiment 3 were duplicated in experiment 3A. However, the lower half of the sand was hydrophobic in order to reverse the effects of capillarity. Figure 7 shows the irregular DNAPL capillary rise (uneven white line) above the DNAPL-water interface. As a result the DNAPL thickness in the sand was difficult to measure.

Coarse gravel was packed in the column for experiment 4 to eliminate the capillary effects. Stringers of PER globules were observed to migrate through the large pores in the gravel. Equilibration of equal PER thicknesses in the well and sand occurred within a few minutes (Figure 8).

The results of experiments 1 through 4 are summarized in Table 2. The results indicate that DNAPL thickness in the wells exceeded

Fig.7 DNAPL capillary rise above the DNAPL level in the well in hydrophobic sand; Experiment 3A. The lowest white line is the DNAPL level in the well. The uneven white line marks the irregular DNAPL-water capillary fringe boundary

Table 2. DNAPL thickness in wells and sands Experiments 1 through 4.

Experiment	DNAPL	Thickness in Sand (cm)	Thickness in Well (cm)
1	PER	-	19.0
1	CAM	-	20.0
2	PER	2.0?	18.0
2	CAM	-	20.5
3	PER	6.0	12.0
3a	PER	4.3	3.9
4	PER	7.0	7.0

- no measurable thickness was evident.

DNAPL thickness in the adjacent sand. The thickness difference was greater in the fine sand than in the coarse sand.

The thickness difference equals the DNAPL-water capillary fringe height which varies with pore size of the media. Pore

Fig.8 PER thickness in the well and gravel during Experiment 4

size is determined by the grain-size distribution of a porous medium and by how it is packed. In general, the smaller the grain size the greater the height of the DNAPL-water capillary fringe.

PER and CAM were retained at the capillary fringe boundary until they could reach the critical capillary pressure and migrate downward. Villaume 1985 suggests that the only way displacement below the water table can occur is for the DNAPL to attain enough mass to overcome the capillary pressure by agglomerating into vertical globule stringers along interconnected pore spaces. These stringers were observed during experiment 4 and in excavating the column packings of experiments 1, 2, and 3. As the grain size of the media decreases, the capillary pressure increases, making it more difficult for downward migration of DNAPL below the water table. Table 3 shows the relationship between the capillarity of the media and its effect upon the DNAPL thickness difference.

The pressure heads at the organic liquid-water interface will determine if the DNAPL achieves equilibrium near the capillary fringe, near the water table, or at various depths below the water table (Cary et al. 1989).

When DNAPL is introduced in a saturated hydrophilic media, it is a non-wetting fluid

Table 3. Capillary rise vs. DNAPL thickness difference.

Experiment	DNAPL	Capillary rise above water level (cm)	DNAPL thickness difference (cm)
1	PER	13.0	19.0
1	CAM	17.4	20.0
2	PER	15.9	16.0
2	CAM	15.3	20.5
3	PER	15.0	6.0
3a	PER	10.0	-0.4
4	PER	0.0	0.0

Note: (-) indicates the DNAPL thickness in the sand > well.

that is immiscible with water (wetting fluid). DNAPLs can only occupy pore spaces in the saturated media by displacing pre-existing water. As grain size decreases, permeability decreases and DNAPL transmission through the small pore spaces becomes more difficult. The well opening offers a less resistant (lower entry pressure) migration pathway. This was seen in Experiment 1 where the wells served as conduits for vertical DNAPL migration in the fine sand. PER entered the wells and migrated down their sides to the bottom.

Cary et al. 1989 suggests that as the DNAPL volume in the well increases, sufficient pressure head may develop to force it out through the bottom of the well. This was seen in Figures 1a and 4a.

4 CONCLUSIONS

In general, the DNAPL-water interface was much higher in the wells than in the adjacent sands, causing DNAPL thickness in the wells to exceed that in the sands. The thickness difference was greater in fine sand than in the coarse sand; there was no difference in the gravel. This thickness difference equals the height of the DNAPL-water capillary fringe, which varies with grain size, packing and DNAPL properties. DNAPL thickness in the hydrophobic sand was greater than in the well, due to DNAPL capillary rise. Water-air-DNAPL thickness relationships observed in the experiments are illustrated in Figure 9.

The observed DNAPL thickness difference between wells and surrounding granular materials is not easy to explain. This result was not expected; the authors expected to see the results of Experiment 4 (equal thicknesses) for all porous materials used due to the influence of gravity upon a DNAPL. However, the force of capillarity also plays a role in finer porous media, which attract water more than DNAPL. The downward pull on water into a DNAPL accumulation results in displacing DNAPL up the well where capillary forces are unimportant. DNAPL thickness in a gravel pack would be expected to be greater than in the surrounding finer media and smaller than in the adjacent well. The thickness difference should be understood as a result of the balance of capillary and gravity forces at equilibrium. These experiments were allowed to continue until equilibrium, up to 110 days in one case (Experiment 2). This permitted adequate time for lateral DNAPL spreading on top of the clay.

We would like to repeat this experiment in a different apparatus. This enclosure would have a greater lateral extent to permit DNAPL spreading. It would also have tanks at each end to maintain constant-head boundary conditions. In these experiments our boundaries were impermeable. DNAPL spill sites in the field would be better approximated by constant-head boundaries. It is possible that our results may change when the boundary conditions change.

This study supported findings from several other studies. The experiments show a DNAPL's migration depends upon its inherent physical properties as well as grain

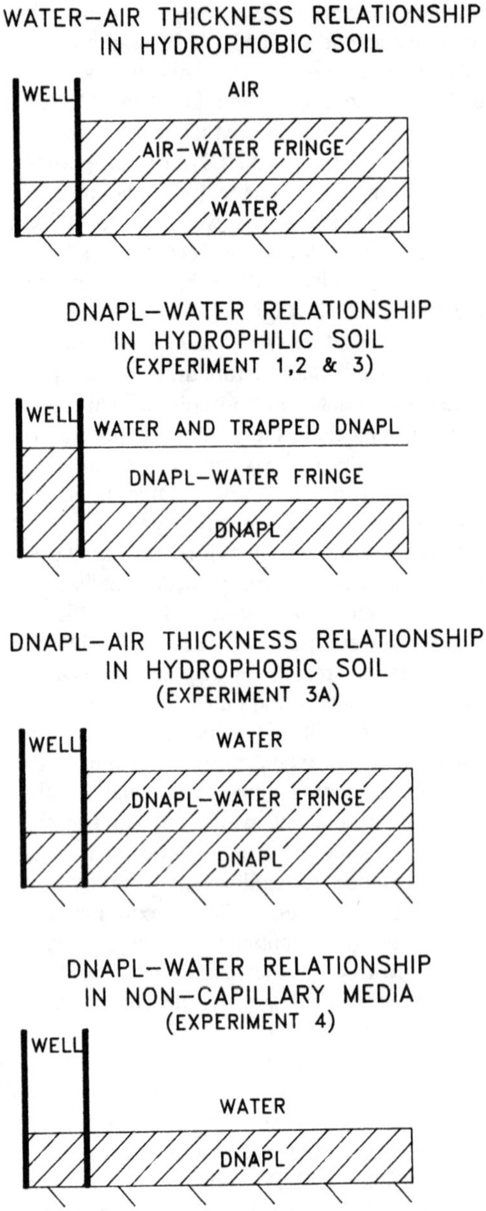

Fig.9 Water-air-DNAPL thickness relationships observed in Experiments 1-4 depending on capillary attraction for water

size of the media. Wells can serve as conduits for vertical DNAPL migration.

ACKNOWLEDGEMENTS

Financial support for this research from the W. E. Upjohn Company, Kalamazoo, MI, and the Research Excellence Fund of the State of Michigan is gratefully acknowledged. We particularly wish to thank Mr. Mel Visser of Upjohn, who originally suggested this project, and Mr. Kevin Walsh for assistance in its implementation.

REFERENCES

Cary, C.R., J.F. McBride, & C.S. Simmons 1989. Trichloroethylene residuals in the capillary fringe as affected by air-entry pressures. J.Envir.Quality 18: 72-77.

Feenstra, S. & J.A. Cherry 1988. Subsurface contamination by dense non-aqueous phase liquid (DNAPL) chemicals. In Proc. Int. Groundwater Symp. of the Int. Assoc. of Hydrogeologists, Halifax, Nova Scotia, May 1-5, 1988, 61-69.

Schwille, F. 1988. Dense chlorinated solvents in porous and fractured media - model experiments. Translated by J.F. Pankow. Chelsea, MI: Lewis Publ. Inc.

Villaume, J.F. 1985. Investigations at sites contaminated with dense non-aqueous phase liquids (NAPLs). Ground Water Monitoring Review. 5(2): 60-74.

4 Numerical and analytical simulation

Subsurface Contamination by Immiscible Fluids, Weyer (ed.) © 1992 Balkema, Rotterdam. ISBN 90 6191 175 3

Modeling flow in three fluid phase porous media with nonwetting fluid entrapment

J.J. Kaluarachchi
Utah State University, Utah, USA

J.C. Parker
Virginia Polytechnic Institute and State University, Va., USA

R.J. Lenhard
Battelle Pacific Northwest Laboratory, Richland, Wash., USA

ABSTRACT: A simplified nonwetting liquid entrapment model for three fluid phase porous media is presented. The model requires one additional parameter in the constitutive relationship, effective residual oil saturation, and can be implemented in a numerical multiphase flow model in a very efficient manner. The validity of the model was verified with good agreement between the predicted and observed saturations. The laboratory column experiment which showed percentage of trapped oil volume under a given set of boundary conditions is very sensitive and varies in a highly nonlinear manner with the effective residual oil saturation.

1 INTRODUCTION

Mathematical models describing the movement of hydrocarbons in the subsurface typically assume nonhysteretic permeability-saturation-pressure (k-S-P_c) relationships (Abriola and Pinder, 1985; Osborne and Sykes, 1986; Kaluarachchi and Parker, 1989). Work by a number of researchers (Killough, 1976; Gillham et al., 1979; Lenhard, et al., 1989) has shown that such nonhysteretic constitutive relationships may predict misleading results in both two phase air-water and three phase air-oil-water systems. Recent work by Parker and Lenhard 1987 and Lenhard and Parker (1987b) has produced a comprehensive hysteresis model to predict multiphase flow in a three fluid phase porous medium. The complexity of this model however limits its application. The objective of this paper is to present a simple and computationally efficient model to predict nonwetting liquid entrapment in a three fluid phase porous medium.

2 MODEL DEVELOPMENT

2.1 Governing Equations for Multiphase Flow

Assuming a three fluid phase incompressible porous medium consisting of water (w), oil (o) and air (a) with water and oil as the mobile phases, the governing equation for mobile p-phase ($p = a, o, w$) continuity can be written as

$$\phi \frac{\partial \rho_p S_p}{\partial t} = -\frac{\partial}{\partial x_i}\left\{\frac{k_{rp} K_{sp_{ij}}}{\eta_{rp}}\left[\frac{\partial h_p}{\partial x_j} + \rho_{rp} e_j\right]\right\} \quad [1]$$

where ϕ is porosity, S_p is the p-phase saturation, x_i (and x_j) are Cartesian spatial coordinates ($i, j = 1, 2$), ρ_p is the density of phase, q_{p_i} is the volumetric flux density along the i direction, k_{rp} is the p-phase relative permeability, K_{swij} is the saturated hydraulic conductivity tensor of water, η_{rp} is the ratio between p-phase and water absolute viscosity, h_p is the p-phase pressure head in equivalent water height of pure water, ρ_{rp} is the density ratio between p-phase and uncontaminated water, $e_j = \partial z/\partial x_j$ is the j-component unit gravitational vector where z is elevation and t is the time.

Solution to the coupled multiphase flow equations given by [1] is described by Kaluarachchi and Parker (1989) and will not be repeated here.

2.2 Saturation-Capillary Pressure Relations

The proposed liquid entrapment model is an extension to the nonhysteretic three phase constitutive relationship of Parker et al. (1987) to accommodate the hysteresis due to nonwetting liquid entrapment. In the present analysis, we disregard gas entrapment in the liquid phases and assume nonhysteretic permeability-saturation relations. The effects of gas entrapment can however be included in the simulations by accommodating the liquid filled porosity instead of the actual porosity and conductivity at field capacity instead of saturated hydraulic conductivity (Kool and Parker, 1987).

Adopting the notation of Parker and Lenhard (1987), effective water and total liquid saturations are defined by

$$\bar{S}_w = \frac{S_w - S_m}{1 - S_m} \quad [2a]$$

$$\bar{S}_t = \frac{S_t - S_m}{1 - S_m} \quad [2b]$$

where $S_t = S_w + S_o$ is the total liquid saturation and S_m is the "irreducible" water saturation. Effective oil saturation is defined such that $\bar{S}_o = \bar{S}_t - \bar{S}_w$, hence

$$\bar{S}_o = \frac{S_o}{1 - S_m} \quad [3]$$

Furthermore, oil saturation is divided into two fractions so that $S_o = S_{of} + S_{ot}$ where subscripts f and t denote "free" and "trapped" oil, respectively. In terms of effective free and trapped oil saturations, defined by analog to [3], we have

$$\bar{S}_o = \bar{S}_{ot} + \bar{S}_{of}. \quad [4]$$

Finally, we define a quantity referred to as apparent water saturation, $\bar{\bar{S}}_w$, which corresponds to the sum of effective saturation of water and trapped oil, i.e.,

$$\bar{\bar{S}}_w = \bar{S}_w + \bar{S}_{ot} \quad [5]$$

Following Fayer and Hillel (1986) and Parker and Lenhard (1987), apparent saturation will be assumed a simple function of capillary pressure.

Prior to the occurrence of oil at a given location, the porous medium is treated as a two-phase air-water system described by the van Genuchten (1980) function

$$\bar{S}_w = [1 + (\alpha\ h_{aw})^n]^{-m} \quad [6]$$

where $\alpha[L^{-1}]$ and $n[L^{\circ}]$ are porous medium parameters and $m = 1 - 1/n$, and $h_{aw} = h_a - h_w$ is the air-water capillary head. Following the occurrence of oil at a given location, the system is described by the three-phase relations

$$\bar{\bar{S}}_w = [1 + (\alpha\ \beta_{ow} h_{ow})^n]^{-m} \quad [7a]$$

$$\bar{S}_t = [1 + (\alpha\ \beta_{ao} h_{ao})^n]^{-m} \quad [7b]$$

where $h_{ow} = h_o - h_w$ and $h_{ao} = h_a - h_o$ are oil-water and air-oil capillary heads, β_{ao} is a scaling coefficient which may be approximated by the ratio of water surface tension to oil surface tension, and β_{ow} is a scaling factor approximated by the ratio of water surface tension to oil-water interfacial tension (Lenhard and Parker, 1987a). To avoid numerical oscillations associated with changes from a two phase air-water system to a three phase air-water-oil system, once a node is classified as a three phase node it should not be allowed to revert to a two phase air-water system (Kaluarachchi and Parker, 1989). The necessary condition to classify a location as a

$$h_o > \frac{\beta_{ao} h_a + \beta_{ow} h_w}{\beta_{ow} + \beta_{ao}} \quad [8]$$

To compute water saturation in the three phase system, an expression for $\bar{S}_{ot}$ is required. Since gas entrapment is disregarded, we assume oil entrapment in a three phase air-oil-water system can be inferred directly from that in a two phase oil-water system. Referring to Figure 1, consider an oil-water system with main drainage and main imbibition curves described by [7a]. On any given scanning curve, the effective water saturation at $h_{ow} = 0$ is less than 1 due to oil entrapment given by $1 - \bar{S}_{or}$ where $\bar{S}_{or}$ is the effective residual oil saturation which may be estimated using an empirical relation given by Land (1968) as

$$\bar{S}_{or} = \frac{1 - \bar{S}_w^{\min}}{1 + R(1 - \bar{S}_w^{\min})} \quad [9a]$$

where

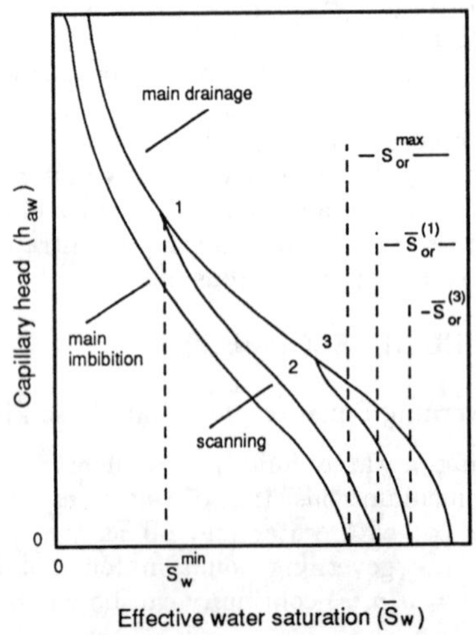

Figure 1 Hypothetical water-oil saturation-capillary head relationship

$$R = \frac{1}{\bar{S}_{or}^{\max}} - 1 \qquad [9b]$$

where $\bar{S}_W^{\min}$ is the minimum effective water saturation corresponding to the reversal from water drainage to imbibition and $\bar{S}_{or}^{max}$ is the effective residual oil saturation for the main imbibition curve. Residual oil saturation corresponds to the trapped oil saturation at zero capillary pressure.

Parker and Lenhard (1987) estimated the trapped saturation at nonzero capillary pressures by linear interpolation with $\bar{S}_{or} = 0$ at $\bar{\bar{S}}_w = \bar{S}_w^{\min}$ and $\bar{S}_{or} = \bar{S}_{or}^{\max}$ at $\bar{\bar{S}}_w = 0$. This method has the advantage of simplicity but is inconsistent with [9]. We have also observed that linear interpolation can lead to numerical problems at capillary heads near zero due to excessive changes in trapped oil saturation with small head changes. A procedure which is fully consistent with [9] involves estimation of trapped oil saturation as the difference between residual oil saturation for the actual scanning curve and that for a curve with a reversal point equal to the saturation on the actual path.

To clarify, consider Point 2 on Figure 1 with apparent water saturation $\bar{\bar{S}}_w$ on a path starting from Point 1 where $\bar{S}_w = \bar{S}_w^{\min}$. The effective trapped oil saturation at Point 2 is computed as the difference between the residual saturation on the path starting from Point 1 and that starting from Point 3. From [9] we have

$$\bar{S}_{ot} = \left\{ \frac{1 - \bar{S}_w^{\min}}{1 + R(1 - \bar{S}_W^{\min})} - \frac{1 - \bar{\bar{S}}_w}{1 + R(1 - \bar{\bar{S}}_w)} \right\} \qquad [10]$$

Equation [10] is valid if $\bar{S}_w > \bar{S}_w^{\min}$, otherwise $\bar{S}_{ot} = 0$. For the three phase air-oil-water system, an additional constraint which must be enforced is that $\bar{S}_{ot}$ at a given node is the maximum of $\bar{S}_{ot}$ computed by [10] and $\bar{S}_o$ at the node—that is, the system cannot trap more oil than is present at the node. Additionally, for the three phase system, $\bar{S}_w^{\min}$ takes on the operational meaning of being the minimum effective water saturation since the first occurrence of oil at the node. The value of $\bar{S}_w^{\min}$ is updated at the end of each time step after convergence and stored for each node.

2.3 Relative Permeability Relationships

At a given water saturation, oil entrapment affects water permeability by displacing water into larger pores. The net effect can be shown theoretically to be small (Lenhard and Parker 1987b) and experimental studies often report no observable hysteresis in wetting phase permeability vs. wetting fluid saturation relations (e.g., Saraf et al., 1982). Therefore, in the present paper, we will assume water relative permeability to be a simple-single valued function of water saturation as described by Parker et al. (1987). Since gas entrapment is not considered in the present analysis, we will assume that gas relative permeability can be described as a unique function of gas saturation also as given by Parker et al. (1987). To describe oil relative permeability, it is assumed that only free oil is a continuous phase subject to convection while trapped oil cannot move. Parker et al. (1987) derived an expression for oil relative permeability in the absence of fluid entrapment which is a function of effective water saturation and effective total liquid saturation. In the presence of trapped oil, the oil-water interface in the porous medium will be displaced to larger pores identified with the apparent water saturation. Replacing effective water saturation by apparent saturation in the previous model should therefore yield a good representation of oil relative permeability.

Phase relative permeabilities in the present model are described by the following relations

$$k_{rw} = \bar{S}_w^{1/2}[1 - \bar{S}_w^{1/m})]^2 \qquad [11a]$$

$$k_{ro} = (\bar{S}_t - \bar{\bar{S}}_w)^{1/2}[(1 - \bar{\bar{S}}_w^{l/m})^m - (1 - \bar{S}_t^{1/m})^m]^2 \qquad [11b]$$

$$k_{ra} = (1 - \bar{S}_t)^{1/2}[1 - \bar{S}_t^{1/2}]^{2m} \qquad [11c]$$

where k_{rw}, k_{ro} and k_{ra} denote relative permeabilities to water, oil and air, $m = 1 - 1/n$ is the van Genuchten parameter and saturations are as previously defined.

3 EXPERIMENTAL VALIDATION

The proposed nonwetting liquid entrapment model was implemented in the numerical multiphase flow and transport code MOFAT. The experimental setup used to validate the liquid entrapment model consists of a 71 cm long soil column (6 cm × 6.5 cm) initially free of oil and the water table located at the upper surface. Fluid saturations along the column were measured using a dual-energy gamma radiation apparatus and liquid phase pressure measurements were taken using ceramic tensiometers connected to pressure transducers. The experiment was conducted with different time dependent boundary conditions that can be summarized as follows:

Stage 1: At $t = 0$, the water table was lowered at a rate of 1 cm per minute until it reached an elevation of 7 cm at $t = 1$ h. The flow system thereafter was allowed to redistribute for a period of 12 h.

Stage 2: At the end of stage 1, a slug of oil containing 90% Soltrol and 10% l-iodoheptane was allowed to infiltrate from the upper surface under a zero water equivalent oil head until the accumulation was 250 cm^3. At the end of the infiltration period, the system was allowed to redistribute under natural hydraulic conditions until $t = 17.45$ h.

Stage 3: The lower surface water table at the end of stage 1 was varied as shown in Figure 2 and the experiment was conducted until $t = 22.1$ h.

Soil hydraulic properties of the medium and the fluid properties of the oil mixture are given in Table 1. The experimental procedure described above was simulated using the nonwetting liquid entrapment model described in this paper. For the simulation, oil head at the upper and lower surfaces were fixed at zero flux condition except during infiltration. The water boundary conditions at the upper surface water head was fixed at zero flux condition during the entire simulation.

Predicted water and oil saturations at 14 and 34 cm depths are shown in Figures 3 and 4 respectively. Predicted water saturations at the 14 cm depth indicate a drop in saturation at approximately 12 h due to the advancing oil front. During the same period, water saturation at the 34 cm depth increased due to the downward movement of water associated with the advancing oil front and decreased thereafter for a short period as the oil front reached this depth. Fluctuations in the water table caused smaller changes in the water saturation at the 34 cm depth compared to the 14 cm depth since during the most period of the simulation, the 34 cm depth remained liquid saturated. Predicted oil

Table 1. Soil hydraulic and bulk fluid properties*

Parameter	Experiment	Hypothetical
ϕ	0.4	0.4
n	2.48	2.5
S_m	0	0
$\alpha[L^{-1}]$	0.05	0.05
$K_{swy}[LT^{-1}]$	121	40
ρ_{ro}	0.83	0.8
η_{ro}	3.62	2
β_{a0}	2.85	2.67
β_{ow}	1.7	1.6
S_{or}^{max}	0.25	0 – 0.4

* Units in cm and hours

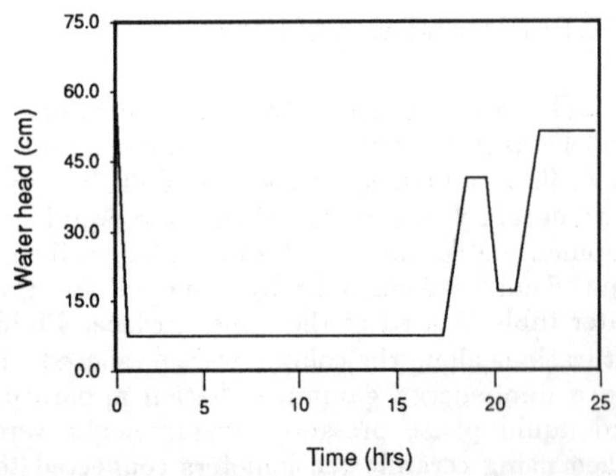

Figure 2 Stipulated water table elevation for column experiment

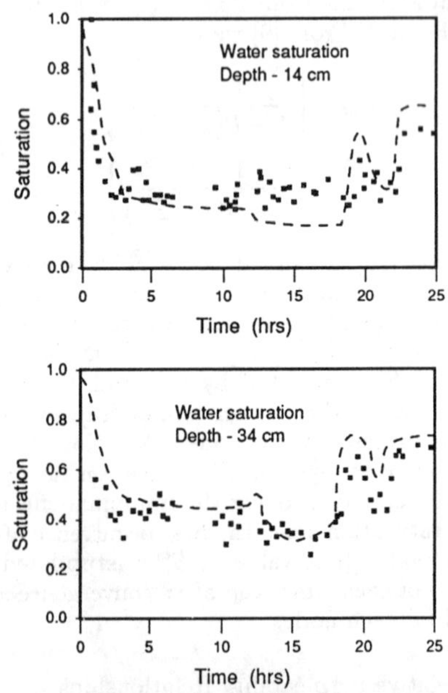

Figure 3 Predicted and observed water saturations for the column experiment; solid line and symbols indicate predicted and observed saturations respectively.

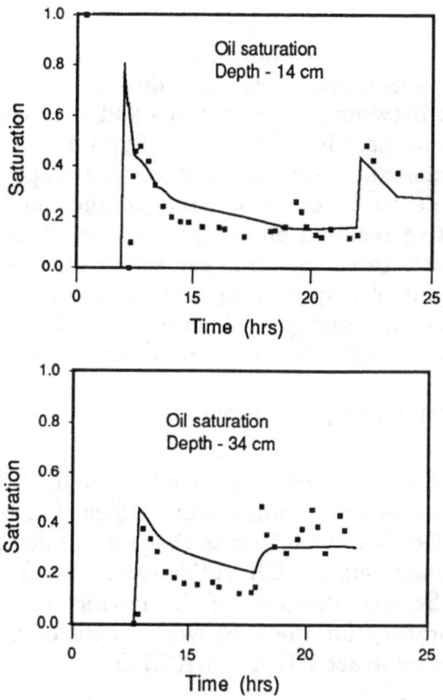

Figure 4 Predicted and observed oil saturations for the column experiment; solid line and symbols indicate predicted and observed saturations respectively.

saturations after 17 h at the 14 cm depth indicate less oil entrapment compared to the 34 cm depth.

The results show good agreement between the observed and predicted fluid saturations. The more pronounced peak in oil saturation as oil moves downward relative to the observed measurements probably reflects the simplifications involved in treating the saturation-capillary pressure relations. In particular, disregarding hysteresis in apparent water saturation may be expected to produce sharper fronts than observed in practice. However, it is noted that these discrepancies diminish markedly further from the source location indicating that ignoring such effects will not have serious practical consequences.

4 HYPOTHETICAL SIMULATION

4.1 Sensitivity of trapped oil percentage effective residual oil saturation

The oil entrapment model described in this paper requires one additional parameter, the maximum residual oil saturation, S_{or}^{max}. The purpose of this example problem is to investigate the sensitivity of the trapped oil volume percentage to this parameter for a set of stipulated boundary conditions. The flow domain is taken to be a 75 cm long soil column which is initially free of oil and at hydrostatic equilibrium with a water table located 10 cm from the lower surface. The soil hydraulic and fluid properties used in the simulation are given in Table 1. A total of seven simulations were performed with S_{or}^{max} values ranging from 0 to 0.4 which is the extreme range which is likely to occur in geologic media. Note that $S_{or}^{max} = 0$ corresponds to no oil entrapment in which case the model reduces to the that described by Parker et al. (1987). Boundary conditions for each of the simulations consist of following stages:

Stage 1: A slug of oil is allowed to infiltrate from the upper surface at a water equivalent head of 1 cm until the total accumulation is 5 cm^3 cm^{-2} while a zero- flux condition was stipulated for the water phase. At the lower boundary, the water head is maintained at the initial value of 10 cm and the oil phase is zero-flux.

Stage 2: Oil is allowed to redistribute for 25 h under zero-flux conditions for both water and oil at the upper surface, while the lower boundary conditions remain the same as in Stage 1.

Stage 3: The water head at the lower boundary is raised from 10 cm to 35 cm over a 1 h period and maintained at this value for 175 h. A zero-flux condition is imposed for water at the upper boundary and for oil at both boundaries.

Predicted oil saturations versus elevation at the end of Stage 3 are shown in Figure 5 for the cases $S_{or}^{max} = 0$ and 0.25. During Stage 3, water saturation increases as the water table rises resulting in oil entrapment when this is considered by the model. As a result, vertical displacement of oil is much less efficient and the zone of oil saturation is much more dispersed when oil entrapment is considered. At the end of the simulation nearly 40% of the oil in the system is trapped for case with $S_{or}^{max} = 0.25$.

Variations in the percentage of added oil which has become trapped at the end of Stage 3 in response to variations in S_{or}^{max} are shown in Figure 6. The results indicate that the fraction of the 5 cm^3 cm^{-2} spill volume which becomes trapped following a 25 cm increase in water table elevation increases in a markedly nonlinear manner with increasing S_{or}^{max}. The results also show that almost 90% of the total spill becomes immobilized under these conditions when $S_{or}^{max} \approx 0.4$. For a given maximum residual saturation, immobilization of the spill would be greater if the total spill volume were smaller or the water table rise greater.

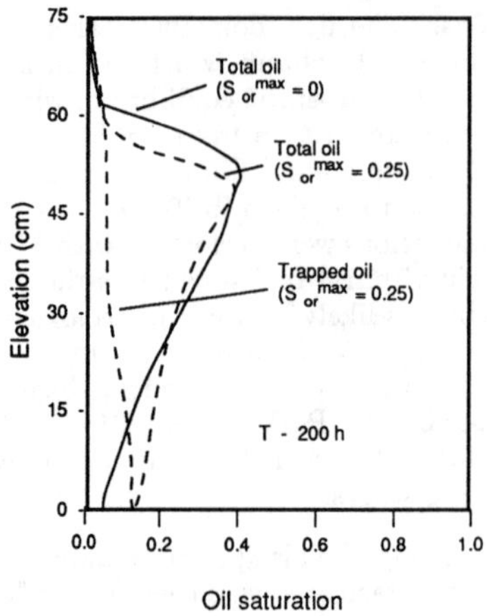

Figure 5 Predicted oil saturation for the hypothetical column simulation at the end of stage 3; initial and final water table elevations are marked.

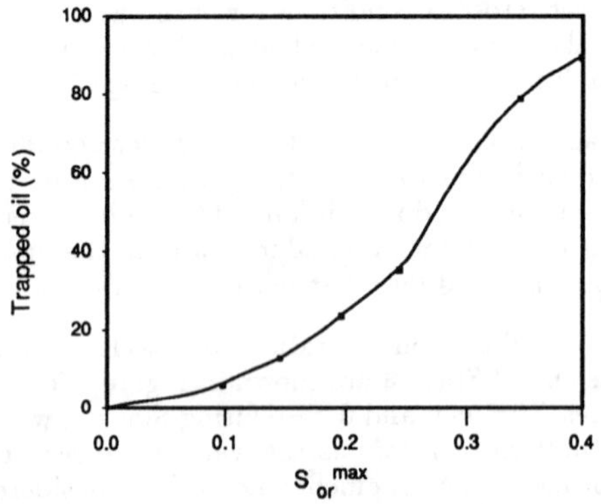

Figure 6 Variation of percentage of trapped oil volume with effective residual oil saturation for the hypothetical simulation.

5 CONCLUSIONS

A simplified procedure has been described to predict effects of oil entrapment on three phase permeability-saturation-capillary pressure relations that can be easily implemented in numerical multiphase flow codes. The fluid entrapment model requires only one additional parameter, the maximum effective residual oil saturation, compared to the corresponding model without fluid entrapment. The model requires minimal additional core storage requirements or computational effort and has been found to be computationally robust. The validity of the model was tested against a set of experimental observations under rapidly changing water table scenarios. Results indicate good agreement between the predictions and the observed saturations. Results from a hypothetical one-dimensional simulation suggest the trapped oil volume to be quite sensitive to the maximum effective residual oil saturation. Small changes in water table elevation can result in large proportions of a spill being immobilized in a manner which is highly nonlinearly dependent on the maximum effective residual oil saturation.

ACKNOWLEDGEMENTS

This work was supported by the U.S. Environmental Protection Agency Office of Research and Development Grants Program under assistance agreement CR-814320-01-0 and Subsurface Science Program of the Pacific Northwest Laboratory for the U.S. Department of Energy under contract DE-AC06-76RLD.

REFERENCES

Abriola, L. M. and G. Pinder. 1985. A multiphase approach to the modeling of porous media contamination by organic compounds, 2. Numerical simulations, Water Resour. Res., 21, 19–26.

Fayer, M. J. and D. Hillel. 1986. Air encapsulation, 1. Measurement in a field soil, Soil Sci. Soc. Am. J., 50, 568–572.

Gillham, R. W., A. Klute and E. Bresler. 1979. A numerical method for estimating infiltration, redistribution, drainage and evaporation of water from soil, Water Resour. Res., 43, 1061–1067.

Kaluarachchi, J. J. and J. C. Parker. 1989. An efficient finite element method for modeling multiphase flow, Water Resour. Res., 25, 43–54.

Killough, J. E. 1976. Reservoir simulation with history-dependent saturation functions, Trans. Am. Inst. Min. Metall. Pet. Eng., 261, 37–48.

Kool, J. B. and J. C. Parker. 1987. Development and evaluation of closed-form expression for hysteretic soil hydraulic properties, Water Resour. Res., 23, 105–114.

Land, C. S. 1968. Calculation of imbibition relative permeability for two- and three-phase flow from rock properties, Trans. Am. Inst. Min. Metall. Pet. Eng., 243, 149–156.

Lenhard, R. J. and J. C. Parker. 1987a. Measurement and prediction of saturation pressure relationships in three-phase porous media sys-

tems, J. Contam. Hydrol., 1, 407–424.
Lenhard, R. J. and J. C. Parker. 1987b. A model for hysteretic constitutive relations governing multiphase flow 2. Permeability-saturation relations, Water Resour. Res., 23, 2197–2206.
Lenhard, R. J., J. C. Parker and J. J. Kaluarachchi. 1989. A model for hysteretic constitutive relations governing multiphase flow 3. Refinements and numerical simulations, Water Resour. Res., 25, 1727–1736.
Osborne, M. and Sykes, J., 1986. Numerical modeling of immiscible organic transport at the Hyde Park Landfill. Water Resour. Res., 22(1), 25–33.
Parker, J. C., R. J. Lenhard and T. Kuppusamy. 1987. A parametric model for constitutive properties governing multiphase flow in porous media, Water Resour. Res., 23, 618–624.
Parker, J. C. and R. J. Lenhard. 1987. A model for hysteretic constitutive relations governing multiphase flow, 1. Saturation-pressure relations, Water Resour. Res., 23, 2187–2196.
Saraf, D. N., J. P. Batycky, C. H. Jackson, and D. B. Fisher. 1982. An experimental investigation of three-phase flow of water-oil-gas mixtures through water-wet sandstones, paper presented at California Regional Meeting, Soc. Pet. Eng., San Francisco, Calif., March 24–26.
Van Genuchten, M. T. 1980. A closed form equation for predicting the hydraulic conductivity of unsaturated soils, Soil Sci. Soc. Am. J., 44, 892–898.

Subsurface Contamination by Immiscible Fluids, Weyer (ed.) © 1992 Balkema, Rotterdam. ISBN 90 6191 175 3

Three-phase simulation of organic contaminations in aquifer systems

H.-W. Dorgarten
Earth Sciences Division, Lawrence Berkeley Laboratory, Berkeley, Calif., USA (Permanently: Aachen University of Technology, Germany)

C.-F. Tsang
Earth Sciences Division, Lawrence Berkeley Laboratory, Berkeley, Calif., USA

ABSTRACT: Depending on soil and fluid parameters, three different paths may dominate the migration of organic contaminants in natural aquifer systems: either the movement as a separate organic phase, or dissolution and subsequent transport of dissolved fractions in water, or volatilization together with transport of volatilized fractions in the gas phase. In order to study such three-phase flow and transport problems, a numerical model (transient two-dimensional finite element model) has been developed and applied. It considers flow of three moving phases, namely water phase, organic phase, and gas phase, as well as water and gas phase transport. The important role of interphase mass transfer processes is taken into account by means of kinetic formulations for dissolution, volatilization, and air-water partitioning. Additionally, transformation of pollutants due to adsorption and biodegradation is included in the model. Simulation results of TCE movement point out the significance of mass transfer kinetics and of density effects within the gas phase.

1 INTRODUCTION

Due to their hydrophobic nature, organic contaminants in aqueous systems primarily exist as a non-aqueous phase. This organic phase usually enters subsurface in the unsaturated zone, where it forms a three-phase system with soil water and soil air. Although immiscible with water, most organic pollutants are soluble in water to an extent far beyond the threshold of drinking water standards. Since these dissolved fractions of the contaminant are much more mobile than the non-aqueous phase itself, their movement will often have the major impact on environmental and water resources concerns. Furthermore, some of the most serious organic pollutants strongly volatilize into soil air. It has become evident that these volatilized fractions can be highly mobile in the soil, transported by diffusion as well as convection.

Depending on soil and fluid parameters, each of these three migration paths might be important in natural aquifer systems. In most cases the different paths are strongly coupled and it is not obvious which path actually dominates. The answer to this question, however, is highly important for the decision about appropriate remediation measures.

Particularly helpful laboratory experiments on the problem of organic contaminations in porous media were performed by Schwille (1971, 1988) and Schiegg (1979, 1980). In their experimental work they investigated a wide range of conditions, in order to gain physical understanding of multiphase flow and transport behavior. For example, they studied the spreading of an organic phase in layered media and in the capillary fringe, the role of residual saturations, microscopic effects, and the role of solubilization and vaporization.

Within the last decade, several numerical models have been developed in order to study the movement of organic pollutants in aquifer systems. However, nearly all of them reduce the physical three-phase problem by some simplifying assumptions to two or even only one phase before solving it. Most common is the assumption of a constant atmospheric pressure in the gas phase, which was used by Abriola (1984), Abriola and Pinder (1985), Faust (1985), Kuppusamy et al. (1987), Kaluarachchi and Parker (1989), Faust et al. (1989), Dorgarten (1989a,b), and others. This assumption, however, limits the applicability of these models to cases where volatilization and gas phase transport are of minor importance.

Other models have been developed especially to simulate volatile organics, but only at the cost of other limitations. Sleep and Sykes (1989) assume the organic phase to be immobile, and simulate a two-phase system of water and gas. In contrast, Falta et al. (1989) assume that water is at residual saturation, and reduce the problem to a two-phase system of organics and air.

Another important limitation of most of these models is that interphase mass transfer is either neglected completely or described by equilibrium approaches. It has become evident, however, that reaction rates may have an important role in phase transfer, especially when flow rates are artificially increased by remedial measures such as pump-and-treat techniques or gas extraction. Only recently, this effect has been taken into account by Sleep and Sykes (1989) and Dorgarten (1989a,b).

The purpose of the present work is to present a three-phase model in order to overcome the limitations that are connected with the assumption of neglecting one or the other phase. Flow of all three phases as well as water and gas phase transport are included. Interphase mass transfer is taken into account by means of kinetic formulations for dissolution, volatilization, and air-water partitioning. Additionally, adsorption and biodegradation reactions are included in the model.

2 THREE-PHASE FLOW

2.1 *Fluid flow equations*

The problem under consideration involves three separate phases, a water or aqueous phase (denoted by w), an organic, oil or non-aqueous phase (denoted by o), and an air or gas phase (denoted by g). Flow of each phase in this system can be described by combining a generalized form of Darcy's Law with the mass continuity equation (e.g. Bear 1970, Peaceman 1977, Aziz and Settari 1979). This results in the fluid flow equations

$$\frac{\partial}{\partial t}(\rho_w \phi S_w) - \frac{\partial}{\partial x_i}\left[\frac{\rho_w}{\mu_w} k_{rw} k_{ij} \left(\frac{\partial p_w}{\partial x_j} + \rho_w g \frac{\partial z}{\partial x_j}\right)\right] + \rho_w Q_w + \Gamma_w = 0 \qquad (1)$$

for the water phase,

$$\frac{\partial}{\partial t}(\rho_o \phi S_o) - \frac{\partial}{\partial x_i}\left[\frac{\rho_o}{\mu_o} k_{ro} k_{ij} \left(\frac{\partial p_o}{\partial x_j} + \rho_o g \frac{\partial z}{\partial x_j}\right)\right] + \rho_o Q_o + \Gamma_o = 0 \qquad (2)$$

for the organic phase, and

$$\frac{\partial}{\partial t}(\rho_g \phi S_g) - \frac{\partial}{\partial x_i}\left[\frac{\rho_g}{\mu_g} k_{rg} k_{ij} \left(\frac{\partial p_g}{\partial x_j} + \rho_g g \frac{\partial z}{\partial x_j}\right)\right] + \rho_g Q_g + \Gamma_g = 0 \qquad (3)$$

for the gas phase. Here, ρ is the density (kg/m^3), ϕ is the porosity of the porous medium (–), S is the phase saturation (–), μ is the dynamic viscosity (Pa·s), k_r is the relative permeability (–), k_{ij} is the intrinsic permeability tensor of the porous medium (m^2), p is the phase pressure (Pa), g is the gravitational constant (m/s^2), and z is the vertical spatial coordinate (m). Q (m^3/m^3·s) represents external sinks and sources, while Γ (kg/m^3·s) represents internal sinks and sources due to reactions and interphase mass transfer. Although this final term may be negligible in most cases, it will be significant, e.g., for highly volatile organics, or when residual amounts of the organic phase are diminished by dissolution or volatilization.

Eqs. (1) to (3) are strongly coupled and nonlinear, due to the dependences of the saturations on the capillary pressures, and of the relative permeabilities on the saturations (cf. section 2.2). Additionally, the fluid flow is coupled with the transport, due to the dependence of the densities on the concentrations (cf. section 2.3), and due to interphase mass transfer (cf. section 3.2).

2.2 *Saturations and relative permeabilities*

Since only little work has been done on capillary pressure-saturation relationships in three-phase systems (Eckberg and Sunada 1984, Lenhard and Parker 1987, 1988), saturations are commonly derived from two-phase data. Assuming that the wettability in a three-phase system decreases from water to the organic phase to gas and that there are no interfaces between water and gas as long as the organic phase is present, the saturations can be calculated from the two-phase relationships between water and organic phase and between organic phase and gas.

For two-phase systems, the relationship that is widely used in recent years was proposed by van Genuchten (1980) and may be written as

$$\overline{S} = \frac{S - S_r}{S_s - S_r} = \left[1 + \left(\alpha \frac{p_c}{\rho g}\right)^n\right]^{-m} \quad (p_c > 0),$$
$$\overline{S} = 1 \quad (p_c \leq 0). \qquad (4)$$

In this formulation, $\overline{S}$ is the effective saturation of the wetting fluid, S is its actual saturation, and S_s and S_r are the saturated and the residual values of S, respectively. p_c is the capillary pressure, i.e. the difference between the phase pressures of nonwetting and wetting fluid. α (m^{-1}) and n (–) are fitting parameters, m is given by $m = 1 - 1/n$.

By combining scaled versions of the van Genuchten model (eq. 4) for the different two-phase systems, Parker et al. (1987) proposed a consistent parametric model for the three-phase system. As long as oil is present in the system, the effective water saturation is given by

$$\overline{S}_w = \left[1 + \left(\beta_{ow}\alpha \frac{p_o - p_w}{\rho_w g}\right)^n\right]^{-m} \quad (p_o > p_w),$$
$$\overline{S}_w = 1 \quad (p_o \leq p_w), \tag{5}$$

and the effective total saturation (water plus organic phase) is given by

$$\overline{S}_t = \left[1 + \left(\beta_{go}\alpha \frac{p_g - p_o}{\rho_w g}\right)^n\right]^{-m} \quad (p_g > p_o), \tag{6}$$
$$\overline{S}_t = 1 \quad (p_g \leq p_o).$$

In the case that the organic phase is absent, the remaining two-phase system of water and gas has the effective water saturation

$$\overline{S}_w = \left[1 + \left(\beta_{gw}\alpha \frac{p_g - p_w}{\rho_w g}\right)^n\right]^{-m} \quad (p_g > p_w),$$
$$\overline{S}_w = 1 \quad (p_g \leq p_w). \tag{7}$$

To achieve consistency when the organic phase enters or leaves the system, the scaling factors β (-) should hold the relationship $1/\beta_{ow} + 1/\beta_{go} = 1/\beta_{gw}$, as was pointed out by Kaluarachchi and Parker (1989). In this case, the organic phase enters a two-phase system of water and gas, when the organic phase pressure p_o exceeds $(\beta_{gw}/\beta_{go})\, p_w + (\beta_{gw}/\beta_{ow})\, p_g$.

In addition to the capillary pressure-saturation relationships, saturation-relative permeability relationships are required in order to describe flow in a multiphase system. Measurements of relative permeabilities in three-phase systems date back to Leverett and Lewis (1941), and a review of their and many further measurements was given by Fayers and Matthews (1984).

In the present work, the relative permeabilities are described by closed-form expressions, based on van Genuchten's (1980) approach and its extension to a three-phase parametric model by Parker et al. (1987):

$$k_{rw} = \overline{S}_w^{\,1/2}\left[1 - \left(1 - \overline{S}_w^{\,1/m}\right)^m\right]^2, \tag{8}$$

$$k_{ro} = (\overline{S}_t - \overline{S}_w)^{1/2}\left[\left(1 - \overline{S}_w^{\,1/m}\right)^m - \left(1 - \overline{S}_t^{\,1/m}\right)^m\right]^2, \tag{9}$$

$$k_{rg} = (1 - \overline{S}_t)^{1/2}\left(1 - \overline{S}_t^{\,1/m}\right)^{2m}, \tag{10}$$

where m is the same parameter as in eq. (4). These relations state that the relative permeability of water is a function only of water saturation, and that the relative permeability of gas is a function only of gas saturation, while the relative permeability of the organic phase is a function of all phase saturations, as indicated by most measurements (Fayers and Matthews 1984). The permeability of the gas phase is additionally influenced by the gas slip or Klinkenberg effect. For coarse material, however, this effect is negligible, while in fine material it may lead to a significant increase of the gas phase permeability (Klinkenberg 1941, Stonestrom and Rubin 1989).

2.3 *Densities and viscosities*

Densities of gases as well as liquids are general functions of composition, pressure, and temperature. Of special interest for the present problem are density changes within the gas phase, since they can get much larger than in the fluid phases, and thus may have significant influence on flow and transport. For a gas mixture the mass density ρ_g (kg/m^3) is given by

$$\rho_g = \sum \frac{c_i M_i}{1000}, \tag{11}$$

where c_i and M_i are molar concentration (mol/m^3) and gram molecular mass (g/mol) of the species i, respectively. The sum of the molar concentrations, c, is related to the gas phase pressure p_g (Pa) and its absolute temperature T (°K) by the ideal gas law

$$\sum c_i = c = \frac{p_g}{RT}, \tag{12}$$

where the universal gas constant R is 8.313 m^3·Pa/mol·°K (Metcalfe and Farquhar 1987). For a mixture of two gases, pure air (denoted by a) and organic vapor (denoted by o), the combination of eqs. (11) and (12) yields

$$\rho_g = \frac{M_a}{1000RT}\, p_g + \frac{M_o - M_a}{1000}\, c_o. \tag{13}$$

If molar concentrations c_i are replaced by mass concentrations $C_i = c_i M_i/1000$ (kg/m^3), the equation of state for the gas phase can be formulated as

$$\rho_g = \frac{M_a}{1000RT}\, p_g + \frac{M_o - M_a}{M_o}\, C_g, \tag{14}$$

where C_g is the mass concentration of organic vapor in the gas phase. Under isothermal conditions, the factors in front of p_g and C_g are constant values, which are equal to the partial derivatives $\partial \rho_g/\partial p_g$ and $\partial \rho_g/\partial C_g$. While the first factor has nearly the same value in any natural aquifer system, the second one changes significantly with the molecular mass of the organic contaminant. If M_o gets smaller than M_a (about 29.0 g/mol), it can even change its sign.

Similarly, the liquid phase densities are represented by $\rho_w = \rho_w(p_w, C_w)$ and $\rho_o = \rho_o(p_o)$, respectively. However, the influence usually is much smaller than in case of the gas phase. Like the densities, the viscosities are general functions of pressure and composition, but their changes have only smaller influence on flow and transport. Thus, they are considered constant, although provisions to include variable viscosities are already implemented into the code.

3 TRANSPORT AND REACTIONS

3.1 *Water and gas phase transport*

Considering ecological and water resources concerns, the major issue of organic contaminations in aquifers is the pollution of groundwater and soil air by dissolved and volatilized fractions of the organic phase, respectively. Thus, transport of those fractions in the other phases is an essential part of investigating and simulating organic contaminations in aquifer systems.

Transport is described by conventional convection-dispersion equations, which may be formulated for partially saturated media as

$$\frac{\partial}{\partial t}(\phi S_w C_w) + \frac{\partial}{\partial x_i}(v_{w_i} C_w) - \frac{\partial}{\partial x_i}\left(\phi S_w D_{w_{ij}} \frac{\partial C_w}{\partial x_j}\right) + Q_w \tilde{C}_w + \Gamma_w = 0 \qquad (15)$$

for the water phase, and

$$\frac{\partial}{\partial t}(\phi S_g C_g) + \frac{\partial}{\partial x_i}(v_{g_i} C_g) - \frac{\partial}{\partial x_i}\left(\phi S_g D_{g_{ij}} \frac{\partial C_g}{\partial x_j}\right) + Q_g \tilde{C}_g + \Gamma_g = 0 \qquad (16)$$

for the gas phase. Here, v_i is the Darcy velocity (m/s), D_{ij} is the tensor of hydrodynamic dispersion (m^2/s), $\tilde{C}$ (kg/m^3) is the concentration of the sink/source flux Q, and all other parameters have been defined previously.

D_{ij} covers both mechanical dispersion and molecular diffusion (Bear 1972). Mechanical dispersion is represented by the conventional approach as defined by Scheidegger (1961), and molecular diffusion is represented by the Millington-Quirk model (Millington 1959, Millington and Quirk 1961). The tensor of hydrodynamic dispersion is then given by

$$\phi S D_{ij} = \delta_{ij}\alpha_t|v| + (\alpha_\ell - \alpha_t)\frac{v_i v_j}{|v|} + \delta_{ij}\phi^{4/3} S^{10/3} D_0 , \qquad (17)$$

where α_ℓ and α_t are longitudinal and transversal dispersivities (m), D_0 is the molecular diffusion (m^2/s) in the pure phase (water or air), and δ_{ij} is the Kronecker delta.

It is important to note that the Γ terms of eqs. (15) and (16) may consist of several components from reactions and interphase mass transfer, which are still undetermined and will be defined later on (cf. sections 3.2 to 3.4). Additionally, it should be pointed out that, in general, the concentrations in the different phases are coupled, so that eqs. (15) and (16) cannot be solved independently.

3.2 *Interphase mass transfer*

For low-solubility organic contaminants, a proper description of water and gas phase transport requires the consideration of interphase mass transfer. Dissolution and volatilization of the organic phase are responsible for the input of the pollutants into groundwater and soil air, and air-water partitioning can significantly affect transport within these phases.

If not ignoring these processes at all, most models treat them as being at local equilibrium rather than being kinetic. This may be appropriate for many groundwater applications, due to the slow velocities of natural groundwater flow and the long time scales. For the specific problem that is under investigation here, however, these conditions are no longer valid. By remedial measures such as pump-and-treat procedures or gas extraction, the flow velocities are strongly increased. Furthermore, with these measures the intention is always to reduce the time scales as much as possible. In such cases the assumption of local equilibrium would lead to an overestimation of transfer rates, and thus to an overestimation of the efficiency of the simulated remedial measure. Therefore, a-priori assumptions of local equilibrium should be relaxed, and reaction rates should be taken into account, as it has been done recently by Sleep and Sykes (1989) and Dorgarten (1989a,b).

Since dissolution defines the conjunction between multiphase flow and water phase transport, its correct characterization is essential for any prediction of groundwater pollution in the present problem. Recent laboratory experiments by Geller and Hunt (1988, 1989) and Mayer et al. (1988) have emphasized that dissolution may be subject to mass transfer limitations, depending on flow velocity, organic phase distribution, and interfacial area between the phases. In a kinetic formulation, the dissolution rate is controlled by the difference between an equilibrium concentration $\bar{C}_w$ and the actual concentration C_w of the organics in the groundwater, and by a rate coefficient κ_{ow} (s^{-1}),

$$\frac{\partial C_w}{\partial t} = \kappa_{ow}(\bar{C}_w - C_w) . \tag{18}$$

For a low-solubility, single-component contaminant, the equilibrium concentration $\bar{C}_w$ is given by its solubility in water, which can be taken from a large number of references (e.g. Pankow 1988). In the case of a multicomponent contaminant, in contrast, the equilibrium concentrations of the single components are generally controlled by selective dissolution and phase partitioning (van der Waarden et al. 1971). The rate coefficient κ_{ow} is a general function of the organic phase saturation, and is controlled by other parameters such as molecular diffusivity of the pollutant in water.

Volatilization is of the same importance for the gas phase as dissolution is for the water phase. According to physical similarity, its mathematical formulation is similar to eq. (18), with gas phase instead of water phase parameters,

$$\frac{\partial C_g}{\partial t} = \kappa_{og}(\bar{C}_g - C_g) . \tag{19}$$

The equilibrium concentration $\bar{C}_g$ is the saturated vapor concentration of the contaminant in air. Its value can be calculated from vapor pressure data, which are given in numerous references (e.g. Pankow 1988). The rate coefficient κ_{og} (s^{-1}) is a function of the organic phase saturation, the molecular diffusivity of the pollutant in air, and other parameters.

Interphase mass transfer by dissolution and volatilization will generally be the dominating phase exchange in regions where a significant amount of the organic phase is present. In other regions, in contrast, mass transfer between water and gas phases is important.

Air-water partitioning is mainly controlled by Henry's Law constant H (-) that expresses the relation between the gas phase concentration C_g and the water phase concentration C_w at equilibrium. Again, the transfer rate is represented by a kinetic formulation, which may be written as

$$\frac{\partial C_g}{\partial t} = \kappa_{wg}(HC_w - C_g) . \tag{20}$$

Values for H are tabulated numerously in the literature, or can be calculated from $H = \bar{C}_g/\bar{C}_w$. It should be pointed out that many different definitions and dimensions have been used for H; here it is used in a dimensionless form in terms of concentration relations. The dependences of the rate coefficient κ_{wg} (s^{-1}) have been studied intensively in the field of wastewater treatment in order to optimize air stripping technologies (McCarty 1987, Fleming 1989).

3.3 *Reactions*

Although other reactions may take place, adsorption and biodegradation seem to be the most important ones for the problem under consideration.

By adsorption processes a fourth, non-moving phase is introduced, namely the solid phase (denoted by s). In the present problem it can be assumed that water is always the wetting fluid, and the solid surface is continuously covered by water. Thus, adsorption can be considered as a process taking place between water and solid phases only, i.e. being independent from the other two phases.

For low-solubility organic contaminants, it is widely accepted that adsorption can be represented by a linear relationship between the concentrations in the solid phase and in the water phase. Considering reaction kinetics, adsorption may be formulated as

$$\frac{\partial C_s}{\partial t} = \kappa_{ws}(K_d C_w - C_s) , \tag{21}$$

where C_s is the mass-related concentration in the solid phase (-), and K_d is the distribution coefficient (m^3/kg). K_d can be calculated as the product of the organic carbon content f_{oc} (-) and the organic carbon partition coefficient K_{oc} (m^3/kg), where K_{oc} may be estimated from the octanol/water partition coefficient of the contaminant (Schwarzenbach and Westall 1981, Karickhoff 1984).

The presence of microorganisms below the root zone and under saturated conditions has become evident during recent years. Thus, there is the potential for biodegradation, depending on the presence of nutrients, substrate, toxicants, and other factors (Lee et al. 1988). Although several, more detailed relationships have been developed in order to describe biodegradation kinetics, such as Monod or Michaelis-Menten equations, biodegradation is mostly taken into account by a first order decay equation,

$$\frac{\partial C}{\partial t} = -\lambda C , \tag{22}$$

which holds for constant chemical and biological conditions. The rate coefficient λ (s^{-1}) depends on the microbial concentration, its utilization rate, and many other factors (e.g. Bouwer and McCarty 1984, Lee et al. 1988).

Most microorganisms that exist in the subsurface are fixed to the surface of the solid soil material (Lee et al. 1988). Thus, only the contaminants that are dissolved in the water phase or adsorbed to the solid phase can be subject to biodegradation. This is taken into account by the respective forms of eq. (22) for those two phases.

3.4 *Equation summary of transfer processes*

In order to take the reactions and transfer processes of eqs. (18) to (22) into account within the multiphase flow and transport problem, they have to be inserted into the Γ terms of eqs. (1) to (3), and of eqs. (15) and (16), respectively. Therefore, the terms of the different processes are summed for each phase, and related to the bulk volume of the porous medium. This yields

$$\begin{aligned}\Gamma_w = &- \phi S_w \kappa_{ow}(\bar{C}_w - C_w) \\ &+ \phi S_g \kappa_{wg}(HC_w - C_g) \\ &+ (1-\phi)\rho_s \kappa_{ws}(K_d C_w - C_s) + \phi S_w \lambda_w C_w \end{aligned} \quad (23)$$

for the water phase,

$$\Gamma_o = \phi S_w \kappa_{ow}(\bar{C}_w - C_w) + \phi S_g \kappa_{og}(\bar{C}_g - C_g) \quad (24)$$

for the organic phase, and

$$\Gamma_g = -\phi S_g \kappa_{og}(\bar{C}_g - C_g) - \phi S_g \kappa_{wg}(HC_w - C_g) \quad (25)$$

for the gas phase. Finally, summation of the solid phase terms yields

$$\frac{\partial C_s}{\partial t} = \kappa_{ws}(K_d C_w - C_s) - \lambda_s C_s . \quad (26)$$

Eq. (26) is needed as an additional equation to describe the transport problem, since C_s is an additional unknown variable, due to the kinetic adsorption approach.

4 NUMERICAL MODEL

The coupled multiphase flow and transport problem as defined in sections 2 and 3 is solved by the Galerkin finite element method, using linear triangular elements within a two-dimensional vertical domain. The unknowns that have been chosen for the numerical solution are summarized in table 1. Flow and transport are completely decoupled. Thus, the influence of changing concentrations on the flow field, mainly by density changes and by the interphase mass transfer terms, is lagged by one timestep.

By choosing fluid pressures instead of saturations, the flow equations can be solved continuously for the saturated and unsaturated zone without switching variables. Mass balance errors are avoided by a mixed formulation that keeps the phase saturations in the equations, similar to the method proposed by Celia et al. (1990) for the simulation of unsaturated flow. Because of the strong coupling the three flow equations are solved simultaneously. The strong nonlinearities are treated with a Newton-Raphson iteration scheme with relaxation.

Table 1. Unknowns of the numerical solution

Equation	(1)	(2)	(3)	(15)	(16)	(26)
Solved for	p_w	p_o	p_g	C_w	C_g	C_s
Matrix Size*	3 * N			2 * N		–

* (N = Number of Nodes)

Due to the kinetic formulation of interphase mass transfer and reactions, the equations for water, gas, and solid phase concentrations are strongly coupled. Therefore, both transport equations for water and gas are solved simultaneously, where the solid phase equation is taken into account by an algorithm described previously (Dorgarten 1989a,b). By means of mass lumping of the reaction terms, the algorithm allows the calculation of the solid phase concentrations very efficiently without additional matrix equations.

Time derivatives, both for flow and transport, are approximated by an implicit finite difference formulation. Timesteps are controlled automatically, depending on the number of Newton-Raphson iterations for the flow field. Additionally, timesteps are limited by an upper bound in order to fulfill the restrictions of the transport problem.

Since most of these numerical features are either standard formulations or similar to those of the previous two-phase model (Dorgarten 1989a,b), they are not described in detail here. Similarly, verification of the numerical model by analytical and quasi-analytical solutions has been completed, but is not presented due to space restrictions.

5 MODEL APPLICATION

After verification, the model is applied to study the movement of trichloroethylene (TCE) in a saturated-unsaturated aquifer system. A schematic representation of the studied domain is shown in figure 1, together with the geometry and the finite element mesh (992 nodes, 1860 elements). The data that have been used to specify the problem are given in table 2. As far as possi-

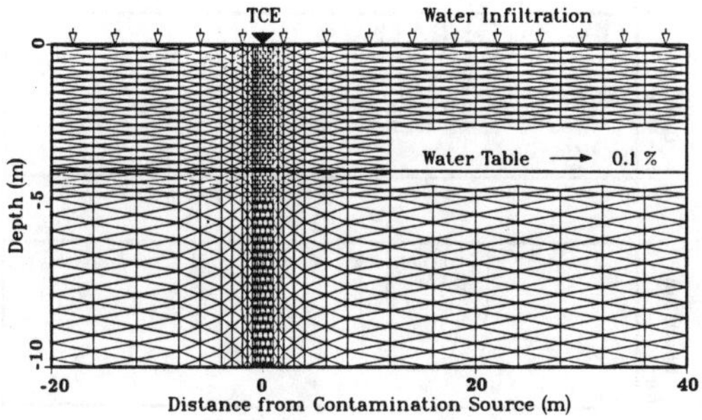

Figure 1. Scetch of the studied system with finite element discretization

ble, they are real data taken from numerous references. Soil properties are chosen to represent a coarse sandy aquifer.

Since the previous model (Dorgarten 1989a,b) was used for simulations of two-phase systems with water and organics, application of the present three-phase model has been focused on gas phase phenomena. Thus, TCE input is not simulated as inflow of a separate organic phase, but by dissolution and volatilization of a non-moving organic phase. Its initial saturation is 0.20 within an area of 0.5 m^2 at the ground surface, resulting in a total TCE mass of 58.4 kg. All simulations cover a period of one year, with time steps increasing from initially one day up to one month at the end of the calculation period.

The first simulations were performed in order to determine the significance of mass transfer kinetics. Apparently, results of these simulations strongly depend on the selection of the rate coefficients. Although mass transfer limitations have been emphasized by many authors, knowledge about rate coefficients is very limited. Nevertheless, the chosen values seem to reflect a reasonable order of magnitude according to the references cited in section 3.2.

Figure 2 shows water and gas phase concentrations after one year for the basic case, i.e. with a kinetic formulation according to the parameters of table 2. In comparison, figure 3 shows the results when the local equilibrium assumption (LEA) is applied. The contrast between these two cases demonstrates that the LEA is apparently not valid for the system studied. (It should be noted that different contour lines are plotted in the two figures.)

To investigate this phenomenon more detailed, further simulations were performed, with only parts of the processes assumed to be at local equilibrium. In table 3, the interphase mass transfer and the final mass distribution of TCE are given for each of these cases. The results indicate that the LEA for dissolution and volatilization has a

Table 2. Data specification

Parameter	Specified Value
Solid Phase	
Permeability	$k_{xx} = 10^{-10}$ m^2
Anisotropy	$k_{zz}/k_{xx} = 1:5$
Porosity	$\phi = 0.40$
Density	$\rho_s = 2650$ kg/m^3
Compressibility	$\alpha_s = 10^{-8}$ Pa^{-1}
Organic Carbon Content	$f_{oc} = 0.001$
Water Phase	
Density	$\rho_w = 1000$ kg/m^3
Compressibility	$\beta_w = 4.4 \cdot 10^{-10}$ Pa^{-1}
Viscosity	$\mu_w = 1.0 \cdot 10^{-3}$ kg/m·s
Infiltration Rate	$q_w = 5 \cdot 10^{-9}$ m^3/m^2·s
Gas Phase	
Density	$\rho_g = 1.20$ kg/m^3
Compressibility	$\partial\rho_g/\partial p_g = 1.19 \cdot 10^{-5}$ kg/m^3·Pa
Viscosity	$\mu_g = 1.8 \cdot 10^{-5}$ kg/m·s
Organic Phase: TCE	
Density	$\rho_o = 1460$ kg/m^3
Compressibility	$\beta_o = 10^{-9}$ Pa^{-1}
Viscosity	$\mu_o = 5.9 \cdot 10^{-4}$ kg/m·s
Solubility in Water	$\bar{C}_w = 1.1$ kg/m^3
Saturated Vapor Conc.	$\bar{C}_g = 0.418$ kg/m^3
Henry's Law Constant	$H = \bar{C}_g/\bar{C}_w = 0.38$
Org. Carbon Part. Coeff.	$K_{oc} = 0.126$ m^3/kg
Density Effect in Water	$\partial\rho_w/\partial C_w = 0.32$
Density Effect in Gas	$\partial\rho_g/\partial C_g = 0.78$
Multiphase Parameters	
van Genuchten (1980)	$n = 2.5$
Model Parameters	$\alpha = 2.0$ m^{-1}
Parker et al. (1987)	$\beta_{gw} = 1.0$
Parametric Model	$\beta_{ow} = 1.5$
Scaling Factors	$\beta_{go} = 3.0$
Residual Saturation	$S_r = 0.20$
Transport Parameters	
Longitudinal Dispersivity	$\alpha_\ell = 0.50$ m
Transversal Dispersivity	$\alpha_t = 0.01$ m
Molecular Diff. in Water	$D_{0w} = 5 \cdot 10^{-10}$ m^2/s
Molecular Diff. in Gas	$D_{0g} = 8 \cdot 10^{-6}$ m^2/s
Rate Coefficients	
Biodegr. Const. in Water	$\lambda_w = 0.0$ s^{-1}
Biodegr. Const. in Soil	$\lambda_s = 0.0$ s^{-1}
Dissolution Rate Coeff.	$\kappa_{ow} = S_o * 10^{-6}$ s^{-1}
Volatilization Rate Coeff.	$\kappa_{og} = S_o * 10^{-6}$ s^{-1}
Air-Water Rate Coeff.	$\kappa_{wg} = 10^{-6}$ s^{-1}
Adsorption Rate Coeff.	$\kappa_{ws} = 10^{-6}$ s^{-1}

strong impact on water and gas phase transport, while the LEA for air-water partitioning and adsorption is of minor importance. This contrast is explained by the fact that the latter processes can take place all over the aquifer, while dissolution and volatilization are limited to the small region where the organic phase is present. Therefore,

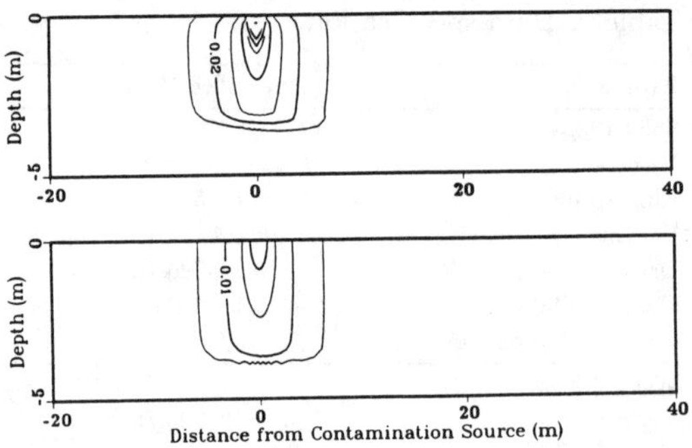

Figure 2. Water (top) and gas phase concentrations of TCE (in kg/m^3) after one year: kinetic formulation, with density effect

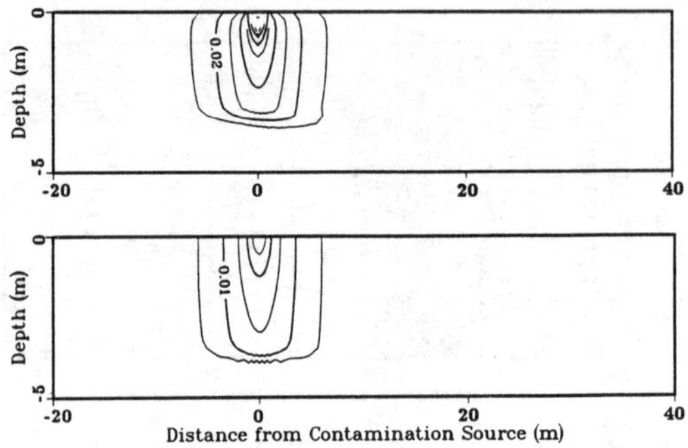

Figure 4. Water (top) and gas phase concentrations of TCE (in kg/m^3) after one year: kinetic formulation, no density effect

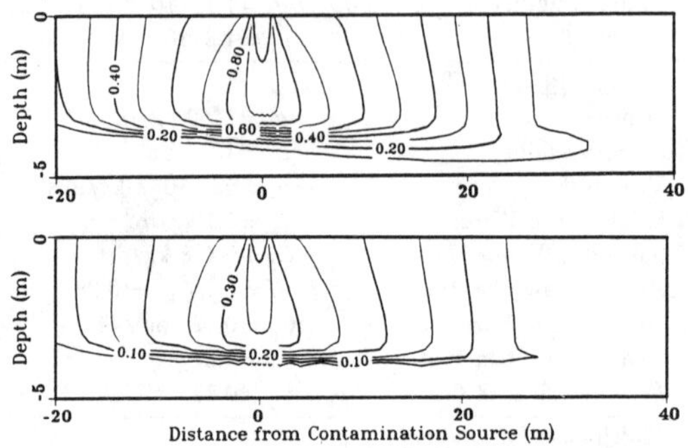

Figure 3. Water (top) and gas phase concentrations of TCE (in kg/m^3) after one year: local equilibrium assumption, with density effect

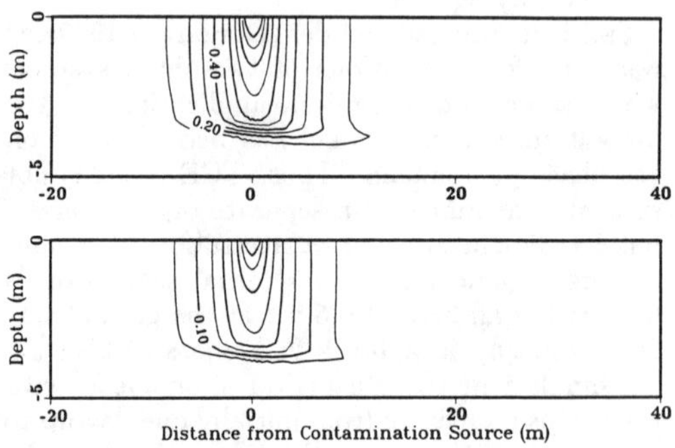

Figure 5. Water (top) and gas phase concentrations of TCE (in kg/m^3) after one year: local equilibrium assumption, no density effect

dissolution and volatilization kinetics do not depend on a global but on a local time scale, i.e. on the duration water and gas need to flow through the region of non-moving organics. Thus, in addition to the rate coefficients, the thickness of the organic region, the vertical permeability, the strength of the density effect, and the water infiltration rate can be identified as controlling parameters, and their influence should be determined in laboratory experiments.

The important role of density-driven flow in the gas phase has been emphasized by several authors (Sleep and Sykes 1989, Falta et al. 1989, Mendoza and Frind 1990). In order to isolate this effect in the present problem, additional simulations were performed with $\partial\rho_w/\partial C_w$ and $\partial\rho_g/\partial C_g$ set equal to zero. In the kinetic case, however, neglecting the density effect causes only slight differences in the results (table 3, figure 4 vs. figure 2). This indicates that gas phase transport is controlled by diffusion rather than density-driven flow. This implication derives from the fact that gas phase concentrations, and consequently density differences in the gas phase, remain small because of mass transfer limitations. In the equilibrium case, in contrast, the impact of neglecting the density effect is significant. Due to the lack of vertical gas flow, water and gas phase transport (figure 5 vs. figure 3) as well as volatilization and air-water partitioning (table 3) are strongly affected. The different influence of the density effect in kinetic and equilibrium systems implies that transport mechanisms in the gas phase might be misinterpreted, if local equilibrium is assumed.

In all the previous simulations, the gas phase path dominates TCE movement in the subsurface, as could be expected for a highly volatile contaminant. Thus, the coupled consideration of volatilization, gas phase transport, air-water partitioning, and water phase transport, in conjunction with multiphase flow, is necessary for the prediction of the groundwater contamination that derives from the non-moving organic phase.

Table 3. Interphase mass transfer of TCE within one year and phase distribution of TCE after one year (in kg)

Simulated Case	Phase Transfer			Final Phase Distribution				
	o → w	o → g	g → w	Org.	Water	Gas	Solid	Loss
Kinetic Formulation of all Processes	0.31	0.37	0.20	57.72	0.22	0.15	0.29	0.02
Local Equilibrium Ass. (LEA) for all Processes	0.50	46.00	35.42	11.90	16.26	8.29	19.07	2.88
LEA only for Dissolution and Volatilization	0.52	43.62	30.66	14.26	13.71	8.73	17.07	4.63
LEA only for Air-Water Part. and Adsorption	0.35	0.37	0.21	57.68	0.25	0.14	0.31	0.02
Kinetic Formulation, no Density Effect	0.31	0.37	0.21	57.72	0.22	0.16	0.30	0.00
Local Equilibrium Ass., no Density Effect	0.53	12.86	10.18	45.01	4.72	2.68	5.89	0.10

6 SUMMARY AND CONCLUSIONS

A finite element model is presented for the simulation of coupled three-phase flow and transport problems. After a detailed discussion of the physical and chemical processes and their mathematical formulation, the model is applied to study the movement of TCE in the subsurface. First results indicate that mass transfer kinetics may be of significant importance for dissolution and volatilization of an organic phase, while it is of minor importance for air-water partitioning and for adsorption. Furthermore, results show that in many cases the role of density-driven flow in the gas phase might be less important than generally assumed, due to mass transfer limitations. To get more evidence for these results, further investigations on the rate coefficients and their dependences are needed.

ACKNOWLEDGEMENTS

This work was made possible by the support of a post-doctorate scholarship, awarded to the first author by the NATO Scientific Committee through the German Academic Exchange Service (DAAD), Bonn, West Germany. Part of the work was jointly supported by the Office of Drinking Water, U.S. Environmental Protection Agency and by the Director, Office of Energy Research, Office of Basic Energy Sciences, Engineering and Geosciences Division of the U.S. Department of Energy under contract No. DE-AC03-76SF00098.

REFERENCES

Abriola, L.M. 1984. *Multiphase migration of organic compounds in a porous medium, a mathematical model*. New York: Springer.

Abriola, L.M. & G.F.Pinder 1985. A multiphase approach to the modeling of porous media contamination by organic compounds, 1. equation development. *WRR* 21: 11-18.

Aziz, K. & A.Settari 1979. *Petroleum reservoir simulation*. New York: Elsevier.

Bear, J. 1970. Two-liquid flows in porous media. In V.T. Chow (ed.), *Advances in Hydroscience* 6: 141-252, New York: Academic Press.

Bear, J. 1972. *Dynamics of fluids in porous media*. New York: Elsevier.

Bouwer, E.J. & P.L.McCarty 1984. Modeling of trace organics biotransformation in the subsurface. *Ground Water* 22: 433-441.

Celia, M.A., E.T.Bouloutas & R.L.Zarba 1990. A general mass-conservative numerical solution for the unsaturated flow equation. *WRR* 26: 1483-1496.

Dorgarten, H.W. 1989a. Das Verhalten hydrophober Stoffe in Boden und Grundwasser. *Mitt. Inst. für Wasserbau und Wasserwirtschaft*, RWTH Aachen, West Germany, No. 73.

Dorgarten, H.W. 1989b. Finite element simulation of immiscible and slightly soluble pollutants in soil and groundwater. In H.E.Kobus & W.Kinzelbach (eds.), *Proc. Int. Symp. on Contaminant Transport in Groundwater*, Stuttgart, West Germany, p. 389-396, Rotterdam: Balkema.

Eckberg D.K. & D.K.Sunada 1984. Nonsteady three-phase immiscible fluid distribution in porous media. *WRR* 20: 1891-1897.

Falta, R.W., I.Javandel, K.Pruess & P.A.Witherspoon 1989. Density-driven flow of gas in the unsaturated zone due to the evaporation of volatile organic compounds. *WRR* 25: 2159-2169.

Faust, C.R. 1985. Transport of immiscible fluids within and below the unsaturated zone: a numerical model. *WRR* 21: 587-596.

Faust, C.R., J.H.Guswa & J.W.Mercer 1989. Simulation of three-dimensional flow of immiscible fluids within and below the unsaturated zone. *WRR* 25: 2449-2464.

Fayers, F.J. & J.D.Matthews 1984. Evaluation of normalized Stone's methods for estimating three-phase relative permeabilities. *Soc. Petr. Eng. J.* 24: 224-232.

Fleming, J.L. 1989. Volatilization technologies

for removing organics from water. *Poll. Techn. Rev.* 164, Park Ridge, NJ: Noyes Data Corp.

Geller, J.T. & J.R.Hunt 1988. Groundwater contamination by immiscible organic liquids: experimental results on dissolution kinetics. *Eos* 69: 1189.

Geller, J.T. & J.R.Hunt 1989. Dissolution of nonaqueous phase organic liquids in porous media. *Eos* 70: 1092.

Kaluarachchi, J.J. & J.C.Parker 1989. An efficient finite element method for modeling multiphase flow. *WRR* 25: 43-54.

Karickhoff, S.W. 1984. Organic pollutant sorption in aquatic systems. *ASCE J. Hydr. Eng.* 110: 707-735.

Klinkenberg, L.J. 1941. The permeability of porous media to liquids and gases. In API (publ.), *Drilling and Production Practice*, p. 200-213, New York.

Kuppusamy, T., J.Sheng, J.C.Parker & R.J.Lenhard 1987. Finite-element analysis of multiphase immiscible flow through soils. *WRR* 23: 625-631.

Lee, M.D., J.M.Thomas, R.C.Borden, P.B.Bedient, J.T. Wilson & C.H.Ward 1988. Biorestoration of aquifers contaminated with organic compounds. *CRC Critical Reviews in Environmental Control* 18: 29-89.

Lenhard, R.J. & J.C.Parker 1987. Measurement and prediction of saturation-pressure relationships in three-phase porous media systems. *J. Contam. Hydrol.* 1: 407-424.

Lenhard, R.J. & J.C.Parker 1988. Experimental validation of the theory of extending two-phase saturation-pressure relations to three-fluid phase systems for monotonic drainage paths. *WRR* 24: 373-380.

Leverett, M.C. & W.B.Lewis 1941. Steady flow of gas-oil-water mixtures through unconsolidated sands. *Trans. AIME* 142: 107-116.

Mayer, A.S., C.T.Miller & M.M.McNeill 1988. Mass transfer from non-aqueous phase liquids to the aqueous phase in groundwater systems. *Eos* 69: 1189.

McCarty, P.L. 1987. Removal of organic substances from water by air stripping. In B.B.Berger (ed.), Control of organic substances in water and wastewater. *Poll. Techn. Rev.* 140: 119-147, Park Ridge, NJ: Noyes Data Corp.

Mendoza, C.A. & E.O.Frind 1990. Advective-dispersive transport of dense organic vapors in the unsaturated zone, 2. sensitivity analysis. *WRR* 26: 388-398.

Metcalfe, D.E. & G.J.Farquhar 1987. Modeling gas migration through unsaturated soils from waste disposal sites. *Water, Air and Soil Poll.* 32: 247-259.

Millington, R.J. 1959. Gas diffusion in porous media. *Science* 130: 100-102.

Millington, R.J. & J.P.Quirk 1961. Permeability of porous solids. *Trans. Faraday Soc.* 57: 1200-1207.

Pankow, J.F. 1988. Translator's appendix: physical and chemical properties of dense solvent compounds. In F.Schwille, *Dense chlorinated solvents in porous and fractured media, model experiments*, Chelsea, MI: Lewis Publ.

Parker, J.C., R.J.Lenhard & T.Kuppusamy 1987. A parametric model for constitutive properties governing multiphase flow in porous media. *WRR* 23: 618-624.

Peaceman, D.W. 1977. *Fundamentals of numerical reservoir simulation*. New York: Elsevier.

Scheidegger, A.E. 1961. General theory of dispersion in porous media. *J. Geophys. Res.* 66: 3273-3278.

Schiegg, H.O. 1979. Verdrängungssimulation dreier nicht mischbarer Fluide in poröser Matrix. *Mitt. Versuchsanstalt für Wasserbau, Hydrologie und Glaziologie*, ETH Zürich, Switzerland, No. 40.

Schiegg, H.O. 1980. Grundlagen, Aufbau und Resultate von Laborexperimenten zur Erforschung von Ölausbreitungen in Grundwasserträgern. *Mitt. Versuchsanstalt für Wasserbau, Hydrologie und Glaziologie*, ETH Zürich, Switzerland, No. 43.

Schwarzenbach, R.P. & J.Westall 1981. Transport of nonpolar compounds from surface water to groundwater, laboratory sorption studies. *Environ. Sci. Technol.* 15: 1360-1367.

Schwille, F. 1971. Die Migration von Mineralöl in porösen Medien. *gwf-Wasser/Abwasser* 112: 307-311, 331-336, 465-472 (3 papers).

Schwille, F. 1988. *Dense chlorinated solvents in porous and fractured media, model experiments*. Chelsea, MI: Lewis Publ.

Sleep, B.E. & J.F.Sykes 1989. Modeling the transport of volatile organics in variably saturated media. *WRR* 25: 81-92.

Stonestrom, D.A. & J.Rubin 1989. Air permeability and trapped-air content in two soils, *WRR* 25: 1959-1969.

van der Waarden, M., A.L.Bridie & W.M.Groenewoud 1971. Transport of mineral oil components to groundwater I. *Water Res.* 5: 213-226.

van Genuchten, M.T. 1980. A closed-form equation for predicting the hydraulic conductivity of unsaturated soils. *Soil Sci. Soc. Am. J.* 44: 892-898.

Subsurface Contamination by Immiscible Fluids, Weyer (ed.) © 1992 Balkema, Rotterdam. ISBN 90 6191 175 3

Modeling multicomponent organic chemical transport

J.C. Parker & Jagath J. Kaluarachchi
Virginia Polytechnic Institute and State University, Blacksburg, Va., USA

ABSTRACT: A two-dimensional finite element model was developed to predict coupled flow and multicomponent transport in a three fluid phase porous medium. The organics can partition between the water, oil, gas and solid phases assuming local equilibrium. Results from a number of simulations suggest that updating of interphase mass transfer rates and phase densities have negligible effects during short periods of highly transient flow but will be very important for long term hydrocarbon migration. Due to low volatility and solubility of most organic liquid compounds, the efficiency of removal by gas phase diffusion and aqueous phase dissolution can be very poor.

1 INTRODUCTION

Contamination of ground water resources due to spills and leakage from underground storage tanks are of great environmental concern due to the hazardous nature of the constituent chemicals. Models developed to predict the movement of hydrocarbons in the subsurface should consider the potential convective and dispersive transport together with mass partitioning between different phases. Recent work by a number of researchers *(Abriola, 1984; Corapcioglu and Baehr, 1987; Reeves and Abriola, 1988; Sleep and Sykes, 1989; among others)* have addressed these issues subjected to various simplifications. The purpose of this paper is to present a robust numerical model to predict multicomponent transport in a three fluid phase porous medium with emphasis on interphase mass transfer and efficiency of remedial schemes in removing contaminants from subsurface.

2 MODEL DEVELOPMENT

2.1 Multiphase liquid flow

Assuming a three fluid phase incompressible porous medium consisting of water (w), oil (o) and air (a) with gas at constant pressure, the governing equation for mobile p-phase (p=o,w) continuity can be written as (*Parker*, 1989)

$$\phi \frac{\partial \rho_p S_p}{\partial t} = - \frac{\partial \rho_p q_{p_i}}{\partial x_i} + R_p \qquad [1]$$

where ϕ is porosity, S_p is the p-phase saturation, x_i (and x_j) are Cartesian spatial coordinates (i,j = 1, 2), ρ_p is the density of phase, q_{p_i} is the volumetric flux density along the i direction, R_p is the net mass transfer per unit porous media volume into (+) or out of (−) phase p, and t is the time.

The volumetric flux of p phase can be described by the Darcy's law written as

$$q_{p_i} = - \frac{k_{rp} K_{sp_{ij}}}{\eta_{rp}} \left(\frac{\partial h_p}{\partial x_j} + \rho_{rp} e_j \right) \qquad [2]$$

where k_{rp} is the p phase relative permeability, $K_{sw_{ij}}$ is the saturated hydraulic conductivity tensor of water, η_{rp} is the ratio between p phase and water absolute viscosity, h_p is the p phase pressure head in equivalent water height of pure water, ρ_{rp} is the density ratio between p phase and uncontaminated water and $e_j = \partial z / \partial x_j$ is the j-component unit gravitational vector where z is elevation.

Combining [1] and [2] leads to the p phase flow equation

$$\phi \frac{\partial \rho_p S_p}{\partial t} - R_p =$$

$$- \frac{\partial}{\partial x_i} \left(\frac{\rho_p k_{rp} K_{sp_{ij}}}{\eta_{rp}} \left[\frac{\partial h_p}{\partial x_j} + \rho_{rp} e_j \right] \right) \qquad [3]$$

Solution to the coupled multiphase flow equations given by [3] is described by *Kaluarachchi and Parker (1989)* and will not be repeated here.

2.2 Multicomponent transport

Mass conservation of each partitionable component α in the p phase can be written as

$$\phi \frac{\partial C_{\alpha p} S_p}{\partial t} = \frac{\partial}{\partial x_i}\left(\phi S_p D_{\alpha p_{ij}} \frac{\partial C_{\alpha p}}{\partial x_j}\right) - \frac{\partial C_{\alpha p} q_{p i}}{\partial x_i} + R_{\alpha p} - \mu_{\alpha p} C_{\alpha p} \qquad [4]$$

where $C_{\alpha p}$ is the concentration of component α in the p phase given as mass per unit p phase volume, $D_{\alpha p_{ij}}$ is the dispersion coefficient of α component in p phase, $R_{\alpha p}$ is the net mass transfer of α component to the p phase and $\mu_{\alpha p}$ is the first order decay coefficient for component α in p phase. The dispersion coefficient is taken as the sum of components due to diffusion and mechanical dispersion. The former is given as the product of the free phase diffusion coefficient, $D^{o}{}_{\alpha p}$, and a tortuosity term. Gas and oil phase diffusion coefficients are estimated from the aqueous diffusion coefficient and phase viscosity. Mechanical dispersion is computed using the conventional formulation in terms of longitudial and transverse dispersities ϵ_l and ϵ_t and phase velocity.

Using the continuity equation given by [1] in [4] and neglecting the time derivative of the density term, transport of noninert component α in p phase can be written as

$$\phi\, S_p \frac{\partial C_{\alpha p}}{\partial t} = \frac{\partial}{\partial x_i}\left(\phi S_p D_{\alpha p_{ij}} \frac{\partial C_{\alpha p}}{\partial x_j}\right) - q_{p i} \frac{\partial C_{\alpha p}}{\partial x_i} + R_{\alpha p} - z\left(\mu_{\alpha p} + \frac{R_p}{\rho_p}\right) C_{\alpha p} \qquad [5]$$

Adsorption of α on the solid phase can be described by an equivalent continuity equation

$$\frac{\partial C_{\alpha s}}{\partial t} = R_{\alpha s} - \mu_{\alpha s}\, C_{\alpha s} \qquad [6]$$

where subscript s refers to the solid phase with $C_{\alpha s}$ defined as the mass of α in the solid phase per unit porous media volume.

2.3 Phase summed transport under local equilibrium

Now invoking the local equilibrium assumption for partitioning of α between water, oil, gas and solid phases, the p phase concentration, $C_{\alpha p}$, can be written in terms of the water phase concentration, $C_{\alpha w}$, as

$$C_{\alpha p} = \Gamma_{\alpha p}\, C_{\alpha w} \qquad [7]$$

Also note

$$R_p = \sum_{\alpha=1}^{n_s} R_{\alpha p} \qquad [8a]$$

and

$$R_{\alpha w} + R_{\alpha o} + R_{\alpha s} + R_{\alpha a} = 0 \qquad [8b]$$

since the total mass of α entering p phase is the sum due to all partionable components, n_s, and net mass transfer of component α among the different phases should be zero. Using the relationship given by [7] and [8] in [5] and [6] that describe the transport of α in water, oil, air and solid phases, and summing up these equations leads to the phase summed transport equation

$$\phi_\alpha^* \frac{\partial C_{\alpha w}}{\partial t} - \frac{\partial}{\partial x_i}[\, D_{\alpha_{ij}}^* \frac{\partial C_{\alpha w}}{\partial x_j}\,] + q_{\alpha_i}^* \frac{\partial C_{\alpha w}}{\partial x_i} + \mu_\alpha^*\, C_{\alpha w} = 0 \qquad [9]$$

where

$$\phi_\alpha^* = \phi\, S_w + \phi S_o \Gamma_{\alpha o} + \phi S_a \Gamma_{\alpha a} + \Gamma_{\alpha s} \qquad [10a]$$

$$D_{\alpha_{ij}}^* = \phi\, S_w\, D_{\alpha w_{ij}} + \phi\, S_o\, D_{\alpha o_{ij}}\, \Gamma_{\alpha o} + \phi S_a\, D_{\alpha a_{ij}}\, \Gamma_{\alpha a} \qquad [10b]$$

$$q_{\alpha_i}^* = q_{\alpha w_i} + q_{\alpha o_i}\, \Gamma_{\alpha o} \qquad [10c]$$

$$\mu_\alpha^* = \mu_w + \mu_o\, \Gamma_{\alpha o} + \mu_a\, \Gamma_{\alpha a} + \mu_{\alpha s}\, \Gamma_{\alpha s} + \frac{R_w}{\rho_w} + \frac{R_o \Gamma_{\alpha o}}{\rho_o} + \frac{R_a\, \Gamma_{\alpha a}}{\rho_a} \qquad [10d]$$

The form of equation [9] is similar to that of single phase transport equation and needs to be solved for each α component. Although gas pressure gradients have been neglected in the present flow analysis, phase summed transport equation will have a contribution from gas phase diffusion.

2.4 Initial and boundary conditions

Stipulating the initial and boundary conditions in terms of the water phase concentration, we have

$$C_{\alpha w}(x_i, t=0) = C_{\alpha w o}(x_i) \text{ on } R \text{ for } t = 0 \quad [11a]$$

$$C_{\alpha w}(x_i,t) = C_{\alpha w_1}(x_i,t) \text{ on } S_1 \text{ for } t > 0 \quad [11b]$$

$$\frac{\partial C_{\alpha w}}{\partial x_n} = 0 \text{ on } S_2 \text{ for } t > 0 \quad [11c]$$

$$q_{on} C_{\alpha o 3} = q_n^* C_{\alpha w} - D_{\alpha ij}^* \frac{\partial C_{\alpha w}}{\partial x_j} /_n \quad \text{on } S_3 \text{ for } t > 0 \quad [11d]$$

In [11a], $C_{\alpha w o}(x_i)$ represents the initial water phase concentration of component α at location x_i. Equation [11b] describes a type-1 boundary condition on boundary segment S_1 with specified α-component concentration $C_{\alpha w_1}(x_i,t)$. Equation [11c] describes type-2 boundaries with zero normal (n) concentration gradient specified on segment S_2. Equation [11d] is a type-3 boundary condition for use on boundary segments, S_3, which have inward fluid fluxes predicated on the assumption that contaminant enters the system solely via inflow of nonaqueous phase liquid. Here, q_{on} is the normal Darcy velocity of the oil phase at the boundary and $C_{\alpha o 3}$ is the concentration of component in the influent oil phase.

In applying the type-3 boundary condition during periods of oil infiltration it is important to ensure consistency in the specified oil phase concentrations such that

$$\sum_{\alpha=1}^{n_s} C_{\alpha o} + C_{no} = \rho_o \quad [12]$$

where C_{no} is the oil phase concentration of the inert component.

The transport equation for a given component is highly dependent on the flow equations but the flow equations are less weakly coupled with the transport analysis. We exploit this weak back coupling of the transport equation with the flow equation and perform the analysis in the following manner; (a) solve the flow equations using time lagged phase densities and interphase mass transfer rates, (b) solve the phase summed transport equation for each noninert component sequentially (c) back calculate and update the phase densities and interphase mass transfer rates from the individual phase transport equations and (d) proceed to the next time step.

2.5 Updating of interphase mass transfer rates

Interphase mass transfer rates play an important role in multicomponent transport since it is the mechanism governing long term removal of nonaqeous contaminants. Back calculation of mass transfer rates are obtained via individual phase transport equations given by [5] and [6]. During highly transient periods of flow however, back calculation could lead to numerical problems especially at sharp solute fronts. This difficulty is further enhanced by the second order derivative terms in the dispersion part of the transport equation.

To avoid such numerical difficulties and unstable mass transfer rates, we propose to compute the interphase mass transfer terms under the condition that oil velocities in the system are lower than a predetermined fraction of the oil saturated hydraulic conductivity. In the present analysis, we take this fraction to be 0.0001. Our experience indicates that this scheme provides numerically stable results.

Phase densities are then updated as follows:

$$\rho_w = \rho_w^o \left(1 - \sum_{\alpha=1}^{n_s} \frac{C_{\alpha w}}{\rho_\alpha} \right) + \sum_{\alpha=1}^{n_s} C_{\alpha w} \quad [13a]$$

$$\rho_o = \rho_o^o \left(1 - \sum_{\alpha=1}^{n_s} \frac{C_{\alpha o}}{\rho_\alpha} \right) + \sum_{\alpha=1}^{n_s} C_{\alpha o} \quad [13b]$$

$$\rho_a = \frac{P_{atm}}{RT} \left(\frac{\rho_a^o + \sum_{\alpha=1} C_{\alpha a}}{\rho_a^o / M_a^o + \sum_{\alpha=1} C_{\alpha a}/M_\alpha} \right) \quad [13c]$$

where ρ_w^o is the density of uncontaminated water, ρ_α is the density of pure α-component, ρ_o^o is the density of "inert" oil components, P_{atm} is the atmospheric pressure, R is the universal gas constant, T is temperture (°k), M_a^o, ρ_a^o are the molecular weight and density of pure air, respectively and M_α is the molecular weight of component α. Density ratios, ρ_{rw} and ρ_{ro}, are computed as the ratio of ρ_w or ρ_o to ρ_w^o.

Other fluid parameters such as the scaling parameters β_{ow} and β_{ao} *(Parker et al., 1987)*, and viscosity ratio, η_{ro}, are assumed to be concentration independent constants and are computed as the weighted averages based on the initial mass fractions of inert and noninert components.

The finite element solution procedure used in solving the phase summed transport equation is described by *Kaluarachchi and*

Parker (1990) and implemented in the original multiphase flow code MOFAT *(Kaluarachchi and Parker, 1989)*. The numerical technique involves an upstream weighted finite element method with linear rectangular elements.

3 HYPOTHETICAL SIMULATIONS

Example I: One-dimensional two-component transport without volatilization

A 75 cm long vertical soil column is assumed to be initially free of oil and at hydrostatic equilibrium with a water table located 50 cm from the top surface. A slug of organic liquid containing 28% mass fraction of benzene, 19% toluene and 53% inert components was introduced at the upper surface under a water equivalent head of 1 cm until a total oil accumulation of 1.5 $cm^3 cm^{-2}$ was achieved. The lower boundary was maintained at the initial head for the water phase and the oil phase was specified as zero flux. Boundary conditions for the transport problem were zero dispersive flux at the bottom boundary at all times and a type-3 condition at the upper surface with $C_{\alpha w3}$ being equal to 0.5 and 0.1 mg cm^{-3} for benzene and toluene, respectively. Soil hydraulic and fluid properties used in the simulation are given in Tables 1 and 2. From the input data, the density of the oil entering the system is 832 mg cm^{-3} and the oil phase concentration of the influent oil slug is 247 mg cm^{-3} for benzene and 167 mg cm^{-3} for toluene.

Following the oil infiltration period, a water flux of 0.12 cm h^{-1} was introduced at the upper surface and a zero-flux condition was maintained for the oil phase. Assuming uncontaminated water enters the system and no evaporation occurs, a type-3 condition was imposed at the upper surface during water infiltration with $C_{\alpha w3}=0$ resulting in zero total component flux across the boundary. A finite element mesh of 55 elements was used in the simulation. The maximum time step during the oil redistribution period was 0.5 h with a time incremental factor of 1.03 and an upstream weight of 0.2. To demonstrate the effects of updating for mass transfer rates between phases, the simulation was performed with and without the updating for mass transfer and phase density.

The time required for 1.5 cm of oil to enter the system was 0.051 h and the predicted mass balance error was 0.3%. Figure 1 shows component masses apparently lost from the system during the simulation due to outflow from the lower surface computed from space integrals of component concentrations. Also shown are the global mass balance errors computed as the differences between mass losses computed from concentration space integrals and computed boundary fluxes. It is seen for both components that the mass loss via concentration space integrals is under predicted when mass transfer updating is disregarded resulting in a continuously increasing global mass balance error. For simulations with mass transfer updating, the global mass balance remains small and fluctuated during the 500 h period. During the period of infiltration, the effects of disregarding mass transfer rates was negligible indicating that updating for mass transfer and phase densities are insignificant during short periods.

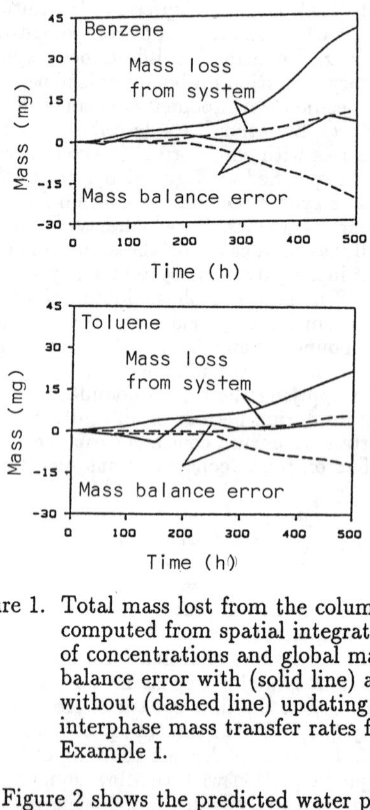

Figure 1. Total mass lost from the column computed from spatial integration of concentrations and global mass balance error with (solid line) and without (dashed line) updating for interphase mass transfer rates for Example I.

Figure 2 shows the predicted water phase benzene and toluene concentrations at the end of the simulation period of 500 h with mass transfer updating. The concentrations are constant at the initially applied concentrations of 0.5 and 0.1 mg cm^{-3} for benzene and toluene respectively over the lower region of the column. Due to the higher solubility of benzene, the extraction front for benzene is approximately 20 cm and corresponds to twice that of toluene.

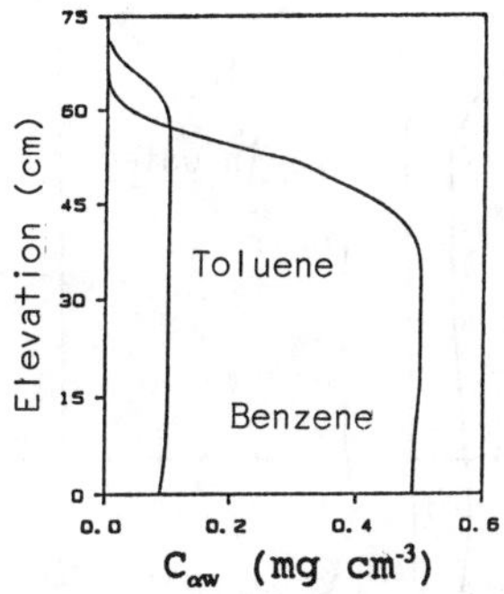

Figure 2 Distribution of water phase concentrations of benzene and toluene at 500 h for Example I with mass transfer updating.

Example II: Two-dimensional transport of dense NAPL with water pumping and volatilization

In the present example, we consider the transport of a spill of pure Trichloroethylene (TCE) during infiltration and redistribution of the organic phase under natural hydraulic conditions followed by removal assisted by water pumping. The flow domain involves a two-dimensional 15 m tall radial section with an inner radius of 0.1 m and an outer radius of 12.1 m. Initial conditions correspond to an air-water system in hydrostatic equilibrium with a horizontal water table located 5 m above the lower surface. Simulation of flow and transport of TCE for this problem was performed in three stages:

1. Infiltration of TCE along a 1.5 m radial section at the center under a water equivalent oil head of 1 cm until total accumulation was 5.9 m^3.
2. Redistribution of TCE for 21 days with zero-flux conditions for oil flow on all boundaries and for the water phase along the unsaturated boundaries.
3. Initiate water pumping on inner radius such that water level in well is maintained at a level 1.5 m below intial water level for a 100 day period.

Boundary conditions for the flow problem are type-1 for the water phase along the water saturated boundary and zero flux elsewhere. For the oil phase, zero flux conditions were stipulated along the entire boundary at all times except along the infiltrating boundary during oil infiltration. A type-3 transport boundary condition was imposed along the infiltrating boundary with $C_{\alpha o3}$ = 1470 mg cm^{-3} and zero thereafter. A zero dispersive flux condition was stipulated along the remaining boundary except along the water saturated right boundary where a type-3 condition was specified with $C_{\alpha w3}$ =0 to allow uncontaminated water to enter the system. The soil hydraulic and fluid properties used in the simulation are given in Tables 1 and 2.

Table 1. Soil hydraulic and bulk fluid properties*

Parameter	Example I	Example II
ϕ	0.4	0.4
n	2.5	2
S_m	0	0
α [L^{-1}]	0.05	0.8
K_{swx} [LT^{-1}]	–	2
K_{swy} [LT^{-1}]	40	2
ρ_{ro}	0.8	1.47
η_{ro}	2	0.58
β_{ao}	2.69	2.34
β_{ow}	1.59	3.02
ϵ_l[L]	0.4	0.2
ϵ_t[L]	–	0.02

* Units for Example I in cm and h and m and d for Example II.

Table 2. Properties of organic components used in the simulations

Parameter	Benzene	Toluene	TCE
$D^o_{\alpha w}$ (cm^{2}h^{-1})	0.04	0.04	0.04
$\Gamma_{\alpha o}$	494	1674	1336
$\Gamma_{\alpha a}$	0.17	0.26	0.38
$\Gamma_{\alpha s}$	0	0	0
ρ_α (mg cm^{-3})	877	862	1470
M_α (g mole^{-1})	78	92	131
η^α_{ro}	2	2	0.58

The predicted results indicates that 5.9 m^3 of TCE entered the system in 1.13 days. The oil saturation contours at 121 d are shown in Figure 3 where it is seen that minimal saturation occurs at the capillary fringe with a lateral spread towards the lower right boundary. Water pumpage from the center well produced an increase in TCE saturation towards the center of the domain. The predicted water and gas phase saturations at 121 d are shown in Figure 4 respectively. The concentrations shown in Figure 4 indicate a decrease in water phase concentration of TCE towards the lower right boundary due to the water entering the domain from the water saturated right boundary under the imposed gradient. Due to the mild impedance caused by the atmospheric boundary layer, a fairly sharp gas phase concentration gradient occurs near the soil surface. Radial spreading of the gas phase concentration of TCE by diffusion alone is little beyond the liquid plume.

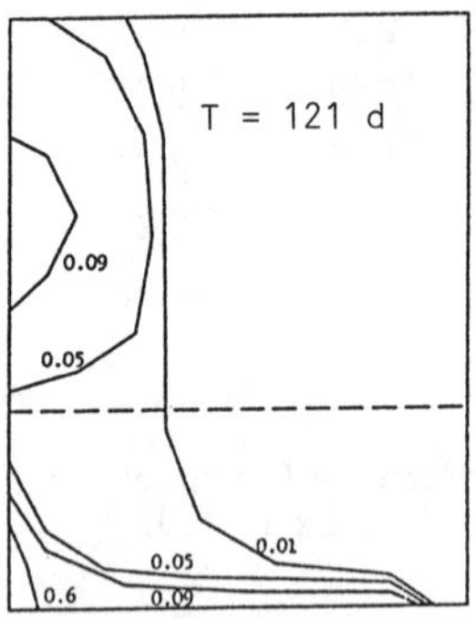

Figure 3 Oil saturations at 121 days for Example II

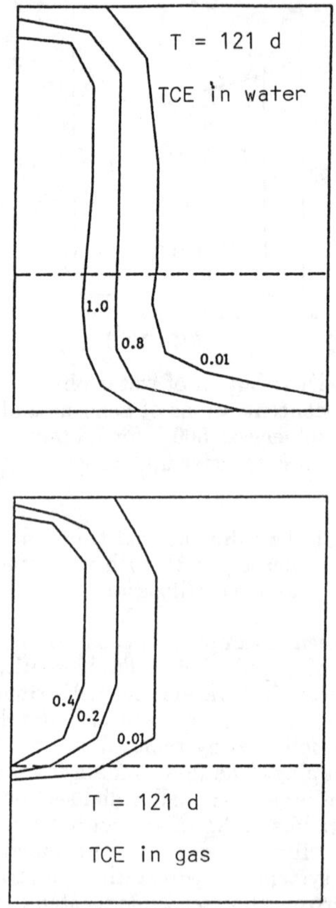

Figure 4 Predicted water and gas phase concentrations (mg cm^{-3}) of TCE at 121 days for Example II.

The cumulative mass of TCE removed from the aqueous and gas phases during the 121 d period are shown in Figure 5. During the early stages of the redistribution, TCE is removed from the gas phase due to volatilization from the upper surface with a maximum rate of 0.18 kg d^{-1} between 20-50 d and gradually decreasing to 0.06 kg d^{-1}. Minimal TCE was removed from the water phase via convective water fluxes during the initial 21 d in the absence of water gradient. The removal rate however changed after 21 d due to the imposed gradient at the well where the removal rate increased to 0.47 kg d^{-1}. The rate was almost constant during the final 100 d since the water phase concentration of TCE was at solubility and the water fluxes were almost constant. The simulation predicts a net removal of TCE from the aqueous and gas phases to be 0.6 and 0.12% respectively over

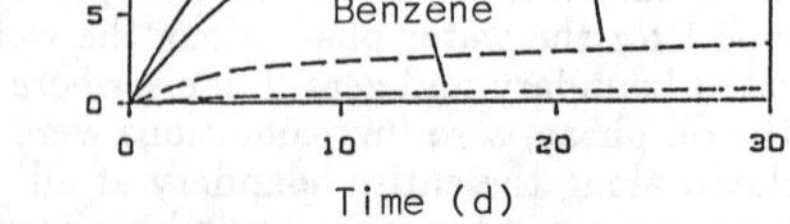

Figure 5 Mass of TCE removed from system via water and gas phases in Example II.

the entire simulation period of 121 days indicating very poor removal efficiency from both phases under the proposed scheme.

4 CONCLUSIONS

A two-dimensional finite element model to predict multicomponent transport in a three fluid phase porous media is presented. The assumption of local equilibrium phase partitioning between water, oil, gas and solid phases reduces the computational effort substantially by reducing the set of transport equations to a single phase summed equation for each partitionable component. Results suggest that updating for interphase mass transfer rates and phase densities play an important role during long term behavior of hydrocarbon transport with negligible importance during periods of highly transient flow. The results also suggest that removal of dense organics from a spill site could be very slow if remediation schemes involve only aqueous phase removal via water pumping and surface volatilization through gas diffusion only.

ACKNOWLEDGEMENTS

This work was sponsored by the U.S. EPA R. S. Kerr Environmental Research Laboratory under cooperative agreement CR-812073-01.

REFERENCES

Abriola, L. M. 1984. Multiphase migration of organic compounds in a porous medium - A mathematical model, Lecture Notes in Engineering, Vol. 8, C. A. Brebbia and D. A. Orszag (eds.), Springer-Verlag, New York, 232p.

Corapcioglu, M. Y. and A. L. Baehr. 1987. A compositional multiphase model for ground water contamination by petroleum products 1. Theoretical considerations, Water Resour. Res. 23:191-200.

Kaluarachchi, J. J. and J. C. Parker. 1989. An efficient finite element method for modeling multiphase flow, Water Resour. Res. 25: 43-54. 1989.

Kaluarachchi, J. J. and J. C. Parker. 1990. Modeling multicomponent organic chemical transport in three fluid phase porous media. in press. J. Cont. Hydrology.

Parker, J. C., R. J. Lenhard and T. Kuppusamy. 1987. A parametric model for constitutive properties governing multiphase fluid flow in porous media. Water Resour. Res. 23:618-624.

Parker, J. C. 1989. Multiphase transport in porous media. Reviews of Geophysics. 27:311-328.

Reeves, H. W. and L. M. Abriola. 1988. A decoupled approach to the simulation of flow and transport of non-aqueous organic phase contaminants through porous media, Proc. VII International Conference on Compuational Methods in Water Resources. MIT. USA.

Sleep, B. E. and J. F. Sykes. 1989. Modeling the transport of volatile organics in variably saturated media, Water Resour. Res. 25: 81-92.

Subsurface Contamination by Immiscible Fluids, Weyer (ed.) © 1992 Balkema, Rotterdam. ISBN 90 6191 175 3

The influence of infiltrating wetting fronts on transport of volatile organic compounds in variably saturated porous media

B.E. Sleep & J.F. Sykes
Department of Civil Engineering, University of Waterloo, Ont., Canada

ABSTRACT: The fate of volatile organic compounds in variably saturated media is examined using a multidimensional finite difference model of transient, multiphase flow and transport in porous media. Air, oil, and water flow are determined using fully coupled equations incorporating capillarity, gravity, and pressure and concentration dependent gas phase density. Partitioning of organic compounds between organic, aqueous and air phases is included. Simulations show the influence of infiltrating wetting fronts on the transport of volatile organic compounds.

1 INTRODUCTION

Residual quantities of volatile organic compounds, immobilized in the vadose zone, may provide long term sources of groundwater contamination. These compounds volatilize into the soil gas phase, and dissolve into the soil water phase. They may then move by advective and diffusive transport in the water and gas phases, and may also move between the water and gas phases by gas-liquid partitioning.

Abriola and Pinder (1985), and Baehr and Corapcioglu (1987) identified the importance of diffusive transport of volatile organic compounds in the vadose zone. The models developed by these authors assumed that gas phase pressures were uniformly atmospheric, and therefore, that there was no advective movement of the gas phase. Sleep and Sykes (1989) demonstrated that significant gas phase advective transport could occur as a result of gas phase density gradients. The effect of a transient water phase on gas phase transport was not examined.

The importance of the gas phase in reducing infiltration rates has been demonstrated by Morel-Seytoux and Billica (1985), and Touma and Vauclin (1986). Conversely, infiltrating wetting fronts may cause significant pressure gradients in the gas phase, resulting in very substantial gas phase advection and subsequent increased subsurface contamination.

In this study a multidimensional transient multiphase flow and transport model is presented for the analysis of the transport of volatile organic compounds in variably saturated media. Air, oil and water flow are determined using fully coupled equations incorporating capillarity, gravity ,and pressure and concentration dependent gas phase density. Partitioning of organic compounds between organic, aqueous and air phases is included.

Results of simulations show the influence of infiltrating wetting fronts on the transport of volatile organic compounds in the subsurface. These results are compared to the steady state water flow case where gas phase transport is caused by diffusion and density gradients only.

2 EQUATIONS OF MULTIPHASE FLOW AND TRANSPORT

The equation describing the movement of a species α, in phase β is

$$\frac{\partial}{\partial t}\left(\theta S_\beta \rho_\beta x_\beta^\alpha\right) + \nabla \cdot \left(\rho_\beta x_\beta^\alpha v_\beta\right) - \nabla \cdot \left(\theta S_\beta \rho_\beta D_\beta^\alpha \nabla x_\beta^\alpha\right) - r_\beta^\alpha - q_\beta^\alpha = 0 \qquad (1)$$

where θ is porosity; S_β is phase saturation; ρ_β is molar density; x_β^α is species mole fraction; v_β is Darcy velocity vector; D_β^α is the dispersion tensor; r_β^α represents interphase transfer of species α to, or from, phase β; and q_β^α represents sinks and sources of species α to phase β.

The Darcy velocity vector is given by

$$v_\beta = -\frac{kk_{r\beta}}{\mu_\beta}\left(\nabla p_\beta + \rho_\beta g \nabla z\right) \tag{2}$$

where k is intrinsic permeability tensor; $k_{r\beta}$ is relative permeability; μ_β is the viscosity; p_β is the phase pressure; g is gravitational acceleration; and z is elevation.

The following constraints apply to equations (1) and (2):

$$\sum_{\beta=1}^{3} S_\beta = 1.0 \tag{3}$$

$$\sum_{\alpha=1}^{N} x_\beta^\alpha = 1.0 \tag{4}$$

$$p_g - p_w = p_{cgw}(S_w) \tag{5}$$

$$p_o - p_w = p_{cow}(S_w) \tag{6}$$

$$p_g - p_o = p_{cgo}(S_t) \tag{7}$$

where $S_t = S_w + S_o$.

Summing the equation for a species α over all the phases gives the species mass balance

$$\sum_{\beta=1}^{3}\left[\frac{\partial}{\partial t}\left(\theta S_\beta \rho_\beta x_\beta^\alpha\right) + \nabla\cdot\left(\rho_\beta x_\beta^\alpha v_\beta\right) - \nabla\cdot\left(\theta S_\beta \rho_\beta D_\beta^\alpha \nabla x_\beta^\alpha\right) - q_\beta^\alpha\right] = 0 \tag{8}$$

Summing the equation for a phase β over all the species gives the phase mass balance

$$\frac{\partial}{\partial t}\left(\theta S_\beta \rho_\beta\right) + \nabla\cdot\left(\rho_\beta v_\beta\right) - \sum_{\alpha=1}^{N}\left[r_\beta^\alpha + q_\beta^\alpha\right] = 0 \tag{9}$$

Interphase mass transfer of organic occurs by dissolution and volatilization of the organic phase and partitioning of organic between the gas and water phases. In a zone containing residual organic phase the mole fraction of organic compound in the vapour phase can be determined from

$$x_g^\alpha = \frac{p^\alpha}{p_g} \tag{10}$$

p^α is the vapour pressure of the pure organic compound.

The mole fraction of organic compound in the water phase can be calculated from the aqueous phase solubility of the compound.

In zones where no organic phase is present Henry's law can be used to relate the organic concentrations in the water and gas phases

$$x_g^\alpha = \frac{H}{p_g} x_w^\alpha \tag{11}$$

Kinetic mass transfer expressions such as those used by Sleep and Sykes (1989) can also be employed to represent interphase mass transfer processes.

Evaporation of water into the gas phase and dissolution of air in the water phase are assumed to be negligible in the present study.

3 THE NUMERICAL MODEL

Conservation equations for water and gas phases and organic species are discretized using block centred finite differences (Aziz and Settari, 1979). The species equation for a block of volume V_i and time increment Δt from $N-1$ to N is

$$\begin{aligned}
&\frac{V_i}{\Delta t}\sum_{\beta=1}^{3}\left[\left(\theta S_\beta \rho_\beta x_\beta^\alpha\right)^N - \left(\theta S_\beta \rho_\beta x_\beta^\alpha\right)^{N-1}\right] \\
&= \frac{V_i}{\Delta x_i}\sum_{\beta=1}^{3}\left[\left(T_\beta \rho_\beta x_\beta^\alpha\right)_{i+1/2}^N \psi_{\beta,i+1/2}^N (1 \right. \\
&\quad - \left(T_\beta \rho_\beta x_\beta^\alpha\right)_{i-1/2}^N \psi_{\beta,i-1/2}^N \\
&\quad + \left(\theta S_\beta \rho_\beta D_\beta^\alpha\right)_{i+1/2}\left(\frac{2}{\Delta x_{i+1} + \Delta x_i}\right)\left(x_{\beta,i+1}^\alpha - x_{\beta,i}^\alpha\right) \\
&\quad \left. - \left(\theta S_\beta \rho_\beta D_\beta^\alpha\right)_{i-1/2}\left(\frac{2}{\Delta x_i + \Delta x_{i-1}}\right)\left(x_{\beta,i}^\alpha - x_{\beta,i-1}^\alpha\right)\right] \\
&\quad + \text{ y and z terms}
\end{aligned}$$

The term $\psi_{\beta,i+1/2}^N$ is given by

$$\psi_{\beta,i+1/2}^N = p_{\beta,i+1}^N - p_{\beta,i}^N + \rho_{\beta,i+1/2}^N g\left(z_{i+1} - z_i\right) \tag{13}$$

The term Δx_i is the x dimension of the ith finite difference block and

$$[T]_{i\pm1/2}^N = \left(\frac{2}{\Delta x_{i\pm1} + \Delta x_i}\right)\left(\frac{k_{i\pm1/2} k_{r\beta}^N}{\mu_\beta}\right) \tag{14}$$

Relative permeabilities are upstream weighted. $k_{i\pm1/2}$ is the harmonic mean of the permeabilities between blocks i and $i\pm1$. Implicit temporal discretization is used to ensure stability of the solution process. Gas phase density is calculated from pressure and composition, assuming ideal gas behavior.

The capillary pressure equations are used to eliminate oil and gas phase pressures, while the gas phase saturation is eliminated by using the saturation constraint equation. In blocks with organic phase present, water and gas phase organic concentrations can be calculated using equilibrium, or kinetic mass transfer relationships. The primary variables solved for are water pressure, water saturation, and total saturation. In blocks with no organic phase present, the primary variables are water pressure, water saturation, and gas phase organic concentration when the gas phase is present, and water phase concentration when the system is water saturated. The gas-water partitioning equations are used to calculate the water phase organic concentrations from gas phase concentrations in the two phase air-water system.

Boundary conditions are specified by using the sink/source formulation of Forsyth (1988). Pressures, saturations, concentrations, or flux rates may be specified. In boundary blocks with prescribed air pressures, air pressure is used in place of water pressure as the primary variable.

The Newton Raphson method is used to solve the system of non-linear coupled equations. This results in a sparse, nonsymmetric system of linear equations with three equations for each finite difference node. The linear equations are solved using block incomplete lower-upper (ILU), second degree factorization with orthomin acceleration (Behie and Vinsome, 1982) An additional reduction of computational effort is achieved by using alternating diagonal ordering (Behie and Forsyth, 1984).

4 MODEL VERIFICATION

Rogers' (1983) solution for vertical displacement of a wetting fluid by a nonwetting fluid was modified to describe non-wetting fluid displacing wetting fluid for model verification.

The relative permeability-saturation relationships used in equation (2) are

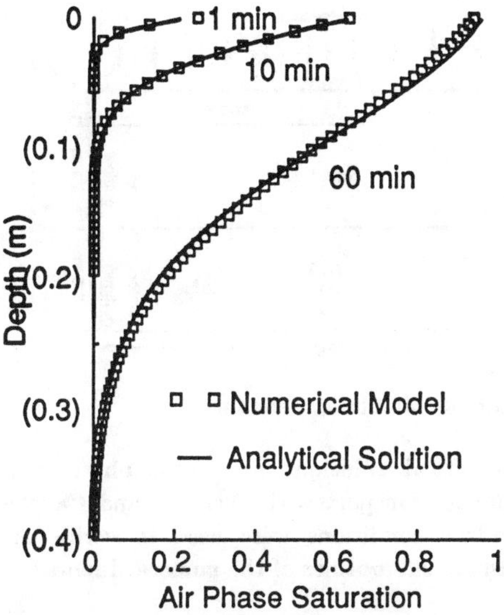

Figure 1: Model verification, air displacing water

$$k_{rw} = S_w - S_{rw} \tag{15}$$

$$k_{rnw} = 1 - S_w - S_{rnw} \tag{16}$$

where k_{nrw} is the relative permeability of the non-wetting phase, and S_{rw} and S_{rnw} are the irreducible saturations of the wetting phase and non-wetting phases. The capillary pressure function, allowing the analytical solution, is of the form

$$p_c(S_w) = a\ ln\Big[(S_w - S_{rw})^b \times (S_w - 1 + S_{rnw})^{-c}(S_w + d)^e\Big] + f \tag{17}$$

where a, b, and f are arbitrary constants and c, d, and e are functions of viscosities and irreducible saturations.

The results of simulations of air displacing water for the numerical and analytical models, in Figure 1, shows the excellent agreement between the solutions. Water and air phase viscosities were 1 and 0.8 cP respectively. The large air phase viscosity was used to meet capillary pressure function constraints of the analytical solution. Fluid densities were 1000 and 1.3 kg/m^3 for the water and air phases. The permeability and infiltration rates were 1x10^{-12} m^2 and 1x10^{-5} m/s. S_{rw} and S_{rnw} were zero.

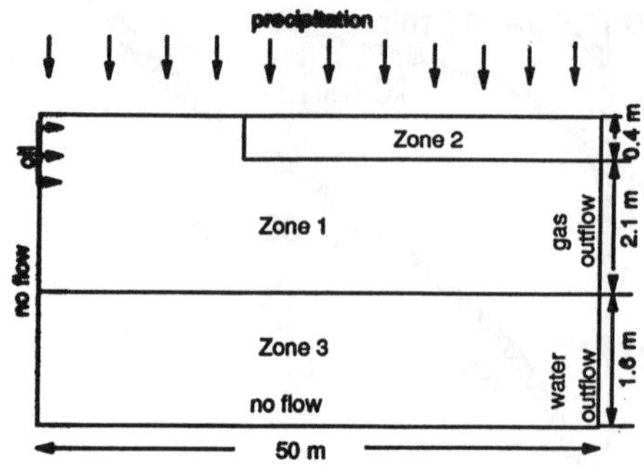

Figure 2: Simulation domain

Analytical solutions for single phase contaminant transport with Dirichlet and Cauchy boundary conditions were used to verify the transport subroutines of the numerical model.

5 TWO-DIMENSIONAL SIMULATIONS

Two-dimensional simulations of the transport of trichloroethylene were performed to demonstrate the effects of infiltrating wetting fronts on the transport of volatile organic compounds.

Trichloroethylene has a molecular weight of 131.4 g/gmole, a vapour pressure is 58 mm. Hg, and an aqueous phase solubility of 1100 ppm. at 20 deg C. The diffusion coefficients in air and water are $6x10^{-6}$ and $1x10^{-9}$ m^2/s respectively.

The simulation domain is illustrated in Figure 2. The domain was divided into three zones. The intrinsic permeabilities in zones 2 and 3 were $10^{-14}m^2$ for all simulations while the intrinsic permeability in zone 1 was varied. The left vertical and lower horizontal boundaries are no flow for all phases. Hydrostatic gas pressures are prescribed on the right vertical boundary in the unsaturated zone, while hydrostatic water pressures are prescribed below the water table.

Trichloroethylene concentrations (equal to 90 % of the saturated vapour phase concentrations at 20 deg C) are prescribed on the upper portion of the left boundary from 0.3 to 0.54 m. depth. At the ground surface, no flux conditions are prescribed for air and trichloroethylene when water infiltration is occurring. For the static water phase simulations, gas phase pressures at the ground surface are prescribed as atmospheric. A Cauchy boundary condition simulating diffusion through a stagnant boundary layer of 0.01 m. thickness (Sleep and Sykes, 1989) is used for trichloroethylene transport across the ground surface boundary.

The nonhysteretic capillary pressure saturation and relative permeability saturation relationships of Parker et. al. (1987) are used. The relationships for capillary pressures are

$$\begin{aligned}\overline{S}_\beta =& \ (S_\beta - S_r) / (1 - S_r) & \\ =& \ [1 + (\alpha\beta_{ij} p_{cij})^n]^{-m} & p_{cij} > 0 \\ =& \ 1 & p_{cij} \leq 0\end{aligned}$$

The relative permeability relationships for water and gas phases are

$$k_{rw} = \overline{S}_w^{1/2}\left[1 - \left(1 - \overline{S}_w^{1/m}\right)^m\right]^2$$

$$k_{rg} = \overline{S}_g^{1/2}\left(1 - \overline{S}_t^{1/m}\right)^{2m}$$

In all cases the values of these parameters were: $\alpha = 2.0$; $n = 2.0$; $S_r = 0.0$; and $\beta_{aw} = 1$.

The Millington-Quirk (Millington and Quirk, 1961) relationship was used to to determine the effective gas phase diffusion coefficient.

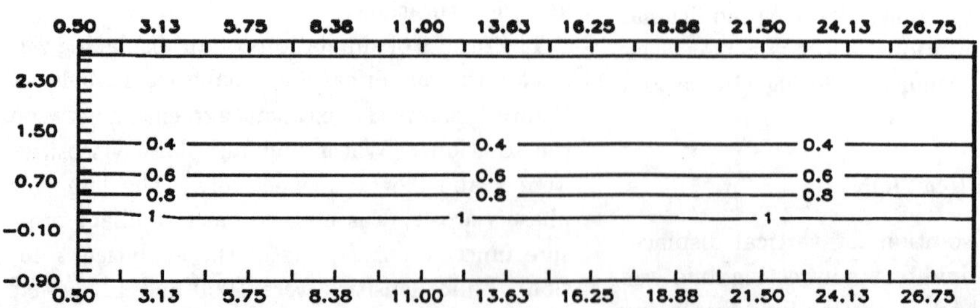

Figure 3: Initial water phase saturations

* author's note: The datum elevation in Figures 3-11 was chosen to coincide with the water table, so that the ground surface was at 3.1 m above datum, and the bottom of the system was 0.9 m below datum.

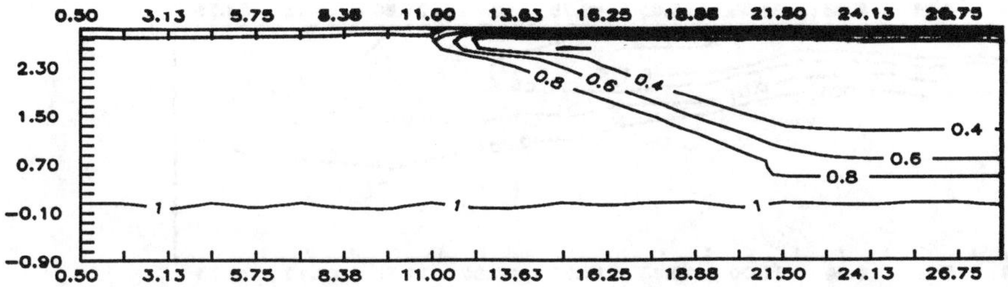

Figure 4: Water phase saturations after 6 days of infiltration ($k = 10^{-11}m^2$)

In the first case the permeability of zone 1 was $10^{-11}m^2$ and the water pressure at the ground surface was prescribed to simulate 1 cm of ponded water across the entire domain. The initial water saturations, and those after 6 days are shown in Figures 3 and 4, respectively (only the left 30 m are shown). The low permeability layer on the right of the domain reduced infiltration leaving lower water saturations. The higher infiltration on the left side of the domain increased the water saturations to near unity. The entrapment of gas by the infiltrating water prevented the attainment of fully saturated conditions, leaving very small gas saturations.

The gas phase concentrations (relative to the source concentration) after 6 days of infiltration are plotted in Figure 5. In the no-infiltration case (density-driven advection only) the 0.001 contour extended laterally to about 3 m. after 6 days (not shown), about half the distance reached in Figure 5. The concentrations after 80 days with no infiltration are shown in Figure 6. The lateral movement of trichloroethylene (TCE) at 80 days was only 1.5 times the movement after 6 days of water infiltration.

As a result of the large aqueous solubility of TCE, relative to its gaseous solubility, transport due to gas phase advection and diffusion is limited at high water saturations, from the retarding effect of dissolution in the stagnant water phase. To further examine gas phase mobilization by infiltrating water the aqueous solubility was re-

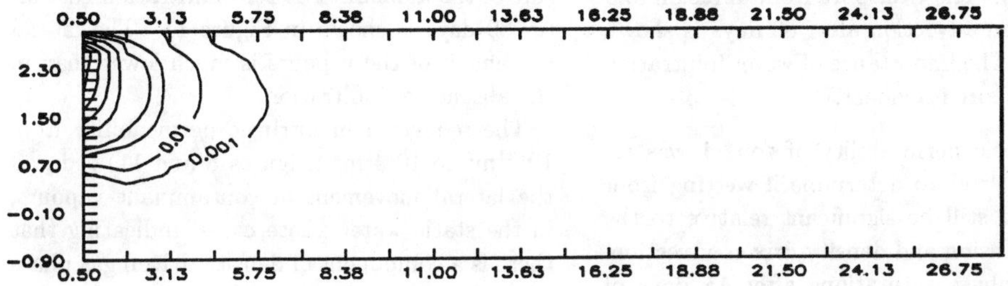

Figure 5: Gas phase concentrations after 6 days of infiltration (aqueous solubility = 1100 ppm, $k = 10^{-11}m^2$)

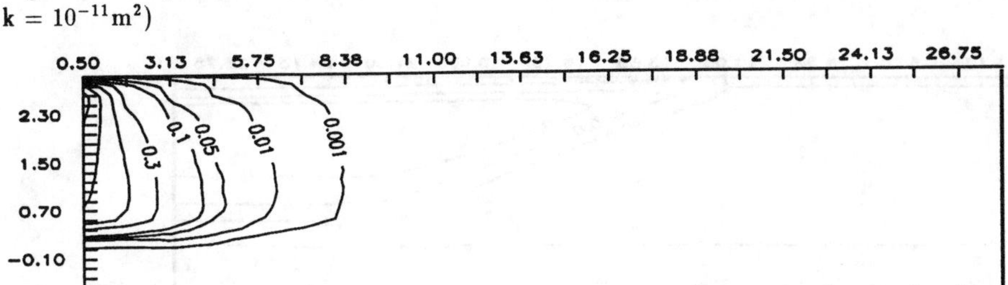

Figure 6: Gas phase concentrations at 80 days, no infiltration (aqueous solubility = 1100 ppm, $k = 10^{-11}m^2$)

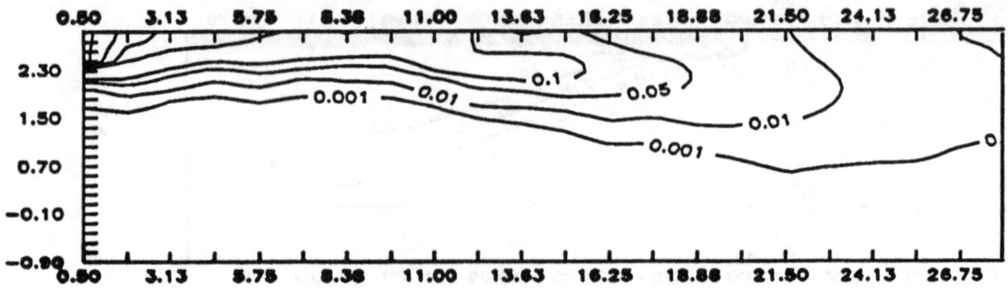

Figure 7: Gas phase concentrations after 36 hours of infiltration (aqueous solubility = 5 ppb, $k = 10^{-11} m^2$)

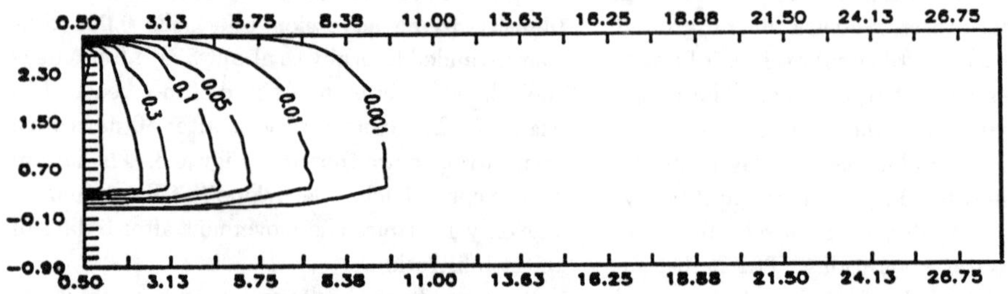

Figure 8: Gas phase concentrations at 80 days, no infiltration (aqueous solubility = 5 ppb, $k = 10^{-11} m^2$)

duced to 5 ppb for subsequent cases to represent less water soluble compounds.

Figure 7 shows gas phase concentrations after 36 hours of water infiltration with reduced aqueous solubility. The transport from diffusion and density driven advection after 80 days is shown in Figure 8. The importance of water infiltration for gas transport is evident.

The intrinsic permeability of zone 1 was reduced to $10^{-12} m^2$ to determine if wetting front effects would still be significant relative to the effects of diffusion and density driven advection. The water phase saturations after 15 days of infiltration are plotted in Figure 9. The gas phase concentrations are in Figure 10. There is still significant lateral movement of the vapours. A small amount of contaminant vapour is also present at depth in the trapped gas on the left side of the domain. The zero infiltration case after 80 days is shown in Figure 11. The lateral movement of the vapours is much lower than in the absence of infiltration.

The reduction in intrinsic permeability from $10^{-11} m^2$ to $10^{-12} m^2$ (Figures 8 and 11) reduced the lateral movement of contaminant vapours, in the static water phase cases, indicating that there is a component of density driven gas phase advection also.

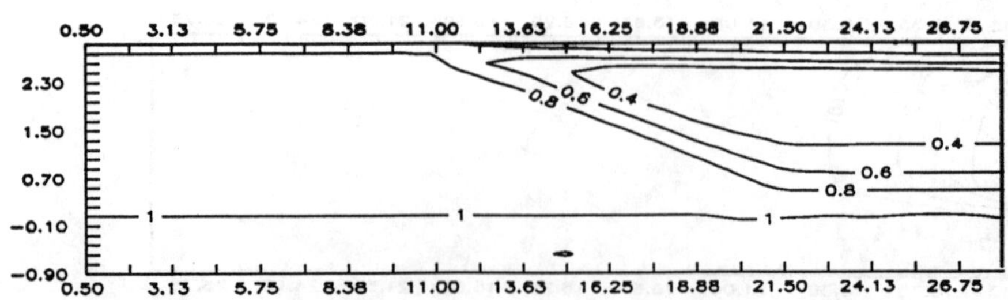

Figure 9: Water phase saturations after 15 days of infiltration ($k = 10^{-12} m^2$)

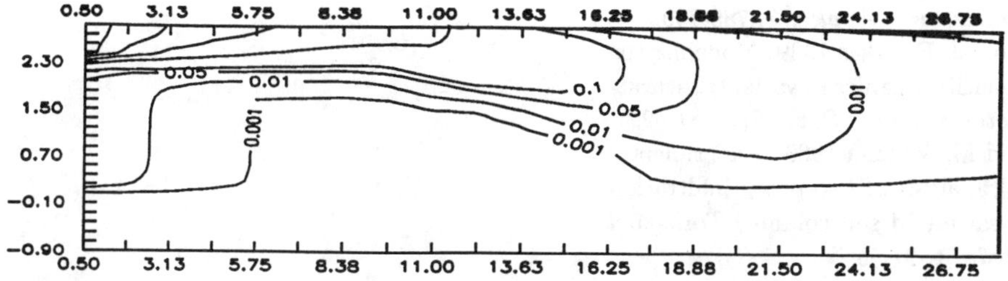

Figure 10: Gas phase concentrations after 15 days of water infiltration (aqueous solubility = 5 ppb, $k = 10^{-12} m^2$)

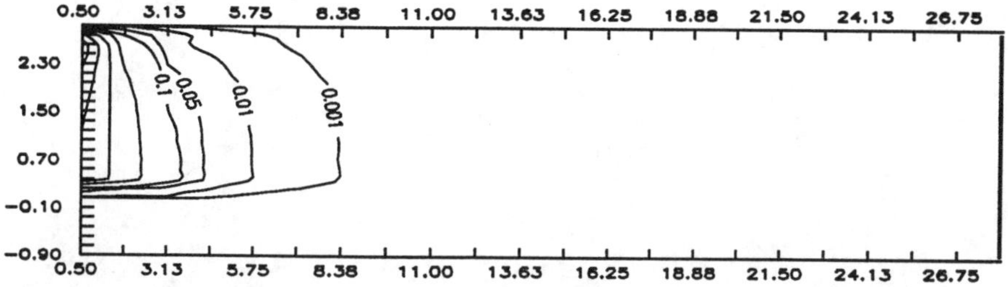

Figure 11: Gas phase concentrations at 80 days, no infiltration (aqueous solubility = 5 ppb, $k = 10^{-12} m^2$)

6 CONCLUSIONS

The infiltration of wetting fronts may cause significant transport of volatile organic compounds in the vadose zone. The water phase plays an important role, not only in causing gas phase advection, but also in absorbing water soluble vapours from the gas phase. The total mobility of a volatile organic compound depends on infiltration rates, water phase saturations, and solubilities of the compound in the water and gas phases.

REFERENCES

Abriola, L. M., and G. F. Pinder 1985. A multiphase approach to the modeling of porous media contamination by organic compounds, 2. Numerical simulation. *Water Resour. Res.* 21(1), 19-26.

Aziz, K. and A. Settari 1979. *Petroleum Reservoir Simulation.* Applied Science, London.

Baehr, A. L., and M. Y. Corapcioglu 1987. A compositional multiphase model for groundwater contamination by petroleum products, 2. numerical solution. *Water Resour. Res.* 23(1),201-213.

Behie, A., and P. A. Forsyth 1984. Incomplete factorization methods for fully implicit simulation of enhanced oil recovery. *SIAM J. Sci. Stat. Comp.* 5, 543-561.

Behie, A., and P. K. W. Vinsome 1982. Block iterative methods for fully implicit reservoir simulation. *Soc. Pet. Eng. J.* 22, 658-668.

Forsyth, P. A. 1988. Simulation of nonaqueous phase groundwater contamination. *Adv. Water Resour.* 11(2), 74-83.

Millington, R. J., and J. M. Quirk 1961. Permeability of porous solids. *Trans. Faraday Soc.*, 57, 1200-1207.

Morel-Seytoux, H. J., and J. A. Billica 1985. A two-phase numerical model for prediction of infiltration: case of an impervious bottom. *Water Resour. Res.* 21(4).

Parker, J. C., R. J. Lenhard, and T. Kuppusamy 1987. A parametric model for constitutuve properties governing multiphase flow in porous media. *Water Resour. Res.* 23(4), 618-624.

Rogers, C., M. P. Stallybrass, and C. L. Clements 1983. On two phase infiltration under gravity and with boundary infiltration: application of a backlund transformation. *Nonlinear Anal.*,

Theory, Methods, and Applic. 7, 785-799.
Sleep, B. E., and J. F. Sykes 1989. Modeling the tranpsort of volatile organics in variably saturated media. *Water Resour. Res.* 25(1), 81-92.
Touma, J., and M. Vauclin 1986. Experimental and numerical analysis of two-phase infiltration in a partially saturated soil column, *Transport in Porous Media* 1, 27-55.

Subsurface Contamination by Immiscible Fluids, Weyer (ed.) © 1992 Balkema, Rotterdam. ISBN 90 6191 175 3

Application of a numerical model and laboratory parameter measurements to the movement of dense, immiscible phase liquids in a heterogeneous sand aquifer

B.H. Kueper
Waterloo Centre for Groundwater Research, University of Waterloo, Ont., Canada (Presently: Department of Civil Engineering, Queen's University, Kingston, Ont., Canada)

E.O. Frind
Waterloo Centre for Groundwater Research, University of Waterloo, Ont., Canada

D.B. McWhorter
Department of Chemical and Agricultural Engineering, Colorado State University, Fort Collins, Colo., USA

ABSTRACT: In this study, a two-dimensional numerical model is utilized to simulate the movement of tetrachloroethylene below the watertable in a heterogeneous sand aquifer. The distribution of permeability is represented as a log-normal, spatially correlated random field for which the statistics have been determined. The spatial distribution of capillary properties is represented by scaling a suite of laboratory-measured tetrachloroethylene-water capillary pressure curves to permeability. Simulations are performed using a grid of sufficient resolution such that the detailed, small-scale variation of aquifer properties is accounted for. Additional simulations are performed where it is assumed that the aquifer is homogeneous and that the aquifer properties can be represented by a single set of laboratory-scale parameters. A comparison of the simulation results demonstrates the errors associated in representing the field-scale aquifer as a homogeneous medium characterized by laboratory-scale data. In particular, the homogeneous simulations greatly underestimate the overall degree of lateral spreading associated with the migrating tetrachloroethylene body, even if the large-scale anisotropy of the saturated hydraulic conductivity field is accounted for.

1 INTRODUCTION

In recent years, there has been an increased awareness of groundwater contamination by denser-than-water non-aqueous phase liquids (NAPLs) such as PCB oils, creosotes, and chlorinated solvents. These liquids have low solubilities in both air and water, and therefore migrate as a separate liquid phase in the subsurface. In partially saturated media, NAPL components partition to the air phase, giving rise to evolving vapour plumes which result in a spread of contamination beyond the region of liquid-phase NAPL alone. Below the watertable, NAPL components partition to flowing groundwater, resulting in the potential for dissolved contaminant migration far from NAPL source areas. In the most general case, the subsurface contamination scenario presented to contaminant hydrogeologists involves the behaviour of an arbitrary number of chemical components distributed amongst the soil, water, air and NAPL phases.

Laboratory and field experimentation involving NAPLs is often hampered by the hazardous nature of these chemicals. In the U.S.A., for example, it is prohibited to release large amounts of NAPL into the subsurface for experimental purposes. As a result, field observations of subsurface NAPL behaviour are usually restricted to indirect evidence from accidental spill and leak sites where little is known regarding the exact volumes, time periods, and manner in which NAPL was introduced into the subsurface. A properly formulated numerical model, however, can provide a tool with which to study specific subsurface processes involving NAPLS in a controlled manner. In particular, sensitivity and uncertainty analyses can be performed which would otherwise be difficult to carry out.

This study focusses on the application of a numerical model to the behaviour of a dense NAPL below the watertable in a heterogeneous sand aquifer. The NAPL is assumed to have only a slight solubility in water such that the dissolution process can be neglected over the time scales at which NAPL migration takes place. There exist at present a variety of models in the contaminant hydrogeology literature capable of simulating two-phase flow through porous media (e.g., Faust 1985, Abriola and Pinder 1985 a,b, Osborne and Sykes 1986, Kuppusamy et al. 1987, Forsyth 1988, Kaluarachchi and Parker 1989, Faust et al. 1989).

In general, these models have not been applied to study specifically the influence of porous media heterogeneity on the movement of NAPL below the watertable.

The application of a multiphase flow model to a heterogeneous field-scale aquifer requires that input data be provided at a scale consistent with that used to discretize the numerical solution domain. If small samples of porous media are collected to determine the aquifer properties governing multiphase flow at the laboratory scale, there is no justification in using such data to represent the aquifer properties at a coarser scale of numerical simulation. To perform coarser-scale simulations, effective large-scale parameters must be defined which represent appropriate averages over scales below that of the numerical discretization. Unfortunately, large-scale parameters governing multiphase flow have not yet been rigorously defined except for selected cases (e.g., Yeh et al. 1985, Mantoglou and Gelhar 1987). As a result, it is tempting to assume that the porous medium of interest is homogeneous, and to utilize a single set of laboratory-scale measurements as input parameters.

In this paper we demonstrate the errors associated by assuming that a field aquifer is homogeneous and represented by a single set of laboratory-scale input data. This is accomplished by comparing the results of three sets of numerical simulations involving the migration of tetrachloroethylene below the watertable. In the first set of simulations, the centimeter-scale variation in permeability and capillary characteristics of the aquifer are accounted for. In the second set of simulations, the aquifer is assumed to be both homogeneous and isotropic. In the third set of simulations, the aquifer is assumed to be homogeneous, but the large-scale anisotropy of the saturated hydraulic conductivity field is accounted for. In the latter two simulations, the aquifer is assumed to be represented by laboratory-scale capillary pressure and relative permeability functions.

2 AQUIFER CHARACTERIZATION

The aquifer of interest is located at Canadian Forces Base Borden near Alliston, Ontario. The portion of the aquifer selected for study is comprised of horizontally bedded, fine to medium grained beach sands, and varies in thickness between 9 - 10 meters. The watertable is generally within 1 - 2 meters below groundsurface, and a relatively uniform hydraulic gradient of 0.004 exists in the north-east direction (MacFarlane et al., 1983).

In general, the aquifer properties that must be characterized with regard to two-phase flow include porosity, permeability, wettability, and the functional relationships between capillary pressure and saturation, and relative permeability and saturation. In this study, we adopt the porosity measurements of Mackay et al. (1986) and the permeability measurements of Sudicky (1986). Mackay et al. (1986) also report that the Borden aquifer exhibits a low organic carbon content (0.02% average). As a result, we will assume that water is perfectly wetting with respect to NAPL at all locations within the aquifer.

To complete the characterization of the aquifer, we complement the above data assemblages by a set of drainage tetrachloroethylene-water capillary pressure - saturation measurements performed on selected samples of Borden sand. The samples were sub-sectioned from a series of vertical cores taken approximately 60 meters east of the core transects taken by Sudicky (1986). Each sectioned sample had a volume of approximately 100 cc, and was selected from within what appeared visually to be a single bed of material void of sub-scale bedding features. In total, seven drainage capillary pressure - saturation curves were measured, each on a sample of different permeability sand. Figure 1 illustrates the seven capillary pressure curves measured, along with the associated sand permeabilities. The figure clearly illustrates the vertical upwards shift of the displacement pressure

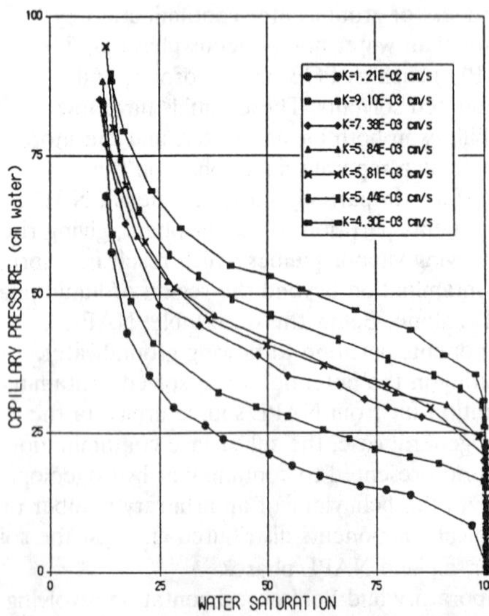

Figure 1 Tetrachloroethylene-water capillary pressure curves measured on selected samples of Borden sand (from Kueper, 1989)

for sands of lower permeability. The details of the capillary pressure measurements are given by Kueper (1989).

The capillary pressure curves illustrated in Figure 1 were scaled to interfacial tension and permeability using a modified form of the function proposed by Leverett (1941):

$$P_{C_D} = \frac{P_C}{\sigma}\left(\frac{k}{\phi}\right)^{\alpha} \tag{1}$$

where P_{CD} is a scaled capillary pressure, P_C is the capillary pressure of interest, σ is the interfacial tension between the liquids of concern, k is the medium permeability, ϕ is the porosity, and α an exponent particular to a given sedimentary environment. The best-fit value of α for the illustrated data was found to be 0.65.

In order to represent the capillary pressure data in smooth functional form, a Brooks-Corey (Brooks and Corey, 1966) capillary pressure - saturation function was fitted to the scaled data. The Brooks-Corey capillary pressure function is given by

$$P_C = P_D\,(S_e)^{\frac{-1}{\lambda}} \tag{2}$$

where P_D is the capillary pressure giving rise to initial non-wetting phase permeability, λ a pore-size distribution index reflecting pore sorting, and S_e an effective wetting phase saturation given by

$$S_e = \frac{S - S_r}{1 - S_r} \tag{3}$$

where S is the wetting phase saturation of interest, and S a residual wetting phase saturation. The best-fit Brooks-Corey curve to the scaled laboratory data was characterized by a pore-size distribution index of 2.48, a scaled residual wetting phase saturation of 0.078, and a scaled displacement pressure of 0.00558. The scaled capillary pressure curve allows determination of a capillary pressure - saturation function given a permeability of interest. Further details of the scaling procedure are given by Kueper (1989).

In addition to the functional relationship between capillary pressure and saturation, the relationship between relative permeability and saturation must be specified for the medium of interest. Various workers have proposed that a relative permeability curve can be constructed from knowledge of capillary pressure data since both relationships are strongly dependent upon pore geometry (e.g., Brooks and Corey 1966, van Genuchten 1980, Parker et al. 1987). In this study, we adopt the relative permeability model of Brooks and Corey (1966), consistent with the choice of their model to represent the relationship between capillary pressure and saturation. Using the Brooks-Corey model, the relative permeability to the wetting phase is given by

$$k_{r,W} = S_e^{\left(\frac{2 + 3\lambda}{\lambda}\right)} \tag{4}$$

and the relative permeability to the non-wetting phase as:

$$k_{r,NW} = (1 - S_e)^2\left(1 - S_e^{\left(\frac{2 + \lambda}{\lambda}\right)}\right) \tag{5}$$

where S_e is an effective wetting phase saturation given by (3).

3 NUMERICAL MODEL

If we assume that the movement of NAPL and water is unaffected by partitioning across phase boundaries, we can write the following equations expressing mass conservation of both the wetting and non-wetting phases:

$$-\frac{\partial}{\partial x_i}(\rho_N q_N) + Q_N = \frac{\partial}{\partial t}(\phi \rho_N S_N) \tag{6}$$

$$-\frac{\partial}{\partial x_i}(\rho_W q_W) + Q_W = \frac{\partial}{\partial t}(\phi \rho_W S_W) \tag{7}$$

where $\frac{\partial}{\partial x_i}$ is the divergence operator, q_W and q_N the wetting and non-wetting phase fluxes respectively, Q_W and Q_N the wetting and non-wetting phase source-sink terms respectively, S_W and S_N the wetting and non-wetting saturations respectively. If we assume that there are no source-sink terms, that the medium and fluid are both incompressible, and that Darcy's law can be extended to multiple fluid systems, we arrive at the following differential equations:

$$\frac{\partial}{\partial x_i}\left[\frac{k_{ij}k_{r,W}}{\mu_W}\left(\frac{\partial P_W}{\partial x_j} + \rho_W\, g \frac{\partial z}{\partial x_j}\right)\right] = \phi\, \frac{\partial S_W}{\partial t} \tag{8}$$

$$\frac{\partial}{\partial x_i}\left[\frac{k_{ij}k_{r,N}}{\mu_N}\left(\frac{\partial P_n}{\partial x_j} + \rho_N g \frac{\partial z}{\partial x_j}\right)\right] = \phi\, \frac{\partial S_N}{\partial t} \tag{9}$$

where k_{ij} is the medium permeability, $k_{r,W}$ and $k_{r,N}$ the relative permeabilities to the wetting and non-wetting phases respectively, μ_W and μ_N the wetting and non-wetting phase viscosities repectively, P_W and P_N the wetting and non-wetting phase pressures respectively, and g gravity. Equations (8) and (9) are coupled through the capillary pressure relation:

$$P_C\,(S_W) = P_N - P_W \tag{10}$$

and are subject to:

$$S_W + S_N = 1 \tag{11}$$

In a typical contamination scenario, there will exist no non-wetting liquid outside the region of the migrating NAPL body. To avoid the complications of having to account for the presence of the non-wetting phase in regions of the solution domain where this phase is absent, we substitute equations (10) and (11) into (8) and (9) to arrive at the following set of equations in terms of the primary variables P_W and S_W:

$$\frac{\partial}{\partial x_i}\left[\frac{k_{ij}k_{r,W}}{\mu_W}\left(\frac{\partial P_W}{\partial x_j} + \rho_W g \frac{\partial z}{\partial x_j}\right)\right] = \phi \frac{\partial S_W}{\partial t} \tag{12}$$

$$\frac{\partial}{\partial x_i}\left[\frac{k_{ij}k_{r,N}}{\mu_N}\left(\frac{\partial (P_C + P_W)}{\partial x_j} + \rho_N\, g \frac{\partial z}{\partial x_j}\right)\right] = -\phi \frac{\partial S_W}{\partial t} \tag{13}$$

The variables P_W and S_W are continuous throughout the solution domain, regardless of whether the non-wetting phase is present. Upon solution for the unknown wetting phase saturation, the non-wetting phase saturation is obtained from (11), and the non-wetting phase pressure from (10).

Equations (12) and (13) are solved in two dimensions using a block-centered finite difference scheme (Peaceman, 1977). The solution scheme is fully implicit, and it is assumed that the coordinate axes are aligned with the principal directions of permeability such that cross terms vanish. The non-linearity of equations (12) and (13) is dealt with using full Newton-Raphson iteration. Full upstream weighting (Aziz and Settari, 1979) is used for all mobility terms. The resulting matrix equation is solved with an iterative solver (Letniowski, 1988) utilizing Orthomin acceleration (Vinsome, 1976) and block Dupont-Kendall-Ratchford pre-conditioning (Dupont et al., 1968). Further details of the numerical solution scheme are given by Kueper (1989). That study also presents the results of model verification against the analytical solution of McWhorter and Sunada (1990), and the results of model validation against the experimental work of Kueper et al. (1989).

4 NUMERICAL SIMULATIONS

In order to investigate the behaviour of a dense NAPL in a field-scale aquifer, numerical simulations were carried out in a two-dimensional solution domain 9.0 meters wide by 3.5 meters high. The top of the solution domain is taken to coincide with the watertable while the bottom boundary is taken to be impermeable. The sides of the solution domain are characterized by a constant wetting-phase pressure representing hydrostatic conditions. The non-wetting fluid is assigned a density of 1600 kg/m^3 and an absolute viscosity of 0.001 Pa·s. An interfacial tension of 0.045 *N/m* is assumed to exist between the two fluid phases. These properties are similar to tetrachloroethylene, a common chlorinated solvent. Initially, the solution domain is completely water saturated and under hydrostatic conditions. The non-wetting fluid is introduced along a 0.40 meter wide source release area in the centre of the top boundary. This source area is characterized by a non-wetting phase saturation of 0.50.

The performed simulations are characterized by first and second-order moments as given by:

$$M_{ij} = \int_{-\infty}^{\infty}\int_{-\infty}^{\infty} x^{i} z^{i}\, V_N \, dxdz \tag{14}$$

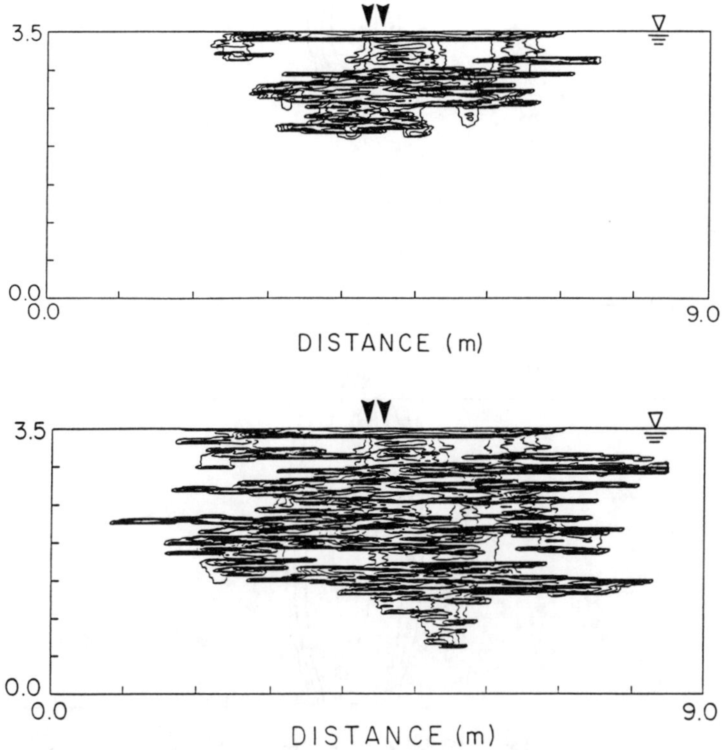

Figure 2 Distribution of tetrachloroethylene in a heterogeneous sand aquifer accounting for small scale variation in aquifer properties. The outermost contour represents 2% non wetting saturation. The contour interval is 10% saturation. (a) t = 3.15 days, (b) t = 7.40 days.

where V_N is the volume of non-wetting fluid. The expression in (14) defines the following moments:

$$X_C = M_{10} / M_{00} \tag{15}$$

$$\sigma^2_{XX} = M_{20} / M_{00} - X_C^2 \tag{16}$$

where X_C is the x-coordinate of the center of mass of the non-wetting body, and $\sigma^2{}_{XX}$ the spatial variance about X_C. These moments will be referred to in order to quantify results of the field-scale simulations.

In the first set of simulations, the solution domain was discretized using a nodal spacing of 0.05 meters in the horizontal direction and 0.025 meters in the vertical direction. This resulted in a total of 25200 nodal points, each with a differing permeability. The permeability field was generated using the Turning Bands method (Mantoglou and Wilson, 1981), with the statistics of the Borden aquifer as determined by Sudicky (1986) used as input. The generated permeability field is characterized by a mean log-transformed hydraulic conductivity (cm/sec) of -4.68 and an overall variance of log-transformed hydraulic conductivity of 0.26. An exponential autocorrelation function was assumed to represent the spatial correlation of permeability. This function was characterized by correlation lengths in the horizontal and vertical directions of 2.8 meters and 0.12 meters respectively. The nodal discretization chosen is sufficiently smaller than these correlation lengths, ensuring that a detailed representation of the aquifer heterogeneity is attained. Capillary pressure and relative permeability curves were assigned to each node using the relationships in (1)-(5).

Figure 2 illustrates the results of the heterogeneous simulation for two selected times. As can be seen, the non-wetting fluid exhibits significant degrees of lateral spreading in response to the spatial distribution of capillary properties. In

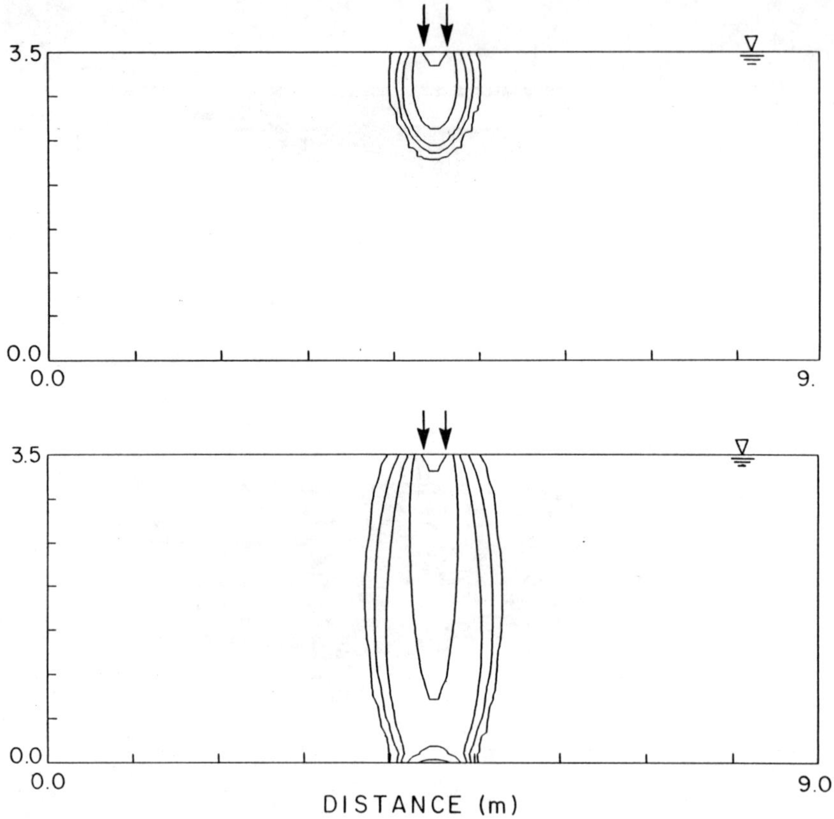

Figure 3 Distribution of tetrachloroethylene in a field-scale aquifer where the medium is assumed to be homogeneous and isotropic. Laboratory-scale parameters are used as input data. The outermost contour represents 2% non-wetting saturation. The contour interval is 10% saturation. (a) t = 0.27 days, (b) t = 1.19 days.

order for non-wetting fluid to penetrate a lower permeability lens, it must accumulate above the lens to bring about a saturation corresponding to a capillary pressure in excess of the displacement pressure of the lens. This accumulation of non-wetting fluid gives rise to lateral pressure gradients, and therefore spreading above the lower permeability material. Once the lens is penetrated, lateral spreading will continue to occur since the lens exhibits a lower relative permeability to the non-wetting fluid than the material immediately above. The interface between varying permeability media is characterized by a continuity in capillary pressure, but a discontinuity in fluid saturation. The overall degree of spreading of the non-wetting body is in general a function of the aquifer structure, as well as the fluid properties and the source release conditions. A sensitivity to fluid properties is demonstrated by Kueper (1989).

In the second set of simulations the Borden aquifer is represented as a homogeneous, isotropic porous medium characterized by a single laboratory-scale capillary pressure and relative permeability function with a pore-size distribution index of 2.48. The permeability is taken to equal $9.46 \times 10^{-12} m^2$. This corresponds to the geometric mean of the permeability distribution representing the Borden aquifer. The single capillary pressure and relative permeability curves were generated using (1)-(5). The identical solution domain, boundary conditions, and fluid properties as were used for the heterogeneous simulation were assigned.

Figure 3 illustrates the results of the homogeneous, isotropic simulation. As shown, the non-wetting body migrates vertically downward in a uniform, well defined manner. The overall shape of the non-wetting body is significantly different

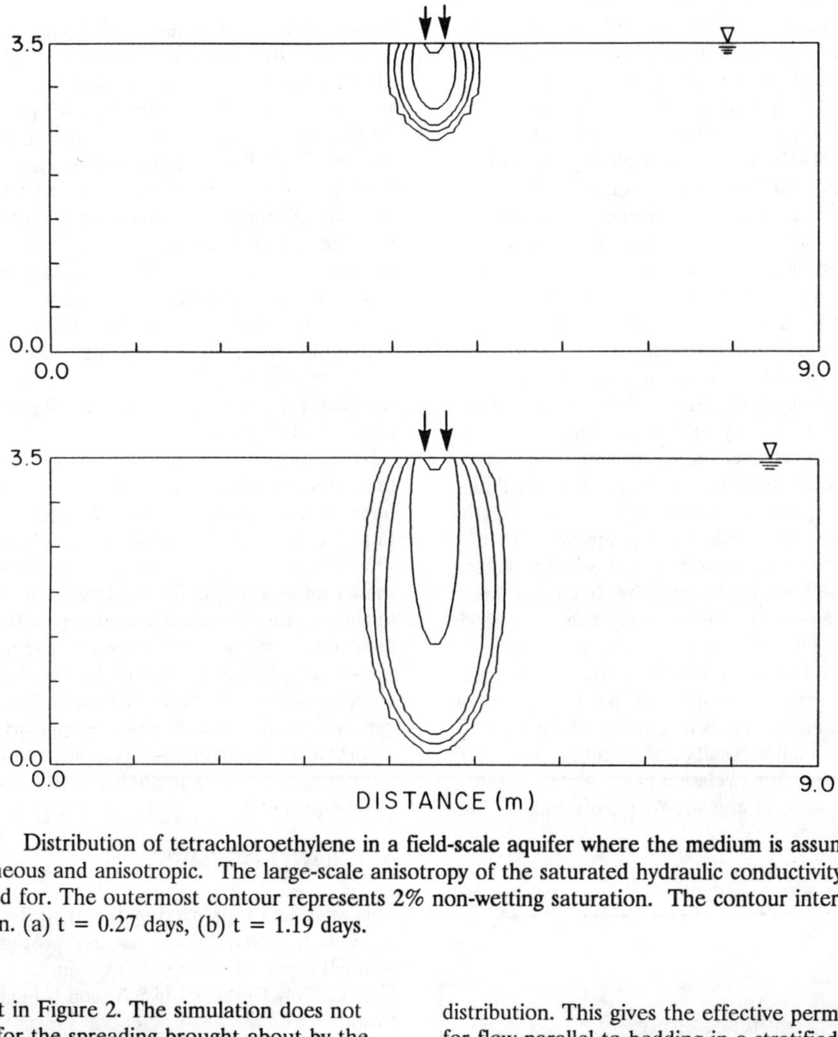

Figure 4 Distribution of tetrachloroethylene in a field-scale aquifer where the medium is assumed to be homogeneous and anisotropic. The large-scale anisotropy of the saturated hydraulic conductivity field is accounted for. The outermost contour represents 2% non-wetting saturation. The contour interval is 10% saturation. (a) t = 0.27 days, (b) t = 1.19 days.

from that in Figure 2. The simulation does not account for the spreading brought about by the spatial variation of permeability and capillary characteristics of the aquifer.

In the third set of simulations the aquifer is again assumed to be homogeneous, but the large-scale anisotropy of the saturated hydraulic conductivity field is accounted for. The identical laboratory-scale capillary pressure and relative permeability curves as employed in the second set of simulations are adopted. The effective vertical permeability is taken to be $8.31 \times 10^{-12} m^2$. This corresponds to the harmonic mean of the Borden permeability distribution, and gives the effective permeability for flow perpendicular to bedding in a stratified aquifer (Freeze and Cherry, 1979). The effective horizontal permeability is taken to be 10.78×10^{-12} m^2, which corresponds to the arithmetic mean of the Borden permeability distribution. This gives the effective permeability for flow parallel to bedding in a stratified aquifer (Freeze and Cherry, 1979). Stochastic-analytic theory (e.g., Gelhar, 1986) allows calculation of these effective permeabilities with adjustment for the correlated structure of the aquifer. In an aquifer such as Borden, however, where the horizontal correlation length is significantly greater than the vertical correlation length, the adjustment is negligible.

Figure 4 illustrates the results of the third set of simulations. As in the isotropic case, the non-wetting body is seen to migrate predominantly vertically downwards in a uniform, well-defined manner. It is clear that accounting for the large-scale anisotropy of the saturated hydraulic conductivity field does not induce the lateral spreading of the non-wetting fluid exhibited in the heterogeneous simulations.

The degree of lateral spreading exhibited by the migrating NAPL body can be characterized by its second moment as calculated with equation (16). This is illustrated in Figure 5 for each set of simulations. As shown, the homogeneous simulations greatly underestimate the overall degree of lateral spreading exhibited by the migrating NAPL body. This is clearly due to the inadequacy of representing a heterogeneous, field-scale aquifer as a homogeneous aquifer characterized by laboratory-scale properties. Accounting for the large-scale anisotropy of the saturated hydraulic conductivity field shows little difference from the isotropic case. This is due primarily to the strong vertical driving force exerted by gravity, and the small degree of large-scale anisotropy exhibited by the saturated hydraulic conductivity field at Borden.

In order to represent a heterogeneous porous medium as an equivalent homogeneous medium, effective large-scale capillary pressure and relative permeability curves need to be employed. The construction of such curves is not a trivial matter, and to date work in this area has been restricted largely to air-water systems (e.g., Yeh et al. 1985, Mantoglou and Gelhar 1987). These studies suggest that the phase relative permeabilities are dependent upon direction, and that the degree of large-scale anisotropy is a function of fluid saturation. Similar results are found by Bear et al. (1987) in examining relative permeability at the laboratory scale in anisotropic porous media.

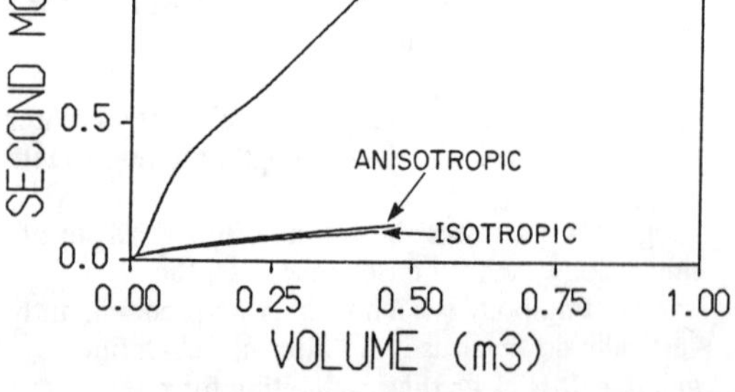

Figure 5 Second moment versus volume. Slopes of the various curves indicate degree of lateral spreading of the migrating non-wetting fluid.

5 SUMMARY AND CONCLUSIONS

In this study, a numerical model is applied to investigate the migration of a dense, non-aqueous phase liquid below the watertable in a heterogeneous sand aquifer. The distribution of permeability is represented as a spatially correlated random field. Tetrachloroethylene-water capillary pressure curves were measured in the laboratory on selected samples of aquifer material, and scaled to permeability such that the spatial distribution of capillary properties is defined. Simulations are carried out in a two-dimensional solution domain accounting for variations in aquifer properties at the centimeter scale. The results of the simulations show the migrating NAPL body to exhibit significant degrees of lateral spreading along lower permeability lenses.

To demonstrate the inadequacy of representing the heterogeneous field aquifer as an equivalent homogeneous porous medium characterized by a single set of laboratory-scale data, additional simulations were carried out. These simulations show that accounting for the large-scale anisotropy of the saturated hydraulic conductivity field alone does not bring about the lateral spreading of non-wetting fluid exhibited in the detailed heterogeneous simulation. It is clear that appropriate effective, large-scale capillary pressure and relative permeability curves need to be defined if a heterogeneous field aquifer is to be modelled as an equivalent homogeneous porous medium.

ACKNOWLEDGEMENTS

This work was supported by the Waterloo Solvents-In-Groundwater research program through financial contributions from Ciba Geigy U.S.A., Dow Chemical U.S.A. and Canada, Eastman Kodak U.S.A., General Electric U.S.A., the province of Ontario University Research Incentive Fund, and the Natural Sciences and Engineering Research Council of Canada.

REFERENCES

Abriola, L.M. and Pinder, G.F., 1985a. A multiphase approach to the modeling of porus media contamination by organic compounds: 1. Equation development. Wat. Resour. Res. 21(1), pp. 11-18.

Abriola, L.M. and Pinder, G.F., 1985b. A multiphase approach to the modeling of porous media contamination by organic compounds: 2: Numerical simulation. Wat. Resour. Res. 21(1), pp. 19-26.

Aziz, K. and Settari, A., 1979. Petroleum Reservoir Simulation. Applied Science, London.

Bear, J., Braester, C., and Menier, P.C., 1987. Effective and Relative Permeabilities of Anisotropic Porous Media. Trans. Por. Med. 2, pp. 301-316.

Brooks, R.H. and Corey, A.T., 1966. Properties of porous media affecting fluid flow. J. Irrig. Drain. Div., Am. Soc. Civ. Eng. 92 (IR2): pp. 61-88.

Dupont, T., Kendall, R.P., and Ratchford, H.H.Jr., 1968. An approximate factorization procedure for solving self-adjoint elliptic difference equations. SIAM J. Numerical Anal. 5: pp. 559-573.

Faust, C.R., Guswa, J.H. and Mercer, J.W., 1989. Simulation of three-dimensional flow of immiscible fluids within and below the unsaturated zone. Wat. Resour. Res. 25(12) pp. 2449-2464.

Faust, C.R., 1985. Transport of immiscible fluids within and below the unsaturated zone: A numerical model. Wat. Resour. Res. 21(4), pp. 587-596.

Forsyth, P.A., 1988. Simulation of nonaqueous phase groundwater contamination. Adv. Wat. Res. 11, pp. 74-83.

Freeze, R.A. and Cherry, J.A., 1979. Groundwater, Prentice Hall.
Gelhar, L.W., 1986. Stochastic subsurface hydrology from theory to applications. Wat. Resour. Res. 22 pp. 135S-145S.

Gelhar, L.W., 1986. Stochastic subsurface hydrology from theory to applications. Water Resources Research., 22(9), pp. 1355-1455.

Kaluarachchi, J.J. and Parker, J.C., 1989. An efficient finite element method for modeling multiphase flow. Wat. Resour. Res. 25(1), pp. 43-54.

Kueper, B.H., 1989. The behaviour of dense, non-aqueous phase liquids in heterogeneous porous media. Ph.D. Dissertation, Department of Earth Sciences, University of Waterloo, Waterloo, Ont., pp. 172.

Kueper, B.H., Abbott, W. and Farquhar, G., 1989. Experimental observations of multiphase flow in heterogeneous porous media. J. Cont. Hyd. 5, pp. 83-95.

Kuppusamy, T., Sheng, J., Parker, J.C. and Lenhard, R.J., 1987. Finite element analysis of multiphase immiscible flow through soils. Wat. Resour. Res. 23(4) pp. 625-631.

Letniowski, F.W., 1988. Faculty of Mathematics, University of Waterloo, Unpublished document.

Leverett, M.C., 1941. Capillary behaviour in porous solids. Trans. Soc. Pet. Eng. AIME 142, pp. 152-169.

MacFarlane, D.S., Cherry, J.A., Gillham, R.W. and Sudicky, E.A., 1983. Migration of contaminants in groundwater at a landfill: A case study 1. Groundwater flow and plume delineation. J. Hydrology, 63: pp. 1-29.

Mackay, D.M., Freyberg, D.L. and Roberts, P.V., 1986. A natural gradient experiment on solute transport in a sand aquifer, 1. Approach and overview of plume movement. Wat. Resour. Res. 22(13), pp. 2017-2029.

Mantoglou, A. and Gelhar, L.W., 1987. Stochastic modeling of large-scale transient unsaturated flow systems. Wat. Resour. Res. 23(1), pp. 37-46.

Mantoglou, A. and Wilson, J.L., 1981. Simulation of random fields with the Turning Bands method. Ralph M. Parsons Lab., Dept. Civil Eng., Mass. Inst. Tech. Report No. 264.

McWhorter, D.B. and Sunada, D.K., 1990. Exact integral solutions for two-phase flow. Wat. Resour. Res. 26(3), pp. 399-413.

Osborne, M. and Sykes, J., 1986. Numerical modelling of immiscible organic transport at the Hyde Park Landfill. Wat. Resour. Res. 22(1), pp. 25-33.

Parker, J.C., Lenhard, R.J. and Kuppusamy, T., 1987. A parametric model for constitutive properties governing multiphase flow in porous media. Wat. Resour. Res. 23(4), pp. 618-624.

Peaceman, D.W., 1977. Fundamentals of numerical reservoir simulation, Elsevier, New York.

Sudicky, E.A., 1986. A natural gradient experiment on solute transport in a sand aquifer: Spatial variability of hydraulic conductivity and its role in the dispersion process. Wat. Resour. Res. 22(13), pp. 2069-2082.

van Genuchten, M.Th., 1980. A closed-form equation for predicting the hydraulic conductivity of unsaturated soils. Soil Sci. Am. J. 44(5), pp. 892-898.

Vinsome, P.K.W., 1976. ORTHOMIN, An iterative method for solving sparse sets of simulatneous linear equations. SPE paper 5729, SPE Sym. Num. Sim. of Res. Per., Los Angeles, Feb. 19-20.

Yeh, T.C., Gelhar, L.W., and Gutjahr, A.L., 1985. Stochastic analysis of unsaturated flow in heterogeneous soils. 2. Statistically anisotropic media with variable .eq alpha . Wat. Resour. Res., 21(4), pp. 457-465.

Subsurface Contamination by Immiscible Fluids, Weyer (ed.) © 1992 Balkema, Rotterdam. ISBN 90 6191 175 3

Numerical simulation of oil entrapment in a sand lens: Implications for organic contamination in the subsurface

B.J. Rostron
Department of Geology, University of Alberta, Edmonton, Alb., Canada

ABSTRACT: Oil migration into a sand lens is controlled by the oil fluid-potential distribution. Numerical simulations show that fluid-potential surfaces are locally modified by: 1) highly permeable lenses; 2) buoyancy forces due to the density difference between oil and water; and 3) capillary pressure differences between the fluid phases and in different media. As saturation distributions change in space and time, fluid-flow directions change because oil fluid-potential distributions are intrinsically linked to saturation distributions through relative-permeability and capillary-pressure curves. It is shown that, in particular situations, buoyancy and capillary forces can lead to oil migration different from, or even opposite to, the water flow direction, and oil being drawn (even downward) into lenses.

1 INTRODUCTION

Prediction of the flow direction of a liquid contaminant within the saturated groundwater zone can be difficult when the contaminant is immiscible with water. This is because each fluid possesses, and acts under, its own fluid-potential field. These fields can generate widely different flow rates, and directions, in their respective fluids. Under certain conditions, vastly different potentiometric surfaces, hence migration directions, can occur.

Correct estimation of contaminant migration directions is essential for proper site assessment and remediation, and this can only occur with a proper understanding of the components that comprise the fluid-potential fields.

Numerical simulations of oil migration and entrapment in a sand lens were conducted to study the evolution of the oil saturation distribution and the oil and water potentiometric surfaces in time and space. Although the results presented here are conducted on larger distance and longer time scales, they are applicable to smaller site-scale contamination problems because the controlling forces are time and length scale-independent.

Analysis of the simulated potentiometric surfaces shows that fluid migration directions depend strongly on buoyancy forces between the fluids; however, capillary-pressure and relative-permeability relationships can also be important.

2 FLUID POTENTIAL ANALYSIS

The flow problem can be considered to consist of two immiscible fluids, oil and water. Before the introduction of the contaminant to the system (in this case oil), the pore space is fully saturated by the water phase. Each phase possesses its own set of properties and hence its own fluid-potential distribution. The fluid potential Φ, for a particular phase f, is given by:

$$\Phi_f = g\,z + \frac{p_f}{\rho_f} = g\,h_f\ , \qquad (1)$$

where: Φ: fluid potential (per unit mass); g: gravitational constant; z: position above a common datum; p: fluid pressure (gage); and ρ: fluid density (assumed constant). Most hydrogeologists work with fluid potential in terms of hydraulic head, h, which is related to fluid potential by the gravitational constant.

The flow rate of a particular fluid is proportional to the gradient of its fluid-potential field, which is expressed by the multiphase form of Darcy's Law:

$$\vec{q_f} = -\frac{k\, k_{rf}\, \rho_f\, g}{\mu_f} \nabla h_f \,, \qquad (2)$$

where: $\vec{q_f}$: specific discharge; k: intrinsic permeability; k_r: relative permeability to each fluid; μ: fluid dynamic viscosity; and ∇h: the gradient of the fluid potential field.

Equations 1 and 2 show that if, for example, the fluids possess different densities, then each will have its own fluid-potential field, hence flow direction and flow rate.

The preceding two equations are coupled via pressure conditions between the fluids at the pore level. At the contact between two immiscible fluids, differences in molecular properties produce interfacial forces, which manifest themselves as a pressure difference between the fluids across their respective boundary. The value of the pressure difference is related to the interfacial tension, the degree of saturation, and the type and distribution of pore-throat radii (Dullien, 1979). Capillary pressure relationships are obtained experimentally for rock samples, with the empirical relationship between fluid saturation and capillary pressure given as (Craig, 1971):

$$p_c = p_{nw} - p_w \,, \qquad (3)$$

where: p_c the capillary pressure difference; p_{nw} : pressure in the non-wetting phase (oil); and p_w : pressure in the wetting phase (water). In this case, the rocks are strongly water-wet, that is, the fluid pressures are higher in the oil phase than in the water phase.

The distribution of fluid-potential within a flow domain is governed by the geometry of the flow domain, controlled by the boundary conditions and modified by the distribution of rock and fluid properties within.

3 NUMERICAL SIMULATION OF OIL ENTRAPMENT

The transient, two-dimensional, multiphase, finite-difference flow model SWANFLOW (Faust and Rumbaugh, 1986) was used for this study. This model was originally developed to study the migration of immiscible contaminants in a shallow groundwater environment and was found to be applicable to petroleum migration problems, such as this, with some simplifying assumptions (Faust and Rumbaugh, 1986; Rostron, 1990).

The simulations were originally conducted to study the long-term migration and accumulation of oil in a commerically-exploitable reservoir; hence, time scales of millions of years, and distance scales of kilometers, apply. Results of these simulations showed that the controlling factors in the migration/entrapment processes were time-independent rock and fluid properties (i.e., capillary pressure and relative permeability curves). These same time-independent flow properties can be found in, and are applicable to, shorter-term contaminant migration problems.

Numerous simulations were conducted using a wide variety of grids; grid geometries; fluid densities; intrinsic permeabilities; capillary-pressure curves; and relative-permeability curves. Space limitations preclude the presentation of all of the simulation results. Instead, the results of one representative case are presented here. This example was selected because it best illustrates the flow-modifying effects most applicable to contaminant migration problems.

3.1 *Simulation data*

A flow domain was constructed using properties representative of the Upper Cretaceous Belly River Formation in west-central Alberta. Selected data were combined into a hypothetical, but presumably realistic, vertical cross-section through a portion of the formation. The domain consists of four units: a highly permeable sandstone lens encased in a less permeable aquifer, which in turn is sandwiched between two aquitards. The finite-difference grid was set up with no-flow boundaries on the top and bottom and constant head boundaries at the upstream and downstream ends. A constant horizontal hydraulic-gradient was set up across the entire grid. The capillary-pressure and relative-permeability curves used in this example are illustrated in Figures 1 and 2. All other simulation data are shown in Table 1.

Each simulation run was conducted in two parts: establishment of the steady-state fluid potentials, and a subsequent transient (migration) phase. The begining of the transient phase (T=0 years) was marked by the introduction of oil (as a line source) along the

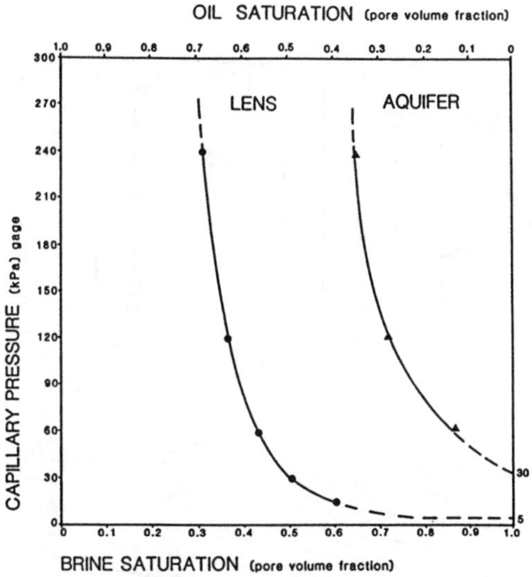

Figure 1. Capillary pressure curves.

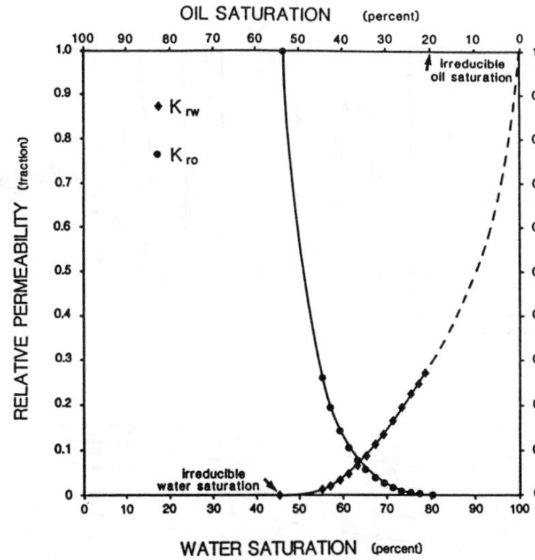

Figure 2. Relative permeability curves.

upstream boundary of the domain. The rate of oil injection remained constant for the duration of the simulation.

3.2 *Steady state fluid-potentials*

Figures 3a and 3b illustrate the water and oil hydraulic-head distributions at T=0.

For water, the fluid potentials are strongly distorted around the upstream and downstream ends of the lens. This effect, termed the "highly-permeable lens effect", is dependent upon lens geometry, lens to aquifer permeability ratio, and regional hydraulic gradient (Tóth and Rakhit, 1988). The flow of water is focused into the upstream end, and expelled out the downstream end of the lens.

Since oil is absent from the flow domain, Figure 3b represents the potential field for oil at zero oil saturation. The potential distribution of oil differs from that of water, in three ways.

First, the magnitude of the oil potential is higher because of two factors: 1) higher pressures in the oil phase relative to the water phase (equation 3); and 2) higher hydraulic heads caused by a lower density (equation 1).

Second, tilted equipotential lines are found outside the lens area. These result from a vector combination of the vertical buoyancy forces

Table 1. Simulation data

Number of grid blocks:	60 x 30	(horizontal : vertical)
Flow domain:	1250 x 300 metres	"
Lens: 10:1 ellipse:	250 x 25 metres	"
Fluid viscosity:	2.0 : 1.0 mPa·s	(oil : water)
Fluid density:	850 : 1000 kg/m^3	(oil : water)
Permeability:	0.1 : 100.0 md	(aquifer : lens)
Porosity:	5 : 20 %	(aquifer : lens)
Formation compressibility:	$1.0{\cdot}10^{-10}$ (kPa^{-1})	
Total fluid injection rate:	$1.89{\cdot}10^{-2}$ m^3/yr	(18.9 l/yr)
Fluid composition:	90 : 10 %	(water : oil)
Regional hydraulic gradient:	0.02 metres/metre	
Timestep length:	500 years	

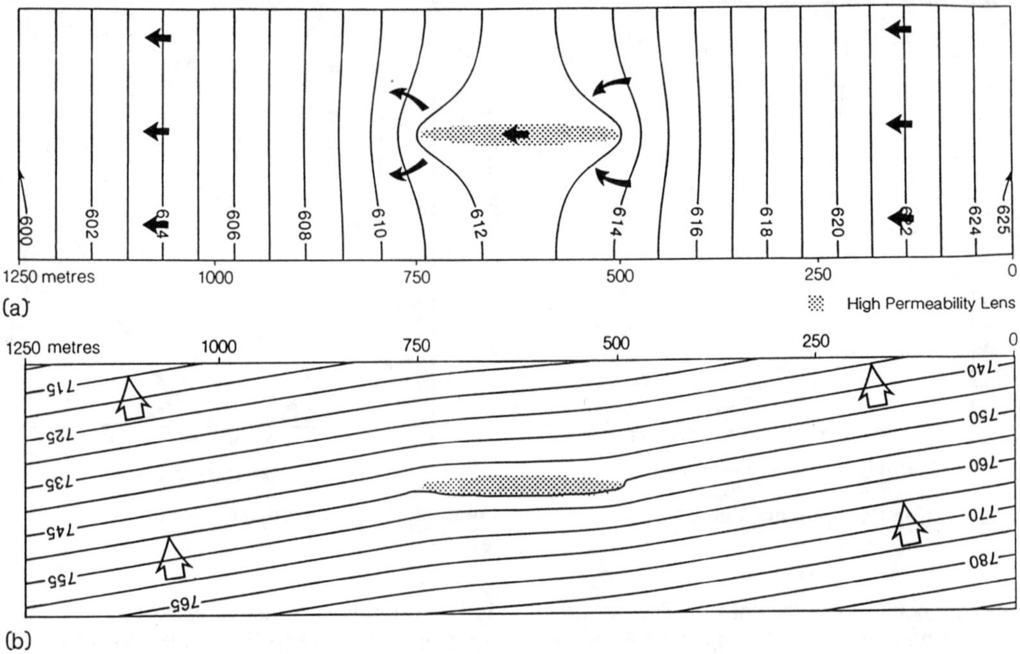

Figure 3. Hydraulic head distribution at steady state: a) water (one metre contour interval); b) oil (five metre contour interval). No vertical exaggeration.

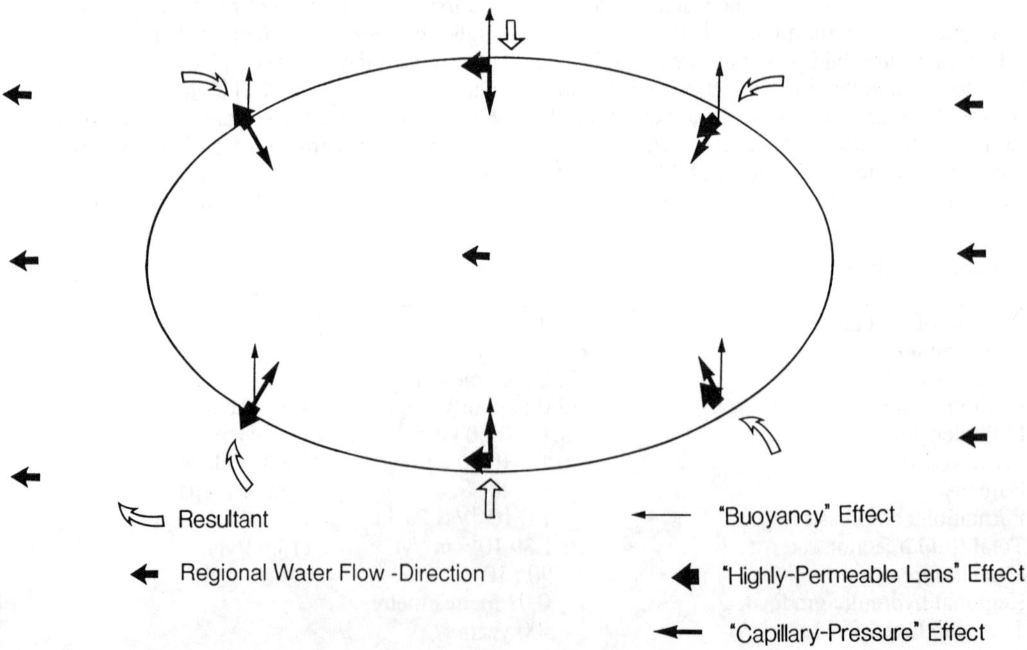

Figure 4. Schematic of forces acting in the area of the lens at steady state conditions.

(caused by fluid density differences) and the horizontal flow forces (caused by the regional water hydraulic-gradient). The tilting of the oil equipotentials is termed the "buoyancy effect".

Third, a relative low-potential region exists inside the lens area (not visible in Figure 3b). This is the result of different capillary pressure curves in the lens and surrounding aquifer. At 100 percent water saturation in both media, the capillary pressure differences between the oil and water phases (theoretically) are 30 kPa and 5 kPa, in the aquifer and lens, respectively (Figure 1). At a saturation just less than 100 percent there is a pressure difference (in the oil phase) of approximately 25 kPa across the boundary of the two media. This pressure difference (hence fluid potential difference) is termed the "capillary pressure effect". In this case, the 25 kPa pressure difference translates to a 2.38 metre head difference, which is not visible with a 5 metre contour interval. The magnitude of this effect depends entirely on the shape of the capillary pressure curves for each lithology within the flow domain.

The combination of the highly permeable lens, buoyancy, and capillary pressure effects on the regional fluid-potential surface produces the fluid potential distribution shown in Figure 3b. A schematic of the interpreted flow paths of oil under these conditions is illustrated as Figure 4. These figures illustrate potential migration into the lens from all directions (even from above). Any particles of oil that enter into the zone of influence of the lens will end up drawn inside. The conditions shown in Figures 3 and 4 persist until the oil saturation distribution changes during migration.

3.3 *Oil saturation distribution*

The oil saturation distribution during the transient portion (T>0) of the simulation is shown in Figure 5.

After one million years of injection and migration, oil concentrations of up to 37 percent (pore volume) can be found, located along the upper boundary of the flow domain (Figure 5a). Oil is primarily migrating upwards with a lesser component of downstream migration and has yet to reach the lens.

After three million years, oil has reached, and is concentrating in the lens (Figure 5b). Oil is drawn downwards into the lens from above. It enters the lens on the upper, upstream "corner", moves downstream, and concentrates at the downstream, upper "corner". Upper regions of the lens have attained 45 percent oil saturation, while oil has not contacted the bottom of the lens. Oil is held within the lens by a fluid-potential barrier between the lens and aquifer.

The oil saturation distribution at five million years is shown in Figure 5c. Oil saturations have reached 55 percent in the upper 5 metres of the lens. The 55 percent saturation is a limiting value imposed by the irreducible water saturation from the relative permeability curves used (Figure 2). Elsewhere within the lens, oil saturations have reached a minimum of 38 percent. Minor amounts of oil have been able to leak from the lens at the downstream end, but there is a continued net oil increase in the lens because the rate of exit is much lower than the rate of influx from upstream.

This simulation was stopped past five million years due to oil reaching and accumulating (spuriously) at the downstream boundary of the flow domain.

3.4 *Final fluid-potential distributions*

Figure 6 illustrates the fluid-potential patterns for water and oil at five million years.

For water (Figure 6a) the potentiometric pattern is similar to the steady state pattern (Figure 3a). Upstream of the lens, the isopotential lines are tightly serried because increased oil saturation in this area reduces the permeability to water. In the lens area, the symmetrical potentiometric distribution has been distorted. Along the downstream end of the lens, the isopotentials have begun to encircle the lens. Once the oil saturations within the lens reach 55 percent, the permeability to water is reduced to zero. Thus, the overall water flow pattern adjusts to flow around that portion of the flow domain, giving rise to the isopotential lines encircling the lens.

The oil potentiometric surface (Figure 6b) mirrors the observed fluid migration pattern. Upstream, oil moves diagonally upwards and downstream. Along the upper boundary of the flow domain, oil flows horizontally downstream. In the lens area, oil moves downwards into the lens upstream end, and collects at its downstream end.

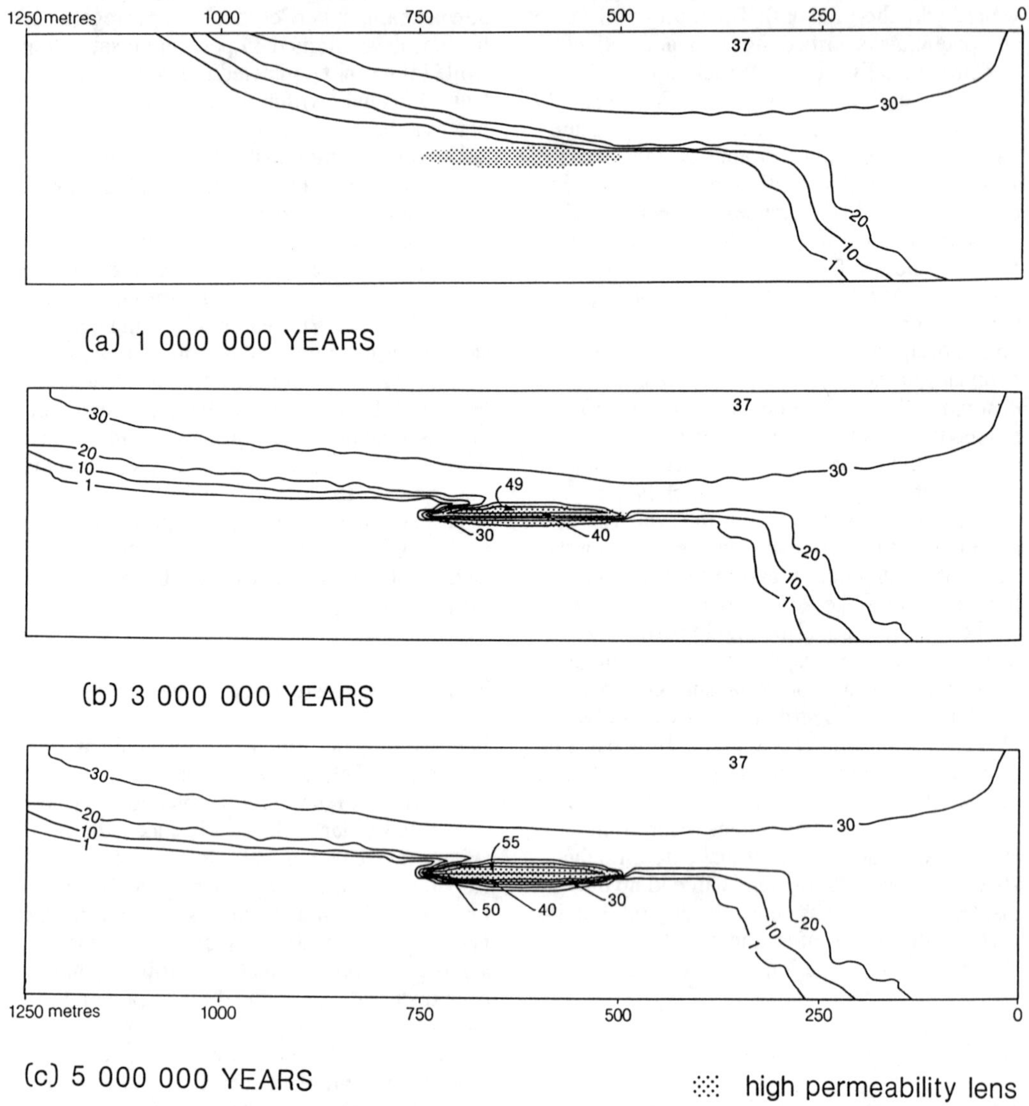

Figure 5. Oil saturation distribution through time (contours in percent oil saturation). No vertical exaggeration.

3.5 *Applicability to contaminant migration*

Results have shown that time-independent rock and fluid properties (capillary-pressures and densities, for example) play an important role in the migration and entrapment of oil. It is unlikely that a contaminant hydrogeologist would worry about a site with a spill rate of approximately 20 litres/year, or be able to wait for thousands of years for any noticeable migration to occur. However, the time and distance scales of the simulations presented (Figures 3-6) can be reduced by suitable manipulations of the intrinsic permeabilities and source rates, such that the same migration patterns are obtained in a time scale that would be important to a contaminant hydrogeologist.

To illustrate this point, a comparison was made between two simulations with different intrinsic permeabilities. The evolution of the 20

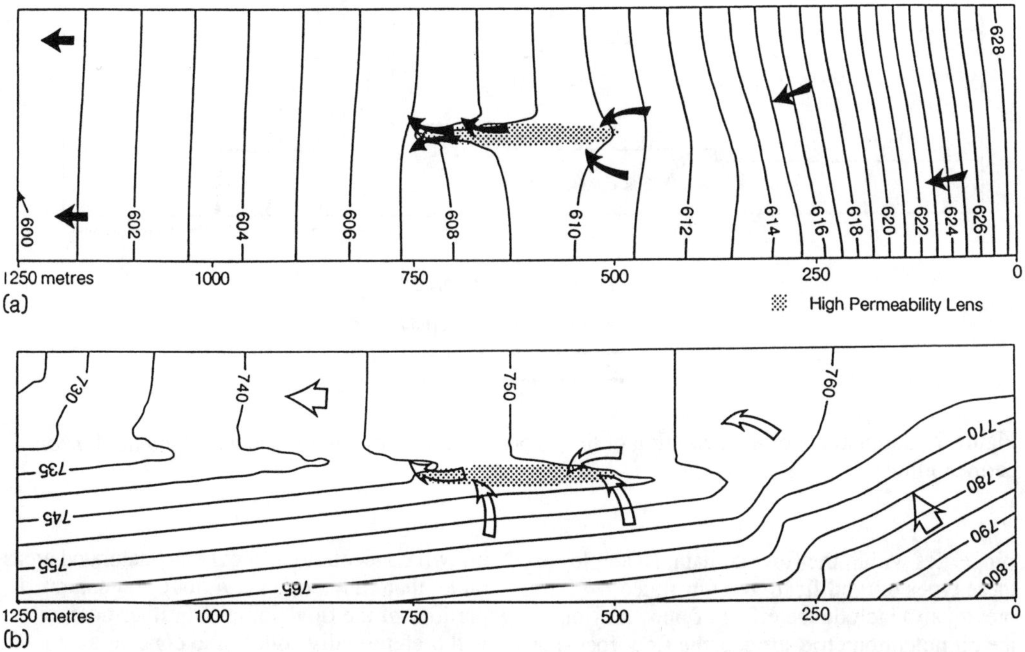

Figure 6. Hydraulic head distribution at five million years: a) water (one metre contour interval); b) oil (five metre contour interval). No vertical exaggeration.

percent saturation contours for two cases where the intrinsic permeabilities differ by 4 orders of magnitude is shown in Figure 7. The saturation contours are congruent at the starting point (termed T=0 years), and essentially congruent at the finishing point (T=50 and 50,000 years). The minor incongruency is an artifact of the contouring package and coarse finite-difference grid, and can be ignored. This figure illustrates that if the permeabilities of the media are increased by 4 orders of magnitude, then there is a corresponding decrease in the amount of time required to reach the same migration distance. In both cases, the shape of the saturation profile is controlled by the oil density, which is not a function of time. Other comparisons were made with increased source rate and hydraulic gradients, with similar results. In every case the most important parameters were oil density, capillary pressure curves, and relative permeability curves.

Therefore, the results from the simulation presented in Figures 3-6 are applicable to contaminant migration problems because the controlling factors are time-independent.

4 IMPLICATIONS FOR ORGANIC CONTAMINATION

Four implications for the flow of immiscible fluids and organic contamination in the subsurface result from this modelling.

First, the presence of one or more immiscible fluids creates multiple fluid-potential fields. Each fluid's migration direction and flow rate is controlled by its own fluid-potential field (given by equations 1 and 2). The potentiometric surfaces are linked together through saturation and pressure relationships (equation 3) but differences in fluid and rock properties can generate completely different fluid-potential fields (Figures 3a and 3b). Although apparently obvious, this has direct implications for contaminant migration assessments, for example, in the estimation of immiscible contaminant flow-direction at a site based on the flow field of water. Water could be flowing in one direction, and the contaminant in another.

The second implication is that the fluid potential fields can be modified, possibly independently, by variations in fluid and rock

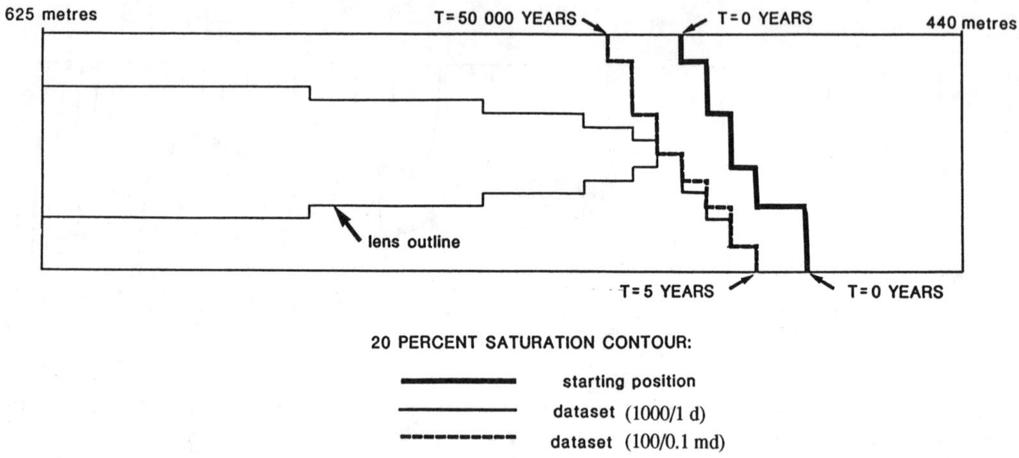

Figure 7. Comparison of the evolution of the 20 percent oil saturation profile with increased media permeabilities.

properties within the flow domain. Examples of these types of modifications (illustrated previously) include the effect of buoyancy on the oil potentiometric surface, the flow-focusing effect of highly permeable lenses, and the capillary pressure generated forces between fluids and in different media. The identification of the presence of any of these factors and their resulting influence on the flow fields is of paramount importance when considering fluid migration pathways. Results show that one or more of these effects can dominate the fluid migration direction. For example in Figure 3, oil is drawn into the lens downward from above, in a flow direction directly opposite to what would be expected for a light oil. This shows the influence the aforementioned factors can have on the regional and local potential fields.

The third implication for organic contamination in the subsurface is that the flow fields generated by the aforementioned process are dependent on fluid saturation distributions. As fluid distributions change in space and time, the potentiometric surfaces, hence flow directions and rates, change also (Figures 3 and 6). For example, migration directions inferred from a site investigation conducted early on could incorrectly predict flow directions at a later time when saturation distributions have changed.

The fourth implication for organic migration is that fluids will collect in areas of local fluid-potential minima. This means, for example, that oil will concentrate in previously saturated areas rather than move into previously non-saturated portions of the flow domain. Furthermore, oil will preferentially collect and concentrate in high permeability material, rather than move downstream in the flow domain.

It is expected that the inclusion of these effects in the analysis of migration problems will improve the understanding and knowledge of immiscible fluid flow problems.

5 CONCLUSIONS

Numerical simulation of oil migration and entrapment in a sand lens has shown that:

1) Fluid potential fields of water and oil govern fluid flow directions and rates, which are modified by: (i) highly-permeable lenses, situated within the flow domain; (ii) buoyancy; (iii) capillary pressures.

2) In certain cases, the fluid flow direction is controlled by one or more of these effects. Opposing flow directions of the separate phases can result.

3) As migration proceeds and saturation distributions change, the fluid-potential distributions change in time and space because the distribution of fluid-potentials is coupled to fluid saturation.

4) Capillary-pressure and relative-permeability relationships for soils and unconsolidated rocks (and rock samples in

general) are not commonly determined, possibly because they are relatively expensive to obtain. These simulations have shown the importance of capillary-pressure curves as controlling factors for migration directions. Therefore, these types of curves should be obtained and utiltzed in organic contamination investigations.

REFERENCES

Craig, F.F. Jr. 1971. *The reservoir engineering aspects of waterflooding*: *SPE Monograph* 5. Dallas: Society of Petroleum Engineers.

Dullien, F.A.L. 1979. *Porous media: fluid transport and pore structure*. New York: Academic Press.

Faust, C.R. and J.O. Rumbaugh III 1986. *SWANFLOW-2D: Simultàneous Water, Air, and Nonaqueous Phase Flow Model in Two Dimensions: Documentation Version 1.0.* Herndon, Virginia, USA: Geotrans, Inc.

Rostron, B.J. 1990. *Numerical simulation of oil migration through a lenticular reservoir.* Edmonton, Canada: University of Alberta, MSc thesis.

Tóth, J. and K. Rakhit 1988. Exploration for reservoir quality rock bodies by mapping and simulation of potentiometric surface anomalies. *Bulletin of Canadian Petroleum Geology* 36, No. 4: 362-378.

Subsurface Contamination by Immiscible Fluids, Weyer (ed.) © 1992 Balkema, Rotterdam. ISBN 90 6191 175 3

Numerical simulation of unsteady organic vapor transport in porous media using the dusty gas model

Linda M. Abriola, Chiu-Shia Fen & Howard W. Reeves
Environmental and Water Resources Engineering, The University of Michigan, Ann Arbor, Mich., USA

ABSTRACT: The transient behavior of organic vapors in unsaturated soils is explored using the Dusty Gas Model. Constitutive equations from this model are employed to develop a coupled system of nonlinear partial differential equations governing component transport in a binary gas system. A finite element simulator is used to investigate the relative importance of various flux components in this model for representative organic chemicals and soil types. To further elucidate transport mechanisms, Dusty Gas Model results are compared with simulations from traditional advective-diffusive transport models.

1 INTRODUCTION

The prevalence of volatile organic pollutant releases in the unsaturated soil zone has resulted in increasing interest in the processes governing multicomponent transport in this region. A number of models have been presented in the hydrology and soil science literature over the past few years to describe organic vapor transport. These models are generally based upon the solution of material balance equations which incorporate Fick's first law to describe diffusive transport. Frequent model assumptions include uniform gas phase pressure and negligible gravity effects (Abriola & Pinder 1985; Baehr 1987; Silka 1986). Some models have treated gas phase density as uniform (Allan 1986; Silka 1986; Baehr 1987). More recently, the influence of applied pressure gradients and gravity have been explored by incorporation of a Darcy flux term (Sleep & Sykes 1989; Mendoza & Frind 1990) .

In the physics literature, a great body of work has been directed towards the description of multicomponent gas transport. General constitutive equations have been derived from the kinetic theory of gases. The so-called "Dusty Gas Model" (hereafter referred to as DGM) which has evolved is fully described in two comprehensive texts (Cunningham & Williams 1980; Mason & Malinauskas 1983). This model has been employed in a variety of applications in physics and nuclear and chemical engineering but has received very little attention in the hydrology literature. In its representation of the diffusive flux, the DGM includes the effects of unequal molar weight components (nonequimolar flux) and Knudsen diffusion as well as the traditional Fickian molecular diffusion.

Quite recently, Thorstenson & Pollock (1989a,b) reviewed the development of the DGM and, through some approximations, explored the implications of this model to the steady state fluxes of various components of multicomponent gas mixtures in the unsaturated soil zone. In some instances, they found significant differences between Fick's law predictions of diffusive fluxes and those predicted by the DGM. A literature survey has uncovered only one soil-related experimental investigation which explored the relative importance of equimolar and nonequimolar fluxes to diffusive transport (Alzaydi 1975; Alzaydi et al. 1978).

This paper presents a preliminary modeling study to examine the significance of various diffusive and advective transport mechanisms to unsteady organic vapor transport in porous media. Material balance equations based upon the DGM are solved for a one-dimensional binary gas system. Simulation results are compared with traditional modeling approaches.

2 THEORY

2.1 Diffusive gas transport

According to the Dusty Gas Model, gas

diffusion in a multicomponent isothermal system is governed by the following equation (Thorstenson & Pollock 1989a):

$$\sum_{i=1,j\neq i,p}^{\upsilon} \frac{X_i N_j^D - X_j N_i^D}{D_{ij}} - \frac{N_i^D}{D_i^k} = \frac{1}{RT} \nabla p_i' \qquad (1)$$

where N_j^D is the total diffusive molar flux of component j (mol L^2 t^{-1}); X_j is the mole fraction of component j; T is the absolute temperature (T); and R is the universal gas constant (ML^2 t^{-2} T^{-1} mol^{-1}). p_i' is the potentiometric head of component i (ML^{-1} t^{-2}) which satisfies:

$$\nabla p_i' = \nabla p_i + \rho_i g \nabla z \qquad (2)$$

Here p_i is the partial pressure of component i ($ML^{-1}t^{-2}$); g is the acceleration of gravity (Lt^{-2}); Z is the vertical (depth) coordinate; and ρ_i is the density of component i (ML^{-3}). The transport coefficients D_{ij} and D_i^k appearing in equation (1) are, respectively, the effective binary molecular diffusivity of component i in j (L^2t^{-1}) and the effective Knudsen diffusivity of component i (L^2t^{-1}) for the porous medium.

Equation (1) represents a system of υ equations where υ is the number of modeled gas components. It assumes ideal gas behavior and treats the soil matrix as a collection of large immobile or "dust" particles (component p). The total diffusive molar flux appearing in the governing equation includes both Knudsen and molecular diffusive transport processes. It can be separated into two components, a Fickian or "equimolar" flux and a so-called "non-equimolar" flux (Cunningham & Williams 1980). One of the goals of the present work is to explore the potential importance of this non-equimolar component to organic vapor transport. It has traditionally been neglected in previous modeling efforts.

The effective diffusivities appearing in equation (1) are presumed herein to be related to the free air diffusivity D_{ij}^o and the pore Knudsen coefficient D_{io}^k according to (Cunningham & Williams 1980):

$$D_{ij} = Q_m D_{ij}^o$$
$$D_i^k = Q_p D_{io}^k \qquad (3)$$

where Q_m and Q_p are coefficients which incorporate the effects of pore geometry and tortuosity (Geankoplis 1972):

$$Q_m = \frac{\varepsilon}{\kappa^2}$$
$$Q_p = \frac{\varepsilon}{\kappa_A^2} \qquad (4)$$

Here ε is the matrix gas-filled porosity and κ and κ_A are the molecular tortuosity and the Knudsen tortuosity, respectively. For the simulations presented in this paper it is assumed that $\kappa = \kappa_A$.

Free air binary diffusivities are estimated herein from the following correlation expression based upon the Chapman-Enskog theory and the Lennard-Jones potential energy function (Geankoplis 1972):

$$D_{AB}^o = 0.0018583 \left(\frac{1}{M_A} + \frac{1}{M_B}\right)^{1/2} \frac{T^{3/2}}{p\sigma_{AB}^2 \Omega_{AB}^D} \qquad (5)$$

where:

M_A	molar weight of gas component A, gm/mole
D_{AB}^o	Molecular diffusivity, cm^2/sec
T	^oK
p	total pressure, atm (=CRT)
σ_{AB}	"collision diameter" , $(=(\sigma_A+\sigma_B)/2)$
Ω_{AB}^D	collision integral based on the Lennard-Jones potential

The pore Knudsen diffusivity is independent of pressure but depends on pore size and component molar weight. It can be estimated from (Geankoplis 1972):

$$D^k_{Ao} = 9.7 \times 10^3\, r \sqrt{\frac{T}{M_A}} \qquad (6)$$

where

r the pore radius, cm

D^k_{Ao} Knudsen diffusivity, cm^2/sec

It is useful to explore the behavior of equation (1) under certain limiting conditions. In very fine grain porous media, the gas molecules will tend to collide only with the solid soil particles. Diffusive flux under such conditions is said to be in the "Knudsen regime." In such a scenario, $D_{ij} >> D^k_i$ and the first term on the left hand side of equation (1) is negligible in comparison with the second term. If gravitational effects are neglected, the following expression may then be obtained for the total diffusive flux of a gas component i:

$$N^D_i = -D^k_i \nabla C_i \qquad (7)$$

Here C_i is the molar concentration of component i (mol L^{-3}) ($C_i = \rho_i/M_i$). Note that equation (7) has a Fickian form.

In contrast, for coarse grain (larger pore) materials ($D_{ij} << D^k_i$) the second term in equation (1) will be negligible. If one can, further, neglect total pressure gradients, equation (1) reduces to the Stefan-Maxwell equation (Thorstenson & Pollock 1989a):

$$\sum_{i=1,j\neq i}^{\upsilon} \frac{X_i N^T_j - X_j N^T_i}{D_{ij}} = \frac{p'}{RT} \nabla X_i \qquad (8)$$

For a general multicomponent mixture, equation (8) will not reduce to a Fickian form.

The work presented herein is confined to the examination of one-dimensional gas transport for two-component mixtures. With some algebraic manipulation, equation (1) can be rearranged to yield the total diffusive molar flux for each of these components (A,B):

$$N^D_A = -\frac{D_{AB} D^k_A C}{D_{AB} + D^k_A(1-\alpha_A X_A)} \nabla X_A - \frac{D^k_A(D_{AB} + D^k_B) X_A}{D_{AB} + D^k_A(1-\alpha_A X_A)} \nabla C - \frac{D^k_A(D_{AB}M_A + D^k_B M) X_A C g}{D_{AB} + D^k_A(1-\alpha_A X_A) RT} \sin\theta \qquad (9)$$

where $\alpha_A = 1-(M_A/M_B)^{1/2}$ and $M = X_A M_A + X_B M_B$. θ is the orientation of the direction of flow with the horizontal. N^D_B, the total diffusive molar flux of component B, may be obtained from equation (9) by interchanging subscripts A and B. Because the Knudsen diffusivities depend directly upon pore size, a more generalized form of (9) has been suggested for a well-graded medium of varying pore diameters. (Feng et al 1973; Alzaydi et al 1978):

$$N^D_A = \sum_{r_{min}}^{r_{max}} \frac{f_v(r)}{\kappa^2} \Big[-D^o_{AB} D^k_{Ao} C \nabla X_A - D^k_{Ao}(D^o_{AB} + D^k_{Bo}) X_A \nabla C - D^k_{Ao}(D^o_{AB} M_A + D^k_{Bo} M) \frac{X_A C g}{RT} \sin\theta \Big] (D^o_{AB} + D^k_{Ao}(1-\alpha_A X_A))^{-1} \qquad (10)$$

Here $f_v(r)$ is the volume void fraction of the medium composed of pores having radius r and D^k_{Ao} is a function of pore radius.

Close examination of equation (9) or (10) reveals that the total diffusive flux will reduce to a Fickian form for this two-component case only under some special conditions. For an isobaric system in which both components have identical molecular weights and when the gravitational effect is small, equation (9) reduces to:

$$N^D_A = -\left(\frac{1}{D^k_A} + \frac{1}{D_{AB}}\right)^{-1} \nabla C_A \qquad (11)$$

In the molecular regime ($D^k_A >> D_{AB}$) equation (11) becomes:

$$N^D_A = - D_{AB} \nabla C_A \qquad (12)$$

This is the familiar Fick's Law commonly employed to model diffusive transport in the unsaturated zone. Note that, although equation (11) reduces to a Fickian form in the Knudsen regime (see equation (7)), gas transport cannot be modeled in this regime by Fick's Law employing an effective molecular diffusivity.

2.2 Viscous flux

For a non-isobaric system in which Knudsen transport is negligible, the molar viscous flux can be expressed by Darcy's Law:

$$N^v = - \frac{B_k p}{\mu RT} \nabla p' \qquad (13)$$

Here B_k is the intrinsic permeability of the medium (L^2) and μ is the dynamic viscosity ($ML^{-1}t^{-1}$). Equation (13) may be expanded in terms of total molar concentration and the gravity component of p' to yield:

$$N^v = - \frac{B_k}{\mu} RTC \nabla C - \frac{B_k}{\mu} C^2 M \, g \sin\theta \qquad (14)$$

It should be noted that, like the nonequimolar component of the diffusive flux, the viscous flux is nonseparative. That is, viscous fluxes of individual species in a multicomponent system are given as:

$$N^v_i = X_i \, N^v \qquad (15)$$

If one examines equation (9) or (10) in conjunction with equation (14) it will be observed that a pressure gradient simultaneously gives rise to a viscous flux and a component of the total diffusive flux of the gas mixture.

2.3 Model governing equations

The total flux of a given component i in a gas mixture is the sum of its diffusive and viscous components:

$$N^T_i = N^D_i + X_i \, N^v \qquad (16)$$

Here N^D_i and N^v are given by equations (10) and (14) respectively. A mass balance equation for the migration of a gas phase component i in an isothermal, rigid porous matrix may thus be written as:

$$\varepsilon \frac{\partial (CX_i)}{\partial t} + \frac{\partial}{\partial x} (N^T_i) = 0 \qquad (17)$$

For a binary gas mixture, the system of equations (17) represents a coupled system of two nonlinear partial differential equations in three unknowns (C, X_A, X_B) where the total diffusive flux is given by equation (9) or (10). Equations (17) assume a single fluid (gas) phase system and implicitly treat this phase as ideal. The coupled equations are subject to a constraint on mole fractions ($X_A = 1 - X_B$) which closes the system.

3 NUMERICAL SIMULATIONS

3.1 Discretization and solution of the governing equations

The system of coupled equations (17) was discretized for a binary mixture using the standard Galerkin method of weighted residuals in space and a backwards difference approximation in time. Linear Chapeau basis functions were employed as trial and weighting functions. Variable coefficients were also expanded in terms of these trial functions. The resulting system of coupled non-linear equations was solved using a Picard (successive substitution) iterative scheme. Convergence of the scheme was presumed when the relative changes in the calculated variables (X_A and C) were less than some specified tolerances (10^{-7} and 10^{-7} mole/cm^3, respectively). Although the model incorporates an algorithm to compute the total diffusive flux according to equation (10), the example simulations presented herein employ the simplified form (9) and assume a porous matrix of uniform pore size.

The model was developed to handle either first or second type boundary conditions. For the example simulations in this paper, first type conditions were employed as follows:

$C = C_o, X_A = X_{Ao}$ @ $x = 0$

$C = C_o, X_A = 0$ @ $x = L$ for pure diffusive flow

$C = C_i, X_A = 0$ @ $x = L$ for combined viscous and diffusive flow

Here L is the domain length.

Based upon the development of equation (11), the system of equations (17) should reduce to the traditional diffusion equation for an isobaric system of equal molecular weight components. The computer simulator was validated for this special case by comparison with an analytical solution for a homogeneous system. A second model validation was conducted by comparing simulations with experimental data and computational results presented in Alzaydi et al (1978).

3.2 Simulation parameters

The goal of this preliminary modeling study is to explore the potential impact of Knudsen transport and nonequimolar flux on predictions of transient organic vapor migration in unsaturated soil columns. To simplify the analysis, soil homogeneity and very low moisture content are assumed and organic vapor partitioning to the solid or aqueous phases is neglected. In the hydrology and soils science literature, vapor transport has traditionally been modeled using a diffusion-advection equation (see, for example, Sleep & Sykes 1989). In terms of molar concentrations, this equation may be written as:

$$\varepsilon \frac{\partial C_i}{\partial t} = \frac{\partial}{\partial x}\left(\varepsilon D^* \frac{\partial C_i}{\partial x}\right) - q \frac{\partial C_i}{\partial x} \qquad (18)$$

where q is the Darcy velocity. D^* is an effective diffusion coefficient usually represented as:

$$D^* = D^o_{AB}\, \tau \qquad (19)$$

Where τ is a tortuosity parameter introduced, as in the Dusty Gas Model, to account for the impedance to the diffusive flux created by the pore structure. Tortuosity is typically represented as a function of air-filled porosity and is presumed to be independent of the molecular weight of the diffusing component.

As has been noted previously, under isobaric conditions and neglecting the effect of gravity, the Dusty Gas Model representation of the diffusive flux reduces to a Fickian form. (first term in equation (9)). The coefficient of this gradient term could be equated with that coefficient appearing in equation (19) to find τ. Note, however, that the resulting "tortuosity" would be a function of the mole fractions of the diffusing components. Only if components A and B have equal molecular weights will a constant τ be obtained. It's form would be as follows:

Knudsen regime

$$\tau \approx \frac{D^k_{Ao} \frac{1}{\kappa^2}}{D^o_{AB}} \qquad D_{AB} \gg D^k_A$$

Molecular regime (20)

$$\tau \approx \frac{1}{\kappa^2} \qquad D_{AB} \ll D^k_A$$

For the simulation comparisons contained herein, tortuosity values for equation (18) and (19) were computed using the appropriate equation (20).

Simulation parameters are given in Tables 1 - 3. The migration of two organic compounds (trichloroethylene (TCE) and methane (CH_4)) are explored for two different soil types. Nitrogen (N_2) is used as a surrogate for air in the binary systems. The simulated domain is a 90 cm column. For all simulations, the soil pore space is initially filled with pure N_2. A constant source of the vapor contaminant is introduced at one end of the column and the far end is presumed uncontaminated. For the pure diffusive transport case, both ends of the column are maintained at a constant (common) total molar concentration. For comparisons with the full diffusion-advection equation, the total molar concentration at each end is set to a constant value consistent with an applied pressure gradient.

Table 1. Soil parameters

	Soil Type		
Para.	Medium Sand	Kaolinite	Reference
ε	0.34	0.7	1
κ	1.05	1.225	1
κ_A	1.05	1.225	1
r	4e-3 cm	4e-7 cm	1
B_k	3e-6 cm^2	1.25 e-12cm^2	2
τ	0.308	0.069	eq'n (20)
D^k_{io}	$\frac{669.79}{\sqrt{M_i}} \frac{cm^2}{sec}$	$\frac{0.06698}{\sqrt{M_i}} \frac{cm^2}{sec}$	eq'n (6)

1 Alzaydi et al 1978
2 Freeze & Cherry 1979

Table 2. Environmental parameters

Parameter	Value
T	298 ^{o}K
p(=CRT)	9.9 e+5 dyne cm^{-2}
R	8.3e+7 cm-dyne/mol ^{o}K
p_L	(9.9 e+5 - 45) dyne cm^{-2}

Table 3. Gas properties

Gas	Viscosity (10^4 g/cm-sec)	Molar Weight (g/mol)	D^o_{AB} (cm^2/sec)
N_2	1.76	28.	$\frac{0.202}{p}$
TCE	0.95	131.	$\frac{0.09657}{p}$
CH_4	1.59	16.	$\frac{0.219}{p}$

p as [atm.]

3.3 Example Simulations

Example simulation results are given in Figures 1-4 to illustrate the importance of different transport mechanisms. In all figures, computed molar fractions of the organic component are shown as a function of distance along the column at various times. An element length of 2 cm was employed for all simulations. In the figures, the Dusty Gas Model is represented as DGM and the traditional advection-diffusion model (equation (18)) is represented as A-D. Simulations were performed in a horizontal domain unless otherwise noted.

Examples for what is herein termed "pure" diffusive transport are shown in Figure 1. Here there is no externally applied pressure gradient. Inspection of the figures reveals that the A-D model appears to adequately predict vapor migration in both large and small pore media at early times if tortuosity has been selected in accordance with equations (20). Nonequimolar effects, however, have an increasingly important influence as travel time (distance) increases. A-D predictions tend to lag those of the DGM in the coarse-grained material (Figure 1(a)). This behavior was observed for compounds of lower molecular weight than air ($\alpha_A < 1$ in Eq.(9)). Alternatively, for $\alpha_A > 1$, the A-D model tends to overpredict diffusive fluxes in large pore media. As pore size decreases ($D_{AB} > D^k_A$), α_A plays a less important role, and the A-D results consistently overestimate the diffusive flux, regardless of molecular weight ratio (Figure 1(b)). The accompanying Figure 2 shows the evolution of pressure along the column during the diffusive processes illustrated in Figure 1. For the coarse-grained soil, pressure deviations from initial conditions are negligible. However, for the fine-grained matrix, one can observe that the presence of compounds of unequal molecular weight gives rise to a pressure gradient within the domain and consequent viscous flux which cannot be predicted by the traditional Fickian model. Similar results were obtained for the saturated TCE vapor case. The importance of this nonequimolar effect in multicomponent systems has been supported

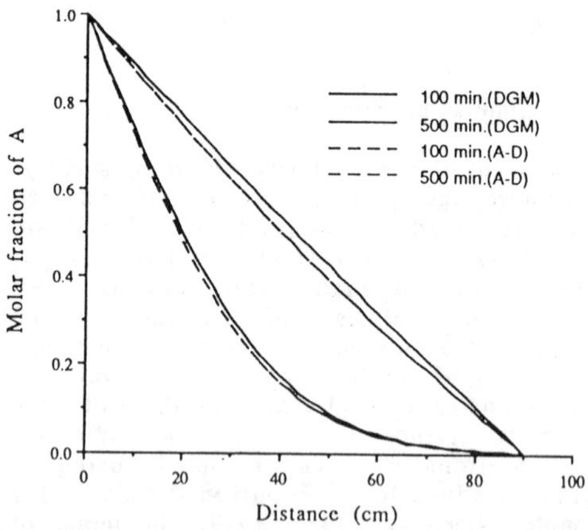

Fig. 1(a) Diffusive transport of methane in sandy soil (Δt=5 min.)

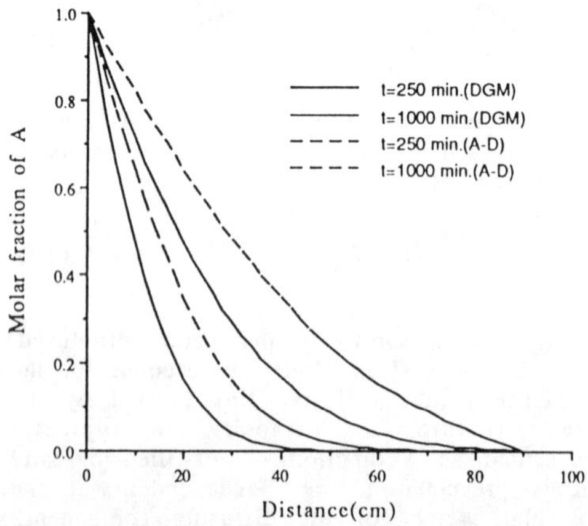

Fig. 1(b) Diffusive transport of methane in clay soil (Δt=5 min.)

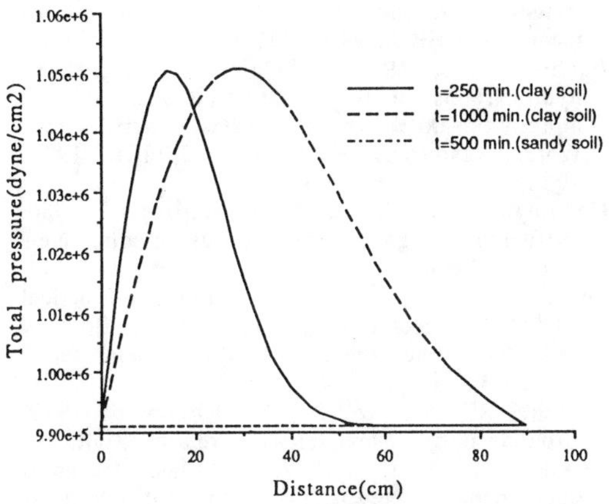

Fig. 2 Variation of total pressure within soil column for diffusive transport of methane (Δt=5 min.)

by the steady-state flux calculations of Thorstenson & Pollock (1989a).

A number of comparisons were made between the DGM and the standard A-D model under application of an externally applied pressure gradient. Even under a small applied pressure gradient, such as that given in Table 2, the two models gave indistinguishable results as long as advection tended to dominate the system (as in the sandy soil). Under an applied gradient in the clay soil, however, diffusive processes still tended to dominate the flux, due to the small intrinsic permeability of the medium. For these conditions, simulation results were quite similar to the pure diffusive case, as illustrated in Figure 1(b).

The effect of gravity on vapor transport is illustrated in Figure 3 for saturated TCE vapor (X_{Ao}= 0.088, Reid et al 1988) in the sandy soil. Here there is no externally applied pressure gradient. Clearly, there is a substantial gravity effect on the transport as is seen by comparison of the horizontal and vertical column results. As anticipated, this gravity effect was found to be more predominant as the difference between component molecular weights increases or as soil intrinsic permeability increases. The contribution of the third term in equation (9) to vapor transport can be seen by comparing curves labeled vertical and vertical*. For the vertical* simulation, this term, which accounts for gravity effects on the total diffusive flux, was dropped from the model. Model results with and without this term are indistinguishable in the figure. All other examined simulations were consistent with this result. Thus, it appears that, although gravity can create substantial advective fluxes under nonequimolar conditions, it has negligible impact upon diffusive fluxes.

In the final figure (Figure 4), three different modeling approaches are compared. The pictured simulations involve purely diffusive horizontal transport of saturated TCE vapor in the sandy soil. Here Model I represents the solution of a coupled system of transport and gas phase flow equations. A detailed description of this model may be found in Reeves and Abriola (1988). It is a Galerkin finite element approach which, through the coupling of flow and transport, can incorporate the influence of the phase density gradient on the mass flux. In this respect it is similar to other gas phase transport models which have appeared recently in the literature (such as Mendoza & Frind 1990). Model I equations are formulated and solved on a mass (rather than a mole)

Fig. 3 Vertical intrusion of saturated TCE vapor in sandy soil (Δt=1 min.)

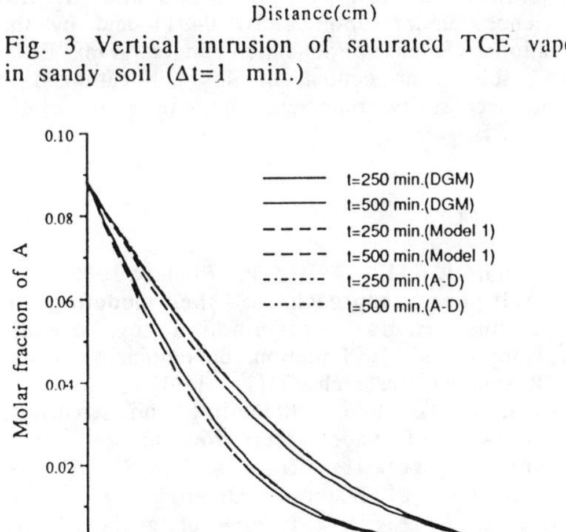

Fig. 4 Diffusive transport of saturated TCE vapor in sandy soil (Δt=5 min.)

basis, as is consistent with the phase balance formulation. Results from the A-D model and the DGM are indistinguishable in the figure. Model I results, however, begin to deviate markedly from the DGM at later times (larger travel distances). This behavior is apparently related to the greater significance of the nonequimolar flux when expressed relative to a mass average velocity.

4. CONCLUSIONS

Based upon the preliminary modeling study described briefly above, the following conclusions may be drawn. It must be noted that these conclusions apply only to binary systems. It is anticipated that deviations of the DGM predictions from the traditional A-D model approach may be more substantial in multicomponent systems, since, for this case, the traditional Fick's law is not valid even under limiting conditions.

1. Nonequimolar fluxes may be significant for "purely diffusive" flow systems. The deviation from Fick's law predictions increases with travel distance. Significant pressure gradients may be induced in finer grained media.
2. Nonequimolar fluxes may not play an important role in advective dominated systems.
3. Deviations of the DGM from the traditional A-D modeling approach appear to be more marked when governing equations are formulated on a mass basis.

ACKNOWLEDGMENTS

This work was supported in part by the National Institute of Environmental Health Sciences under Grant #ESO 04911 and by the National Science Foundation under grant ECE-8451469. The content of this publication does not necessarily represent the views of either agency.

REFERENCES

Abriola, L. M., & G. F. Pinder 1985. A multiphase approach to the modeling of porous media contamination by organic compounds, 1. Equation development. Water Resources Research. 21(1): 11-18.

Allan, R. E. 1986. Modelling and sensitivity analysis of vapor migration in soil from spilled volatile liquids, M.S. thesis, University of Waterloo, Ontario.

Alzaydi, A. A. 1975. Flow of gases through porous media. Ph.D. Thesis. Ohio State University.

Alzaydi, A. A., C. A. Moore, & I. S. Rai 1978. Combined pressure and diffusional transition region flow of gases in porous media. AICHE J.. 24 (1): 35-43.

Baehr, A. L. 1987. Selective transport of hydrocarbons in the unsaturated zone due to aqueous and vapor phase partitioning. Water Resources Research. 23(10): 1926-1938.

Cunningham, R. E. & R. J. Williams 1980. Diffusion in gases and porous media. New York: Plenum.

Feng, C., & W. E. Stewart 1973. Practical models for isothermal diffusion and flow of gases in porous solids. Ind. Eng. Chem. Fund. 12(2): 143-147.

Freeze, R. A., & J. A. Cherry 1979. Groundwater. New Jersey: Prentice Hall.

Geankoplis, C. J. 1972. Mass transport phenomena. New York: Holt, Rinehart & Winston.

Mason, E. A. and A. P. Malinauskas 1983. Gas transport in porous media : The dusty-gas model. Chem. Eng. Monograph, 17. New York: Elsevier.

Mendoza, C. A. & E. O. Frind 1990. Advective-dispersive transport of dense organic vapors in the unsaturated zone, 1. Model development. Water Resources Research. 26(3): 379-389.

Reeves, H. W., & L. M. Abriola 1988. A decoupled approach to the simulation of flow and transport of non-aqueous organic phase contaminants through porous media. Computational Methods in Water Resources, Vol. 1 Modeling Surface and Subsurface Flows. New York: Elsevier. 147-152.

Reid, R. C., J.M. Pransnitz and B.E. Poling 1988. The properties of gases & liquids. New York: McGraw-Hill.

Sleep, B. E. & J. F. Sykes 1989. Modeling the transport of volatile organics in variably saturated media. Water Resources Research. 25(1): 81-92.

Silka, L. R 1986. Simulation of the movement of volatile organic vapor through the unsaturated zone as it pertains to soil-gas surveys. Petroleum hydrocarbons and organic chemicals in groundwater. Worthington, Ohio: National Water Well Association. 204-226.

Thorstenson, D. C., and D. W. Pollock 1989a. Gas transport in unsaturated zones: Multicomponent systems and the adequacy of Fick's Law. Water Resources Research. 25(3): 477-507.

Thorstenson, D. C. & D. W. Pollock 1989b. Gas transport in unsaturated porous media: the adequacy of Fick's Law. Reviews of Geophysics. 27(1): 61-78.

Subsurface Contamination by Immiscible Fluids, Weyer (ed.) © 1992 Balkema, Rotterdam. ISBN 90 6191 175 3

Evaluation of methods for improving the efficiency and robustness of multiphase flow equations

A.K.Katyal, J.J.Kaluarachchi & J.C.Parker
Virginia Polytechnic Institute and State University, Blacksburg, Va., USA

ABSTRACT An adaptive solution domain (ASD) finite element model for simulating hydrocarbon spills has been developed that is computationally more efficient than conventional numerical methods. Coupled flow of water and oil with an air phase at constant pressure is considered. Several formulations of the governing equations are evaluated for use with the ASD method. The most robust formulation involved flow equations written in terms of water and oil pressures with mass storage terms as time derivatives of fluid contents treated as nonlinear functions of pressures. In the ASD formulation, the solution domain for water and oil flow equations is restricted by eliminating elements from the global matrix assembly which are not experiencing significant changes in fluid saturations or pressures. When any nodes of an element exhibits changes in fluid pressures more than specified tolerances during the current time step, it is labelled active and included in the computations for the next iteration.

1 INTRODUCTION

Contamination of groundwater by organic chemicals poses a serious threat to subsurface water resources. Many of these contaminants exhibit low water solubility and will move within unsaturated and saturated subsurface zones as separate liquid phases when introduced as a free liquid phase. Numerical analyses of multiphase flow have been described by a number of researchers in recent years. Due to the strong coupling and high degree of nonlinearity, simultaneous solution of the multiphase flow equations using rather fine grids and small time steps is necessary. In problems involving analysis of nonaqueous phase liquid (NAPL) spills, the assumption of a negligible gas phase pressure gradient is commonly invoked, thus leaving only the NAPL and water flow equations to solve. A variety of formulations of the governing equations for NAPL and water flow have been taken as a starting point for numerical model development. Abriola and Pinder (1985a,b), Osborne and Sykes (1986), Kuppusamy et al. (1987) and Kaluarachchi and Parker (1989) used flow equations in terms of fluid pressures with mass storage terms expressed as the product of time derivative of fluid pressure and saturation-pressure derivatives. The formulations reported by Faust (1985) and Forsyth (1988) have oil pressure and water saturation as the dependent variables. These solutions are based on the fully implicit formulations of the governing equations throughout the spatial domain using conventional spatial discretization procedures. In most practical problems, large changes in fluid pressures and saturations do not occur throughout the spatial domain at a given time step. Computational effort is thereby inefficiently spent solving equations in areas where little activity occurs rather than concentrating effort in the more active zones. However, since the location of active zones changes with time, schemes designed to take advantage of the these variations must be capable of automatically adapting to the current conditions. One method for doing so involves adaptive grid refinement in which the numerical mesh is automatically refined in active zones and expanded in inactive zones. Another general approach which employs a fixed grid varies the solution approach within the domain to gain efficiency. Adaptive implicit methods have been described by Thomas and Thurnau (1983), Forsyth and Sammon (1986) and Forsyth (1988) that involve solving implicit formulations of the governing equations in parts of the domain where changes in the primary variables are large, while in the rest of the domain a computationally less intensive but less robust implicit pressure-explicit saturation (IMPES) formulation is employed.

In the present paper, we describe a new fixed-grid adaptive solution approach designed to reduce the computational effort required in solving multiphase flow problems. The proposed Adaptive Solution Domain (ASD) method, employed with the finite element method for discretizing the governing equations, reduces the solution domain by selectively eliminating elements from the global matrix assembly step if current changes in nodal fluid pressures or saturations are below stipulated tolerances.

2 FORMULATION OF GOVERNING EQUATIONS

Method 1: Pressure formulation with saturation time derivatives

Assuming an incompressible porous medium and incompressible and compositionally uniform liquids, the governing equations for flow in a three fluid phase porous medium may be written as (Parker, 1989)

$$\Phi \frac{\partial S_w}{\partial t} = \frac{\partial}{\partial x_i}\left(K_{w_{ij}}\left[\frac{\partial h_w}{\partial x_j} + u_j\right]\right) \quad [1a]$$

$$\Phi \frac{\partial S_o}{\partial t} = \frac{\partial}{\partial x_i}\left(K_{o_{ij}}\left[\frac{\partial h_o}{\partial x_j} + \rho_{ro} u_j\right]\right) \quad [1b]$$

where Φ is porosity; S_p is the p-phase fluid saturation for $p = o$ (oil) or w (water); t is time; x_i is the i-direction coordinate; K_{pij} is the p-phase conductivity tensor (i,j = 1,2,3) for phase p; $h_p = P_p/\rho_w g$ is the water height-equivalent pressure head of the p-phase where P_p is the p-phase pressure, ρ_w is the density of water, and g is gravitational acceleration; ρ_{rp} is the specific gravity of the p-phase ($\rho_{rw} \equiv 1$); and $u_j = \partial z/\partial x_j$ is the unit gravitational vector where z is elevation. For brevity, we will refer to the nonaqueous phase liquid simply as "oil." Phase conductivities are described by

$$K_{wij} = k_{rw}\, K_{swij} \quad [2a]$$

$$K_{oij} = k_{ro}\, K_{swij}/\, \eta_{ro} \quad [2b]$$

where K_{swij} is the saturated conductivity to water, k_{rp} is the relative permeability of phase p, and η_{ro} is the ratio of oil to water viscosity. We shall assume that the coordinate axes are parallel to the principal direction of the hydraulic conductivity tensor such that $K_{swij} = 0$ for $i \neq j$. Initial and type 1 boundary conditions are specified in terms of water and oil heads.

The unknowns in Method 1 are taken to be water and oil head, h_w and h_o, which will be treated here using a finite element model with linear basis functions. Time derivatives will be handled using a finite difference approximation. Water and oil relative permeabilities, k_{rw} and k_{ro}, and saturations, S_w and S_o, are treated as nonlinear functions of phase pressures which are updated iteratively for the current time step.

Method 2: Pressure formulation with split time derivatives

Method 2 also involves oil and water pressure heads as unknowns, but in this case saturation time derivatives are split into two terms each written as pressure time derivatives. Noting that $S_p = f(h_w, h_o)$ and expanding the time dervatives of [1] yields

$$C_{ww}\frac{\partial h_w}{\partial t} + C_{wo}\frac{\partial h_o}{\partial t} = \frac{\partial}{\partial x_i}\left(K_{w_{ij}}\left[\frac{\partial h_w}{\partial x_j} + u_j\right]\right) \quad [3a]$$

$$C_{ow}\frac{\partial h_w}{\partial t} + C_{oo}\frac{\partial h_o}{\partial t} = \frac{\partial}{\partial x_i}\left(K_{o_{ij}}\left[\frac{\partial h_o}{\partial x_j} + \rho_{ro} u_j\right]\right) \quad [3b]$$

where pq-phase fluid capacities are defined by

$$C_{pq}=\phi\,\frac{\partial S_p}{\partial h_q} \qquad p,\ q = o,\ w \quad [3c]$$

In Method 2, water and oil relative permeabilities, k_{rw} and k_{ro}, and fluid capacities, C_{pq}, are treated as nonlinear functions of phase pressures which are updated iteratively.

Method 3: Capillary pressure formulation

In Method 3, fluid pressure heads are replaced by capillary heads defined for fluid pair pq by $h_{pq} = h_p - h_q$. Substituting air-oil and oil-water capillary heads into [1] and noting that in the present analysis $\partial h_a/\partial x_i = 0$ by assumption yields

$$\Phi \frac{\partial S_w}{\partial t} = \frac{\partial}{\partial x_i}\left(K_{w_{ij}}\left[u_j - \frac{\partial h_{ao}}{\partial x_j} - \frac{\partial h_{ow}}{\partial x_j}\right]\right) \quad [4a]$$

$$\Phi \frac{\partial S_o}{\partial t} = \frac{\partial}{\partial x_i}\left(K_{o_{ij}}\left[\rho_{ro} u_j - \frac{\partial h_{ao}}{\partial x_j}\right]\right) \quad [4b]$$

In this method, air-oil and oil-water capillary heads are taken as the primary variables and phase saturations and relative permeabilities are treated as nonlinear functions of the capillary heads. Initial and type 1 boundary conditions for the model are specified in terms of air-oil and oil-water capillary pressures.

Method 4 Semi-diffusive oil pressure-water saturation formulation

Diffusive forms of mass flow equations may be written by expanding pressure derivative terms into saturation gradients and pressure-saturation derivatives yielding homogeneous equations in terms of fluid saturation. Here, we consider a "semi-diffusive" two phase model which is formulated in terms of water saturation and oil pressure. The basic form of the governing equations in this formulation is the same as used in Method 1. However, the unknowns are now taken to be water saturation, S_w, and oil pressure head, h_o. Other quantities in the equations are treated as nonlinear functions of the primary variables. Water relative permeability is expressed as a function of water saturation, and oil relative permeability is expressed as a function of water saturation and oil head via the constitutive relations described in the Appendix. Oil saturation is computed from the equality: $S_o=S_t-S_w$ where total liquid saturation, S_t, is computed as a function of oil pressure head. Initial and type 1 boundary conditions for Method 4 are prescribed in terms of water saturation and oil pressure.

Method 5: Semi-diffusive water pressure-liquid saturation formulation

Method 5 is a semi-diffusive scheme in terms of total liquid saturation, $S_t=S_o+S_w$, and water pressure head, h_w. The governing equations for this method are identical to those in Method 1. The transformation of the governing equations to obtain S_t and h_w as the primary variables is achieved by treating all other variables in the equations as nonlinear functions of the primary variables. Oil pressure head, h_o, is expressed as a function of S_t via the constitutive model described in the Appendix and S_w is expressed as a function of h_w and S_t. Water and the oil relative permeabilities, k_{rw} and k_{ro}, are expressed in terms of h_w and S_t also via the constiutive relations. Initial and type 1 boundary conditions for Method 5 are specified in terms of total liquid saturation and water pressure.

3 SOLUTION PROCEDURE

The governing equations for each of the above formulations have been solved using a finite element method. Spatial derivative terms in the flow equations are weighted by unsymmetric upstream-weighting functions developed by Huyakorn and Nilkuha (1979), while the remaining terms are handled using linear basis functions. Nonlinearity is handled with the Newton-Raphson procedure. Details of the numerical model are described by Katyal & Parker (1990) and will not be repeated here.

In the ASD method, elements are categorized as either "active" or "inactive" at a given iteration. If the element is classified as active, then it is included in the global matrix assembly indicated by [9], otherwise it is excluded. The criteria for changing a node from inactive to active is based on the changes in water pressure head, h_w, oil pressure head, h_o, oil saturation, S_o, and water saturation, S_w at connected nodes from the last converged time step to the current iteration. The criteria for switching an active element to an inactive status is based on the difference in h_w, h_o, S_w and S_o at the connected nodes between the converged values at the previous and the current time step. The specific criteria for changing the classification of an element is as follows:

Inactive → active if any nodes meet the conditions:

$$\Delta h_w > \tau_1 \qquad [5a]$$

or

$$\Delta h_o > \tau_1 \qquad [5b]$$

or

$$\Delta S_w > \tau_2 \qquad [5c]$$

or

$$\Delta S_o > \tau_2 \qquad [5d]$$

or

$$\sum h_w > \tau_3 \qquad [5e]$$

Active → inactive if all nodes meet the conditions:

$$\Delta h_w < \tau_4 \qquad [5f]$$

and

$$\Delta h_o < \tau_4 \qquad [5g]$$

and

$$\Delta S_w < \tau_5 \qquad [5h]$$

and

$$\Delta S_o < \tau_5 \qquad [5i]$$

where Δh_w, Δh_o, ΔS_w and ΔS_o in [5a-5d] are the changes in h_w, h_o, S_w and S_o, respectively, between the previous converged time step and the current iteration, while in [5f-5i] these refer to the changes between the converged values at the end of the current and the previous time step, and $\sum h_w$ is the cumulative change in h_w from the last time the node met criterion [5a-5d] to the present time.

To ensure an accurate solution, a bias is imposed towards maintaining active elements by making $\tau_4 < \tau_1$ and $\tau_5 < \tau_2$. Preliminary simulations for a number of cases indicated good results using $\tau_3 = 2\ cm$ and $\tau_4=\tau_1/100$ subject to the constraint that $0.001 \leq \tau_4 \leq 0.0001\ cm$ and $\tau_5=\tau_2/10$ subject to the constraint that $0.0001 \leq \tau_5 \leq 0.00001$. The foregoing values will be employed in the present study. To avoid erratic switching between active and inactive subdomains, an active element is labelled inactive only after conditions [5f-5i] are satisfied for three consecutive time steps. This allows a smooth transition between active and inactive elements while achieving high computational efficiency.

It may be noted that complete elimination of a node from the global matrix only occurs if all elements associated with the node are classified as inactive. Equations for nodes on the active-inactive frontier will have contributions only from active elements and the boundary will shift naturally as imbibition or drainage "fronts" pass through the domain.

4 NUMERICAL RESULTS

Five example problems involving infiltration and redistribution of organic liquids in partially water saturated soils will be investigated. The first three examples are used to compare the computational efficiency, mass balance error and robustness of the different equation formulations. Examples 4 and 5 will be described to illustrate application of the ASD algorithm and to compare its performance with a conventional full domain finite element approach for. Assumed fluid and soil parameters in the constitutive model for permeability-saturations-capillary pressure relations described in the Katyal and Parker (1990) are given in Table 1.

Table 1. Soil and fluid properties used in simulations.

Parameter	Example 1	2 & 4	3	5
n	3.00	3.00	3.00	2.70
α, cm^{-1}	0.05	0.05	0.05	0.02
S_m	0.00	0.00	0.00	0.0
K_{sw}, $cm\ h^{-1}$	20.00	20.00	50.00	2.10
β_{ao}	3.20	3.20	3.20	3.20
β_{ow}	1.455	1.455	1.455	1.455
ρ_{ro}	0.80	1.20	0.80	0.8
η_{ro}	0.80	0.80	2.00	2.0
ϕ	0.40	0.40	0.40	0.4

Example 1

This problem involves vertical infiltration and redistribution of light oil in a 200 cm long column initially free of organic liquid and in equilibrium with a water table at a depth of 100 cm. For Methods 1, 2 and 4, oil infiltration was permitted under zero oil pressure, while in Method 3, air-oil capillary pressure was fixed at zero. In Method 5, the top boundary was simulated by maintaining the total liquid saturation, S_t, at unity. In all simulations, infiltration was continued until a total volume of 3.95 $cm^3\ cm^{-2}$ was absorbed. Following the infiltration period, fluid redistribution was allowed with zero oil flux was specified at the upper boundary.

At the lower boundary, zero oil flux and a fixed water head of +100 cm was assumed. This was achieved directly for Methods 1, 2 and 5 by specifying zero oil flux and fixing h_w at 100 cm. For Methods 3 - 4, there is no way to directly specify the desired boundary conditions so surrogate approaches were employed. For Method 3, the capillary pressures were fixed at the lower boundary in such a way as to impose the correct water head and to force zero oil saturation. The air-oil capillary head was assigned a value $h_{ao} = h_a - h_o^{cr}$ where $h_a = 0$ (atmospheric gas pressure) and h_o^{cr} is the oil head at which $S_o \rightarrow 0$, which for the soil and fluid properties in this problem is equal to 31.25 cm (see Appendix). Thus, $h_{ao} = -31.25$ cm at the lower boundary and the oil-water capillary head, $h_{ow} = h_o - h_w$, is $100 - 31.25 = 68.75$ cm. For Method 4, with dependent variables S_w and h_o, the lower boundary conditions were specified by setting $S_w = 1$ and $h_o = 100$ cm.

Example 2

This case is the same as the previous example except that the organic liquid is denser than water (Table 1). Initial and boundary conditions imposed on this problem are taken identical to those of Example 1. The maximum mass balance error for all five methods is given in Table 2 for the redistribution period.

Table 2. Maximum percent mass balance error for oil.

Example	Method 1	2	3	4	5
1	0.50	1.87	0.47	NC	(a)
2	0.34	3.70	(b)	NC	(c)
3	0.75	(d)	0.71	0.10	0.07

NC = no convergence during infiltration

(a) Mass balance error of 0.6% prior to oil front reaching water table; thereafter, mass balance error became greater than 10%
(b) Mass balance error of 0.34% prior to oil front reaching lower boundary; thereafter, oil mass lost by flow through boundary
(c) Mass balance error of 0.1% prior to oil front reaching water table; thereafter, mass balance error became greater than 10%
(d) Mass balance error gradually increasing to greater than 10%

Example 3

In this example, we investigate a situation involving a domain in which water pressure is everywhere negative. Column dimensions, spatial discretization and time-integration parameters are the same as in Example 1. Fluid and porous media propoerties are similar to those of Example 1, except for a higher organic liquid viscosity and higher hydraulic conductivity (Table 1). Initial conditions correspond in this case to equilibrium with a water table at the bottom of the column (200 cm depth) and the water head is fixed throughout the simulation at the lower boundary at $h_w = 0$.

The maximum mass balance error for all five methods is given in Table 2 for the redistribution period. For this problem Methods 1 and 3 yield accuracy as before while Method 2 produces excessive mass balance error. Methods 4 and 5 provide very good accuracy for this problem illustrating the superiority of semi-diffusive schemes for conditions when all phases are at negative pressures. Overall, Method 1 appears to yield the most consistently accurate results for all problem. Therefore, in the subsequent examples we will consider only Method 1 for investigating the ASD method.

Example 4

This problem involves vertical oil infiltration and redistribution into a 225 cm long column initially free of hydrocarbon and in equilibrium with a water table at a depth of 200 cm. A dense oil was added at the top of the column under a constant head of $h_o = 0$ over a period of 0.14 h until the cumulative oil infiltration reached 6.92 cm. At the end of the infiltration period, the upper boundary condition for oil was specified as zero flux. Fluid redistribution was permitted for a period of 250 h after the end of the infiltration stage. A zero flux condition for water was imposed at the upper boundary at all times. At the lower boundary, water pressure was maintained continuously at $h_w = 25$ cm and zero oil flux was imposed. Lateral boundaries were zero flux for both fluids. Spatial discretization for the problem was achieved using 45 rectangular elements each 5 cm wide and spaced 5 cm apart in the vertical direction yielding a total of 92 nodes. All simulations were started with an initial time step of 0.0002 h and a time acceleration factor of $F = 1.03$ was used to increase the time step size up to a maximum time step of 0.5 h.

Simulations were carried out using the ASD method with various values of τ_1 and τ_2 and using a conventional full-domain finite element solution (CFEM) based on the same numerical formulation employed for the ASD solution but with no element exclusion. Simulations were performed in two stages corresponding to the period of oil infiltration and the subsequent redistribution period. Table 3 shows the CPU time required for infiltration and redistribution periods using different ASD tolerances and for the CFEM. The maximum percentage mass balance error for the oil phase (MBE) during the redistribution period is also given in Table 3. Water and total liquid saturation distributions predicted by the ASD method and the CFEM are virtually identical. Small differences among the various solutions were evident in the mass balance error (Table 3). The ASD cases with $\tau_1 \geq 0.01$ cm yielded nearly the same mass balance error as the CFEM, while the ASD case with $\tau_1 = 0.002$ cm exhibited substantially higher, although still tolerable, mass balance error. We attribute the poorer behavior of the latter case to erratic changes in the number of active elements for consecutive iterations of the same time step which result when imposed ASD tolerances undercut the precision in the numerical solution.

Table 3. CPU time and mass balance error on IBM 3090 for Example 4.

τ_1 (cm)	τ_2 (-)	CPU time (s) infiltration	CPU time (s) redistribution	MBE (%)
Conventional FEM		30	127	0.33
0.002	0.00005	9	92	0.66
0.010	0.00010	6	77	0.37
0.050	0.00050	6	76	0.43
0.100	0.00050	6	73	0.46

Marked savings in CPU time are obtained with the ASD method (Table 3). For the $\tau_1 \geq 0.01$ cm cases, the ASD method yields an overall speed-up factor (CPU time for CFEM/CPU time for ASD method) of 5 relative to the CFEM for the infiltration period. For the entire redistribution period, the overall speed-up factor diminishes to 1.7.

Example 5

A two-dimensional domain is considered in this example involving a vertical slice though an unconfined aquifer which is 7.50 m wide and 6.38 m in height (Figure 1). The system intially contains no oil and is in equilibrium with a water table at an elevation of 2.55 m from the bottom boundary. A low density oil is added on a 1.1 m strip (segment A-B, Figure 1) at a constant pressure of $h_o = -0.1$ m for a period of 3.43 d yielding a cumulative oil infiltration of 0.32 m^3 m^{-1}. Thereafter, segment A-B is a zero-flux boundary for oil. All other boundaries are zero-flux for oil at all times. Redistribution was permitted out for a period of 45 days following the end of infiltration. Water pressure is fixed at the intial condition on boundary segment C-D and all other boundaries are zero-flux for water at all times. A total of 294 rectangular elements was used to discretize the domain yielding 330 nodes. All simulations were started with an initial time step of 0.002 d and a factor of F=1.03 was used to increase the time step size up to a maximum time step of 0.2 d.

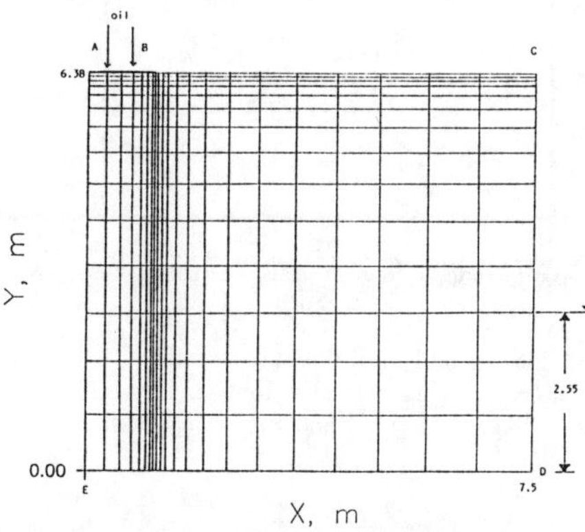

Figure 1. Geometry of Example 5

Simulations were carried out with the ASD method using different tolerances and using CFEM. A comparison of water and total liquid saturation distributions on vertical sections at $x = 0$ and 1.2 m are shown in Figures 2 and 3. It is observed that at $x = 0$ the CFEM results slightly undershoots the ASD solution and slightly overshoots at $x = 1.2$ m. This reflects a slight oscillation in the CFEM solution which could be observed in the y-direction. Such behavior was not evident in the ASD results. The superior stability of the ASD method may also be evidenced by the more rapid convergence behavior of the nonlinear iterations which averaged 4.0 iterations per time step for the CFEM and 3.5 for the ASD method for the $\tau_1 = 0.1$ cm case. The greater accuracy of the ASD method is confirmed by the consistently lower mass balance error for all ASD solutions (Table 4).

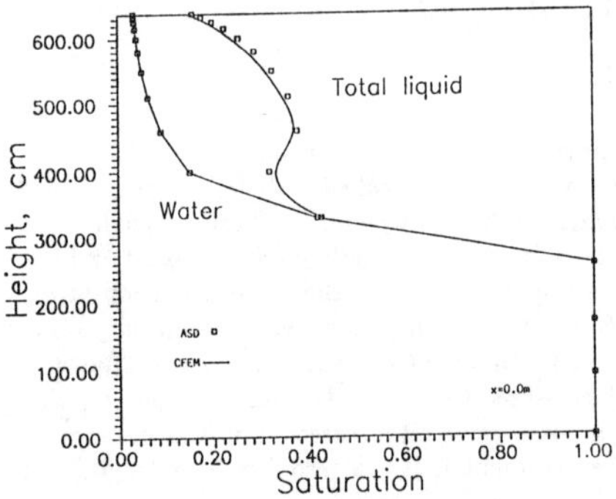

Figure 2. Water and total liquid saturation distributions at the end of redistribution at x=0.0 $m$for CFEM and ASD method with τ_1 =0.1 cm for Example 5.

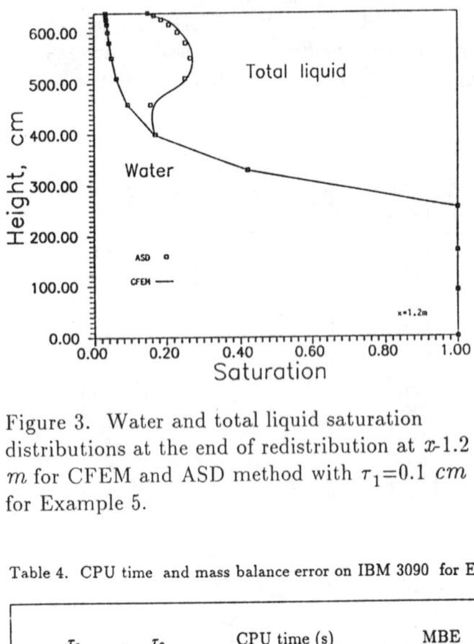

Figure 3. Water and total liquid saturation distributions at the end of redistribution at x-1.2 m for CFEM and ASD method with τ_1=0.1 cm for Example 5.

Table 4. CPU time and mass balance error on IBM 3090 for Example 5.

τ_1 (cm)	τ_2 (-)	CPU time (s) infiltration	redistribution	MBE (%)
Conventional FEM		511	871	0.68
0.05	0.0001	86	371	0.39
0.10	0.0001	84	349	0.12
0.15	0.0001	84	323	0.14

Total CPU times for infiltration and redistribution periods showed even more marked contrasts between CFEM and ASD results than were observed for the one-dimensional problem. Overall speed-up factors for the present case were 6 for the infiltration period and 2.5 for the redistribution period. This greater speed-up reflects the larger proportion savings as dimensionality increases.

5 SUMMARY AND CONCLUSIONS

This paper compares five formulations of the governing equations for multiphase flow for computational efficiency, mass balance accuracy and robustness. Method 1, which formulates the equations with water and oil pressures as primary variables retaining saturation-time derivatives, appeared to be most robust and produced consistently good mass balance results with reasonable efficiency. Method 2, in terms of two fluid pressures but with split time derivative terms, exhibited consistently higher mass balance errors than Method 1 and also exhibited convergence problems in the last case which involved more severe conditions. The problems in Method 2 reflect errors in approximating mass storage terms in the governing equations caused by splitting the saturation-time derivatives into saturation-pressure derivatives and pressure-time derivatives. Method 3, with capillary pressures as primary variables, performed well for all problems, although the results at larger times for Example 2 could not be compared because the formulation did not permit stipulation of the correct lower boundary condition. The capillary pressure formulation does not allow direct stipulation of water pressure at the boundary independently of the oil pressure.

The semi-diffusive schemes performed exceptionally well under dry conditions which were the most difficult for other methods to handle. However, when part of the domain was water saturated, Method 4 failed to converge at all and Method 5 produced high mass balance errors after the oil front reached the water table. The problems that arise with these methods appear to be due to the nonunique specification of the fluid pressure which is not treated as a dependent variable in the formulation.

Considerable savings in computational effort is achieved by the adaptive solution domain (ASD) technique implemented in this code by temporarily excluding elements for which changes in water pressure, oil pressure, water saturation or oil saturation are below prescribed tolerances. A comparison of the computional effort required for a conventional finite element method (CFEM) and for the same code with the ASD method for the two examples included in this paper shows that savings are greatest during the initial stages of a front propogation when the active domain is limited. Savings, furthermore, will generally increase with the dimensionality of the problem due to the tendency for a larger fraction of inactive elements. For Example 5, the ASD method yielded superior performance to the CFEM as evidenced by lower mass balance error, fewer iterations per time step and lack of any oscillatory behavior seen in the CFEM.

Good results for the ASD method were obtained using τ_1= 0.01 to 0.15 cm and τ_2= 0.0001 to 0.0005. In examples 4 and 5 it was observed that the mass balance error increased for extreme high or low tolerances.

ACKNOWLEDGEMENTS

This work was sponsored by the U.S. Environmental Protection Agency joint funding from R. S. Kerr Laboratory and the Water Physics and Chemistry Grants Program as cooperative agreement CR-814320.

REFERENCES

Abriola, L. M. and Pinder, G. F. 1985a. A multiphase approach to the modeling of porous media contamination by organic compounds, 1. Equation development, Water Resour. Res., 21, 11-18.

Abriola, L. M. and Pinder, G. F. 1985b. A multiphase approach to the modeling of porous media contamination by organic compounds, 2. Numerical simulation, Water Resour. Res., 21, 19-26.

Faust, C. R. 1985. Transport of immiscible fluids within and below the unsaturated zone - A numerical model, Water Resour. Res., 21, 587-596.

Forsyth, P. A. 1988. Simulation of nonaqueous phase groundwater contamination, Adv. Water Resour., 11, 74-83.

Forsyth, P. A. and Sammon, P. H. 1986. Practical considerations for adaptive implicit methods in reservoir simulation, J. Comp. Phys., 62, 265-281.

Huyakorn, P. S. and Nilkuha, K. 1979. Solution of transient transport equation using an upstream finite element scheme, Appl. Math. Modeling, 3, 7-17.

Kaluarachchi, J. J and Parker, J. C. 1989. An efficient finite element method for modeling multiphase flow in porous media., Water Resour. Res., 25, 43-54.

Katyal, A. K. and Parker, J. C. 1990. An adaptive solution domain algorithm for solving multiphase flow problems., under revision for Computers and Geosciences.

Kuppusamy, T., Sheng, Parker, J. C. and Lenhard, R. J. 1987. Finite element analysis of multiphase immiscible flow through soils, Water Resour. Res., 23, 625-631.

Osborne M. and Sykes, J. 1986. Numerical modeling of immiscible organics transport at the Hyde Park landfill, Water Resour. Res., 22, 25-33, 1986.

Parker, J. C. 1989. Multiphase flow and transport in porous media, Reviews of Geophysics, 27, 311-328, 1989.

Thomas, G. W. and Thurnau, D. H. 1983. Reservoir simulation using an adaptive implicit method, Soc. Pet. Eng. J., 23, 759-768.

APPENDIX A: Three phase constitutive relations used in simulations

Saturation-capillary pressure relations are described by the model of Parker et al. (1987). Oil-free nodes at which $h_o \leq h_o^{cr}$ are described by

$$\bar{S}_w = [\,1+(\,\alpha\ h_{aw})^n\,]^{-m} \qquad [A1]$$

$$\bar{S}_t = 0 \qquad [A2]$$

and nodes with oil for which $h_o > h_o^{cr}$ are described by

$$\bar{S}_w = [\,1+(\,\alpha\,\beta_{ow}\,h_{ow})^n\,]^{-m} \qquad [A3]$$

$$\bar{S}_t = [\,1+(\,\alpha\,\beta_{ao}\,h_{ao})^n\,]^{-m} \qquad [A4]$$

where

$$\bar{S}_w = (S_w - S_m)/(1 - S_m) \qquad [A5]$$

$$\bar{S}_t = (S_w + S_o - S_m)/(1 - S_m) \qquad [A6]$$

and

$$h_{aw} = h_a - h_w \qquad [A7]$$

$$h_{ow} = h_o - h_w \qquad [A8]$$

$$h_{ao} = h_a - h_o \qquad [A9]$$

$$h_o^{cr} = \frac{\beta_{ao}h_a + \beta_{ow}h_w}{\beta_{ow} + \beta_{ao}} \qquad [A10]$$

Relative permeability relations for water and oil are given by

$$k_{rw} = \bar{S}_w^{\,1/2}\left(1 - \{\,1 - \bar{S}_w^{\,1/m}\,\}^m\right)^2 \qquad [A11]$$

$$k_{ro} = (\bar{S}_t - \bar{S}_w)^{1/2}\left(\{1-\bar{S}_w^{\,1/m}\}^m - \{1-\bar{S}_t^{\,1/m}\}^m\right)^2 \qquad [A12]$$

The parameters S_m, α, n and $m = 1\text{-}1/n$ are regarded as soil properties while β_{ao} and β_{ow} are properties of the fluids.

5 Comparison of numerical simulation to field and laboratory experiments

Subsurface Contamination by Immiscible Fluids, Weyer (ed.) © 1992 Balkema, Rotterdam. ISBN 90 6191 175 3

Organic liquid infiltration into unsaturated porous media

C.S.Simmons, J.F. McBride, J.W.Cary & R.J.Lenhard
Pacific Northwest Laboratory, Richland, Wash., USA

ABSTRACT: Infiltration experiments with immiscible organic liquids (Soltrol [Philips Petroleum Co. ,Bartlesville, OK], mineral oil, and transmission oil) were performed in boxes packed uniformly with sand or glass beads to test a multiphase flow model. Porous media were slightly wet to mimic a vadose zone. Resulting three-dimensional oil plumes and water contents were measured from core samples and photographs of vertical cross sections. Soltrol and transmission oil, with a zero and positive spreading coefficient, respectively, gave regularly shaped plumes similar to those predicted by the model. However, mineral oil with a negative spreading coefficient (i.e., repelled by water) exhibited fingering behavior that was not accurately simulated by the continuum model. The implication is that constitutive theory may not predict well the flow of a nonspreading oil in a vadose zone.

1 INTRODUCTION

Petroleum products and chlorinated hydrocarbons spilled into the subsurface can constitute a serious environmental contamination problem. These organic liquids are usually immiscible with water and flow as a distinct phase. Soluble components of these organic liquids become contamination sources transported by ground water. Multiphase flow models are needed to help select appropriate remediation strategies for cleaning up subsurface contamination caused by such organic immiscible liquids (oils). To be accurate, the models must be based on a confirmed fundamental understanding of how oils behave in the subsurface. The flow behavior of oils in unsaturated zones is especially important because this is where entry into the subsurface first occurs. Therefore, this research involves using three-dimensional organic liquid infiltration experiments designed to test three-phase flow models like those described by Parker (1989).

We are studying how well multiphase flow theory, based on the continuum concept in conjunction with Darcy's law, predicts organic liquid movement in an unsaturated porous medium. Movement of an organic immiscible liquid (oil) in a vadose zone involves complicated coupled simultaneous flow of air, water and oil phases governed by gravitational and capillary forces. The capillary forces cause both the retention and flow of all the fluids in a porous medium. To model immiscible flows, it is necessary to first discover the constitutive relations that comprise the basic controlling parameters. The needed constitutive relations consist of retention relations expressing the functional connection between fluid saturations and capillary pressures (between pairs of contacting phases) and the permeability equations for each fluid involved. This research involves testing how well constitutive relations can be derived from basic physical parameters such as interfacial tension and viscosity.

1.1 Scaling hypothesis

A central scientific hypothesis being tested is called fluid-media scaling. The mathematical statement of this hypothesis describes how to construct three-phase flow constitutive relations from just the retention and permeability equations for air and water in a specific porous medium (Parker 1989, Cary et al. 1989b). In the hypothesis, it is proposed that scale factors representing ratios of water surface tension to interfacial tension for each fluid interface can be used to construct the three-phase retention relations in the following way.

Let Sw =f(haw) be the retention equation for an air-water system where Sw is effective water saturation and "haw" is capillary pressure head. Then

$$Sw = f(Bow\ how) \quad (1)$$

$$So + Sw = f(Bao\ hao) \quad (2)$$

determine the three-phase saturations Sw, So, Sa = 1 - So - Sw, for water, oil, and air where "Bow" and "Bao" are the scale factors. Here "how" and "hao" are the capillary pressure heads for the oil-water and air-oil fluid interfaces. Scaling includes the assumption that water is the wetting phase and oil is nonwetting. In three-phase systems, air is nonwetting and oil is the intermediate wetting phase. Approximately, without taking into account wetting angle effects, "Bow" and "Bao" are the water surface tension divided by the oil-water and air-oil interfacial tensions, respectively. The fluid permeabilities are derived from a capillary pore model, and the oil conductivity is scaled by multiplying saturated water conductivity by the ratio of oil to water viscosity (Parker 1989).

To test the constitutive relations, their associated equations are incorporated in a multiphase flow code and simulation results compared with observed experimental behavior. We have used the MOFHYS code developed by Kaluarachchi and Parker (1989) to perform simulations of the infiltration experiments discussed here. This code that includes a hysteresis modeling capability (Lenhard et al. 1989) is an extension of MOFAT described by Kaluarachchi and Parker (1989).

1.2 Spreading coefficient

The spreading coefficient, which is the difference between the air-water and the sum of the air-oil and oil-water interfacial tensions, quantifies the tendency of an oil to spontaneously spread as a film over a water surface. Using Young's equation, it is easily shown that an oil with a negative spreading coefficient will form a lens on a water surface with a definite contact angle that is between zero and 180 degrees. If the contact angle is greater than 90 degrees, then the oil is nonwetting on a water surface. An oil with a zero spreading coefficient spreads slowly over a water surface and no contact angle exists, whereas an oil with a positive value spreads spontaneously. If a relative spreading coefficient, denoted as S*, is defined by dividing by the water surface tension, then S* can be related to the scale factors as follows:

$$S^* = 1 - 1/Bow - 1/Bao \quad (3)$$

Thus, a positive or zero spreading coefficient is equivalent to the sum, 1/Bow + 1/Bao, being less than or equal to unity, respectively.

2 EXPERIMENTAL DESIGN

In the experiments, an example oil infiltrates from nearly a point source (1.5-cm-diameter ring) into a box filled with moist (water unsaturated) sand or glass beads. A diagram of the experimental setup is shown in Figure 1.

The open-top acrylic box shown in Figure 1 has removable side panels and a 0.5-bar ceramic porous plate at the bottom. The size of the porous media cube held by the box is about 24 cm along an edge. Before oil infiltrates, the initial water content distribution is set by draining to equilibrium the satiated porous medium by applying -110 mbars pressure to the plate. An average initial volumetric water content of 0.07 in the sand was established. A 500 mL quantity of oil was applied through the ring with minimal ponding maintained for each experiment. Boundary effects are minimized because the oil does not reach the side walls during infiltration.

After infiltration ceased and one of the side panels was removed, the porous medium cube was sliced vertically in 2-cm increments, parallel to the box's removed side. The oil plumes were visualized by taking photographs and by contouring liquid saturation measurements obtained from a network of core samples (3-cm diameter by 2 cm deep) removed from exposed faces. Transmission oil, naturally red, and mineral oil, dyed red with Sudan IV, were easily seen in white the glass beads. In sand, the plume was seen as dark brown after evaporating water from an exposed face using a hot air gun. Organic liquid and water contents of the cores were determined gravimetrically by an assay method that uses a porous polyethylene strip to extract oil from a sample (Cary et al. 1990).

A line of electro-optical organic liquid sensors was also used in some experiments to track the front advance in sand (Figure 1). Cary et al. (1989a) describe the design and operation of these probes.

The oils selected for study, based on contrast in their viscosity and interfacial tension, are Soltrol, mineral oil,

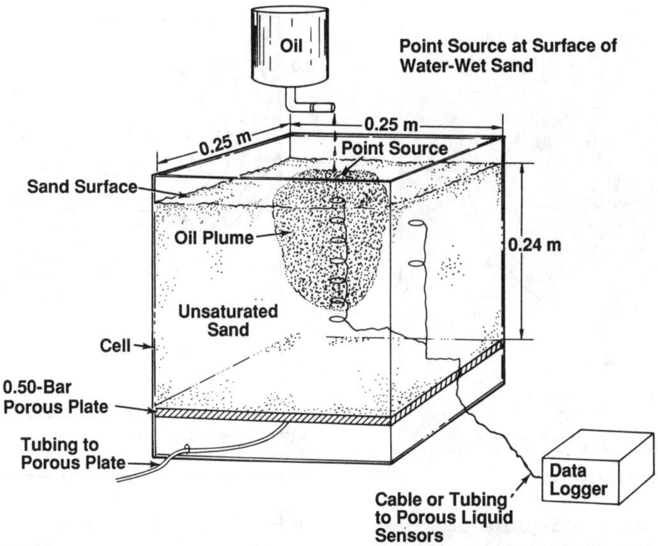

Figure 1. Design of experimental cell for study of oil infiltration into unsaturated porous media (glass beads or sand). Initial volumetric water content is 0.07.

and transmission oil. Oil viscosities relative to water are 4.7, 77, and 71, respectively. Interfacial tensions with air are 26, 31, and 31 dyne/cm and with water are 49, 56, and 16 dyne/cm, respectively. Oil density ranges between 0.8 and 0.85 gm/mL. Transmission oil, with a positive spreading coefficient (24), spreads strongly over a water surface, whereas mineral oil with a negative coefficient (-16) forms oil droplets and is repelled by water. Soltrol has a spreading coefficient nearly equal to zero after contaminating a contacted water surface. Organic components of Soltrol eventually reduce the oil-water interfacial tension after prolonged contact and cause it to exhibit spreading. A tendency of some oils to spread on a water surface is connected with the action of a spreading pressure as discussed by McBride et al. (1989).

3 EXPERIMENTAL OBSERVATIONS

Although transmission and mineral oils have nearly the same viscosity, they behave quite differently during infiltration into moist glass beads or sand. The visualization experiments showed that transmission oil infiltrated uniformly into glass beads (Figure 2), whereas mineral oil displayed channeling behavior (Figure 3). Similar behavior was found for sand, although the channeling was more horizontally pronounced with a more feather like or dendritic pattern. We hypothesize that the channeling behavior is caused by mineral oil's negative spreading coefficient acting in conjunction with pore size variations of the coarse porous media that was used. Oils with a positive or zero spreading coefficient, however, spread evenly through the pore space. Apparently, the flow of mineral oil in sand detects pore scale variations (e.g.,stratifications) that do not affect the flow of a spreading oil.

As oil infiltrates a region, water is expelled as a consequence of the change in interfacial conditions, and this water loss further enhances oil conduction into the increasingly available pore space volume. Measured distributions of Soltrol and transmission oil in water-wet sand are shown in Figures 4a and 5a. Corresponding water content distributions are shown in Figures 4b and 5b. In these cases, the least water is displaced by Soltrol, which has relatively low viscosity and flows more rapidly compared with the other two oils. Transmission oil, however, displaced a substantial amount of water from the plume, as shown in Figure 5b. The initial volumetric water content of 0.07 was reduced nearly to zero in the plume. As a result of much lower viscosity, the Soltrol plume exhibited sinking into the profile before the core samples could be collected. Transmission oil, being much more viscous, remained near the surface.

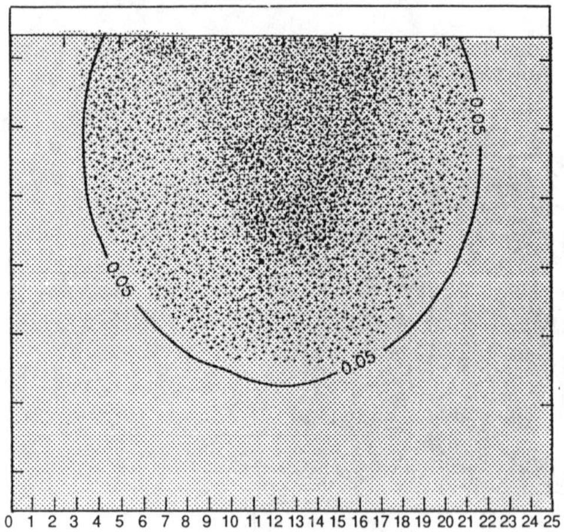

Figure 2. Transmission oil plume observed in glass beads at the central plane after 15 hours infiltration of 490 mL. Solid line is based on simulation of oil front location.

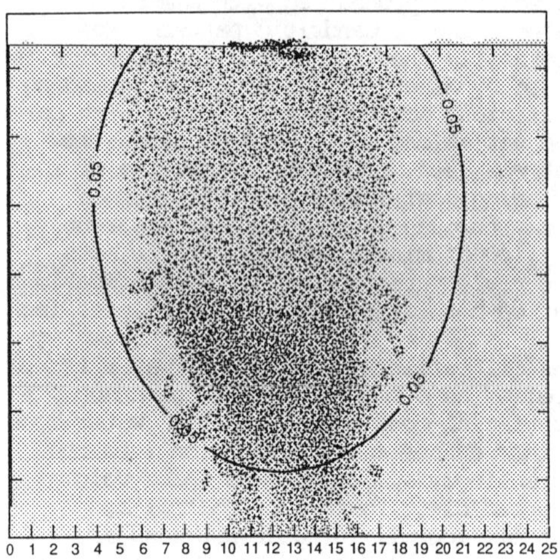

Figure 3. Mineral oil plume observed in glass beads at the central plane after 16.3 hours infiltration of 470 mL. Solid line is based on simulation of oil front location.

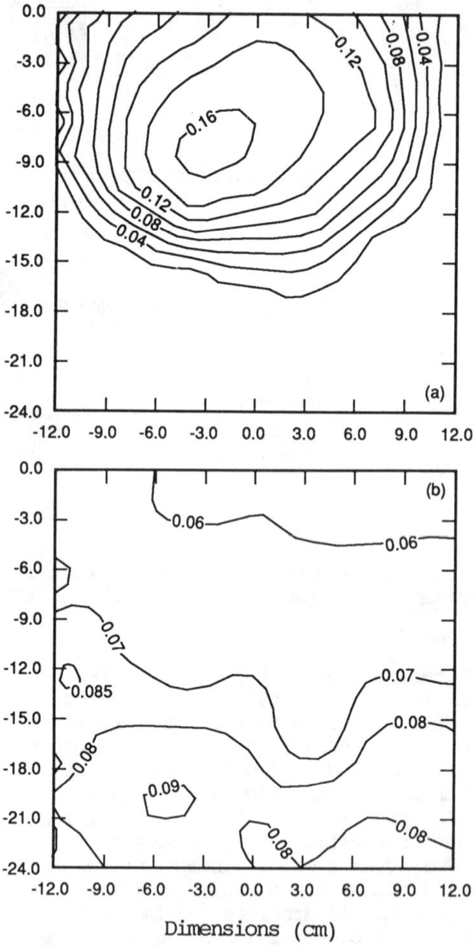

Figure 4. Soltrol infiltration in moist sand a) Volumetric Soltrol content b) Water content distribution after 1.1 hours.

Although the pattern of infiltration was different because of channeling, Mineral oil (Figure 6a) also displaced water substantially (Figure 6b), but less than transmission oil. Figure 6a shows that the mineral oil, as measured by core samples, reached deeper into the box and dispersed more than transmission oil. Thus, oils with either positive or negative spreading coefficient caused displacement of water. In fact, it can be shown using the scaled retention relations that a pressure gradient in the water phase always prevails from the oil-contacted region into an adjacent pristine air-water region. The greater the absolute value of S*, the greater is the gradient causing water to flow out of an oil-contacted region. Soltrol, being least viscous and with S* nearly zero, apparently moves rapidly into the porous medium before water has time to be displaced. Transmission and mineral oils with high viscosity, however, move slowly enough for water to be expelled by the capillary pressure gradient associated with interfacial changes within the plume region.

The 500 mL of Soltrol entered the sand in 50 minutes, and measurements for the distribution in Figure 5a were collected about 15 minutes later, as time passed while the box was sectioned. It took 760 minutes for transmission oil to infiltrate. Scaled to the Soltrol infiltration time by using a viscosity ratio of 71/4.7 (disregarding a small density difference), the estimated entry time for transmission oil is 755 minutes, which is nearly equal to that observed. Mineral oil entered in 680 minutes. The shorter time may be associated with the channeling, which produced a larger and deeper plume.

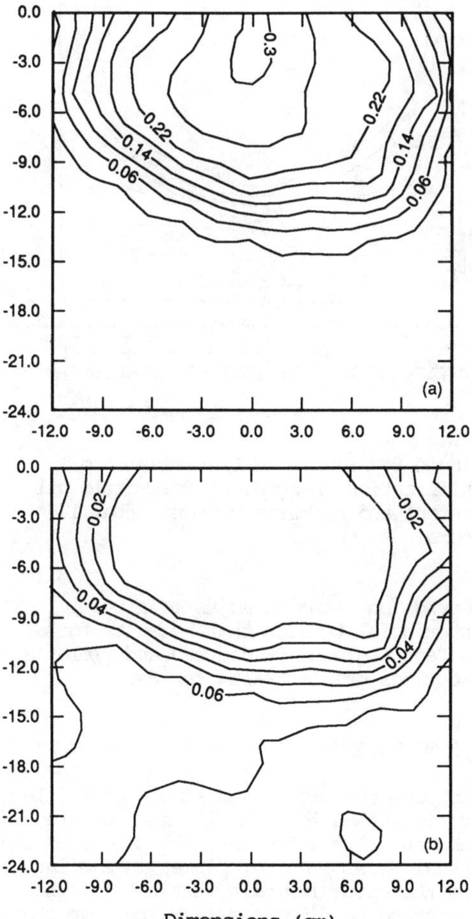

Figure 5. Transmission oil infiltration in moist sand a) Volumetric oil content b) Water content distribution after 13 hours.

4 SIMULATIONS

Oil infiltration was simulated with the MOFHYS code (Kaluarachchi and Parker 1989) in radial coordinates by using the van Genuchten equation for the retention relation, f(haw). Only the imbibition path parameters were used for the sand. For sand, the alpha and n parameters are 0.039 and 3.7, and for glass beads, they are 0.035 and 7.9. A drainage path value for alpha equal to 0.02 was used to model hysteretic flow in the glass beads. For sand, saturated water conductivity had the value of 60 cm/h when measured with 0.04 air content entrapped for a water content of 0.36. Saturated conductivity of the glass beads is 70 cm/h for porosity equal

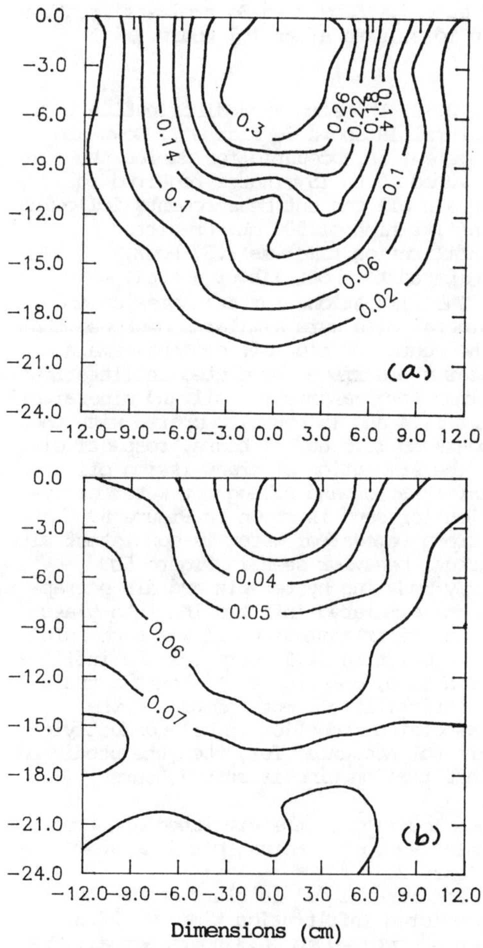

Figure 6. Mineral oil infiltration in moist sand a) Volumetric oil content b) Water content distribution after 12.7 hours.

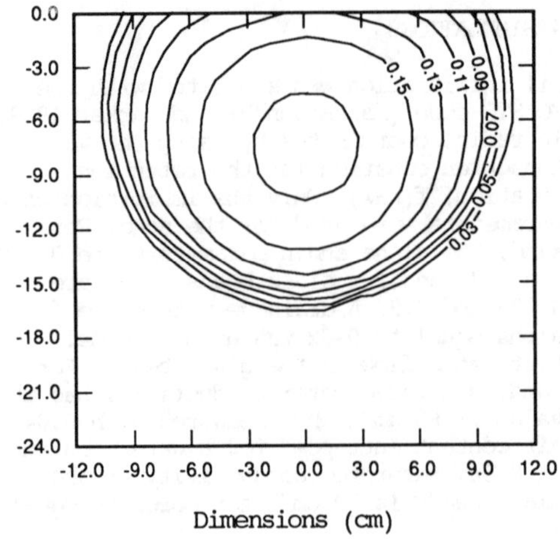

Figure 7. Simulated Soltrol infiltration in moist sand after 1.7 hours.

to 0.42. The simulated distribution for Soltrol is shown in Figure 7. Downward movement is accounted for during the assumed 10 to 15 minutes required to obtain oil content measurements following infiltration of 500 mL. Predicted infiltration time was 1.57 hour, compared to about 1 hour actual.

The simulations for transmission and mineral oils were similar and showed that the center of high oil content remained near the surface. Predicted infiltration times for transmission oil and mineral oil are 17.4 and 19 hours, compared with the measured 12.5 and 12 hours, respectively.

The simulation of transmission oil infiltration and subsequent water exclusion for sand is shown in Figure 8. Strong removal of water is consistent with actual behavior seen in Figure 5b.

By modeling hysteresis and air entrapment, simulated infiltration into glass beads of transmission oil was successful in predicting 14.5 hours for the infiltration time, compared to 15 hours actual. In contrast, if air entrapment and the associated reduction in oil conductivity are not accounted for, then the predicted infiltration time is only 7 hours.

The location of the simulated oil front relative to the actual plume is shown in Figure 2.

For mineral oil in glass beads the predicted infiltration time was 23.5 hours compared to 16.3 hours actual. The simulated oil front shown in Figure 3 is more elongated than that of transmission oil, although both oils have nearly the same viscosity and density. This result

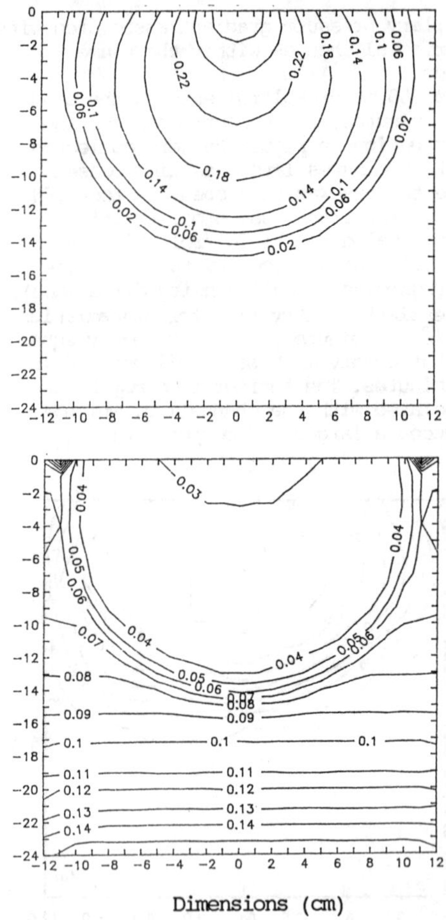

Figure 8. Simulated transmission oil infiltration in sand, a) Volumetric oil content and b) Water content after 17.4 hours.

depends on using an adjusted air-oil interfacial tension that accounts for a contact angle of 60 degrees and yields a two times greater value for Bao.

5 CONCLUSIONS

Although the actual and simulated infiltration times for the oils in sand are substantially different, the predicted and actual development of plumes is similar for the spreading oils, but not for the nonspreading mineral oil. We believe that controlling parameters, especially saturated conductivity, did not quite reflect the actual values appropriate in the sand-packed box. In particular, the independently determined 60 cm/h for saturated conductivity was probably too low and reflected air entrapment that was not accounted for in the nonhysteretic simulations. However, simulations of transmission oil infiltrating glass beads while accounting for hysteresis and air entrapment gave good agreement for both the infiltration time and location of the oil front, whereas disregarding air entrapment did not. Simulation of mineral oil infiltration required accounting for the effect of wetting angle. Observed channeling behavior of mineral oil, however, indicated a definite limitation of the continuum multiphase flow model. Indeed, the channeling mineral oil penetrated to much greater depth than predicted. Such immiscible flow of a non-spreading oil is likely driven by an interfacial phenomenon that has not been appreciated before. The exact mechanism of this phenomenon as manifested in particular porous media is unexplained and requires further study.

ACKNOWLEDGEMENTS

Research was supported as part of the Subsurface Science Program of the Office of Health and Environmental Research, Ecological Research Division of the U.S. Department of Energy under Contract DE-AC06-76RLO 1830.

REFERENCES

Cary, J.W., J.F. McBride, and C.S. Simmons 1989a. Electro-optic detection of liquid in translucent porous materia. Soil Science Society of America 53:1591-1595.

Cary, J.W., C.S. Simmons and J.F. McBride 1989b. Predicting oil infiltration and redistribution in unsaturated soils. Soil Science Society of America 53:335-342.

Cary, J.W., J.F. McBride, and C.S. Simmons 1990. "Assay of Organic Contents in Predominantly Water-Wet Unconsolidated Porous Media." Journal of Contaminant Hydrology. In press.

Kaluarachchi, J.J. and J.C. Parker 1989. An efficient finite element method for modeling multiphase flow. Water Resources Research 25:43-54.

Lenhard, R.J., J.C. Parker and J.J. Kaluarachchi 1989. Model for hysteretic constitutive relations governing multiphase flow 3. Refinements and numerical simulations. Water Resources Research 25:1727-1736.

McBride, J.F., J.W. Cary and C.S. Simmons 1989. Direct measurement of the spreading pressure of organic liquids on water. Journal of Colloid and Interface Science 132:363-372.

Parker, J.C. 1989. Multiphase flow and transport in porous media. Reviews of Geophysics 27:311-328.

Subsurface Contamination by Immiscible Fluids, Weyer (ed.) © 1992 Balkema, Rotterdam. ISBN 90 6191 175 3

Transport of trichloroethylene vapours in the unsaturated zone: Numerical analysis of a field experiment

Carl A. Mendoza, Bernadette M. Hughes & Emil O. Frind
Waterloo Centre for Groundwater Research, University of Waterloo, Ont., Canada

ABSTRACT: A finite-element model that simulates advective-dispersive transport of vapours in a domain with spatial temperature variations is used to simulate two field experiments, one performed in the winter and one in the summer, involving the migration of trichloroethylene vapours from a residual source in the unsaturated zone. Sensitivity analyses are performed to obtain insight into the importance of the various parameters and transport processes that play a role in the migration of the vapours. Density-induced advection is seen to be more significant in the summer experiment than in the winter experiment. This is due to higher temperatures causing higher source concentrations, and hence stronger density gradients within the domain. Increased temperatures also cause more rapid migration of the vapours due to decreases in the retardation factor. An open ground surface allows mass to leave the system but does not significantly affect the mass distribution below the higher organic content layer in the upper part of the section.

1 INTRODUCTION

The presence of dissolved chlorinated solvents in groundwater is of increasing environmental concern due to the potential health hazard posed by these chemicals in drinking water supplies. As a result, much work is currently being directed at determining how solvents find their way into groundwater. One possible pathway is through vapour transport from a residual solvent source in the unsaturated zone.

Soon after a discrete solvent release near the ground surface, the free-product stops migrating and forms an immobile residual source in the unsaturated zone. The compound evaporates and its vapours migrate by diffusion through the unsaturated zone. Density-induced advection may also occur because many solvents have high molecular weights and vapour pressures, and so their saturated vapours are commonly denser than uncontaminated air. As the vapours migrate, they dissolve into the soil moisture and sorb onto the soil solids. Although these partitioning processes retard the vapour migration rate, they also make contaminants readily available for transport to groundwater through infiltration of rain water or watertable fluctuations. An open ground surface may allow the passive loss of contaminants to the atmosphere.

Several numerical models have been developed to simulate vapour transport in the unsaturated zone. Baehr (1987) simulated the migration of hydrocarbons by vapour-phase diffusion in equilibrium with water infiltrating at steady state. Sleep and Sykes (1989) investigated density-dependent, advective-dispersive transport of vapours from a residual TCE source in the presence of infiltrating water. Their assumed mass transfer coefficients describing phase partitioning resulted in relative phase concentrations close to equilibrium. Both Falta et al. (1989) and Mendoza and Frind (1990a,b) examined the case of density-induced, advective-dispersive vapour transport from a residual source, in the presence of an immobile water phase, under the assumption of equilibrium partitioning between phases. Falta et al. (1989) performed a sensitivity analysis using carbon tetrachloride and toluene gases. Mendoza and Frind (1990b), using generic organic compounds, evaluated the relative importance of different vapour transport mechanisms, including density flow and advection due to source vapourization. Mendoza and McAlary (1990) used decoupled vapour transport and groundwater transport models to illustrate possible groundwater contamination effects of TCE migration from a residual source.

These studies show that an uncovered ground surface may contribute to subsurface remediation by allowing contaminants to escape to the atmosphere, and that groundwater contamination may be significantly increased as a result of vapour transport. They also show that phase partitioning may be a significant control on transport and that, under certain conditions, density-dependent advection may be an important transport process.

Because all of these studies have been done on a theoretical basis, there is an urgent need for validation by application of the models to field problems. Furthermore, the above models all assume isothermal conditions. This may limit their applicability to real problems since some important vapour transport parameters are highly temperature dependent, and spatial and temporal temperature variations may be large in the unsaturated zone.

This study uses a vapour transport model that incorporates effects of temperature- and density-dependence of vapour flow to examine the results of two detailed field experiments reported by Hughes et al. (1990). The investigation focusses on the effects of density-induced transport, ground-surface boundary conditions, temperature variations, and soil organic content on vapour transport.

2 NUMERICAL MODEL

The numerical model is based on that developed by Mendoza and Frind (1990a). Finite elements in an axisymmetric coordinate system are used to represent vapour phase diffusion and advection from a localized residual solvent source in the presence of an immobile soil moisture phase.

Several modifications have been made to the model to more fully represent the field scenario. One characteristic of the experiments was that the subsurface temperature varied significantly between experiments and spatially within each experiment (Figure 1). Because the heat capacity of soil gas is low, we assume that the migration of the gases does not affect the subsurface temperature distribution and that the soil gas is in thermal equilibrium with the surrounding porous medium. In addition, the temperature distribution is assumed constant in time over the duration of a simulation. With these assumptions, the subsurface temperature and the temperature-dependent compound properties can be specified *a priori* for each point in the system. The compound parameters most sensitive to temperature, Henry's constant and vapour pressure, both increase with increasing temperature. Diffusion coefficients and viscosities are also slightly increased with higher temperatures.

To adequately accommodate the temperature dependence, the governing equation for vapour flow is written as:

$$\frac{1}{r}\frac{\partial}{\partial x_i}\left[r\,k_r\,k_{ij}\,\frac{\rho_0 g}{\mu}\rho\left(\frac{\partial h^*}{\partial x_j}+\frac{\rho-\rho_0}{\rho_0}\frac{\partial z}{\partial x_j}\right)\right]$$

$$= \theta_a\,\rho_0\,g\,\frac{\rho}{P}\frac{\partial h^*}{\partial t} \quad (1)$$

where k_r and k_{ij} are the relative and intrinsic permeabilities, ρ is the fluid density, ρ_0 is the density of uncontaminated soil gas, g is the gravitational constant, and μ is the viscosity of the gas mixture. The equivalent head of the gas mixture is defined as $h^* = P/\rho_0 g + z$, with P being the fluid pressure. The air-filled porosity is represented by θ_a.

The density of the gas phase is determined from the ideal gas law and incorporates the effects of composition, temperature, and pressure on density.

The transport equation is written as:

$$\frac{1}{r}\frac{\partial}{\partial x_i}\left[r\,\theta_a\,D_{ij}\,\rho\frac{\partial X}{\partial x_j}\right]-\frac{\partial}{\partial x_i}\left[\theta_a v_i\,\rho X\right]$$

$$= \theta_a\,R\frac{\partial}{\partial t}(\rho X) \quad (2)$$

where D_{ij} is the dispersion tensor which includes diffusion, v_i is the velocity, and X is the mass fraction of contaminant. The velocity is calculated from the equivalent head distribution using a modified form of Darcy's Law.

The retardation factor describing equilibrium phase partitioning is expressed as:

$$R = 1+\frac{\theta_w}{\theta_a}\frac{1}{H}+\frac{\rho_b}{\theta_a}\frac{K_{oc}f_{oc}}{H} \quad (3)$$

where θ_w is the soil moisture content, ρ_b is the soil bulk density, H is the dimensionless Henry's constant, K_{oc} is the compound's organic carbon partitioning coefficient, and f_{oc} is the soil organic content.

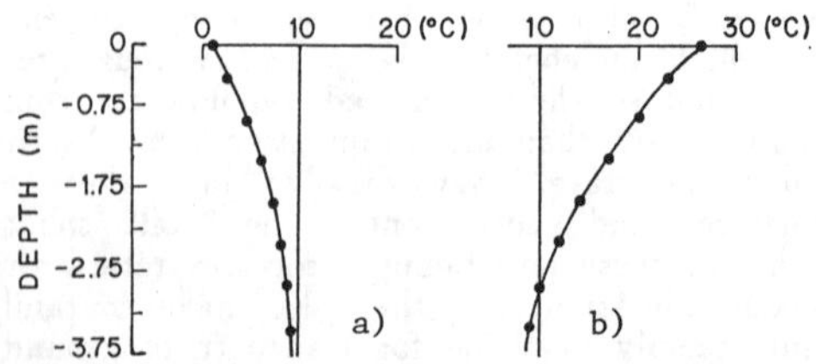

Figure 1 Subsurface temperature profiles: (a) Experiment 1 (winter), (b) Experiment 2 (summer).

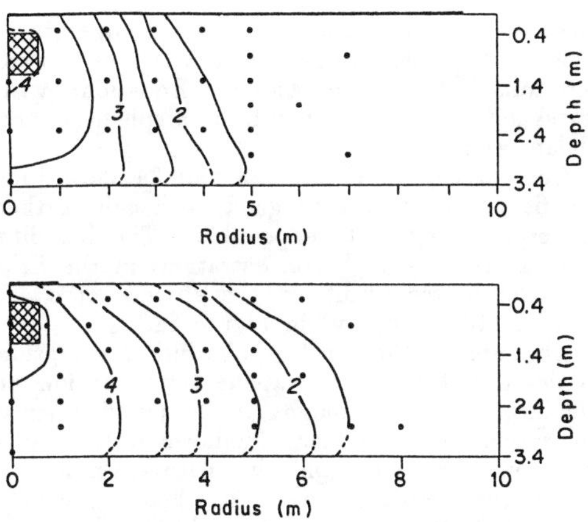

Figure 2 Observed vapour concentrations at 18 days: (a) Experiment 1 (winter), (b) Experiment 2 (summer). Contours represent logarithms of concentration in ppmv.

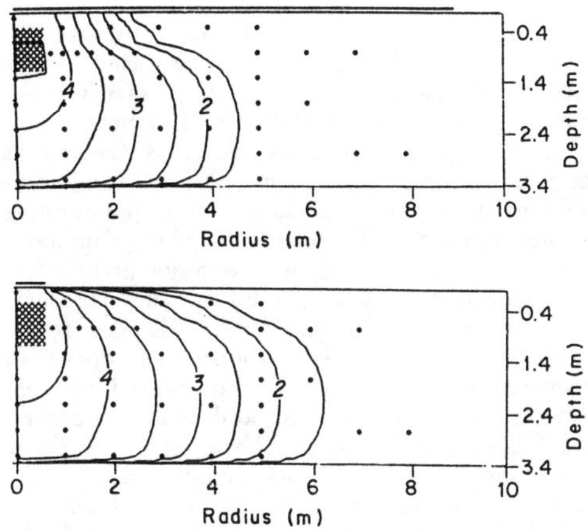

Figure 3 Simulated vapour concentrations at 18 days: (a) Experiment 1, (b) Experiment 2. Contours represent logarithms of concentration in ppmv. Results are contoured on the full finite-element grid (651 nodes).

3 FIELD EXPERIMENTS

The two related field experiments discussed here were performed at the Borden field site. The experiments are documented in detail by Hughes (1991) and summarized by Hughes et al. (1990).

The unsaturated zone consists of fine- to medium-grained sands with interbedded silty-organic layers in the upper section. Besides the thin layer of organic-rich material at the ground surface, there is a predominant layer with higher organic content above 1 m depth over much of the site. The sands were at, or near, residual moisture saturation throughout the experiments. The average permeability is 1.3×10^{-11} m^2. This is in the range where density effects may become important (Mendoza and Frind, 1990b).

For each experiment, approximately 60 kg of TCE were mixed with sand and placed in the ground to represent a residual solvent source. This source was about 0.6 m in radius, 0.8 m high, and the top was located at 0.35 m depth.

Vapour samples were obtained at regular intervals from multilevel sampling nests along a profile intersecting the source. TCE concentrations were determined using a portable gas chromatograph. The numerical model was used to assist in designing the sampling grid and the monitoring schedule.

In Experiment 1, performed in November-December 1988, the ground surface was covered to a radius of 9 m from the source to prevent infiltration and vapour losses to the atmosphere. The temperature profile was fairly constant over the duration of the experiment, varying from 10 °C at the watertable to about 0 °C, or less, near the ground surface (Figure 1). The watertable rose 8 cm over the 26 day duration of the experiment.

In Experiment 2, performed in July-August 1989, only a very small cover (0.9 m radius) was placed directly over the source; the remainder of the ground surface was left open to the atmosphere. There was no infiltration for the first 21 days. The temperature was again relatively constant over time, varying from 9 °C at the watertable to around 26 °C near the surface (Figure 1). There was background contamination of the soil gas of between 10 and 15 parts per million by volume (ppmv) as a result of the first experiment. The watertable dropped 44 cm over the 42 day experiment.

Concentration contours for the two experiments at 18 days are shown in Figures 2a and b respectively. Contours correspond to the logarithm of vapour concentration in ppmv, at half-order of magnitude increments. Actual concentrations contoured range from 31.6 ppmv to 31,600 ppmv.

4 SIMULATION OF FIELD EXPERIMENTS

An average depth of 3.45 m to the top of the saturated capillary fringe was used for the finite-element domain. Nodes were placed to coincide with field samplers and elements were chosen to represent the geology and temperature profiles. The vertical node spacing varied from 10 to 25 cm, while the horizontal spacing varied from 15 to 100 cm. Initial time steps of between 1.0 and 2.4 hours were used. The spatial and temporal discretization satisfies the criteria for numerical stability and accuracy.

The top of the capillary fringe is treated as being impermeable to both gas flow and contaminant transport. In fact, some partitioning of contaminant will occur at this boundary; however, it has been shown to have little effect on vapour distributions (Mendoza and McAlary, 1990). A covered ground surface is also represented as being impermeable. An open, or uncovered ground surface is represented by constant atmospheric pressure and zero concentration. This neglects atmospheric pressure variations and assumes that contaminants are removed instantaneously at the ground surface. Lateral exit boundaries are represented by hydrostatic pressure and zero dispersive flux. A fixed zero-concentration boundary condition applied at the lateral exit boundaries was found to produce essentially the same results. The source is represented as being at constant concentration.

The two temperature profiles shown in Figure 1 were used as a basis for the simulations. Henry's constants and diffusion coefficients corresponding to these temperature profiles were assigned by elemental layer thoughout the domain. Effective diffusion coefficients were calculated from the relationship of Millington and Quirk (1961).

Physical parameters, including permeability, soil moisture content, and organic content, were measured in two cores from the site (Hughes et al., 1990). Smoothed, average functions of these parameters were used in the simulations. The model structure thus corresponds well to the observed soil structure at the site, which is characterized by two relatively organic-rich layers near the ground surface and sands of low organic content at depth.

Simulation results corresponding to the two field experiments at 18 days are shown in Figures 3a and b respectively. The model fit was evaluated by visual comparison of the contoured simulation results to the field measurements. Breakthrough curves at sampling points (not shown) were also used in the comparison.

The source temperature in Experiment 1 was between 2 and 5 °C, while the surface temperature was generally below 0 °C. The actual source concentration in the first experiment could not be measured properly because saturated vapours would condense upon being transported to a lower-temperature environment. Vapour concentrations measured at the source were extremely variable with a maximum of about 22,000 ppmv which corresponds to a temperature of about −3 °C. Because initial simulations using these concentrations gave consistently low results, an equilibrium concentration of between 30,000 and 35,000 ppmv was used in the simulations. These temperature-dependent equilibrium concentrations were determined from the Antoine Equation, with regression constants taken from Stephenson and Malanowski (1987).

Comparisons of Figures 2a and 3a show that the fit obtained is very good, although further refinement may still be possible. The inability to measure very high concentrations in the field near the source results in the absence of a 4.0 contour (10,000 ppmv) in Figure 2a.

The use of the local equilibrium assumption to describe the source vapour concentration at the large scale is thought to be justified since the emplaced source had a uniform residual distribution. For a spill in natural material, where the residual might be heterogeneously distributed, this assumption might not be justified.

Mass fate calculations for Experiment 1 show that after 18 days, of the original 60 kg of TCE placed in the source, 0.55 kg are in the vapour phase, 1.2 kg are in the aqueous phase, and 5.2 kg are sorbed onto the solids. Much of the sorbed phase resides in the two organic-rich layers near the ground surface. Essentially no mass is lost to the atmosphere.

In Experiment 2, which took place in the summer, the source temperature ranged from 23 to 19 °C, and maximum measured concentration was about 88,000 ppmv, which corresponds to a temperature of almost 23 °C. Source equilibrium concentrations of between 73,500 and 90,000 ppmv were used in the simulations. Since there was no precipitation during this period of the experiment, the assumption of an immobile soil moisture profile is likely valid. Comparison of Figures 2b and 3b shows that the fit is also very good.

After 18 days, of the original 60 kg emplaced for Experiment 2, 1.75 kg are in the gas phase, 2.28 kg are in the aqueous phase, and 4.65 kg are sorbed. The model calculates that 6.0 kg have been lost to the atmosphere.

There are several sources of error that can lead to disagreement between the experimental and simulation results. One is the heterogeneity of the medium, particularly in the upper zone, which is not represented in the model. Variations in the organic content of the soil would result in notable variations in the retardation coefficient, which can be expected to affect the concentration distribution (see Section 5.4). The model uses an organic content distribution that is based on cores taken at only two locations (Hughes et al., 1990); however, in the field, the sediments are observed to be heterogeneous both laterally and vertically. Another source of error is the variations in watertable elevation that are neglected in the simulations. Nevertheless, the results indicate that the main transport processes are represented accurately in the model.

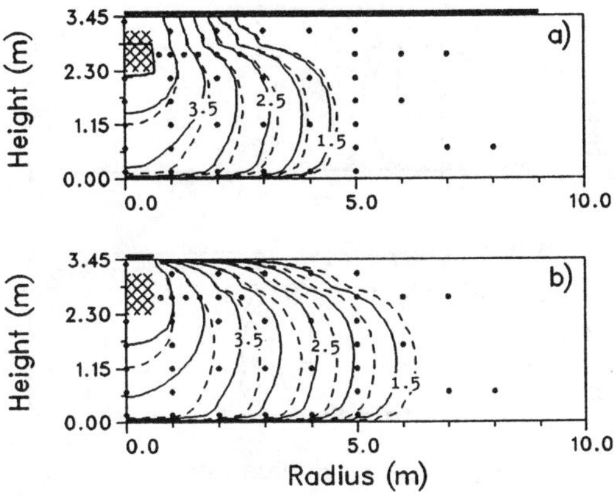

Figure 4 Effect of vapour density: (a) Experiment 1, (b) Experiment 2. Contours correspond to purely-diffusive transport (solid lines), and advective- diffusive transport with density effect (dashed lines).

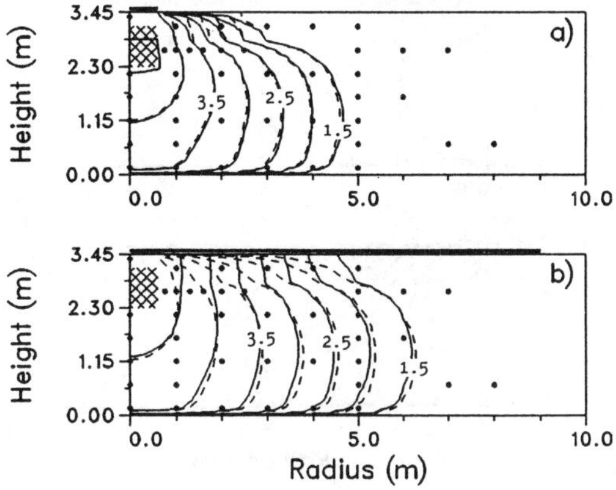

Figure 5 Effect of surface cover: (a) Experiment 1, (b) Experiment 2. Contours correspond to surface cover as shown (solid lines), and surface cover as in original experiments (dashed lines).

5 SENSITIVITY ANALYSES

In order to gain insight into the processes and parameters influencing the field results, we performed several sensitivity analyses to address the effects of density-induced advection, the ground surface cover, the temperature distribution, and the organic content distribution. For comparison, the original simulation at 18 days is shown by dashed lines on each contour plot (Figures 4 to 7).

5.1 *Effect of Density*

From Figure 3, it appears that density-dependent advection may play a role in the vapour transport process observed in both experiments. The relative importance of density-dependent transport may be investigated by comparing the simulation results with those obtained on the basis of a purely diffusive transport mechanism.

Figures 4a and b give this comparison for the two experiments. The difference in concentrations, remembering that the contour intervals represent half an order of magnitude, is fairly large for Experiment 2. In each case, the full transport mechanism, which includes density effects, (dashed lines in Fig. 4), produce a greater advance of the vapours in the radial direction. This incremental advance is due to the downward velocities in the source area (about 15 cm/day for Experiment 1 and 50 cm/day for Experiment 2), which carry more vapours away from the source area in any given time period than under the pure-diffusion scenario. The larger advective contribution in Experiment 2 arises because the maximum density relative to uncontaminated air is 1.32 compared to 1.13 in Experiment 1. These differences in density are primarily a result of the different source concentration arising from the different source temperatures. The degree of density-dependent advection agrees with the generic compound simulations of Mendoza and Frind (1990b).

5.2 *Effect of Surface Cover*

The primary differences between the two experiments were the soil temperature profiles and the ground surface boundary. In order to provide insight into the individual effects of these two variables, we reverse the ground cover arrangement for the two experiments.

Figure 5a shows a simulation under the same conditions as in Experiment 1 (winter), except that the surface boundary is changed from full to partial cover. Likewise, Figure 5b shows a simulation under the same conditions as in Experiment 2 (summer), except that the cover now extends over the full area.

For winter conditions, the change in the surface cover has a significant effect only in the immediate vicinity of the boundary. For summer conditions, the effect is again most noticeable near the surface, where the closed boundary reduces mass losses to the atmosphere. The closed boundary also causes a slight reduction in the frontal advance of the vapour phase at depth; this is due to the decrease in

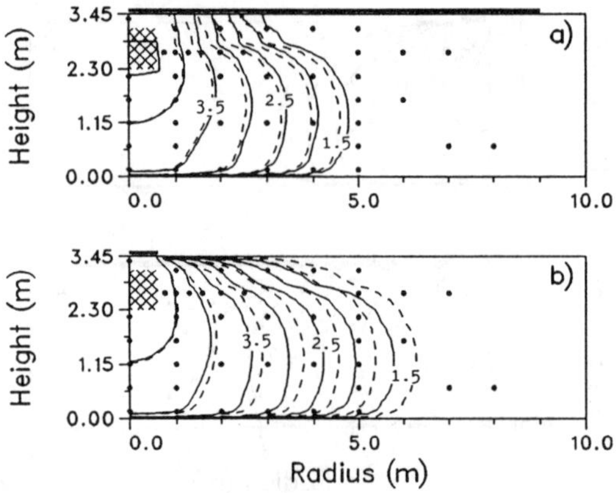

Figure 6 Effect of temperature-dependent parameters: (a) Experiment 1, (b) Experiment 2. Contours correspond to transport parameters based on a uniform temperature of 9 °C (solid lines), and parameters based on temperature profiles (dashed lines).

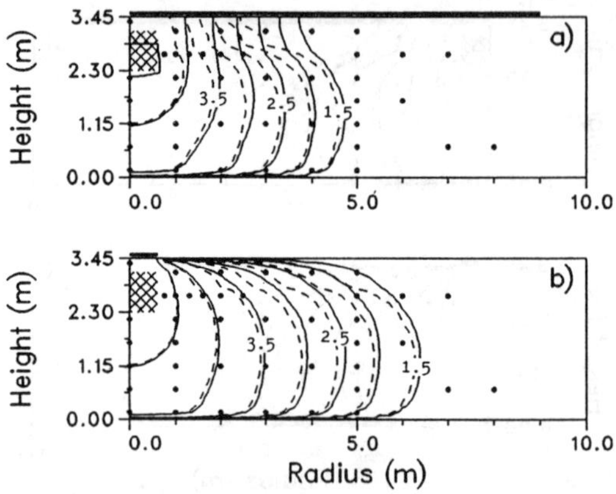

Figure 7 Effect of soil organic content: (a) Experiment 1, (b) Experiment 2. Contours correspond to uniform carbon content of 0.2 % to a depth of about 1 m (solid lines), and variable carbon content up to 1.9 % (dashed lines).

advective air circulation through the source.

Compared to earlier simulations with a homogeneous medium (Mendoza and Frind, 1990b) the effect of the ground cover seems to be slight. This relative lack of response is due to the controlling influence of the organic-rich layer which limits transport to the exposed ground surface.

5.3 *Effect of Temperature on Parameters*

Temperature variations affect the vapour transport processes both through the source vapour concentrations and through the temperature-dependence of the transport parameters. The source vapour concentrations determine the absolute vapour concentrations throughout the domain, while the temperature-dependent parameters control the migration rates. These parameters include Henry's constant, which in turn affects the retardation coefficient (equation 3), as well as the diffusion coefficient and the viscosity. The sensitivity of the retardation coefficient on temperature, through Henry's constant, can be large; for example, the near-surface organic layer has a retardation factor of 120 in the winter but only 43 in the summer, while the lower sand has an average retardation factor of 3.3 in the winter and 2.5 in the summer.

In order to determine the effect of temperature dependence, we performed a simulation with parameters corresponding to a uniform 9 °C background temperature and compared the results to those obtained with parameters corresponding to the actual temperature profiles observed in the experiments (Figure 1).

The comparison is given in Figure 6. The source concentration in each case is that used in the simulation of the original experiment (30,000 to 35,000 ppmv for Experiment 1 and 73,500 to 90,000 for Experiment 2). The results show that under winter conditions, the vapour front is retarded relative to average conditions, while under summer conditions, the vapour front is advanced. In each case, this effect is primarily due to the change in the retardation coefficient that results from the temperature change.

5.4 *Effect of Organic Content*

The vapour transport process is also sensitive with respect to the organic content of the soil. To demonstrate this sensitivity, simulations were performed where the near-surface organic-rich layer (organic content varying between 0.2 and 1.9 %) was replaced with a layer having a uniform 0.2 % organic content.

The results are shown in Figure 7. The change (solid lines) causes an increased frontal advance due to the lower retardation coefficient in both the winter and summer scenarios. This increased advance is more noticeable in Experiment 1 due to the closed surface boundary which prevents losses to the atmosphere. In the lower part of the system, changes are small.

The mass fate is also affected by the greatly reduced capacity for sorption onto the solids. For the summer case, there is about 1.9 kg in the vapour phase, 2.4 kg in the aqueous phase, and only 2.9 kg sorbed onto the solids. Atmospheric losses increase to 7.2 kg from the 6.0 kg in the experiment. This increase in passive venting occurs primarily because the predominant organic layer no longer exerts a confining effect on vapour transport.

6 CONCLUSIONS

This work closes a gap in our understanding of solvent vapour processes in the subsurface by providing validation of a numerical model on the basis of a controlled field experiment. The good agreement obtained between the simulations and field observations provides confidence in the numerical model.

The effect of temperature variations in vapour transport processes is shown to be substantial, resulting in different behaviour under summer and winter conditions. The temperature effect manifests itself in two ways, first through the vapourization of the residual solvent, which results in dramatically different vapour concentrations at the source and throughout the unsaturated zone, and second through the temperature-dependence of the transport parameters, which affects vapour migration rates. For a given source concentration, the most sensitive transport parameter is Henry's constant, which affects the retardation factor and hence the rate of advance of the vapour front, as well as the mass fraction stored in the aqueous and solid phases.

Density-induced advection is also affected by temperature through the source concentrations. Increased source vapour concentrations cause greater vapour densities which in turn cause stronger advective gas currents through the source as well as wider spreading of the vapour plume.

A ground surface cover prevents significant losses to the atmosphere. In the two experiments, the effect of atmospheric losses seems to be important mainly in the upper zone of the system, while the overall spreading of the plume is only slightly affected. This response is specific to the present system, where the presence of layers with high organic content near the ground surface restricts vapour migration. For a homogeneous system, the effect of the surface cover would be greater. The organic content also affects transport rates and mass fate within the domain. At lower organic content, retardation is decreased and less mass is stored on the solid phase.

ACKNOWLEDGEMENTS

We would like to thank Dr. R.W. Gillham for his guidance and assistance throughout this project. This work was funded in part by the University Consortium for Solvents-in-Groundwater Research which is sponsored by: the Ciba-Geigy, Dow Chemical, Eastman Kodak, and General Electric companies; the Ontario University Research Incentive Fund; and, the Natural Sciences and Engineering Research Council (NSERC). Computing services were supplied by the Waterloo Centre for Groundwater Research, an Ontario Centre of Excellence. Funding was also supplied by NSERC in the form of post-graduate scholarships.

REFERENCES

Baehr, A.L. 1987. Selective transport of hydrocarbons in the unsaturated zone due to aqueous and vapor phase partitioning. *Water Resour. Res.* 23(10): 1926-1938.

Falta R.W., I. Javandel, K. Pruess & P.A. Witherspoon 1989. Density-driven flow of gas in the unsaturated zone due to the evaporation of volatile organic compounds. *Water Resour. Res.* 25(10): 2159-2169.

Hughes, B.M. 1991. Vapour transport of trichloroethylene in the unsaturated zone: A field experiment. MSc thesis, Dept. Earth Sciences, University of Waterloo.

Hughes, B.M., R.W. Gillham & C.A. Mendoza 1990. Transport of trichloroethylene vapours in the unsaturated zone: A field experiment. This volume.

Mendoza, C.A. & E.O. Frind 1990a. Advective-dispersive transport of dense organic vapors in the unsaturated zone 1. Model development. *Water Resour. Res.* 26(3): 379-387.

Mendoza, C.A. & E.O. Frind 1990b. Advective-dispersive transport of dense organic vapors in the unsaturated zone 2. Sensitivity analysis. *Water Resour. Res.* 26(3): 388-398.

Mendoza, C.A. & T.A. McAlary 1990. Modeling of ground-water contamination caused by organic solvent vapors. *Ground Water* 28(2): 199-206.

Millington, R.J. & J.P. Quirk 1961. Permeability of porous solids. *Trans. Faraday Soc.* 57: 1200-1207.

Sleep, B.E. & J.F. Sykes 1989. Modeling the transport of volatile organics in variably saturated media. *Water Resour. Res.* 25(1): 81-92.

Stephenson, R.M. & S. Malanowski 1987. *Handbook of the Thermodynamics of Organic Compounds.* Elsevier, New York.

Subsurface Contamination by Immiscible Fluids, Weyer (ed.) © 1992 Balkema, Rotterdam. ISBN 90 6191 175 3

Quantitative comparison of computational and experimental models for three-phase immiscible transport

Wendy E. Soll
Massachusetts Institute of Technology, Cambridge, Mass., USA

Michael A. Celia
Princeton University, N.J., USA

John L. Wilson
New Mexico Institute of Mining and Technology, Socorro, N.Mex., USA

ABSTRACT: This paper presents a study of three-phase pore-scale interactions through a combination of computational modeling and experimental analysis. The computational model simulates capillary driven movement of two or three concurrently mobile fluids in a two- or three-dimensional pore space. The model uses a network-type description of the pore space, and an extended form of percolation theory to describe the fluid transport through the system. Various aspects of laboratory pressure-saturation tests can be simulated, including drainage/imbibition cycles, hysteresis, residual saturation, and wettability effects. The experimental set-up was designed to run pressure-saturation tests using micromodels to simulate the pore space. Saturations at each pressure level were computed from photographic images taken at equilibrium states. Observations from the experiments have shown that many of the physical processes occurring in two-phase systems are also the dominant processes in three-phase systems; however, comparisons of model and experimental results have identified the need to account for transport phenomena other than those of simple capillary-based displacement.

1 INTRODUCTION

Typical problems for the groundwater analyst involve prediction of contaminant movement or removal at the field scale. To solve these problems, the system of interest must somehow be characterized through description of the solid(s), fluids, and interactions between them. When multiple fluids are present in the porous medium, the governing equations are highly non-linear and the characterization of the system is complex. In some cases there may be three mutually mobile fluid phases: a gas, an aqueous liquid, and a non-aqueous liquid, which adds significantly to the complexity of the governing equations. The constitutive parameters enter into the transport equations through pressure-saturation relationships, permeability-saturation relationships, and conductivity. These parameters are implemented in these equations in one of two ways: 1) using empirically derived functions or 2) using laboratory data from core samples. Both methods of arriving at the material relations have severe limitations - laboratory tests are expensive and time consuming, while functional forms sidestep the physical principles and may not account for such important features as hysteresis. In either case, the three-phase information used in models is generally some combination of two-phase data (Stone 1973), due to the inherent difficulty in obtaining true three-phase values. The difficulty in applying material relations in this way is that there is very little understanding of the physics governing systems having more than two fluids. There is disagreement about the extent to which each fluid affects the others (Baker 1988), and there is no proof that combining two-phase data gives valid three-phase results.

The work presented here undertakes to increase understanding of three-phase physics and to use this knowledge to develop a numerical model for constitutive behavior in porous systems. Experiments were used to gain an understanding of the physics dominating three-phase transport as well as to obtain quantitative data to use in model validation. This paper presents a computational model for capillary pressure - saturation (P_c-S) relations based on pore-scale processes, and the experimental technique used to study the interaction between two or three coexistant fluids. Experimental and computational results are analyzed and compared, and important aspects of the transport process are identified.

2 COMPUTATIONAL MODEL

A computational model has been developed to simulate capillary pressure - saturation relationships for two or three fluid phases in a porous medium. The process of interest is quasi-static displacement of immiscible fluids in a porous solid. For quasi-static transport interfacial forces are dominant. The governing equation is the Laplace equation, relating the pressure difference across an interface to the curvature of the interface and the interfacial tension between the fluids and solid:

$$P_c = \frac{\sigma}{\rho} \tag{1}$$

where ρ is the radius of curvature at the interface, and σ is the surface tension.

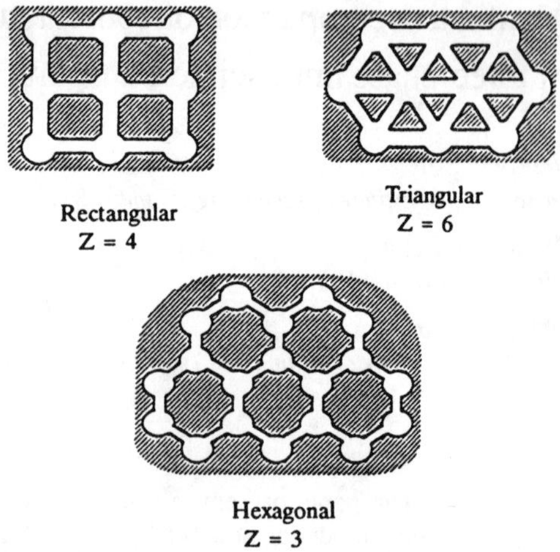

Fig 1: Lattices of various connectivity

2.1 Model Basis

The computational model combines the techniques of network modeling as a description of the porous solid, with a modified form of percolation theory to govern the movement of the fluids through the system. The network model was first applied to flow in porous media when Fatt (1956) used a network of interconnected tubes to study pore-size distributions and capillary pressure curves. Since that time, a number of studies have used networks to describe the pore space (i.e. Chatzis and Dullien 1985, Bear et al. 1987). A general network might consist of a lattice of large pore spaces (bodies, sites) interconnected by narrower spaces (throats, bonds). Material characteristics, such as size, wettability, etc., can be assigned to the spaces making up the network. The lattice is characterized by its connectivity (coordination number), which may be defined as the average number of adjacent sites to which a given site is directly connected by bonds (see Figure 1). The connectivity of each site is not necessarily a constant, especially in an actual porous material, but is generally modeled as such for computational simplicity. The benefit of these networks is that they describe, at least in part, the interconnected and branching nature of a porous medium, for which other models, such as the bundle of (non-intersecting) tubes or the lattice tree, are not able to account.

Percolation theory is used to describe deterministic transport through a random system (Broadbent and Hammersley 1957, Shante and Kirkpatrick, 1971). The percolation process utilizes a lattice which carries the random character of the medium, and a set of rules (governing equations) that the transport follows. These rules are applied to each element to determine whether that element would be active for the given system conditions. Percolation is a purely binary process in that the result of applying the rules is a yes or no answer: either the element is conducting or not. Once the characteristics have been assigned to the lattice, transport is entirely deterministic, and the entire system state is determined by repeat application of the percolation equation to each element in the lattice.

The governing equation for capillary dominated transport is the Laplace equation. For a simplified model of the pore space, the equation of state becomes:

$$P_c = \frac{2\sigma_{ij}\cos\theta}{R} \quad \text{(cylinder)} \tag{2a}$$

$$P_c = \frac{\sigma_{ij}\cos\theta}{R} \quad \text{(sphere)} \tag{2b}$$

where θ is the angle the advancing fluid makes with the solid, σ_{ij} is the surface tension between fluid i and j, and R is the radius of the pore space. If the lattice of interest is a network of pores each of which has been assigned some radius, then the percolation process is a matter of applying equations (2) at each interface and comparing against the bulk fluid pressure drop at that position. As an example, in drainage, if the bulk pressure of the less-wetting fluid minus the bulk pressure of the more-wetting fluid [P(lw)-P(mw)] is greater than P_c (as computed from (2)), then the pore fills with the non-wetting fluid, otherwise the wetting fluid remains. After a sweep through the system, all pores with radius large enough to drain will be filled with the non-wetting fluid, while all other pores will contain wetting fluid.

The basic percolation process has undergone a number of modifications to make it more applicable to modeling flow in porous media. In basic percolation theory the state of a given element is strictly a function of its characteristics with respect to the factors determining transport (the governing equation). The state of the element is neither a function of its position in space nor the state of its neighbors. This is not realistic for modeling flow in porous media, where some path or history allows a fluid to invade a given pore space. The first modification to standard percolation theory is, therefore, to incorporate an accessibility rule requiring a continuous path of fluid from its source to any given site which it invades. This modified form of percolation theory has been applied by a number of researchers (i.e. Chatzis and Dullien, 1985; Chandler et al. 1982; Larson et al. 1981) to study issues such as relative permeabilities, pressure-saturation behavior, and effects of flood velocities. A further modification makes it possible to account for fluids which become isolated from their source. Immobilization of discontinuous regions of fluid has been supported in very low flow rate studies of two-phase systems (Chatzis and Morrow 1984); preliminary observations of three-phase systems suggest this to be generally acceptable as well. To prevent these fluids from being displaced in a percolation model it is necessary to recognize elements having no path along which to move. Such elements can then be bypassed. Identification of isolated fluid is an important enhancement of percolation theory which allows a more accurate representation of the fluid distribution in a porous solid. It is required to produce the non-zero residual saturations which are observed in real systems.

The basic appropriateness of a percolation model to two-fluid systems is evident, with one fluid being denoted as "on" while the other is "off". Application of percolation based transport to systems with three mobile fluids is not as straightforward. When three fluids coexist, and elements in the system may have any of three different values, the principles of percolation theory must be extended to adequately handle the new options. In the three-phase model presented below a number of assumptions about fluid behavior have been made to maintain a binary model structure. The primary assumptions require that any given interface involves only two fluids, thus reducing it to a two-phase system locally, and that only one fluid advances into the network at a time, although all three phases may be simultaneously mobile.

2.2 Model Description

In this section the form and operation of the computational model is described. The model pore space is a two- or three-dimensional network of pore bodies interconnected by pore throats. The overall lattice is either rectangular (2D) or rectangular parallelepiped (3D). The number of sites in any given row is the same, though not necessarily the same number as the other direction(s) (i.e. the system need not be square or cubic). The system size is unrestricted, but is typically on the order of 10000 sites. The connectivity of the lattice is also unrestricted, however to date only systems of connectivity 4, 6, or 8 have been implemented. The cases presented herein are all two-dimensional with connectivity four, to reflect the experimental system.

Figure 2 illustrates the configuration of the 2-D model system. The boundary conditions on all sides are either open-flow, from which a fluid may freely exit the system at any time, or no-flow, which acts like an impenetrable barrier. Three of the domain boundaries contain fluid reservoirs. These reservoirs may span any portion of the side,

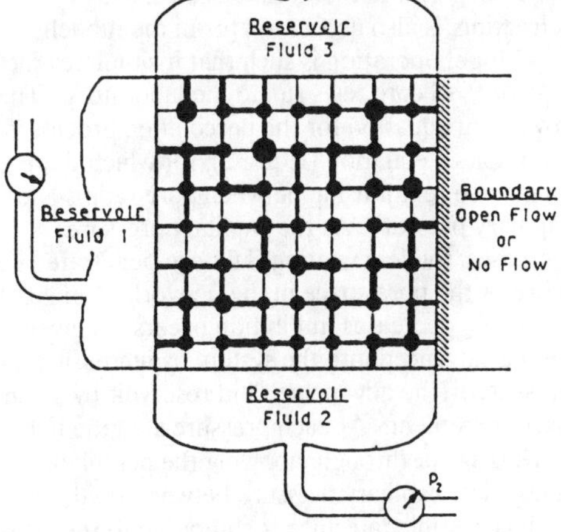

Fig 2: Configuration of the 2-D system for the computational model

as specified by the user. Although a border with a reservoir may be defined as a no-flow boundary, the reservoir is always open to its own fluid.

The controlling characteristic in capillary dominated flow is the size of the pores. The choice of pore size distribution (PSD) is a complicated matter for which no clear criterion has yet been established. Due to the complex and irregular shape of the pore space it is not possible to measure the PSD of a porous material. Experience indicates that the choice of pore-size distribution governs the overall shape of the P_c-S curve, and influences what percent of the total pore volume is contributed by the throats. The model assigns a radius to each pore space, usually from an internally generated probability distribution. There are a number of distributions available, including the commonly used log-normal, uniform, and Weibull (Bear et al., 1987, Chatzis and Dullien 1977). Pore body and pore throat radii are assigned from two independent distributions. The only restriction on pore size is that any pore body must have a radius that is at least as large as that of the largest connected throat. This requirement is reasonable in terms of the basic definition of bodies versus throats. The overall network is statistically homogeneous and isotro-

Another important material characteristic featured in equation (2) is the relative wettability of the porous solid, and of each fluid phase with respect to each other and the solid. Relative wettability provides a distinction between behavior of the different phases. Wetting is accounted for through the contact angle, θ, in the Laplace equation, and may range from 0° to 180°. Contact angle hysteresis, the change in contact angle depending on whether a fluid is advancing or retreating, is also accounted for in the model.

Model operation is such that it simulates the type of P_c-S core tests run in the laboratory. The governing equation for the percolation process is the Laplace equation (Eqns. 2), for which there is an inverse relationship between pore radius and capillary pressure. As the capillary pressure increases, the less wetting fluid can penetrate more of the pore space in the network (drainage), and as P_c decreases imbibition occurs. A given fluid is advanced into the system by increasing the pressure in the advancing fluid reservoir by some fixed increment. At each pressure step the network is swept through, applying the percolation rules. The capillary pressure between fluids at each accessible interface is compared to the critical pressure at that interface. If the comparison favors advancement (i.e. $P_c > 2\sigma\cos\theta/R$ for drainage) then the advancing fluid displaces the resident fluid; otherwise, the resident fluid remains. The cycle of changing pressure and advancing fluid is continued until some predetermined (user specified) capillary pressure is reached (this pressure may or may not correspond to residual saturation of the displaced phase(s)).

The model includes the ability to identify immobilized fluids. The mode of trapping incorporated in the model is fluid bypass, wherein an area of resident fluid becomes discontinuous when a line of advancing fluid moves around it and intersects another region of advancing fluid. An element, or cluster of elements, is identified as immobile if there is no continuous path from it to either the external reservoir or an open-flow border. Immobile fluids may not be displaced; however, they may be remobilized if a continuous path of the same fluid reconnects those sites.

The effect of gravity is included for modeling sloped or vertical domains (Wilkinson 1984). Gravity is accounted for through a hydrostatic correction to equation (2):

$$P_c = \frac{2\sigma\cos\theta}{R} - (\rho_{mw} - \rho_{lw})gz + K \qquad (3)$$

where K is a correction which accounts for the fluid in the vertical reservoir. The fluid pressure in the vertical reservoir (fluid 1 in Fig. 2) is assumed to be hydrostatic. In the tests presented in this paper (both experimental and computational) all systems were run horizontally and so gravity is neglected.

A number of simplifying assumptions have been made to make the displacement rules computationally tractable. The first is that although all three fluid phases may be concurrently mobile, only one fluid is advancing into the system at any given pressure step. The other two fluids move as a result of being "pushed back". The retreating fluids are displaced into their respective reservoirs or escape through open-flow boundaries. A further restriction of the computational model is that only one fluid is allowed to occupy any given pore at any given time. The implications of this assumption will be discussed below.

Simulations have been run on both a MicroVax II and a Sun4/330 SparcStation. CPU time required for two advancing fluid cycles (i.e. one complete imbibition/drainage cycle for two-phase, or some combination for the three-phase system) is approximately 8 seconds on the Sun and 60 seconds on the MicroVax for a 50x50 network (the size of the experimental system). For a 100x100 network those times increase to 55 seconds and 20 minutes respectively. The three-dimensional model runs a 30x30x30 network in 4 minutes on the Sun, and a 60x60x60 network in 150 minutes.

3 EXPERIMENTAL MODEL

The experimental purpose was twofold: 1) to observe the interaction of three fluid phases in a porous medium so that important physical aspects of the transport process could be identified, and 2) to obtain quantitative data against which to test the computational model. In fulfilling these purposes, both two- and three-phase experiments were run. These appear to be the first two-phase experiments that have been designed for quantitative analysis at equilibrium states, and the first experiments of their kind using three mutually mobile fluid phases.

3.1 Experimental Set-up

A two-dimensional system was chosen for the experiments because it is a system with minimal complexities which lends itself to full visual examination. The experimental model was designed to reflect the computational system as closely as possible to make comparisons more meaningful. The "porous medium" chosen for these experiments was a glass plate "micromodel", which consists of glass plates that have been chemically etched to produce a pattern of channels in the glass, then fused to create a sealed space between the plates. The technique for making micromodels is well established (McKellar and Wardlaw 1982; KTI Publication, 1989). This type of plate has been used in a number of projects, for example to identify quasi-static and dynamic displacement processes (Chen 1986), and to study blob mobilization (Wardlaw 1982).

The pattern used to make the micromodel was a 50x50 network created by reproducing the pore space from a run of the computational model onto paper. The network was composed of five sizes each of pores and throats taken from uniform distributions. The requirement of a quasi-static system meant that the fluids had to be prevented from breaking through into the other fluid reservoirs. To accomplish this, each reservoir was designed with capillary barriers etched into the plate. This meant that each reservoir was specific to one fluid, and the relative positions of the fluids in the reservoirs could not be switched. Two and three phase tests were run using different plates with the appropriate number of reservoirs. The reservoirs spanned only sixty percent of the total length of the model in an effort to reduce boundary effects and to minimize the amount of interference between reservoirs on adjacent sides of the plate. For three-phase tests the two liquids were always at either end of the model, while the air was on the side.

The fluids chosen for the experiment were air (least-wetting fluid), distilled and deionized water dyed with blue food coloring (Schillings) (most-wetting fluid), and Soltrol 130 (Phillips Petroleum) dyed with Oil Red O (intermediate-wetting fluid). The choice of fluids was based on ease and safety in handling, relative mutual immiscibility, and availability. Fluids were injected into the plates through holes drilled in each reservoir. Fluid lines were attached at the holes using C-clamps to hold the connectors. The connectors were fitted with O-rings to create a positive seal with the plate and eliminate possible fluid leaks.

The experiments were designed to replicate a standard laboratory P_c-S core test and were thus pressure driven (refer to Figure 3 for a schematic of the laboratory set-up). Each reservoir pressure was independently controlled. The liquid reservoirs were controlled via head from a large volume buret on a stand of adjustable height. Pressure of the liquid phases could be controlled to within 0.0625" H_2O. The air pressure was controlled through multiple regulators with a pressure range of vacuum to 20 PSI. The regulator adjustment was sensitive to approximately 0.125" H_2O.

The system was initialized with one fluid at approximately 100% saturation. The initial resident fluid was primarily H_2O, due to the

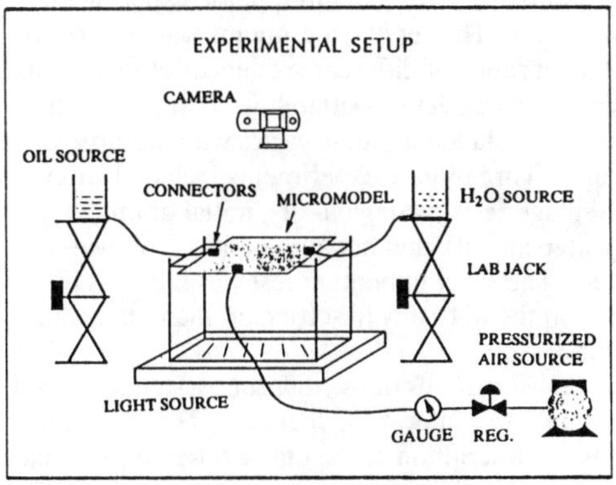

Fig 3: Schematic of the experimental set-up

difficulty in setting up an initial condition with anything but the most-wetting fluid. Some of the experiments were run with oil as the initial resident fluid, but it was not possible to begin with air. After the network was saturated, the other reservoirs were filled with their respective fluids. It was almost inevitable that at least some of the "non-resident" fluid(s) would enter into the model during the set-up of the initial condition. A small quantity of such invasion was allowed as being unavoidable (generally less than 5% saturation).

The procedure for advancing the fluids through the micromodel was the same as that described for the computational model. One fluid was advanced into the system at any given time by increasing the pressure of that fluid by small increments, typically 0.125" H_2O. The system was then allowed to reach equilibrium. At that time a photograph was taken of the system state and the procedure was repeated. In most cases the displacement cycle was continued until the resident fluids reached residual saturation. After the first fluid had advanced into the system the cycle was repeated with the next advancing fluid. Three-phase experiments were also run where the initial resident fluid was not displaced to residual. In these cases there were two mutually mobile phases in the system when the third fluid entered.

Saturations of each of the fluids in the system were computed at each equilibrium state. The saturations presented here are site fractions, computed from the images of the system taken at equilibrium. All comparisons between the experimental and numerical systems presented here were made using the site fraction information.

3.2 Experimental Observations

A number of two- and three-phase experiments were run. The set of experiments was selected to study a range of different sequences of fluid invasion with respect to wettability. Two-phase experiments included multiple cycles with all three fluid pairs. Three-phase experiments included strict drainage tests (water-oil-air), initial drainage (water-air-oil), and initial imbibition (oil-water-air). The strict imbibition test was not possible due to the difficulty in setting up the initial condition.

Observations of the different systems showed a number of interesting features. The following is a brief description of the major observations made during the experiments:

1. Wedge (thick film) flow is an important mode of fluid transport during imbibition. Wedge flow describes the fluid transport occurring in the surface roughness and corners of the pore space. Corners are formed in micromodels at the seam where the plates are fused because of the inexactness of the etching process. These tight spaces become preferred flow paths for the more wetting phase. Wedge flow is manifested during imbibition steps by the more wetting fluid often appearing in "inaccessible" pores.
2. Fluid topology has a strong influence on whether or not a given pore will fill. By "fluid topology" we refer to pores and throats that are filled with the fluid of interest, and, in particular, those throats leading into a pore are filled with the same fluid as the pore itself. The influence of fluid topology on pore filling in a system of connectivity four was studied in the early 1980's (Lenormand et al. 1983). That work defined different types of imbibition based on the fluid topology and identified approximate pressure levels at which each occurred. The filling patterns observed in certain individual pores in our experiments were similar to those described in the above mentioned work. The influence of fluid topology manifests itself as partially filled pores occurring during some imbibition cycles. Pore bodies shared by two apparently continuous phases were observed in experiments where oil was imbibing against air. It is not clear whether this type of displacement actually affected the transport process; however it did serve to identify that the tranport process was not purely capillary dominated.
3. Although the micromodel material was glass, the intrinsic wettability was not obviously water wet. During cycles of the tests in which water was "imbibing" against oil, negative capillary pressures ($P(H_2O) > P(oil)$) were always required for displacement of the major quantity of oil. Repeated occurrences of this phenomenon, regardless of the particular micromodel used or the treatment of the plate prior to running the test, suggested that the etched glass was not strongly wetted by either H_2O or oil; however, there was evidence of wedge-type transport of the H_2O phase against the oil, which suggests that water is mildly preferred.
4. The order in which the fluids enter the pore space has a strong affect on the transport

behavior of the different phases. The three-phase experiments run included tests in which the displacement cycle was $H_2O \rightarrow$ Oil $\rightarrow$ Air and Oil $\rightarrow H_2O \rightarrow$ Air. There was a marked difference in the behavior of the air phase between these two orders. In the first case (W→O→A), the air phase produced long thin fingers in the system and had none of the frontlike qualities expected from capillary displacement. In the second case (O→W→A), the air displaced the other fluids in the expected frontlike manner. Furthermore, tests in which displacement was not to residual showed the air invading in fingers only in pores that had H_2O which was displaced by oil. The difference in behavior of the air phase suggests that in the first case the interface occurred between the three fluid phases while in the second case the interface was between the two fluids and the glass.

5. The experiments were limited by both boundary effects and effects resulting from the form of the capillary barriers. The size of the system we could create was limited by the physical pattern that could be printed in conjunction with a plate size that could be handled during the process of creating the micromodel. The design of the capillary barriers was limited by the existence of three differently-wetting fluids. It was particularly difficult to design a barrier that would block both the most- and least-wetting fluids yet allow the intermediate-wetting fluid to pass. The resulting barrier design created large spaces bordering the oil reservoir that were preferred by the air phase, thus filling immediately on contact with it, subsequently isolating any oil in the network. Further, these capillary barriers could withstand only a limited capillary pressure, which would not be high enough to cause measurable fluid displacement through film flow.

Most of the above observations regarding displacement rules occurred predominantly or exclusively during imbibition. Displacement during drainage appeared to occur more through capillary displacement processes.

4 RESULTS

In this section the computational model results will be compared to the experimental data. The computer simulations run used the same network as was used to generate the micromodel with one modification. It was found that the model was better able to match experimental data using ten classes of pore body and throat size, rather than the five classes from which the artwork was generated. The pore sizes in the micromodel actually span a wide range with five evident clusters as a result of the inexactness of the chemical etching process. An increase to ten sizes in the computational model appeared to be sufficient to capture the experimental smearing. The actual pore sizes used in the numerical model were based on dimensions measured from the micromodels. It was possible to measure pore dimensions only in the plane of the glass. The pore depth is generally much shallower (due to limitations in etching glass) and thus a very influential term in the Laplace equation. Typical measurements (Campbell and Orr 1985) have shown pore depth to be in the range of 100-200 μm. Using this range and knowledge of the pressure ranges required in the experiments, a value of 150 μm was selected for the numerical tests.

The other parameter which is not known exactly in the experiments is the value of the numerator in equation (2), $\sigma\cos\theta$. Contact angles could not be measured in such a confined system, nor was it possible to determine the extent of contact angle hysteresis. Simulations presented here assume no hysteresis. The contact angles themselves are not separated from the surface tension, so that $\sigma\cos\theta$ is one parameter. Measured and published values of surface tensions result in a range of values for each of the three fluid pairs. Furthermore, surface tensions vary depending on fluid contamination and the presence of a third phase. The values assigned to the term $\sigma\cos\theta$ were thus based on the experimental data, and restricted to lying within the published range. For the three-phase simulations, values of this term were determined using the H_2O-oil-air tests (Fig. 6) and remained unchanged for the other three-phase tests. The values of $\sigma\cos\theta$ used in the two-phase simulations were chosen independently from the three-phase runs, and varied somewhat from those of the respective fluid pairs in the three-phase simulations. The difference in surface tension values can be explained by the existence of an additional fluid in the system. The values used for $\sigma\cos\theta$ are noted in the figures.

Figures 4-5 are comparisons of computational and experimental results for two-fluid tests. The first two-phase relationship illustrated here is for water and air. It can be seen that the model has simulated the experimental data well overall, how-

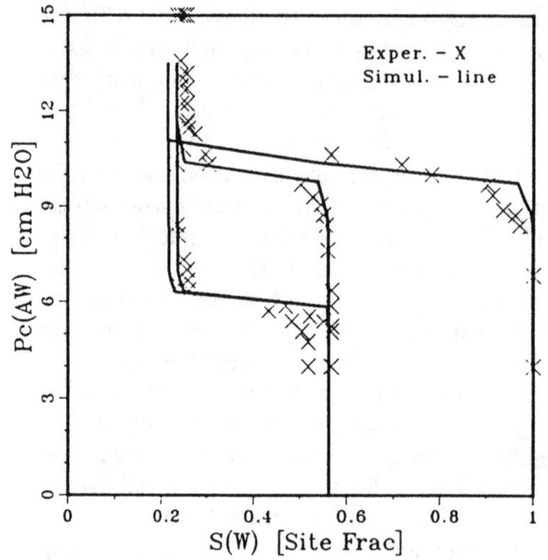

Fig 4: P_c-S relation for a water-air system, computational vs. experimental values ($\sigma_{aw}\cos\theta$ = 0.074 dyn/cm^2)

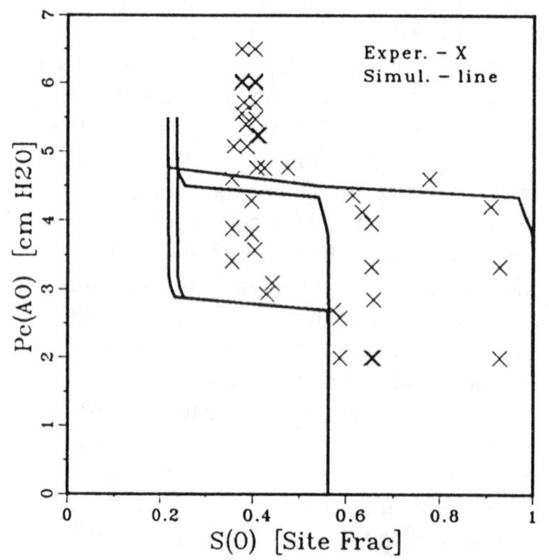

Fig 5: P_c-S relation for an oil-air system, computational vs. experimental values ($\sigma_{ao}\cos\theta$ = 0.034 dyn/cm^2)

ever, the model was not able to reproduce the small changes in saturation at the extreme pressures. A better discretization of the pore sizes may result in a closer fit. The second relationship (fig. 5) is for oil and air. In this case the model has predicted a much lower residual saturation than was obtained in the experiments. In fact, the saturations for the water-air and oil-air models are the same. The only difference between the computational results for these two tests is in the pressure, which differs by a factor of $\sigma\cos\theta(aw)/\sigma\cos\theta(oa)$. This is a result of fluid movement in the model being controlled solely by capillary rules. Pores fill when the capillary pressure is sufficient to displace the fluid resident in them. This is determined strictly by pore size, therefore the pores fill in the same order and at the same relative time, just at different absolute values of the capillary pressure. The mechanisms involved in producing different saturations for the different fluid combinations in the experiments is not clear but will be studied in greater detail in the future.

Comparison of experimental and numerical results for three-phase systems are shown in figures 6 and 7. Two different orders of fluid invasion are shown in these figures: water → oil → air and water → air → oil, respectively. These two runs illustrate the difference between modeling pure drainage and modeling imbibition. The water-oil-air data can be matched fairly well by the numerical model, but given the same parameters, the other test is not as well matched. In fact, the initial displacement in the water-air-oil experiment shows a relatively good match, but the comparison deteriorates during the second cycle.

Recall that at this point the computational model is not able to account for many of the observed processes active in the experimental system. Points (1) - (3) of the previous section refer to physical processes which affect fluid movement. Of these three, the computational model accounts only for the third, relative wettability, however the matters of wedge flow and fluid topology are important, based on physical observations. The continuation of the modeling process is taking on the task of accounting for each of the other processes affecting fluid displacement.

We have not yet looked at the positional distribution of the fluids, but have only looked at bulk saturations. The benefit of having images of the system at various points during the experiments is that there is the possibility of comparing consecutive steps and exactly identifying where the changes are occurring. This enables us to identify which processes are likely to be controlling the transport at a given time. Also, it is useful to compare the positional distributions of the computational and experimental systems to verify that the saturations obtained by the computer model are a result of the same behavior that established the fluid distribution in the experiments.

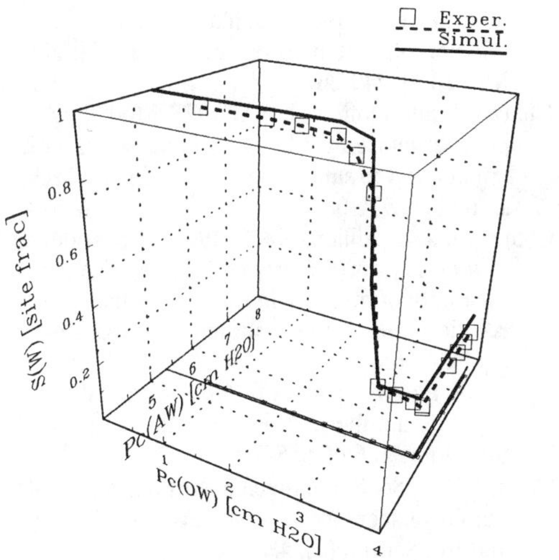

Fig 6a: P_c-S(H_2O) relation for Water→Oil→Air test, computational vs. experimental. Projections of the curves in the x-y plane are also shown. ($\sigma_{aw}\cos\theta$ = 0.060, $\sigma_{ao}\cos\theta$ = 0.027, $\sigma_{ow}\cos\theta$ = 0.024) [units are dyn/cm^2]

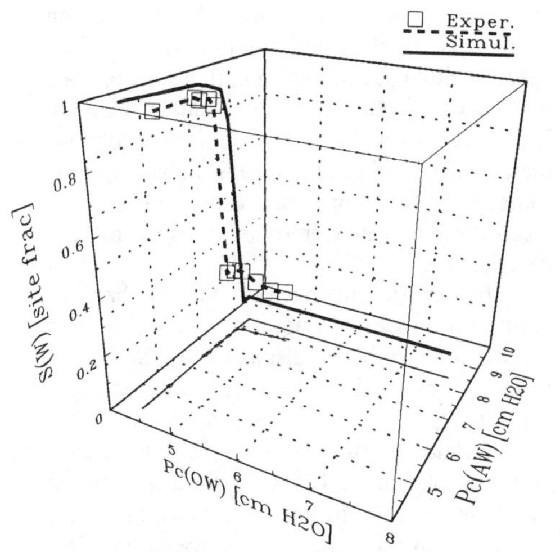

Fig 7a: P_c-S(H_2O) relation for Water→Air→Oil test, computational vs. experimental. Projections of the curves in the x-y plane are also shown. ($\sigma_{aw}\cos\theta$ = 0.060, $\sigma_{ao}\cos\theta$ = 0.027, $\sigma_{ow}\cos\theta$ = 0.024) [units are dyn/cm^2]

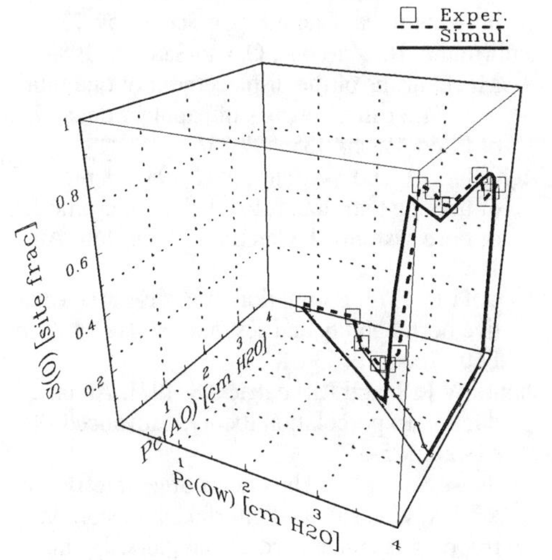

Fig 6b: P_c-S(Oil) relation for Water→Oil→Air test, computational vs. experimental. Projections of the curves in the x-y plane are also shown. ($\sigma_{aw}\cos\theta$ = 0.060, $\sigma_{ao}\cos\theta$ = 0.027, $\sigma_{ow}\cos\theta$ = 0.024) [units are dyn/cm^2]

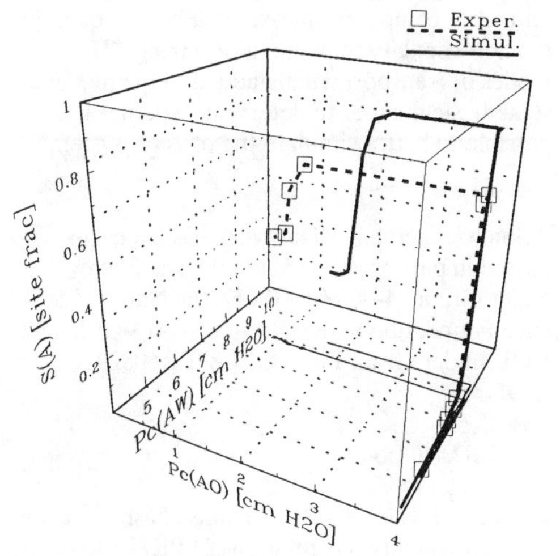

Fig 7b: P_c-S(Air) relation for Water→Air→Oil test, computational vs. experimental. Projections of the curves in the x-y plane are also shown. ($\sigma_{aw}\cos\theta$ = 0.060, $\sigma_{ao}\cos\theta$ = 0.027, $\sigma_{ow}\cos\theta$ = 0.024) [units are dyn/cm^2]

5. CONCLUSIONS

A computational model for two- and three-phase capillary pressure - saturation relations has been presented. This model is based on quasi-static, pore-scale processes. In conjunction, an experimental technique for testing the model and developing understanding of the system processes has been discussed. The model has the capability of reproducing the dominant behavior in a P_c-S curve, including: imbibition and drainage, scanning curves, hysteresis, and residual saturations. It has been seen that the model is able to match the experimental data to a reasonable extent for some of the fluid combinations and orders of infiltration; however, in other cases the match deteriorates.

The experiments have shown that there are a number of processes operating in the system, not simply capillary-based displacement. Some of the behavior observed includes: 1) more-wetting-phase transport through wedges, 2) two mutually mobile fluids sharing a given pore space, 3) strong influence of fluid topology on displacement, and 4) influence of the order of fluid infiltration into the system as a function of relative wettability. The extent to which these processes influence the system has not yet been quantified. This is the next step in the modeling process. More elaborate saturation computations, as well as detailed analysis of specific order of filling, and changes occurring between pressure steps will be made using a high resolution image analysis system. The computer analyses will help to quantify the non-capillary movement occurring. These modes of transport will be added to the model in some basic manner to determine whether it is possible to better simulate the physical system.

Acknowledgements: This work has been supported in part by the U.S. Geological Survey under Grant 14-08-001-G1747, the National Science Foundation under Grant 8657419-CES, and DOE under Grant DE-FG04-89ER60829

REFERENCES

Baker, L.E. 1988. Three-phase relative permeability correlations. SPE/DOE paper #17369.

Bear, J., Braester, C. and Menier, P. 1987. Effective and relative permeabilities of anisotropic porous media. Transport in Porous Media 2: 301-316.

Broadbent, S.R. and Hammersley, J.M. 1957. Percolation processes I: crystals and mazes. Proceedings of the Cambridge Phil. Soc. 53: 629-641

Campbell, B.T. and Orr, F.M. Jr. 1985. Flow visualization for CO_2/crude-oil displacements. SPEJ Oct: 665-678.

Chandler, R. et al. 1982. Capillary displacement and percolation in porous media. J. of Fluid Mech 119: 249-267.

Chatzis, I. and Dullien, F.A.L. 1977. Modelling pore structure by 2-D and 3-D networks with application to sandstones. J. Can. Petr. Tech. Jan-Mar: 97-108.

Chatzis, I. and Dullien, F.A.L. 1985. The modeling of mercury porosimetry and the relative permeability of mercury in sandstones using percolation theory. International Chem Eng'g 25(1): 47-65.

Chatzis, I. and Morrow, N.R. 1984. Correlation of capillary number relationships for sandstone. SPE Journal Oct: 555-556.

Chen, J-D. 1986. Some mechanisms of immiscible fluid displacement in small networks. J. Coll. and Int. Sci. 110(2): 488-503.

Fatt, I. 1956. The network model of porous media. Petr. Trans. of AIME 207: 144-163.

KTI Chemicals Inc. 1989. Photofabrication methods with KTI macro photoresists. Macro Guide-89-03-2000.

Larson, R.G., Scriven, L.E. and Davis, H.T. 1981. Percolation theory of two phase flow in porous media. Chem Eng'g Sci. 36: 57-73.

Lenormand, R., Zarcone, C. and Saar, A. 1983. Mechanisms of the displacement of one fluid by another in a network of capillary ducts. J. of Fluid Mech. 135: 337-353

McKellar, M. and Wardlaw, N.C. 1982. A method of making two-dimensional glass micromodels of pore systems. J. Can. Petr. Tech. July-Aug: 1982.

Stone, H.L. 1973. Estimation of three-phase relative permeability and residual oil data. J. Can. Petr. Tech. 12: 53-61.

Shante, V.K.S. and Kirkpatrick, S. 1971. An introduction to percolation theory. Advanced Physics 20: 325-357.

Wardlaw, N.C. 1982. The effects of geometry, wettability, viscosity, and interfacial tension on trapping in single pore-throat pairs. J. Can. Petr. Tech. 21(3): 21-27.

Wilkinson, D. 1984. Percolation model of immiscible displacement in the presence of bouyancy forces. Phys. Rev. A 30(1): 520-531.

Subsurface Contamination by Immiscible Fluids, Weyer (ed.) © 1992 Balkema, Rotterdam. ISBN 90 6191 175 3

Field measurements and numerical simulation of pressure drop during air stripping in the vadose zone

J.Croisé
Institut für Wasserbau, Lehrstuhl für Hydraulik und Grundwasser, Stuttgart, Germany
V.Kaleris
Department of Civil Engineering, Division of Hydraulic and Geotechnical Engineering, University Campus, Patras, Greece

Abstract: Prediction of the steady-state air flow field around gas extraction wells can be made by means of numerical simulations. In this study, field measurements conducted around a single gas extraction well are presented. The observed steady-state pressure field around the well was not radially symmetric. Several possible causes are given for the lack of symmetry, including variability of the permeability and the water content of the soil, are given. Furthermore, the validity of using Darcy's law to describe the steady-state air flow field around a gas extraction well is discussed critically. Initial results of 3-D numerical simulations are presented, showing the impact of an asymmetric skin around the well on the pressure field.

1. INTRODUCTION

Numerical methods are required to design subsurface venting systems used to remove volatile organic compounds (VOC's) from the unsaturated zone of the soil. The design goal is to maximize the flux of pollutants to the gas extraction wells. Numerical methods for the simulation and optimization of the design of subsurface venting systems, as they exist for groundwater clean-up measures, are still to be developed. For practical field cases, predictions of contaminant fluxes are generally not possible since the extent of the contaminated area is not sufficiently well known. Furthermore, the mass transfer processes occuring between the air, water and non-aqueous liquid phases, are still not sufficiently understood (Sleep and Sykes, 1989).
A meaningful approach to the optimization problem is by numerical investigations of the steady-state air flow. First results are presented in (Johnson et al., 1988), (Baehr et al., 1989), (Croisé et al., 1989), (Sabadell et al.), 1989.
In this paper, field experiments are presented, which were conducted in order to test several techniques used for the measurement of the air flow field produced by a single gas extraction well and to provide information about the validity of assumptions made in modelling.

2. TEST SITE AND HYDROGEOLOGICAL SITUATION

The test site is located in the Neckar valley near Tübingen (F.R.G) on a meadow in proximity to a transformer station (fig. 1). The gas venting installation comprises an extraction well (inner diameter: 600 mm, depth: 6,15 m, see fig. 2) and an air blower (type Elektror HRD 2/5, rated output: 3,1 kW, maximal capacity 1680 m^3/h) directly coupled to the well head.

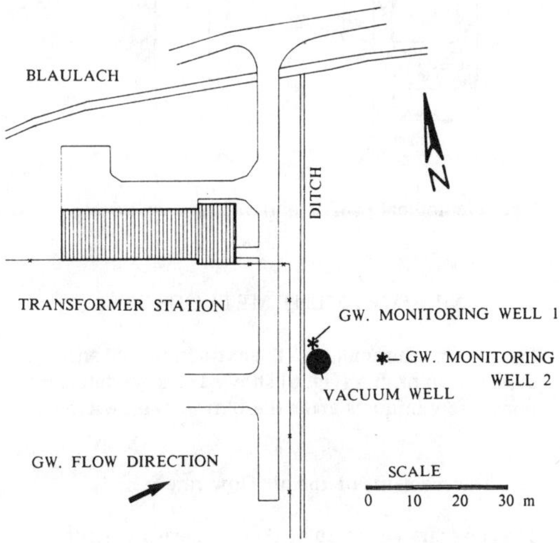

Fig.1 Test site location

The extraction well was already designed before the research project started, so no influence on the design has been possible.
Information about the structure of the unsaturated and saturated zone of the soil is provided through 2 boreholes and several split spoon samples (fig. 6). Three distinct horizontal layers of the soil are observed: clay, silt-sand and gravel (fig. 2). During the investigation period the groundwater level varied between 3.2 m and 3.9 m below land surface (BLS). In Fig. 2, the water level given and measured 13 February 1989, is the groundwater water level representative for the following investigations.

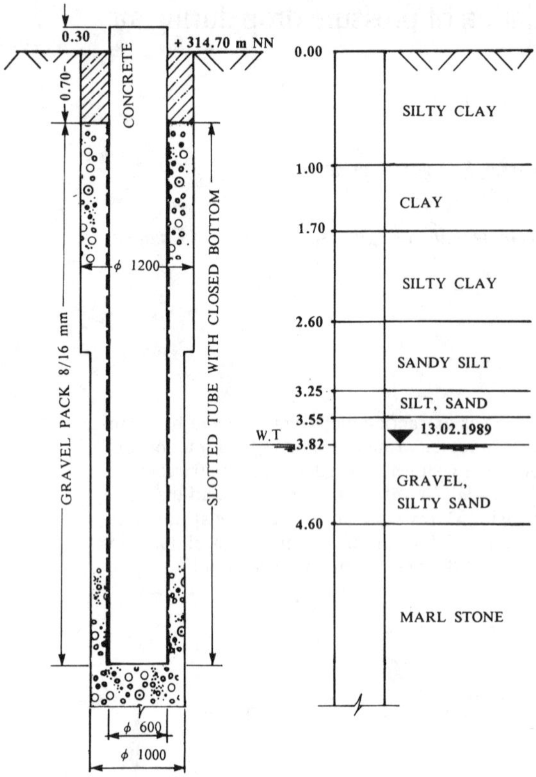

Fig.2 Geological profile and vacuum well design

3. MEASUREMENT METHODS

Pressure measurements in the extraction well and in the observation probes (fig. 6) showed steady-state conditions a few minutes after the blower was switched on.

3.1 Measurements of the air flow rate

The flow rate was calculated using measurements of the dynamic pressure taken at the end of a 2 m long pipe (cf. Kaleris et al., 1989). A flow straightening baffle was placed at the outlet of the air blower in order to remove the spin of the air flow (fig. 3). By measuring the dynamic pressure along a radius of the measuring section of the pipe, the air flow rate to the well can be determined with high precision. However, this type of measurement takes a very long time. Alternatively, a propeller air flowmeter with the same diameter as the air blower outlet pipe can be used. According to the manufacturer the systematic error is on the order of about 2 % of the measurement range, which is acceptable under the given field conditions.

3.2 Vacuum measurement setup in the extraction well

The pressure in the well was directly measured at the well head by means of a differential pressure gauge. All pressure values presented in this paper are negative values relative to atmospheric pressure.
In order to control the pressure measurement, the water-upconing in the well was observed with a water level meter for different operating conditions. Figure 4 shows the systematic difference between the pressure at well head and the water table upconing (converted to equivalent mbar for direct comparison with the vacuum in the well) in the well during the venting.
The observed difference may have been caused by the pressure sampling port being located at the outer edge of the well where flow is strongly deflected (fig. 3). As the water table is more difficult to measure than the pressure in the well, the pressure data obtained at well head was corrected using the graph shown in figure 4.

3.3 Measurement of the pressure in the soil

Two types of pipes were used to measure the pressure in the soil (fig. 5):

- -Type (a) is a steel pipe with a drive-point at its lower end, which is pushed away after the pipe has been driven in.
- -Type (b) is a thin plastic pipe which is introduced into a borehole with diameter of 50 mm, placed on a sand filter of about 20 cm length and sealed (to the atmosphere) by a clay seal (clay pellets).

The upper ends of the measuring pipes were connected to a pressure gauge using gastight seals. The pressure

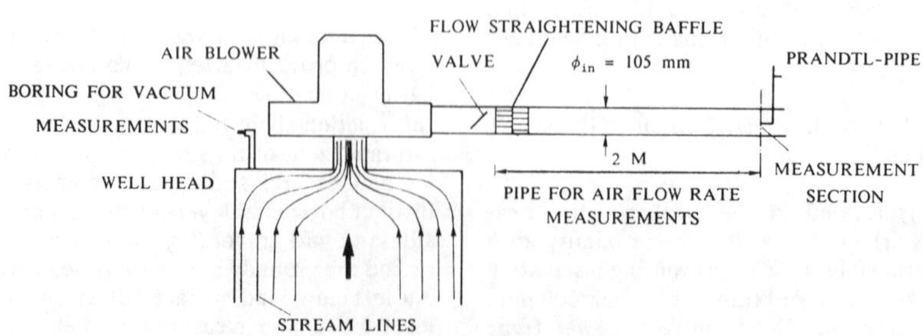

Fig.3 Venting system

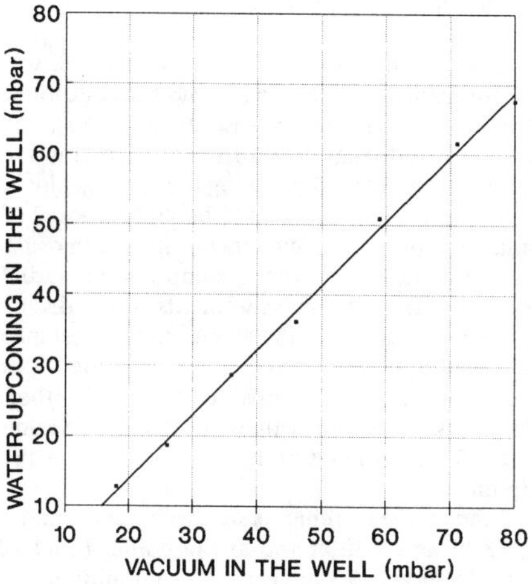

Fig.4 Watertable-upconing versus vacuum at the well head

values greater than 10 mbar were measured with an accuracy of ±2 mbar and those below this value with an accuracy of ±0.06 mbar. The reproducibility of the measurements was within the range of error given by the error bars on the plot of pressure values. The measurements with the two types of pipes were compared at 5 different locations (1,2), (8,9), (11,12), (13,14), (14,15). Fig. 6 presents the location of the observation probes set (drive-point system/ drill-pipe system). All measuring depths are about 3.2 m BLS. Fig. 7 shows the pressure measurements as obtained from the drive-point and the drill-pipe system. The agreement is very good only for set number (11,12), situated the farthest away from the well (6.3 m). For the other probes, the observed differences between the measurement of the two kinds are significant. Several reasons for the observed differences may be given:

1. In the drive-point systems, the soil permeability near

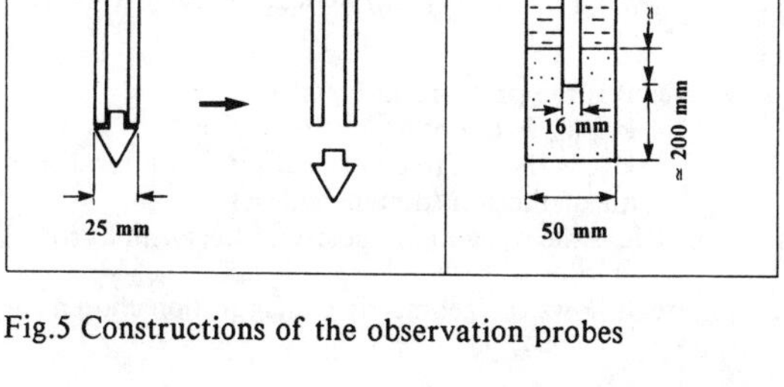

Fig.5 Constructions of the observation probes

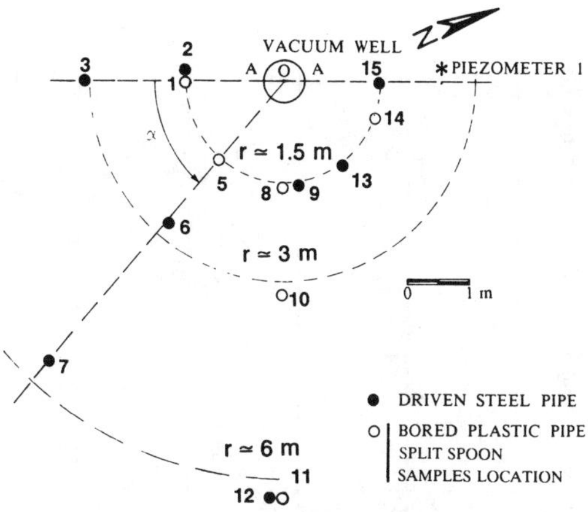

Fig.6 Location of the observation probes

the inlet end may be reduced due to densification of the soil during driving of the probe. The densification effect should play an important role, especially at locations where the silt content of the silty-sand layer is high. According to this explanation, Fig. 7 shows that except for pair 14-15 the measured pressures in the drill-point system probes are always greater than those of the drive-point systems.

2. In drive-point systems, the measurements are point measurements whereas in drill-pipe systems, they represent an averaged value of the pressure along the sand filter (20 cm).

Moreover, the geological units of figure 2 may demonstrate an unknown spacial variability in permeability and water content, which may give a random component to the air permeability of the soil: even the same type of system at given radial distances to the well provides values which are strongly differing from each other (s. part 4.2).

4. SHAPE OF THE AIR FLOW FIELD

4.1 Top layer sealing

Pressure measurements made during the investigation period at depths of 0.5 m and 2.0 m BLS as well as laboratory air permeability measurements of undisturbed samples, showed that the silty clay top layer of the soil is impervious to air. Assuming that the concrete seal at the top of the well is airtight, the underground air flow would be confined to the horizontal silt- sand and gravel layers (fig.2). In this particular geological situation (top layer of the soil is airtight), there was no point in using any system to apply the vacuum only on the part of the screen physically in contact with the silt-sand and gravel layers, therefore the entire well screen was put under vacuum.

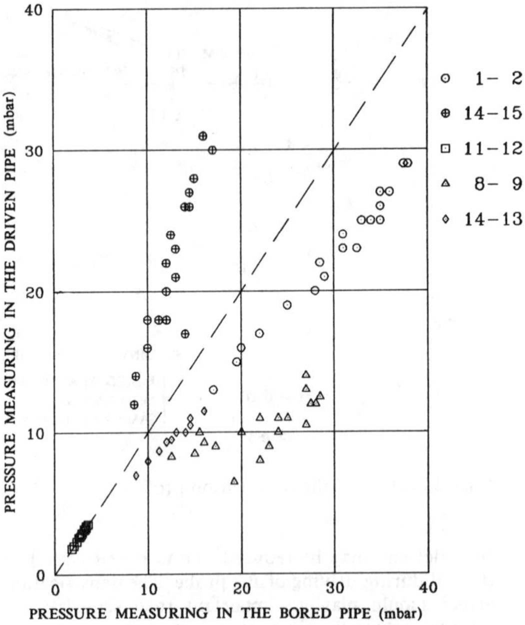

Fig.7 Comparison between pressures measured at the two kinds of observation probes

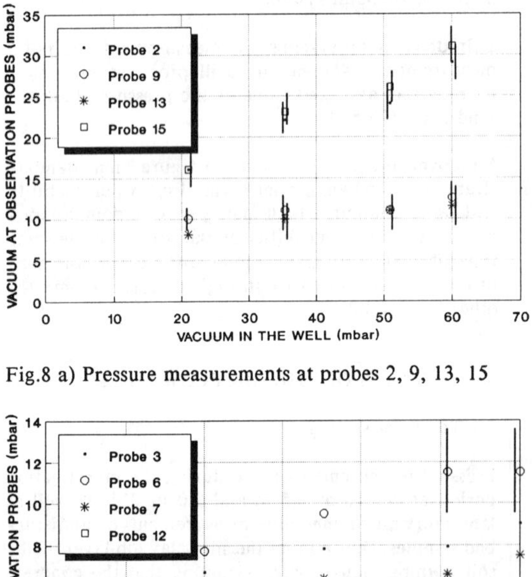

Fig.8 a) Pressure measurements at probes 2, 9, 13, 15

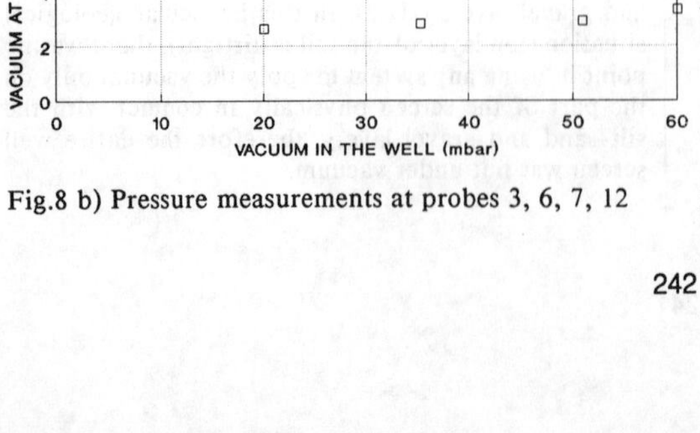

Fig.8 b) Pressure measurements at probes 3, 6, 7, 12

4.2 Symmetry of the air flow field

In previous work, the air flow field around a well was calculated for horizontally homogeneous soil conditions (see Baehr et al., 1989 and Croisé et al., 1989), the assumption was made that the air flow to the extraction well is radially symmetric. Following the given geological data (fig. 2), the soil is assumed being composed of 3 horizontal and globally homogeneous layers. In order to check the validity of the assumption of radial symmetry at the test site, measurements in the drive-point systems situated at the same radial distance around the well and approximately the same depth in the soil (3.2 m BLS.) are compared with each other. Figure 8a shows pressure values at the observation probe 2, 9, 13, 15 situated at 1.55 - 1.6 m from the well midpoint;

Figure 8b shows measurements at the points 3 and 6 located at 3.15 and 2.80 m and at the points 7 and 12 located at 5.65 m and 6.3 m from the well midpoint. The observed pressure differences in the probes located at a radius between 1.55 and 1.6 m from the well demonstrate clearly that the air flow close to the well is not radially symmetric. At greater radii (probes 3 and 6, 7 and 12) the observed pressure differences are also very large. Obviously, this shows that the radial symmetry of the flow field in the depth of 3.2 m (silt-sand layer) is disturbed by some heterogeneities. Further work will help to investigate this more precisely.

5. COMPUTATION OF THE AIR PERMEABILITY OF THE SOIL LAYERS

A first estimation of air permeability of the soil layers was computed using an axisymmetric finite element flow model (Croisé, 1987). The following assumptions were made:

- The air behaves like an ideal gas
- Darcy's law is valid
- The one-phase air flow is radially symmetric around the well
- Gas slippage effects are negligible.
- The soil is at field capacity and the soil moisture is constant in time.

The governing equation for a radially symmetric air flow is obtained from the continuity equation, Darcy's law and from the ideal gas law. Expressed in cylindrical coordinates (r,z) it reads,

$$\partial/\partial r\,[(k_{rr}/\mu)\cdot k_{rea}\cdot \partial/\partial r(P^2)] + (1/r)\cdot[(k_{rr}/\mu)\cdot k_{rea}\cdot \partial/\partial r(P^2)] + \partial/\partial z\,[(k_{zz}/\mu)\cdot k_{rea}\cdot \partial/\partial z(P^2)] = 0 \quad (1)$$

where: P is the pressure in N/m^2

k_{rr}, k_{zz} is the permeability of the soil in m^2

k_{rea} is the relative permeability with respect to air of the soil (dimensionless)

μ is the dynamic viscosity of the air in kg/(m.s)

Figure 9 shows the geometric configuration chosen for

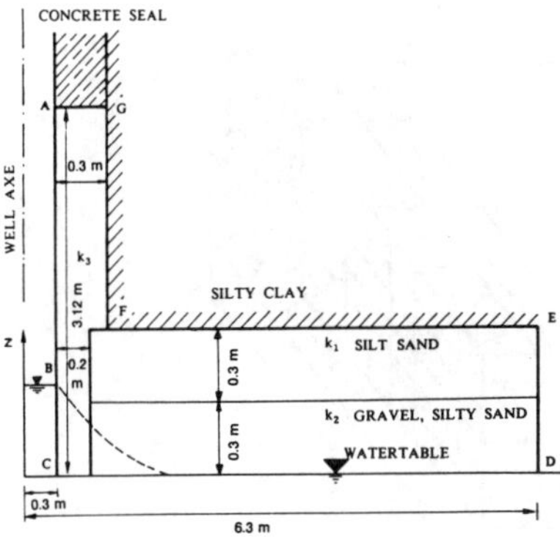

Fig.9 Flow configuration

the flow computation. The location of the two layers having different air permeabilities are obtained from geological survey. For the flow computation, the sides AB and ED of the model are handled as first type boundaries (Dirichlet). The left boundary (side AB) is set to the vacuum value measured in the well, and the right boundary (side ED) is set to the pressure value measured in the observation probe 12. The watertable rise in the well was taken into consideration by setting the boundary side BC to zero flow condition. For all other boundaries, zero flow conditions were also assumed.

For each computation, the air flow rate in the well was calculated using Darcy's law:

$$q = (k \cdot k_{rea})/\mu \cdot \text{grad } P \qquad (3)$$

The hydraulic conductivity of the gravel pack of the well was set to $k_{water} = 5.10^{-1}$ m/s according to the particle size distribution. As it is described below, this arbitrary choice of the hydraulic conductivity of the gravel pack, will not have any influence on the pressure field because the geological formations are several order of magnitude less pervious than the gravel pack. The air conductivity k_{airo} of a dry soil is given by:

$$k_{airo} = k_{water} \; (\nu_{water} \; / \; \nu_{air})$$

where:

k_{water} : hydraulic conductivity of the porous material in m/s

ν: kinematic viscosity of the fluid (air or water) in m^2/s

For a temperature of 15 °C a value of:

$$k_{airo} = 3.9x10^{-2} \text{ m/s.}$$

is obtained.

For field conditions, the influence of the soil water content on the air conductivity has to be taken into account. This can be done to a first approximation by the following way:

$$k_{air} = k_{airo} \cdot k_{rea}$$

where:

k_{rea} is the relative air permeability of the soil at field capacity saturation.

The ratio of the air conductivities of the layers 1 (k_{air1}) (silt-sand) and 2 (k_{air2}) (gravel, silty sand) was set to 1:100 according to the values given by Busch und Luckner, 1972.

The model was calibrated for 4 different venting experiments (4 different air flow rates). Model calibration was done by comparing the measured and computed values of the total air flow rate. The calibrated air conductivities are listed in table 1 together with the results of air permeability tests made in the laboratory on field samples taken while augering groundwater monitoring well 2, (fig.1) at field capacity (s. Brauns et al., 1989).

Tab.1

Pressure in well [mbar]	Pressure probe 12 [mbar]	Air flow rate [m^3/h]	k_{air2} comput. [m/s]	permeametry measured [m/s]
				$k_{air1} \simeq 1.10^{-7}$
-78	3.0	146	$4.5.10^{-5}$	after a week
-52	3.0	92	$3.7.10^{-5}$	k_{air2} =
-38	2.6	84	$4.6.10^{-5}$	$1.3.10^{-5}$ to
-24	2.0	40	$7.1.10^{-5}$	$5.2.10^{-5}$
Mean value of k_{air2} :			5.10^{-5}	

The differences between the measured (in lab.) and computed values are less than a factor 3 or 4 for the gravel. This is satisfactory considering the high uncertainty related with the measurement of the air conductivity by permeametry in the disturbed samples of gravel. Because of the big difference between the air conductivities of both layers (from permeability test we see that it could be also 1:1000 or more depending on the water content of the silty sand), the calibration of the model made by adjusting the air flow rate, is unsensitive to changes of the conductivity of the less permeable layer. This means that only the sandy gravel layer has been actually calibrated by this procedure.

6. 3-D MODELING OF THE AIR FLOW FIELD

In order to compute the pressure distribution around the gas extraction well, a 3-dimensional finite-difference groundwater model was modified to allow for computation of steady-state flow of ideal gases.

6.1 3-D finite-difference flow model

The governing equation of the one-phase steady-state 3-dimensional flow of an ideal gas with the same assumptions made in 5. (except that of the symmetry of the flow field), is given by:

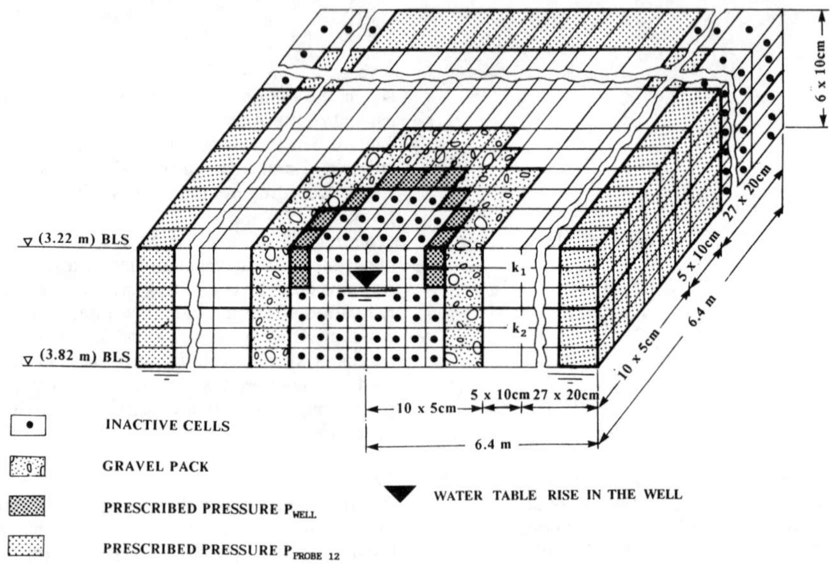

Fig.10 Model geometry

$$\partial/\partial x\,[(k_{xx}/\mu)\bullet k_{xxrea}\bullet\ \partial/\partial x(P^2)] +$$

$$+\ \partial/\partial y\,[(k_{yy}/\mu)\bullet k_{yyrea}\bullet\ \partial/\partial y(P^2)] +$$

$$+\ \partial/\partial z\,[(k_{zz}/\mu)\bullet k_{zzrea}\bullet\ \partial/\partial z(P^2)] = 0 \quad (2)$$

To solve the differential equation (2) the finite-difference groundwater model MODFLOW (Mc Donald and Harbaugh, 1988) was used. The modifications of the program needed to treat the steady-state ideal gas flow are as follows: the hydraulic conductivities (k_f) is replaced by $[(k/\mu)\bullet k_{rea}]$ and the piezometric head (h) is replaced by the pressure squared (P^2). After computing the flow field, P is calculated before volumetric overall and cell-by-cell budget are calculated. The fact that the compressibility of the ideal gas was taken into account (the equation is not written in P but in P^2), causes a difference between out- and inflow volumetric rates of up to 5%.

6.2 Model geometry and model discretization

The domain of study is shown in figure 6. Only the last 60 cm of the unsaturated zone (silt-sand, gravel) were taken into account in the computation. The upper part (silty clay) was not considered in the computation. Discretization of the domain and boundary conditions are shown in figure 10.

The air conductivities were set equal to the mean values obtained in the calibration of the axisymmetric model.

6.3 Results and interpretation

Figure 11 shows the measured and calculated pressure

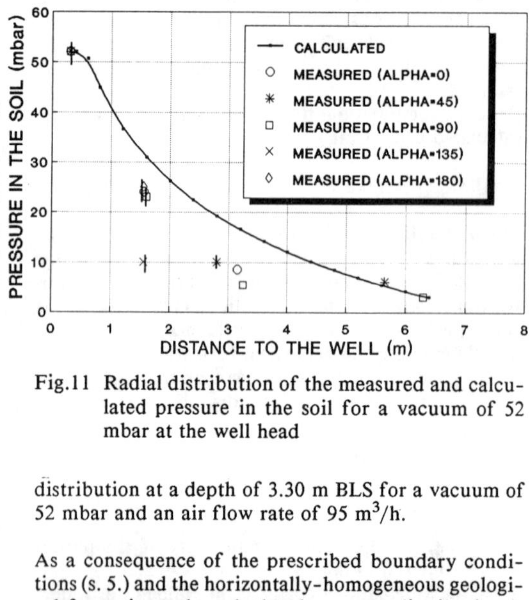

Fig.11 Radial distribution of the measured and calculated pressure in the soil for a vacuum of 52 mbar at the well head

distribution at a depth of 3.30 m BLS for a vacuum of 52 mbar and an air flow rate of 95 m^3/h.

As a consequence of the prescribed boundary conditions (s. 5.) and the horizontally-homogeneous geological formations, the calculated pressure distribution is vertically constant and radially symmetric around the well. Figure 11 shows that the pressure distribution is overestimated by the model. The difference becomes lower, for lower vacuums and lower air flow rates to the well (figures 12 and 13).

As the geological survey indicates that the two soil layers (silt-sand and gravel) are horizontally homogeneous other factors have to be mentioned which may cause the discrepancy between measured and calculated pressures:

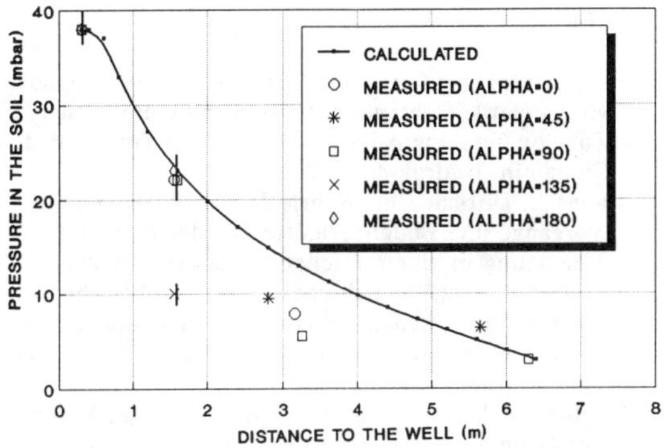

Fig.12 Radial distribution of the measured and calculated pressure in the soil for a vacuum of 38 mbar at the well head (ALPHA=α, s. fig.6)

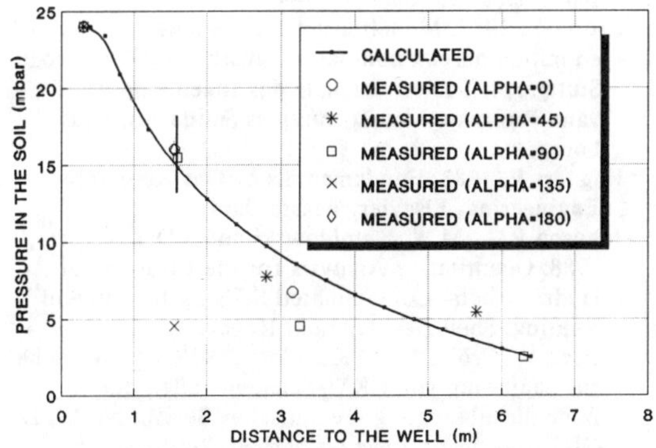

Fig.13 Radial distribution of the calculated and measured pressure in the soil for a vacuum of 24 mbar at the well head

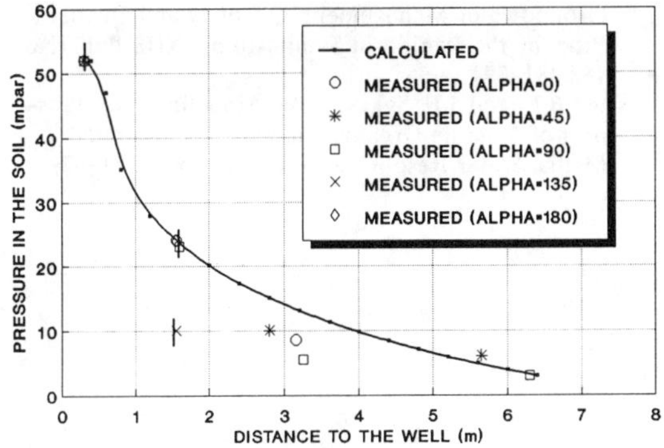

Fig.14 Radial distribution of the measured and calculated pressure in the soil for a vacuum of 52 mbar at the well head and an homogeneous skin around the gravel pack of the well

- Clogging of the well gravel pack by fine soil particles.
- Non-Darcy flow (energy dissipation) in the immediate vicinity of the well due to inertial forces caused by the locally-high velocities of the air flow (Hagoort, 1988).

In order to show the possible influence of these effects on the pressure distribution in the soil, the air conductivities corresponding to the computational cells surrounding the gravel pack were reduced to 1/10 of the value of the gravel formation. The results of the simulation for the highest flow rate (95 m^3/h) are shown in figure 14.
As expected, the skin of lower permeability causes a higher pressure drop around the well. The values calculated give in this case a better match to the measured ones. At the same time the skin also reduces air flow rate by approximatly 25%. In order to obtain the same air flow rate as before, the air conductivity has to be increased. Such an adjustement of the air conductivity is admissible as long as it does not exeed the values of the air permeabilities given in item 5.
In order to explain the differences between the measured values at observation probes located 1.6 m from the well (low pressure values were observed in probes 9 and 13, fig. 8a and fig. 6), an attempt was made to represent an asymmetric skin around the gravel pack. The air conductivity of the skin was modified only in the region defined by α= 100° and α=170° corresponding to the direction of probe 9 and 13 (figure 6). The corresponding value was set to 1/10 of the old value. Figure 15 shows the horizontal distribution of pressure. Figure 16 shows the radial pressure drop in section 1 and 2 of figure 15. The pressure distribution in the soil is no longer radially symmetric around the well midpoint. At distances higher then 1.5 m from the well, the flow is affected only in the form of a shift along the axis AOA of the well, causing a reduction of the pressure in the region between α=90° and α=180° and a better match between calculated and measured values in this area. On the other hand the differences between calculated and measured values, observed in the region between α=0° and α=90°, are still non-negligible.
At this point of the study a final explanation of the apparent asymmetry of the measured flow field is still to be found.
For this reason, more observation probes will be installed around the well to permit a better description of the flow field. In addition further field experiments e.g. gas tracer tests and longtime measurements (several days or weeks) are planned.

7. CONCLUSIONS

In this study, measurements of the steady-state air flow field around a gas extraction well were carried out. Pressure observation probes of 2 kinds were used and the differences in their results were analysed. Differences between measurements made at the same radial distance around the well showed that the flow field was not radially symmetric. A calibration of the air conduc-

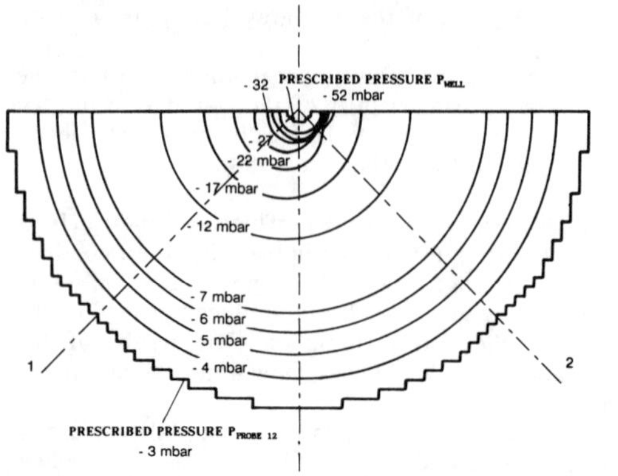

Fig.15 Horizontal pressure distribution in the soil for the case of an asymmetric skin around the gravel pack of the well

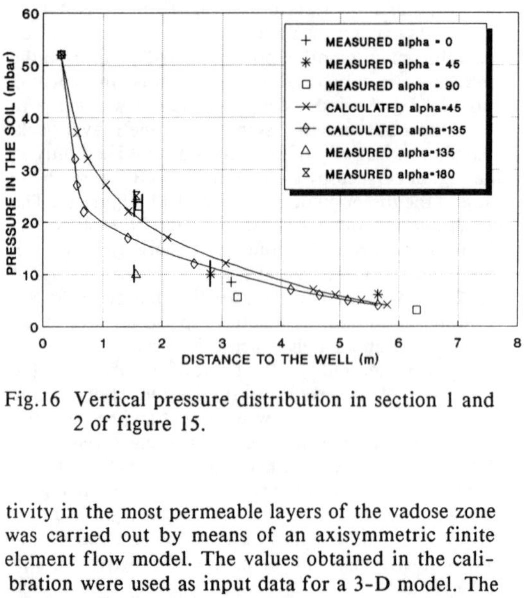

Fig.16 Vertical pressure distribution in section 1 and 2 of figure 15.

tivity in the most permeable layers of the vadose zone was carried out by means of an axisymmetric finite element flow model. The values obtained in the calibration were used as input data for a 3-D model. The pressure distribution computed with the 3-D finite-difference flow model was compared with the field measurements. Different possible causes for the discrepancies between measurements and calculations were discussed. The study will be completed by further numerical and field experiments.

ACKNOWLEDGMENTS

We thank Jörg Kopold for his assistance in this study. The financial support of the Landesanstalt für Umweltschutz of the state Baden-Württemberg is gratefully acknowledged.

REFERENCES

Baehr A.L., Hoag G.E. and Marley M.C., 1989. Removing Volatile Contaminants from the Unsaturated Zone by Inducing Advective Air-Phase Transport. J. Contamin. Hydrol. 4: 1 - 26.

Brauns J., Dittmar Ch., Wehrle K., 1989. Strömungsvorgänge und Bodenverhalten bei der Bodenluftabsaugung in einem flachen bedeckten Lockergesteinskörper (EVS-Tübingen). Abschlußbericht zum Forschungsvorhaben, im Auftrag der Landesanstalt für Umweltschutz Baden-Württemberg, Karlsruhe (unpublished)

Busch K.F., Luckner L., 1972. Geohydraulik, Verlag für Grundstoffindustrie, Leipzig.

Croisé J., Kinzelbach W. and Schmolke J., 1989. Computation of air flows induced in the zone of aeration during in situ remediation of volatile hydrocarbon spills, in Proc. International Symposium on "Contaminant Transport in Groundwater", Stuttgart, Balkema, Rotterdam: 437-444.

Croisé J., 1987. Modélisation d'un écoulement gazeux en milieu poreux non-saturé, Institut für Wasserbau Stuttgart - Ecole Nationale des Ingénieurs des Travaux Ruraux et des Techniques Sanitaires, Strasbourg.

Hagoort J., 1988. Fundamentals of Gas Reservoir Engineering, Elsevier, Amsterdam.

Johnson P.C., M.W. Kemblowski and J.D. Colthart, 1988. Quantitative Analysis for the Clean-up of Hydrocarbon- Contaminated Soils by In-Situ Soil Venting, Shell Development Report.

Kaleris V., Croisé J. and Kopold J., 1989. Feldversuche zur Sanierung von CKW-Schadensfällen durch Bodenluftabsaugung. Technischer Bericht Nr. 89/12 (HG 103). Institut für Wasserbau, Stuttgart.

Mc Donald M.G. and A. W. Harbaugh, 1988. A Modular Three-dimensional Finite-Difference Groundwater Flow Model. USGS-Report 83-875

Sabadell G.P., D.K. Sunada and N.S. Grigg,1989. A Numerical Model for Simulating the In-Situ Volatilization of hydrocarbon contaminants in Soils. Groundwater Management: Quantity and Quality . Proc. of the Benidorm Symposium. IAHS Publ. No. 188:381-391

Sleep B.E. and J.F. Sykes, 1989. Modeling the Transport of Volatile Organics in Variably Saturated Media. Water Resources Research, 25 (1): 81-92.

Subsurface Contamination by Immiscible Fluids, Weyer (ed.) © 1992 Balkema, Rotterdam. ISBN 90 6191 175 3

Transport of volatile chlorinated hydrocarbons by suction of subsoil air: Simulations and laboratory experiments

Dieter Weth

Prof. Mull & Partner GmbH, Garbsen, Germany (Formerly: Institute for Water Resources, Hydrology and Agricultural Hydraulic Engineering (IWH), University of Hannover, Germany)

Rolf Mull & Martina Pielke

Institute for Water Resources, Hydrology and Agricultural Hydraulic Engineering (IWH), University of Hannover, Germany

ABSTRACT: Suction of subsoil air can be used for the removal of volatile organics from the unsaturated zone. A numerical model is presented which describes the air flow, the phase transfer processes of the volatile compounds and the transport of the gaseous organics through a porous medium. The decrease of liquid contaminants in the soil caused by phase transfer processes liquid-gas has been calculated with the help of a reaction constant, which is dependent on the pore geometry and the apparent vapor diffusion coefficient. The work has been done in two steps, a theoretical and an experimental part. The results of the program system were compared with the results of the laboratory experiments.

1 INTRODUCTION

The contamination of soil and groundwater with chemical compounds e.g. volatile chlorinated hydrocarbons (CHC) demands effective methods of decontamination like soil venting. During this process air flows from the soil surface through the contaminated zone to a well. The liquid organic substances evaporate into this air stream corresponding to their vapor pressures.

Until now this technique is based only on empirical knowledge (Störk 1986, Bruckner & Kugele 1985). Investigations of Schwille et al. (1984) showed the behaviour of CHC, infiltrating into soil and groundwater. Numerical analysis of multiphase contaminant flow in the porous medium is the objective in a number of research projects (e.g. Coats et al., 1967; Corapciogly and Baehr, 1987; Parker et al., 1987; Abriola, 1984). Mathematical calculations of air flow in soil are given by Croisé et al. (1989). Experimental studies of vapor phase diffusion in soil are investigated by Sallam et al. (1984) and Karimi et al. (1987).

The purpose of our paper is to present a mathematical model describing the essential physical and chemical processes determining the progress of decontamination by soil venting. Therefore the following has to be considered:

1. Calculating with a numerical model the pressure distribution and the air flow in a soil being influenced by the
- porosities
- absolute and relative permeabilities
- boundary conditions.

2. Determining the process of phase transfer liquid-gas for special CHC with respect to the
- vapor pressure of CHC
- interfaces between the gas-filled pore volume and the liquid CHC-drops
- geometry of the pore volume
- velocity of the gas stream

3. Building up a transport model describing the movement of particles (molecules) in the gas phase by means of the air stream.

4. Coupling the above listed modules to a program system and verifying this system by comparing it with results from the laboratory experiments.

2 THEORY

2.1 Air Flow

The mathematical model describing the pressure distribution in a porous medium is based on the following differential equation (1).

The solution of equation (1) gives pressure heads as a function of space and time. As a result of a three dimensional pressure distribution, surfaces of equal pressure can be interpolated. Air flows

$$\frac{n_{am}}{p} * \frac{\delta p}{\delta t} = - \frac{k_0 * k_{rg}}{p * \mu_g} * (\text{grad } p)^2 + \\ + \text{div}\left[\frac{k_0 * k_{rg}}{\mu_g} \text{grad } p\right] \quad (1)$$

with

p : pressure
k_0 : intrinsic permeability
k_{rg} : rel. permeability of the gas
μ_g : dynamic viscosity of the gas
n_{am} : effective, airfilled porosity
t : time

perpendicular to these isosurfaces, and the Darcy law expresses the filter velocity of the air flow $\vec{v}_{fg}$:

$$\vec{v}_{fg} = - \frac{k_0 * k_{rg}}{\mu_g} * \text{grad } p \quad (2)$$

The steady-state form of equation (1) is solved numerically with the method of Finite Differences.

2.2 Phase Transfer Liquid-Gas

Because water always forms the wetting and CHC the non wetting phase, the surface of the soil particles and the small pores in between are occupied by water (n_w, see Fig. 1). Only the larger interstices (n_l) can be filled up by later infiltrating CHC, until the largest pores, which are able to hold the CHC against the force of gravity, are filled. The remaining pore space (n_{am} and n_{ai}) contains soil air.

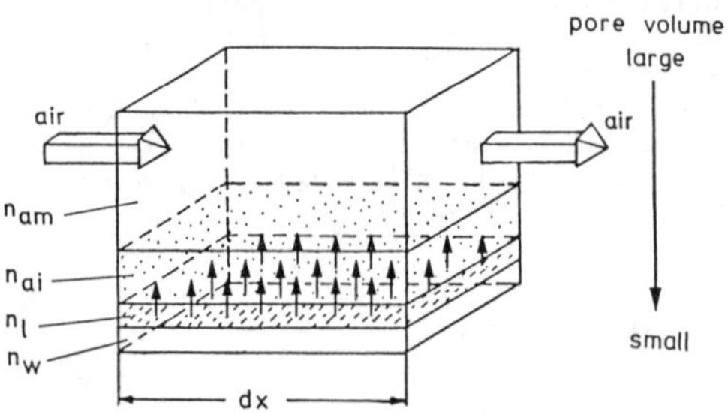

n_{am} effective, air-filled porosity
n_{ai} non-effective, air-filled porosity
n_l CHC(l)-filled porosity
n_w water-filled porosity

Fig.1 Phase transfer liquid-gas during soil venting

Because of their high vapor pressure CHC evaporate quickly and distribute into the airfilled pores.

If an air stream passes through the contaminated area, the CHC are transported within this stream in the effective airfilled porespace n_{am}. Because of the thermodynamic equilibrium the decreasing CHC-concentration in the gas phase causes a compensation to reach the saturated vapor pressure. This occurs as long as there is CHC in liquid phase present. Fig. 1 shows this process, schematically.

Two processes cause the increase of CHC-concentration in the gas phase to reach saturated vapor pressure. Firstly, CHC-molecules are delivered from the liquid drops into the double layer, which surrounds these liquid CHC-drops. Secondly, the particles are transported by diffusion from this double layer into the airfilled pore space. The last process is slower than the first one and therefore this is the process which controls the velocity of phase transfer.

The increase of CHC-concentration in the gas phase $C_{g,diff}$ caused by phase transfer liquid-gas can be expressed as:

$$C_{g,diff} = (C_{g0} - C_{g1}) * (1 - e^{-\Gamma * \Delta t}) \quad (3)$$

with

C_{g0} : vapor saturation density
C_{g1} : initial vapor concentration
Γ : reaction constant
t : time

The reaction constant Γ [s^{-1}] is dependent on the pore geometry and apparent vapor diffusion coefficient:

$$\Gamma = \frac{D_g^* * A}{V_g * \delta'} \quad (4)$$

where

D_g^* : vapor diffusion coefficient
A : surface of the electric double layer
V_g : air-filled porosity
δ' : distance of diffusion path

The increase in vapor concentration of CHC results in a decrease in mass of liquid CHC. This process proceeds as long as there are liquid CHC in the pore volume. A mass balance model gives the temporary and local distribution of liquid CHC in soil. The number of molcules, transfered to the gas phase, calculated with the ideal gas law and Avogadro's number is:

$$N = n' * N_A \qquad (5)$$

where

N : number of particles
N_A : Avogadro's number
n' : amount of substance
p : vapor pressure
V : volume
R : gas constant
T : temperature

with

$$n' = \frac{p * V}{R * T} \qquad (6)$$

This number of particles must be subtracted from the remaining number of particles in the liquid phase:

$$N = \frac{V * \sigma}{M_{mol}} * N_A \qquad (7)$$

where

σ : density
M_{mol} : molar mass

A decrease in the mass of liquid CHC remaining in soil has an effect on the phase transfer processes. Because of longer pathways for the particles in the pores and the apparent decrease of the vapor diffusion coefficients the reaction constant Γ will get smaller. This effect is considered by a linear function between Γ and the remaining number of particles in soil (Fig. 2). By means of column experiments the order of magnitude of the reaction constant Γ could be defined as ca. 10^{-4} - 10^{-2} [s^{-1}] comparing the measured with the calculated results of a simple mixed-cell model.

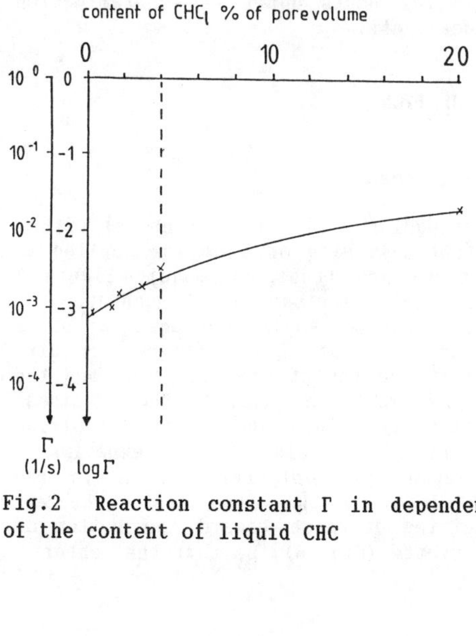

Fig.2 Reaction constant Γ in dependency of the content of liquid CHC

2.3 Vapor Phase Transport

The mass transport of CHC in the vapor phase carried by the air stream is described by the following equation:

$$\frac{\delta C_g}{\delta t} = \text{div } D * (\text{grad } C_g - \frac{\vec{v}_{fg}}{n_{am}} * C_g) + \\ + \Gamma * (C_{0g} - C_g) - \frac{\delta S}{n_{am} * \delta t} \qquad (8)$$

where

D : dispersion coefficient
S : concentration of the sorbed phase

The sink/source term of equation (8) includes the change of mass caused by adsorption, and phase transfer processes. Equation (8) is also solved with the method of finite differences.

2.4 Operation of the Program System

The first step is the calculation of the steady state pressure distribution under the given boundary conditions. Then the velocity of the air flow is determined by Darcy's law (equation (2)) with respect to the variation of relative permeability in dependency of the water content. Knowing the velocity of the air flow and its direction it is possible to solve the transport equation (8). The calculation of the respective phase transfer liquid-gas is carried out within every time step t. A mass balance module computes the remaining number of CHC-particles in the liquid phase for every cell and time step, simultaneously.

3 LABORATORY EXPERIMENTS

Laboratory experiments were carried out to investigate the transport parameters and to verify the computer simulation.

3.1 Experimental Set-up

The experimental set-up consists of cuvettes, which allow a two dimensional consideration of mass transport.

The cuvettes are filled with dry soil homogeneously. Then an air stream is passed through and the distribution of pressure is measured. The pressure measuring devices are placed at equal spacing along the cuvettes. Variations of the position and the kind of air inlet (point and/or line source) are possible.

For the transport experiments the pressure measuring devices are replaced by tensiometers to measure the water content and septa to take soil air samples out of the cuvettes. An analysis of the air samples by gas chromatography allows to determine the distribution of vapor phase CHC-concentration.

Two different soils are used, a sand and a silt. Adsorption of CHC on the sand has not to take into account, because it has a loss of ignition of 0.1%. The utilized CHC are Trichloromethane ($CHCl_3$, Chloroform), 1.1.1-Trichloroethane ($C_2H_3Cl_3$) and Trichloroethene (Tri, C_2HCl_3).

3.2 Sampling and Laboratory Analysis

A certain volume of soil air is sucked through a small glas column (Fig. 3) filled with a non-polar synthetic resin, XAD-4 (Rohm and Haas), either by a syringe or a vacuum pump. The CHC are adsorbed by the resin.

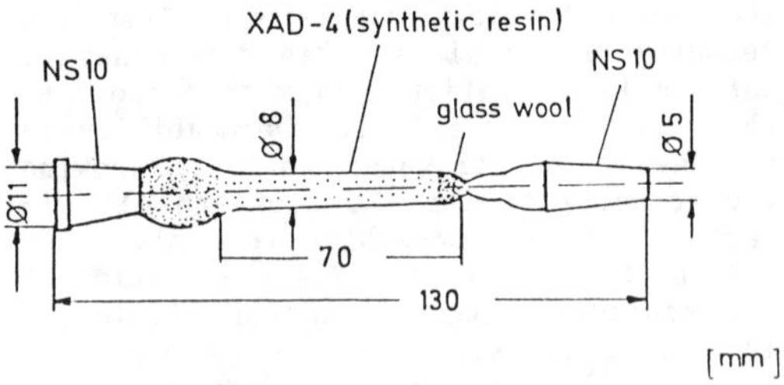

Fig.3 Air sampling with adsorption in test tubes

The resin is filled in a Headspace-vial and the quantitative analysis is done by Gas Chromatographic Headspace analysis. The Gas Chromatograph is a GC8500 Perkin-Elmer with a HS 101 (Headspace Injector), an Electron Capture Detector and capillar column. Parameters of the GC-method are listed in table 1.

Table 1. Parameters of the GC-method

oven temperature	65 °C
time (isotherm)	10 min
injector temp.	115 °C
detector temp.	300 °C
carrier gas	N_2
calibration	ext. standard

3.3 Experimental Procedure

Fig. 4 shows the arrangement of one of these cuvette-experiments.

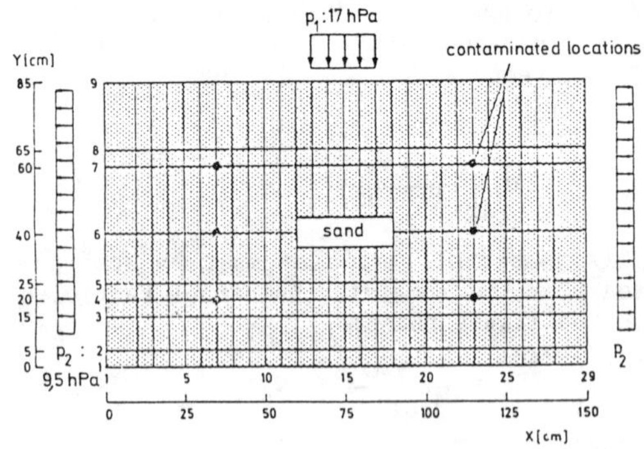

Fig.4 Arrangement of the experiment

The cuvette is filled with sand packed homogeneously. The air inlet sits on top of the cuvette right in the middle. The outflow occurs on both sides. As a result of the pathway of the air stream the flow velocities decrease from the top to the bottom of the cuvette. The contami-nated locations are also shown in Fig. 4. The contaminant is introduced there by injection with needles.

After contamination with CHC a period of 24 hours is necessary to allow the air-filled pore volume becoming saturated with vapor phase CHC by diffusion processes. Then the air stream is passed through the system and samples are taken over a period of several hours until the contamination is sucked off.

4 SIMULATION

4.1 Air Flow

A rectangular grid (31 X 18 nodes) with a constant mesh size of 5 cm was applied to the flow- and transport model. Figure 5 shows the distribution of contures of equal pressure as a consequence of a pressure gradient of 7.5 hPa between air-inlet at the top of the soil body and the two air-outlets at the sides. Additionally, Fig.5 shows the resulting symmetric flow-field of the experiment. The larger air velocities occur in the area close to the air inlet. The air velocities decrease toward the bottom of the cuvette (Fig. 6), so that the centers

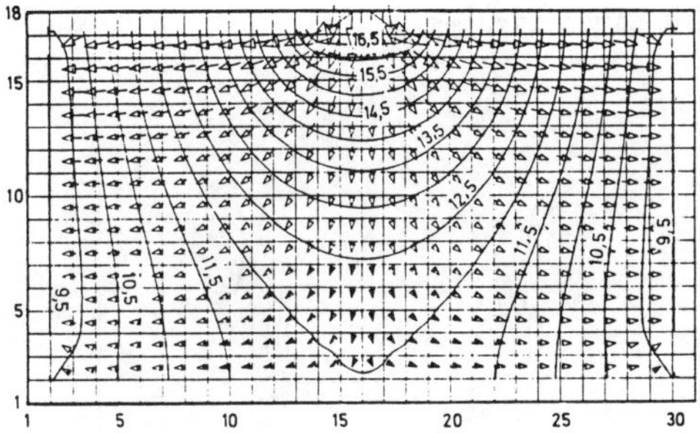

Fig.5 Pressure distribution and flow-field

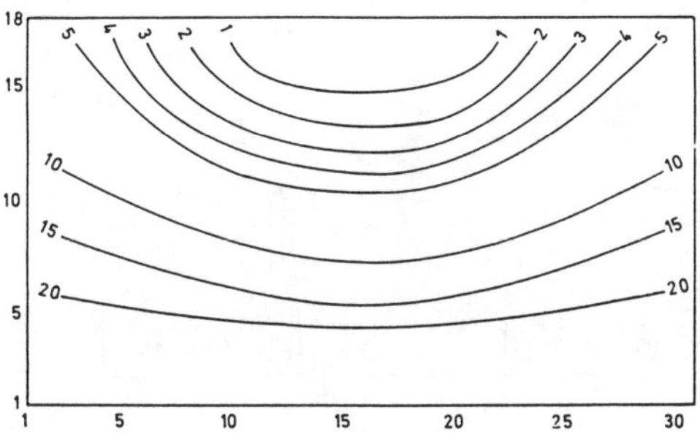

Fig.6 Lines of equal travel times

of contamination by CHC are located in areas with different air flow velocities. Depending on the height above the bottom of the cuvette they vary between 0.04 cm/s and 0.20 cm/s in the contaminated areas having a total flow rate of 5.52 l/min through the cuvette.

4.2 CHC Transport

During the model calibration with respect to the non-steady state progress of vapor phase concentration in soil the permeabilities, dispersivities, and the reaction constant are varied predominantly. The comparison between the experimental and simulated results shows that the distributions of vapor phase CHC during the transport processes can be predicted with good accuracy. Fig. 7 shows measured and calculated distributions of CHC at three different times.

Fig. 7a illustrates a point of time shortly after the beginning of decontamination where the vapor phase concentrations change rapidly. During this period (in the following named 'starting phase') the gaseous CHC, having previously expanded due to vapor diffusion, were carried away by the air stream. The starting phase ended ca. 35 min after the beginning of decontamination followed by a period with hardly any alterations of the distribution of vapor phase CHC-concentration. Because of phase transfer processes a thermodynamic equilibrium is built up during this period of time, which we named 'evaporation phase'. The length of this period becomes relatively large in contrast to the starting phase. In this experiment it reduces to approximatly 2 h 45' caused by a small storage of liquid CHC in soil. Fig. 7b shows CHC-concentrations at the beginning of the evaporation phase. The 'end phase' begins as soon as the storage of liquid CHC is exhausted at any location of the contaminated area. Now CHC-concentrations decrease and approach a low level, rapidly. Fig. 7c shows one distribution of CHC-concentrations during this period. The end phase lasted ca. 1 h 15'.

The reduction of $CHC_{(l)}$ in the pores proceeds from the upper areas down to areas at the bottom of the cuvette, corresponding to various air flow velocities.

At three different locations in the contaminated area breakthrough curves have been measured and were compared with calculated breakthrough curves (see Fig. 8).

The average vapor phase CHC-concentration during the evaporation phase and the length of time of this period are mainly dependent on the following aspects:

1) The order of magnitude of the reaction constant Γ.

2) The amount of liquid CHC.

3) The length of the pathway of the air stream through the contaminated area up to the considered locations.

4) The velocity of the air stream along the pathway through the contaminated area.

The air velocity at point 1 is faster than at point 2 and 3. Additionally, the length of the pathway through the contaminated area up to point 1 is shorter than the one to points 2 and 3. The result is a lower level of CHC-concentration during the evaporation phase and a earlier beginning of the end phase. The different velocities between point 2 and 3 cause small differences concerning the concentration level, but the slower velocity at point 3 generates a higher CHC-concentration due to the longer residence time for evaporation. The pathways of air through the contaminated area up to the considered points 2 and 3 have about the same length. Therefore the longer evaporation phase of point 3 can be explained by the different air velocities.

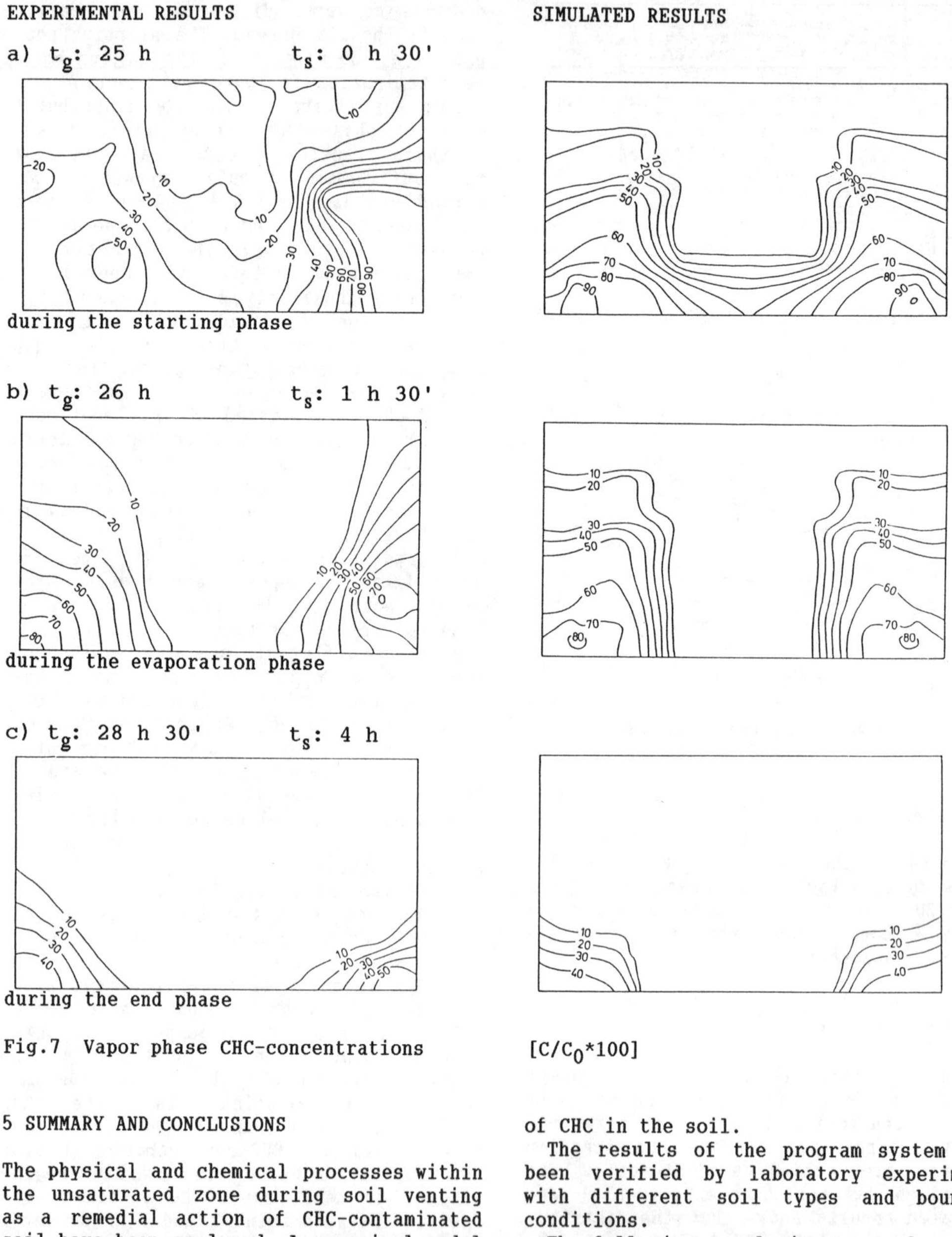

Fig.7 Vapor phase CHC-concentrations [C/C_0*100]

5 SUMMARY AND CONCLUSIONS

The physical and chemical processes within the unsaturated zone during soil venting as a remedial action of CHC-contaminated soil have been analyzed. A numerical model system has been built to describe the flow- and transport processes with a special interest on the phase transfer liquid-gas. This work contains:

1) A flow-model to calculate pressure distributions and flow-fields.

2) A mixed-cell model to compute the order of magnitude of the reaction constant Γ.

3) A transport model, which describes the non-steady state vapor phase transport of CHC in the soil.

The results of the program system have been verified by laboratory experiments with different soil types and boundary conditions.

The following conclusions are of special importance. Different intervals during the decontamination period need to be considered. The whole process can be divided into the starting phase, the evaporation phase and the end phase. The velocity of the air stream has to be optimized to get the best success of decontamination. If the velocity is too slow, saturation takes place within the contaminated area. The reaction constant Γ gives the maximum number of particles evaporated into the pore

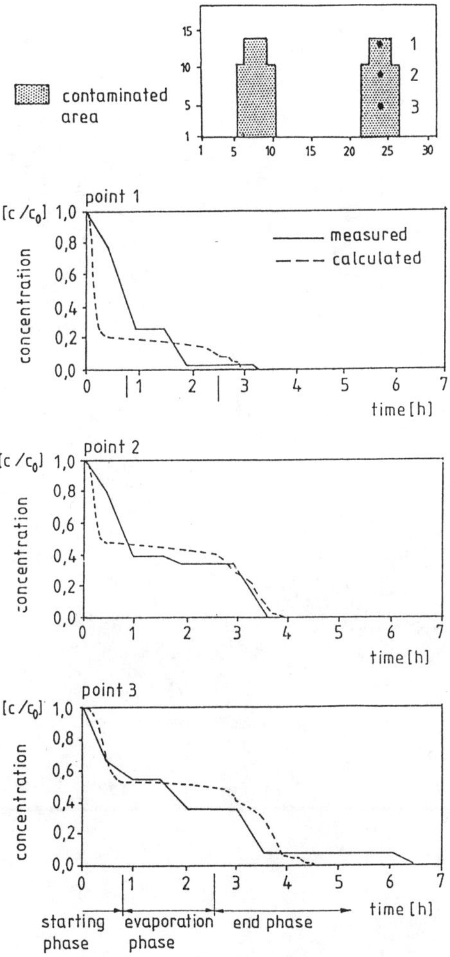

Fig.8 Breakthrough curves at three different locations

volume within one time step. For this reason a higher velocity can hardly accelerate the decontamination process.

The comparison between measured and calculated vapor phase concentration distributions for different soil types and boundary conditions shows that the transport processes within the air stream and the phase transfer processes liquid-gas of these compounds can be predicted with good accuracy.

The simulations were carried out on a PC-386/25 MHZ and 4 MB Memory. The next step is the transfer to field scale.

To accomplish this the soil and boundary conditions (free groundwater surface as impervious boundary, impervious covers caused by buildings etc.) must be evaluated precisely. As the local soil properties influence the phase transfer and transport processes, the transfer from the laboratory to field scale must be tested and verified by measurements. For this kind of simulations it will be necessary to use computers with higher capacities.

REFERENCES

Abriola, L. 1984. Multiphase Migration of Organic Compounds in a Porous Medium -A Mathematical Model-. Berlin, Heidelberg, New York, Tokyo: Springer-Verlag.

Bruckner, F. & H. Kugele 1985. Sanierung von mit leichtflüchtigen Kohlenwasserstoffen kontaminierten Böden mittels Absaugverfahren der Bodenluft. Korrespondenz Abwasser, 10(32):863-866.

Coats, K.H., R.L. Nielsen, M.H. Terhune and A.G. Weber 1967. Simulation of Three Dimensional, Two Phase Flow in Oil and Gas Reservoirs. Soc. Pet. Eng. J., 7(4): 377-388.

Corapciogly, M.Y. & A.L. Baehr 1987. A Compositional Multiphase Model for Groundwater Contaminations by Petroleum Products. 1. Theoretical Considerations. Water Resources Research, 23(1):191-200.

Croisé, J., Kinzelbach, W. & J. Schmolke 1989. Computation of air flows induced in the zone of aeration during in situ remediation of volatile hydrocarbon spills. Proccedings International Symposium on Contaminant Transport in Groundwater, IAHR-AIRH, Stuttgart, H.E. Kobus & W. Kinzelbach (Editors):437-444. Rotterdam: Balkema.

Karimi, A.A., Farmer, W.J. & M.M. Cliath 1987. Vapor-phase Diffusion of Benzene in Soil. J. Environ. Qual., 16(1):38-44.

Parker, J.C., Lenhard, R.J. & T. Kuppusamy 1987. A Parametric Model for Constitutive Properties Governing Multiphase Flow in Porous Media. Water Resources Research, 23(4):618-624.

Sallam, K, Jury, W.A. & J. Letey 1984. Measurement of Gas Diffusion under relatively Low Air-filled Porosity. Soil Sci. Soc. Am. J., 48:3-6.

Schwille, F., W. Bertsch, R. Linke, W.R. Reif and S. Zaüter 1984. Leichtflüchtige Chlorkohlenwasserstoffe in porösen Medien. Besondere Mitteilungen zum Duetschen Gewässerkundlichen Jahrbuch, Nr.46. Herausgeber: Bundesanstalt für Gewässerkunde, Koblenz.

Störk, K. 1986. Anwendung des Bodenluftabsaugverfahrens zur Sanierung von Schadensfällen mit leichtflüchtigen halogenierten Kohlenwasserstoffen (HKW) und als präventive Schutzvorkehrung bei HKW-Anwendern. DVWK-Schriften 78.

Subsurface Contamination by Immiscible Fluids, Weyer (ed.) © 1992 Balkema, Rotterdam. ISBN 90 6191 175 3

Laboratory studies of vapor extraction for remediation of contaminated soil

Martin B. Bloes
Converse Environmental West, Pasadena, Calif., USA

Klaus M. Rathfelder
University of Michigan, Ann Arbor, Mich., USA

Douglas M. Mackay
University of Waterloo, Ont., Canada

ABSTRACT: Column experiments were conducted to investigate the venting of organic liquid (NAPL) from dry porous media (glass beads or sand). The medium was charged with NAPL to desired saturation, then vented by passing dry nitrogen upwards through the column at constant flow rate. The NAPL was either a single component fluid (TCE, benzene or toluene) or a multicomponent mixture of all three. The effluent gas was sampled and analyzed by gas chromatography to yield relatively detailed venting curves. The results were compared to simulations using the model of Rathfelder (1989). The model was found to simulate the experimental data for the single and multicomponent experiments with glass beads or with sand quite well. This finding suggests that the model assumptions are valid for venting of organic liquids from dry porous media, i.e. that vapor-liquid partitioning may be assumed to be at equilibrium and the vapor pressures of individual components above a multicomponent liquid are well-described by Raoult's law. Some low-level tailing of vapor concentrations was observed in experiments with the sand which may be due to rate-limited desorption.

1 INTRODUCTION

Many computer models have been developed to simulate the behavior of vapors in the unsaturated zone (Baehr, 1984; Massmann 1987; Metcalfe and Farquhar, 1987; Silka, 1988; Sleep and Sykes, 1989; Rathfelder, 1989; Mendoza and McAlary, 1990). There have been relatively few laboratory experiments reported on vapor extraction of organic liquids from porous media. Most experiments (e.g. Texas Research Institute, 1980; Hoag et al., 1984; Baehr et al., 1989) have investigated the venting of gasoline from soil columns. Hoag et al. (1984) collected detailed data on the venting of the individual components of gasoline from a gasoline contaminated column, which was modeled by Baehr et al. (1989). However, neither paper discussed the behavior of the individual components in detail; instead they reported and modeled the total mass of components removed as a function of time. In the work reported here and by Bloes (1990), controlled laboratory experiments were conducted to investigate individual component behavior during vapor extraction of multi-component organic liquids from porous media. The results have been compared to predictions of a one dimensional computer model developed by Rathfelder (1989). The model simulates the extracted soil gas composition over time for soil venting of organic liquids. The laboratory studies reported here explore the suitability of various model assumptions, namely 1) that partitioning between the liquid and vapor phases is at equilibrium at gas flow rates of typical concern, and 2) that Raoult's law adequately describes the vapor pressures of individual components in equilibrium with a multicomponent organic liquid trapped in porous media.

2 EXPERIMENTAL METHODS

Column studies were conducted using a glass column (2.5 cm diameter, 30.0 cm length; Kontes Products); the column ends were machined aluminum (Figure 1). Glass wool was used to create a small porous plug in the inlet and outlet to retain the column contents, i.e. the porous medium. The internal volume of the column occupied by the porous medium was approximately 147 cm^3. The column was filled with either idealized porous medium (glass beads) or sandy aquifer material (Borden sands). Measured and estimated properties of these media are listed in Table I. The glass beads ranged in diameter from 0.5 to 1.0 mm, roughly equivalent in size to coarse sand. Samples of sandy aquifer material from the Borden site were obtained from D. McClellan of the University of Waterloo, Ontario, Canada, who has conducted a field study of vapor extraction of TCE (McClellan and Gillham, 1990). The sand, received moist, was subsequently dried for two days at ambient temperature in a laboratory fume hood. Particles in the range 0.25 to 0.5 mm in diameter were used in the column experiments.

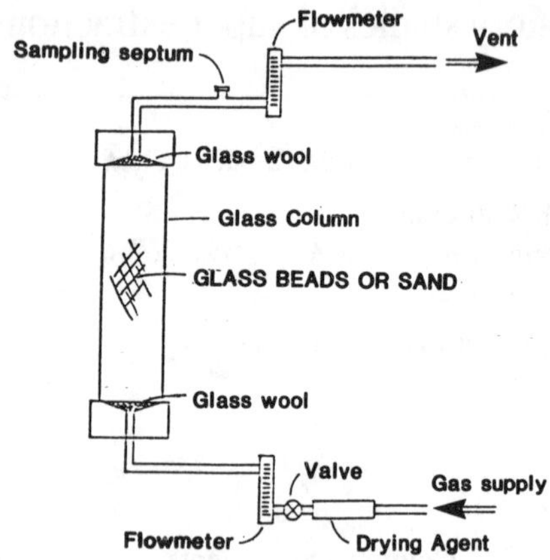

Figure 1. Experimental Apparatus

Table I. Properties of the porous media.

	Glass beads	Sand
Diam. (mm)	0.50-1.00	0.25-0.50
Bulk density (g/cm^3)	1.535	1.650
Solid density (g/cm^3)	2.449	2.705
Porosity (-)	0.373	0.390

Dry nitrogen gas was conveyed via 3.2 mm (1/8 inch) copper tubing, a flow control valve, a flowmeter and fittings to the bottom of the column; the effluent was conveyed through similar tubing past a sampling port, through a flowmeter, to a vent in a fume hood. Gas flow (ml/min) was measured with a "Mini-Buck Calibrator" (A.P. Buck Inc. #M5). The measured gas flow rates, q (ml/min), were divided by the total pore volume in the column (ml) to yield flow rates in units of pore volumes per minute.

The gas flow rates varied between 0.53 to 0.58 pore volumes per minute. Using manometers, the pressure drop across the column was found to be on the order of 7×10^{-4} and 4×10^{-4} atm (0.03 and 0.15 inches of water) for these flow rates through dry glass beads and dry sand, respectively. This finding is consistent with that of Hoag et al. (1984), who noted no significant pressure drop across a column packed with fine to coarse sands at flow rates up to 7 pore volumes/min. For calculations in this work, the small pressure gradient was ignored and the pressure assumed to be one atmosphere throughout the system.

The organic chemicals used in the experiments were benzene, toluene and trichloroethylene (TCE). All were at least 99.5% pure and obtained from Fisher Scientific. The vapor pressures of the chemicals at 23°C (the temperature of the experiments) were calculated from data provided by Perry (1984) to be 0.118 atm, 0.092 atm, and 0.0289 atm for benzene, TCE and toluene, respectively. The saturated vapor concentrations at 23°C were calculated from the ideal gas law to be 379 mg/l, 497 mg/l and 117 mg/l, respectively.

"Single component" experiments investigated the venting of each organic liquid separately. "Multicomponent" experiments were conducted by charging the column with various total masses of a mixture composed of the components.

In the experiments with glass beads, the column containing the glass beads was filled with the organic liquid, excluding all air, and then allowed to drain via gravity. Glass wool was used at each end of the column to contain the beads in the column during this process. To ensure that all the organic liquid was evenly distributed within the column, the column was tilted many times over and the contaminant allowed to drain in both directions. Before the initiation of each experiment the glass wool at each end of the column was changed to ensure that all the organic liquid in the column was contained within the porous medium and not on the glass wool. After the glass wool was changed the column containing the organic liquid was weighed. The column was weighed again after the venting experiment to determine, by difference, the mass of organic liquid present initially. This procedure resulted in the organic liquid initially occupying approximately 8-10% of the pore volume.

The same method was attempted for the single component (TCE) experiment with the Borden sands, but this resulted in TCE content well above residual saturation (TCE occupying 31% of pore volume); TCE appeared to fully saturate some portions of the sand at the bottom of the column. Consequently in the multicomponent experiment with the sand, an alternate approach was used. An appropriate amount of organic liquid was mixed with the sand in a wide mouth glass jar; the jar was subsequently stored 2 days (run M-S-2) in the refrigerator to allow some time for the liquid to penetrate the solids. The solids were then handpacked into the column under a fume hood.

During the venting experiments, gas samples of column effluent were taken periodically with a syringe and analyzed using a gas chromatograph (SRI Model 8610) equipped with a flame ionization detector and capillary column. Depending on the experiment, 25 to 100 μl gas samples were analyzed. The analysis was achieved with an isothermal oven temperature of 60°C for single component experiments and 45°C for multicomponent experiments. Sample concentrations were estimated by comparing their chromatogram peak areas to the peak area resulting from analysis of a standard (the vapor in equilibrium with pure organic liquid at 23°C (room temperature). The peak area obtained from the injection of the standard was assumed to represent the saturated vapor concentration at room temperature. In preliminary work, the response of the detector had been shown to be linear at this concentration and below. The mass of each compound vented from the column was calculated from the measured effluent concentration histories and the measured flow rates.

The one-dimensional model presented by Rathfelder et al. (1991) was used to simulate the experiments. The reader is referred to that paper or Rathfelder (1989) for detail; only a brief summary of the modeling approach is provided here. Transport equations are solved for each organic component, coupled with the vapor phase flow equations. The model assumes equilibrium phase partitioning, which is defined by the component vapor pressures at the temperature of the experiment and the mole fractions of the components in the organic liquid. The model input includes the column volume and porosity, the gas flow rate, the initial saturation of the organic liquid (calculated from the initial mass, composition and density of the organic liquid), and the vapor pressures of the components at the experimental temperature. The organic liquid is assumed to be distributed uniformly through the column. Boundary conditions for pressure are defined by Darcy's Law, the measured constant influent flux and atmospheric conditions at the effluent end of the system. Conditions for organic vapor concentrations are zero influent and zero gradient at the effluent end.

3 RESULTS

3.1 Single component experiments

The results of the single component venting experiments are shown in Figures 2 through 5. Figures 2 through 4 show the results for venting of benzene, TCE and toluene, respectively, from glass beads at gas flow rates of 0.53 to 0.58 pore volumes per minute. Figure 5 shows the results for venting of TCE from Borden sand at a flow rate of 0.53 pore volumes per minute. The figures also present the model simulations. Various raw data and calculated results are listed in Table II for the single component venting experiments.

In all single component experiments, there is considerable scatter in the initial "plateau" concentrations, then the concentrations drop off abruptly. Prior to the decline, the observations vary about the simulated plateau value, the saturated vapor concentration at the temperature of the experiment (23°C) calculated from tabulated data (Perry, 1984). As shown in Table II, the calculated saturated vapor concentrations fall within one standard deviation of the measured means in all cases. This implies that the gas passing through the column reached equilibrium with the organic liquid prior to exiting the column. Whether equilibrium was reached at the vaporization front cannot be determined from the plateau values.

There is relatively close agreement between the mass determined by weighing the column before and after the experiment and the mass calculated from the measured effluent vapor concentrations and the gas flow rate; in the

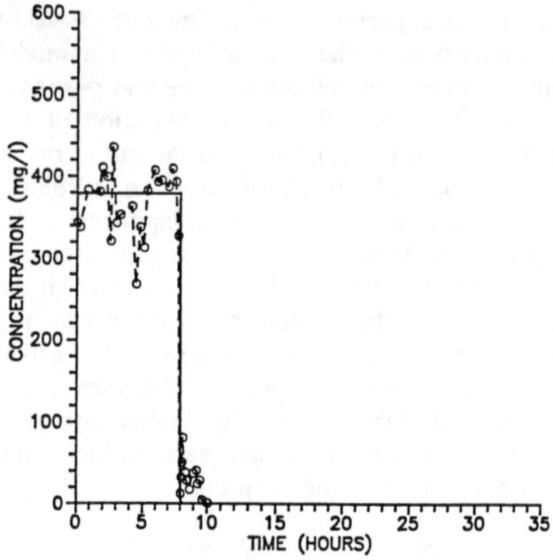

Figure 2. Venting of benzene from glass beads (Run S-G-1). See Table II for quantitative comparison of simulation (solid line) to observations (dashed line).

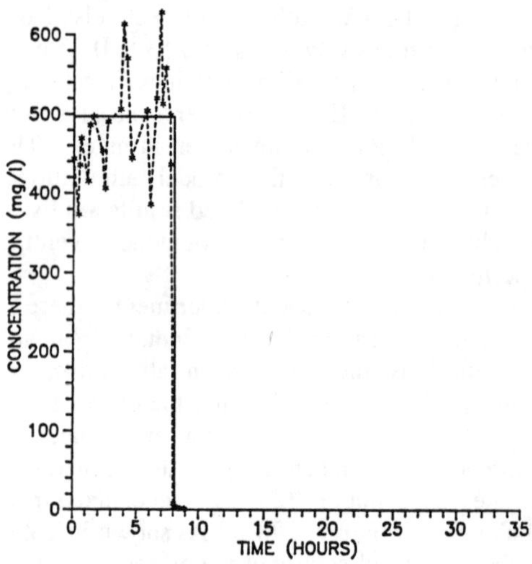

Figure 3. Venting of TCE from glass beads (Run S-G-2). See Table II for quantitative comparison of simulation (solid line) to observations (dashed line).

experiment with benzene the discrepancy is 14%, in the other experiments the discrepancies are 7% or lower. Given the variability in the analytical results, these mass balances appear to be relatively good. Rathfelder et al. (1991) state that the simulations of the experiments generally

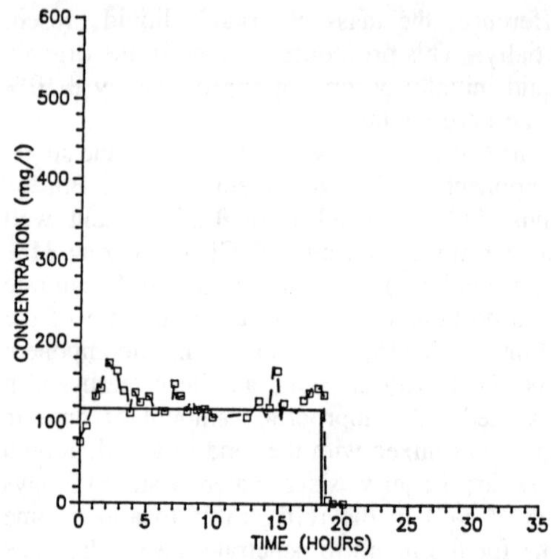

Figure 4. Venting of toluene from glass beads (Run S-G-3). See Table II for quantitative comparison of simulation (solid line) to observations (dashed line).

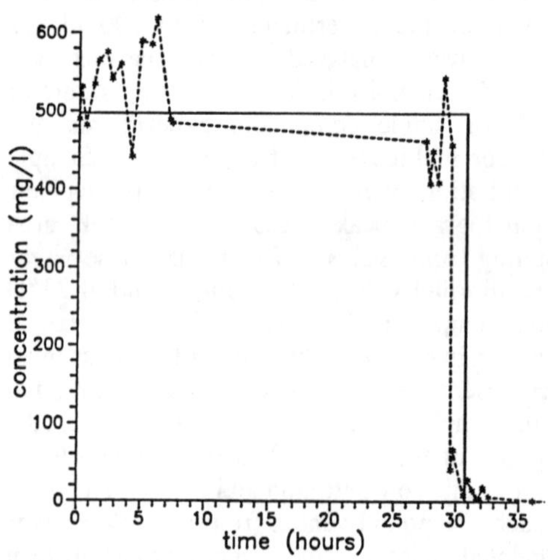

Figure 5. Venting of TCE from Borden sand (Run S-S-1). See Table II for quantitative comparison of simulation (solid line) to observations (symbols and dashed line).

had a mass balance error on the order of 1-2% for each component.

An apparent tail in the concentration history is observed for benzene vented from glass beads (Figure 2, 8-10 hrs). Since there is little reason to expect such a tail (and since the mass estimat-

Table II. Results of single component venting experiments.

MEDIUM Run No., Compound	Vented Mass (g) Gravimetric	 Calculated*	Vapor concentration (mg/l) literature**	 measured***
GLASS BEADS				
S-G-1, Benzene	4.7	5.38	379	367 ± 41
S-G-2, TCE	7.4	7.31	497	497 ± 73
S-G-3, Toluene	4.5	4.21	117	126 ± 22
SAND				
S-S-1, TCE	25.0	26.1	497	506 ± 67

* Mass calculated from measured concentration histories and measured flow rates.

** Saturated vapor concentrations calculated from tabulated vapor pressures (Perry, 1984), corrected to temperature of experiments (23°C).

*** Mean ± standard deviation.

ed from the vapor measurements is somewhat high), we suspect that this tail is an artifact, due possibly to liquid benzene contamination of parts of the apparatus downstream of the column. This could have occurred during filling of the column and/or during venting, if aerosols were blown from the column. Other explanations cannot be discounted, however. Slight tailing is also seen for venting of TCE from Borden sand (Figure 5); the tailing represents approximately 0.8% of the initial TCE mass. This tailing may be due to rate-limited desorption of TCE sorbed to the sand, as discussed later.

3.2 Multicomponent experiments

The results of two multicomponent venting experiments are presented in Figures 6 and 7. Table III lists various raw data and calculated values for the experiments. For all of the multicomponent experiments, an attempt was made to saturate the media with a liquid composed of equal mass fractions of the components (mole fractions of 0.410, 0.243, and 0.347 for benzene, TCE and toluene, respectively). However, Table III indicates that the masses of the components vented from the experiments imply a different initial composition than intended. The handling and storage of the liquid and the saturated media apparently resulted in changes to the composition of the liquid before the venting experiments began; this is not surprising since much of the handling and transfer of the liquids and liquid-solid mixtures was conducted in a fume hood, during which time volatilization of the organic chemicals was likely to have been occurring. Thus the mole fractions calculated from the vented masses (Table III) were assumed to be the most reliable description of the initial composition of the multicomponent organic liquid and were used as input to the model.

Figure 6 shows the results for venting of a three component liquid from glass beads at a gas flow rate of 0.53 pore volumes per minute; also depicted are the concentration histories simulated by the model. Note that the y-axis in these figures covers a smaller concentration range than the previous figures, and that the initial vapor concentrations of the components are significantly lower than their saturated vapor concentrations (Table II), as a consequence of Raoult's law. As predicted by the model, the most volatile contaminant (benzene) is removed first, a peak appears in the TCE concentration at the same time as the benzene concentration drops (because of the rapid change in the mole fraction and vapor pressure of the remaining components), and the toluene elutes last with a large peak in concentration when the other two components have been removed. Figure 7 presents the results of venting of the multicomponent liquid from Borden sand. Again the model appears to fit the data quite well.

In both Figures 6 and 7, it is evident that some tailing of concentrations of benzene and TCE occurs, at least until the toluene is removed from the column. There is some evidence in

Table III. Results of multicomponent venting experiments.

MEDIUM, RUN NO. Compound	Vented Mass (g) Grav.	Calc.	Xi, Mole fraction	Initial vapor concentration (mg/l) Calculated*	Measured**
GLASS BEADS, M-G-2					
Benzene		1.82	0.467	177	186 ± 15
TCE		1.82	0.277	138	145 ± 11
Toluene		1.18	0.257	30	39 ± 6
Total	4.5	4.82	1.001		
SAND, M-S-2					
Benzene		1.64	0.432	164	158 ± 19
TCE		1.63	0.255	127	114 ± 17
Toluene		1.40	0.313	37	43 ± 2
Total	4.8	4.67	1.000		

* Initial vapor concentration calculated from the estimated mole fraction and the vapor pressure of the component at the temperature of the experiment (23°C).

** Mean ± standard deviation.

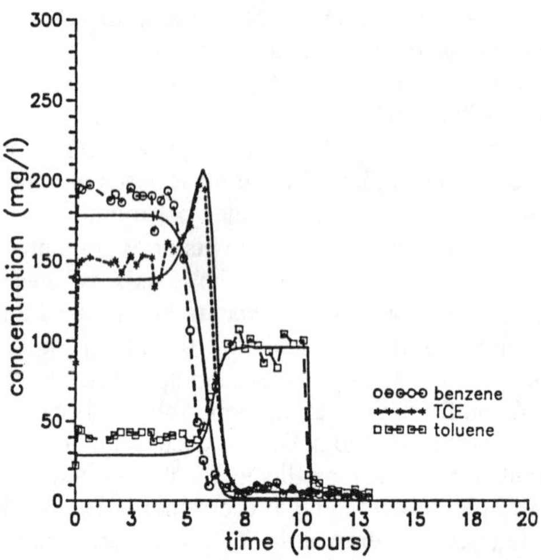

Figure 6. Venting of multicomponent organic liquid from glass beads (Run M-G-2). See Table III for quantitative comparison of simulations (solid lines) to observations (dashed lines).

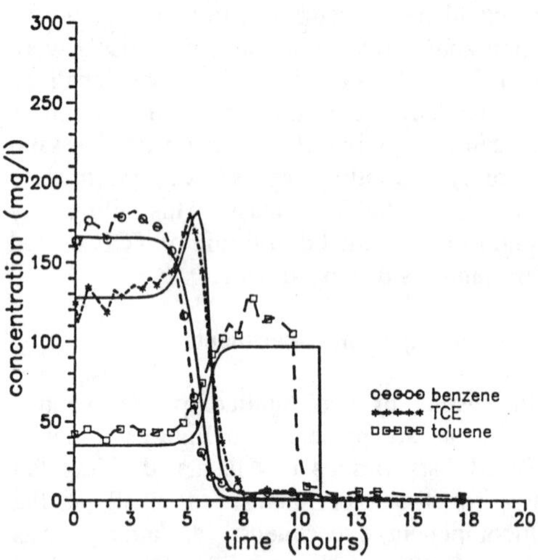

Figure 7. Venting of multicomponent organic liquid from Borden sand (Run M-S-2). See Table III for quantitative comparison of simulations (solid lines) to observations (dashed lines).

both multicomponent experiments that tailing of all three components continues for a few hours after the bulk of the organic liquids have been removed (i.e. after 10-11 hr). Furthermore, in the multicomponent experiment with Borden sand (Figure 7), the tailing of toluene decreases slowly from approximately 10 mg/l at 11 hours to 1 mg/l at 20 hours and 0.2 mg/l at 30 hours.

After over 30 hours of venting, analysis of a methylene chloride extract of the sand indicated that toluene was present at low levels (14 ng/g in a 20 g sample of the sand). It appears likely that this toluene was sorbed to the sands. Chiou and Shoup (1985) found that vapors of benzene and various chlorobenzenes sorbed relatively strongly on dry soil. Since Ball et al. (1990) have shown that the Borden sands contain on the order of 1-5% intragranular porosity, it is possible that some or all of the toluene was sorbed to intragranular surfaces. Mass transfer from such pores to a mobile vapor phase might well be rate-limited. If, in fact, the desorption of the sorbed vapors is rate-limited, then tailing in the venting data would be expected. Further work is required to confirm that the dry sands sorb organic vapors, to determine whether significant fractions of the sorbed organic mass are present in the intragranular porosity, and to investigate the rates of and limitations to desorption to the vapor phase.

4 CONCLUSIONS

The experimental data for venting of multicomponent organic liquids from porous media illustrate that the gas composition changed in complex ways as a result of the changing composition of the residual organic liquid. Some tailing was observed, perhaps as a result of slow desorption from the dry sands, but the vapor concentration was very low. Comparisons of measured and predicted effluent vapor concentrations suggest that equilibrium partitioning from organic liquid to vapor is a reasonable assumption for modeling a water-free system with idealized porous media (glass beads) or real geologic media (Borden sands). Considering that the gas velocities in the laboratory studies are on the same order or greater than encountered in field settings (except near extraction vents), the results of this work suggest that equilibrium partitioning may be a reasonable assumption in modeling field scale vapor extraction systems when the geologic media are dry. Further experiments are clearly warranted to investigate the effect of moisture on the venting process. The model (Rathfelder et al. 1991) and field data (McClellan and Gillham, 1990) suggest that tailing is more significant when the medium is moist because of mass transfer limitations from the sorbed and dissolved phases.

5 ACKNOWLEDGEMENTS

This work was supported in part by the UCLA Engineering Research Center for Hazardous Substance Control and in part by the University of Waterloo Solvents-in-Groundwater Research Program.

6 REFERENCES

Baehr, A.L. 1984. Ph.D. Dissertation, University of Delaware.

Baehr, A.L., G.E.Hoag & M.C.Marley 1989. Removing volatile contaminants from the unsaturated zone by inducing advective air-phase transport. *Contam. Hydrol.* 4:1-26.

Ball, W. P., C. Buehler, T.C. Harmon, D. M. Mackay & P.V.Roberts 1990. Characterization of a sandy aquifer material at the grain scale. Contam. Hydrol. 5:253-295.

Bloes, M. 1990. *Laboratory studies of vapor extraction for remediation of contaminated soil. M.S. thesis. UCLA School of Public Health.*

Chiou, C.T. & T.D.Shoup 1985. Soil sorption of organic vapors and effects of humidity on sorptive mechanisms and capacity. Environ. Sci. Technol. 19:1196-1200.

Hoag, G.E., C.J.Bruell & M.C.Marley 1984. The study of the mechanisms controlling gasoline hydrocarbon partitioning and transport in groundwater systems, NTIS Publication No. PB85-242907.

Massmann, J. 1987. Applying ground-water flow models in vapor extraction system design. *J. Env. Engr. ASCE* 115(1):129-149.

McClellan, D. & R.W.Gillham 1990. Vacuum extraction of trichloroethylene from the vadose zone. Paper presented at the IAH conference on Subsurface Contamination by Immiscible Fluids, Apr. 18-20, Calgary, Alberta, Canada.

Mendoza, C. A. & T. A.McAlary 1990. Modeling of ground-water contamination caused by organic solvent vapors. *Ground Water* 28(2): 199-206.

Metcalfe, D.E. & G.J.Farquhar 1987. Modeling gas migration through unsaturated soils from waste disposal sites. *Water, Air & Soil Polln.* 32:247-259.

Perry, R.H. 1984. *Perry's chemical engineer's handbook.* McGraw-Hill, Inc.

Rathfelder, K.M. 1989. Ph.D. Dissertation, UCLA Department of Civil Engineering.

Rathfelder, K., W. Yeh & D. Mackay 1991. Mathematical simulation of soil vapor extraction systems: model development and

numerical examples. *Contaminant Hydrology* (in press).
Silka, L.R. 1988. Simulation of vapor transport through the unsaturated zone - interpretation of soil gas survey. *Ground Water Mon. Rev.*, Spr., 115-123.
Sleep, B.E. & J.F.Sykes 1989. Modeling of transport of volatile organics in variably saturated media. *Water Resour. Res.* 25:81-92.
Texas Research Institute 1980. *Examination of venting for removal of gasoline vapors from contaminated soil.* American Petroleum Institute, Wash. D.C.

6 Investigative techniques

Subsurface Contamination by Immiscible Fluids, Weyer (ed.) © 1992 Balkema, Rotterdam. ISBN 90 6191 175 3

Practical aspects of multiphase equilibria in evaluating the degree of contamination

Nicholas Sitar, James R. Hunt & Jil T. Geller
Department of Civil Engineering, University of California, Berkeley, Calif., USA

ABSTRACT: One of the major problems in dealing with subsurface contamination by immiscible fluids is the evaluation of the degree and the extent of contamination. Typical data collected during a site investigation consist of soil, water, and gas samples and the problem is in evaluating whether the measured concentrations of contaminants represent equilibrium conditions, and whether they accurately reflect the extent and type of contamination. Data from field sites together with laboratory experimental results are used to demonstrate the need for consistent data evaluation based on principles of multiphase equilibria. The results illustrate that at sites contaminated by immiscible fluids it is necessary to collect more than one type of sample in order to accurately assess the degree of contamination.

1 INTRODUCTION

In the last few years, groundwater contamination by nonaqueous liquids, organic solvents and gasoline, has become recognized as a significant problem from the point of aquifer restoration and cleanup. The main difficulty is that once a spill of a nonaqueous phase liquid (NAPL) occurs, the NAPL tends to remain trapped as discontinous drops and lenses either above or below the water table. These drops and lenses can be displaced only partially by groundwater flushing and they tend to slowly dissolve, serving as long term sources of groundwater contamination. Thus, the identification of the presence of the separate phase is of paramount importance during a site investigation. A major complication is that pure component liquids are rarely released to the subsurface. The waste liquids that become trapped in saturated and unsaturated porous media, gasoline for example, have many components with variable chemical properties. The existence of a multicomponent liquid has major consequences in the recovery of specific contaminants because of the reduction in aqueous or gas phase solubility and the changing trapped liquid composition as it weathers (Hunt et al., 1988a). This weathering process substantially alters the rates of recovery for components initially present in the trapped liquid.

The purpose of this paper is to examine the factors controlling the equilibrium between the aqueous and gas phases and a trapped multicomponent liquid, and to evaluate the role of soil, groundwater, and soil gas sampling in delineating the extent and severity of subsurface contamination by NAPLs.

2 MULTICOMPONENT CHEMICAL EQUILIBRIUM

For the typical organic solvents and petroleum hydrocarbon contaminants present in trapped liquids in the subsurface environment, the organic compounds are individually only sparingly soluble in water with a pure phase solubility indicated by C_S. If the trapped liquid contains more than one component, then there is a reduction in equilibrium aqueous solubility, C, predicted by

$$C = X_L \gamma_L C_S \qquad (1)$$

where X_L is the mole fraction of the component of interest in the trapped liquid and γ_L is the activity coefficient for the particular contaminant in the trapped liquid (Mackay and

Shiu, 1981). The activity coefficient corrects for nonideal behavior of organic chemicals with dissimilar structure. For a two-component mixture of benzene (B) and toluene (T), their nearly identical aromatic structure makes them behave as an ideal liquid such that equation (1) becomes

$$C_B = X_{BL}\, C_{BS} \qquad (2)$$

$$C_T = X_{TL}\, C_{TS} \qquad (3)$$

For a trapped liquid composed of benzene and toluene in equal molar amounts, the mole fraction would be 0.5 for each component in the liquid. Water in equilibrium with this liquid would have benzene at about 900 g m^{-3} and toluene at 270 g m^{-3}, half their respective aqueous solubilities of 1800 g m^{-3} and 540 g m^{-3}.

Gasoline presents a more complicated picture as shown in Table 1. The solubility of gasoline is on the order of 100 g m^{-3}. Yet, benzene, toluene, and xylene (BTX) which represent only about 21% of the total gasoline mixture, account for almost 50% of the dissolved hydrocarbons in equilibrium with gasoline. Most importantly, however, these concentrations are well below the solubility of the individual pure phase compounds. Since most typical sites involve multicomponent NAPLs, this simple example illustrates the problem in using groundwater concentration measurements as evidence for the presence of the NAPL as a separate phase. The problem is essentially identical when using soil gas samples.

Table 1. Typical aqueous concentrations of BTX in equilibrium with gasoline (Hunt et al., 1988c)

Compound	% in Gasoline	Aqueous Solubility (g/m^3)	
		Pure Phase	With Gasoline
Benzene	2.3	1800	11
Toluene	8.3	540	24
Xylene	10.3	185	12

3 TIME EVOLUTION OF EQUILIBRIUM GROUNDWATER COMPOSITION

Given a trapped liquid with components of variable water solubility, the liquid is expected to change in composition with time as the more water soluble components are dissolved away. Considering only a two component liquid containing benzene and toluene, the rate of change in component mass present in the liquid is controlled by the aqueous concentration and the flow rate of water. Assuming that a flow rate of Q equilibrates with the trapped liquid, the mass balance expressions for benzene and toluene mass, M_B and M_T, become

$$\frac{dM_B}{dt} = -QC_B \qquad (4)$$

$$\frac{dM_T}{dt} = -QC_T \qquad (5)$$

where aqueous phase benzene and toluene concentrations at equilibrium are given in equations (2) and (3). These four equations are not complete because they involve six unknowns, M_B, M_T, C_B, C_T, X_{BL}, and X_{TL} that are functions of time. The final two equations needed for solution relate masses in the trapped liquid to the mole fractions through

$$X_{BL} = \frac{M_B/MW_B}{M_B/MW_B + M_T/MW_T} \qquad (6)$$

$$X_{TL} = \frac{M_T/MW_T}{M_B/MW_B + M_T/MW_T} \qquad (7)$$

where MW_B and MW_T are the molecular weights of the benzene and toluene, respectively. Assuming initial masses of benzene and toluene in the trapped liquid, equations (2) through (7) can be solved numerically to predict the aqueous concentration leaving a site and the changing composition of the trapped liquid.

To test the applicability of this model, the model predictions are compared with experimental data reported by Hunt et al. (1988c) on the dissolution of a benzene-toluene mixture. The experimental sand column had a pore volume of 716 cm^3 and received 7.94 g of benzene and 7.83 g of toluene (9 mL each) injected as a liquid into the center of the water-saturated sand pack. Water was pumped through the sand pack at an approach velocity of 15 m d^{-1}. A previous experiment with a single component liquid of trichloroethylene showed that the effluent water reached chemical equilibrium with the trapped liquid after the displacement of one pore volume.

Figure 1 compares the predicted and observed evolution of the benzene and toluene concentration in the water as a function of pore volume of water displaced from the sand pack.

The actual experiment was terminated at 11 pore volumes, and the comparison is reasonably good between measurements and predictions. Beyond 11 pore volumes it is possible that the model assumption of chemical equilibrium between the liquid and the flowing water will not hold because the surface area available for dissolution is likely decreasing. Figure 2 plots the predicted evolution of the mole fractions of benzene and toluene in the trapped liquid. As the liquid ages, benzene is depleted and toluene becomes more concentrated.

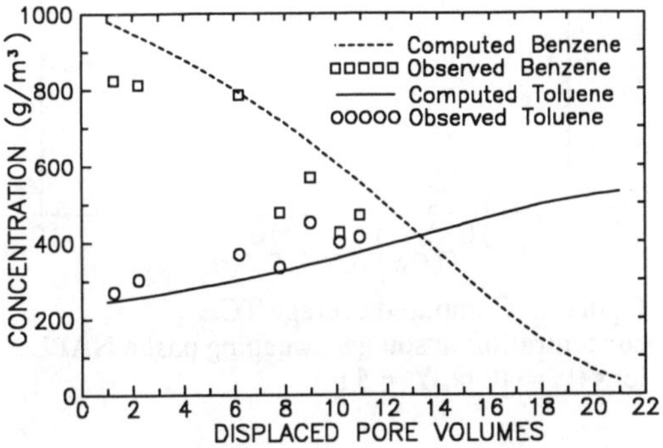

Figure 1. Predicted and observed evolution of aqueous concentration of benzene and toluene in equilibrium with separate phase.

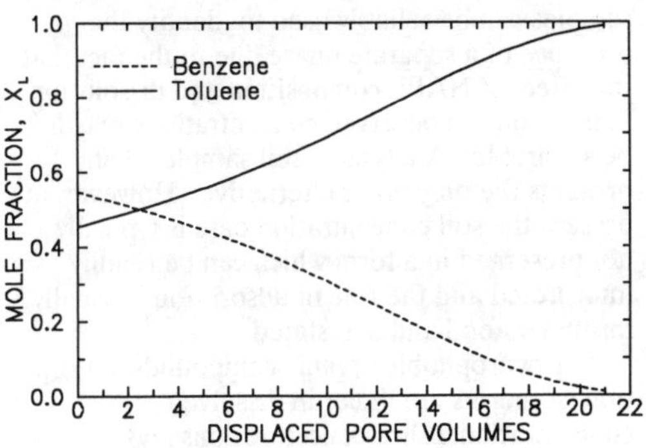

Figure 2. Evolution of the trapped liquid composition.

Current research is finding that chemical equilibrium is not always achieved between flowing water and trapped liquids at low residual concentrations. Under these circumstances mass transfer limits dissolution and experimental results are providing models to predict the changing composition of multi-component liquids along the flow path and over time. The analysis is practically identical for equilibrium between the liquid and a gas phase.

4 DISSOLUTION KINETICS

Dissolution kinetics is an additional factor which has to be considered in evaluating the aqueous and/or soil gas effluent concentration both during site investigation and during cleanup. The problem can be illustrated by considering an idealized geometry of a lens of NAPL in contact with the pore fluid as shown in Figure 3. Assuming parallel flow of pore fluid with velocity U, over the length of the lens, L, and the thickness of the sampled zone given by Y, the expression for average concentration C(L,Y) leaving the lens is given by (Hunt et al., 1988b):

$$\frac{\bar{C}(L,Y)}{C_S} = \frac{1 - e^{-w^2}}{w\sqrt{\pi}} + \mathrm{erfc}(w) \qquad (8)$$

where

$$w = \frac{Y}{2(DL/U)^{\frac{1}{2}}} \qquad (9)$$

and

$$D = D^* + \alpha_t U \qquad (10)$$

in which D^* is the effective diffusion coefficient and α_t is the transverse dispersivity.

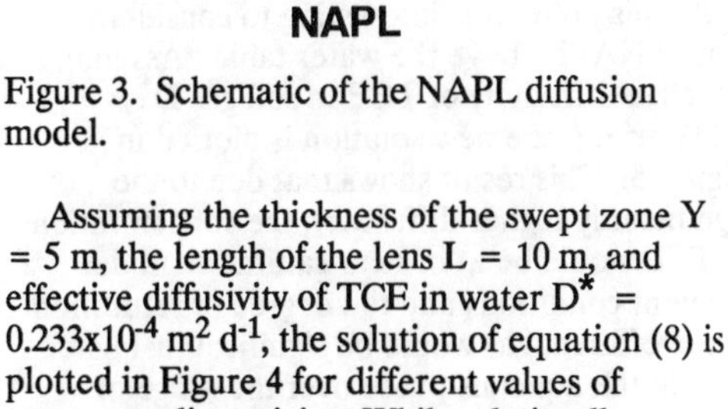

Figure 3. Schematic of the NAPL diffusion model.

Assuming the thickness of the swept zone Y = 5 m, the length of the lens L = 10 m, and effective diffusivity of TCE in water D^* = 0.233×10^{-4} m^2 d^{-1}, the solution of equation (8) is plotted in Figure 4 for different values of transverse dispersivity. While admittedly oversimplified, the solution is useful in that it shows that the average concentration of TCE leaving the lens would be about 30% of its solubility even at very low sweep velocities. Of course, the average concentration would increase with decreasing thickness of the swept zone. However, while decreasing the thickness of the sampling interval will obviously help in getting a better resolution, the location of the sampled interval relative to the lens becomes of great importance.

Possibly much more significant is the well defined trend of decreasing average concentration with increasing sweep velocity. In this respect, the result is equivalent to the evolution of the effluent concentration after beginning of pumping. As the groundwater velocity increases, the concentration of the dissolved compound in the effluent decreases exponentially, a trend commonly noted during pump-and-treat operations. As can be seen, this decrease in concentration does not mean that the source concentration is in fact decreasing, which is a commonly made but erroneous conclusion.

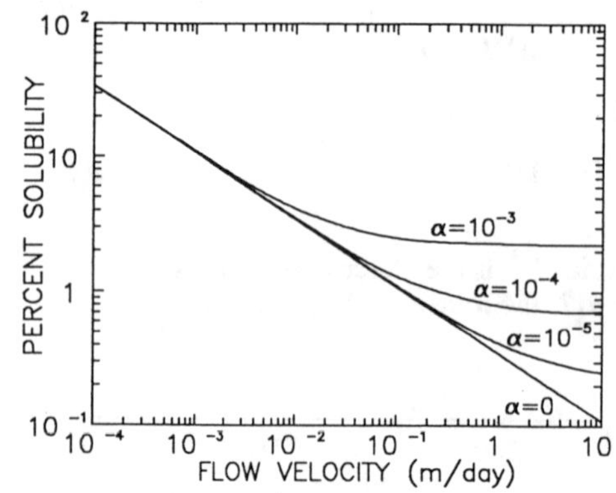

Figure 4. Computed average TCE concentration in groundwater sweeping past a NAPL lens (L = 10 m, Y = 5 m).

At this point, it is interesting to consider a lens of NAPL above the water table. Assuming effective diffusivity of TCE in soil gas $D^* = 0.233\ m^2\ d^{-1}$, the new solution is plotted in Figure 5. This result shows that due to the significantly higher diffusivity, the concentration of TCE would be at or near saturation under ambient conditions and the degree of saturation in the effluent gas would be significantly higher than in the aqueous phase over the range of typical sweep velocities. This result suggests that soil gas sampling, when possible, should be a more reliable indicator of the degree of contamination, as long as the separate phase is in direct contact with the soil gas. However, it is important to note again the exponential decrease in concentration in the effluent gas with increasing sweep velocity, as is typically observed in the field.

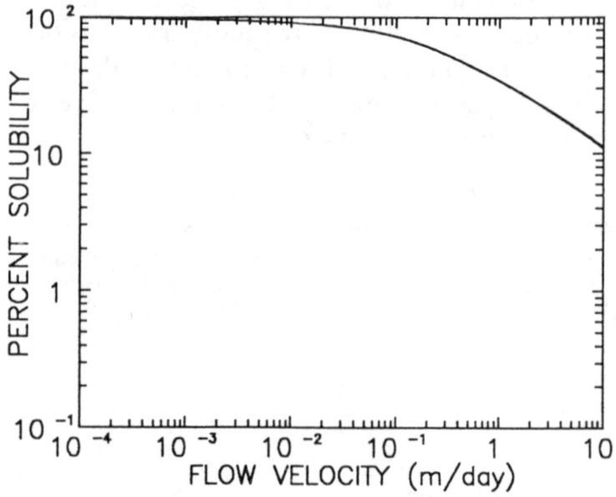

Figure 5. Computed average TCE concentration in soil gas sweeping past a NAPL lens (L = 10 m, Y = 5 m)

5 SOIL SAMPLE ANALYSIS

From the previous discussion it should be apparent that neither groundwater nor soil gas samples can be reliably used to identify the presence of a separate phase due to the fact that the effect of NAPL composition and dissolution kinetics on the observed concentrations may not be separable. Analysis of soil samples then presents the only other alternative. However, at present the soil concentration data is typically not presented in a form which can be readily interpreted and the role of adsorption is usually misunderstood and overstated.

For hydrophobic organic compounds, adsorption isotherms are linear in dissolved concentration at low concentrations and represented as

$$x/m = K_d C \qquad (11)$$

where x/m is the mass adsorbed per mass of solid phase (mg kg^{-1}), K_d is the slope of the adsorption isotherm and C is the equilibrium concentration of the contaminant dissolved in water (g m^{-3}). Karickhoff et al. (1979) have shown that K_d can be linearly correlated with the fraction of organic carbon in the solid phase, f_{oc}, and the octanol-water partition coefficient,

K_{ow}, which is a measure of the compound's hydrophobicity. For sediment particles less than 50μm they observed that

$$K_d \approx 0.6\, f_{oc}\, K_{ow} \qquad (12)$$

Thus, knowing the chemical characteristics of the NAPL it is a relatively simple matter to compute the adsorbed mass in equilibrium with the observed aqueous concentration. Sitar et al. (1987) reviewed data from a number of NAPL contaminated sites in Santa Clara Valley, CA, and found that the reported concentrations in soil samples typically exceeded the amounts which could be adsorbed, assuming pore fluid at saturation with respect to the specific compounds.

The problem can be illustrated by considering water and soil concentrations of BTX obtained from the same depth interval at a site contaminated with gasoline (Table 2). As can be seen, the aqueous concentrations of BTX are at saturation with respect to a typical gasoline (Table 1) with benzene concentration somewhat higher than expected. Considering typically reported K_{ow} values of 135, 490, and 1300 for benzene, toluene, and xylene, respectively, and assuming a conservative value of f_{oc} = 0.01, the predicted adsorbed concentrations in equilibrium with the reported aqueous concentrations should be 22, 76, and 109 mg kg^{-1} for benzene, toluene, and xylene, respectively. Thus, the observed soil concentrations are clearly well above the adsorption limit and in fact indicate the presence of free gasoline.

Table 2. Measured hydrocarbon concentrations in soil and water from the same borehole at a gasoline contaminated site (data from Dresen et al., 1986)

Compound	Measured Concentration	
	in water [g m^{-3}]	in soil [mg kg^{-1}]
Benzene	27	270
Toluene	26	1100
Xylene	14	1100
THC	N.A.	9400

An important consequence of this observation is that the reported concentrations of NAPLs in soil, which are typically reported in mg or μg per kg of either dry or moist soil, should be more correctly interpreted in terms of concentration of the NAPL in the pore water. For example, if a saturated soil sample has a water content of 25% by weight (40% by volume), then the actual concentration of the contaminant in the pore water is 4 times greater than its reported concentration in dry soil, assuming pore fluid density of 1.0. If the reported concentration is given per unit weight of wet soil, the actual concentration in pore water is 5 times greater. Consequently, reported NAPL concentrations in soil samples, which are as much as 4-5 times lower than the corresponding aqueous solubility, are indicative of the presence of separate phase. In case of gasoline, to continue with our example, this would correspond to soil concentrations on the order of 20 - 25 mg kg^{-1}.

Since the recognition of the presence of separate phase is crucial to accurate assessment of the potential degree of contamination and potential cleanup options, the method used for reporting soil concentration data must be revised. Specifically, the measurement of water content of the analyzed samples should become a standard part of the laboratory routine and concentrations should be reported relative to both the solid and the water.

6 CONCLUSIONS

The data and the analyses presented herein illustrate the problem in evaluating the degree of contamination at NAPL contaminated sites. Specifically, it is shown that equilibrium water concentrations depend on the mole fractions of individual components of a NAPL and, therefore, in absence of a complete knowledge of the NAPL components the interpretation of water concentration data may be difficult, if not misleading. The same caveat applies to soil gas concentrations.

Dissolution kinetics also has an important influence on the measured effluent concentrations, particularly in the aqueous phase. In this respect, because diffusion coefficients in air are 4-5 orders of magnitude greater soil gas measurements should be more effective in assessing the degree of contamination as long as the NAPL is in direct contact with the soil gas.

Finally, measurement of NAPL concentrations in soil samples is the surest way for obtaining a clear understanding of the distribution of the NAPLs. However, the present methods of reporting soil concentration data make it difficult to recognize the presence of separate phase and should be changed to account for the moisture content, so that the NAPL concentrations in the pore water can be computed.

ACKNOWLEDGMENT

The work presented herein was supported in part by the National Institute of Environmental Health Sciences Superfund Program, Grant No. 3 P42 E504705-0281. The opinions and conclusions presented are those of the authors and do not in any way represent the opinions of NIEHS or the University of California.

REFERENCES

Dresen, M.D., F. Hoffman, S. Lovejoy Jr. 1986. *Subsurface distribution of hydrocarbons in the building 403 area at LLNL*. Report No. UCID-20787, Lawrence Livermore Nat. Lab., Livermore, CA.

Hunt, J.R., J.T. Geller, N. Sitar, K.S. Udell 1988a. Subsurface Transport Processes for Gasoline Components. *Proc. 1988 Joint CSCE-ASCE Nat. Conf. on Env. Engrg.*, Vancouver, B.C.

Hunt, J.R., N. Sitar, and K.S. Udell 1988b. Nonaqueous phase liquid transport and cleanup: 1. Analysis of mechanisms. *Water Resour. Res.* 24(8): 1247-1258.

Hunt, J.R., N. Sitar, and K.S. Udell 1988c. Nonaqueous phase liquid transport and cleanup: 2. Experimental studies. *Water Resour. Res.* 24(8): 1259-1269.

Karickhoff, S. W., D. S. Brown, and T. A. Scott 1979. Sorption of hydrophobic pollutants on natural sediments. *Water Res.* 13(3), 241-248.

Mackay, D., and W.Y. Shiu 1981. A critical review of Henry's law constants for chemicals of environmental interest. *J. Phys. Chem. Ref. Data* 10(4), 1175-1199.

Sitar, N., J.R. Hunt, and K.S. Udell 1987. Movement of nonaqueous liquids in groundwater. *Proc. Geotech. Practice for Waste Disposal '87*, ASCE, pp. 205-223.

Subsurface Contamination by Immiscible Fluids, Weyer (ed.) © 1992 Balkema, Rotterdam. ISBN 90 6191 175 3

Comparison of active versus passive soil gas techniques

Bud E. Smith
Union Carbide Chemicals and Plastics Company Inc., South Charleston, W.Va., USA

James C. Hickey & Ronald E. McCullen
Northeast Research Institute, Inc., Farmington, Conn., USA

ABSTRACT: A comparison study of two predominant types of soil gas study methods was conducted at a chemical manufacturing facility in New Jersey. One technique involved passive sampling coupled with analysis by desorption-mass spectrometry and the other utilized active sampling and analysis in the field using a portable gas chromatograph. Following data analysis and interpretation, the results from each system's respective surveys were compared to identify the advantages and disadvantages of each system. A checklist stipulating the conditions under which both systems are best applied was developed.

1 INTRODUCTION

Environmental managers are faced with the task of identifying contamination problems caused by an extensive list of volatile and semivolatile organic compounds (VOCs and SVOCs). Early assessments of potentially contaminated sites have typically been conducted using standard soil/ground water sampling techniques and analytical protocols whose development centered on use for enforcement and litigation actions. These regulatory protocol investigations have proven to be very costly and, in some cases, have led to incomplete source evaluation analyses and spatial delineation of subsurface contaminants. During the mid to late 1980's, emphasis has been placed on identifying lower cost, initial site assessment tools that would establish a better database from which the environmental manager can make more informed decisions about subsurface contamination. Researchers and practitioners from industry, academia, and the government began to investigate the feasibility of using soil gas surveying as a reconnaissance tool to meet the demand for site screening (Devitt 1987, Eklund 1985).

Soil gas sampling and analytical techniques were developed as a resource exploration tool dating back to the late 1920's (Hickey 1986). Over the ensuing years, several different approaches to sampling and analysis emerged including: active soil gas sampling, desorption of sorbed and occluded gases from soils, and use of a passive collection medium. During this development, the analytical procedures progressed from simple combustion and manometry to advanced instrumental techniques such as gas chromatography and mas spectrometry (Hickey 1986).

Adaptations of two of these methodologies, (1) active soil gas sampling combined with gas chromatographic analysis and (2) passive (integrative) soil gas sampling coupled with desorption mass spectrometric analysis, have emerged as the principal soil gas techniques for initial environmental site assessment (Devitt 1987, Eklund 1985).

Active soil gas sampling methods obtain aliquots of soil gas by extracting gases through a probe using either a pump to create a vacuum followed by sampling of the gas stream with a gas-tight syringe or using a high volume gas-tight syringe to extract the sample directly from the probe (Thompson & Mann 1987, Karably & Babcock 1987). Probe configuration varies in construction from 1" PVC with a length of 2 feet and open end at depth to facilitate gas sampling to use of a heavy-walled hollow steel tube that is screened at the sampling end and can be increased to approximately 15 feet in length (Thompson & Mann 1987, Karably & Babcock 1987). Analysis of the aliquot is performed by a gas chromatograph either on site via a mobile van or in a laboratory.

Passive soil gas sampling devices are constructed using two basic components: an adsorption material and a support apparatus. The adsorption material possesses a significant affinity for volatile organic compounds (VOCs) and semivolatile organic compounds (SVOCs), such as most chlorinated solvents and petroleum hydrocarbons. The support apparatus does not limit sorbate/sorbent interactions (Hickey 1986, Voorhees, Hickey & Klusman 1984) or contribute to a blank background. Investigators have utilized various sorbent materials including charcoal, carbon black, Tenax GC, silica gel, molecular sieves, etc., depending on the nature of the targeted (pathfinder) gases. Support apparati have ranged from glass culture tubes supporting a sorbent coated filament, to canisters of sorbent placed inside a can with the sorbent being held by various devices inside the unit (Hickey 1986, Voorhees, Hickey & Klusman 1984).

This paper presents the results of a comparison study made between two soil gas study methods.

2 OBJECTIVES

The objective of the survey was to compare the active and passive soil gas methods as they are routinely implemented by consulting contractors. Consequently, special optimization adjustments to routine procedures were not made to accommodate existing geological or site conditions such as low permeable soils, asphalted areas, sewer lines, varying water levels or weather conditions. The performance of the two methods was judged on the ability to identify and delineate subsurface contamination, time and ease of implementation, cost of survey, appropriate sensitivity of the measurement and total information available for environmental management decisions. An additional objective was to develop a check list which could be used by environmental managers to determine when each of these technologies could be effectively applied.

The survey was not intended to be a research project which promoted original research nor compared accuracy and precision between two methods.

3 SITE DESCRIPTION

The site is located at a chemical manufacturing plant in central New Jersey. A thin deposit of unconsolidated deposits formed by the residuum of weathered bedrock and fill materials comprises the upper 2 to 5 feet. The Triassic Brunswick Formation consisting of red shales interbedded with massive siltstones and occasional layers of arkose and conglomerates underlies the surficial materials. The Brunswick formation is several hundred feet thick and comprises the first aquifer beneath the site. The depth to the water table ranges from 5 to 20+ feet across the site.

4 SOIL GAS SURVEY DESIGN

The passive system survey design was developed by Northeast Research Institute, Inc. (Petrex). The survey design consisted of 250 sample locations deployed in a regular 50 foot grid pattern over the site of an old chemical process unit area. The active method was performed by an environmental consultant at identical locations to 100% duplication of two subareas: one believed to be contaminated and one believed to be background. The two subareas in which the active method was performed comprised about 25% of the total study area. The passive method was completed first, then the active method was completed within the next two weeks. The sample locations of the two methods were within one foot of each other.

5 PASSIVE METHOD ANALYSIS

The Petrex monitor consists of a ferromagnetic wire coated with an activated adsorbent placed inside a one-inch diameter glass tube (Figure 1). Each tube was placed in a two-inch diameter hole that was 18 to 24 inches deep, then the hole was backfilled. The tubes were retrieved three weeks later. The samples were then thermally desorbed by heating of the Curie Point wire resulting in the gases being directly injected into a mass spectrometer. The total ion counts were recorded for the individual masses for the compounds present.

The hydrocarbon compounds were identified by their characteristic peak occurrences including molecular ions, fragment ions, and carbon isotope ions. The molecular ions for the aromatic compounds, such as benzene, toluene, and xylene isomers, or ethyl benzene (BTX), were utilized to help identify and subsequently report relative abundances of these compounds. There was little or no interference of the specific mass fragments for the target compounds that were chosen for mapping. The ion count value of the indicator peak was used to report the relative concentration for that compound.

It should be noted that indicator peak intensity is not an absolute measure of concentration but rather a measurement of the total amount of a compound collected at a given location during the exposure period. Sample locations are shown with the corresponding relative ion count levels which relate directly to varying compound concentrations in the subsurface. Only relative differences in subsurface compound concentrations of a site can be initially inferred from the relative ion count values for a given compound. Also, the relative ion count values for

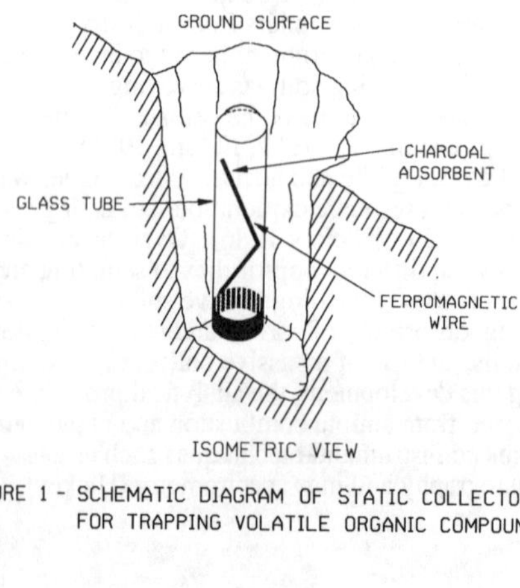

FIGURE 1 - SCHEMATIC DIAGRAM OF STATIC COLLECTOR USED FOR TRAPPING VOLATILE ORGANIC COMPOUNDS

one compound cannot directly be quantitatively compared to the values for a different compound. It is sometimes possible at specific sites to use results from conventional ground water analysis of well samples as a calibration of surface ion flux levels observed by the soil gas measurement. This has to be done with caution because of varying subsurface conditions that may not be obvious.

6 ACTIVE METHOD ANALYSIS

One-inch diameter borings were drilled at each sample location. The holes were an approximate depth of 24 to 36 inches. The depth varied due to the thickness of man made material overlying the native soil. The design of the hole was to be 24 inches deep into the native soil. A soil gas probe, designed as depicted in Figure 2, was then inserted into the boring and the annulus at the surface sealed with cuttings from the boring. The fit between the probe and the boring was considered sufficiently tight to insure that soil gas vapors entered only through the drilled holes and the bottom of the soil gas probe.

The pump of a Foxboro Model 128 Organic Vapor Analyzer (OVA) with a flame ionization detector (FID) was used to draw gas vapors from the soil and the maximum total volatile organic compound (VOC) reading recorded for each boring. In addition to the maximum reading being measured, a stabilized VOC reading was recorded. The OVA was calibrated to a methane standard and results recorded in total parts per million (ppm) as methane. Prior to taking a total VOC measurement, background VOCs were measured in the vicinity of the survey location in the upwind direction and the instrument zeroed to prevent interference.

Soil gas was collected for GC analysis by inserting a gas-tight syringe through the septum port

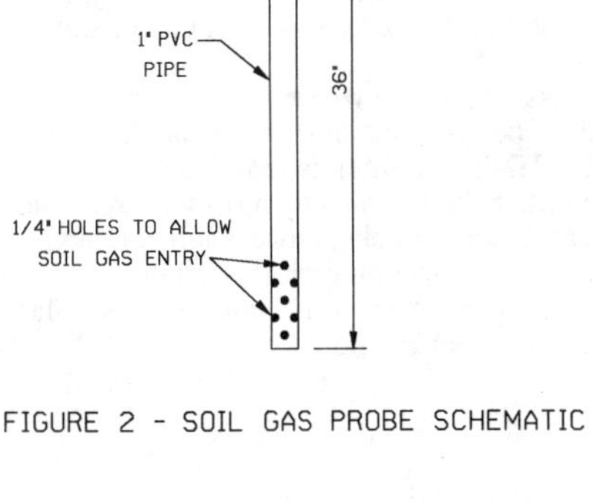

FIGURE 2 - SOIL GAS PROBE SCHEMATIC

of the soil gas probe at the time the maximum total VOC reading was reached. An approximate volume (based on the VOC reading) of the soil gas was then withdrawn from the probe and injected onto the column of a photovac 10S50 Photoionization Detector (PID) portable gas chromatograph (GC). The GC was equipped with a 4 foot CSP-20M column to provide separation of benzene, toluene and xylene (ortho, meta and para isomers). Prior to initial use, the GC was calibrated with 0.5 ppm level standard of each compound. Injection volumes were adjusted for the standards and soil gas samples to provide the best achievable sensitivity, typically not detected at 50 parts per billion (ppb).

Standard gases were used to generate example chromatograms of benzene, toluene, and xylene for the field chromatograph. Response factors were inferred from the injection of known concentrations of these compounds.

7 COMPARISON OF RESULTS

The comparison of the two methods was made in accordance with the stated objectives previously described. A summary of the comparison is listed below in two categories: common characteristics and differing characteristics.

<u>Common Characteristics</u>

1. These soil gas study methods are relatively inexpensive "scanning" tools to determine the presence or extent of subsurface contamination. The results of such studies can be used to direct more detailed studies in a cost effective manner (e.g., better location of borings and wells; less wells required).

2. Neither method can differentiate between soil and ground water contamination unless actual samples of each are tested separately. Such testing usually involves additional work generally beyond the initial scope of work of the soil gas study.

3. While the concentrations from the passive method (ion counts) and the active method (mg/l) are in different units of measurement, neither quantification can be directly correlated to a contaminant concentration in the soil or ground water.

<u>Differing Characteristics</u>

1. The passive method yields a more detailed and refined description of a contaminant plume(s). For example, the Petrex contours show detailed definition of the limits of specific organic compounds (Figure 3) to reveal several distinct sources and relative contaminant concentrations within the area rather than two contours and many non detects (ND's) as resulted from the active method (Figure 4). The passive analysis can identify the presence of organic compounds below the detection limit of the active method.

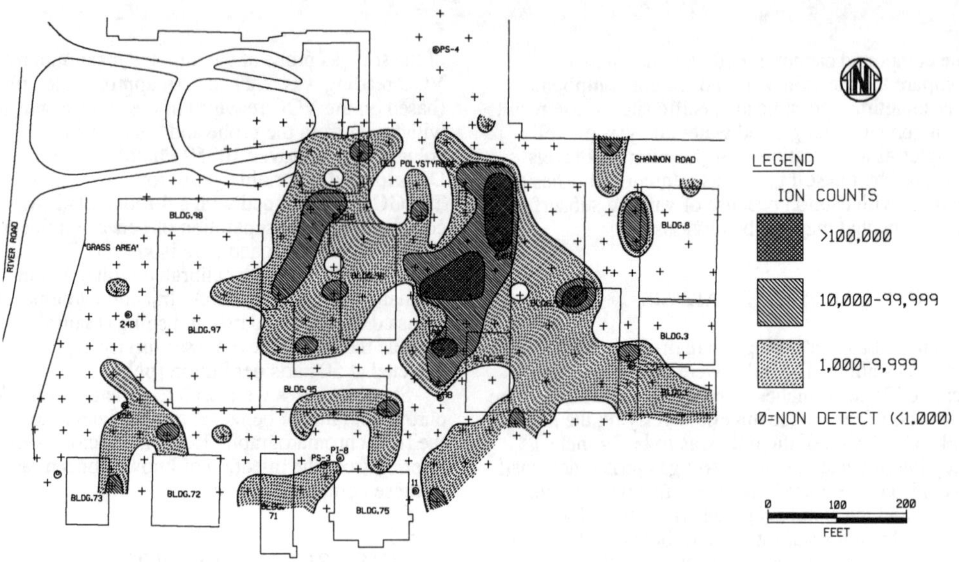

FIGURE 3 - PASSIVE METHOD-RELATIVE FLUX STYRENE

2. The active method provides an average vapor phase target compound concentration of the soil gas but will not always be able to differentiate specific organic compounds. For example, it did not identify styrene in the old Polystyrene Unit area because of the constraints of detection limits and short-term sampling variability. This lead to an "ND" being shown on a final drawing, whereas the Petrex drawing showed the presence of styrene and gave a more detailed indication of the concentration trend. In addition, the active method resulted in a false positive as a result of the presence of naturally occurring compounds, such as methanol or terpene.

3. A cost comparison for the two methods was made by dividing the total contract invoiced price by the total number of samples analyzed. The cost per sample for the passive method was $144 compared with $265 for the active method. The costs included supplying all sampling equipment, sample procurement (tube installation and retrieval for Petrex), analyses, and written report.

8 RECOMMENDATIONS

The following recommendations were identified during the course of the comparison study at the test site, but they are probably applicable to most any site.

1. As part of a soil gas study, collect several soil samples with depth and analyze by headspace MS and/or laboratory GC/MS to identify sources of soil gas detected by Petrex tubes or GC probes and to establish actual concentrations. Some of these samples could come from the cuttings from the shallow holes drilled for the soil gas study, but additional samples should be taken at depth, preferably to the ground water table.

2. Collect several ground water samples (from extended borings recommended above) and analyze by head space MS or GC/MS analysis to help define the source of the soil gas and establish actual concentrations. Highly volatile, but soluble organic compounds may not be detected by soil gas detectors once they are dissolved in ground water.

3. Develop a site history to define old unit areas, spill areas, and other suspected sources of contamination. Use interviews with long time plant personnel and review old aerial photographs to help with the interpretation of the results of the soil gas study.

4. Augment active studies with some MS analyses to provide definition of compounds to prevent false positives and other impacts caused by interferences.

5. Review the properties of the suspected compounds, subsurface conditions, climatic conditions and time constraints before selecting a specific soil gas study method.

6. Use the following checklist as a guide for selection of the type of soil gas study method.

Use passive method when:

a. Specific contaminants are not known.

b. High resolution is required (e.g., to differentiate between many nearby sources, such as sewers, loading docks, process units, tanks).

c. Low detection limit is required.

d. Organic compounds range in molecular weight between 29 and 240.

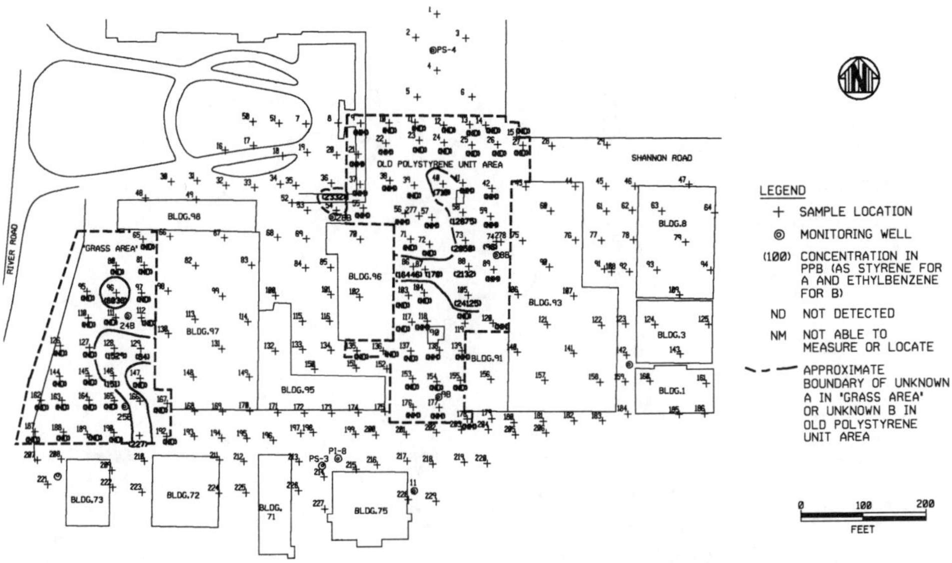

FIGURE 4 - ACTIVE METHOD-UNKNOWN "A" CONCENTRATIONS

e. High quality graphic display of data is required.

f. High ground water level precludes soil gas extraction.

g. Project schedule allows minimum of one month.

Use active method when:

a. Contaminant compound(s) is known.

b. Immediate (real time) data are required (e.g., spill detection).

9 CONCLUSIONS

Soil gas studies comprise a relatively inexpensive investigative tool for scanning sites for potential soil and ground water contamination. The passive method usually requires more time to implement but yields a more detailed description of the presence of subsurface contaminants.

REFERENCES

Devitt, D. A. 1987. Soil Gas Sensing for Detection and Mapping of Volatile Organics. EPA Report # EPA/-600/8-87/036.

Eklund, B. 1985. Detection of Hydrocarbons in Groundwater by Analysis of Shallow Soil Gas/-Vapor. API Publication #4394.

Hickey, J. C. 1986. Preliminary Investigations of an Integrative Gas Geochemical Technique for Petroleum Exploration. Unpublished Master's Thesis. Colorado School of Mines.

Karably, L. S. & K. B. Babcock. 1987. Effect of Environmental Variables on Soil Gas Surveys. Proceedings of the 19874 Superfund Conference. HMCRI.

Thompson, G. M. & D. L. Mann. 1987. Soil Gas Contaminant Investigation - A Dynamic Approach. Ground Water Monitoring Review. Vol. VII, No. 3, p. 88.

Voorhees, K. J., J. C. Hickey & Klusmann. 1984. Analysis of Groundwater Contamination Contamination by a Near Surface Static Trapping/Mass Spectrometry Technique. Analytical Chemistry. v. 56 #13.

Subsurface Contamination by Immiscible Fluids, Weyer (ed.) © 1992 Balkema, Rotterdam. ISBN 90 6191 175 3

Limitations of sampling subsurface contamination by volatile hydrocarbons in unconsolidated, fractured and karstic rocks: 1. Practical experiences from 220 cases

Benedikt B.Toussaint
Hessian Agency of the Environment, Wiesbaden, Germany

K.Udo Weyer
WDA Consultants Inc., Calgary, Alb., Canada & WKC Weyer Consultants, Düsseldorf, Germany

ABSTRACT: Volatile hydrocarbons are ubiquitous pollutants in soil, soil gas and groundwater of industrialized countries. The decision to investigate polluted sites and the extent of field investigations and subsequent remediation in Germany are based on contaminant concentrations found during preliminary examination of soil and groundwater.

Soil samples are usually taken from cores of small diameter sounding drills; groundwater is sampled from existing or newly installed wells and piezometers. Reference is then made to limiting concentration values which have been rather arbitrarily set by the state regulatory agencies. The reasoning and pitfalls in setting reference values are not the topic of this paper. Instead focus is on the reasons why concentration values obtained for pollutants are often grossly misrepresenting the actual field conditions. The results of chemical analyses for volatile hydrocarbons are therefore often unsuited as a basis for informed decision making.

The choice of sampling procedures for soil, soil gas and groundwater and their interaction with the geologic substratum and the local hydrodynamics determine the quality of results. A second paper contains the evaluation of three typical case histories (Toussaint & Weyer, this volume).

1 INTRODUCTION

Volatile halogenated hydrocarbons are contaminating groundwater worldwide to an ever increasing degree. In Germany, organic compounds are used in about 10 % of industrial plants and service shops as solvent, extraction and cleaning agents. By way of spills and leakage from pipes, tanks, and sewer systems immiscible fluids find their way into soil and groundwater.

The very dangerous potential of halogenated hydrocarbons as pollutants and the total extent of their invasion of soil and groundwater systems are only now becoming fully known. Methods which can recognize their presence in nature and determine the levels of contamination are still in development. They are by no means foolproof. The procedures for sampling volatile hydrocarbons are still more an art than routine techniques.

In laboratories, stringent quality control is necessary to ensure reliable and reproducible results. In Germany, procedures for the analysis of heavy hydrocarbons are laid down in the industrial standards DIN 38407-F4 and F11 as well as DIN 38413-P2. The hydrocarbons are separated by gas chromatographic methods and then analyzed with electron capture detectors and/or flame ionization detectors (Hagendorf et al. 1987a, 1987b, Kanitz 1986). The chemists are quite certain that these methods are reliable.

The real cause of inaccuracies in concentration results would then lie in the sampling techniques, conservation of the sample, and its transportation to the laboratory. Procedures dealing with volatile hydrocarbons need to take into account fully their physical properties and the hydrologic characteristics of the geologic substratum.

Nevertheless, in spite of all the uncertainties involved, most states of the Federal Republic of Germany, including Hessia, opted for a schematic decision- making framework of reference values for deciding whether to investigate and subsequently to cleanup contaminated soil and groundwater (Table 1). One may assume that uncertainties in the concentration values obtained are not significant for the preservation of a healthy environment. This is not the case, however, mainly because of three reasons: (1) the chemicals involved pose a high potential health risk for humans even at low concentrations, (2) the concentration determination may be in error by orders of magnitude, and (3) the decision to investigate and subsequently to clean up is governed by the analytical values obtained.

Table 1. Reference concentrations of volatile halogenated hydrocarbon contamination for decision making by regulatory agencies in Hessia.

Investigation of groundwater	
> 10 µg/l	decision to start investigation
25 µg/l	outer limit of plume
> 100 µg/l	decision to start remediation
Remediation of groundwater	
disposal of processed water into aquifer	
< 25 µg/l	within depression cone at start of clean-up
≤ 1 µg/l	within depression cone at end of clean-up
≤ 1 µg/l	reinfiltration outside of depression cone
disposal of processed water outside of aquifer	
≤ 25 µg/l	use for technical processes and then disposal in sewer
< 50 µg/l	disposal in sewer, if water has been cleaned to < 5 % of original contamination
< 50 µg/l	disposal in surface water not passing water protection zone
≤ 1 µg/l	disposal in surface water passing water protection zone
≤ 10 µg/l	decision to finish remediation
Remediation of soil	
≤ 5 mg/kg	decision to start remediation
≤ 1 mg/kg	decision to finish remediation
Disposal of soil gas	
> 20 mg/m³	filtering through active coal required
≤0.5 mg/m³	decision to finish soil gas stripping (stripping to be repeated at least five times)
Disposal of soil and waste in landfills	
<detect. limits	landfills for soil and rubble
< 100 mg/kg	sanitary landfills
≥ 100 mg/kg	toxic waste disposal sites

This paper presents the basic principles and likely pitfalls encountered in sampling for volatile hydrocarbons. It does not attempt to discuss in detail the transportation of samples to the laboratory or their preparation for analysis. Three case histories are presented in a second paper (Toussaint & Weyer, this volume), illustrating procedures commonly used at spill sites.

2 SAMPLING TECHNIQUES

The physical and chemical behaviour of organic compounds exert an important influence on procedures suited for sampling. The characteristic hydrodynamic behaviour of immiscible fluids heavier than water has been explained by the work of Schwille et al.(1984) and Schwille (1988). Increases in permeability of clay, due to the dielectric behaviour of immiscible liquids, have been known to exist since 1954 (Michaels & Lin 1954, Mesri & Olsen 1971, Anderson & Brown 1981, Fernandez & Quigley 1985). Table 2 summarizes some of the physicochemical properties of halogenated hydrocarbons, such as boiling point, solubility, vapour pressure and others.

The basic migration principles are summarized in Fig. 1 showing the typical migration paths of the immiscible liquid phase, the plume of dissolved hydrocarbons, and the vapour phase. Hydraulic gradients in the groundwater system cannot be taken as indicators for the direction of migration of the liquid phase due to its immiscible behaviour. Rather, the flow direction is determined by geologic structures and heterogeneities in the permeability system.

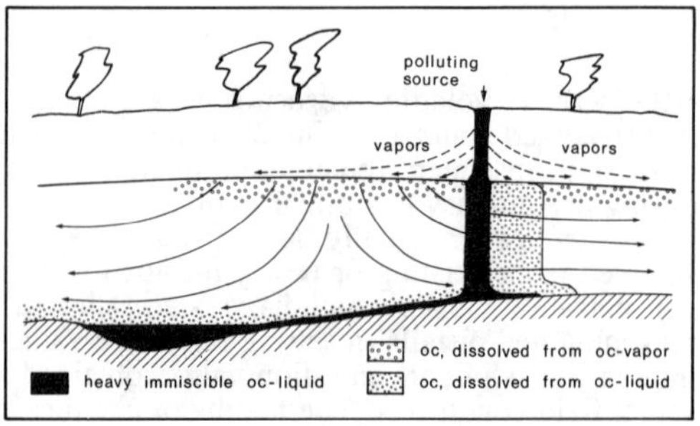

Figure 1. Schematic diagram of flow of heavy, immiscible and volatile organic compounds (oc) and resulting groundwater contamination. Long arrows: flow lines for groundwater; dashed arrows: flow lines for vapours (from Weyer 1987).

The sinking immiscible liquid migrates towards the deepest location along its flow path. While migrating it leaves behind residual saturation of liquid phase in the pore and fracture space. This material then acts as a source for additional plumes of dissolved hydrocarbons. It is this behaviour which makes it so difficult to detect the full extent of groundwater pollution by immiscible fluids heavier than water (Weyer 1987). Schwille (1988) points out that heavy organic liquids move faster in fractured rocks than in unconsolidated rocks.

Dissolved multi-component concentrations of hydrocarbons in groundwater could exceed 20 g/l or more (compare Table 2: solubility). In this case the increase in density can be sufficient to modify flow directions as compared with uncontaminated water. Highly concentrated dissolved plumes would move farther down within the aquifer flow system. This behaviour needs to be considered in the strategy for monitoring and sampling.

Normally only very expensive geologic drilling and possibly surface-geophysical investigations allow the early detection of the free-liquid phase in the groundwater system. These are not

Table 2. Material properties of selected volatile halogenated hydrocarbons (modified from Toussaint 1989b).

SUBSTANCE									
IUPAC nomenclature	tetrachloro–methane	trichloro–methane	dichloro–methane	1,1,1–tri–chloro–ethane	tetrachloro–ethene	trichloro–ethene	cis 1,2–dichloro–ethene	chloro–ethene	hydrogen–oxide
common name	carbon tetrachloride	chloroform	methylene chloride	methylene chloroform	perchloro–ethylene	trichloro–ethylene	cis 1,2–dichloro–ethylene	vinyl chloride	water
CAS number	56–23–5	67–66–3	75–09–2	71–55–6	127–18–4	91–01–6	540–59–0 (trans)	75–01–4	7732–18–5
chemical formula	CCl_4	$CHCl_3$	CH_2Cl_2	$C_2H_3Cl_3$	C_2Cl_4	C_2HCl_3	$C_2H_2Cl_2$	C_2H_3Cl	H_2O
molecular mass [g/mol]	153.82	119.38	84.93	133.41	165.83	131.29	96.94	62.50	18.02
density at 20°C [g/cm³]	1.59	1.48	1.33	1.34	1.62	1.46	1.28	0.91*	0.99
solubility in water at 20°C [mg/l]	1,160	7,300	20,000	1,300	160	1,090	800	1,100**	–
vapor pressure at 20°C [Pa]	12,000	20,600	47,300	13,300	1,860	7,800	23,400	309,200	2,340
boiling point at 1013 hPa [°C]	76.5	61.7	40.6	74.1	121.2	86.7	60.3	–13.4	100.0
diffusion coefficient in air at 25°C and 1013 hPa [cm²/s]	0.0820	0.0888	0.1037	0.0794	0.0797	0.0875	0.0911	0.1071	–
dielectric constant at 20°C	2.4	4.9	8.9	7.1	2.3	3.4	9.3	6.2	81.6
partition coefficient water/air at 20°C	0.97	7.69	8.10	1.68	1.22	2.74	0.51	0.39	–
partition coefficient n–octanol/water at 20°C	525	93.3	17.8	309	398	195	30.2	4.0	–
log P(ow)	2.73	1.97	1.25	2.49	2.60	2.29	1.48	0.60	–
surface tension of liquid phase at 20°C [mN/m]	27.0	27.1	26.5	25.6	31.7	29.3	28.0	20.9***	73.0
dynamic viscosity at 20°C [mPa*s]	0.97	0.58	0.43	0.86	0.54	0.57	0.47	0.21	0.89

* under pressure ** at 1013 hPa *** at –10°C

routine investigations; they are based on trial and error. As a result none have been undertaken so far in Hessia. In Germany the search for the free-liquid phase is not part of investigations for heavy hydrocarbons within groundwater systems, except very close to the surface.

Instead, emphasis is put on the investigation of soil at the point of pollution, and the dissolved plume of contamination in the groundwater system. Investigations in the unsaturated zone at pollution centres are usually undertaken to provide proof of cause for the courts.

Subsurface samples are taken from the soil, the soil gas and the groundwater system. They are discussed below.

2.1 Soil sampling

Most of the soil sampling is done in the unsaturated zone, some of it in aquifers as well. The intent of soil sampling often is to determine the volume of polluted material for excavation or air stripping and its degree of contamination. The information gained also helps in estimating active and future groundwater pollution by the contaminated soil.

Soil samples are taken from boreholes or excavations, with sampling from boreholes normally preferred.

Only drilling techniques without circulation of air, water or mud can be applied without loss of volatile pollutants from the sample. In German practice, soil samples are generally taken with small diameter (1" or 2") sounding drills which usually are rammed into the ground by motor driven hammers.

This method has the advantage of low cost and high speed. There are, however, grave disadvantages. Compression of core material often disturbs the geologic log significantly and makes an exact depth determination difficult. Within the drillstem, polluted material may spread significantly beyond its original position leading to cross contamination within the borehole. Cross contamination between boreholes can be avoided by cleaning the drillstem with acetone. The small volume of the samples encourages losses by degassing due to its unfavorable ratio between volume and surface area. During the ramming action the temperature at the tip of the drill may reach up to 60°C thus enhancing degassing. Vibrating effects can cause separation of liquid phase and matrix and lead to subsequent migration and loss.

In our experience, soil sampling by ram sounding drills only provides qualitative results. In regard to quantity of pollution the results could be completely unreliable, for the above reasons.

Using different drilling methods can lead to significantly different quantitative results. For example results of a few μg/kg pollution received with ram sounding drills often compare to pollution values of a few hundred to thousand $\mu g/kg$ found with methods such as split-spoon sampling.

Best drilling results are obtained with a method similar to split-spoon sampling (Hagendorf et al. 1987b). The core barrel used is lined with PVC or better yet, teflon. Its length is 0.5 to 3 m with an inner diameter of up to 10 cm. Upon drawing of the drillstem, the PVC or Teflon pipe is immediately sealed at both ends. It is then delivered to the laboratory in its entirety. The disadvantage of this method is the high cost.

Samples taken from the walls of trenches or other excavations usually represent the natural conditions better if the excavations are fresh. Often, however, the conditions at sampling time are not representative for the subsurface condition before the opening of the excavation due to losses caused by changes in temperature, by changes in the flow pattern of the soil gas, and by dissolution of hydrocarbons in precipitation water hitting the walls of the excavation. Samples are usually taken by Shelby tubes pushed into the walls.

An advantage of excavations can be the view of the near-surface geology. The high costs for excavating often prove to be a serious disadvantage. Also, this method may not be applicable underneath or near buildings.

Soil samples should never be mixed and quartered. Both would lead to significant losses of volatile hydrocarbons. The field sample used to be stored in wide-mouth bottles in as large a quantity as possible (roughly 500 g as a minimum). In the laboratory, the outer layers of the sample are then cut away before analysis. Analytical work should only be done with the inner part of the sample. In the outer fringes degassing losses and adsorption effects on surfaces can be significant (Remmler 1990). Wide-mouth bottles made of glass with polished glass plugs are still often used but do not perform well enough.

Use of headspace containers in the field minimizes losses caused by degassing. Hagendorf et al. (1987a, 1987b) consider a soil sample size of 5 g as optimal for headspace containers.

By its very nature, soil sampling returns only point values for contamination. These point values are , however, often not representative for the surrounding soil volume, due to its heterogeneity. Matrix heterogeneities of the soil, as well as time and climate (air pressure, temperature, precipitation) dependent changes in the dynamic equilibrium between organic molecules and the surface forces at the matrix particles of the soil, make spatial and time related concentration variations within the soil unpredictable.

Detailed knowledge of the local geology helps, however, in estimating the true conditions. For illustrative purposes Figure 2 shows a schematic block diagram of typical small-scale heterogeneous soil and permeability variations. Should peat be present it would act as a sink with high concentrations of heavy hydrocarbons in free-liquid phase while there may be none in the sand layer nearby.

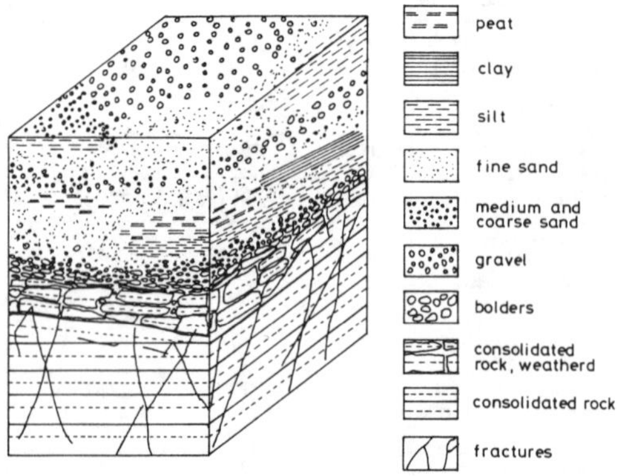

Figure 2. Schematic block diagram of unconsolidated and consolidated rocks.

In summation, the results of soil investigations in Hessia only allow observations on the presence or absence of pollution. The concentration values analyzed are not representative for the degree of pollution. In consequence singular analyses values for soil are not of much use. Even a close sampling grid may miss the most polluted areas. It would need an extraordinary number of additional samples, hydrogeologic tests and evaluation procedures to overcome this disadvantage. Such efforts are usually not made in this type of investigation.

2.2 Soil gas sampling

Low costs make soil-gas sampling an attractive investigation method. The method is seemingly simple and results are obtained quickly. A variety of methods exists and has been described by various authors (Neumayr, 1983; Swallow & Gschwend, 1983; Voorhees et al., 1984; Kanitz, 1986; Marrin & Kerfoot, 1988). General principles have been discussed by Neumayr (1983), Nerger et al. (1988), and Peterson et al.(1988).

Boiling point and vapour pressure of the pollutant determine the flux of pollutants from soil or water pollution into the soil air. From the point of pollution the soil air migrates by convection and diffusion. This flow of polluted air is captured for concentration measurements.

In Germany, active sampling by a syringe (direct method) or by a piston pump (enrichment method) is preferred over long-term passive sampling with in-situ adsorbing of hydrocarbon from soil gas on buried activated carbon or resin. Passive sampling returns concentration values integrated over time and contaminant species. That may be an advantage for certain types of investigations such as the survey of landfills.

Singular samples are not of great value; a large number of samples is necessary for meaningful interpretation. The spatial interpretation of soil gas concentrations is hampered by the same limitations as soil sampling.

Direct sampling with syringes can produce actual point values allowing pinpointing of the contaminant source, and disposable syringes avoid cross contamination. Because of the small volumes taken, concentration values reflect local conditions well. The borehole need not be sealed and concentration values are not integrated over time and space.

Sampling by piston pumps typically captures the contaminants from 20 l of soil gas on activated carbon or resin. Thus contaminated soil gas is sampled from a larger volume of soil than by the syringe method. This method leads to comparatively fast delineation of major contamination sites. In turn, the concentration values are integrated over time and space. The error margin is considerably larger than that of the syringe method, though, and problems are often encountered with the sealing at the top of the borehole.

In Germany, background values of organic pollutants in soil air without any particular local soil contamination are up to 500 $\mu g/m^3$ in industrial areas and 50 - 100$\mu g/m^3$ elsewhere. These concentration ranges need to be sampled by piston pumping with enrichment.

Soils with pollution by liquid phase typically have soil-gas values in the mg/m^3 to g/m^3 range. Besides using the above methods in conjunction with GC analysis, such high concentrations are sometimes measured in boreholes using testing tubes with a suction ball. A colour indicator in the tip of the tubes gives the approximate degree of contamination, integrating several species of contaminants.

Depth profiling of soil-gas pollution helps to identify the type of pollution sources involved. Depth profiling can be achieved by installation of separate boreholes to different depth or by analysis of air from core samples. Packer systems are difficult to install and unreliable because of the small diameter of boreholes.

The shape of the concentration-versus- depth curve helps separate sources within the vadose zone from sources within the saturated domain. Sources in the unsaturated zone would show up with peaks along the vertical sampling line (Fig.3: g, h). This would be the case for pollution from sewers, from underground storage tanks and for stratified soil pollution.

Should degassing of unconfined groundwater be the source of soil-gas contamination, then concentrations would increase with depth. Here, the shape of the concentration versus depth curve is mainly determined by two parameters: (1) reduction of concentration caused by diffusive or convective processes and trapping of volatiles by soil moisture (Fig 3 a) and (2) the effects of geologic and lithologic pattern in the unsaturated zone. These processes determine the type of modification of the simple vertical concentration curve. Type curves have been schematically summarized in Figure 3 (diagrams b to e). Diagram b shows a concentration curve under an inhibiting surface layer, diagrams (c) and (d) a modification by a strongly inhibiting layer and a series of partially inhibiting lenses. Diagram (e) refers to two groundwater sources, one of them in a perched system.

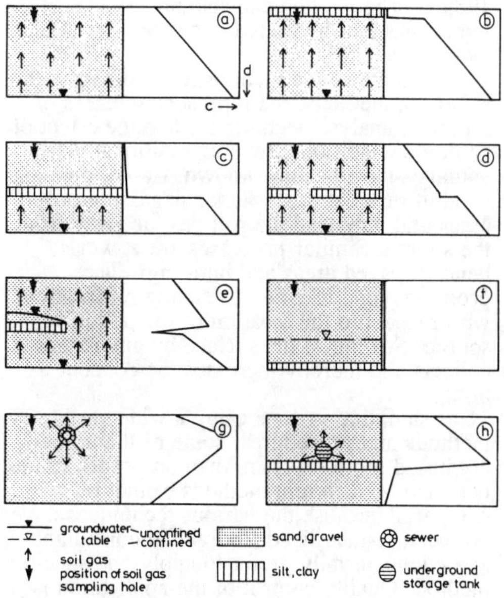

Figure 3. Concentration of soil gas in dependence of hydrogeology and source of pollution (left half: hydrogeology; right half: concentration of halogenated hydrocarbons).
a - f: concentration of soil gas over a polluted plume of groundwater under differing geologic and hydrodynamic conditions (unconfined: a-d, f; confined: e).
g, h: effect of a source of pollution in the unsaturated zone.

If the source of pollution is within a confined aquifer (diagram f) then the chance of detecting this pollution in the soil gas of the overlying unsaturated zone is practically non-existent.

In humid climates, as in Hessia, the distance between groundwater table and sampling points should not exceed 12 to 15 m and highly permeable layers, such as sand and gravel, should prevail (Neumayr 1983). Low-permeability material tends to inhibit the migration of gas. Perched groundwater and high water content in the unsaturated soil successively strip contamination from the soil gas. Under all conditions the point of sampling has to be located below the often loamy and humic surface layers.

It has been stated (Grathwohl 1989) that groundwater and soil pollution may be calculated from soil gas values. In practice, however, this cannot be done readily because necessary data such as porosity distribution, amount and type of organic substance, and moisture contents of the layers penetrated are usually not available and cannot easily be determined (Hagendorf et al. 1987a, Peterson et al. 1988, Swallow & Gschwend 1983, Toussaint 1986, 1989b). The method can also not be used if the liquid phase of hydrocarbons is present (Hahn 1990).

Besides the above mentioned problems with sampling methods, the representativeness of samples, analysis techniques, and the effect of geologic conditions, several additional disadvantages have become known. Cable, pipe and conduit installations enhance degassing processes and may lead the soil gas far away from the source. Similar processes are at work beneath paved areas and buildings. These processes tend to give misleading information with respect to the location of the pollution source. Sealing of the surface by any means collects and thereby increases the concentrations.

Gas sampling is done using a wide variety of methods and procedures, some of them possibly based more on claim than on proof. Values obtained by different methods cannot be compared because the various techniques have not been standardized. Individual consultants and clients usually prefer to apply one specific method. Quality control of the application is almost impossible.

Due to changing weather conditions, sample series taken in the same area at different times cannot be readily compared and are often impossible to interpret. Precipitation causes lower concentration values in soil air as increasing water content within the pore space reduces the liquid and volatile content of volatile hydrocarbons. Low temperatures and especially freezing of soil tend to trap volatile hydrocarbons by condensation (Hagendorf et al. 1987b, Neumayr 1983).

Variation of concentrations is also dependent on pumping history and pumping stress. Areas supposedly cleaned by continuous pumping often show significant increases in the concentration of pollution during non-pumping periods, often even back up to original concentration values. The cause for this behaviour is often residual saturation and slow diffusion of pollutants from lower permeable areas into higher permeable pathways.

Besides the physico-chemical properties and processes involved, the migration of soil gas is modified by geological conditions of the subsurface. The geology is usually not known in enough detail. Low permeability material reduces or even inhibits transport of volatiles to the surface. Significant vertical and horizontal variation in lithology of unconsolidated and consolidated rocks and their fracture patterns cause great variations in concentration in soil air.

Nevertheless, soil gas sampling can deliver valuable insights into subsurface pollution. The sampling strategy for an investigation of soil pollution should use intersecting sampling lines. An investigation for groundwater pollution should favour singular lines across the suspected direction of the plume of pollution.

In conclusion, the soil gas sampling is suited to optimize the selection of piezometers when supplemented with proper drilling and groundwater sampling programs. It does, however, not replace groundwater installations and investigations.

2.3 Groundwater sampling

The main objectives for groundwater investigations are to determine the hydrodynamical properties of the rocks in general, the flow directions of the groundwater in particular, and the type, degree and source of pollution. The results determine the human health risk and, independent of that, the need for remediation. To this end a number of monitoring approaches are available.

Spring discharge is ideal for non-destructive and non-disturbing sampling of conditions of the groundwater body close-by. Most contaminations occur, however, away from spring areas. There, sampling through wells or piezometers is necessary.

In a first step one might want to make use of previously installed wells and piezometers. This is not adequate for sampling volatile chlorinated hydrocarbons. Large diameter or suction-lift well arrangements encourage significant degassing. Wells and piezometers with very shallow or very deep screen positions often miss the pollution plume. Wells and piezometers with screens in two or more aquifers are totally inadequate. They cause cross-contamination between aquifers and should not be used for sampling. Also, they do not allow any meaningful measurement of potentials needed for the determination of flow directions.

Existing installations often are defective in three ways (Toussaint 1989a, Toussaint and Pütz 1987):

1. In the past, and sometimes even today, wells and piezometers have been sealed above the screen with only a simple cement seal. In many groundwater systems waters are aggressive because of the presence of dissolved CO_2 and of sulfate. Under these conditions normal cement seals usually fail within a few years after installation.

2. Another method of sealing uses air- dried "bentonite" pellets. It has been shown by unpublished laboratory tests with uranin tracers (Schwille 1990) that these seals can fail to function immediately after installation. In these tests swelling was insufficient to close the pore space between the bentonite pellets completely. There are also some indications that "bentonite" pellets commonly used in the past may have been subject to disintegration. A new brand of sealing clay with embedded gamma ray emitters now makes it possible to determine the exact location and extent of the seal, and to estimate its continuity.

3. Installation of piezometers made from polymeric materials such as PVC also has its pitfalls. Using television inspection and geophysical borehole logging, Toussaint (1989a) and Toussaint & Pütz (1987) showed many cases where open cracks had developed during installation of PVC pipes by forceful or insufficient connection of the couplings. The weight of the pipes themselves (in Germany usually 4" or even 6" diameter) hanging during the process of installation tends to damage the thread and thus open a leakage path for contaminants. The critical weight was reached by installations exceeding 60 to 70 m in length.

The older couplings were standard thread connections with sockets. The manufacturer now states that a new type of stronger pipes and couplings with double sockets and embedded sealing elements are waterproof up to several hundred metres depth. Nevertheless, in Hessia the additional use of outside shrinking sealing sleeves is required by the responsible government agency.

Installation of new piezometers also has its limitations. Drilling can mobilize and open pathways for migration of liquid phase or dissolved contaminants between aquifers.

Groundwater flow system configurations require depth-dependent monitoring and sampling. Dissolved organic compounds migrate along the flow path of groundwater and are not subject to density flow. Hence screen positions need to be well selected within an aquifer with heterogeneous permeability. Screening the total length of a thick aquifer is not good practice, not even for a first assessment of the presence of contamination by organic compounds. Installation of a complete filter often changes the flow field of an aquifer significantly. Within nests, piezometers should be installed in individual boreholes.

More recently, multi-port piezometers have been used in increasing numbers. Often the annulus has been filled with sand, regardless of the lithology and permeability of the surrounding material. In these cases vertical crossflow between ports cannot be excluded. Analyses of water samples taken this way may lead to erroneous conclusions. We only recommend multi-port piezometers as additional installations if a functional seal is installed between ports and the annulus around a port is filled with highly permeable sand.

The material of the piezometer and the sampling equipment can adsorb hydrocarbons (Remmler 1990). Sampling material, that will be exposed to groundwater for any prolonged time should, therefore, consist of teflon only. Piezometer casings and screens in low-permeability areas should also consist of teflon. In highly permeable rocks the contact time between water and PVC can be minimized by longtime pumping such that adsorption effects are negligible.

A particular German custom is the use of filters with filter sand and gravel glued to the filter pipe with organic resin. Many of these filters are routinely installed for investigations of contaminations, including organic contaminations. We know that certain hydrocarbons dissolve these resins, sometimes liberating phenol. In fact phenol extracted this way led some investigators on a wild-goose chase for non-existing former coking facilities or coal gasification facilities.

In Germany, most piezometers by regulation are between 4" and 6" in diameter. Use of 2" or 3" piezometers is not recommended. It is widely believed that smaller diameter piezometers are not suited to chemical sampling, probably because in the past 2" piezometers in Germany used to be installed as well points only. In spite of the much higher costs involved it will need some convincing to change this often unjustified view.

In regard to the sampling itself it is the general view that pumping is better than bailing because:

1. results are easier to reproduce,

2. bailing can cause cross-contamination through the bailer itself and by disturbing surficial layers of hydrocarbons in the piezometer,

3. bailing does not allow constant measurement of electrical conductivity to determine quasi-stable conditions (normally three pipe volumes should be enough to reach this condition, but sometimes it is not),

4. pumping creates a cone of depression which tends to draw nearby pollution into the sampling point.

Pumping should only be done with submersible pumps or other pressure increasing systems, not with suction pumps. Use of suction pumps would enhance degassing of water.

There are, however, two situations where the application of a bailer is advantageous:

1. Small bailers can detect a pollution maximum at the groundwater table. This has helped in several cases to detect nearby soil pollution above the groundwater table which by pumping would have escaped detection (maximum of concentration depth profile is near the groundwater table). 2. The comparison of bailer results with results of later pumping at the same piezometer can help in the interpretation in the following manner: (a) if the results are the same the piezometer is likely in the centre of dissolved pollution; (b) if the bailer result is higher than the pump result then the piezometer is positioned close to the edge of the plume or draws less polluted water through more permeable zones; (c) if the bailer result is lower than that of the pump then the piezometer draws pollution from more contaminated areas. In this way bailing can help in determining the configuration of a plume of dissolved contaminants.

There is a difference between sampling piezometers and sampling pumping wells. Piezometers collect a sample from a relative small flow tube in the flow field, while pumping wells sample a complete capture zone. Also piezometers ideally are screened only in a part of the aquifer, whereas pumping wells usually capture all permeable layers of an aquifer.

The choice of sampling procedure depends also on local geologic and well conditions. Inflow from an aquifer into a well is governed by the pumping gradients applied, the duration of pumping, the length of the filter and the permeability distribution across the aquifer. No cookbook procedure exists. In practice, procedures and strategies need to be optimized based on the site conditions. Because of the above differences, the measured concentration values of contamination samples from neighbouring piezometers and pumping wells can be orders of magnitude apart though they are sampling the same contamination.

Finally, sampling requires careful handling of the water itself. It should be placed into the sample container without flow turbulence and in such a manner that degassing is prevented as much as possible. Ten or 20 cm^3 glass bottles with headspace and teflon septum are often used as sample containers. Air from the headspace is then analyzed and the results are back calculated for pollutant concentration in water, using partition coefficients for water/air systems (Table 2). The accuracy of this method has been found superior to others (Hagendorf et al. 1987b).

3 CONCLUSIONS

Investigation of subsurface pollution by heavy hydrocarbons is complex and difficult. There are no routine procedures.

There is no sure way to quantify contamination by heavy hydrocarbon in the subsurface. All sampling methods methodically lead to losses of the volatile components. Groundwater sampling methods come closest to reality. The volume and distribution of the free and residual phases are usually not known. Hence, time estimates for remediation by soil air suction, microbiological methods and groundwater pumping are often far too low and therefore unrealistic.

Uncertainties in sampling and heterogeneous distribution of pollutants in soil make results of analyses only suited to the determination of the order of magnitude of the pollution. For the same reason disposal criteria for unsaturated soil are often unrealistic. Attempts to determine iso-concentration lines in soil do not represent reality.

In the unsaturated zone, soil air investigations are much less expensive and faster than soil sampling. They are, therefore, increasingly substituted for soil sampling. Values obtained can only lead to statements about "more" or "less" pollution. The values returned cannot be transformed into concentrations in soil and groundwater. In principle, soil-gas sampling has inherent disadvantages similar to soil sampling. But it is less expensive and faster. It has the advantage of allowing one to come to some conclusions on pollution within the groundwater. This helps in designing a groundwater monitoring network.

Groundwater samples are necessary to show the extent of subsurface pollution. The limitations and intricacies of sampling, transporting and analyzing groundwater are best understood. Chemically, water standards allow back calculation, while uncertainties with soil samples and soil-gas samples do not.

Experiences from Hessia indicate that the main problems in groundwater sampling originate from:

1. defective installations of piezometers, and
2. improper lateral and vertical positioning of piezometer screens without due regard of geology and migration behaviour of hydrocarbons.

Proper positioning of piezometers is easier in unconsolidated rocks than in fractured or karstic rocks. Pollution plumes within groundwater systems are usually more homogenized in porous media than in fractured media or karstic rocks.

Today's knowledge and experience do not allow investigations without some involvement of trial and error procedures. More and better investigation tools and sampling procedures need to be developed and, even more important, to be applied in the field.

The total spectrum of investigative procedures needs to be considered for soil, soil gas and - the most important part of any investigation - for groundwater, to yield acceptable results. It is our practice to consider the totality of all analyses for all materials together with the results of proper geologic and hydrogeologic investigations as the proper basis for decision making.

REFERENCES

Anderson, D.C. & K.W.Brown 1981. Organic leachate effects on the permeability of clay liners. Proc. 7th Annual Res.Symp., Ft. Mitchell, KY, EPA-600/9-81-002b: 119-130.

Fernandez, F. & R.M.Quigley 1985. Hydraulic conductivity of natural clays per-meated with simple liquid hydrocarbons. Can. Geotech. J. 22: 205-214.

Grathwohl, P. 1989. Verteilung unpolarer organischer Verbindungen in der wasserungesättigten Bodenzone am Beispiel leichtflüchtiger aliphatischer Chlorkohlenwasserstoffe (Modellversuche). TGA C1: 102p.

Hagendorf, U., R.Lechber, G.Milde & M.Nerger 1987a. Erkennung, Ortung und Bewertung von Untergrundkontaminationen durch leichtflüchtige Chlorkohlenwasserstoffe. WaBoLu-Hefte 1986(7):47 p.

Hagendorf, U., R.Lechber, M.Nerger & W.Rotard 1987b. Bestimmung leichtflüchtiger chlorierter Kohlenwasserstoffe in Bodenproben. Fresenius Z. Anal. Chem. 326:33-39.

Hahn, R. 1990. Aussagekraft der Analysenergebnisse. In: P.Bock, H.Hötzl, and M.Nahold (eds.): Untergrundsanierung mittels Bodenluftabsaugung und In-Situ-Stripping. Schr.Angew.Geol. Karlsruhe 9:69-79.

Kanitz, J. 1986. Bodenluftuntersuchungen, Möglichkeiten und Grenzen. LWA - Materialien 86(2): 43-55.

Marrin, D.L. & H.B.Kerfoot 1988. Soil-gas surveying techniques. Environ. Sci. Technol. 22: 740-745.

Mesri, G. & R.E.Olsen 1971. Mechanisms controlling the permeability of clays. Clay and Clay Minerals 19: 151-158.

Michaels, A.S. & C.S.Lin 1954. Permeability of kaolinite. Industrial and Engineering Chemistry 46: 1239-1246.

Nerger, M., U.Hagendorf & R.Mergler 1988. Zur Erkennung von Kontaminationen in der ungesättigten Zone durch leichtflüchtige Chlorkohlenwasserstoffe (LCKW) mit unterschiedlichen Aufschlußtechniken. Wasser und Boden 40: 564-569.

Neumayr, V. 1983. Möglichkeiten und Grenzen zur Erfassung von Untergrundverunreinigungen durch halogenierte Kohlenwasserstoffe. DVGW-Schriftenr. Wasser 36: 35-63.

Peterson, M.S., L.W.Lion & C.A.Shoemaker 1988. Influence of vapor-phase sorption and diffusion on the fate of trichloro-ethylene in an unsaturated aquifer system. Environ. Sci. Technol. 22: 571-578.

Remmler, F. 1990. Einflüsse von Meßstellenausbau und Pumpenmaterialien auf die Beschaffenheit einer Wasserprobe. DVWK-Mitteilungen 20:152p.

Schwille, F. 1988. Fluid-mechanical aspects of the migration of chemicals in fractured media. In: Custodio,E. et al.(eds.): Groundwater flow and quality modeling: 515-537. D.Reidel Publishing Company.

Schwille, F. 1990. Oral communication to K.U.Weyer, January.

Schwille, F., W.Bertsch, R.Linke, W.Reif & S.Zauter 1984. Leichtflüchtige Chlorkohlenwasserstoffe in porösen und klüftigen Medien - Modellversuche. Bes. Mitt. Dt. Gewässerkdl. Jb. 46: 270 p.

Swallow, J.A. & P.M.Gschwend 1983. Volatilization of organic compounds from unconfined aquifers. NWWA, Proc. 3rd National Symp. Aquifer Restoration and Groundwater Monitoring: 327-333.

Toussaint, B.B. 1986. Verunreinigung des Grundwassers in Hessen durch leichtflüchtige chlorierte Kohlenwasserstoffe - CKW-Studie Hessen, Stand 31.12.1985. Umweltplanung und Umweltschutz 36: 506 p.

Toussaint, B.B. 1989a. Anforderungen an den Bau von Grundwassermeßstellen aus hydrogeologischer Sicht. Oberrhein. geol. Abh. 35: 111-128.

Toussaint, B.B. 1989b. Möglichkeiten und Grenzen der Aussagefähigkeit von Boden-, Bodenluft- und Grundwasserproben im Zusammenhang mit CKW-Schadensfällen. Dt. gewässerkdl. Mitt. 33: 150-160.

Toussaint, B.B. & W.Pütz 1987. Erfahrungen mit Eignungsprüfungen von Meßstellen zur Überwachung der Grundwasserbeschaffenheit. bbr 38: 341-343.

Toussaint, B.B. & K.U.Weyer 1992. Limitations in sampling subsurface contamination by volatile hydrocarbons in unconsolidated, fractured and karstic rocks: 2. Evaluation of three case histories. This volume.

Voorhees, K.J, J.C.Hickey & R.W.Klusman, 1984. Analysis of groundwater contamination by a new surface static trapping/mass spectrometry technique. Analytical Chemistry, 56:2602-2604.
Weyer, K.U. 1987. Contaminant hydrogeology and the role of geology and fluid dynamics in the investigation of subsurface pollutions. Proc. Intern. Conf. Vulnerability of soil and groundwater to pollutants, The Hague: TNO Committee on Hydrological Research, Proceedings and Information 38: 869-883.

Subsurface Contamination by Immiscible Fluids, Weyer (ed.) © 1992 Balkema, Rotterdam. ISBN 90 6191 175 3

Limitations of sampling subsurface contamination by volatile hydrocarbons in unconsolidated, fractured and karstic rocks: 2. Evaluation of three case histories

Benedikt B.Toussaint
Hessian Agency of Environment, Wiesbaden, Germany

K.Udo Weyer
WDA Consultants Inc., Calgary, Alb., Canada & WKC Weyer Consultants, Düsseldorf, Germany

ABSTRACT: A first paper (Toussaint & Weyer, this volume) discussed the basic principles and likely pitfalls encountered when sampling for volatile hydrocarbons in soil, soil gas and groundwater. Geologic substratum and multiphase flow dynamics determine the choice of sampling procedures and investigation tools. This paper presents three case histories illustrating some of the actual difficulties faced in practical work.

1 INTRODUCTION

A first paper (Toussaint & Weyer, this volume) discussed the basic principles and likely pitfalls encountered when sampling for volatile hydrocarbons in soil, soil gas and groundwater.

Volatile hydrocarbons are ubiquitous pollutants in soil, soil gas and groundwater of industrialized countries. Soil samples are usually taken from cores of small diameter sounding drills; groundwater is sampled from existing and newly installed wells and piezometers. Due to inherent difficulties, concentration values for volatile hydrocarbons encountered in soil, soil gas and groundwater are often grossly misrepresenting the actual field conditions.

During soil sampling substantial degassing losses from the mobile and immobile phase liquid cannot be avoided. There are no rational means to reconstruct reliably the magnitude of the losses. The results of chemical analyses of soil samples usually are not representative for the field conditions, due to the prevailing heterogeneity of the soil matrix. Values received are point values only.

Because of the above difficulties and much lower costs involved, soil gas sampling has been widely adopted as the panacea. Another advantage is seen in the ability to draw air, and thereby the trace of pollution, from a greater volume. This overcomes the point limitation of soil sampling to some degree. There are disadvantages which often invalidate the method. The sampling techniques do not follow a common procedure and therefore the results cannot be easily compared. Temperature, temperature gradients and soil moisture can play havoc with the results. Hydrocarbon contamination in low-permeability media may not release enough soil gas to be reflected in a measurement. The implicit lack of detailed hydrogeologic information often causes misinterpretation of groundwater pollution, in particular if lenses of clay are present above or in the upper part of the groundwater body.

When sampling groundwater, the problem often lies with the well and piezometer installations. Because of perceived cost savings, available wells and piezometers are often used as the first and only source of groundwater samples, although they have not been placed and constructed for these investigations. In the placement and construction of dedicated piezometers and boreholes inadequacies are common- place as well. In most of the cases the location and extent of the free phase and of the plume of polluted groundwater are not fully known. There is also a general tendency to ignore basic facts of hydrodynamics of water and free liquid phase flow, as well as general hydrogeological parameters.

This paper presents three case histories illustrating some of the actual difficulties faced in practical work.

2 EXAMPLES OF CASE HISTORIES

In 1989 about 220 cases of groundwater pollution by halogenated hydrocarbons were known to the regulatory agencies of Hessia. It is not unrealistic to estimate the total number of cases of groundwater pollution recognized in the state, but unknown to the government, to be probably three to four times as high. Most of these cases are located in and around Frankfurt. About 50 % have probably been caused by metal processing. At best 10 to 15 % of the cases have been investigated to some degree. Not one has been satisfactorily cleaned up so far.

Three examples will be discussed below. All have been investigated to some detail and remedial action has been started. Usually there is a time span of 2 to 5 years between the start of an investigation and the beginning of the remedial action. The examples selected are situated in unconsolidated rocks at the Frankfurt Airport, in consolidated, fractured rocks at the City of Dreieich between Frankfurt and Darmstadt, and in limestone karst at the City of Hadamar north of Limburg (Fig. 1). Details of the three cases are summarized in Tables 1 to 3. Some data for two of these sites have previously been published by Toussaint (1986).

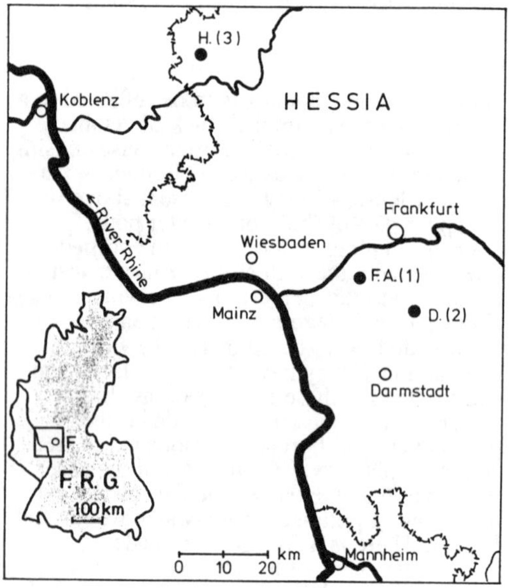

Figure 1. Location of case histories discussed in detail.

2.1 Unconsolidated rocks: Frankfurt Airport

History and geology

This case of contamination at the Frankfurt Airport was the first case recognized as pollution by halogenated hydrocarbons in Hessia. It is now one of five pollution cases at the airport alone. Contamination is mainly by tetrachloroethene and also by 1.1.1-trichloroethane and trichloroethene. Cis 1.2-dichloroethene occurs as a intermediate product of biodegradation especially in the front part of the plume (Toussaint 1986).

Table 1. Synoptic summary of subsurface pollution by halogenated hydrocarbons in unconsolidated rocks at the Frankfurt Airport.

CASE	Unconsolidated rocks Frankfurt Airport – major airline
Sources of contamination	sewage canal in airplane repair facility, car repair facility, storage of barrels with solvents, underground tank with solvents, test facility for airplane engines
Type of halogenated hydrocarbons	tetrachloroethane, trichloroethene, cis 1,2–dichloroethene, 1,1,1–trichloroethane
Start of leakage	possibly since 1936
Geology	two aquifers: upper: 35–50m coarse sand and gravel (Quaternary); lower: 110–140m fine and medium sand (Tertiary); dividing layer: 5m silt and clay with windows
Depth to groundwater	10 – 15 m
Permiabilty	upper aquifer: 0.001–0.002; lower aquifer: 0.0004–0.00006 (from pump tests at water wells; in m/s)
Groundwater velocity	0.5 – 1 m/d in upper aquifer
Dispersion	ratio long/transv: 5–10/1
Type of investigation	soil gas: 495 samples groundwater: 43 piezometers installed
Degree of polution	soil: not measured soil gas: max. 3.7 g/m³ groundwater: max. 80 mg/l
Spread of polution	
liquid phase	not known
plume	upper aquifer: length 3.5 km, width 0.8 km, thickness 30m lower aquifer: not known
Remediation	infiltration of cleaned water of River Main since 1981 to divert plume; withdrawal and cleaning of groundwater in steps since 1983; suction and clean up of the soil gas in steps since 1989

The contamination was recognized in 1977 when organic compounds were found in drinking water from wells of the Hinkelstein gallery of the City of Frankfurt, located about 500 m from the airport. Sources were the sewer discharge from the hangar of a major airline, a car repair shop, a storage area for barrels with solvents, an underground storage tank for solvents, careless handling of solvents in the parking area for airplanes, and, in particular and most important, a testing facility for airplane engines. Estimates of the amount of these liquids entering the soil have been attempted but did not lead to any satisfactory results. The airport opened in 1936 and first contamination may have started shortly thereafter. The bulk of contamination probably occurred after the war. The amount of contamination could well exceed 100 tons of hydrocarbons.

The subsurface consists of 30 to 50 m of coarse sand and gravel underlain by 110 to 140 m of fine to medium sand. Between the two aquifers, a dividing layer of silt and clay exists

Table 2. Synoptic summary of subsurface pollution by halogenated hydrocarbons in fractured consolidated rocks at the City of Dreieich.

CASE	Consolidated, fractured rocks City of Dreieich plant for medical equipment
Sources of contamination	barrel storage area for solvents, solvent recycling process
Type of halogenated hydrocarbons	tetrachloroethene, trichloroethene
Start of leakage	possibly since 1965
Geology	sequence os sandstone, siltstone, conglomerate and thin, non-karstic limestone layers (Permian) with intrusions of diorite and basalt; located on the edge of the Upper Rhine Graben
Depth to groundwater	3 – 5 m
Permiabilty	not known
Groundwater velocity	not known
Dispersion	mainly longitudinal
Type of investigation	soil: 1 cored borehole soil gas: 10 samples groundwater: 5 piezometers installed
Degree of polution	soil: max. 2.6 g/kg soil gas: max. 5.7 g/m³ groundwater: max. 41 mg/l
Spread of polution	
liquid phase	not known
plume	not known
Remediation	withdrawal and cleaning of groundwater since 1988; suction and clean up of soil gas since 1988

in places. The lower aquifer is underlain by more than 100 m of silt and clay. The groundwater flows in NW direction (Fig. 2). Its table is at 10 to 15 m depth; permeability of the upper aquifer is between 10^{-3} and 10^{-4} m/s. The hydraulic conductivity of the lower aquifer is probably between 10^{-4} and 10^{-5} m/s.

Investigation

In 1978 groundwater and soil gas investigations were started. Groundwater investigation was first done by using about 25 existing wells and piezometers. Construction details were often not known. From 1978 to 1989 an additional 43 piezometers (4" to 6" diameter) dedicated for sampling were installed step by step with the goal to determine the location of the front and the shape of the pollution plume. In 1988 and 1989, 25 additional piezometers of 2" diameter were installed near four remediation wells. These piezometers are for hydraulic observation only. As pointed out in Toussaint & Weyer (this volume) German regulations recommend against chemical sampling from 2" piezometers.

Table 3. Synoptic summary of subsurface pollution by halogenated hydrocarbons in karstic rocks at the City of Hadamar.

CASE	Karstic rocks City of Hadamar – textile plant
Sources of contamination	cleaning and distillation facilities, leaks in pipes, barrel storage area for solvents
Type of halogenated hydrocarbons	tetrachloroethene
Start of leakage	possibly since 1966
Geology	karstic limestone and diabase tuff (Devonian), overlain by clay (Tertiary), gravel and loess (Quaternary)
Depth to groundwater	upper aquifer in gravel: 7 – 8 m main aquifer in limestone: appr.30m
Permiabilty	not known
Groundwater velocity	not known
Dispersion	longitudinal
Type of investigation	soil: 8 cored boreholes in loess soil gas: 30 samples in loess groundwater: 1 piezometer in limestone
Degree of polution	soil: max. 2.9 g/kg soil gas: max. 2.9 g/m³ groundwater: max. 0.6 mg/l
Spread of polution	
liquid phase	not known
plume	length at least 0.5 km (exact extent not known)
Remediation	cleaning of drinking water since 1986; suction and cleanup of soil gas in 1986 and 1987

Within the plume, dissolved concentrations close to the source were originally >80 mg/l. Today it is about 5 mg/l because of active remediation. Along the centre line of today's plume concentration values range from about 1 mg/l about 1 km downstream of the source to about 0.1 mg/l close to the front.

By now the plume (concentration values >25 μg/l) is about 3.5 km long and 850 m wide within the upper aquifer (Fig.2). Some pollution in the μg/l range has reached the lower aquifer. Samples taken in December 1989 suggest that the plume in the upper aquifer may have migrated under the nearby River Main to the other side. This would indicate that here the River Main may not function as a hydraulic barrier and recipient of groundwater flow.

In its course the plume also passed through a field of six industrial wells, five of which were completed in the lower aquifer. The groundwater supply well-gallery Hattersheim of the City of Frankfurt is located about 1.5 km from the opposite bank of the river.

Some 495 samples of soil gas were taken over the length of the investigation in blocks in May 1978 (about 55 samples), in 1986 (about 150 samples), in 1987 (about 240 samples), in 1988 (25 samples) and in 1989 (25 samples). Sampling and analyses were often done by different

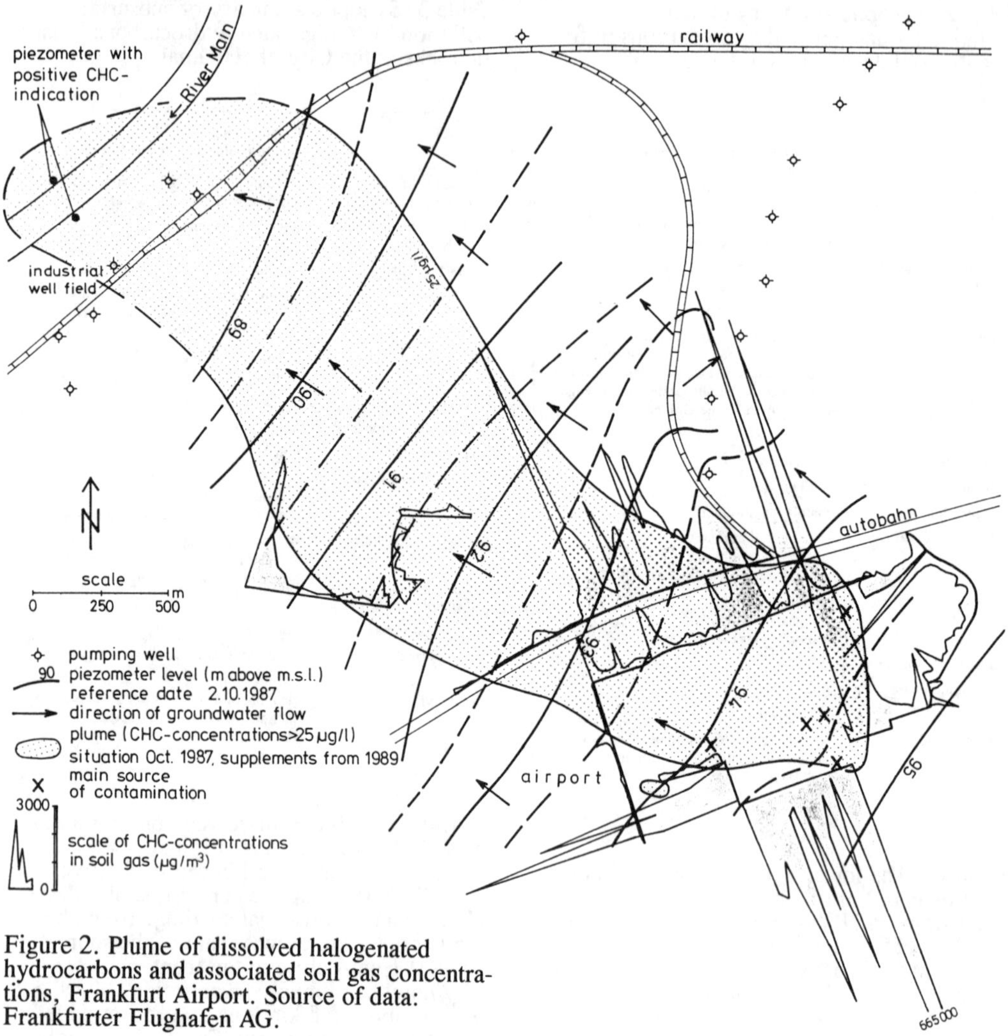

Figure 2. Plume of dissolved halogenated hydrocarbons and associated soil gas concentrations, Frankfurt Airport. Source of data: Frankfurter Flughafen AG.

consultants and laboratories at differing sampling sites. Accordingly, results for the different sampling events cannot meaningfully be compared.

Soil gas sampling was intended to outline the areas of source pollution in soil and the plumes of contamination in groundwater. Concentration values in soil gas were between 3.7 g/m^3 near the source areas to some mg/m^3 over the polluted groundwater about 500 m downstream of the sources. Based on soil gas values the plume of pollution in the groundwater near the source areas consists of five arms originating at differing sources of pollution. The downstream migration path of these arms was not investigated in detail. Instead the perceived outer fringe of the zone of pollution was targeted.

The distribution of soil gas concentrations along sampling lines disagrees somewhat with the interpretation of groundwater data. Soil gas values are documented in the small shaded curve areas in Figure 2. Immediately apparent are disagreements just south of the source area and west of the plume. The reasons for the peaks are not known. It appears, however, that additional unknown sources of pollution may exist south of the area discussed. The lack of high values of groundwater pollution in this area may be caused by clay layers in the vadose zone protecting the groundwater from contamination from above.

Sampling strategy and sampling methods applied so far do not allow to understand the distribution and spreading of pollution adequately.

Remediation

Of immediate concern in 1978 was the protection of existing water supply wells about 300 to 500 m north of the pollution sources. The southernmost of the wells was abandoned in 1978. By 1981 five water infiltration wells were installed to the south and southwest of the line of water supply wells to hydraulically divert the moving plume of pollutants to the west. The operation was successful.

Remediation of the source areas has been undertaken since 1983 with groundwater pumping (about 10 m^3/h) in one well situated under the source area. This well evacuated 2.2 tons of hydrocarbons between April 1983 and December 1989.

Since 1988 three additional wells were operated in the lower half of the plume where higher contamination values were encountered by chance. They produced 0.64 tons of hydrocarbons between July 1988 and June 1989. Late in 1990 three further wells will start pumping in a line across the plume at the northern boundary of the airport.

In-situ soil decontamination by air stripping was started in August 1989 with 4 suction wells distributed over two of five pollution centres within the unsaturated zone. The screens of the suction wells were positioned in the centre of soil gas pollution, three at 9 m depth and one at 6 m depth. The amount of hydrocarbon extracted was about 8 kg by the end of January 1990.

So far, the costs of investigation and remediation are estimated to exceed DM 50 million (about US$ 30 million). The total amount of hydrocarbons extracted by remediation has been less than 4 tons.

Comments

It appears that the investigation was not done with as much vigor as one may have expected. No investigation has been done to illuminate the fate of the liquid phase and, thus, no remedial action has been started in this regard. A systematic removal of soil with the immobile liquid phase under the source areas was not attempted. The available concentration data do not allow to draw a detailed map of concentration distribution with confidence, neither for the past nor for today's conditions. Even today the exact configuration and extent of the plume of contamination is not known.

A general strategy for the groundwater investigation is still missing. Until now the approach has been piecemeal and at no time was there an educated attempt made to understand all the implications of the case. Also remedial measures did not go to the core of the problem. At the source, clean-up efforts have been insufficient and were started too late. The same applies for the plume. At one time it was thought that injections at the water well site would also reduce the pollutant's concentration by mixing. Since January 1990 it is known that a number of Frankfurt water supply wells at the Hattersheim gallery north of the River Main may be threatened by the moving plume. The hydrodynamics of groundwater flow in the area are not understood adequately.

2.2 Fractured, consolidated rocks: City of Dreieich

History and geology

A precautionary investigation of a medical needle factory at the City of Dreieich by a regulatory agency in 1986 indicated that contamination by chlorinated hydrocarbons could have occurred in two areas. The courtyard held a storage area with barrels containing used and distilled solvents, without any protection against seepage into the subsurface. In the basement, the solvent distillation and recycling plant had a capture basin for leakage made of concrete only, without a protective coating. A first investigation of the soil gas under the two suspected source areas by the regulatory agency revealed pollution values of up to 5.7 g/m^3. A detailed investigation was then ordered by the agency.

Estimates of the amount of solvent leakage into the ground are difficult because of the unknown, but certainly significant degree of volatilization. After refit of the distillation and recycling plants, the need for replenishment of hydrocarbons was reduced dramatically to less than 20%. From that one may conclude that tons of hydrocarbons must have seeped into the subsurface.

Built in 1965,the factory is located on soil above fractured red beds consisting of claystone, siltstone, sandstone, conglomerates, marl and non-karstic limestone (Fig. 3). Intrusions by basalt and diorite are present close by. The upper 4 - 5 m of rocks are weathered.

Hydraulic conductivities for the different rocks are not known at the site. The pattern of jointing and faulting also has not been investigated. The nearby positioned edge of the Rhine valley rift suggests occurrence of SW-NE faulting parallel to the edge of the graben at that location. The slope of the groundwater table is to the NW. Over fractured rocks it occurs at 3 to 4 m depth, at the edge of the graben, in unconsolidated rocks and it is said to drop to 17 m. The investigating engineering consultant concluded, from short-time pumping in holes 1 and 2 in the red beds and 5, 6 and 7 in the basalt, that permeability of the red beds was very low and that of the basalt much higher. At

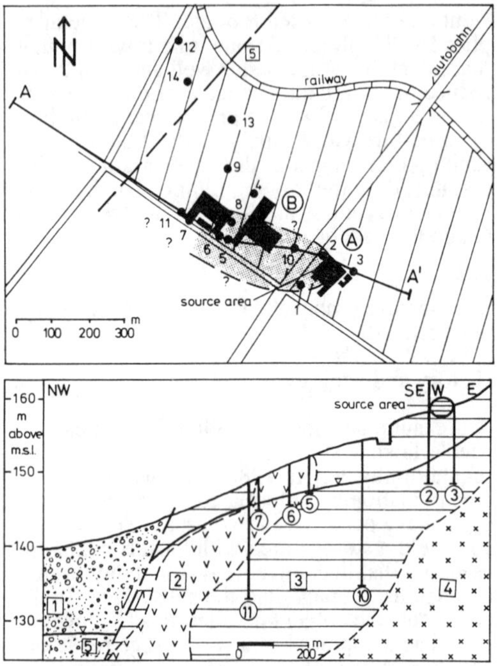

Figure 3. Migration of dissolved halogenated hydrocarbons in fractured consolidated rocks, City of Dreieich. Upper part: birds eye view; lower part: cross-section.

a pumping rate of 5 l/min in red beds the hole was dry in 15 minutes. In basalt a pumping rate of 30 - 40 l/min could be sustained for several hours. Permeability values were not given.

Investigations

In spring 1987 holes 1, 2, and 3 were drilled for the investigation and instrumented with piezometers. Holes 4 to 8 were drilled independently in 1988 by another consultant for a different client during investigation before purchase of a property. They were abandoned and sealed after one time sampling of groundwater in open holes. Holes 10, 11, 13, and 14 were drilled in 1989 for the factory by the previous consultant. Holes 9 and 12 are wells existing since decades. Only some technical data are known for these holes.

Contamination in groundwater was found to be between not detectable and 41 mg/l (Table 4)

The spatial distribution of the values suggests a complex pattern of migration. Considering the components of the contamination at the different boreholes, the picture becomes even more confusing. Just down-gradient from the

Table 4. Groundwater contamination

wells	contamination [μg/l]
1, 5	100 - 200
2	41
3	50 - 70
4, 9,10, 11	< 10
6, 7, 8	$\approx$ 50
12, 14	not detected
13	300

needle factory holes 2 and 3 show predominance of tetrachloroethene over trichloroethene in a ratio of 80 to 20. Hole 3, technically just groundwater upgradient from the factory, shows predominance of trichloroethene over tetrachloroethene in the same ratio. The factory claims to use no trichloroethene at all.

Most of the holes show a slight predominance of tetrachloroethene over trichloroethene. Hole 13 has a very strong predominance of 1.1.1-trichloroethane. The source of this substance may be an abandoned landfill nearby.

A line of soil gas sampling points was put across the suspected contamination plume down-gradient from the factory with 10 boreholes sampled in 1987. The maximum contaminant content of about 1 mg/m^3 was unexpectedly found at the southwestern end of the sampling line with tetrachloroethene dominating. The concentration in all other holes was only the background value for soil air of 200 μg/m^3. Hence the results of this investigation increased the confusion.

Shallow soil sampling at the barrel storage area of the needle factory in 1987 showed tetrachloroethene contaminations of up to 2.6 g/kg.

In 1988 a third source area was located by soil gas sampling. The pollution there was caused by collection of heavy vapours at the deepest point of the basement and subsequent seepage of heavy volatiles through the concrete floor into the soil by gravity drainage.

In 1989 two soil gas samples were taken from new holes in a rubber factory nearby (B in Figure 3). Only background values were found. In spite of these results further investigations are planned because of suspicions that sampling procedures applied may not have been adequate.

Remediation

Starting in March 1988 vapour extraction methods have been applied for clean-up of the unsaturated zone under the three source areas. By December of 1989, 335 kg hydrocarbons had been removed.

Remediation of groundwater started in May 1988 with intermittent pumping in holes 1 and 2 at a rate of less than 10 l/min. Much of the time the holes are dry. So far about 2 kg hydrocarbons have been removed.

Total costs were DM 180,000 (about US$ 100,000) so far.

Comments

The interpretation of the data suffers from a lack of detailed investigations. For example, no observation holes were installed in the upper 4 m of weathered rock close to the site of pollution. There was no dedicated attempt made to find the liquid phase. There is a lack of geologic investigations in regard to jointing and faulting patterns. The hydraulics of the groundwater flow system(s) are not clearly known.

It was assumed by the consultant that liquid phase and dissolved contamination did not travel significantly into the red beds. In our opinion there is no real reason for this assumption. We feel that the few boreholes installed in all likelihood missed highly permeable vertical fractures. In these fractures, liquid phase and dissolved plumes could spread unnoticed. The same could happen in the zone of weathering. Holes 10 and 11, low in contamination, may have missed transport pathways because the screens may be positioned too deep and below these pathways. Furthermore, no boreholes were installed in the source areas.

2.3 Karstic rocks: City of Hadamar

History and Geology

When drinking-water standards were changed, an investigation of the City of Hadamar water supply (Fig. 4, hole 4) in 1983, showed a small content (<25 μg/l) of tetrachloroethene. In a nearby spring the content of hydrocarbons was found to be higher (up to 40 μg/l). Further investigations revealed that a textile company 200 m away was the source. The company operated from 1966 to 1986.

The sources of pollution were: cleaning and distillation facilities; leakage from solvent pipelines; a barrel storage area; and an area where textiles were softened with tetrachloroethene for welding. At the main point of pollution the liquid phase had moved into a cable duct and from there directly into the soil. Detailed estimates of the total amount of leakage where attempted by the Solicitor General without success. From general deductions the losses can be estimated to be between 50 and 150 tons of liquid.

The plant is located in hilly terrain on 5 to 8 m of loess above 2 to 3 m of gravel, with clay

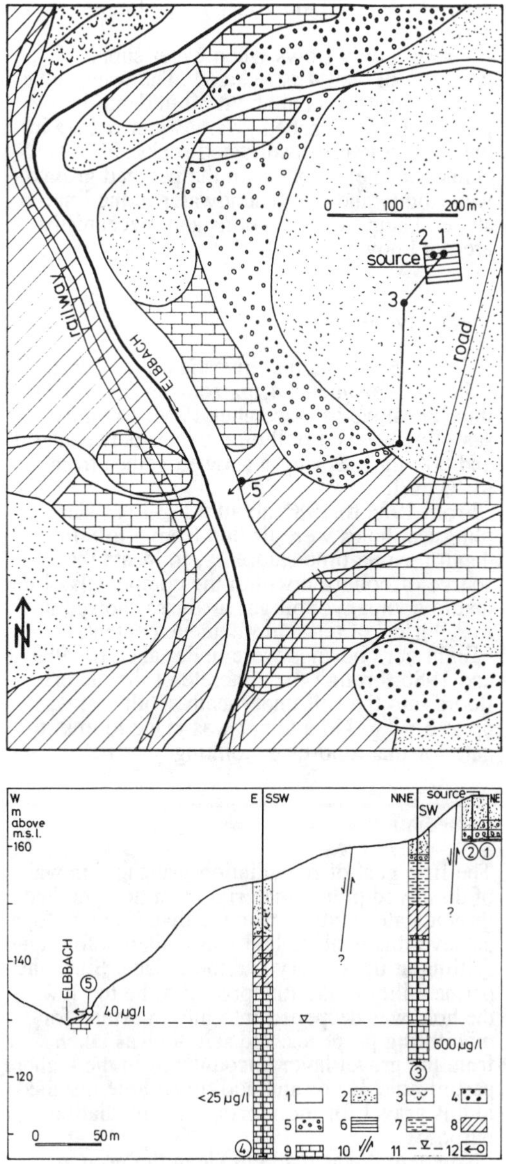

Figure 4. Contamination by halogenated hydrocarbons in karstic rocks, City of Hadamar. Upper part: geological map; lower part: cross-section.

underneath (Fig. 4, holes 1, 2). Clay was not penetrated by drilling to prevent leakage of liquid phase into the deeper subsurface. Karstic limestone is, however, known to exist underneath. The lithologic profile in the nearby hole 3 consists of 7.5 m of silt and 6 m of diabase tuff over limestone. Based on the occurrence of a river terrace at different elevations the existence of vertical fault zones in the limestone is certain.

Spring 5 only flows when water supply pumping in well 4 is reduced. The water table in the limestone is at about 30 m depth in piezometer 3. Perched groundwater occurs in the overlying unconsolidated rocks. Flow connections exist between the perched groundwater body and the groundwater in the limestone. Their exact locations and mechanisms are not known.

Investigation

Following the original investigation of groundwater in piezometer 4, piezometer 3 was drilled into limestone in December 1984. Monthly sampling showed pollution values between 200 and 600 μg/l. Spring 5 showed values from 20 to 40 μg/l.

From 1984 to 1986 about 30 soil gas sampling test holes were drilled from 5 to 8 m depth for one-time sampling, to locate the source of pollution within the compounds of the textile company. Soil gas analyses showed up to 2.9 g/m^3 of tetrachloroethene content.

Eight core borings were undertaken in 1985 and 1986 through the concrete of the factory floors into the soil underneath. Pollution was up to 3 g/kg. No attempt was made to investigate for macropores or jointing.

Remediation

The first goal of remediation was the removal of dissolved plume material from the perched groundwater body within the diabase tuff. To achieve this goal, hole 3 was drilled before any drilling at the factory site took place. Since the permeability in the tuff proved to be too low, the hole was deepened into the limestone for monitoring purposes. No sample was taken from the gravel layer encountered in the higher part of hole 3, nor any additional hole installed in this gravel for monitoring or remediation purposes.

No attempt was made to clean the groundwater near the factory, neither in the perched groundwater body nor in the limestone. Instead, air stripping has been applied since May 1986 to the water pumped from the water supply well 4 (pollution <25 μg/l). No record is kept of the amount of hydrocarbon extracted there.

Vacuum extraction by six wells was undertaken from August 1986 to July 1987. By August 1987 the soil gas values were low. The regulatory agencies therefore allowed the vacuum extraction to be permanently discontinued. About 70 kg of tetrachloroethene had been extracted. Originally, excavation of the polluted soil was ordered by the regulatory agency. Excavation of the estimated volume of 4800 m^3 of polluted material would have cost about 4.5 million DM (US$ 2.7 million). Bankruptcy lawyers, however, successfully argued in court that excavation should be replaced by the much less expensive vacuum extraction.

Total costs for investigation were DM 100,000 (US$ 60,000). Because it could not be clearly shown that the groundwater pollution was caused by the company, 50 % of the costs were carried by the state of Hessia. For the same reason the City of Hadamar carries the total cost for clean up of the water supply. By 1990 these costs exceeded DM 100,000 (US$ 60,000).

Comments

In the eyes of the regulatory agency this case has been properly investigated and cleaned up. It is clear, however, that the flow path and fate of the liquid phase are unknown. The observed reduction of pollution in continuous suction of soil air is normal and does not mean the soil has been cleaned. Instead, intermittent pumping would be required to extract additional volatiles migrating from the fine into the coarser pores.

The extent and even location of the plume of contamination within the groundwater is still not known. None of the known geophysical methods for locating water-filled karstic systems have been applied, nor any tracer test to find other exits from the groundwater system than the known spring 5 and the pumped borehole 4.

3 CONCLUSIONS

Geologic drilling is the most important tool to understand the hydrogeologic framework and to discover the position of deeply penetrating liquid phase. All our experiences suggest that leadership by an experienced geologist is necessary for successful investigation and subsequent remediation. Proper process engineering is important for the technical side of the clean-up.

The costs for discovery of the liquid phase are high because hit-or-miss drilling is involved. Geophysical methods can be of help. Removal of the liquid phase of organic compounds is the basis for any clean-up. To remediate plumes of dissolved organic compounds is also very expensive and time consuming. Time requirements for pump-and-treat technology can reach 1000 years and more due to residual concentrations of liquid phase (Hall 1989, Schmelling this volume).

Today German law protects the water, not the soil. The soil, in turn, needs to be cleaned to prevent further pollution of groundwater. Groundwater itself is protected totally. This means no pollution at all is allowed by law. In reality, nobody tries to enforce this law to the letter (see above: case 3). In practice, the less stringent rules for drinking water are applied, allowing pollution by heavy hydrocarbons of up to 25 μg/l. A system of guide values is also used. By law, any polluter has the right to remediate with the most cost-effective methods he can convince the court to accept. In the past it was often believed that both vacuum extraction and pumping of polluted groundwater lead to fast clean-up. Experience has shown that neither is the case.

Proper investigations form the necessary basis for any realistic remediation. They are also a necessary condition for cost reduction if serious remediation efforts are intended. Remediation without timely and exhaustive preceding investigations increase costs by orders of magnitude. To date this does not seem to be generally apparent to polluters and their insurance companies. They often do not see the seriousness of the contamination and tend to try to minimize initial costs. Regulatory agencies sometimes also select this approach.

From the above statements one may conclude that the investigative end of things may be particularly bad in Hessia or Germany. Weyer (1987) showed these conditions are not unique to Hessia or Germany and occur in other jurisdictions.

REFERENCES

Hall, C. 1989. Practical limits to pump and treat technology for aquifer remediation. Conference on Prevention & Treatment of Groundwater & Soil Contamination in Petroleum Exploration & Production, May 9-11, 1989, Calgary, Alberta, Canada, Proceedings: 29.0-29.7.

Schmelling, S., 1992. Technical consideration for pump and treat in groundwater remediation. This volume.

Toussaint, B.B. 1986. Verunreinigung des Grundwassers in Hessen durch leichtflüchtige chlorierte Kohlenwasserstoffe. Stand 31.12.1985. Umweltplanung und Umweltschutz 36:506 p., Wiesbaden.

Toussaint, B.B. & K.U.Weyer 1992. Limitations in sampling subsurface contamination by volatile hydrocarbons in unconsolidated, fractured and karstic rocks: 1. Practical experiences from 220 cases. This volume.

Weyer, K.U. 1987. Contaminant hydrogeology and the role of geology and fluid dynamics in the investigation of subsurface pollutions. Proc. Intern. Conf. Vulnerability of soil and groundwater to pollutants, The Hague: TNO Committee on Hydrological Research, Proceedings and Information 38: 869-883.

Subsurface Contamination by Immiscible Fluids, Weyer (ed.) © 1992 Balkema, Rotterdam. ISBN 90 6191 175 3

A reconnaissance method for determining the nature and extent of deeply penetrating DNAPLs

Ken W.Campbell & David B.Holmes
Hydro-Search, Inc., Milwaukee, Wis., USA
Joe R.Storrs
Avganic Industries, Cottage Grove, Wis., USA

ABSTRACT: This paper describes a cost effective screening method for determining the nature and extent of groundwater contamination at the site of a release of chlorinated and non-chlorinated organic compounds.

The "hydrogeological profiling" procedure consists of a carefully packaged combination of methodologies including rock coring, packer installation, in-situ hydraulic conductivity testing, and groundwater sampling. The method provides a nearly continuous "profile" of rock core, hydraulic conductivity and groundwater chemistry. The methodology was utilized at a solvent reclaiming facility in the U. S. Midwest where up to 150,000 litres of organic solvents may have been released from leaking drums over a period of several years. A total of 26 packer tests were performed in three boreholes at depths ranging from 23 to 85-m below ground surface. A zone of dense non-aqueous phase liquids (DNAPL) accumulation was identified in one of the boreholes at a depth of 41 to 50-m below ground surface. Through careful use of temporary and permanent casing strings, it was possible to drill through the base of the DNAPL zone, continue profiling, and install a monitor well in a non-impacted formation directly below the DNAPL zone. An analysis of a groundwater sample subsequently collected from the monitor well reported non-detectable concentrations for all volatile organic compounds (VOCs) tested.

Although the methodology is intended as a screening tool, the data obtained have allowed distinct organic plumes to be identified in the bedrock as well as variations in the percentages of compounds present.

Refinements to the methodology and careful execution of the "profiling" program should make this methodology applicable to a wide variety of geological settings.

1 THE PROBLEM

The mechanisms that control the subsurface migration of soluble organic and inorganic compounds and to some extent "light non-aqueous phase" compounds are relatively well understood. The migration, or potential migration, of these compounds are predictable to the extent that earth scientists with an insight into the nature of the compounds can execute field programs using convention investigative techniques to evaluate the nature and extent of subsurface impacts.

The hydrogeological understanding of the mechanisms that influence the migration of DNAPLs in the subsurface are less well understood by the hydrogeological community. Although multiphase flow in porous media has been intensively studied by petroleum engineers and geologists for many years, a resurgence in scientific interest in the behavior or DNAPLs was stimulated by an increasing number of "environmental" incidents in which groundwater had become contaminated by organic chemicals. The state of our understanding of DNAPL migration in 1990 is perhaps most clearly summarized by Friedreich Schwille (1988) who states that the forces dominating the migration of DNAPL are gravitational in nature rather than hydraulic. Simply put, DNAPLs will sink through a water-saturated medium until the body no longer holds together as a single related continuous phase, but rather lies in isolated residual globules in the so-called condition of residual saturation.

As is apparent from Dr. Schwille's experiments, the migration pathways of the DNAPLs are dominated by gravitational forces and modified by what may appear to be subtle changes in the magnitude and nature of the hydraulic conductivity. The result is a migration pathway that may appear both tortuous and difficult to rationalize. From a practical standpoint, in instances where the volume of released DNAPL is small, the geological setting is relatively simple (2 or 3 layers) and the water table is high (<10-m), the methodology that is used to determine the nature and extent of impacts in the subsurface may not be dramatically different from a program designed to evaluate soluble organics or inorganics. The situation changes, however, in scenarios where the geological setting is complex, the volume of the DNAPL release is high, and a significant period of time has elapsed since the release occurred.

In such complex and potentially high risk settings, the cost of conducting a hydrogeological investigation can be extremely expensive to execute with the inherent safety and quality assurance/quality control (QA/QC) protocols. Nonetheless, in order to develop an effective remedial program, the nature and extent of the potential impacts must be evaluated in sufficient detail to allow for the design of the remedial measures.

2 THE PROCEDURE

The methodology discussed in this paper was developed and applied at a Resource Conservation and Recovery Act (RCRA) permitted solvent reclaiming facility in the U. S. Midwest where up to 150,000 litres of organic solvents may have been released from leaking drums over a period of several years. The contaminants released include at least 12 high density chlorinated VOCs and 8 low density VOCs including aromatic hydrocarbons, alcohols, and ketones. Six phases of investigation using conventional investigation techniques had been performed at the site since 1982 to define soil and groundwater impacts. Based on these investigations, it was believed that vertical and horizontal migration of the DNAPL was limited. Vertical migration of the DNAPL beneath the facility was believed to be restricted by the bedrock surface at a depth of approximately 25-m. Although soluble organic impacts were measurable several hundred metres downgradient from the site, these soluble impacts were viewed as being manageable. Field work performed in July, 1988 resulted in the discovery of DNAPL contaminants in monitor well P38 (Figure 1) at a depth of 55-m below ground level at a distance some 100-m from the facility boundary.

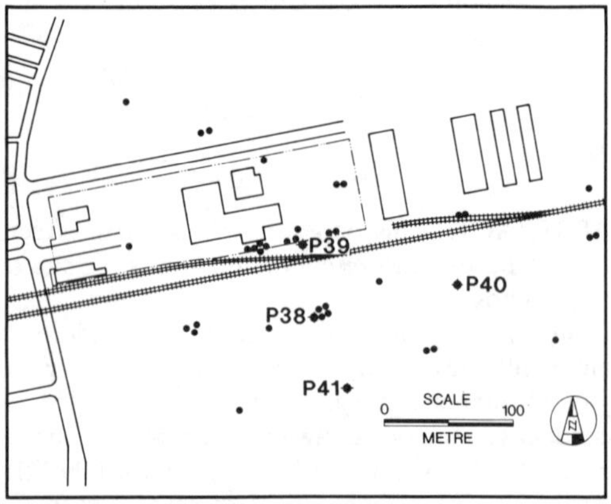

Figure 1. Site Location

With the discovery of the DNAPL in the P38 monitor well, it became apparent that the distribution of the DNAPL had not been adequately determined and that significant additional investigatory work was necessary before final remedial measures could be selected, designed, and installed.

The overall purpose of this investigation was to provide additional subsurface data needed to "focus" the scope of a RCRA Facility Investigation (RFI) work plan that was being prepared under a RCRA Corrective Action Plan.

The specific objectives of this investigation were to:

1. further define the nature and extent of DNAPL impacts at the site,
2. further define the bedrock stratigraphy and structure, and
3. characterize the hydraulic properties and assess groundwater impacts in previously untested bedrock units.

In order to generate the maximum understanding of the impacts and hydrogeological setting of the deeper bedrock units, a method was developed that provides a nearly continuous profile of the stratigraphy, hydraulic conductivity, and groundwater chemistry.

The scope of the investigation consisted of packer

testing of three deep boreholes (P39, P40, P41) ranging in depth from 69-m to 85-m. Borehole P39 was located on site in the area of highest contaminant concentrations in the unconsolidated deposits. Borehole P40 was located 100-m southeast of the site in the apparent migration direction for impacted groundwater. Borehole P41 was located approximately 122-m south of the site, in alignment with previous locations in which DNAPLs had been found. The work performed at each borehole location included collection of continuous bedrock core samples and at selected intervals, purging and collection of groundwater and DNAPL samples, and performance of hydraulic tests.

2.1 Coring

A total of 159-m of bedrock was cored as part of this investigation. Bedrock coring was performed using a Mobil B-57 drill rig equipped with a Longyear HQ wire-line system with a 3-m long core barrel. The diamond coring bit produced a 96 mm diameter borehole and 63.5 mm diameter core.

During coring, the approximate volumes of drilling fluid lost to the borehole were estimated and recorded. Photoionization detector (PID) measurements of rock core were also recorded.

Temporary and permanent 254, 152, and 102-mm diameter steel casings were installed in the boreholes during drilling to seal off multiple impacted, highly transmissive, or incompetent zones present in each borehole (Figure 2). Permanent 254-mm diameter steel casings were installed in all three boreholes prior to the start of packer testing to seal off the incompetent sand and gravel surficial deposits and upper 1.5 to 3.0-m of bedrock.

The permanent casings were cemented in place by pumping a 15:1 cement (Portland Type I)/ bentonite (Aquagel Goldseal) grout through a tremie pipe into the annular space between the casing and the borehole wall.

The temporary 102-mm diameter casing was fitted with a reaming shoe, which allowed it to be advanced by rotating the 102-mm casing and reaming out a thin rim of rock at the edge of the 96-mm diameter core hole.

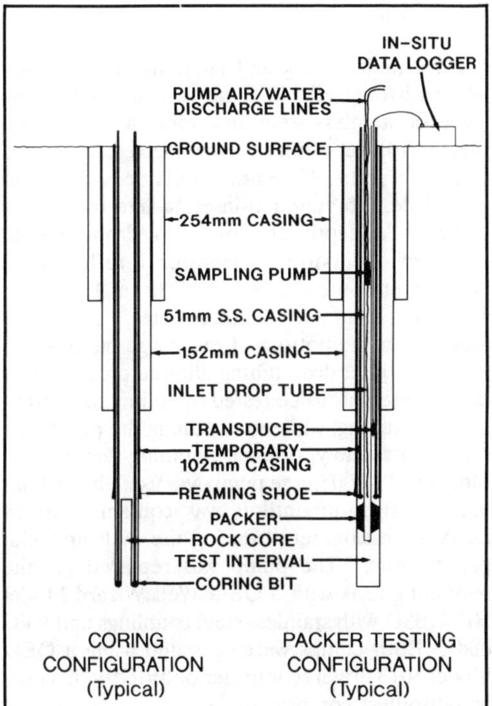

Figure 2. Testing Configuration

2.2 Packer String

After removal of the core from a test interval, the drilling rod was raised to expose the borehole walls within the test interval. An inflatable packer was attached by a stainless steel nipple to 3.0-m sections of stainless steel pipe which were inserted and lowered through the center of the drill rod to the precise top of the planned test interval. The pipe was 508-mm inner diameter, 314 stainless steel with flush threaded joints and the packer was a Tigre Tierra Model 34B-60-1.0 with stainless steel heads and center tube. The packer had a maximum inflated diameter of 12.7 mm and an uninflated diameter of 54.1 mm. The center tube had a 25.4-mm inner diameter.

After lowering the packer to the desired depth, the packer inflation line was attached to a pressure regulator on a nitrogen tank and the packer inflated to a pressure of between 1,380 to 1,517 kPa. Test intervals were typically approximately 6-m long.

2.3 *Hydraulic Testing*

For hydraulic testing and sampling, a QED Well Wizard Model HR-4200 gas displacement pump made of stainless steel and teflon and equipped with a drop tube inlet was connected to bonded 19-mm and 12.7-mm outside diameter polyethylene tubing and inserted into the packer string to a depth of 17-m. A drop tube of 12.7-mm outside diameter teflon-lined polyethylene tubing, attached to a 1-m long, 12.7-mm diameter stainless steel pipe was attached to the bottom of the purge pump. The use of the inlet drop tubing allowed progressively deeper intervals to be tested by adding sections to the drop tubing, while maintaining the pump at a relatively shallow, optimal pumping depth. The Model HR-4100 purge pump was used throughout the investigation until it was contaminated by DNAPLs encountered during the 20th and 21st packer tests. The pump was replaced for the remaining tests with a QED Well Wizard Model HR-4105D with stainless steel couplings and PVC body. Both pumps were operated using a QED Model 3013 driver/controller outfitted with a high output oilless compressor.

A 345 kPa pressure transducer was inserted into the stainless steel riser pipe at a depth of approximately 0.5-m above the top of the purge pump. The transducer was connected to an In Situ SE 1000B data logger. The data logger was used to record water level changes in the test interval during the purging interval and during the recovery period following purging. The recovery data were used to calculate the hydraulic conductivity of the test intervals using the methods of Cooper and Jacob, 1946, and Cooper et. al, 1967.

2.4 *Groundwater and DNAPL Sampling*

As the test interval was purged, field parameters including pH, specific conductance, temperature, and photoionizable constituent concentrations were measured in the purged water. Photoionizable constituents were measured using an HNu Model PI-101 PID equipped with an 11.7 eV probe.

At a minimum, three test interval volumes were removed during purging. In addition, an attempt was made to remove an additional volume equal to the estimated volume of drilling fluid lost during coring of the test interval.

Groundwater samples collected during testing were initially analyzed for VOCs immediately after collection by the site laboratory using a modified version of SW846 Method 3810 (United States Environmental Protection Agency, 1986). SW846 Method 3810 is a static headspace technique for extracting VOCs from samples. The method was modified by equilibrating the samples at 80 °C instead of 90 °C, and withdrawing 1 ml of headspace gas instead of 2 ml.

Duplicate groundwater samples were analyzed by a contract laboratory for VOCs using the SW846 Method 5030 for sample preparation and extraction, and SW846 Methods 8010, 8015, 8020, and 8030 for determination of the compound concentrations (U. S. EPA, 1986).

Packer test equipment, including the packer, packer inflation tubing, stainless steel riser pipe, purge pump, and pump tubing were decontaminated by pressure washing and selected steam cleaning. This procedure was generally adequate until DNAPLs were exposed to the system during the 20th and 21st tests performed. Pump blanks were collected prior to subsequent tests to evaluate the adequacy of decontamination procedures. These samples were immediately submitted to the site lab for VOC analysis.

2.5 *Installation of Monitor Well*

After completion of the final packer test in each borehole, the temporary 102-mm casing was removed and the borehole was reamed using a 149-mm diameter tricone bit.

Monitor wells were constructed according to Wisconsin state protocol using 50.8-mm I.D. schedule 80 PVC riser pipe with flush-threaded joints and 3-m long PVC screens.

3 RESULTS

3.1 *Geology*

The bedrock at the site consists of a series of Ordovician and Cambrian-age sandstone, siltstone, shale and dolomite formations (Ostrom, 1978). In descending order, these units include:

1. The Tonti Member of the St. Peter Formation, a well sorted, medium-grained sandstone,

2. The Readstown Member of the St. Peter Formation (a series of deformed breccia, sandstone, and shale layers,
3. The Black Earth Member of the St. Lawrence Formation (a dolomite),
4. The Lone Rock Formation (siltstone, sandstone, and shale), and
5. The Wonewoc Formation (a medium-grained well sorted sandstone).

The Readstown Member of the St. Peter Formation was further differentiated at the site into five informal stratigraphic subunits:
1. An upper sandstone unit,
2. An upper brecciated unit of mixed lithologies,
3. A siltstone, mudstone shale unit,
4. A lower brecciated unit of mixed lithologies,
5. A lower sandstone unit.

3.2 Hydraulic Properties

Prior to this investigation, the hydraulic properties of the bedrock at the site had been defined only to a limited extent as a result of seven short-term pump/recovery tests performed following development of the previously installed monitor wells. The packer test methodology provided a substantial amount of additional data as a result of the hydraulic tests performed on 16 bedrock test intervals.

The hydraulic conductivity values calculated from these data ranged from 1.38×10^{-5} cm/sec to 4.36×10^{-2} cm/sec. In general, the values calculated from the packer test data were consistent with other hydraulic tests performed in similar bedrock lithologic units, and with tests performed in bedrock monitor wells on the site. For example, packer tests performed on the Wonewoc Formation showed hydraulic conductivity values of 6.37×10^{-5} cm/sec, 3.18×10^{-4} cm/sec, 3.31×10^{-4} cm/sec, and 3.68×10^{-4} cm/sec. Hydraulic tests performed on the two monitor wells completed in the Wonewoc Formation had values of 2.90×10^{-4} cm/sec and 4.92×10^{-4} cm/sec.

Hydraulic conductivity values were variable within the Lone Rock Formation, which has interbedded layers ranging from coarse-grained sandstone to shale. The calculated hydraulic conductivity ranged from 1.38×10^{-5} cm/sec to 4.36×10^{-2} cm/sec (which were also the lowest and highest values, respectively, calculated from any of the packer tests). The water level data from the packer tests were also used to calculate potentiometric elevations for each test interval. These values were estimated by projecting the late recovery data in a similar manner to a Horner plot (Horner, 1951).

Additional insights into the hydraulic properties of the bedrock were also grained through monitoring of drilling fluid losses during coring. Based on the occurrence of 100 % drilling fluid losses at the same stratigraphic horizon in two or more boreholes, two laterally extensive, thin, high-permeability zones were identified within the Lone Rock Formation. These high-permeability zones are interpreted to be the result of dissolution along bedding planes. The upper zone coincided with the completion interval of a monitor well and a packer test interval from which DNAPLs were recovered, and represents a very significant migration pathway for both DNAPLs and DNAPL-impacted groundwater.

Ground water flow conditions at the site are illustrated on Figure 3.

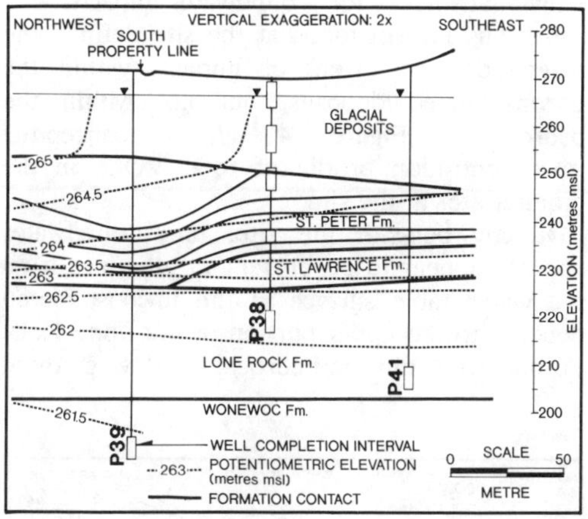

Figure 3. Flow Regime

Within the glacial deposits and shallow bedrock, the horizontal component of flow is oriented southeast, parallel to the line of the cross section. Vertical gradients are slightly downward within the glacial deposits, but steepen within the bedrock, and are downward to at least the base of the current site monitoring network. The downward gradients are believed to be caused by regional pumping of a major sandstone aquifer underlying the Wonewoc Formation, which provides the water supply for approximately 350,000 people, and has a pumping center located approximately 15 kilometres west of the site. The direction of flow within bedrock units below the St. Peter Formation has not been documented, due to interference from a local municipal water supply well approximately 400 metres northwest of the site, completed at elevations between 210 and 150 metres msl.

3.3 Groundwater Impacts

Impacted groundwater at the site has a complex chemistry due to the large number of contaminant compounds released. Within the packer test groundwater samples collected during this program, 21 VOCs were detected, including 13 halogenated hydrocarbon compounds, 4 aromatic hydrocarbon compounds, 3 ketones, and 1 alcohol. In addition to the complex chemistry, the geometry of the impacted groundwater is also complex due to the occurrence of numerous high permeability zones within the bedrock and multiple zones of DNAPL potentially acting as individual sources for groundwater impacts.

Previous investigations at the site defined the magnitude and extent of impacts within the unconsolidated deposits, but not within the bedrock. Figure 4 is a conceptual isoconcentration profile of total VOCs in the groundwater at the site.

Directly beneath the site, the groundwater impacts appear to be present in all zones from the water table surface in the unconsolidated deposits to the upper portion of the Lone Rock Formation within the bedrock. In downgradient areas, the impacted groundwater appears to form three distinct plumes which occur mainly within the upper and lower sandstone units of the Readstown Member of the St. Peter Formation and the upper portion of the Lone Rock Formation.

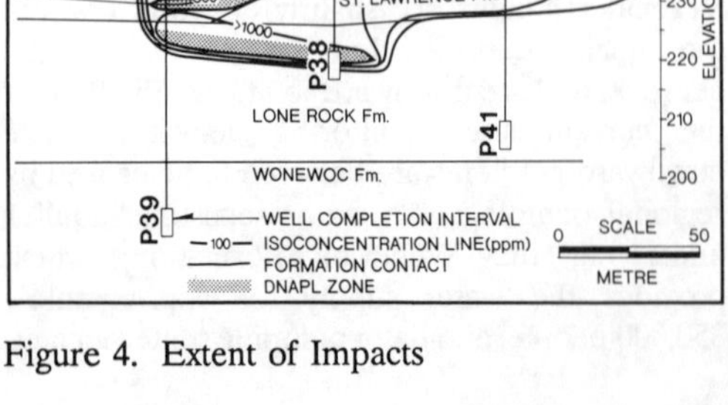

Figure 4. Extent of Impacts

The upper bedrock plume straddles the contact between the unconsolidated depth and the upper sandstone bed in the Readstown Member of the St. Peter Formation. The middle bedrock plume occurs within the lower sandstone unit in the Readstown Member of the St. Peter Formation, and is separated from the upper bedrock plume by a low permeability siltstone, mudstone, and shale unit. The plumes were differentiated based on relative contaminant concentrations and stratigraphic position. In boreholes P39 and P41, the position of the packer test intervals did not allow precise differentiation of the plumes; however, in the other borehole (P40), the plumes were clearly demonstrated to be separate by the occurrence of an intervening zone with a total VOC concentration of only 0.001 mg/l.

The deep bedrock plume occurs within the upper 6 to 10-m of the Lone Rock Formation. Groundwater impacts in the upper Lone Rock were present in two of the boreholes. In P40, the groundwater sample from the upper Lone Rock Formation contained a total VOC concentration of 75 mg/l, the highest concentration present in tests performed at the borehole. In P39, the groundwater sample from the upper Lone Rock Formation had a total VOC concentration of 1,619 mg/l and contained up to 20% by volume of DNAPLs. In both the P39 and P40 boreholes, groundwater samples from the subsequent deep packer test intervals contained total VOC concentrations approximately 150 to 300 times less than those in the upper Lone Rock Formation. These low level concentrations were attributable to contamination of the deeper zones as a result of drilling, sampling, and decontamination procedures. This interpretation was supported by PID measurements on the purge water which, in most of the deep zones, showed continuous decreases in measured VOC concentrations during purging. However, even with contamination from the procedures, the packer test methodology still was able to demonstrate the existence of an extremely sharp concentration gradient between the upper and middle portions of the Lone Rock Formation.

3.4 Extent of DNAPL

Prior to this investigation, the extent of DNAPLs at the site was very inadequately defined. It was known that DNAPL was present in at least three site wells, two of which were completed within the base of unconsolidated deposits, and the other within the upper portion of the Lone Rock Formation. The occurrences indicated only that DNAPL was present in some form at one or more horizons that intersected with the well completion interval. The exact elevation at which the DNAPL entered the well and the nature of the DNAPL occurrence (i.e, isolated ganglia versus continuous bodies of DNAPL in "pools") were unknown.

Use of the packer test methodology significantly increased the understanding of the nature of the DNAPL impacts. Valuable insight was gained by screening the bedrock core samples from the test intervals with a PID. The PID measurements for the majority of the core samples were relatively low (< 1 units); however, very high readings (10 to 50 units) were measured on several fracture surfaces and thin beds of medium-grained sandstone within the two test intervals from which DNAPLs were recovered. These high readings confirm the conclusion that DNAPLs may be concentrated at discrete horizons.

Additional insights into the nature of the DNAPL occurrences were gained through careful observation and monitoring of the DNAPLs as they were recovered during the packer tests. The DNAPL recovered during test P39-4 occurred in the form of extremely small dispersed droplets, which were initially invisible, but which coalesced after five to ten minutes t form a thin layer of large droplets on the bottom of the sample jar, representing approximately 1 to 2% of the total sample volume. This suggests that the DNAPL in the P39-4 test interval may have occurred in the bedrock in the form of small droplets dispersed in intergranular pore spaces.

In contrast, the DNAPL recovered during test P39-5 occurred as large globules of DNAPL which were readily visible and settled immediately to the bottom of the sample jar. The DNAPL initially represented as much as 20% of the total sample volume. An extended pump test was performed at P39-5 to determine the rate at which the proportion of DNAPL (as a percent of total fluid pumped) would decrease over time. The test was performed for five hours, during which the volume of DNAPL decreased steadily from 20% to 1.2% of the total fluid volume. An estimated total volume of 83 litres of DNAPL were recovered.

The volume of DNAPL recovered and its occurrence in the form of large globules, suggests that the DNAPL in the P39-5 test interval may occur as a continuous body of DNAPL concentrated within open fractures or dissolution cavities along bedding planes. A likely location for such a fracture or dissolution zone would be at an elevation of 221.3-m where a 100% loss of drilling fluid occurred during coring of the P39-5 test interval. It is very significant that a 100% drilling fluid loss occurred at the same elevation in a second site borehole (P41). This suggests the existence of a continuous high permeability horizon in the upper portion of the Lone Rock Formation with a lateral extent of at least 122-m. This same elevation also coincides with the test interval in which the greatest impacts were present in the third borehole (P-40). Unfortunately, this fracture or dissolution zone could not be distinguished in the core samples from other breaks in the core along bedding planes generated during coring.

The maximum lateral and vertical extent of DNAPLs within the upper Lone Rock Formation at the site is unknown. Based on the known occurrence within the Lone Rock Formation of interbedded high and low permeability rock units that are nearly horizontal and laterally continuous over wide areas, it is possible that DNAPLs have spread outward in several directions within one or more high permeability horizons in the upper Lone Rock Formation. The maximum vertical extent of DNAPLs within the Lone Rock Formation is also unknown.

Two DNAPL samples (both from the upper Lone Rock Formation) have been analyzed for VOCs by an outside laboratory. By weight, halogenated hydrocarbons as a group account for 83.3% and 90.1% of the total organic concentrations in the two DNAPL samples, versus 16.7% and 9.9% for the aromatic hydrocarbons. The concentrations of individual DNAPL compounds, as a percentage by weight of total detects, vary greatly. The DNAPL is composed, in order from greatest to least percentage by weight of total detects, of the following compounds:

1. 1,1,1-trichloroethene (40.8% and 47.0%
2. Trichloroethene (22.2% and 23.9%)
3. Tetrachloroethene (11.8% and 13.9%)
4. Toluene (5.2% and 7.4%)

5. 1,2-dichloroethane (5.3% and 6.1%)
6. Xylenes (3.6% and 7.6%)
7. Ethylbenzene (1.1% and 1.7%)
8. Methylene chloride (0.8% and 0.9%).

Several new insights into the migration and dissolution dynamics of DNAPLs at the site were gained from evaluation of the DNAPL and groundwater chemistry data. For each compound, the concentrations measured in the DNAPL samples (in units of mg/l) were converted to units of moles/l, based on the molecular weight of each compound. The mole concentration of each compound was divided by the total molar concentration of all the compounds in each sample to calculate the mole fraction of each compound of the total organics in each DNAPL sample. The mole fractions were then multiplied by the solubility in water for each compound to derive the estimated concentration for each compound for groundwater saturated with respect to the DNAPL. These predicted values can be compared with the measured values in groundwater samples collected from each DNAPL zone.

The validity of this analytical method can be demonstrated in DNAPL and groundwater samples by the relatively close correlation between the predicted and observed concentrations of various compounds, in particular for the aromatic compounds. The only anomalous values are for 1,2-dichloroethane, for which the observed concentrations are approximately 30 times lower than the predicted concentrations.

The application of this procedure provides several significant insights. First, the relative concentrations of individual compounds present in the groundwater may vary greatly from their proportion in the DNAPL. Methylene chloride forms only about 1% of the DNAPL, but about 10% to 25% of the total organic concentrations in the groundwater. In contrast, ethylbenzene, which also forms about 1% to 2% of the DNAPL, forms only about 0.3% of the measured concentrations in the groundwater.

One implication of this relationship is that removal of methylene chloride will occur far more rapidly than removal of ethylbenzene or other lower solubility compounds. A second implication is that, as the DNAPL is depleted over time, the ratio of the concentrations of high and low solubility compounds should change, as the high solubility compounds are depleted at a higher rate. These changes could potentially be monitored and used to estimate the rate of DNAPL removal over time during remediation.

The evaluation of the distribution of organic compounds in groundwater samples can also potentially be used to provide insights into the location and nature of DNAPL sources based on concentrations measured in downgradient monitor wells. For example, the mole fractions of the compounds in the P40-4 sample are virtually identical to the fractions in the P38 groundwater sample, and the concentrations are approximately 10% of those observed in the P38 groundwater sample. This suggests that the source of the contaminated groundwater plume at P40-4 is DNAPL with a composition identical to the P38 DNAPL sample, and a relatively near source.

4 CONCLUSIONS

Using the described "profiling" methodology, it was possible to accurately sample groundwater and DNAPL where present.

The distribution of soluble organics in groundwater and DNAPLs at the site is complex and it is not possible to cost-effectively evaluate or monitor the distribution of these contaminants using conventionally constructed monitor wells. It has been recommended that further site evaluations be undertaken using the "profiling" technique coupled with the installation of multiport monitor wells. Protocols and procedures for implementing this work is presently being reviewed by staff of the EPA Region V.

Using this technique, varying levels of QA/QC may be applied. A limitation of the sampling is that when the packer string and pump are installed, they must be lowered through contaminated drilling fluid. The problem of contaminated delivery lines can be overcome by replacing the air and sample delivery lines between sampling events.

Hydraulic testing is functional, particularly for low permeability zones. In zones of higher conductivity, testing is restricted by the capacity of the pump. Where high hydraulic conductivities are anticipated, it may be advisable to execute the project using larger diameter core and packer equipment to allow the use of an electrical submersible pump during hydraulic testing.

Although not tested during this program, it is possible to use a straddle packer to isolate and conduct hydraulic tests on specific intervals within any cored interval.

The success of this program was possible in large part due to the quick turnaround and analysis of lab samples. Certainly not every site will have an organics lab in such close proximity; however, with the increasing reliability of field portable GC equipment, similar turnarounds are possible at most sites.

Valuable insight into subsurface conditions can be gained by closely monitoring the drilling fluids. The presence of highly impacted zones is possible by monitoring the drilling fluid with a PID. Similarly, zones of high hydraulic conductivity can be identified during coring operations by closely monitoring drilling fluid losses.

This program certainly confirms that DNAPLs can, in fact, be pumped from a bedrock formation. However, it is apparent that much is yet to be learned about the effective management of DNAPL pumping systems.

REFERENCES

Cooper, H. H., Jr. & C.E. Jacob, 1946. A generalized graphical method for evaluating formation constants and summarizing well field history. Transactions of the American geophysical union 27: 526-534.

Cooper, H. H., Jr., J. D. Bredehoeft & I. S. Papadopoulos, 1967. Response of a finite-diameter well to an instantaneous charge of water. *Water resources research 3*: 263-269.

Horner, D.R., 1951. Pressure build-up in wells. Proceedings of Third World Petroleum Congress. Section II: 503-521.

Ostrom, M. E., 1978. Stratigraphic relations of lower paleozoic rocks of Wisconsin, in Ostrom, M. E. (ed), Lithostratigraphy, petrology, and sedimentology of late cambrian-early ordovician rocks near Madison, Wisconsin. *Wisconsin geological and natural history survey, field trip guidebook No. 3*: 142 p.

Schwille, F., 1988. Dense chlorinated solvents in porous and fractured media, model experiments (english language edition): 146 p. Lewis Publishers: Chelsea, Michigan

U. S. EPA, 1986. *Test methods for evaluating solid waste, physical/chemical methods, SW846, 3rd Edition*

Subsurface Contamination by Immiscible Fluids, Weyer (ed.) © 1992 Balkema, Rotterdam. ISBN 90 6191 175 3

Potential applications of stable isotopes to studies of immiscible groundwater contaminants

H.R.Krouse
Department of Physics and Astronomy, The University of Calgary, Alb., Canada
S.Pickering & B.Batts
School of Chemistry, Macquarie University, N.S.W., Australia

ABSTRACT: Stable isotope data have at least three potential applications in the study of immiscible groundwater contaminants: (1) tracing of groundwater flow paths, (2) tracing and characterization of soluble and insoluble contaminants, and (3) studies of redox reactions leading to products such as sulphate, H_2S, CO_2, and CH_4. Illustrative examples from field and laboratory studies are given.

1 INTRODUCTION

In recent years, stable isotope techniques have been increasingly applied to the problem of distinguishing the products of pollution from compounds of natural origin in the environment. Such research is not straightforward since pollutants may indirectly affect natural conversions, e.g. alter microbial populations.

Isotopes are forms of an element which differ in the their masses because of different numbers of neutrons in their nuclei. Many chemical and physical processes are mass dependent so that isotopic species participate at different rates. When processes are incomplete, the remaining reactant has a different isotope composition than the product, i.e. the isotopes have been "fractionated".

For unidirectional processes, the term "kinetic isotope effect" is used. This strictly refers to "the rate of a chemical reaction being sensitive to atomic mass at a particular position in one of the reacting species" (Hayes, 1982). More generally, the term has been used to express the overall isotopic selectivity in chemical conversions or even physical phenomena, e.g. evaporation.

Isotope exchange reactions may occur among different chemical compounds, e.g.

$$Pb^{34}S + Zn^{32}S \rightleftharpoons Pb^{32}S + Zn^{34}S \qquad (1)$$

or different phases or states of one compound, e.g.

$$H_2{}^{18}O_{vapour} + H_2{}^{16}O_{liquid} \rightleftharpoons$$

$$H_2{}^{16}O_{vapour} + H_2{}^{18}{}_{liquid}O \qquad (2)$$

Isotope equilibrium may be obtained (but not always) in which case, a temperature dependent equilibrium constant can be measured.

There are two aspects of stable isotope studies of pollutants. One is the use of isotope compositions as tracers and the other is the use of isotope fractionation to better understand physical and chemical conversions. Clearly, if fractionation is extensive, then a tracer loses its identity. Information may not be lost if the system is closed since depletion of an isotope in one component results in its enrichment in another. The isotopic composition of a compound derived from two different sources can be used to deduce the proportions derived from each, provided that the isotope compositions of the sources are known.

Absolute isotope abundances are difficult to measure whereas the isotope compositions of two samples can be compared very precisely. Therefore, stable isotope data are expressed on δ- (del or delta) scales defined as the $^{o}/oo$ (per mil or parts per thousand) deviation in the abundance ratio of two isotopes in a sample relative to that of a standard, i.e.

$$\delta\ (^{o}/oo) = \left(\frac{R_{sample} - R_{standard}}{R_{standard}}\right) x\ 1000$$

where R is the abundance ratio of the heavy to the light isotope, e.g. $^{13}C/^{12}C$, $^{18}O/^{16}O$, $^{34}S/^{32}S$. The δ-value is designated in terms of the heavier isotope, e.g. $\delta^{13}C$, $\delta^{18}O$, $\delta^{34}S$. International standards have been chosen.

Relative stable isotope abundances are usually measured with mass spectrometers which feature simultaneous collection of ion beams (corresponding to different isotopes) and alternating admissions of standard and unknown gases to the source. The measurement procedures are consistent with the δ-scale designation and realize precisions of better than $\pm 0.1\ ^{o}/oo$.

For a given element, one or two processes account for the bulk of the terrestrial isotope variation. For hydrogen and oxygen, fractionation occurs in the meteorologic cycle since isotope species of water have different vapour pressures. For carbon isotopes, fractionation during photosynthesis is dominant although large kinetic isotope effects are realized during methanogenesis. Natural variations in sulphur isotope abundances are due mainly to kinetic isotope effects during bacterial sulphate reduction.

In studies of subsurface contamination by immiscible fluids, there are three possible applications of stable isotope data;
(1) tracing of water, (2) tracing of soluble and insoluble components, and (3) studies of degradation of the contaminants.

2 TRACING OF GROUNDWATER FLOW

The world acts as a complex distillation system for isotopic species of water. The H- and O-isotope composition of precipitation and surface waters vary with latitude, altitude, temperature, amount of precipitation, and storm paths (e.g. Dansgaard, 1964; Friedman et al., 1964; Yonge et al., 1989). A defined standard, SMOW (Standard Mean Ocean Water), has been used historically and more recently, a close approximation, V-SMOW has been distributed by the International Atomic Energy Agency in Vienna. Meteoric waters have negative δD and $\delta^{18}O$ values with lower values at higher latitudes. Isotope data have proved useful for tracing groundwater flow paths (e.g Weyer et al., 1979; Fontes, 1980) and detecting meteoric water in deeper formations (Clayton et al., 1966; Hitchon and Friedman, 1969; Bailey et al., 1973).

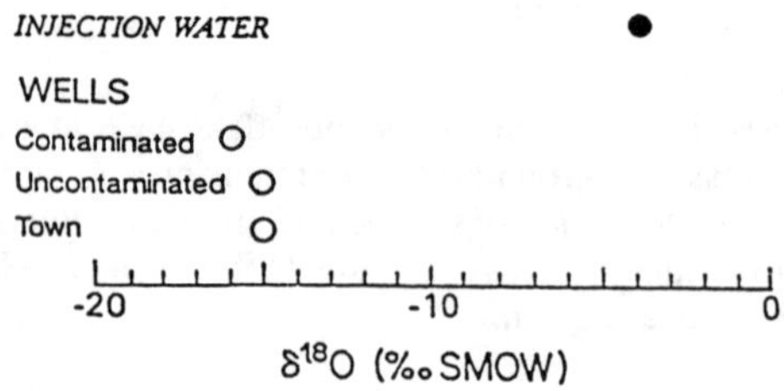

Figure 1 Oxygen isotope data indicating that a water injection operation for enhanced oil recovery was not responsible for introducing hydrocarbon contaminants into a nearby domestic water well.

Data for a groundwater contamination study are shown in Fig. 1. A farmer complained of an odour in his well water which he attributed to the use of water injection for enhanced recovery in a nearby oil well. As it turned out, the company had used water pumped up from a deep formation. It was not as depleted in the heavy isotopes of hydrogen and oxygen as local meteoric water. In contrast, the farmer's well, labelled "contaminated" in Fig. 1, had more negative δD and $\delta^{18}O$ values than either an adjacent farm well judged to be uncontaminated or the nearby town water supply. This clearly meteoric water isotope signature ruled out the water injection operation as the cause. In fact, other indicators such as nitrate in the water suggested that possibly a nearby cattle feed lot operation was a more likely source of the organic contamination. Whereas this problem could have been solved by many techniques including organic chemical analyses, $\delta^{18}O$ or δD data constituted the easiest approach. However, if meteoric water had been used for injection, isotopic analyses of water would not have been successful.

3 LEACHING EXPERIMENTS

Understanding organic chemistry of subsurface water is important for monitoring migration and degradation of immiscible hydrocarbons. For example, Barth et al., (1987) provided evidence that organic acids in groundwaters may affect the hydrocarbon wettability behaviour. Colin (1975) suggested that low organic acid levels (50 mg/ℓ) or about 10 to 20 mg/ℓ of dissolved organic carbon can solubilize and mobilize significant quantities of immiscible hydrocarbons. It is therefore important to study water leaching as a mechanism of releasing organic acids into subsurface water systems.

Leaching experiments were conducted from 22°C up to 300°C. This range was chosen to provide data relevant to shallow and deep natural conditions, as well as in situ subsurface industrial operations such as steam injection or combustion drives for bitumen recovery.

Data were obtained for three peat samples and one soil. Used motor oil was also mixed with a portion of the soil sample and leaching carried out at 22°C, 50°C, and 95°C with the mixture and with oil alone.

The Macquarie Island Green Gorge Peat Mound (54° 37' 57"N; 158° 54' 10"E) is 8500 years old and formed from an herb and grass swamp under a cool hyperoceanic subantarctic climate. One sample each from the unweathered interior and weathered surface were taken. The Tomaga sand-dune peat (32° 30'S; 151° 48'E) is 50,000 years old and derived from a woody environment. A peat sample was selected at 1 to 2 m burial below the water table and a podzol soil taken from another site at 2 m depth above the water table. The soil contained 2.2 percent carbon whereas the peat samples ranged from 47.3 to 48.7 percent carbon.

Five gram samples of peat and 25 gram samples of soil were leached by 100 ml of distilled H_2O in an evacuated, sealed, rocking steel bomb for 72 hours. The leachates were cooled to ambient temperature prior to removal from the bomb. The water was decanted and centrifuged at 8000 rpm for 30 min. at 10°C. Organic matter retained in the water after passage through a 0.45 μ Duropore Hydrophylic membrane filter was classified as dissolved organic matter (Thurmann, 1985).

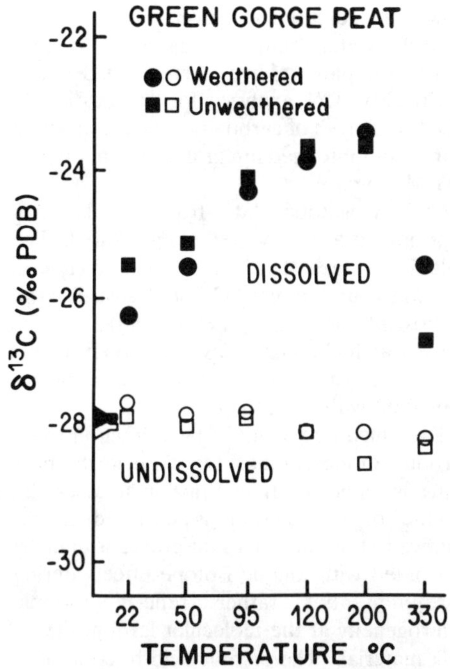

Figure 2 Carbon isotope data for leaching experiments with Macquarie Island Green Gorge peat. Values for the starting materials are given by the symbol on the y-axis.

Detailed discussion of the molecular composition of the dissolved organic matter is beyond the scope of this paper. Acid yields ranged from 0.2 mg/g carbon for the 22°C experiments up to greater than 250 mg/g carbon at the higher temperatures. Humic/fulvic acids dominated in the 22°C runs. At 50°C, monocarboxylic acids (C_1-C_5) dominated in the Tomago soil leachate whereas humic/fulvic acids dominated in the peat leachate. Humic/fulvic acids were less than 0.001 mg/g carbon in all samples leached at higher temperatures. At 95°C and 120°C, dicarboxylic acid (C_2-C_4) dominated in the Tomago soil extract whereas monoacids dominated in the peats. At 200°C and 330°C, the monoacids dominated in the soil and peat samples. The dominance of humic/fulvic acid in the low temperature leachates is consistent with the nature of shallow groundwaters associated with wetlands/swamps (Thurmann, 1985). The higher temperature leachates seemed to have embraced thermal degradation as well as increased solvation.

After water removal, dissolved organic matter samples (~1 mg) were loaded into a Carlo Erba CN analyzer for combustion to CO_2 for analysis of carbon isotope composition with a computerized integrated system built by VG Micromass.

Carbon isotope data from the leaching experiments are presented in Figs. 2 to 4. The bulk $\delta^{13}C$ values of -27 to -28 ‰ are typical of plants undergoing C_3 or Calvin-Benson photosynthesis (e.g. O'Leary, 1981). Since grasses at higher latitudes and woody plants are in this category, the C-isotope data are consistent with the source.

The much larger undissolved fraction has a carbon isotope composition close to the bulk value as expected. In contrast, in all cases, the soluble organic matter is enriched in ^{13}C relative to the bulk. This difference is unlikely associated with kinetic isotope effects during dissolution but rather reflects isotopic heterogeneity at the molecular level in the bulk material. The organic acid fraction of in some C_3 plants has been shown to be enriched in ^{13}C by about 5 ‰ as compared to total leaf (Whelan et al., 1970; Deines, 1980). In Green Gorge peat, the difference increased steadily with temperature of leaching up to 200°C and decreased to about 2 ‰ at 330°C (Fig. 2). With the exception of the experiment at 95°C, a similar but smaller magnitude trend is noted with the Tomago peat sample (Fig. 3). A similar trend was found for the Tomago soil sample (Fig. 4). Discussion of how these trends possibly relate to the chemical and isotopic heterogeneity of the original material and subsequent conversions with losses of volatiles is beyond the scope of this paper.

Of more relevance to the problem of immiscible fluids in subsurface waters, are the data obtained by leaching the soil-oil mixture. At 20°C, 50°C, and 95°C, the mass of recovered dissolved organic matter (order of mg/g C of total starting materials) was about double that obtained in experiments with soil alone. Further, the $\delta^{13}C$ value of the dissolved material tended to approach that of the bulk oil with greater yields at higher temperatures, as is strikingly evident in the 95°C experiment

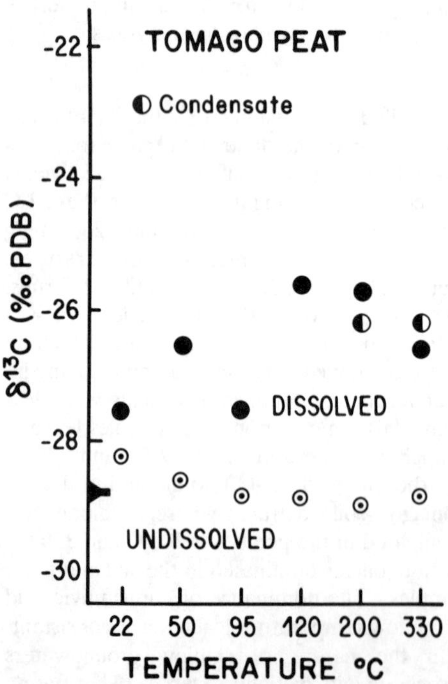

Figure 3 Carbon isotope data for leaching experiments with Tomago Peat.

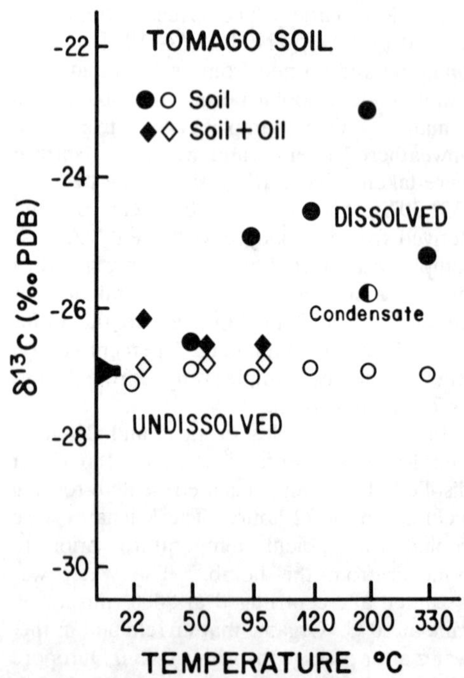

Figure 4 Carbon isotope data for leaching experiments with Tomago soil and a soil-used motor oil mixture.

(Fig. 4). In contrast, in leaching experiments with oil alone, the amount of dissolved organic matter was too small (order of μg/g) to permit isotope analyses. Therefore, the isotope data of Fig. 4 support previous evidence that organic acids can promote solubilization and mobilization of otherwise immiscible hydrocarbons (Colin, 1975). The mechanisms require elucidation and the phenomenon probably relates to the physical definition of soluble organic matter as being that which will pass through a 0.45 μ filter.

During higher temperature leaching, steam vented from the reactor was condensed and dissolved organic matter recovered (labelled "condensate" in Figs. 3 and 4). For Tomago soil leached at 200°C, the condensate was 3 $^{o}/oo$ depleted in ^{13}C as compared to dissolved material in the reactor. This reflects the more volatile molecules being relatively depleted in ^{13}C and the tendency for the lighter isotopic species of a given compound to favour the vapour phase. In the Tomago peat experiment, the isotopic difference was in the same direction but smaller, although there was a small effect in the opposite direction for the extractions at 330°C.

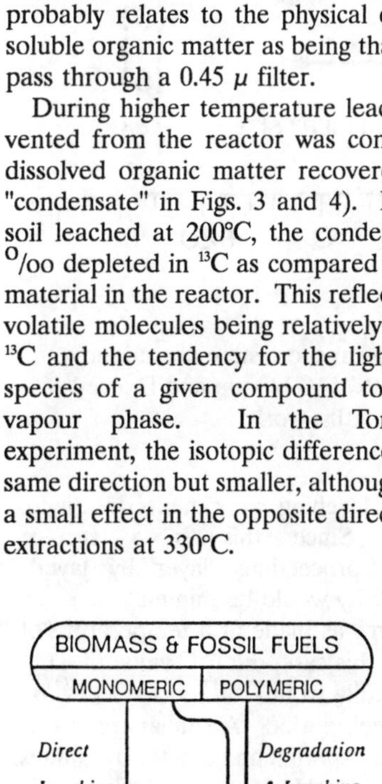

Figure 5 Simplified scheme of natural degradation of organic materials.

4 ORGANIC MATTER DEGRADATION

Whereas the previous section considered leachates and undissolved organic matter, this section will address the isotopic composition of the products of degradation. One can consider two basic types of reactions: (1) specific organic molecules consumed in processes such as sulphate reduction and (2) degradation of organic matter by a number of processes, producing CH_4 and CO_2. A simplified diagram of the latter is presented in Fig. 5.

A well-established isotopic indicator of sulphate reduction is product sulphide that is significantly depleted in the heavier ^{34}S with respect to reactant sulphate. If the reaction is extensive in a closed system, the unreacted sulphates may become very enriched in ^{34}S because of preferential reduction of ^{32}S-containing sulphate. This criterion was used along with chromatographic evidence of depletion of high molecular weight paraffins as evidence of biodegradation of oil in the Mission Canyon formation of Saskatchewan (Bailey et al., 1973). Further, O-isotope data for the formation waters showed that degradation was most extensive where waters were dominantly meteoric. It is unlikely that the immiscible crude oil served as the carbon source for the sulphate reducers, but rather products of its degradation by aerobic bacteria associated with greater meteoric water influx.

Where organic sulphur compounds have been oxidized, S- and O-isotope analyses of product sulphate prove informative. The S-isotope composition is close to that of the organic sulphur. Oxygen in the sulphate is derived from both O_2 and H_2O. The former has quite positive $\delta^{18}O$-values (+23 $^{o}/oo$ in the atmosphere) whereas the latter is usually meteoric water with negative $\delta^{18}O$ values.

Carbon isotope data for CH_4 and CO_2 associated with decaying organic matter in different environments, are shown in Fig. 6. In most cases, the CO_2 and CH_4 are significantly enriched and depleted respectively in ^{13}C with respect to co-existing organic matter. The basic question is whether the isotope data can be used to identify the source. The immediate answer is that the task is very difficult. This arises because CO_2 and CH_4 can be generated from a variety of organic molecules, e.g. alcohols, volatile fatty acids and reaction of CO with water. Carbon dioxide can be reduced to methane and methane oxidized to carbon dioxide. The latter would be curtailed in anaerobic environments. CO_2 might also be converted to acetate as demonstrated in a sludge digester (Chynoweth and Mah, 1971).

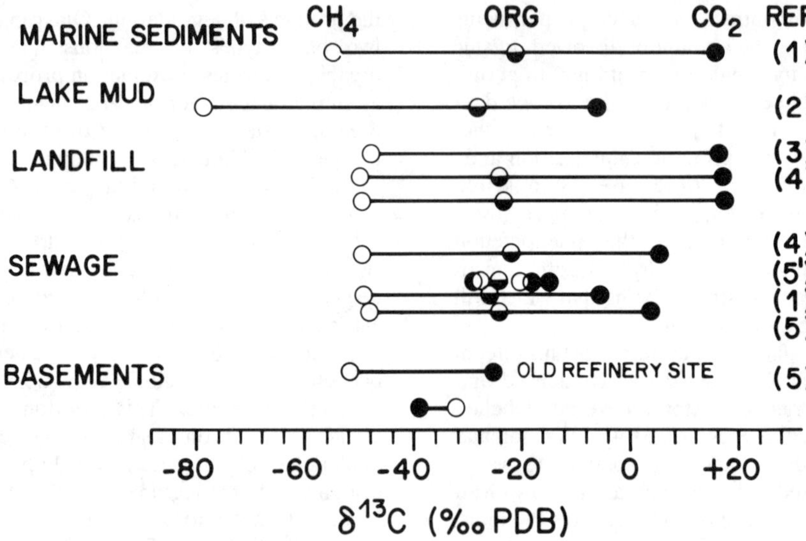

Figure 6 Carbon isotope compositions of methane and carbon dioxide at sites where degradation of organic matter is occurring. References (1) Nissenbaum et al., 1972 (2) Oana and Deevey, 1960 (3) Games and Hayes, 1974 (4) Games and Hayes, 1976 (5,5′) this work.

When the results of several studies are combined, it is found that CH_4 in soils, glacial drift, marshes, lake muds, and marine sediments has a wide range of $\delta^{13}C$ values, -48 to -86 ^{O}/oo. The data for the three landfill sites in Fig. 6, are quite consistent, but a larger data base might show otherwise.

Similarly, data from sewage sludge reactors show less spread but this is probably the consequence of operating at higher temperatures (35°C or above). Isotope fractionation, whether kinetic isotope effects or exchange, decreases with temperature. In our study, gases produced at the Bonnybrook treatment plant in Calgary, were examined and found to be in agreement with other reported studies. A suspension of the sewage was set up in the laboratory at ambient temperature. It was constantly purged with N_2 while CH_4 and CO_2 products were collected. The isotope data differed from the higher temperature data obtained at the processing plant. The $\delta^{13}C$ values for both CH_4 and CO_2 were close to that of the raw sewage (5' in Fig. 6). This means that a step in the process with little isotope selectivity controlled the rate. Since the organic matter was dominantly suspended material, dissolution was probably much slower at the cooler temperature, consistent with the findings of the leaching experiments described previously. Since this is a surface phenomenon, proceeding "layer by layer", isotope selectivity would be minimal.

Another run was made at a temperature of 36 ± 1°C. In this case, the $\delta^{13}C$ value of CO_2 increased steadily from -18 ^{O}/oo to +12 ^{O}/oo over a two week period. A similar trend was found using an experimental reactor by Games and Hayes (1976). However, the $\delta^{13}C$ value of the methane did not decrease much below -30 ^{O}/oo.

When CH_4 and CO_2 production waned in the room temperature experiment, Na_2SO_4 was added to see if sulphate reducing bacteria were present. H_2S was barely detectable immediately, but after one week was produced in copious quantities. It was depleted in ^{34}S by up to 20 ^{O}/oo with respect to the added sulphate. CO_2 production increased significantly and its $\delta^{13}C$ value was slightly more negative than previously. Interestingly, methane production also continued with about the same carbon isotope composition as the suspended organic matter. The implication is that biologically relevant inorganic ions such as sulphate can significantly affect the degradation of immiscible hydrocarbons.

A practical problem that often arises is

determining the source of gas entering through cracks in basements. One source is the gas heating supply but there are usually other possibilities. Two such investigations are shown at the bottom of Fig. 6. In both cases, the CO_2 content was about three times that of the methane.

In the lowermost case, the gas mains had a $\delta^{13}C$ value near -49 $^{o}/oo$ and so it was ruled out. Another possibility was a vacant lot where once stood a gas service station. The soil contains large quantities of disposed used motor oil, some of which has resided there for four decades. Since the CO_2 had a $\delta^{13}C$ value which is about 8 $^{o}/oo$ lower than methane, it was most likely generated by methane oxidation. In this process, the lighter ^{12}C methane oxidizes faster whether under biological or pure chemical influence (Barker and Fritz, 1981; Kiyosu and Krouse, 1989). The relatively small isotopic difference between the CH_4 and CO_2 is consistent with a high percentage conversion and hence the observed dominance of CO_2.

The second study was a greenhouse operation on a former refinery site. In this case, a hole was drilled through the basement wall to sample soil gas with concentrations of 17,000 and 51,000 ppm for CH_4 and CO_2 respectively. The $\delta^{13}C$ value for the methane was fairly close to that of the mains. One theory advanced was that an underground leak in the gas mains was mobilizing the immiscible hydrocarbons. The methane carbon isotope composition would tend to support this. However, ethane was barely detectable in the soil gas, but constituted 11 percent of the mains supply. Further the hydrogen isotope compositions of the soil and mains methane were quite different. In combination with the dominance of CO_2, the isotope data suggest that degradation of immiscible hydrocarbons was occurring. The greenhouse operation used CO_2. However, its $\delta^{13}C$ value of -70 $^{o}/oo$ and the physical location of the supply made it an unlikely contributor to the soil gas.

5 CONCLUSIONS

Stable isotope techniques have been used quite successfully to elucidate fairly complicated processes, particularly where the starting materials are known. However, deducing processes and starting materials from products found in suspected polluted sites is more difficult and the success has been variable. Some problems such as tracing of the water carrier flow path or evidence of sulphate reduction are more readily pursued.

There are additional ways in which stable isotopes can be used to trace immiscible liquids and follow their conversions. Most involve further differentiation of the sample, even down to individual compounds or structures in a given compound. In the case of sulphur, one approach has been to pyrolyze the sample while raising the temperature at a steady rate (~2°/min). Gases, including H_2S are evolved as peaks over discrete temperature ranges. Some of the peaks are believed to correspond to different sulphur structures in the original sample and their sulphur isotope compositions have been measured (Krouse et al., 1987). Samples that have been degraded (e.g. bacterial oxidation of thiophenic sulphur) are expected to show different H_2S evolution and isotopic behaviours.

Degradation of immiscible fluids and the effects of dissolved organic acids on leaching and degradation should be more thoroughly studied in the laboratory and evidence for such interactions sought in the field situations. The potential certainly exists for stable isotope data to answer many questions about subsurface immiscible fluids, but in most cases, an intensive study is required in coordination with other techniques. With complex systems, sophisticated techniques are required and restriction to one approach often assures failure.

ACKNOWLEDGMENTS

Support for the Stable Isotope Laboratory at the University of Calgary from the Natural Sciences and Engineering Research Council of Canada is gratefully acknowledged.

REFERENCES

Bailey, N.J.L., H.R. Krouse, C.R. Evans, and M.A. Rogers 1973. Alteration of crude oil by waters and bacteria. Evidence from geochemical and isotope studies. *Bull. Amer. Assoc. Petrol. Geol.* 57: 1276-1290.

Barker, J.F. and P. Fritz 1981. Carbon isotope fractionation during microbial methane oxidation. *Nature* 293: 289-291.

Barth, T., A. Borgund, A. Hopland, and A. Graue 1987. Volatile organic acids produced during kerogen maturation-amount, composition and role in migration of oil. *Adv. in Organic Geochemistry* 3: Nos. 1-3.

Chynoweth, D.P. and R.A. Mah 1971. Volatile acid formation in sludge digestion. In R.F. Gould (ed.). *Adv. in Chem. Ser.* 105: 41-54 Washington: American Chemical Society.

Clayton, R.N., I. Friedman, D.L. Graf, T.K Mayeda, W.F. Meents, and N.F. Shimp 1966. The origin of saline formation waters, I. Isotopic composition. *J. Geophys. Res.* 71: 3869-3882.

Colin, G. 1975. Geochemistry of oil field waters - *Developments in Petroleum Science I.* Amsterdam: Elsevier.

Dansgaard, W. 1964. Stable isotopes in precipitation. *Tellus* 16: 436-467.

Deines, P. 1980. The isotope composition of reduced organic carbon. In P. Fritz and J. Ch. Fontes (eds.), *Handbook of environmental isotope geochemistry 1. The terrestrial environment*, A, p. 329-406. Amsterdam: Elsevier.

Fontes, J. Ch. 1980. Environmental isotopes in groundwater hydrology. In P. Fritz and J. Ch. Fontes (eds.), *Handbook of environmental isotope geochemistry 1. The terrestrial environment*, A, p. 75-140. Amsterdam: Elsevier.

Friedman, I., A.C. Redfield, B. Schoen, and J. Harris 1964. The variation in the deuterium content of natural waters in the hydrologic cycle. *Rev. Geophys.* 2: 177-224.

Games, L.M. and J.M. Hayes 1974. Carbon in groundwater at the Columbus, Indiana landfill. In D.B. Waldrip and R.V. Ruhe (eds.). *Solid waste disposal by land burial in Southern Indiana*, p. 81-110. Bloomington: Indiana Univ. Water Resources Res. Center.

Games, L.M. and J.M. Hayes 1976. On the natural anaerobic environments. In J.O. Nriagu (ed.). *Environmental biogeochemistry*, 1. Carbon, nitrogen, phosphorous, sulfur and selenium cycles, p. 51-73. Ann Arbor Science.

Hayes, J.M. 1982. Fractionation, et al. An introduction to isotopic measurements and terminology. *Spectra* 8: 3-8.

Hitchon, B. and I. Friedman 1969. Geochemistry and origin of formation waters in the western Canada sedimentary basin, I. Stable isotopes of hydrogen and oxygen. *Geochim. Cosmochim. Acta* 33: 1321-1349.

Kiyosu, Y. and H.R. Krouse 1989. Carbon isotope effect during abiogenic oxidation of methane. *Earth Planet. Sci. Lett.* 95: 302-306.

Krouse, H.R., R.G.S. Ritchie, and R.S. Roche 1987. Sulphur isotope composition of H_2S evolved during the non-isothermal pyrolysis of sulphur-containing materials. *J. Anal. Appl. Pyrolysis* 12: 19-29.

Nissenbaum, A., B.J. Presley, and I.R. Kaplan 1972. Early diagenesis in a reducing fjord, Saanich-Inlet, British Columbia - I. Chemical and isotopic changes in major components of interstitial water. *Geochim. Cosmochim. Acta.* 36, 1007-1027.

Oana, S. and E.S. Deevey 1960. Carbon 13 in lake waters, and its possible bearing on paleoliminology. *Amer. J. Sci.* 258A: 253-272.

O'Leary, M. 1981. Carbon isotope fractionation in plants. Phytochemistry 20: 553-567.

Thurmann, E. 1985. *Organic geochemistry of natural waters.* Martinus Nijhoff. W. Junk. Member Kluwer Academic Publishing Group. Netherlands, Boston.

Weyer, K.U., H.R. Krouse, and W.C. Horwood 1979. Investigation of regional geohydrology south of Great Slave Lake, Canada, utilizing natural sulphur and hydrogen isotope variations. In *Isotope Hydrology* 1978, 1, p. 251-264. Vienna: IAEA.

Whelan, T., W.M. Sackett, and C.R. Benedict 1970. Carbon isotope discrimination in a plant possessing the C_4 dicarboxylic acid pathway. *Biochem. Biophys. Res. Comm.* 41: 1205-1210.

Yonge, C., L. Goldenberg, and H.R. Krouse 1989. An isotope study of water bodies along a transverse of southwestern Canada. *J. Hydrol.* 106: 245-255.

Subsurface Contamination by Immiscible Fluids, Weyer (ed.) © 1992 Balkema, Rotterdam. ISBN 90 6191 175 3

A detailed geophysical investigation of a shallow sandy aquifer

P.D. Bauman
Piteau Engineering Ltd, Calgary, Alb., Canada
J.P. Greenhouse & J.D. Redman
Waterloo Centre for Groundwater Research, University of Waterloo, Ont., Canada

ABSTRACT: This study explores the capability of various borehole, surface, and laboratory geophysical methods to distinguish small scale and subtle (i.e. significant but not indicative of a distinct change in material) changes in stratigraphy and physical properties in a relatively homogeneous sandy aquifer. Detailed radar profiles suggest subtle stratigraphic features which may have a controlling influence on the migration of spilled solvents. Conductivity, density, and porosity logging were also able to recognize subtle changes in physical properties within the sandy aquifer. Measurements showed that, in fact, considerable variation exists both spatially and with depth. Laboratory measurements on split cores are capable of precisely identifying changes in physical properties on a scale of centimetres, the same scale at which soil heterogeneities are visually observed. The immediate relevance of this work is a better understanding and description of heterogeneities in the Borden aquifer which may influence the path of chlorinated organic solvents in a slow release spill.

1 INTRODUCTION

The hydrogeologic problems associated with contamination by chlorinated organic solvents, also known as DNAPLs (dense nonaqueous phase liquids) stem from the physical properties of these organic solvents. As the acronym implies, DNAPLs are denser than water and have very low solubilities in water. Additionally, DNAPLs are less viscous than water and are therefore highly mobile. Spilled in a sandy aquifer, they will tend to migrate downward as a separate phase, regardless of groundwater flow direction. Because of their low solubility, solvent may exist for decades in the saturated zone trapped by capillary forces as immobilized globs. These residual volumes may exceed 15 percent of pore space in fine grained sands (Feenstra and Cherry, 1988).

While the migration downward of DNAPLs into a porous medium is largely unaffected by the groundwater flow, laboratory investigations have shown that soil heterogeneities may control the migration path and residual volume of free product (Kueper et al., 1989; Wilson et al., 1988; Schwille, 1988; Kueper and Frind, 1988). Understanding the subsurface movement of DNAPLs in a porous medium has become, to a large extent, a problem of understanding the influence and distribution of heterogeneities in porous media.

From a geophysical perspective, the prospect of remote detection of organic solvents from either the surface of-or boreholes within-a sandy aquifer, are not encouraging. The direct detection of organics depends on contrasts of physical properties with the pore fluid which is displaced. Fresh water and organic solvents are similarly very resistive, reducing the efficacy of geoelectric methods. Solvent densities and hydrogen content, while different from water, are not so anomalous as to be accounted for by small variations in porosity.

Most organic solvents are relatively non-polar; that is, they possess very low dielectric constants. Water, however, is an extremely polar substance having a dielectric constant up to 35 times greater than that of many common solvents. Ground penetrating radar (GPR), an instrument which responds to changes in the dielectric constant, has therefore been considered as a possible tool for direct detection.

Another outstanding feature of GPR is its ability to image the subsurface in fine detail. Even in the visually homogeneous Borden sands, GPR profiles reveal a plethora of subtle and not so subtle features.

Spectral induced polarization (SIP) is another technique which in theory may be capable of direct detection of organic contamination. The premise of using SIP to detect organic contaminants in a sandy aquifer depends entirely on the existence of a significant induced polarization (IP) effect that is altered by clay-organic reactions in the porous medium (Krumenacher and Taylor, 1988; Sadowski, 1988; Olhoeft, 1986)

In this project, GPR and other geophysical methods are being used to map the physical properties of a small volume of a shallow sandy aquifer, with the following goals: 1) to identify fine scale stratigraphic features and spatial changes in physical properties within the aquifer; and 2) to investigate techniques which may recognize very subtle anomalies created by the introduction of a DNAPL into a sandy aquifer.

2 SITE DESCRIPTION

The geophysical investigation sites include two 15 X 15 m field "cells" proposed for use by other researchers as controlled solvent spill locations. Surrounding each of these cells is a gridded 25 X 25 m geophysical cell (Figure 1). Geophysical data from these cells will be discussed in this paper. The cells are situated in an abandoned sand pit on Canadian Forces Base Borden, about 80 km northwest of Toronto, Ontario. The cell areas are flat and grass covered.

The stratigraphy of the sand pit is a result of a prograding foreshore sequence resulting from deglaciation and subsequent creation of the main stage of Lake Algonquin between 11,500 and 10,500 years B.P. (Bolha, 1986). The sand pit is excavated from reworked and redeposited deltaic sediments from meltwater channels terminating to the north.

Previous coring suggested that the top 1.5 m of the aquifer consisted of structureless, homogeneous, windblown sands. Below this is a 0.3 m thick peaty layer which is laterally extensive but discontinuous. Bedded beach deposits extend from the peat layer to a silty clay layer at about 7 m. Between 8 m and 10 m, a soft, gray clay is reached. This clay is the top of an 8 m thick aquitard.

3 METHODOLOGY

Detailed seismic refraction, streaming potential, electromagnetic, galvanic resistivity, magnetometry, SIP, borehole logging and GPR surveys were performed within geophysical cells 1 and 2. Borehole porosity, density, natural radioactivity, and electromagnetic logs were performed in four 7.5 m deep PVC cased boreholes from within the cells. Cores from 2 of the boreholes were removed for electrical conductivity profiling, time domain reflectometry (TDR) profiling, hydraulic conductivity profiling, visual description, matrix density determination, and grain size analysis. An in situ and permanent electrical profiling device (DEL) was installed in the vicinity of one of the boreholes (Figure 1). Discussed below are those methods which were found to be most applicable to the study.

3.1 Ground Penetrating Radar (GPR)

GPR is an electromagnetic field technique in which high frequency pulses are transmitted into the ground and returned to a receiving antenna via subsurface reflectors believed to be interfaces of contrasting dielectric permittivity. The resulting images visually and conceptually resemble seismic reflection records.

Water has a much higher dielectric permittivity than the soil matrix. Within the saturated zone, changes in porosity will result in changes in the dielectric permittivity. Although it has yet to be unequivocally

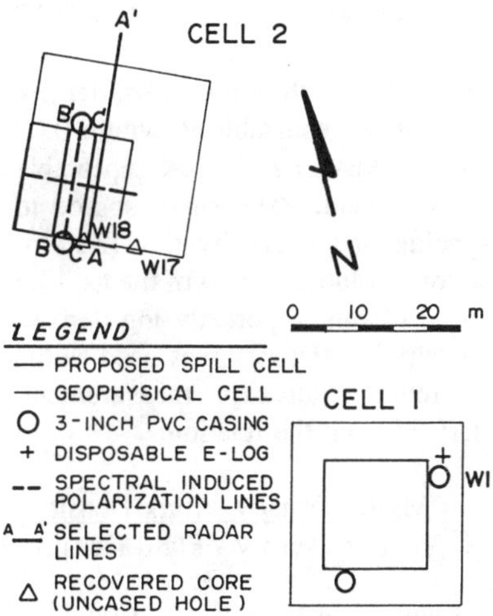

Fig. 1 Investigation site layout and location of radar lines discussed in paper

proven, the source of radar reflections in a shallow sandy aquifer is generally taken to be changes in water content, even as small as 1 percent (Cosgrave, 1987). Cosgrave has further suggested that stratigraphically detailed reflection packages found in radar records gathered over glaciofluvial and glaciolacustrine environments are, in fact, imaging the stratigraphic characteristic of that particular depositional environment.

A total of almost 0.5 km of GPR profiling was performed in the two geophysical cells. For all the surveys, the 100 MHz transmitting and receiving antennas were spaced 1 m apart, and moved together along the surface in discrete steps of 0.2 m or 0.25 m. Sixty-four traces were stacked and averaged for each trace recorded. This is an enormous amount of detailed data gathered from a very small area. This detailed geophysical survey performed on such a small scale is in itself unique, and for this reason alone of great interest. The radar records not only provide high resolution images of the aquifer, but serve as records for comparison to intended postspill surveys.

The principal limitation of GPR in Borden is a maximum depth of penetration of 6 m due to conductive attenuation. While lower frequency antennas might increase the depth of investigation, a loss of vertical resolution would offset this gain.

3.2 Spectral Induced Polarization (SIP)

SIP measures subsurface changes in apparent resistivity and phase over a broad range of frequencies. The detection of organic contaminants with SIP depends upon the muting of significant IP effects observed in background surveys. Two perpendicular lines were surveyed in Cell 2. A double-dipole configuration of 9 electrodes, a dipole spacing of 2 m, and a maximum number of receiver pairs of 6 was used. The length of each line from end to end was 16 m.

The principal limitation of applying SIP to a prespill background survey was the low clay content in the Borden sands. Klein and Sill (1983) found maximum polarization effects in sandstone with clay contents of between 5 and 9 percent. The Borden sands are noted for being free of all mineralogical clay except chlorite (Roberts and Mackay, 1986).

3.3 Vertical Electrical Conductivity Determination

Three methods of measuring downhole electrical conductivity were applied, each possessing a different vertical resolution of measurement. Inductive conductivity logging in the 4 PVC cased holes was performed with the Geonics EM39 which has a vertical resolution of 50 cm. An in situ electrical profiling device known as a disposable E-log (DEL) (Greenhouse and Pehme, 1990) was installed 1 m from borehole W11. The DEL is essentially a continuous downhole Wenner array with an MN spacing and approximate vertical resolution of 10 cm. Electrical conductivity profiling in the lab along split cores from W11 and W17 allowed conductivity measurements to be performed every 2 cm.

The purpose of implementing the three conductivity devices of various resolving capabilities was to gain some understanding of the scale of heterogeneity of electrical conductivity. By measuring the electrical conductivity profile on recovered cores, it was hoped that a correlation could be made between any measured electrical heterogeneities and observed stratigraphic heterogeneities. EM39 logging of borehole W11, electrical profiling of 7.65 m of split core from W11, and a DEL 1 m distant provided an ideal situation for comparative purposes.

3.4 Time Domain Reflectometry (TDR) Profiling

TDR is an electromagnetic technique which measures the dielectric constant of soils at frequencies between 10 MHz and 1 GHz. Essentially, it is "a one-dimensional linear radar system" (Topp and Davis, 1985). Topp et al. (1980) developed a simple relationship between dielectric constant and volumetric water content of a soil. Therefore in theory, TDR seems to be an ideal method for establishing vertical control for radar sections, and as a nondestructive method of profiling the porosity along a core. TDR measurements were taken every 5 cm along 9.1 m of split core.

The principal limitation in this TDR profiling method, when applied to split core, is the relatively poor accuracy of the measured dielectric permittivity and the inferred volumetric water content. The technique is capable of resolving the gross features but not the subtle variations that are of particular interest. The measurement inaccuracies are due to difficulties in ensuring that the core is saturated and disturbance of the core during the measurement process. The measurement accuracy could be improved substantially if applied to core that is not split.

3.5 Neutron (Porosity) and Gamma-Gamma (Density) Logging

Neutron logging determines the change in hydrogen atom concentration, and thus water content, with depth. Gamma-gamma logging determines the change in electron density, and therefore bulk density, with depth. Logging at slow speeds with a digital data acquisition system was performed in each of the four PVC cased holes.

At any particular depth in the wells, density measurements were repeatable to within 0.02 g/cc. Porosity measurements were repeatable to within 0.5 percent. The small source to detector spacing of the density tool (25 cm) offers superior resolution to that of the neutron tool. Because of this, a porosity log derived from the density log (DPHI), was used as the primary borehole porosity measurement. DPHI is derived from the relation:

$$\text{DPHI} = \frac{\text{Matrix Density - Bulk Density}}{\text{Matrix Density - Fluid Density}}$$

The fluid density is that of water, 1.00 g/cc, the bulk density is taken directly from the calibrated density logs, and the matrix density is that calculated by helium pycnometry and indirect calculation from the determined mineral composition of aquifer solids (Roberts and Mackay, 1986).

4 RESULTS

Discussed below are those results which were most relevant to satisfying the goals of the study.

4.1 Ground Penetrating Radar (GPR)

In the Borden sand pit, the GPR produced images are replete with stratigraphic detail (Figure 2). While it has not been conclusively shown that the radar stratigraphy of cells 1 and 2 are representative of real stratigraphy, the circumstantial evidence is very strong. The angled bedding of Cell 2 is very similar in strike and dip to the angled bedding of an excavated unsaturated zone site 70 m distant. Cores from Cell 2 have similar stratigraphy to cores from the same unsaturated zone site.

GPR's ability to image stratigraphic packages of high angle, low angle, and flat-lying bedding may be more important than its ability to delineate any single outstanding feature. Poulsen and Kueper (personal communication, 1989) found that for a slow release of tetra-

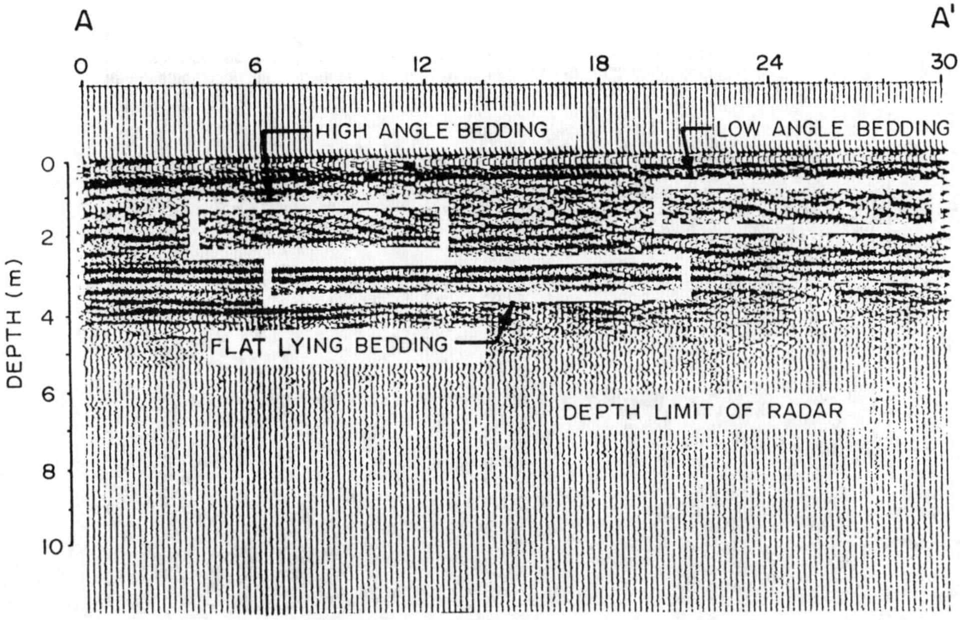

Fig. 2 Radar line A-A' (Figure 1) run across Cell 2

chloroethene in the unsaturated zone site, capillary forces control the migration of the solvent throughout the section. The evidence for this was that the dyed solvents, upon excavation, had observably migrated along angled bedding rather than overcoming the capillary entry pressures and shooting vertically downward. The solvents travelled through the coarse quartz lamina which lie above the finer grained heavy mineral lamina. The significance of this observation is that, at least in the case of Borden, subtle heterogeneities such as those observed in the radar records can determine the migration path of spilled solvent.

Furthermore, while all radar reflectors cannot be accounted for by ground-truthing or geophysical logs, the outstanding ones can be. A 20 cm thick dipping peat layer in Cell 2 (Figure 3) was identified in core, by porosity and density logs, and by TDR profiling. Reflectors within a 1 m thick peat layer in Cell 1 were correlated to variable porosity layers by porosity and density logs. A strong flat-lying reflector extending across Cell 2 was correlated to a 12 cm thick unusually coarse

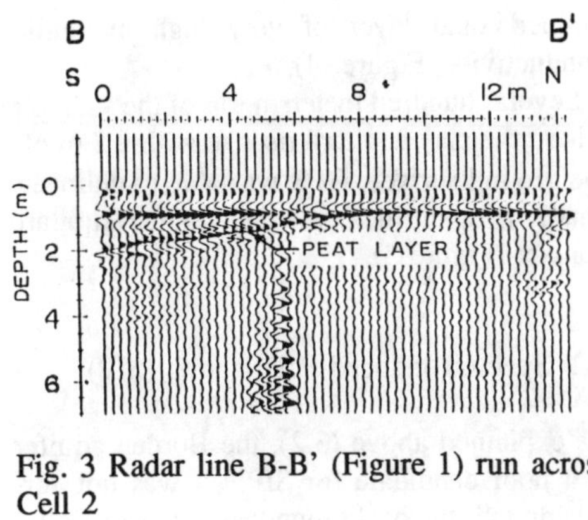

Fig. 3 Radar line B-B' (Figure 1) run across Cell 2

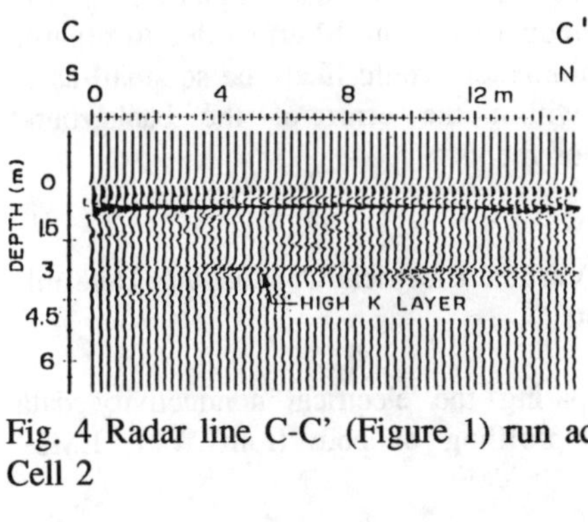

Fig. 4 Radar line C-C' (Figure 1) run across Cell 2

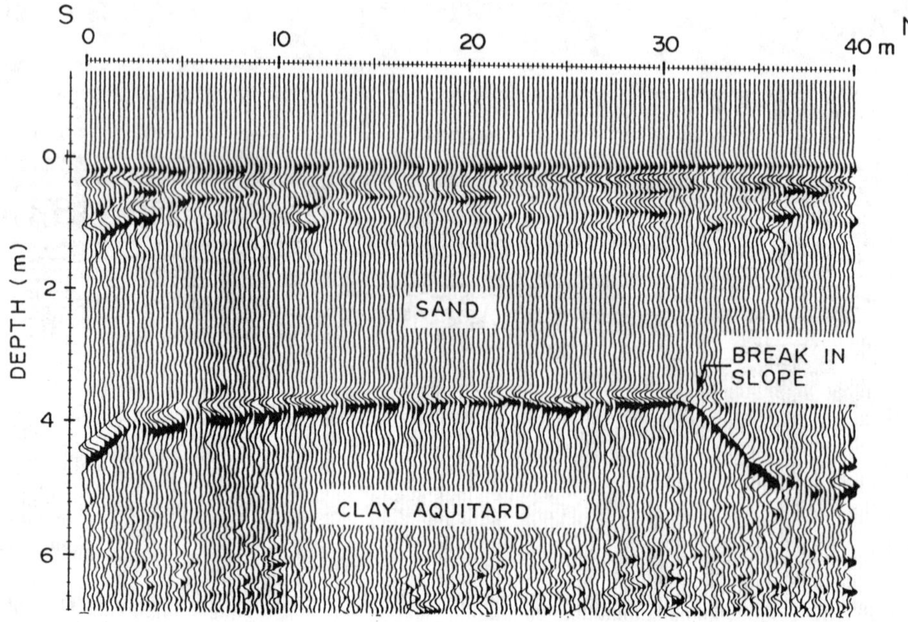

Fig. 5 Radar line surveyed north of the sand pit where the aquitard rises closer to surface

grained sand layer of very high hydraulic conductivity (Figure 4).

Several hundred meters north of the sand pit where the clay aquitard rises to within 5 m of the ground surface, GPR was able to delineate small irregularities in the aquifer/aquitard contact (Figure 5).

4.2 Spectral Induced Polarization (SIP)

As explained above (3.2), the Borden aquifer is a poor candidate for SIP. It was not surprising that the background survey over Cell 2 showed a small and scattered phase response. Any reduction in the IP effect due to organic contamination would likely be so small as to fall within the error of the background measurements.

4.3 Vertical Electrical Conductivity Determination

Comparing the electrical conductivity data from profiling of core from W11, EM39 logging of W11, and DEL logging very near to W11 gave good correlations, but also manifested many subtle and not so subtle variations in bulk conductivity as determined by the more detailed laboratory core measurements (Figure 6). What appears to be a large degree of scatter in the profiled conductivity data is actually real and repeatable variations correlatable with visually observable fine scale changes (on the order of centimetres) in grain size along the core.

Electrical conductivity variations in a saturated soil with no significant clay content are controlled by changes in pore water conductivity, porosity and the tortuosity of the conduction path. In soil conductivity models, the effects of porosity and tortuosity are usually grouped together in a single parameter, the formation factor.

In order to determine whether these conductivity variations were due to changes in pore water conductivity or changes in the formation factor, a 1.5 m section of core was desiccated and saturated with a high conductivity (0.86 S/m) solution of KCl. The core was then electrically profiled. Dividing the observed

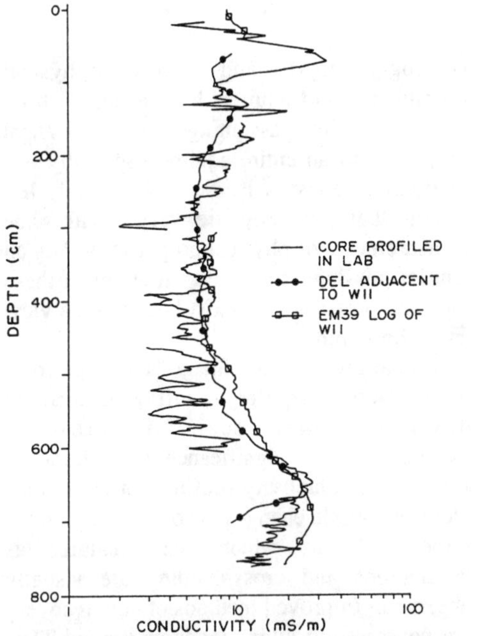

Fig. 6 Electrical conductivity data from profiling of core from W11, EM39 logging of borehole W11, and DEL logging 1 m from W11

saturated bulk soil conductivities by the known pore water conductivity yields the formation factor. Multiplying the formation factor by the original (before drying) measurements of bulk soil conductivity yields values for the pore water conductivity along the entire core section. Results showed a constant formation factor along the core (Figure 7). This indicates that the fine to small scale changes in conductivity are due to changes in pore water conductivity.

While these conductivity variations are real, the cause for some of the fine scale changes is probably differential drainage inside the core tubes from zones of larger grain size and wider capillary throats, to zones of smaller grain size and narrower capillary throats. This assumption is supported by the observations that: 1) the cores are not completely saturated when split; 2) the greatest fluctuations in conductivity are seen in coarse grained zones where conductivity minima are observed (eg. 2.95 m), and 3) in the sand from 2.5 m to 6

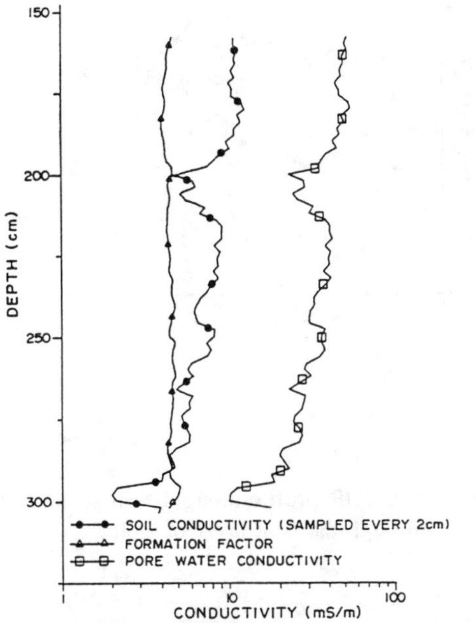

Fig. 7 Formation factor determination along 1.5 m section of core from borehole W11

m, where fluctuations in the sand are most noticeable, even the profiled conductivity peaks lie below the DEL and EM39 conductivities, suggesting pore water has been lost. Conductivity variations occurring on the scale of 10's of centimetres, appearing as continuous trends, and correlating well with EM39 measurements probably indicate small shifts in the conductivity of the original pore water.

4.4 Time Domain Reflectometry (TDR) Profiling

The limited accuracy of the measurements and an inability to ensure complete saturation of the split cores at conditions identical to the in situ aquifer removed any chance of a good correlation of the TDR profiles with the radar records. A TDR profile through the peat layer delineated by the radar section of Figure 4, however, shows that the method can readily pick out small zones with a relatively large contrast in water content (Figure 8).

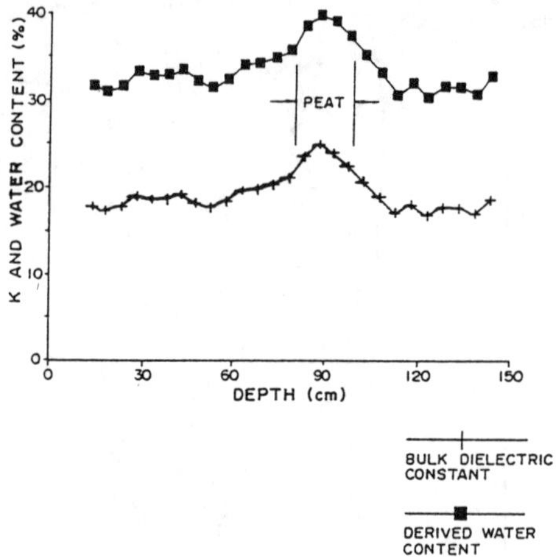

Fig. 8 A TDR profile along 1.5 m section of core through the peat layer of Cell 2

4.5 Porosity and Density Logging

Porosity and density logs indicated very subtle changes in physical properties could be recognized within the sand. Discrete changes in porosity of up to 3 percent over a few centimetres were observed and found to be repeatable.

The precision of bulk density measurements indicates that density logging may recognize DNAPL on a repeat survey basis. Given a scenario of a spill of tetrachloroethene (density 1.63 g/cc) in the Borden aquifer, let us assume a lower bound residual saturation of 5% of pore space, and an upper bound of 20% of pore space (Bernard Kueper, personal communication). Neutron logging has determined porosity in the Borden sand to be about 33 percent. Therefore, Borden sands at residual saturation may exhibit increases in density from a minimum of 0.01 g/cc to a maximum of 0.04 g/cc. Denser solvents, larger residual saturations, higher porosities, unsaturated zones, and pooled solvent will all further increase the change in measured bulk density of a given porous medium. While such variations in density would normally be unrecognizable on a first survey basis, they could be apparent in the context of a monitoring program.

5 CONCLUSIONS

Hydrogeological and hydrogeophysical investigations of aquifers have generally been satisfied with assigning bulk physical properties to an entire aquifer, especially one as homogeneous as Borden. This study has shown that not only are very small scale measurements of physical properties within an aquifer possible, but subtle variations of these physical properties do exist both in plan view and with depth.

GPR allows one to "image" the upper 6 m of the Borden aquifer in sufficient detail to describe small scale features of hydrogeological significance. Laboratory electrical conductivity profiling of split cores identifies subtle changes in conductivity on the same scale as which such features as laminations and cross-bedding are visually observed. Improved methods of core recovery are necessary to allow conductivity and TDR laboratory profiling data to be accurately compared to in situ measurements.

Without an actual controlled spill and a comparison of before and after measurements, nothing definitive can be said of the ability of the various geophysical methods used to detect DNAPLs in a sandy aquifer. While in theory, GPR has the greatest likelihood to directly detect spilled solvent, radar's depth of investigation is limited to 6 m in Borden. SIP is an unsuitable technique due to the almost complete absence of mineralogical clay and consequent low IP effect. Density logging at slow speeds may be able to detect DNAPL on a monitoring basis.

REFERENCES

Bolha, J., Jr. 1986. A sedimentological investigation of a progradational foreshore sequence: C.F.B. Borden. Unpublished M.Sc. thesis. University of Waterloo, Ontario.

Cosgrave, T.M. 1987. An investigation of shallow stratigraphic reflections from ground penetrating radar. Unpublished M.Sc. thesis. University of Waterloo, Waterloo, Ontario.

Feenstra, S. and Cherry, J.A. 1988. Subsurface contamination by dense non-aqueous phase liquid (DNAPL) chemicals. Proceedings of the International Groundwater Symposium, Halifax, Nova Scotia.

Greenhouse, J.P. and Pehme, P.E. 1990. A disposable E-log. In submittal (preprint available from first author). Department of Earth Sciences, University of Waterloo, Waterloo, Ontario.

Klein, J.D. and Sill, W.R. 1983. On the electrical properties of clay-bearing sandstone. Geophysics 48:62-75.

Krumenacher, M.J. and Taylor, R.W. 1988. Innovative application of induced polarization for detecting organic groundwater contamination. Proceedings of Petroleum Hydrocarbons and Organic Chemicals in Ground Water: Prevention, Detection and Restoration, Houston, Texas.

Kueper, Bernard, H. 1989. Personal Communication.

Kueper, Bernard H., McWhorter, D.B., and Frind, E.O. 1989. The behaviour of dense nonaqueous phase liquid contaminants in heterogeneous porous media. Proceedings of the International Symposium on Contaminant Transport in Groundwater, April 1989, Stuttgart, Germany.

Kueper, B.H., Frind, E.O. 1988. An overview of immiscible fingering in porous media. Proceedings of the International Symposium on Contaminant Transport in Groundwater, Stuttgart, Germany.

Olhoeft, G.R. 1986. Direct detection of hydrocarbon and organic chemicals with ground penetrating radar and complex resistivity. Proceedings of the National Water Well Association Conference on Petroleum Hydrocarbons and Organic Chemicals in Ground Water, Houston, Texas.

Poulsen, Mette and Kueper, Bernard. 1989. Personal Communication.

Roberts, P.V. and Mackay, D.M., eds. 1986. A natural gradient experiment on solute transport in a sand aquifer. Technical report no. 2. Department of Civil Engineering, Stanford University, Palo Alto, California.

Sadowski, R.M. 1988. Clay organic interactions. Ph.D. dissertation. Colorado School of Mines, Golden, Colorado.

Schwille, F. 1988. Dense Chlorinated Solvents in Porous and Fractured Media. Chelsea, Michigan: Lewis Publishers.

Topp, G.C. and Davis, J.L. 1985. Measurement of soil water content using time-domain reflectometry (TDR): a field evaluation. Soil Sci. Soc. Am. J. 49:672-678.

Topp, G.C., Davis, J.L. and Annan, A.P. 1980. Electromagnetic determination of soil water content: measurements in coaxial transmission lines. Water Resources Research 16(3):574-582.

Wilson, J.L., Conrad, S.H., Hagan, E., Mason, W.R. and Peplinski, W. 1988. The pore level spatial distribution and saturation of organic liquids in porous media. Proceedings of Petroleum Hydrocarbons and Organic Chemicals in Ground Water: Prevention, Detection and Restoration, Houston, Texas.

7 Remediation techniques

Subsurface Contamination by Immiscible Fluids, Weyer (ed.) © 1992 Balkema, Rotterdam. ISBN 90 6191 175 3

Combined steam injection and vacuum extraction for aquifer cleanup

K.S.Udell & L.D.Stewart
University of California at Berkeley, USA

ABSTRACT: One-dimensional, two-dimensional, and pilot-scale experiments have been performed to assess the effectiveness of a combined steam injection and vacuum extraction process for the removal of contaminants from aquifer environments. One-dimensional experiments have been performed in the laboratory using soil samples taken from several contaminated sites. Excellent recovery effectiveness was achieved in samples containing solvents, gasoline and diesel. The steam displacement/vacuum extraction process effectively removed a pool of xylene floating on a water table in a scaled two-dimensional experiment. A pilot-scale study was performed on a proposed EPA Superfund site contaminated by a mixture of solvents and illustrates the effect of heterogeneities on the cleanup process. As a result of the experiments, several mechanisms have been identified which account for the observed removal of the contaminants. These are vaporization of components with low boiling points, enhanced evaporation rates of semi-volatile components, physical displacement of low viscosity liquids, dilution and displacement of aqueous contaminants, and removal of residual contaminants by vacuum drying.

1 INTRODUCTION

Soil and groundwater contamination by organic solvents, petroleum products, agricultural chemicals, and radioactive compounds pose problems of major national and international concern. While above-ground techniques for soil cleaning can be applied to small, near-surface spills, effective in situ technologies are needed for sites where the contaminants are found at depth or where excavation would be impractical. In any remediation project, the first task is the removal of the source of contamination, commonly a leaking underground storage tank or associated piping which contains a non-aqueous phase liquid (NAPL). The next step is to carefully characterize the extent of the contamination. Initially, the contamination flows rapidly downward until some barrier such as a clay layer or the water table is reached or until enough small, discrete ganglia of the contaminant are formed and held within the pore spaces by capillary and adhesion forces to contain the total volume of contaminant spilled. Spreading continues by mass diffusion into the soil gas and into flowing groundwater. The final task is to evaluate, select and implement various remedial actions to remove the contaminant from this complex system. In situ thermal techniques have been shown to be particularly effective in removing the trapped lenses or ganglia, primarily due to higher mass transfer rates from the contaminant phase to the flowing gas phase as driven by increasing contaminant vapor pressures with increasing temperature.

Thermal techniques for the recovery of volatile and semi-volatile liquids from porous media are not new. A large body of research on steam injection for enhanced oil recovery by viscosity reduction and distillation can be found in the petroleum literature (i.e., Mandl and Volek, 1969, Volek and Pryor, 1972, and Konopnicki et al., 1979). However, the use of thermal processes for the in situ recovery of contaminants has a very short history. Early exploratory field work on steam injection was carried out in the Netherlands in the early 1980's (Hilberts, 1985). Radio frequency heating has been tested (Dev, 1986) as has in situ vitrification by resistance heating (Fitzpatrick et al., 1986). Most recently, steam injection/vacuum recovery (patent pending) for solvent recovery was pilot tested by Udell and Stewart (1989).

More definitive laboratory experiments on steam injection for volatile and semi-volatile liquid contaminant recovery have been reported. Lord et al. (1987) reported kerosene removal efficiencies as functions of time of steam contact and soil type. Hunt, Sitar and Udell (1988) reported removal of pure trichloroethene, a mixture of benzene and toluene, and gasoline as separate liquid phases flowing ahead of the steam condensation front. Stewart and Udell (1988) developed three phase flow solutions to the problem of steam displacement of volatile contaminants and compared those results to experimental data with good agreement.

The research reported here adds to this body of literature through the presentation of new experimental data and further interpretation of previously reported data. The experimental data aid in defining the mechanisms which are responsible for the surprising success of combined steam injection and vacuum extraction for soil and groundwater decontamination.

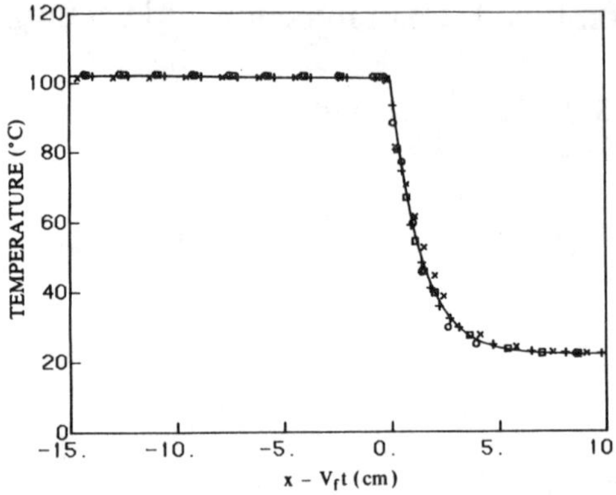

Fig. 1 Temperature distribution near steam condensation front

2 MECHANISMS

When saturated steam is injected into an initially cool porous medium, the steam condenses, giving up the latent heat of vaporization to heat the porous solid and interstitial fluids. With continued injection three distinct zones will develop: an early isothermal steam zone, a variable temperature zone, and an isothermal zone at the initial medium temperature saturated with water and mobilized contaminant (see Figure 1). The rate of expansion of the steam zone is directly proportional to the injected steam enthalpy flux if the length of the variable temperature zone remains constant with time (Menegus and Udell, 1985, Hunt, Sitar and Udell, 1988, and Stewart and Udell, 1988). The interface between the steam zone and the variable temperature zone is essentially the location of steam condensation if the temperature gradient in the steam zone is not significant. For injection pressures near atmospheric and soil thermal properties and temperatures characteristic of field conditions, approximately 350 pore volumes of steam will be required to condense at the condensation front to expand the steam zone an additional unit volume. Thus, steam injection is characterized by high vapor volumetric flow rates in the steam zone. Within this thermal and fluid distribution context, contaminant removal mechanisms can be described.

2.1 Vaporization of low boiling point contaminants

For non-aqueous phase contaminants with boiling points lower than that of steam at pressures near atmospheric, vaporization will occur in the variable temperature zone ahead of the steam condensation front (point A in Figure 1). The resultant contaminant vapor would occupy a region extending to the left of point A into the steam zone. Yet, because of the high flux of steam in the steam zone, the contaminant vapor will be convected forward and displaced into the variable temperature zone ahead of point A where it must condense. Therefore, the low boiling point liquid will be mobilized ahead of the steam condensation front in spite of capillary forces which would otherwise trap the second phase liquid. As evidenced by the experimental observation of second phase contaminants flowing from the exit of the experiments reported by Hunt, Sitar and Udell (1988), it is expected that the contaminant vapor either condenses upon the second phase liquid or that any produced emulsions coalesce to the second phase liquid bank rather than flow as contaminant-in-water emulsions. Without doubt, pure separate liquid phase contaminants with boiling points less than that of water will be completely removed from the steam zone except for the small amount which is adsorbed to solid surfaces or is dissolved in liquid water which may be present in the steam zone.

With the accumulation of distributed, high volatility second phase contaminants into a coherent bank ahead of the steam condensation front, buoyancy forces reenter the displacement process. Hydrostatic forces in a connected contaminant bank can lead to percolation of the contaminant either upward or downward depending on whether the density is less than or greater than that of water, respectively.

2.2 Enhancement of contaminant evaporation rates

In most remediation projects, the contaminant will consist of many chemical components of varying thermodynamic properties. It is inevitable that the low vapor pressure components of the separate phase contaminant will remain in the pore spaces after the passage of the steam condensation front. However, these components will be contacted by the

Table 1 Physical properties of various contaminants

Compound	Molecular Weight (g/g-mol)	Liquid Density (g/cm^3)	Melting Point (°C)	Boiling Point (°C)	Water Solubility at 20°C (mg/L)	MW x Vapor Press. at 20°C (g-kPa/g-mol)	MW x Vapor Press at 100°C (g-kPa/g-mol)	Evaporation Factor
Benzene	78.1	0.88	5.5	80.1	700	781	Vapor	NA
Toluene	92.1	0.87	-95	110.6	540	267	6,630	25
Ethylbenzene	106.2	0.87	-95	136.2	140	99	3,720	38
o-Xylene	106.2	0.88	-25	144.4	185	88	3,290	37
Naphthalene	128.2	0.97	80	218	32	1.4	389	278
Dodecane	170.3	0.75	-9.6	216.5	0.0034	2.7	346	128
Phenanthrene	178.2	0.98	101	340	1.18	0.005	11	2200
Hexadecane	226.5	0.77	18	287	0.0063	0.2	29	145
Acetone	58.1	0.79	-94	56.3	∞	1,409	Vapor	NA
2-Butanone	72.1	0.81	-86	79.6	27,000	683	Vapor	NA
Tetrachloroethene	165.9	1.63	-19	121.0	140	376	8,847	24
1,1,1-TCA	133.4	1.31	-30	74.1	720	1,707	Vapor	NA
Trichloroethene	131.4	1.46	-73	86.8	1,100	1,016	Vapor	NA

steam that is flowing toward the condensation front. The evaporation of these components will occur due to their significant vapor pressures at the steam temperature. While the rate of this evaporation process will be mass transfer limited as with isothermal vacuum extraction procedures, the evaporation rate of each component will be proportional to its gaseous phase concentration at equilibrium with the multicomponent liquid. This concentration is proportional to the vapor pressure multiplied by the constituent molecular weight, corrected for its molar fraction in the multicomponent liquid. By Raoult's law, this correction factor is equal to the component's molar fraction for ideal mixtures. By increasing the temperature from initial conditions to the steam temperature, significant increases in the vapor pressures of most contaminants will result. Table 1 lists the vapor pressures at 20 °C and 100 °C of several contaminants of interest. The evaporation increase factor is simply the ratio of the vapor pressure at 100 °C to that at 20 °C and is representative of the increase for equal vapor flowrates. This evaporation increase factor translates roughly to the decrease in the remediation timescale. As illustrated in the table, decreases of cleanup time scales over ambient temperature vacuum extraction alone of 25 to 40 are possible for components such as tetrachloroethene or ethylbenzene. Furthermore, for longer chained "non-volatile" components such as naphthalene, evaporation can occur at significant rates only if the temperature is increased. Therefore, for less volatile petroleum hydrocarbons such as diesel or lubricating oils, steam injection appears to be particularly applicable.

The fate of the vaporized low volatility components is similar to that of the highly volatile constituents, namely condensation near the steam condensation front and coalescence into a bank of second phase liquid flowing ahead of the steam condensation front. Such banks of second phase contaminant have been experimentally observed in the laboratory for gasoline, xylene, a mixture of benzene and toluene, and diesel.

2.3 Displacement of NAPLs at the steam condensation front

The displacement of second phase liquids at the steam condensation front has been studied from a theoretical standpoint (Stewart and Udell, 1988) and has been visually observed during steam displacement of o-xylene in two dimensional experiments. A distinguishing feature of steam displacement is the relatively high pressure gradient in the steam zone due to the high steam velocities necessary for the propagation of the steam condensation front. Theoretical pressure profiles for steam displacement of water and a NAPL are shown in Figure 2. The pressures shown are referenced to the water pressure at the steam condensation front and nondimensionalized by dividing by the steam-water interfacial tension and multiplied by the square root of the ratio of the permeability to porosity. The high pressure gradients in the steam zone give rise to two primary effects. First, the water remaining in the steam zone is driven to the steam front, which reduces the water content to values represented by the zero mobility

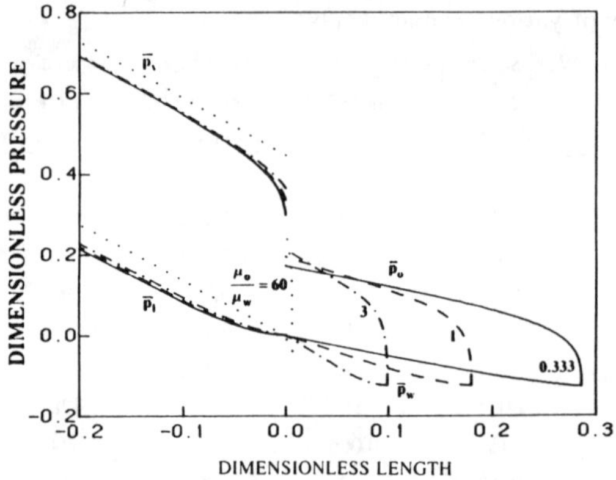

Fig. 2 Pressure distributions near steam condensation front for various viscosity ratios (from Stewart and Udell, 1988)

limit. Second, the high pressure gradient is stabilizing with respect to displacement stability, as determined by the Saffman and Taylor (1958) stability criterion of a higher pressure gradient in the displacing phase than in the displaced phase. Indeed, Stewart and Udell (1988) showed that stable displacement of second phase liquids with viscosities greater than water can occur. Also, steam is capable of mobilizing any length of NAPL bank for a sufficiently low NAPL viscosity.

2.4 Dilution of initial pore water

The reduction of soil contaminant concentrations can occur through steam displacement of the contaminant in solution in the aqueous phase. As mentioned in the discussion of the displacement effectiveness of steam injection, all but the immobile water is removed from the pores. Therefore, if 80% of the contaminated water is displaced from the soil, the soil concentration, referenced to the mass of contaminant per dry mass of soil, would be expected to be reduced by at least 80% if all contaminant is in solution. However, a bank of distilled water representing the steam condensate is formed ahead of the steam condensation front. Through mixing, this steam condensate dilutes the initial water. Therefore, the water remaining in the steam zone would not necessarily contain the same contaminant concentration as the initial water.

To determine the reduction of aqueous phse contamination by steam displacement and dilution, experiments were performed in the same experimental apparatus as used in Hunt et al. (1988) and Stewart and Udell (1988). The solute chosen was sodium chloride, because of its low adsorption on quartz and zero vapor pressure. A dilute saline solution of 14.0 g/kg concentration was injected into a sandpack until the effluent fluid electrical resistivity matched that of the injected fluid. The 5.04-cm diameter sandpack was then subjected to a steam flowrate of 2 mg/min. A steady steam condensation front velocity was observed. The effluent was collected in a graduated cylinder. To maintain one-dimensional conditions in the sandpack, a series of heater tapes were sequentially turned on and adjusted in accordance with a predetermined calibration of heat loss versus sandpack/room temperature difference. After steam was observed in the outlet tubing, noted as time of steam breakthrough, injection continued until the effluent mass flux equaled the steam injection rate. The water remaining in the sandpack was thus considered to be immobile and at the irreducible saturation, which was determined from a mass balance to be 12.9%. A measured volume of distilled water was then injected into the steam zone as the column was allowed to cool. At the end of this procedure, the sandpack was completely saturated with water. The system was allowed to rest for 48 hours, allowing the salt remaining in the sandpack to diffuse into the injected water. The water was then allowed to drain from the sandpack through the injection port by gravity. All effluent solutes were analysed for salt concentrations by atomic adsorption spectroscopy.

Table 2 Water volumes and salt masses during various sequences of steam injection

	Water Volume (ml)	Salt Mass in Solution (g)
Initial State	724	10.0
Steam Injection (1.97 ml/min)		
- before breakthrough	188	0
- after breakthrough	147	0
- total	335	0
Effluent		
- before breakthrough	803	9.80
- after breakthrough	163	0.0133
- total	966	9.81
Residual Water		
- before breakthrough	109	0.191
- final	93	0.177

A summary of the material balances of water and salt is presented in Table 2. Of the 10.1 grams of salt initially in solution in the 0.724 liters of initial pore water, 9.8 grams were recovered in the 0.803 liters of effluent water at the time of breakthrough. An additional 0.0133 grams of salt were removed after breakthrough accounting for the remaining mobile water which flowed from the sandpack from the steam zone. From the measured volumes and salt concentrations of the drained water recovered after the waterflood and resting period, a total salt mass of 0.177 grams was determined to have remained in the sandpack after steam injection. Therefore, while 87.1% of the saline water was displaced by the steam, a decrease in salt concentration of 98.45% was observed as would be calculated on a dry sand mass basis. Thus, the residual water remaining in the steam zone was diluted on average by a factor of 7.3. The sensitivity of this dilution factor on the length of the sandpack, the flowrate of the steam, the pore dimension, and the pore size distribution is not known.

2.5 Desorption of contaminants

When fluids containing a contaminant come into contact with a solid such as a porous matrix, there is a tendency for a fraction of the contaminant to collect on the solid surface. This surface excess of contamination arises because intermolecular forces result in the formation of a monomolecular layer of contaminant on the solid surface which in turn may cause the formation of other layers due to the attractive forces among the contaminant molecules. In soil systems the attractive forces between organic carbon and organic contaminants are particularly large. For an isothermal system, an increase in pressure will result in an increase in the contaminant adsorption as predicted by the Langmuir adsorption isotherm. Yet, for this adsorption to occur a considerable amount of heat must be evolved which is called the latent heat of adsorption. Likewise, for desorption to occur a large amount of energy must be supplied to the system.

The energy necessary for desorption to occur in contaminated soils can be provided by steam condensation. Although raising the temperature of the contaminated soil increases the vapor pressure of contaminants, the intermolecular forces required for a surface to hold a molecule are significantly increased. The net result in most instances is a decrease in the adsorbed fraction of vaporized or dissolved contaminant. In fact, the use of low pressure steam to regenerate activated carbon used in solvent recovery systems has been common in industry since early in this century (Scheflan and Jacobs, 1953).

In the context of the steam injection/vacuum extraction process, desorption is expected to occur in the steam zone in a manner analogous to that of second phase contaminant evaporation. Furthermore, desorption in the region ahead of the condensation front is expected due to the dilution of the interstitial water by steam condensate as well as the increase in temperature within the variable temperature zone. The latter effects would be most prevalent in the use of steam injection to remediate aquifers containing aqueous phase contaminants.

2.6 Boiling of interstitial water

The thermodynamic system consisting of steam, water, semi-volatile contaminant, and porous matrix found within the steam zone is complex and unusual. Because of capillary forces stemming from the radius of curvature of the water-vapor interface, the water is in a stable superheated state in equilibrium with a slightly superheated vapor (Udell, 1983). This situation produces an effective liquid to vapor phase change mechanism where a decrease in the local vapor pressure will promote evaporation of the interstitial water at a rate determined primarily by heat transfer from adjacent solids. From an analysis of an adiabatic control volume containing steam, water and a porous matrix, it is evident that the removal of a given mass of steam of large enthalpy, will require a decrease of the system internal energy as well as a decrease in the interstitial water mass. The decrease in the porous matrix internal energy provides the necessary heat for the vaporization of the liquid water. With continued steam withdrawal, the system temperature will continue to drop until the liquid water content is reduced to zero. The significance of this effect is that the vacuum withdrawal of steam from the steam zone can produce a vapor flux from microscopic and macroscopic regions that would not be otherwise contacted by flowing steam. This mechanism can be brought to bear in a field process by simply stopping steam injection to a contaminated region while maintaining a vacuum withdrawal of the vapor phase.

The enhanced contaminant recovery rates obtained by this mechanism have been determined in a series of experiments that were performed on soil samples taken from sites contaminated by diesel. A schematic of the experimental apparatus is shown in Figure 3. For each run, the sample was packed into a cylindrical stainless steel holder of 3.8-cm inner diameter and 21.6-cm length. Approximately 2.5 cm of Ottawa sand (100-115 mesh) was packed on either end of the sample, and screens were placed between the sand and the end plates. Once packed, the holder was wrapped with a thin

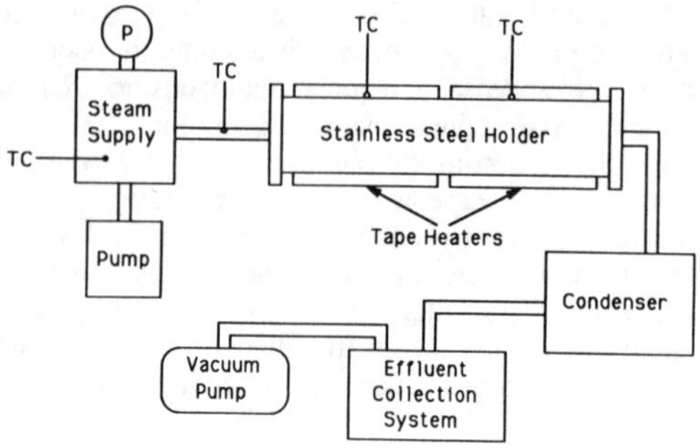

Fig. 3 Experimental apparatus for steam injection into contaminated soils

ceramic tape. Two tape heaters were then applied along the length of the holder and covered the entire radial surface area. The heater tapes were used to maintain adiabatic conditions at the surface during steaming. Finally, the apparatus was placed in a 2.5-cm thick ceramic blanket. The end plates were covered with additional insulation.

The steam generator consisted of copper tubing wound around electrical resistance heating plates and surrounded by an insulating blanket. A constant displacement pump supplied deionized, deaerated water at ambient temperature to the steam generator at a constant rate. During injection, the heater tapes were turned on sequentially as the steam front propagated through the core. A flexible 3-mm teflon exit line was run through an ice bath over a length of approximately 0.5 meters. The effluent was collected in 8- and 40-ml vials which were also held in an ice bath. After steam breakthrough, the steam injection was continued for varying lengths of time before termination.

The steam injection was followed by a period of vacuum extraction at 711 mm Hg. A vacuum pump was connected to the exit of the sample holder and the vacuum was applied for approximately five seconds to clear the exit line of steam injection condensate. The vacuum was then applied for some period of time and a sample of the condensate was collected. With the apparatus under vacuum, the samples cooled quickly due to the removal of energy from the hot matrix which was used to vaporize the remaining contaminated interstitial water.

A summary of the experimental conditions are given in Tables 3 and 4. The flow rate refers to the rate of steam injection into the sample, the maximum injection pressure is the pressure recorded at the steam generator at steam breakthrough, breakthrough corresponds to the time elapsed between the initiation of steam and the appearance of steam at the exit, steaming period refers to the time for which steam injection continued beyond breakthrough, and

Table 3 Diesel contaminated soil samples used in experiments of combined steam injection/vacuum extraction

Expt #	Description of Sample
1	Silty clay with gravel
2	Silty clay, uniform
3	Clayey gravel with sand
4	Clayey gravel, dry
5	Fine sand, dry
6	Fine sand, dry

Table 4 Experimental conditions

Expt #	Flow Rate (ml/min)	Max. Inj. Pressure (kPa)	Breakthrough (min)	Steam Period (min)	Vacuum Period (min)
1	0.50	<7	105	125	100
2	0.67	34.5	100	45	87
3	0.80	10	64	30	100
4	0.80	<7	66	34	90
5	0.40	1.4	184	31	96
6	0.60	1.4	105	17	97

Table 5 Summary of concentrations in soils and effluent liquids

Expt #	Original Soil TPH Concentration (mg/kg)	Processed Soil TPH Concentration (mg/kg)	TPH Concentration in Steam Condensate (mg/L)	TPH Concentration in Vacuum Condensate (mg/L)
1	1500	19	110	21
2	120	<10	82	48
3	110	45	92	49
4	440	<10	<10	578
5	710	220	25	180
6	<10	<10	5	230

vacuum period refers to the elapsed time during which the sample was held under vacuum. The steam injection rate, the time periods of application of each phase and the soil type can be correlated to the levels of remediation achieved. The soil and water samples were analyzed for total extractable hydrocarbons (TEH) according to EPA Method 3550 for soils and EPA Method 3510 and 8015 for water. A summary of the results of the chemical analyses are presented in Table 5. As well as providing data on the importance of processing conditions and soil heterogeneities on decontamination rates, the data in Table 5 clearly illustrates the importance of the vacuum drying process on contaminant recovery rates. The solubility of diesel in water at room temperature is approximately 25 ppm and therefore the concentrations above 25 ppm observed in the condensed steam are evidence of removal of the diesel as a separate phase. The initial steam condensate from runs 4, 5 and 6 contained very little diesel compared to the separate phase of diesel which was recovered during the first three runs. However, the large amount of diesel recovered as a separate phase during the application of the vacuum indicates the importance of this phase of the process.

3 FIELD RESULTS

Given the extraordinary success of laboratory studies of the combined steam injection and vacuum extraxction technique, a field study was undertaken. The site chosen for the study has been used as a solvent recycling and acid treatment facility since 1973. Soil contamination at the site resulted from leaking underground storage tanks and surface spills associated with the handling of these industrial wastes. Twenty organic volatile compounds and six non-volatile compounds have been identified at the site. Fourteen of the organic compounds have been detected in concentrations exceeding 1.5 ppm. The principal contaminants of the soil are xylenes, ethylbenzene, 1,2-dichlorobenzene, 1,1,1-trichloroethane and acetone which, along with trichloroethene and tetrachloroethene, occur with maximum concentrations greater than 500 ppm. Further details of the site description can be found in Udell and Stewart (1989).

The location of the pilot study was in the region of highest contaminant concentrations. A clay barrier is located approximately twenty feet below the surface which inhibited the further vertical migration of the contaminants. In characterizing the extent of soil contamination, the complications associated with mixtures were apparent. High concentrations of acetone were found throughout the site which is not surprizing given its high volatility and complete miscibility in water. Most of the other contaminants at this site are miscible or highly soluble in acetone. Thus, the acetone provided a mechanism for contaminants with low water solubilities and mobilities to migrate with the water.

The pattern of injection and recovery wells used for the pilot study was selected as a prototype for a design to remediate a small-scale site. The system consisted of a central recovery well surrounded by six injection wells evenly spaced on a ten-foot diameter circle. Also, 12 temperature observation wells were placed inside the pattern, 3 on the edge, and 2 outside the pattern. The recovery well was completed throughout and extended into the clay layer. Liquids were removed from the recovery well via a jack pump and vapors were drawn out through application of subatmospheric pressure provided by

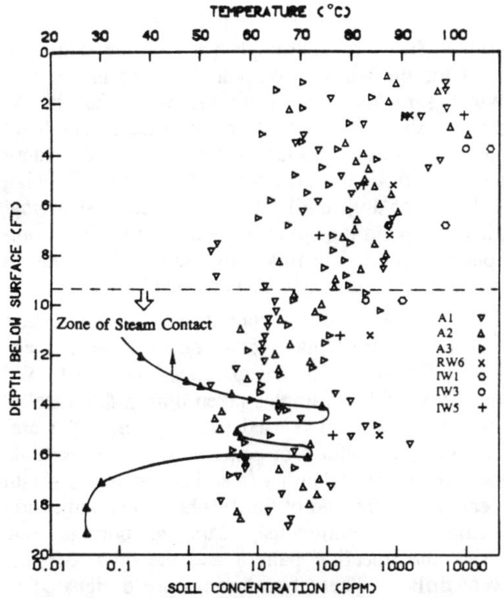

Fig. 4 Total soil contaminant concentrations (less acetone) and temperature profile before breakthrough (from Udell and Stewart, 1989)

a vacuum pump placed on an extraction line downstream of a condenser. The injection wells were drilled to 19 feet and completed over only the bottom three feet.

Baseline permeability, gas composition, and mass transfer properties of the contaminated zone surrounded by the injection wells, were obtained during a 40-hour vacuum extraction test. Recovered gases were saturated with the contaminant vapors with concentrations of approximately 60 grams per meter cubed (gm/m^3). Gas flow rates remained nearly constant at 0.9 standard cubic meters per minute (scmm). Steam was injected into the six wells surrounding the recovery well at maximum pressures of 41 kPa. Total flow rates of approximately 110 kg/hr of steam were maintained for most of the duration of the 140-hour steam injection period. After 32 hours of steam injection, the recovery well temperature began to increase dramatically, indicating that steam had broken through. Thereafter, noncondensible gas flow rates decreased although gaseous phase contaminant concentrations were indicative of saturated conditions. Yet, the concentrations of compounds with lower vapor pressures and higher boiling points increased and the concentration of the compounds with higher vapor pressures and lower boiling points decreased. The more volatile compounds were removed first and the less volatile compounds later, and in greater proportions, after steam had heated the formation and was flowing through the soil.

The water pumped out of the recovery well prior to steam breakthrough was near the saturation limit for the various compounds found in the soil. Water samples taken from the well after breakthrough showed a sharp decrease in the contaminant concentrations. Contamination in the vapor removed from the steamed soil was initially high and did not drop until after 100 hours of steaming. Sharp drops in the concentration of individual components occurred at times inversely proportional to the vapor pressures of each compound.

The total soil concentrations of contaminant less acetone are plotted versus depth below the surface in Figure 4. The designations RW6, IW1, IW3 and IW5 refer to samples taken during the installation of the wells and used to characterize the presteaming distribution of contamination. These data show the large variability in soil concentrations with location. At the end of the 140-hour steam injection period, two continuously sampled borings, one inside the injection pattern and the other outside, were drilled. These two borings are designated as A1 and A2. At the time which these borings were recovered, temperature observation wells nearest these locations indicated an isothermal steam zone between 9 and 17 feet below the ground surface. This is indicated in Figure 4. A third boring was recovered after a short period of steam and vacuum cycling and is referred to as A3. Also plotted in Figure 4 is the vertical temperature profile at a location midway between the injection and recovery wells after 11 hours of steam injection. The temperature profile is shown to indicate evidence of a low permeability region at a depth of about 14.5 feet which apparently was not contacted by the injected steam. As evidenced by the results of this extensive soil analysis, the levels of soil contaminate concentrations decreased nearly two orders of magnitude in the steam zone. Also, apparent is the scatter of the soil concentration data which is presumed to be due to the heterogeneous nature of actual geologic media.

Vacuum cycling did produce contaminated vapors which were qualitatively observed to be of higher contaminant concentrations than those measured during active steam injection. The effectiveness of this phase of the steam injection/vacuum extraction process for recovering contaminants in the low permeability regions could not be adequately tested because of concerns regarding the possibility of drawing contaminants in from outside the test pattern.

4 CONCLUSIONS

Several mechanisms have been identified which account for the observed removal of separate phase liquid contaminants from porous media by combined steam injection and vacuum extraction. These are:

1. Vaporization of low boiling point components.
2. Accelerated evaporation of semi-volatile components.
3. Displacement of low viscosity liquids.
4. Dilution and displacement of the original contaminated water by steam and its condensate.
5. A possible decrease in the fraction of contaminant adsorbed to the solids.
6. Removal of residual contaminated water from the steam zone by drying using vacuum extraction alone.

ACKNOWLEDGMENTS

Support for this work was provided by NIEHS Award No. 3 P42ESO4705-0251 and EPA Award No. R-815247-01-0. Additional funding for Dr. Stewart's work was provided by the Switzer Foundation. Mr. Lawrence Moffett is gratefully acknowledged for performing the salt displacement experiments.

REFERENCES

Dev, H., 1986, "Radio Frequency Enhanced *In Situ* Decontamination of Soils Contaminated with Halogenated Hydrocarbons," *Proc. of EPA Conf. on Land Disposal, Remedial Action, Incineration and Treatment of Hazardous Waste*, Cincinnati, Ohio, April 1986, published by EPA Haz. Waste Engin. Res. Lab., Cincinnati, Ohio.

Fitzpatrick, V.F., Timmerman, C.L., and Buelt, J.L., 1986, "*In Situ* Vitrification - A Candidate Process for *In Situ* Destruction of Hazardous Waste," *Proceedings, Seventh Superfund Conference*, Washington, DC, 1986, published by HMCRI.

Hilberts, B., 1985, "*In Situ* Steam Stripping," Assink, J.W., Van Den Brink, W.J., Eds., *Contaminated Soil*, Proc. of First Intern. TNO Conf. on Contaminated Soil, Utrecht, The Netherlands, November 11-15, 1985, pp. 680-687.

Hunt, J.R., Sitar, N. and Udell, K.S., 1988, "Nonaqueous Phase Liquid Transport and Cleanup 2. Experimental Studies," *Water Resources Research*, Vol. 24, pp. 1259-1269.

Konopnicki, D.T., Traverse, E.F., Brown, A., and Deibert, A.D., 1979, "Design and Evaluation of the Shiells Canyon Field Steam-Distillation Drive Pilot Project," *J. Pet. Tech.*, Vol. 31, pp. 546-552.

Lord, A.E., Jr., Koerner, R.M., Murphy, V.P. and Brugger, J.E., 1987, "Vacuum-Assisted, *In Situ* Steam Stripping to Remove Pollutants from Contaminated Soil," *Proc. of EPA Conf. on Land Disposal, Remedial Action, Incineration and Treatment of Hazardous Waste*, Cincinnati, Ohio, May 6-8, 1987, p. 511.

Mandl, G. and Volek, C.W., 1969, "Heat and Mass Transport in Steam-Drive Processes," *Soc. Pet. Eng. J.*, Vol. 9, pp. 59-79.

Menegus, D.K. and Udell, K.S., 1985, "A Study of Steam Injection Into Water Saturated Capillary Porous Media," *Heat Transfer in Porous Media and Particulate Flows*, ASME HTD, **46**.

Saffman, P.G., and Taylor, G.I., 1958, "The Penetration of a Fluid into a Porous Medium or Hele-Shaw Cell Containing a More Viscous Liquid," *Proc. Roy. Soc.* (London), Series A, Vol. 245, pp. 312-329.

Scheflan, L., and Jacobs, M.B., 1953, *The Handbook of Solvents*, D. Van Nostrand Co., New York.

Stewart, L.D., and Udell, K.S., 1988, "Mechanisms of Residual Oil Displacement by Steam Injection," *SPE Reservoir Engineering*, Vol. 3, pp. 1233-1242.

Udell, K.S., and Stewart, L.D., 1989, "Field Study of *In Situ* Steam Injection and Vacuum Extraction for Recovery of Volatile Organic Solvents," Sanitary Engineering and Environmental Health Research Laboratory, University of California Berkeley, UCB-SEEHRL Report No. 89-2.

Udell, K.S., 1983, "Heat Transfer in Porous Media Heated from Above with Evaporation, Condensation and Capillary Effects," *Journal of Heat Transfer*, Vol. 105, pp. 485-492.

Volek, C.W. and Pryor, J.A., 1972, "Steam Distillation Drive - Brea Field, California," *J. Pet. Tech.*, Vol. 24, pp. 899-906.

Subsurface Contamination by Immiscible Fluids, Weyer (ed.) © 1992 Balkema, Rotterdam. ISBN 90 6191 175 3

Contamination of aquifers by hydrocarbons and restoration with surfactants

J. Ducreux, C. Bocard, C. Gatellier & L. Minssieux
Institut Français du Pétrole, Rueil-Malmaison, France

ABSTRACT : During movements of the water table the contamination of groundwater can result from leaching trapped oil at residual saturation in the vadose zone. Solubilization of aromatics from gasoline and fuel oil # 2 is simulated by sand column experiments. Total and individual aromatics are analysed by high resolution gas chromatography. Most of the light aromatics are quickly solubilized compared with C9-C10 monoaromatics and diaromatics, accountable for a long-term contamination.
The use of surfactants is suggested to restore quickly an oil contaminated soil. Physical modelling of injection parameters will lead to design a process without any groundwater contamination by surfactants. Different other parameters are taken into account for surfactant selection : low interfacial tension, drainage efficiency, retention on porous matrix, distribution between oil and water phases. In order to mitigate the adsorption of anionic surfactant, a sodium chloride preflush is suggested. Even when implemented at concentration lower than required for moving trapped oil, surfactants enhance aerobic biodegradation by improving the oil and nutrients contact and increasing the microbial oil colonization.
Field experiments are in progess to demonstrate the operational feasibility of soil clean up by in-situ surfactant flushing.

1 INTRODUCTION

Extensive field experience has provided a large knowledge on the underground transport of oil pollution and especially, has revealed the significance of the unsaturated zone as important in the transfer of pollutants.

Simulation experiments in columns has led to measure the individual fate of aromatics in the water phase. To avoid the long-term effects of low water soluble and non-volatile aromatic hydrocarbons, immobilized in the soil matrix, a treatment has to be undertaken. Experiences of the authors in oil secondary recovery and in petroleum microbiology has led to evaluate the chance of surfactants to enhance both oil displacement by in-situ flushing and natural biodegradation of residual hydrocarbons.

2 EXPERIMENTS

2.1 Column test used in solubilization of aromatic hydrocarbons

A plexiglass column, 2.5 cm i.d. x 15 cm long was packed with sand containing no organic matter in order to avoid the adsorption interference. Its grain-size distribution was in a relatively narrow range (79 % between 297 and 420 μm). The water filling was performed in order to obtain a residual air phase (i.e. dry filling), followed by a drainage step to simulate water table fluctuation in the capillary fringe zone. The accessible porous volume (Vp_{acc}) was 25 ± 0.5 milliliters. The water saturation was about 85 %. Water spiked with a biocide (1 mg/l NaN_2) was pumped through the column at a constant rate (21 ml/h). A leaded gasoline and fuel oil # 2 (FO2) were used as non-aqueous phase liquid (NAPL). Two Vp_{acc} were injected downward into the column, then the water flow was introduced at counterflow to displace a part of the oil. Behind the flow, oil was entrapped at residual saturation. Column effluent was sampled intermittengly and analyzed for dissolved hydrocarbons. The gas chromatograph was fitted with non polar capillary column (DB1, 30 m x 0.32 mm i.d., J&W.) and a flame ionization detector (FID). The quantitative results were processed by peak area integration on HP1000 computer.

The GC analyses allowed to identify and quantify up to 52 individual aromatic hydrocarbons which are found in a water soluble fraction (WSF). Whatever the pollutant is (fuel-oil or gasoline), these aromatic hydrocarbons are found in the WSF's, but their concentrations may differ. Only data corresponding to the most representative components are presented in this study. A characteristic chromatogram is shown in Figure 1.

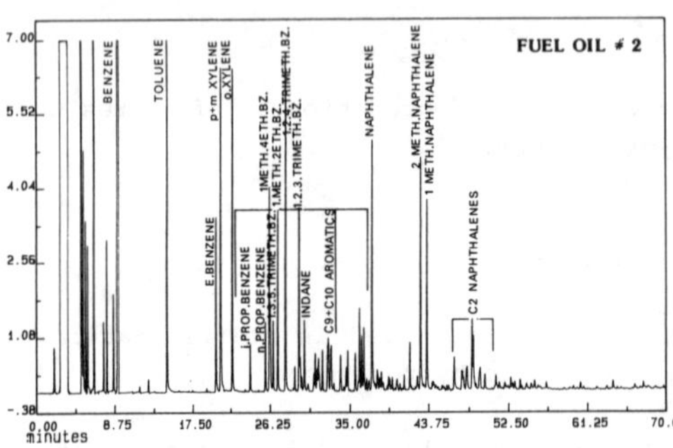

Figure 1. Water soluble aromatics (experiment with fuel oil # 2)

2.2 Determination of interfacial tensions

The interfacial tensions (< 2 mN/m) were measured at 25°C using a Krüss spinning - drop tensiometer. The interfacial tension is given by :

$$\gamma\,(\mathrm{mN/m}) = \pi^2 r^3 \Delta\rho\, n^2$$

where r is the drop radius, n the frequency of rotation and $\Delta\rho$ is the difference in density between the liquids (Aveyard, et al. 1985).

2.3 Retention of surfactants on silt by adsorption and/or precipitation

A silt "Motelle" was provided by INRA (Institut National de Recherche Agronomique - France). Its granulometric and chemical characteristics have been previously described (Ducreux et al. 1990).

The batch experiments were performed in centrifugation cells, where 25 ml surfactant solution, with increasing concentrations (100 to 3000 mg/l), were in contact with 2.5 g silt. The mixture was shaken for 24 hours at 20°C. The residual surfactant concentration Cr was determined on the supernatant after centrifugation.

The amounts of surfactant fixed per unit mass of silt A_{ads} (mg/g) were calculated from the following equation :

$$A_{ads} = (Co - Cr)\,\frac{V}{P}$$

where Co and Cr are respectively the initial and residual surfactant concentrations, V/P is the ratio of the surfactant solution volume to the weight of silt.

2.4 Efficiency tests : oil recovery by using surfactant

In the efficiency test (20°C), the column was filled with the sand used in solubilization experiments. It was saturated by water under physical and hydraulic parameters already described (Ducreux et al. 1990). Before surfactant flushing, gasoline or FO2 remained entrapped at their residual saturation. Following, aqueous surfactant solution washdowns (10 Vp_{acc}) at a 21 ml/h flowrate, the surfactant contents of effluent fraction were analysed (Anionics : NF T73-258, 1972 ; nonionics : Boyer et al. 1977). The oil recovered as a separated liquid phase was measured according to a gravimetric method, as well as the emulsified oil present in the aqueous effluent by the organic carbon content analysis related to the oil (Dohrman, DC 80.). For a given surfactant, the efficiency is defined by the ratio of total recovered oil (both separated and emulsified) to the residual entrapped oil. It is obvious that the aim of this remedial process is to recover the most oil as a separated phase rather than an emulsion.

2.5 Simulating of surfactant flushing - (Physical modelling)

The surfactant injection was simulated on a large 2D model homogeneous porous medium (250 x 50 x 12 cm).

The side walls of the model were constructed of 10 mm thick plexiglass. The model was filled layer by layer with calibrated sand (95 % between 600 and 840 μm). The thickness of packed sand was 46 cm. The gradient (i) was produced by tipping the model. The experiments could be followed visually and recorded photographically after addition of a dye (Bromophenol red) (figure 2). It has been proved that surfactant solution at 0.1 % concentration behaved as the dyed water. The model was fitted by multiple sampling ports system. The injections were performed into the unsaturated zone, close to the surface, by using a fritted glass filter (i.d. 4.7 cm).

The permeability (K) of the porous medium was caculated = 2.0×10^{-3} m/s.

Figure 2. Photograh of the large 2D physical model

2.6 Aerobic biodegration in column

A glass column, 4.5 cm i.d. x 17.0 cm long was packed with sand, the type of which has been described in modelling experiments. In order to simulate an unsaturated zone, the dry sand was wetted by up and down movement of water till a water residual saturation of about 16 %. A mixed microbial culture obtained by isolating strains from soil contaminated by FO2, was injected into the column both at the top and the bottom. A nutrient solution, containing inorganic nitrogen, phosphorus and tryptycase broth, was flushed up and down through the porous medium in order to obtain a good microbial colonization (on average : 4.10^8 bact./g). After that, the oil residual saturation (on average : 27 %) was established by successive flushings (from the column bottom) of FO2 and nutrient aqueous solution. Thus, to simulate the effects of infiltration in the unsaturated zone, an aqueous solution containing nutrients and surfactant (alkylphenol polyglycol ether (APPE 1) concentration : 0.1 %) was percolated, during 50 days, by up and down movement through the sand, giving sufficient oxygen supply (> 1 mg/l) for non-limited aerobic metabolism (wetting / drainage cycle : 3 h./5 h. and 3 cycles per 24 hours). The sand was finally extracted by n-hexane in a soxhlet extractor, to operate a liquid chromatographic separation on activated silica, of aliphatics, aromatics (both further analysed by GC), and polar compounds.

3 RESULTS AND DISCUSSION

3.1 Solubilization of individual and total aromatics : nature of residual oil contamination in case of leaded gasoline and fuel oil # 2

The study of the total polluting potential requires to measure the persistence of contamination and to evaluate its nature. Our experiments simulated the mass exchange mechanisms which operate in the porous medium subjected to the effects of water table fluctuation. The hydrocarbons entrapped after an oil spill constitute the source for dissolved aromatics in the shallow aquifer system (Pfannkuch 1982, 1984). Generally they represent a high percentage of the WSF of gasoline or even

F02 (i.e. approximately 95 % by weight or greater ; Coleman et al. 1984). As is well known, small dissolved quantities of aromatics can render water supply unuseable. Figure 3 displays the variation of total hydrocarbon contents in water (LnCw) analysed during water flushing of the contaminated column.

However, after 1000 relative pore volumes of water flushed through the column, the total hydrocarbon content in effluent from gasoline remains higher (2.4 mg/l) than that from F02 (0.8 mg/l). It is due to a higher content of soluble aromatic hydrocarbons in gasoline.

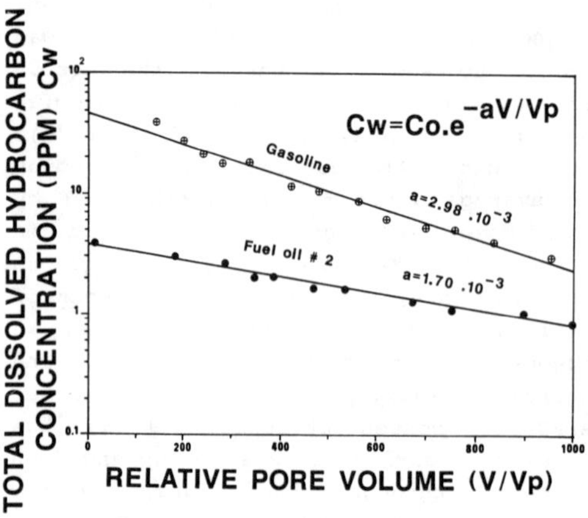

Figure 3. Log total concentration of dissolved aromatic hydrocarbons vs relative pore volume of water through column from experiments with gasoline and fuel-oil # 2

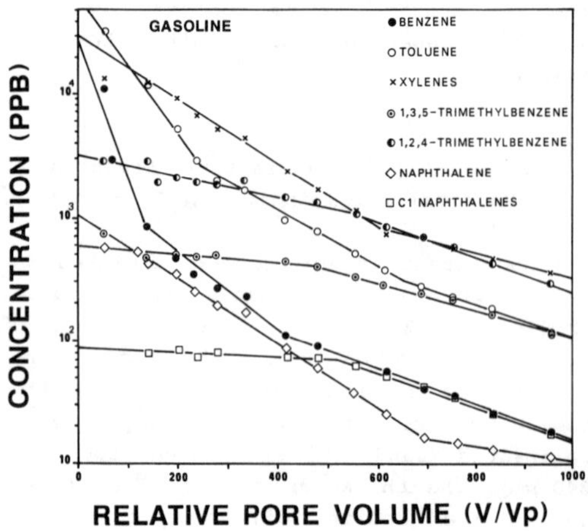

Figure 4. Log concentration of dissolved aromatic hydrocarbons vs relative pore volume of water through column from an experiment with gasoline

The log concentrations of total aromatics are inversely proportional to the discharged water volume. It suggests that the solubilization of total aromatics is an exponential law which may be defined as :

$$C_w = Co\ e^{-a\ V/Vp}$$

where,

C_w : concentration of total dissolved aromatic hydrocarbons after a certain pore volume injected,

Co : initial aqueous concentration

V/Vp : relative pore volume.

Pfannkuch (1984) suggested a proportionnal relationship between the absolute amount of mass exchanged in a given time period and the contact aera.

The decrease observed with gasoline is faster than that with F02.

The extraction of individual aromatics is shown in Figures 4 and 5. In the two cases a linear decrease of individual aromatic contents was observed. However, some changes in the shape of the curves were noted. These phenomena occur whatever the oil phase is. These may be explained by modification of the water/NAPL interfacial area (Pfannkuch 1984). Moreover, it is known that the equilibrium aqueous phase concentration of a component of a complex mixture as a petroleum fraction is a function of both its mole fraction in the NAPL and its specific aqueous solubility. In these experiments the composition of the NAPL is changing as far as each component is dissolving at a different rate ; the equilibrium aqueous concentration of each compound consequently depends upon dissolution (Geller and Hunt 1989). Ramanantsoa et al. (1986) observed earlier only on pure hydrocarbons, that the leaching of an oil body is selective and progessive.

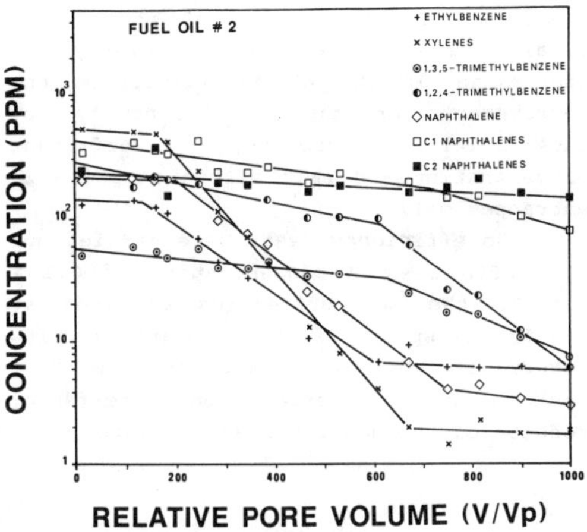

Figure 5. Log concentration dissolved aromatic hydrocarbons vs relative pore column of water through column from an experiment with fuel oil # 2

In these experiments, the concentration of the more soluble compounds (BTEX : Benzene, Toluene, Ethylbenzene, Xylenes) decreased faster than the polysubstituted alkylbenzenes and naphthalenic compounds. However, a particular behaviour was observed with naphthalene. Its solubilization kinetic was similar to those of BTEX's. That is in agreement with the results obtained by Sutton and Calder (1975) in oil/water equilibrium experiments.

Table 1. Typical aromatic composition of WSF (Percent of the total extracted WSF at each sampling time.)

Oil	V/Vp	BTEX (%)	Identified C9 + C10 Aromatics (%)	Identified Naphthalenic Compounds (%)
Gasoline	58.0	80.8	14.7	0.2
	202.7	64.4	26.2	0.3
	481.9	29.4	46.2	0.7
	957.0	18.6	48.0	0.6
Fuel oil#2	16.4	25.4	28.2	14.8
	164.3	18.3	44.2	15.5
	471.8	1.8	53.4	21.7
	1022.0	0.9	35.5	24.9

Table 1 summarizes the relative proportions of characteristic groups of aromatic hydrocarbons found in a WSF. It illustrates the relatively fast elimination of BTEX compared with the other more persistent compounds. Whatever the petroleum fraction is, it appears that the long-term contamination will be mainly due to C9 and C10 aromatics (poly-substituted alkylbenzenes). They represent respectively 48 and 35.5 % in gasoline and FO2 tests. Furthermore in the FO2 case, the other persistent contaminants were the naphthalenic hydrocarbons (about 25 % of WSF). The solubility of the aliphatic hydrocarbons is too low to allow their detection in water. Therefore, these compounds affect potability less than the aromatics and are lesser health risks.

Duffy et al. (1977) found in percolation experiments of soils, that the equivalent of 120 years rainfall applied to hydrocarbon saturated soil only reduced leachate hydrocarbon concentration from 1500 to 400 ppb. An API study, Engineering Science Inc. (1986) showed a long time period (20 years) to restore a site by using only water injection or reinjection technologies. Thus, we propose an in-situ remediation technique, by using surfactant flushing to clean up quickly the oil-contaminated porous medium.

3.2 In-situ surfactant flushing

3.2.1 Surfactant selection

It has been recognized in laboratory tests that aqueous surfactant solutions could be used for in-situ washing of slightly hydrophilic (water soluble) and hydrophilic organics from soils (Texas Research Institute (TRI) 1979, 1985 ; Ellis et al. 1985 ; Nash 1987 ; Ducreux et al. 1990). The enhanced oil recovery (EOR) techniques displace residual oil by reducing the interfacial energy between the oil and water phases and minimize the viscosity of oil-water emulsions. Both of these characteristics are needed to promote flow of residual oil trapped in the vadose zone and saturated zone.

A significant range of commercial products are available. The surfactant selection was based on its ability to lower the interfacial tension within a given system and to displace oil trapped in sand at residual saturation. Moreover, the adsorption parameter will be examined.

Nineteen non ionic and anionic commercial available surfactants were investigated and only the most representative surfactants are presented in this study.

Considering the too high values of interfacial tensions of sulfates (> 0.7 mN/m), due to their hydrophilic character, no efficiency test was carried out. Table 2 summarizes the results obtained with sulfonates and alkylphenol polyglycol ether (APPE). The values of interfacial tensions and oil-recovery efficiency (in the range 0.5 to 83 %) illustrate the improvement brought simultaneously by interfacial property, surfactant concentration and salt effect to mobilize entrapped oil.

No efficiency tests were carried out with APPE 1 at a concentration of 0.1 % due to the low effiencies obtained at 0.5 % in spite of the relatively low interfacial tensions. Moreover, the salt has no or low effect on interfacial tension of the non ionic surfactants.

Table 2. Interfacial tensions and oil recovery efficiencies on sand column of some commercial available surfactants in experiments with gasoline and fuel oil # 2 (NaCl concentration = 2 g/l).

Oil	Surfactant type	Active Surfactant Conc. (5%)	Interfacial Tension (mN.m-1)	Recovery Efficiency (%)	SO/EO (**)
GASOLINE	SULFONATE 1	0.5*	0.202	49.2	3.7
		0.5	0.085	48.8	2.7
		0.1	0.065	22.1	23.8
	SULFONATE 2	0.5	0.015	41.3	11.4
		0.1	0.025	13.3	15.8
	SULFONATE 3	0.5	0.358	4.9	0
		0.1	0.409	18.3	14.8
	APPE 1	0.5*	0.091	14.5	3.1
		0.1*	0.230	-	-
		0.1	0.620	-	-
FUEL OIL # 2	SULFONATE 1	0.5*	0.265	83.4	3.9
		0.5	0.173	4.0	-
		0.1	0.300	0.5	3.8
	SULFONATE 2	0.5*	1.413	77.7	4.3
		0.5	0.085	83.1	13.6
		0.1	0.100	7.8	6.2
	SULFONATE 3	0.5	0.745	4.4	3.8
		0.1	0.550	3.3	2.8
	APPE 1	0.5*	0.130	8.8	5.3
		0.1*	0.270	-	-
		0.1	0.310	-	-

* : no salt test

** : SO/EO : ratio of separated oil to emulsified oil

\- : not analyzed

Whatever the anionic surfactant is, it is easier to displace FO2 (> 75 %) than gasoline in spite of higher interfacial tensions. For the latter, the best efficiency is close to 50 %. By using APPE 1, the efficiency is better with gasoline although its value remains low (14.5 %). In all cases an admixture of sodium chloride (2 g/l) lowered the interfacial tensions and increase the recovery rates.

Thus, the 2 g/l salt concentration seems to be a convenient content in the field experiments. Some experiments were carried out with higher salt concentration (10 g/l) but no significant effects on interfacial tension lowering were observed. Later, this NaCl concentration will only be used as preflush treatment on silt.

Generally, 0.5 % active surfactant concentrations were more effective than 0.1 %. That corresponds to values higher than critical micellar concentration(CMC). The most effective surfactant to displace FO2 was the sulfonate 2 (in salt solution and a 0.5 % concentration of active surfactant). In gasoline experiments, its behaviour was comparable with sulfonate 1. However, the sulfonate 2 enhanced a displacement of oil as a separated phase (high ratio SO/EO), whereas the sulfonate 1 led to solubilization of the oil in emulsion. Low efficiencies were obtained with the APPE 1 in spite of relatively low interfacial tensions (about 0.1 mN/m). The low recovery rates obtained with the sulfonate 3 suggest that useable surfactants will have to provide interfacial tensions less than 0.1 mN/m in a system oil/salt surfactant solution. As a general rule, the effectiveness of in-situ remediation by using surfactant flushing appeared limited by soil adsorption due to the presence of specific materials (clay, silt, ...).

The retention of surfactants on silt, C_{ads} (absorption and/or precipitation) is plotted in Figure 6 as a function of the useable surfactant concentration (initial concentration). All experiments show an increasing retention with an increase of initial surfactant concentration up to a maximum followed by a decrease. This adsorption phenomenon may be explained by the formation of electrostatic interactions between the solid phase and surfactant, and/or precipitation of surfactant by ion-exchange process taking place between multivalent metal ion of adsorbing materials such as Ca^{2+}, Mg^{2+}, Al^{3+} and ion of ionic surfactant.

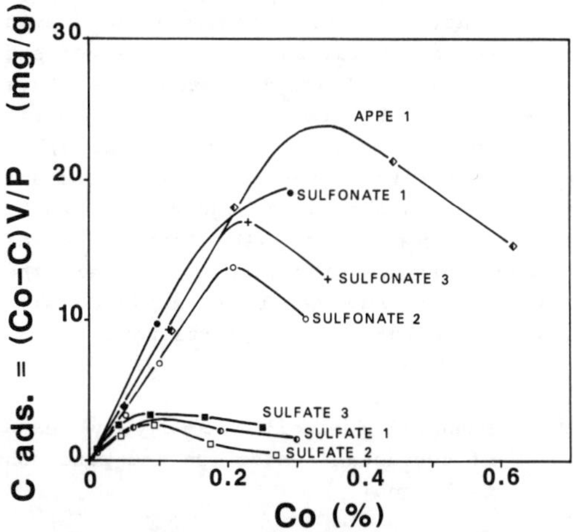

Figure 6. Surfactant adsorption on silt vs initial concentration

Sulfonates were more adsorbed on silt than sulfates, suggesting a direct relationship between their retention and the hydrophilic lipophilic balance (HLB) on considered groups (ethoxylated + 0.33, sulfonate + 11, sulfate + 38.7, LIN and MARSZALL, 1978).

From these data, it appears that sulfate groups are more hydrophilic than sulfonate and ethoxylated groups. It is in agreement with a previous study on adsorption of alkylsulfostearate which showed a close relation between the retention and hydrophily (Bazin and Eisenzimmer, 1988). HLB group numbers may also explain the strongest retention of the APPE 1 compared with anionic surfactants. The shape of diagrams, prior to the maximum, could be due to formation of surface aggregates which leads to the saturation of accessible and active sites (Harwell et al. 1984). The maximum value being related to the CMC, no more absorption of micelles can take place.

In that case, the high negative charges cause a repulsion at the mineral surface. Further, the accessibility to the clay active sites, already saturated, is less easy. It is the "micellar exclusion".

Several workers (Celik et al. 1982, Zundel and Siffert 1984) showed other interaction types leading to precipitation of surfactants, a phenomenon adding to adsorption. Indeed, multivalent ions, such as Ca^{2+} may cause precipitation followed by redissolution at sufficiently high concentrations. With sulfonates, even complete redissolution can be achieved and several solubilization mechanisms have been proposed (Celik et al. 1979). According to the data of the Figure 6, it appears that the useable active surfactant concentration should be higher than 0.3 %. This is compatible with the data of recovery efficiencies (Table 2), where the best results were generally achieved with a concentration of 0.5 % surfactant.

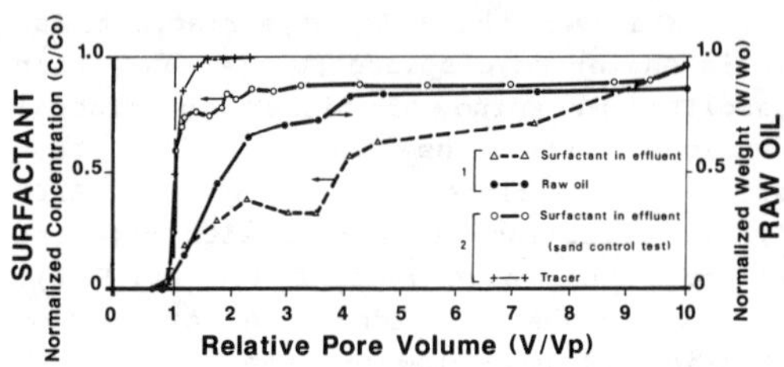

Figure 7. Surfactant effectiveness on the fuel oil # 2 recovery. Effect of the surfactant partitioning between water and oil. (1) sand column with residual oil flushed by a surfactant solution (2) control test without oil : surfactant breakthrough curve.

3.2.2 Enhanced fuel-oil recovery by means of surfactant injection and preflush technique

(i) Surfactant injection tests in clean sand columns

Simultaneously a tracer and a 0.5 % anionic surfactant solution in tap water with 2 g/l NaCl, were injected in a clean sand column as a no oil control test (Ducreux et al. 1990). Then, the same solution was injected in a column containing a residual FO2 phase, at irreducible concentration (Wo), left after water-flooding. In this case, a water flushing without surfactant has no effect on raw oil recovery. In each experiment, effluents were collected and analysed. in Figure 7 (1) are given the cumulative FO2 recovered (Wc/Wo) as a separated phase, and surfactant normalized concentration (C/Co) in effluent, as a function of relative pore volumes (V/Vp) of injected fluid.

The surfactant elution took place after one pore volume injected, simultaneously with the tracer [Figure 7(2)], so the adsorption of surfactant (sulfonate 2, previously shown) appeared negligible in the sand used. However, the stabilized value of C/Co is less than unity, indicating a low surfactant retention in the sand column.

Considering the calcium content in tap water (100 ppm detected), this remanent loss of additive was attributed to retention of calcium salt precipitate of anionic surfactant.

The oil recovery test in Figure 7 (1) shows a good oil mobilization, with a final recovery of 86.7 % of residual oil in place. Moreover the oil produced in emulsified form was only 4.6 % on overall mobilized oil. In order to clarify the apparent loss of surfactant in the presence of oil, complementary determinations of oil/water equilibrium were performed in a sample tube. Both oil and aqueous phases were gently shaken for one day and stored two weeks at 20° C. Afterwards, surfactant and calcium contents were determined in each phase. Table 3 summarizes the values of surfactant (sulfonate 2) oil/water partitioning observed under the indicated conditions.

In conclusion, a large part of surfactant is solubilized in oil as calcium salt. Partitioning coefficient amounts to 5.6 in the 2 g/l NaCl injected solution. The salinity effect upon oil/water partitioning of the anionic surfactant is clearly shown. The drastic effect of calcium ion contained in tap water is also included in the table, by comparison with a 4 g/l NaCl solution prepared in deionized water.

Table 3. Sulfonate 2 oil/water partioning coefficient vs NaCl concentration

NaCl conc. (g/l)	Surfactant partioning coefficient C oil/C water	Ca^{2+}concentration in aqueous phase (ppm)
2	5.6	48
4	9.8	48
4*	0.18	-
5	76.0	48
10	∞	69
5**	114	195

* : in deionized water
** : with silt (2 g)

(ii) Surfactant injection tests in oil-contaminated sand columns

A series of oil recovery tests were performed in sand mixed with 5 % (w/w) of silt. In order to modify ion exchange on clays and reduce the calcium and magnesium content of the silt material, a preflush technique with 41 pore volumes (i.e. 500 ml) NaCl solution (2 and 10 g/l) as operated before surfactant injection (Figure 8).

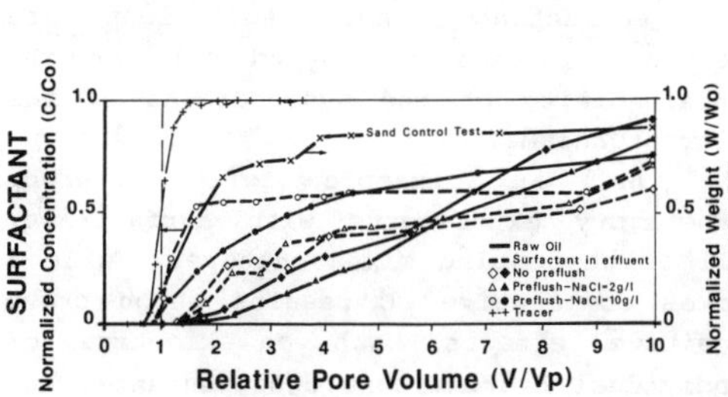

Figure 8. Influence of a salt preflush on the enhanced fuel oil and surfactant recoveries.

In the case of experiments without preflush or with a 2 g/l NaCl preflush, oil production is delayed as well as surfactant breakthrough, and the retained additive amount is increased.

These effects result mainly from a high adsorption on clays containing Ca^{2+} ions. Consequently a good oil recovery needs much more relative pore volume injection of surfactant.

The technique of salt preflush is a means to improve surfactant efficiency in oil-contaminated sand when provided salinity is high enough. With 10 g/l NaCl preflush, breakthrough of surfactant takes place much earlier and oil production is made faster.

In field application, such a technique will allow to reduce the amount of surfactant and will decrease the cost of treatment. Additive transfer into immobilized residual oil is noticeable, even in the case of a 10 g/l NaCl salinity preflush. Presumably a 10 g/l NaCl preflush is not high enough to exchange the whole calcium. This hypothesis is supported by the observation, that an addition of native clay material leads to a clear release of calcium in both oil and water phases at equilibrium, as well as a higher surfactant oil partitioning (Table 3).

3.2.3 Simulating of surfactant flushing - Physical modelling

Following and controlling the migration of surfactant in a soil demands to simulate the remedial system in a large 2D physical model (Figure 2). Efforts have been focused mainly on surfactant migration in the vadose zone especially near the top of the water table where high oil concentrations result from the residual oil saturation in the soil after the up and down movement of the floating oil phase. In certain conditions there is a possibility that surfactant and solubilized organic compounds may escape capture by a recovery well and be widely disseminated through the aquifer. This present work shows that such an event could be avoided by controlling some parameters, especially the flushing flowrate related to the hydraulic gradient. In pratice, the water table can be lowered locally by a pumping well with a magnitude depending upon time and permeability (CONCAWE 1979).

In our experiments, this technique was simulated by tipping the model more and less. The influence of injection flowrate on penetration in capillary fringe (Δh) at increasing hydraulic gradient is shown in Figure 9. The extrapolation at zero flowrate provides the capillary fringe heigh (9 cm). The higher the hydraulic gradient the higher the injection flowrate that can be used without contamination of groundwater. Thus, under our experimental conditions, at the higher gradient (i = 0.049), the injection flowrate should be less than 2.5 l/h. Then, aqueous surfactant solution will migrate horizontally to the water table increasing the water saturation in places. However, under such conditions, it is noticeable that the suction forces existing in the capillary fringe are not disturbed by the surfactant migration.

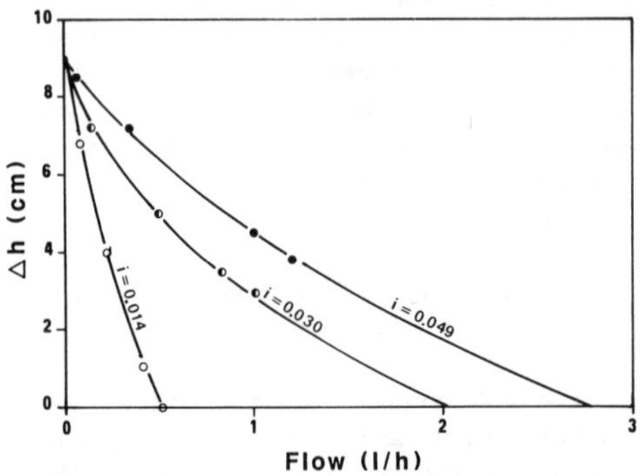

Figure 9. Capillary fringe penetration by surfactant vs injection flowrate and hydraulic gradient

Only when very high flowrates are applied (penetration into groundwater) an increase of capillary pressure is observed, due to the modification of local water saturation leading to the creation of new capillary fringe. Thus penetration depth was consistent with the magnitude of the head pressure. Moreover the lower the gradient is, the more downwards oriented the resultant of applied forces. Unlike the field experiments, the observed penetration effect may be considered as excessive, considering the limited lateral spreading in this physical model.

The interest of this study has been to show the possibility for limiting the surfactant contamination of groundwater by control of injection flowrates depending upon the permeability of the unsaturated zone, its water content and capillary forces, and at the same time, by increasing the hydraulic gradient.

4 AEOROBIC BIODEGRADATION ENHANCED BY SURFACTANT

Hydrocarbons entering an aquatic ecosystem provide a source of carbon and energy for microbial growth (Watkinson 1978). Many authors found that aeration plus addition of water-soluble nitrogen and phosphorus compounds to a groundwater reservoir contaminated with petroleum products promoted microbial activity and accelerated removal of the pollution (Atlas 1984).

The interfacial area of oil in contact with water can also limit microbial degradation of oil, and use of emulsifiers in clean-up treatment was proposed to enhance natural biodegradation of oil. In-situ bioremediation could use indigeneous microbial populations to carry out the ultimate mineralisation of aliphatic and aromatic hydrocarbons to cardon dioxyde and water via biotransformation in oxydized organic compounds present in the carbon cycle in nature. The aim of this study was to measure the effect of low concentrations of surfactant (not sufficient to mechanicaly recover trapped oil) on the attainability of the hydrocarbons by the microorganisms.

As an example of different laboratory experiments with surfactants, all of them using mixed cultures, Table 4 gives comparative GC results which prove positive effects both on kinetics of individual hydrocarbons disappearance, and on apparent yield of total biodegradation. It is notable that the biochemical processes, observed in this study, are coherent with the generally accepted scenario of microbial petroleum degradation (Walker et al. 1976, 1977) : after an initial fast disappearance of n-alkanes, the biodegradation affects both branched-chain alkanes and aromatic hydrocarbons (alkylbenzenes and alkyl-naphthalenes). Figure 10 shows clearly the disappearance of simply structured hydrocarbons by the decreasing number of identified peaks. As a refined petroleum

Table 4. Aerobic biodegradation of fuel # 2 enhanced by surfactant. Modification of residual oil composition in sand compared with original pollutant composition.

		Initial fuel oil # 2	Control test without surfactant	Test with surfactant (APPE1 0.1 %)
	TOTAL	61.5	59.5	51.5
ALIPHATIC	n-alkanes	16.0	4.2	0.7
HYDROCARBONS	nC17/Pr* ratio	2.65	0.54	0.18
%	nC18/Ph* ratio	2.39	0.54	0.19
	TOTAL	35.0	39.7	31.5
	Ethyl benzene	0.021	0.002	-
	Xylenes	0.141	0.010	0.005
AROMATIC HYDROCARBONS (%)	1,3,5+1,2,4 Trimethyl Benzenes	0.148	0.024	-
	Naphthalenes	0.123	0.089	0.004
	C1 Naphthalenes	0.606	0.222	0.020
	C2 Naphthalenes	1.448	1.036	0.396
POLAR COMPOUNDS (%)	TOTAL	3.5	1.2	17.00
ALIPHATICS / AROMATICS ratio		1.76	1.49	1.60
% DISAPPEARED**		-	53.1	98.8

* : Pr = Pristane, Ph = Phytane } Isoprenoïds

** : % calculated from ratio between residual oil in sand after and before biodegradation experiment

product such as FO2 may contain more than 5000 different compounds, it is often convenient to observe in the residual oil the branched compounds (isoprenoïds) which are recognized as persistent compounds (Kator, 1973). The decreasing ratio of adjacent alkanes to pristane and phytane measures the importance of biodegradation processes. At the end of experiments only unresolved complex mixtures can be detected, but only after increasing the GC response level.

Many hydrocarbon-using micro-organisms produce emulsifying agents and in practice emulsification of FO2 is taken as evidence for biodeterioration. In our experiments, such a bacterial activity could produce a sufficient amount of surfactants to develop mechanical clean-up responsible for the high value of disappearance. In these experiments, no inhibitory effect, due to last metabolites, could be shown. Further experiments will put the accent on more precise mass balance by analyzing micro-emulsified hydrocarbons washed off by the high volume of percolating water.

5 CASE HISTORY

An oil spillage gave us the opportunity to test the in-situ surfactant flushing to mobilize the trapped oil still in place after a mechanical traditional recovery. Unhappily, the unknown amount of spilled

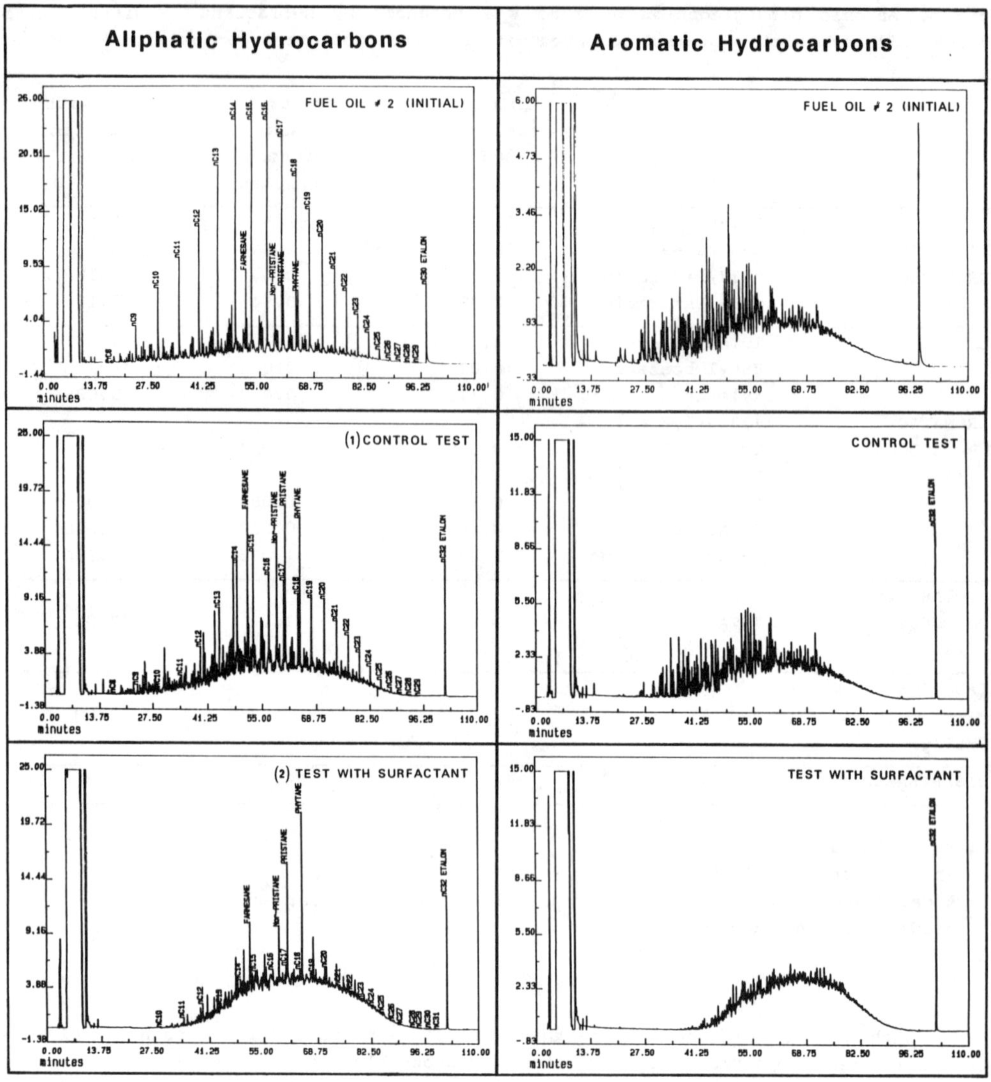

Figure 10. Aerobic biodegradation enhanced by surfactant : chromatograms of aliphatic and aromatic hydrocarbons

(1) Control test = biodegradation without surfactant
(2) Test with surfactant = biodegradation with APPE 1 0.1 %

petroleum products (a mixture of spirit and naphthalenic oil) did not give the opportunity to carry out an exact mass balance (Béraud et al. 1989). A simple tap water flushing mobilized less than 100 liters of products even with a very large amount of flushing water and after a long period which led only to the extraction of soluble hydrocarbons (low level) from the soil matrix. By using a nonionic surfactant (APPE 1), which can be considered today as medium efficient (Table 2), a new oil recovery occurred and in the same time emulsified oil was washed away with the flushing water. Another trial is in progress with a more efficient surfactant, selected by the laboratory method above described. It will let us check in the field the surfactant injection parameters as defined in the physical model.

6 CONCLUSIONS

After using classical mechanical methods to recover oil as separated phase, a relatively important amount of residual hydrocarbons remains underground entrapped. We demonstrate that such a residual saturation is an important source of contamination mainly in the vadose and fluctuation zones. We emphasize a fast solubilization of lighter aromatics (BTEX) responsible for a high level contamination, followed by a longer term pollution related to soluble organics (C9 and C10 aromatics and naphthalenic compounds) present at a sufficient level to render the tap water unfit for human consumption. It is suggested to operate an in-situ clean up treatment by surfactant flushing using some commercial products selected to be efficient to remobilize the residual oil. Use of such compounds requires to control their migration through the porous medium in order to avoid any further groundwater contamination. A large 2D physical model allowed us to define injection parameters linked to hydraulic gradient. Moreover, it is shown not only that these additives are required for moving trapped oil but also, at much lower concentrations they enhance the aerobic biodegradation of aliphatic and aromatic hydrocarbons, the interfacial area of which is now more accessible by active microorganisms.

REFERENCES

Atlas, R.M. 1984. Petroleum microbiology, Mac Millan : 692.

Aveyard, R., Binks, B.P., Lawless, T.A. and J. Mead 1985. Interfacial tension minima in oil - water - surfactant systems. J. Chem. Soc., Faraday trans. 1, 81:2155-2168.

Bazin, B. and A. Eisenzimmer 1988. Adsorption de tensio-actifs utilisables dans des conditions de fortes salinités. 2ème Congrès mondial des agents de surface, 24-27 mai, Paris.

Beraud, J.F., Ducreux, J. and C. Gatellier 1989. Use of soil-aquifer treatment in oil pollution control of underground waters. Proceedings Oil Spill Conference (Prevention, Behavior, Control, Clean up) February 13-16, San Antonio, Texas : 53-59

Boyer, L.S., Guin, R.C., Kelly, R.M., Mausner, L.M., Robinson, F.H., Setzkorn 1977. Analytical method for nonionic surfactants in laboratory biodegradation and environmental studies. Environ. Sci. Technol. 11, No. 13:1167-1171.

Celik, M., Goyal, A., Manev, E. and P. Somasundaran 1979. The role of surfactant precipitation and redissolution in the adsorption of sulfonate on minerals. SPE paper 8263, 54th Annual SPE Meeting, Las Vegas.

Celik, M.S., Manen, E.D. and P. Somasundaran 1982. Sulfonate precipitation-redissolution-reprecipitation in inorganic electrolytes. Interfacial phenomena in enhanced oil recovery. AICHE Symposium series, No. 212, Vol.78:86-96.

Coleman, W.E., Munch, J.W., Streicher, R.P., Ringland, H.P. and F.C. Kopler 1984. The identification and measurement of components in gasoline, kerosene and No. 2 fuel oil that partition into the aqueous phase after mixing. Arch. Environm. Contam. toxicol.,13:171-178.

CONCAWE Report. Protection of groundwater from oil pollution 1979. Report No 3/79.

Ducreux, J., Bocard, C., Muntzer, P., Razakarisoa, O. and L. Zilliox 1990. Mobility of soluble and non-soluble hydrocarbons in contaminated aquifer. Proceedings of international symposium on processes governing the movement and fate of contaminants in the subsurface environment. Stanford, Wat. Sci. Tech. Vol. 22.No.6,27-36.

Duffy, J.J., Mohtadi, M.F. and E. Peake 1977. Subsurface persistence of crude oil spilled and land and its transport in groundwater. Proceedings Oil Spill Conference (Prevention, Behaviour, Control, Clean up), march 8-10, New-orleans, Louisiana : 475-478.

Ellis, W.E., Payne, J.R. and G.D. Mc Nabb 1985. Treatment of contaminated soils with aqueous surfactants, U.S. EPA

Report, No EPA/600/2085/129, PB 86-122 561.
Engineering-Science, Inc. 1986. Cost model for selected technologies for remocal of gasoline components from groundwater. Health and Environmental Sciences Department. API Publication, No. 4422.
Geller, J.L., Hunt, J.R. 1990. Non-aqueous phase organic liquids in the subsurface : Dissolution kinetics in the saturated zone. Proceedings of international symposium on processes governing the movement and fate of contaminants in the subsurface environment. Stanford.
Harwell, J.H., Schechter, R. and W.H. Wade 1984. Recent developements in the theory of surfactant adsorption on oxides. Colloque bilan - Nancy, France.
Kator, H. 1973. Utilization of crude oil hydrocarbons by mixed cultures of marine bacteria. D.G. Ahearn and S.P. Meyers (eds), Microbial degradation of oil pollutants. Louisiana State Univ. Center for Wetland Resources, Pub.No. LSU-SG-73-01.
Lin, I.J. and L. Marszall 1978. Partition coefficent, HLB and effective chain length of surface-active agents. Prog. Colloid Polym. Sci. 63:99-104.
Nash, J.H. 1987. Field studies of in-situ soil washing. U.S. EPA Report, No. EPA/600/2-87/110, PB 88-146 808.
Pfannkuch, H.O. 1982. Problems of monitoring network design to detect unanticipated contamination. Groundwater Monitoring Review. Vol.2, No. 1 : 67-76.
Pfannkuch, H.O. 1984. Mass exchange processes at the petroleum-water interface. Groundwater contamination by crude oil at the Bemidji-Minesota Research Site. U.S. Geological Survey Toxic Waste Technical Meeting, Tucson, Arizona. Hult, M.F. (Editor) U.S. Geological Survey Water Resource Investigation Report. 84-4188:23-48.
Ramanantsoa, B., Muntzer, P. and L. Zilliox 1986. Dissolution sélective d'un mélange d'hydrocarbures par l'eau en milieu poreux saturé. Application à la pollution des eaux souterraines par les produits pétroliers. Sciences de l'eau 5:149-168.
Sutton, C and J. Calder 1975. Solubility of alkylbenzenes in distilled water and seawater at 25°C, J. Chem. Eng. Data 20 : 320-322.
Texas Research Institute, Inc. 1979. Underground movement of gasoline on groundwater and enhanced recovery by surfactants. API Report No. 7743 (1-3)-F.
Texas Research Institute, Inc. 1985. Test results of surfactant enhanced gasoline recovery in a large-scale model aquifer. API Publication, No. 4390.
Walker, J.D. and Colwell, R.R. 1976. Biodegradation rates of components of petroleum. Canadian Journal of Microbiology 22(8) : 1209-1213.
Walker, J.D. and Colwell, R.R. 1977. Role of autochthonous bacteria in the removal of spilled oil from sediment. Environ. Pollut (12): 51-56
Watkinson, R.J. 1978. Developments in biodegradation of hydrocarbons. Applied Science: 232.
Zundel, J.P. and B. Siffert 1984. Mécanisme de rétention de l'octylbenzène sulfonate de sodium sur les minéraux argileux. Solid-Liquid interactions in porous media, Colloque-Bilan, Nancy, Ed. Technip, Paris : 447-462.
NF T73-258, Association Francaise de Normalisation (AFNOR), 1990. Agents de surface, détergents. Détermination de la teneur en matière active anionique selon une méthode manuelle ou mécanique par titrage direct dans deux phases. Janvier.

Subsurface Contamination by Immiscible Fluids, Weyer (ed.) © 1992 Balkema, Rotterdam. ISBN 90 6191 175 3

In-situ remediation of volatile contaminants in groundwater by a new system of 'Vacuum-Vaporizer-Wells'

B. Herrling, W. Buermann & J. Stamm
Institute of Hydromechanics, University of Karlsruhe, Germany

ABSTRACT: A new hydraulic system for in-situ remediation of volatile contaminants (e.g. chlorinated hydrocarbons) in groundwater is presented. The contaminated groundwater is stripped by air in a vacuum pressure field in a special screened well. The vertical well discharge initiates a circulation flow in the region surrounding the well. The operation of the remediation well is explained and the sphere of influence, or the capture zone, for the case of natural groundwater flow is discussed. Further, measurement data from two investigation sites are presented which demonstrate the action of the well.

1 INTRODUCTION

The contamination of groundwater by volatile substances, particularly by volatile chlorinated hydrocarbons, is a significant problem in all industrial countries. As an alternative to conventional hydraulic remediation measures (pumping, off-site cleaning and reinfiltration of the groundwater), a new method for in-situ remediation of groundwater is currently being used at several locations in Germany (Herrling and Buermann 1990a, Herrling et al. 1990). The contaminated groundwater is stripped by air in a below atmospheric pressure field in a so called "Vacuum-Vaporizer-Well" (German: Unterdruck-Verdampfer-Brunnen, abbreviation: UVB), see acknowledgement. The used air, charged with volatile contaminants, is cleaned employing activated carbon.

2 IN-SITU REMEDIATION OF VOLATILE CONTAMINANTS BY THE UVB-METHOD

The Vacuum-Vaporizer-Well (UVB) helps to remove volatile substances from the groundwater and from the unsaturated zone. When using the UVB-method a special well with two screen sections is employed, one at the aquifer bottom and one at the groundwater surface (see figure 1) or at the top of a confined aquifer. The borehole reach between the two screen sections should be made impermeable. One well should be used to remediate only one aquifer (phreatic or confined) and should not connect separate aquifers.

The upper, closed part of the well has a pressure less than atmospheric which is generated by a ventilator. The fresh air for the upper part of the well casing is introduced through the fresh air pipe whose upper end is open to the atmosphere and which ends beneath the water surface in a pinhole plate. The height of the pinhole plate is adjusted such that the water pressure is lower there than the atmospheric pressure. Therefore, the fresh air is sucked into the system. The reach between the pinhole plate and the water surface in the well casing is the stripping zone, in which an air bubble flow is developed. The rising air bubbles produce a pumping effect which moves the water up and causes a suction effect at the well bottom. In recent wells a separating plate and an additional pump (figure 1) are used to support the pumping effect of the air bubbles.

Additionally, soil air is sucked from the surrounding contaminated unsaturated zone at many sites. Stripped air and possibly soil air pass through the ventilator and across activated carbon where the contaminants are adsorbed. Thus, only cleaned air escapes into the atmosphere. The cleaning effect of the well is based on one hand on the reduced pressure which supports the escape of volatile contaminants out of the water. As a result of the air intermixing, the cleaning effect is also based on the considerable surface area of the air bubbles and on the concentration gradient. In this sense the permanent vibration, caused by the air bubbles, is beneficial to the escaping process of the contaminants. This vibration is transmitted as compression and shear waves into sediment and fluid and influences presumably the mobility of contaminants even outside the well.

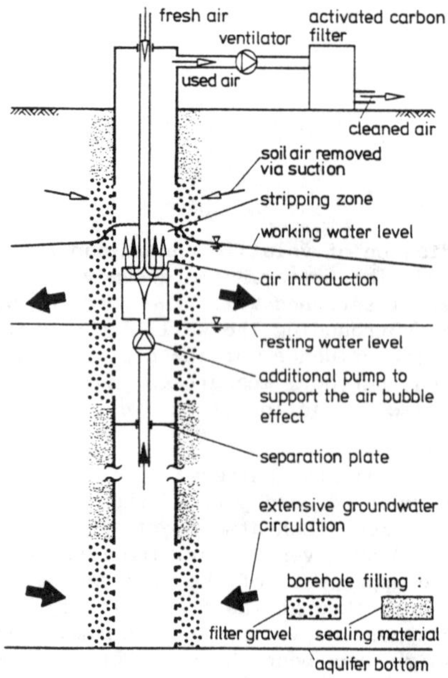

Fig.1 Vacuum-Vaporizer-Well, called UVB

The upward streaming, stripped groundwater leaves the well casing through the upper screen at the groundwater surface, which is lifted in a phreatic aquifer by the explained pumping processes and by the below atmospheric pressure. The groundwater returns in an extensive circulation to the well bottom. In this way the groundwater surrounding the well is also remediated. The expansion of the groundwater circulation is positively influenced by the anisotropy existing in each natural aquifer with higher horizontal hydraulic conductivities than vertical ones. The artificial groundwater circulation determines the sphere of influence of a well and is overlapped with the natural groundwater flow (see below).

The pinhole plate and all the installations within the well casing are designed as a float, so they can adjust to changing groundwater levels.

The above-ground installations, from an investigation site in Mannheim, can be viewed in figure 2. In the right foreground, the upper end of the well casing is visible. To the right of it can be seen one of two measuring pipes which are situated in the bore-hole and reach either the lower or upper screen sections of the well. The fresh air pipe, open against the atmosphere, leads into the well casing. The used air pipe leaves the casing at its left side and leads into the ventilator. From there the used air flows to four barrels filled with activated carbon. The cleaned air leaves the barrels at the bottom. At the far left, the switch cabinet for the electricity supply of the ventilator and the additional pump is visible.

Fig. 2 View of the above-ground installations at Mannheim-Kaefertal

When using the UVB in the implementation of figure 1, with additional pump and separation plate, precipitations of iron, manganese and calcium may occur depending on the water quality at a remediation site. In general, there is no oxygen in the water when it enters the well. When it leaves the well after the stripping zone, full saturation of oxygen is reached with the well known effects. The supersaturation with carbon dioxide at the well influx is decreased in the stripping zone by the exchange with the air. The pH-value increases and calcium precipitation occurs. To avoid or to reduce these effects, which will finally seal the upper screen section, the closed UVB-method has been developed (see figure 3): The cleaned used air is recirculated. Thus after only a few circulations water and air will reach an equilibrium and the unwanted precipitation effect will not occur. Only those contaminants, which are sorbed by activated carbon, are removed from the air circulation. Volatile substances not sorbed remain in the groundwater and will not contaminate

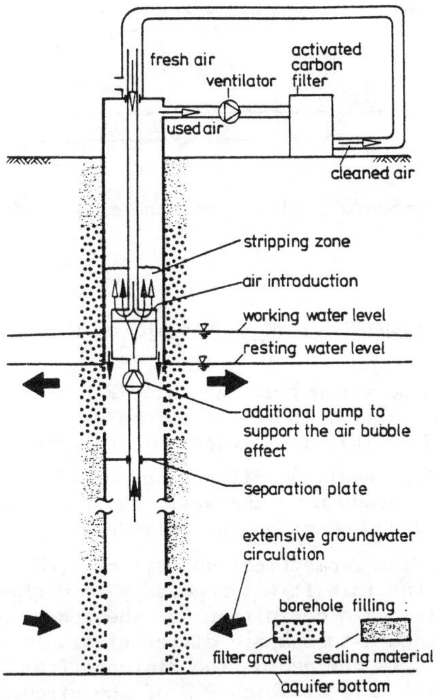

Fig. 3 UVB-method with closed air circulation

the atmosphere. But special cleaning methods, incorporated in the air circulation, are possible.

Using the closed UVB-method the soil air must be separately removed via suction when required at a site. Further, special effort must be made to seal the upper well casing and the pipe system for the air circulation so that no fresh air can enter the closed system.

At many remediation sites the UVB is used without an additional pump and separating plate, see figure 4. In this case a circulation flow occures within the well casing which is produced by the pumping effect of the rising air bubbles. For the most part, the stripped water follows the path of least resistance and flows down to the end of the suction pipe. Thus, in the entire well casing a water of uniform temperature and oxygen content appears. The water temperature is influenced by the withdrawn evaporation heat in the stripping zone and by the temperature of the fresh air. Depending on the groundwater temperature around the well the water leaves the well casing through the upper screen section and contaminated water enters the UVB at the lower screen section.

This occurs when the groundwater is colder than the circulation water in the well casing. On the other hand, when the water in the well is colder than the surrounding groundwater, an outer circulation occurs which is opposite to that shown in figure 4: the water leaves the well at the lower screen section and enters it at the upper. Both cases, influenced by density differences of the involved water bodies, have been observed at different sites.

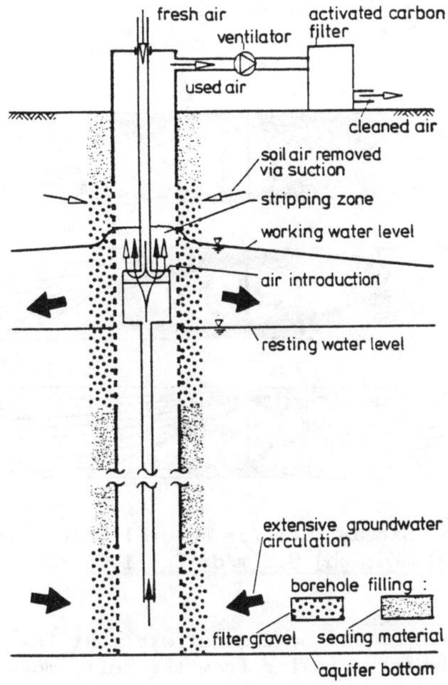

Fig. 4 UVB-method, driven by the air bubble effect

3 SPHERE OF INFLUENCE AND CAPTURE ZONE OF AN UVB

The extended circulation field outside the well is of special interest. In this section only UVBs with additional pump and separating plate will be discussed (figure 1 or 3). In principle two different cases have to be considered: (1) when there is no, or only negligible, small natural groundwater flow the sphere of influence, or the range R, of an UVB is of interest and; (2) when natural groundwater flow must be included, the extent of the capture zone has to be determined to set up the installations for a remediation site.

Before going into more detail, the complex flow field near an individual UVB is explained for a vertical cross section in the direction of the natural groundwater flow (symmetry plane of the flow problem). In figure 5 the streamlines of three case studies are demonstrated with Darcy velocities of natural groundwater flow of 0.0 m/d , 0.3 m/d and 1 m/d. All other parameters remain constant: the discharge through the well casing is 20.16 m^3/h, the thickness of the aquifer is 10 m, the anisotropic hydraulic conductivities are 0.001 m/s (horizontal) and 0.0001 m/s (vertical) and the lengths of the screen sections are 1.2 m (at the bottom) and 2.1 m (at the top).

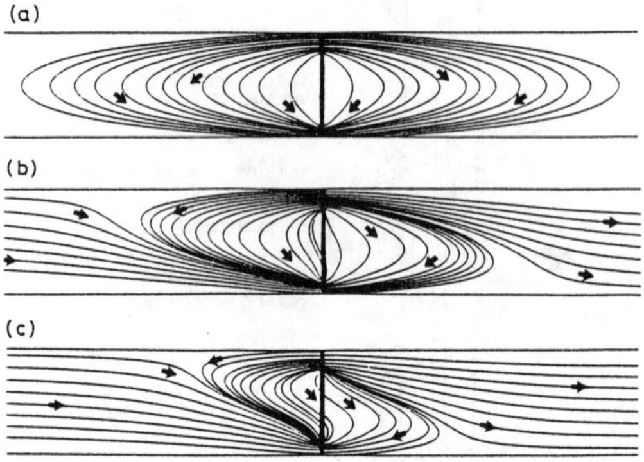

Fig. 5 Streamlines for natural velocities (a) 0.0 m/d; (b) 0.3 m/d; (c) 1.0 m/d

Figure 5b and 5c demonstrate that the groundwater, flowing from the left, moves downward to the lower screen section and is transported upward in the well casing, and that the cleaned water flows out in all directions at the upper screen section. The flow situation can only be calculated and plotted in such a simple way in this cross section. Otherwise the complex three-dimensional flow field has to be viewed.

Without natural groundwater flow and for a confined aquifer the sphere of influence R is dependent on the discharge Q through the UVB, on the thickness H of the aquifer, on the horizontal and vertical hydraulic conductivities K_H and K_V and on the upper and lower screen lengths a_1 and a_2 (figure 6):

$$R = f(Q, H, K_H, K_V, a_1, a_2) \qquad (1)$$

The density has been neglected which seems to be permissible for an UVB with separating plate and additional pump.

Using the dimension analysis, equation (1) is transformed to

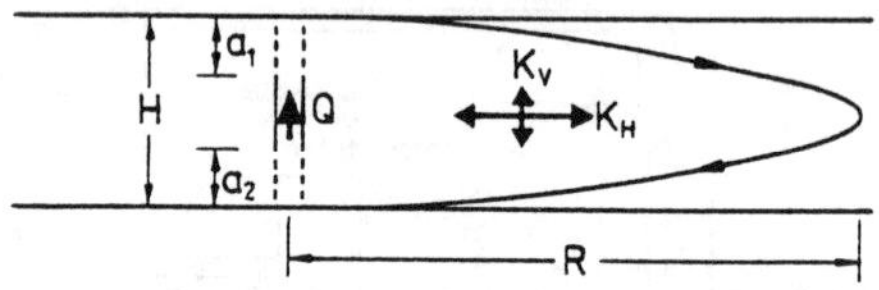

Fig. 6 Sketch, clarifying the substantial variables

$$R/H = f(Q/(H^2K_H), K_H/K_V, a_1/H, a_2/H) \qquad (2)$$

It can be shown that the ratio R/H is independent from the first parameter $Q/(H^2K_H)$. This is significant in connection with resulting differences of the hydraulic heads near the well. The ratio R/H is only dependent on the anisotropy K_H/K_V and on the geometrical quantities a_1/H and a_2/H. For this it was necessary to decide on a suitable definition for the sphere of influence R to compare different simulations: R is defined as the distance from the well axis at which 98% of the circulating water has dived downward and flows towards the lower screen section. From figure 7 the relation

$$R/H = c\ (K_H/K_V)^{1/2} \qquad (3)$$

with $c = f(a_1/H, a_2/H)$ is derived.

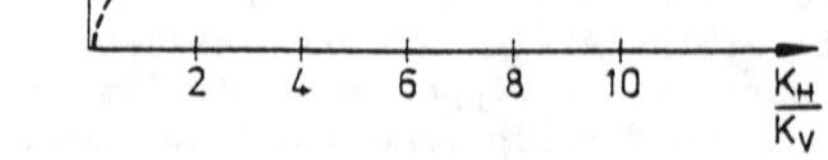

Fig. 7 Dependency of the ratio R/H on the anisotropy for a confined aquifer without natural groundwater flow

At most of the UVB-installation sites the natural groundwater flow rate cannot be ignored. Here the relationship (3) to estimate the sphere of influence cannot be used. It is convenient to employ in this case the term capture zone of an UVB.

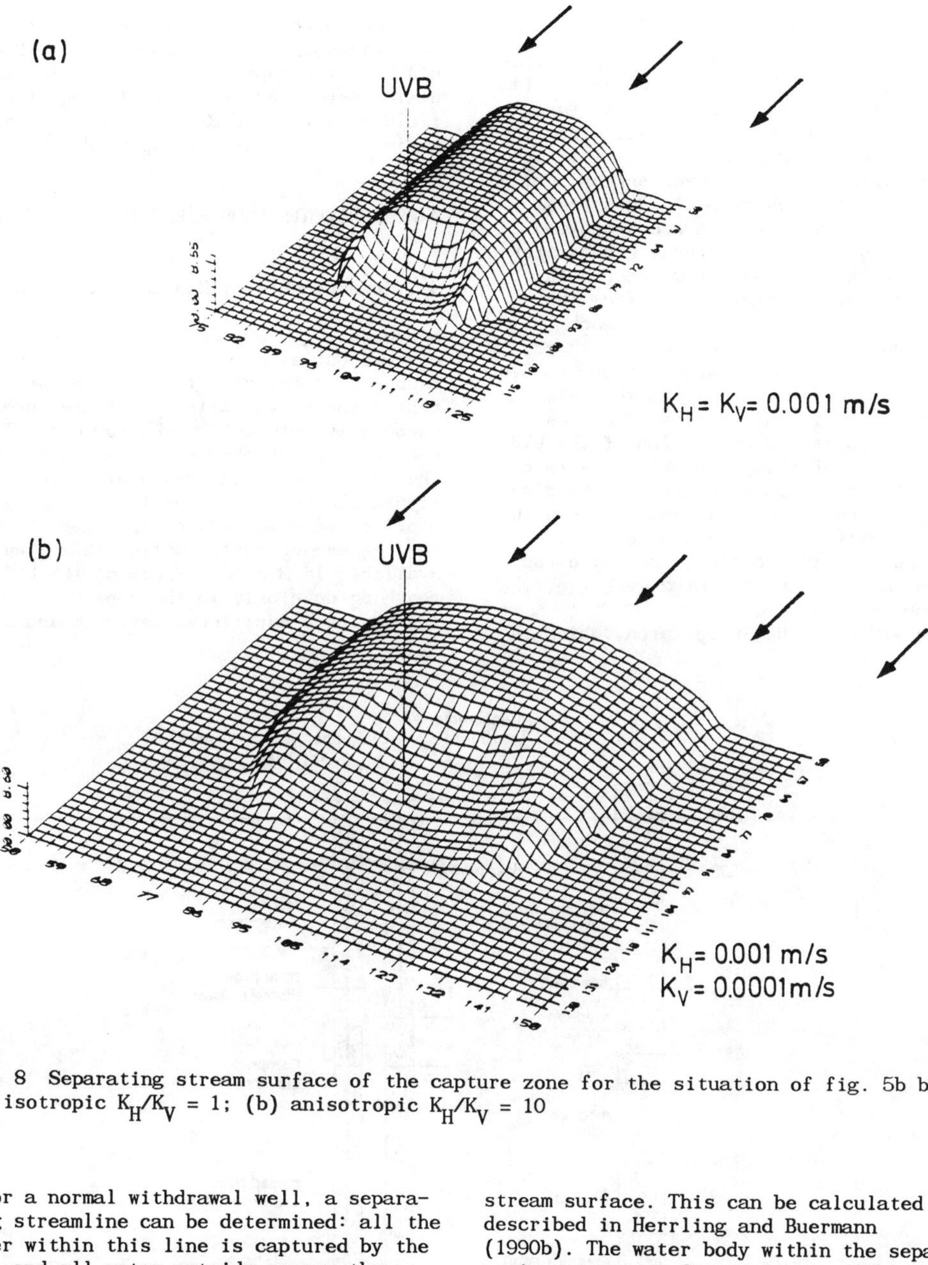

Fig. 8 Separating stream surface of the capture zone for the situation of fig. 5b but (a) isotropic $K_H/K_V = 1$; (b) anisotropic $K_H/K_V = 10$

For a normal withdrawal well, a separating streamline can be determined: all the water within this line is captured by the well and all water outside passes the well. In principle the situation is the same when using an UVB. But in contrast to a normal withdrawal well, where the flow can be considered horizontally, the flow around an UVB must be regarded as three-dimensional. Thus the water body, flowing to the UVB from upstream and being captured by the lower screen section, cannot be delimited by a simple separating streamline but only by a curved separating stream surface. This can be calculated as described in Herrling and Buermann (1990b). The water body within the separating stream surface is captured by an UVB and that outside of it, which flows from upstream, passes the well. In figure 8 the outer surface of the capture zone, calculated numerically, is plotted for two cases (the natural groundwater flows from the background at the right side to the UVB, see the vectors). Figure 8a and 8b were calculated for the situation described for figure 5b; the only difference is that for figure 8a the vertical hydraulic conductivity is K_V = 0.001 m/s which means the calculation is based on isotropic conditions. The captured water is cleaned within the well and leaves it through the upper screen section in all directions (not shown in figure 8). Parts of it are again captured by the lower screen section and the rest flows directly downstream.

In a vertical upstream cross-section perpendicular to the natural groundwater flow, and so far away from the UVB that the circulating water coming from the upper screen section does not reach this location (see figure 5b and 5c), there exists an inflow area which stretches over the full aquifer height. All the water flowing through this inflow area flows into the lower screen section of the UVB. The contour of this area or the capture zone of the UVB is dependent on the discharge through the well casing, the natural velocity of the groundwater, the thickness of the aquifer, the anisotropy of the aquifer and the arrangement of the screen sections.

The width of the inflow area, which increases to the bottom of the aquifer, characterizes the capture zone of an UVB. The width of the capture zone is an important design parameter for dimensioning of a field of several UVBs if a broad contamination plume should be captured.

4 MEASUREMENTS FROM FIELD INVESTIGATIONS

4.1. Measurements at Mannheim-Kaefertal

Since January '89 measurements have been carried out at an experimental well at Mannheim-Kaefertal. At this site the concentration of volatile chlorinated hydrocarbons is between 100 and 200 µg/l. The construction of the well, the placing of the two accompanying measurement locations within the bore-hole and the geological profile are demonstrated in figure 9. One further measurement location (B225) was available 15.3 m downstream of the UVB, reaching down only to the upper part of the aquifer. Additional nearby measure-

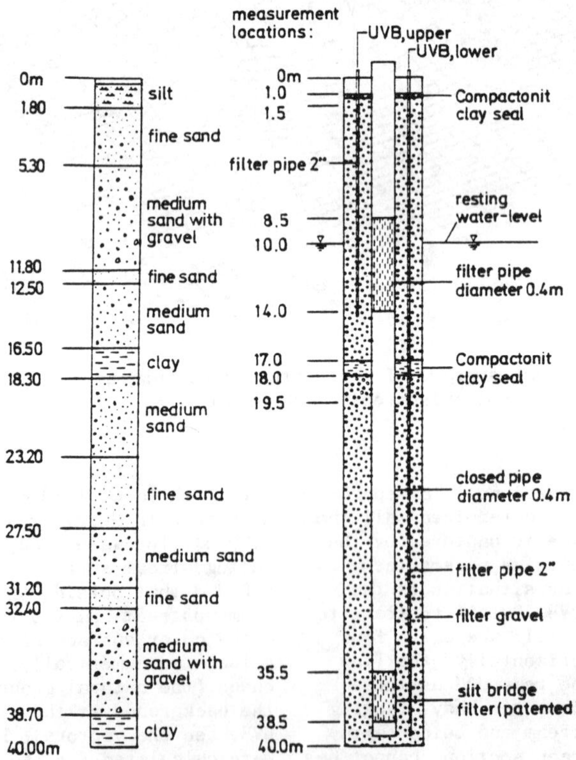

Fig. 9 UVB-site at Mannheim Kaefertal

ment locations could not be built finally for financial reasons. The claybank, 17 m below the surface, is assumed to be only locally because the upper and lower part of the aquifer is contaminated and no differences of the hydraulic heads could be ascertained.

The UVB has been run under two different operation conditions during '89. Until June 13, '89 an UVB only driven by the air bubble effect (figure 4) was used and after that date one with a separating plate and an additional pump (figure 1) has been employed. Figure 10 demonstrates the measured hydraulic heads at the above mentioned measurement locations. Until June 13 the hydraulic heads at the lower measurement location of the UVB are higher than at the upper one. The water circulated outside the well from the lower screen section to the upper one, which means opposite to the flow direction demonstrated in figure 4. This was caused by density effects.After June 13 the water flows outside the well from the upper to the lower screen section. This was forced by the separation plate and the additional pump. The difference of the hydraulic heads decreased from 0.8 m to small values because the upper screen section became sealed in consequence of precipitation effects after some time. An increasing clogging of the screen section yielded a decreasing well discharge. The hydraulic head at the upper part of the well increased but it decreased at the upper measurement location which is behind the sealed screen section.

Figure 11 demonstrates the concentration histories of PCE at the three mentioned measurement locations. The measurement locations within the bore-hole of the UVB show that clean water leaves the well, depending on the operation of the UVB, at lower or upper screen section, respectively. At the measurement location B225, in the upper part of the aquifer, a decrease of concentration takes place from 130 µg/l to 30 µg/l in the second phase.

The cumulative curve of the withdrawn chlorinated hydrocarbons is shown in figure 12 and demonstrates that the UVB is effective during both kinds of operation. At the end of the second phase the upper screen section is sealed.

During field measurements on August 25, '89 a relative humidity of only 33% of the used air was measured between the ventilator and the activated carbon. The same values are found at other UVB-installations. This means that the used air need not be dried additionally. Further, a vacuum pressure of $49 \cdot 10^2$ Pa less than atmospheric pressure was measured at the top of the well casing. It was determined that the quantities of fresh air, soil air and used air were 180 m^3/h, 320 m^3/h and 500 m^3/h, repectively.

The water discharge through the well with the additional pump and separating plate was measured at 18.1 m^3/h (with impeller) on June 28 and at 13.6 m^3/h (with wing wheel) on October 2 after cleaning of the sealed screen section.

4.2 Tracer experiment at Untersulmetingen

At Untersulmetingen near Ulm a tracer experiment was performed to demonstrate the effectiveness of an UVB (Kaess and Strayle 1990). At this remediation site the aquifer thickness is only 4.5 m. Below 1.1 m of medow loam there is a partly unconfined aquifer of Pleistocene gravel (hydraulic conductivity 0.005 m/s). Clayey silt of the upper fresh-water molasse is found be-

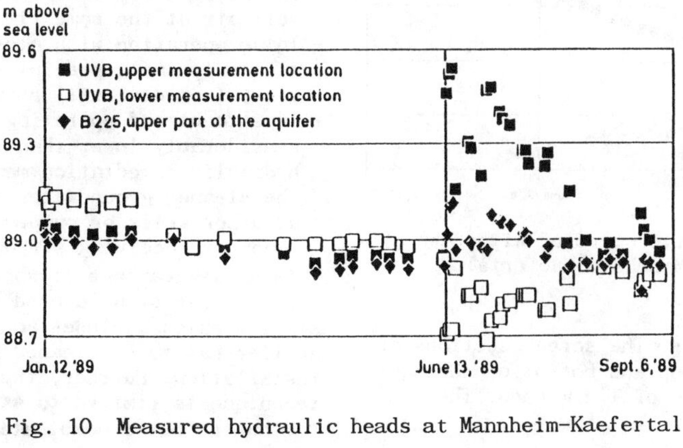

Fig. 10 Measured hydraulic heads at Mannheim-Kaefertal

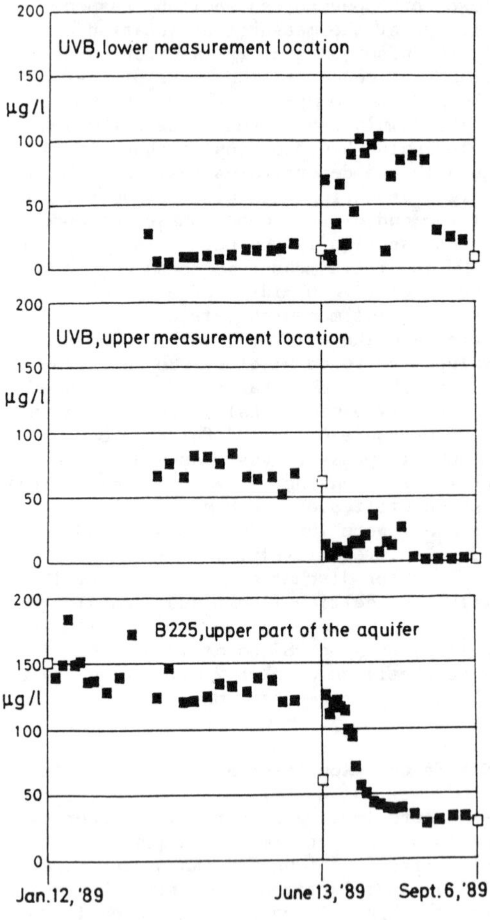

Fig. 11 Concentration history of PCE at Mannheim-Kaefertal

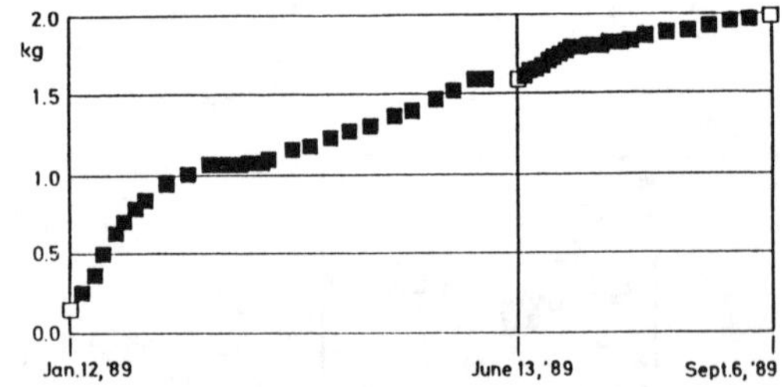

Fig. 12 Cumulative curve of withdrawn contaminants at Mannheim-Kaefertal

neath the aquifer. The screen sections of the UVB at the top and bottom of the aquifer have a length of 1.2 m each. The groundwater table is 1.5 m beneath the ground surface.

There are 24 measurement locations, each 12 were screened at the height of the upper and lower screen section of the UVB. The extent of the field is 27 m upstream and 14.5 m downstream of the UVB. The transport velocity of the natural groundwater flow is 5.7 m/d (Darcy velocity 1 m/d).

During the tracer experiment with sulphorhodamine B an UVB with additional pump and separating plate (figure 1) was used. The propagation of the tracer is demonstrated in figure 13, the dotted lines mark measurements at the lower locations. The results demonstrate clearly that, especially at the upper part of the aquifer, the circulation flow reached clearly to the upstream side as shown in figure 4b or 4c. Because of the small extent of the measurement field on the downstream side, no statements can be made about the concentrations in this region.

5 ASSESSMENT OF THE UVB-METHOD

In the previous sections the operation of the UVB was demonstrated. Theoretical results concerning the sphere of influence and the capture zone were presented along with measurements from field investigations. In conclusion a few further advantages of the UVB-method, which has been used at more than 60 locations, will be listed:

- Lowering of groundwater levels is not to be expected.
- Within the sphere of influence, groundwater flows vertically through less permeable, horizontal layers.
- The remediation of the groundwater is effected down to the aquifer bottom.
- Even at low well capacity a continuous remediation operation is ensured because no pumping off of the well occurs.
- The possibility exists to remediate the soil air at the same time.
- In an operation with closed air circulation only small precipitation of calcium, iron etc. will result.
- Investment and operating costs will be considerably lower than for conventional hydraulic remediation measures for which the cleaned groundwater is reinfiltrated at other wells or pumped to a sewage plant or receiving stream.

As a disadvantage it should be mentioned that in case of a layered subsoil with stories which includes aquitards, each aquifer has to be cleaned by a UVB-installation. Further, the described technique is limited to strippable contaminants, but a combination with in situ

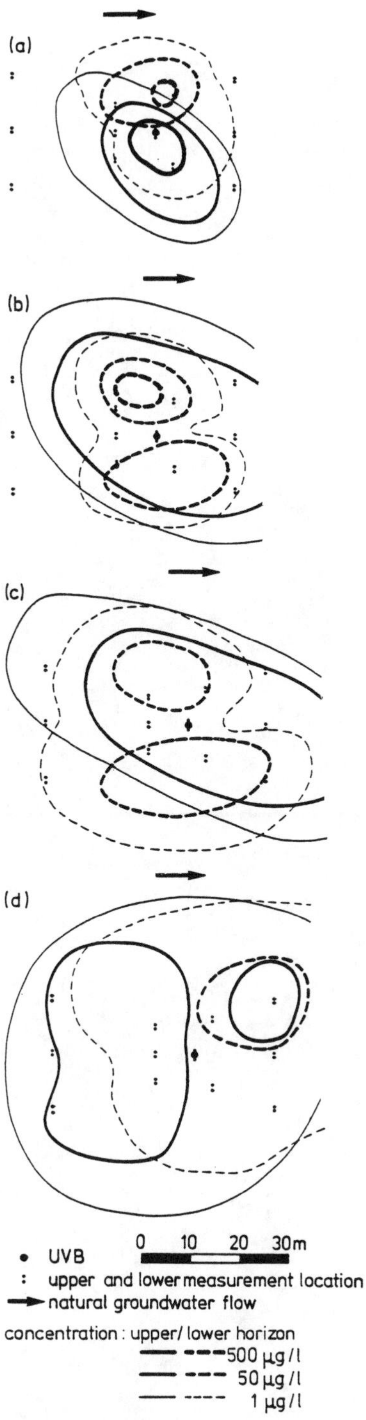

Fig. 13 Propagation of the tracer after (a) 8 h; (b) 32 h; (c) 78 h; and after (d) 750 h (modified from Kaess and Strayle, 1990)

bioreclamation is possible and has been used.

6 ACKNOWLEDGEMENT

The authors greatfully acknowledge B. Bernhardt, IEG mbH, inventor and patent holder of the UVB-method, for the valuable discussions, the cooperation and financial support in studying and in developing the UVB-technique. E.J. Alesi, GfS mbH; P. Brinnel, HYDRODATA GmbH; W. Kaess; and H.J. Lochte, UTB mbH, are thanked for many helpful discussions.

REFERENCES

Herrling, B., Buermann, W. 1990a. UVB-Verfahren - Grundprinzip und Messungen. In P. Bock et al. : Untergrundsanierung mittels Bodenluftabsaugung und In-Situ-Strippen. Schriftenr. Angew. Geologie, Uni. Karlsruhe, Bd. 9, pp. 275 - 290.

Herrling, B., Buermann, W. 1990b. A new method for in-situ remediation of volatile contaminants in groundwater - numerical simulation of the flow regime. In G. Gambolati et al.: Computational methods in subsurface hydrology. Springer, Berlin, pp. 299 - 304.

Herrling, B., Buermann, W., Stamm, J. 1990. In-Situ-Beseitigung leichtflüchtiger Schadstoffe aus dem Grundwasserbereich mit dem UVB-Verfahren. In H.-P. Luehr et al.: Neuer Stand der Sanierungstechniken von Altlasten. IWS - Schriftenreihe, Bd.10, E. Schmidt publisher, Berlin, pp. 71 - 99.

Kaess, W., Strayle, G. 1990. UVB-Verfahren aus der Sicht des Anwenders - Ergebnisse von Markierungsversuchen. In P. Bock et al.: Untergrundsanierung mittels Bodenluftabsaugung und In-Situ-Strippen. Schriftenr. Angew. Geologie, Uni. Karlsruhe, Bd. 9, pp. 291 - 304.

Subsurface Contamination by Immiscible Fluids, Weyer (ed.) © 1992 Balkema, Rotterdam. ISBN 90 6191 175 3

Vacuum extraction induced biodegradation of petroleum hydrocarbons

R.E.Hinchee & S.K.Ong
Battelle Memorial Institute, Columbus, Ohio, USA

R.N.Miller
USAF Center for Environmental Excellence, Brooks AFB, Tex., USA

C.Vogel
USAF Civil Engineering Support Agency, Tyndall AFB, Fla., USA

D.C.Downey
Engineering-Science, Inc., Denver, Colo., USA

ABSTRACT: Bioventing is the process of aerating subsurface soils to stimulate in-situ bioremediation. Air forced into the ground provides an oxygen source for microorganisms. This stimulates the soil-indigenous organisms to aerobically metabolize hydrocarbons, such as jet fuel, in unsaturated soils. Volatile compounds also may be simultaneously removed. For contaminated sites near buildings, air withdrawal may be necessary to prevent the buildup of volatile compounds and CO_2 in subsurface structures. Bioremediation differs from soil venting, which maximizes volatilization of low-molecular-weight compounds, in that bioventing maximizes biodegradation of any aerobically biodegradable compound, regardless of molecular weight. Based on field applications studies, addition of nutrients was found to be ineffective in enhancing biodegradation. In some instances, it may be necessary to dewater the soil to eliminate saturated conditions and to increase air/contaminated contact and zone of remediation. Soil temperature was observed to have a significant effect on biodegradation rates.

1 INTRODUCTION

Bioventing is the process of aerating subsurface soils to stimulate in-situ bioremediation. It is related to the soil venting (aka soil vacuum extraction, soil gas extraction, in-situ soil stripping) process. The significant difference is that soil venting is designed and operated to maximize volatilization of low-molecular-weight compounds. In most soil venting remediations, some biodegradation occurs. In contrast bioventing is designed to maximize biodegradation of any aerobically biodegradable compound, regardless of molecular weight. In many bioventing remediations, some volatilization will occur. The significant difference in the technologies is that the objective of soil venting is volatilization, and the objective of bioventing is biodegradation. Although both technologies involve venting of air through the subsurface the differences in objectives result in different design and operation of the remedial systems.

2 DISCUSSION

Petroleum distillate hydrocarbons such as JP-4 (jet fuel) are generally biodegradable if naturally occurring microorganisms are provided an adequate supply of oxygen and basic nutrients (Atlas, 1986). Natural biodegradation does occur and, at many sites, may eventually mineralize most fuel contamination. However, the process is dependent upon natural oxygen diffusion rates (Ostendorf and Kambell, 1989) and, as a result, is frequently too slow to prevent the spread of contamination. Such sites may require remediation of the contaminant source to protect sensitive aquifers. An acceleration or enhancement of the natural biodegradation process may prove the most effective remediation at these sites.

An understanding of the distribution of contaminants is important in any in-situ remediation. Much of the residue of hydrocarbons at a fuel-contaminated site is found in the unsaturated zone soils, in the capillary fringe, and immediately below the water table. Seasonal water table fluctuations typically spread residues in the area immediately above and below the water table. Any successful bioremediation effort must treat these areas. Bioventing can provide oxygen to unsaturated zone soils.

2.1 Conventional enhanced biodegradation

Over the past two decades the practice of enhanced biodegradation has increased, particularly for treating soluble fuel components in groundwater (Lee et al., 1988). Less emphasis has been given to enhancing biodegradation in the unsaturated zone. The current conventional enhanced bioreclamation process uses water to carry the oxygen or an alternative electron acceptor to the contamination, whether it occurs in the groundwater or unsaturated zone.

A recent field experiment at a jet fuel-contaminated site used infiltration galleries and spray irrigation to introduce oxygen (as hydrogen peroxide), nitrogen, and phosphorus to unsaturated sandy soils. It was unsuccessful because rapid hydrogen peroxide decomposition resulted in poor oxygen distribution (Hinchee et al., 1989). A study being conducted by the U.S. Environmental Protection Agency (USEPA) and the U.S. Coast Guard at Traverse City, Michigan, uses deep well injection to raise the water table in order to supply oxygen-enriched water to the contaminated soils. Pure oxygen and hydrogen peroxide have been used as oxygen sources, and recently nitrate has been added as an alternative to oxygen. Although results indicate better hydrogen peroxide stability than achieved by Hinchee et al. (1989), it was concluded that much of the hydrogen peroxide decomposed rapidly and was lost as bubbles (Huling et al., 1990). Some degradation of aromatic hydrocarbons appears to have occurred; however, no change in total hydrocarbon contamination levels was detected in the soils (Ward, 1988).

Oxygen is the limiting factor for biodegradation in most cases where water is used as the oxygen carrier. If pure oxygen is utilized and 40 mg/L of dissolved oxygen is achieved, approximately 80,000 kg of water must be delivered to the formation to degrade a single kilogram of hydrocarbon. If 500 mg/L of hydrogen peroxide is successfully delivered, then approximately 13,000 kg of water are necessary. As a result, even if hydrogen peroxide can be successfully used, substantial volumes of water must be pumped through the contaminated formation to deliver sufficient oxygen.

2.2 Bioventing

A system engineered to increase the microbial biodegradation of fuel hydrocarbons in the vadose zone using forced air as the oxygen source is a cost-effective alternative to conventional systems. This process stimulates soil-indigenous microorganisms to aerobically metabolize fuel hydrocarbons in unsaturated soils. Volatile compounds may be simultaneously removed from contaminated soils depending on airflow rates.

By using air as an oxygen source, the minimum ratio (based on stoichiometry) of air to hydrocarbon on a mass basis is approximately 13 to 1. This compares with over 10,000 to 1, water to hydrocarbon, for a conventional waterborne enhanced bioreclamation process. An additional advantage of using an airborne process is that gases have greater diffusivity than liquids. At many sites, geological heterogeneities present an added problem with a waterborne oxygen source because fluid pumped through the formation is channeled into the more permeable pathways. For example, in an alluvial soil with interbedded sand and clay, initially all of the fluid flow will take place in the sand. As a result, oxygen must be delivered to less permeable clay lenses through the diffusion. In a gaseous system (as is found in unsaturated soils) this diffusion can be expected to take place at a rate several orders of magnitude greater than in a liquid system (as is found in saturated soils). Although it is not realistic to expect diffusion to aid significantly in water-based bioreclamation, in an air-based application, diffusion may be a significant mechanism for oxygen delivery to less permeable zones.

To the authors' knowledge the first documented evidence of unsaturated zone biodegradation resulting from forced aeration was reported by the Texas Research Institute, Inc., in a study for the American Petroleum Institute. A large-scale model experiment was conducted to test the effectiveness of a surfactant treatment to enhance recovery of spilled gasoline. The experiment accounted for only 30 L of the 250 L originally spilled and raised questions about the fate of the gasoline. A subsequent column study was conducted to determine a diffusion coefficient for soil venting. This column study evolved into a biodegradation study in which it was concluded that as much as 38% of the fuel hydrocarbon was biologically mineralized. Researchers concluded that venting would not only remove gasoline by physical means, but also could enhance microbial activity (Texas Research Institute, 1980; 1984).

The authors believe that the first field-scale bioventing experiments were conducted by Jack van Eyk working for

Shell Oil. At his direction, Delft Geotechnics in the Netherlands initiated a series of experiments in 1983 to investigate the effectiveness of bioventing for treatment of gasoline and contaminated soils. These studies are reported in a series of papers (Anonymous 1986, Staatsuitgeverij 1986, van Eyk and Vreeken 1988, and van Eyk and Vreeken 1989a and b).

Wilson and Ward (1986) suggested that using air as a carrier for oxygen could be 1,000 times more efficient than transferring it to water, especially in deep, hard-to-flood unsaturated zones. They made the connection between soil venting and biodegradation by observing that "soil venting uses the same principle to remove volatile components of the hydrocarbon." In a general overview of the soil venting process, Bennedsen et al. (1987) concluded that soil venting provides large quantities of oxygen to the vadose zone, possibly stimulating aerobic degradation. He suggested that water and nutrients would also be required for significant degradation and encouraged additional investigation in this area.

Biodegradation enhanced by soil venting has been observed at several field sites. Investigators at a soil venting site for remediation of gasoline-contaminated soil claim significant biodegradation as measured by a rise in temperature when air was supplied. Investigators pulsed pumped air through a pile of excavated soil and observed a consistent rise in temperature that they attributed to biodegradation. They claimed that the pile was cleaned up during the summer primarily by biodegradation (Conner, 1988). However, they did not control for natural volatilization from the aboveground pile, and not enough data were published to critically review the biodegradation claim.

Researchers at Traverse City measured toluene concentration in vadose zone soil gas as an indicator of fuel contamination in the vadose zone. They assumed absence of advection and attributed toluene loss to biodegradation. The investigators concluded that because toluene concentrations decayed near the oxygenated ground surface, soil venting is an attractive remediation alternative for biodegrading light volatile hydrocarbon spills (Ostendorf and Kambell, 1989).

Ely and Heffner (1988) of the Chevron Research Company patented a process for the in-situ biodegradation of spilled hydrocarbons using soil venting. Detailed experimental design and data are not provided, but findings are presented graphically. At a gasoline- and diesel oil-contaminated site, a slightly higher removal rate was observed through biodegradation than through evaporation. At a gasoline-contaminated site, results indicated that about two-thirds of the hydrocarbon removal was by volatilization and one-third by biodegradation. At a site containing only fuel oils, approximately 75 L per well per day were biodegraded, while vapor pressures were too low for removal by volatilization. Ely and Heffner claimed that the process is more advantageous than strict soil venting because removal is not dependent only on vapor pressure. In the examples stated in the patent, CO_2 was maintained between 6.8% and 11% and O_2 between 2.3% and 11% in vented air. The patent suggests that the addition of water and nutrients may not be acceptable because of flushing to the water table, but nutrient addition is claimed as part of the patent. The patent recommends flow rates between 50 and 420 m^3/min per well and states that airflows higher than those required for volatilization may be optimum for biodegradation.

The U.S. Air Force initiated its research and development program in bioventing in 1988 with a study at Hill Air Force Base (AFB), near Ogden, Utah. In the course of this study, it became apparent that bioventing had great potential for remediation of JP-4-contaminated soils. It was also apparent that additional research was needed before the technology could be routinely applied in the field. The work was initially supported by the U.S. Air Force Civil Engineering Support Agency (AFCESA), previously known as the Air Force Engineering and Services Center. Subsequently they were joined by the U.S. Air Force Center for Environmental Excellence (AFCEE) and later Hill and Eielson AFBs in research and development support of this technology. Following the Hill AFB study, a more controlled bioventing study was completed at Tyndall AFB in Florida. Currently the Air Force is supporting a number of field programs to further improve and demonstrate the technology. At Hill AFB after completion of the initial site, a low-intensity bioreclamation research program at another site was initiated in late 1989. At Eielson AFB near Fairbanks, Alaska, a field demonstration of bioventing in a subarctic environment was initiated in the summer of 1991. This study includes a soil heating experiment to attempt to improve biodegradation rates. The USEPA Risk Reduction Engineering Laboratory (RREL) has become interested in the Air Force's program and has jointly funded, and technically supported the work at both Hill and Eielson AFBs. Additionally, the AFCESA is supporting a well-documented bioventing demonstration at a cold-weather site, where field work will be initiated during the summer of 1992.

2.3 Applications

The use of an air-based oxygen supply for enhancing biodegradation relies on airflow through the hydrocarbon-contaminated soils at rates and configurations that will ensure adequate oxygenation of aerobic biodegradation and minimize or eliminate the production of a hydrocarbon-contaminated off-gas. The addition of nutrients and moisture may be desirable to increase biodegradation rates; however, field research to date does not support this (Dupont et al., 1991 and Miller et al., 1991). Dewatering may at times be necessary depending upon the distribution of contaminants relative to the water table. However, it is required at many fuel hydrocarbon-contaminated sites. A key feature is the screened soil gas monitoring points that sample only a short vertical section of the soil. These points are required to determine local oxygen concentrations. Measurements of oxygen levels in the vent well is not representative of local conditions. Nutrient and moisture addition typically may take any of a variety of configurations.

The use of a conventional soil venting installation where air is drawn from a vent well in the area of greatest contamination is possible. The advantage of the configuration is that it allows straightforward monitoring of the off-gases; the disadvantage is that hydrocarbon off-gas concentration is probably maximized, and may require permitting and treatment, and all of the capillary fringe contamination may not be treated. Figure 1 involves air injection only. Figure 2 illustrates this configuration coupled with dewatering to allow more effective treatment of deeper soils. Optional soil moisture and nutrient addition is also illustrated. This is the lowest cost configuration; however, careful consideration must be given to the fate of injected air. The objective is for most if not all of the hydrocarbons to be degraded, and simply result in CO_2 emissions at some distance from the injection point. If a building or subsurface structure exists within the radius of influence hydrocarbon vapors may be forced into that structure. Protection of subsurface structures may be required. Figure 3 illustrates a configuration in which air is injected (the injection may also be by passive well) into the contaminated zone and withdrawn from clean soils. This configuration allows the more volatile hydrocarbons to degrade prior to being withdrawn and thereby eliminating contaminated off-gases. This configuration typically does not require air emission permitting (site specific exceptions may apply). Figure 4 illustrates a configuration which may alleviate the threat to subsurface structures while achieving the same basic effect as air injection alone. In this configuration soil gas is extracted near the structure of concern and reinjected at a safe distance. If necessary make-up air can be added before injection. The optimal configuration for any given site will, of course, depend upon site-specific conditions and remedial objectives.

The significant features of this technology include the following:

- Optimizing airflow to reduce volatilization while maintaining aerobic conditions for biodegradation
- Monitoring local soil gas conditions to assure aerobic conditions, not just monitoring vent gas composition
- Adding moisture and nutrients as required to increase biodegradation rates although, as stated earlier, it appears based upon field studies that this may not be necessary at many if not most sites
- Manipulating the water table as required for air/contaminant contact.

2.4 Hill AFB site

A spill of approximately 100,000 L of JP-4 occurred when an automatic overflow device failed at Hill AFB. The contamination is primarily found in the upper 20 m of a delta outwash of the Weber River. This surficial formation extends the surface to a depth of approximately 20 m and is composed of mixed sand and gravel with occasional clay stingers. Depth to regional groundwater is approximately 200 m, however, water may occasionally be found in discontinuous perched zones. Soil moisture averaged less than 6% in the contaminated soils.

Soil samples that were collected indicated JP-4 concentrations up to 20,000 mg/kg, with an average concentration of approximately 400 mg/kg (Oak Ridge National Laboratories, 1989). Contaminants were unevenly distributed to a depth of 20 m. Vent wells were drilled to

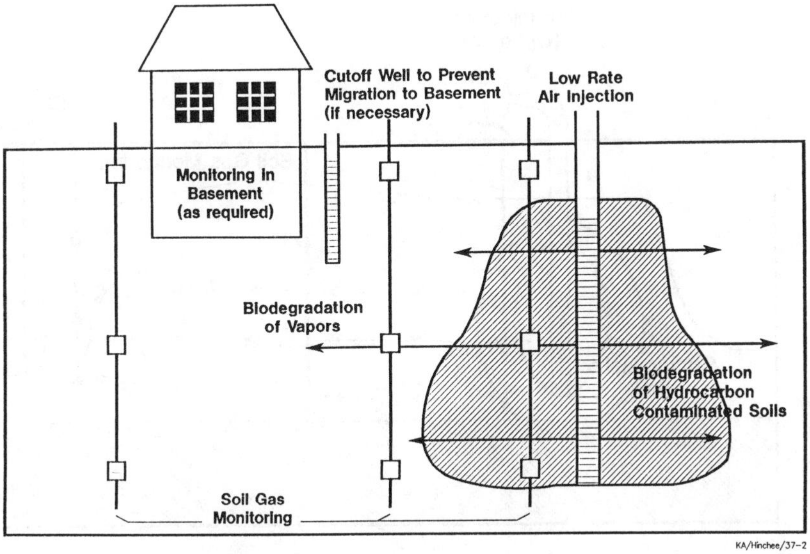

Figure 1. Conceptual Layout of Bioventing Process with Air Injection Only

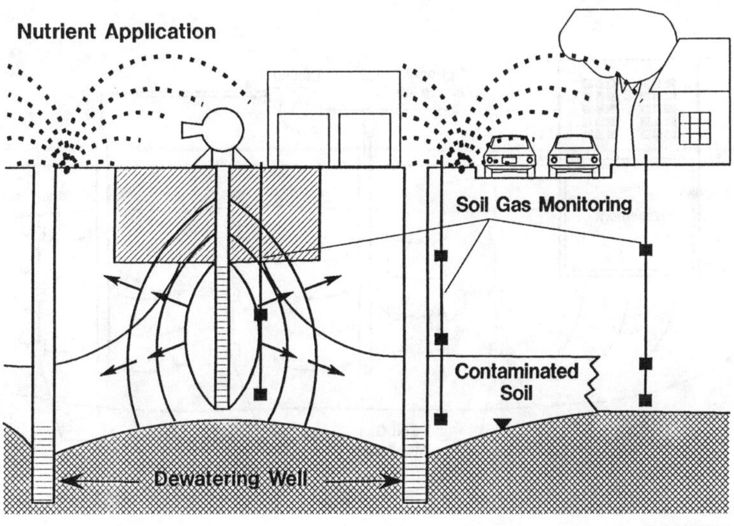

Figure 2. Conceptual Layout of Bioventing Process Coupled with Dewatering and Nutrient and Moisture Addition

approximately 20 m below the ground surface and screened from 3 to 18 m below the surface. A background vent was installed in an uncontaminated location in the same geological formation approximately 200 m north of the site.

Venting was initiated in December 1988 at a rate of approximately 45 m^3/hr. The off-gas was treated by catalytic incineration, and it was initially necessary to dilute the highly concentrated gas to remain below explosive limits and within the incinerator's hydrocarbon operating limits. The venting rate was gradually increased to approximately 2,500 m^3/hr as hydrocarbon concentration levels dropped. Between December 1988 and November 1990, more than 10,000,000 m^3 of soil gas were

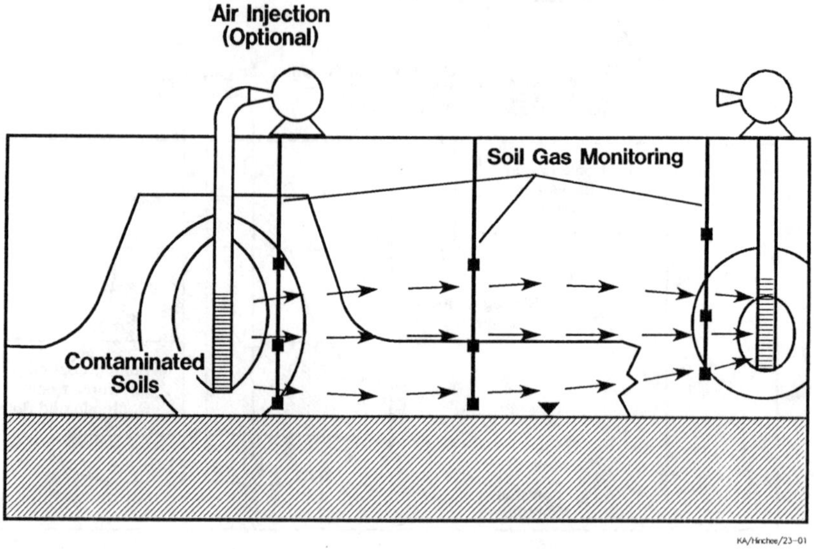

Figure 3. Conceptual Layout of Bioventing Process with Air Withdrawn from Clean Soil

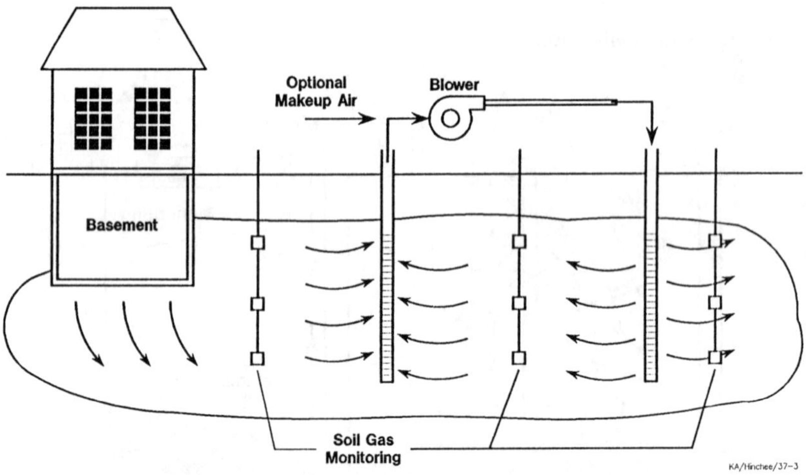

Figure 4. Conceptual Layout of Bioventing Process with Soil Gas Reinjection

extracted from the site. In November 1989, ventilation rates were reduced to approximately 500 to 1,000 m^3/hr to provide aeration for bioremediation but to reduce volatilization. This resulted in removal of the catalytic incinerator, saving approximately $6,000 per month. Oxygen and hydrocarbon concentrations were measured in the off-gas during extraction. To quantify the extent of biodegradation at the site, the oxygen was converted to an equivalent basis. This was based upon the stoichiometric oxygen requirement for hexane mineralization. JP-4 was determined based on direct readings of a total hydrocarbon analyzer calibrated to hexane. Based upon these calculations, the mass of JP-4 as carbon removed and degraded was approximately 53,000 kg volatilized and 42,000 kg biodegraded. Figures 5 and 6 illustrate these results.

Hinchee and Arthur (1991) conducted bench-scale studies utilizing soils from Hill AFB and found that, in the labora-

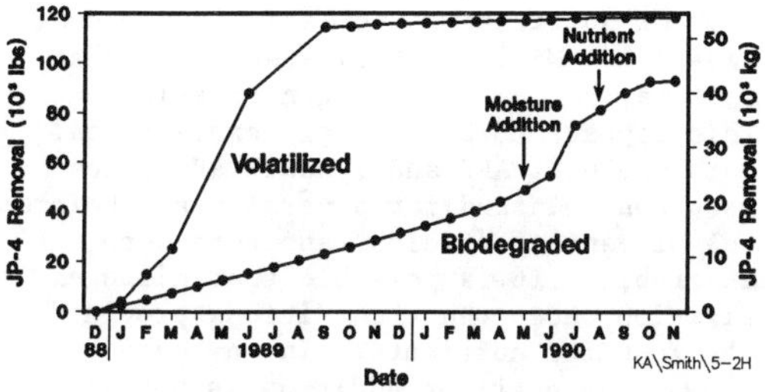

Figure 5. Cumulative Hydrocarbon Removal from the Hill AFB Building 914 Soil Venting Site

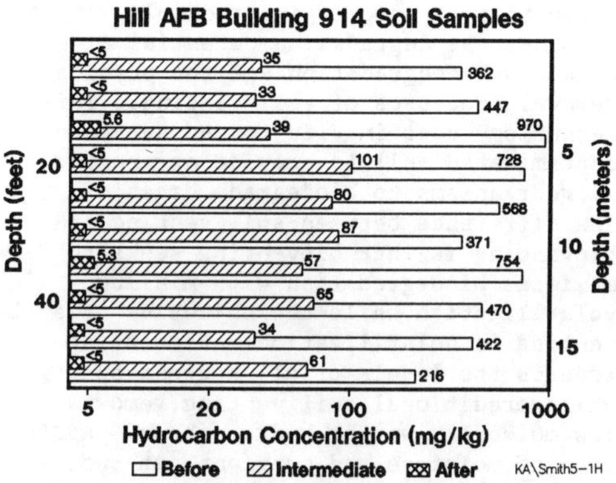

Figure 6. Results of Soil Analysis at Hill AFB Before and After Venting. Each Bar is the Average of 14 or More Samples. (Dupont et al., 1991)

tory, both moisture and nutrients became limiting after aerobic conditions were achieved. This lead to first the addition of moisture and then nutrients in the field. The results of these field additions can be seen in Figure 5. Moisture addition clearly stimulated biodegradation, nutrient addition did not. The failure to observe an effect of nutrient addition could be the result of a number of factors including: failure of nutrients to move in soils, this is particularly a problem for ammonia and phosphorus (Aggarwal et al., 1991); remediation of the site was entering its final phase and the nutrient addition may have been too late to result in an observed change; or nutrients may simply have not been limiting.

2.5 Tyndall AFB site

As a follow-up to the Hill AFB research, a more controlled study was designed. The experimental area of Tyndall AFB was located at a site where past JP-4 storage had resulted in contaminated soils. The nature and volume of fuel spilled or leaked is unknown. Initial site characterization indicated soil hydrocarbon levels in the 5,000 to 20,000 mg/kg range. The site soils are a fine- to medium-grained quartz sand and depth to groundwater is 0.5 to 1.0 m. Four test cells were constructed to allow control of gas flow, water flow, and nutrient addition. Two of the test cells were installed in the hydrocarbon-contaminated zone and two in uncontaminated soils. The contaminated area was dewatered and hydraulic control was maintained to keep the depth to groundwater at approximately 1.6 m. This exposed more of the contaminated soil to aeration. During normal operation, air-flow rates were maintained at approximately one air-filled void volume per day. Biodegradation and volatilization rates were much higher than those observed at Hill AFB; this was likely due to higher average levels of contamination, warmer temperatures, and moisture conditions. After 200 days of aeration, hydrocarbon reduction of approximately one-third was observed.

This represents a reduction in total hydrocarbons of approximately 20%. The study stopped because the process monitoring objectives had been met, biodegradation was still vigorous. Although the total petroleum hydrocarbons had only been

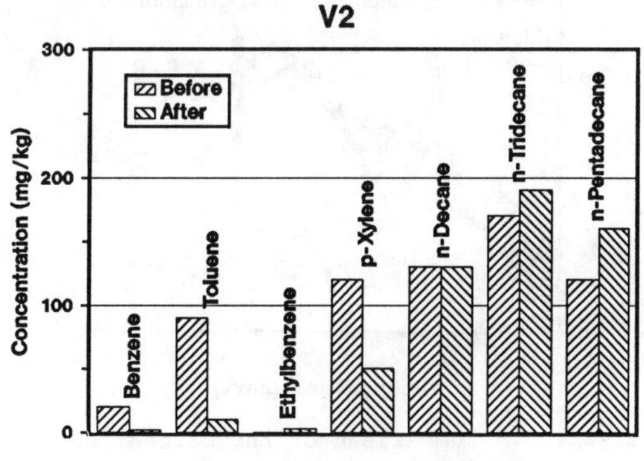

Figure 7. Results of Soil Analysis at Tyndall AFB Before and After Venting (data from Miller, 1990)

reduced by 20%, the low-molecular-weight aromatics, benzene, toluene, the xylenes, and ethylbenzene (BTXE) were reduced by more than 90% (see Figure 7). It appears that the bioventing process more rapidly removes the BTXE compounds than other JP-4 constituents.

Another important observation of this study is the effect of temperature on biodegradation rate. Miller (1990) found that the van Hoff-Arrhenius equation provided an excellent model of temperature effects. In the Tyndall AFB study soil temperature varied by only approximately 7 C and yet biodegradation rates were approximately twice as high as 25 C as at 18 C.

In the Tyndall AFB study, the effect of moisture and nutrients were observed in a field test. Two side by side plots were identically treated except that one (V2) received both moisture and nutrients. From the outset, the other plot (V1) received neither for 8 weeks then moisture only for 14 weeks, followed by both moisture and nutrients for 7 weeks. As illustrated in Figure 8 no significant effect of moisture or nutrients was observed. The lack of moisture effect contrasts with the Hill AFB findings, but is most likely the result of contrasting climatic and hydrogeological conditions. Hill AFB is located on a high-elevation desert with a very deep water table. Tyndall AFB is located in a moist subtropical environment and, at the site studied, the water table was maintained at a depth of less than 2 m.

The nutrient findings support field observations at Hill AFB, that the addition of nutrients do not stimulate biodegradation. Miller (1990) speculates that this may be due to nitrogen fixation. This appears to be a likely explanation. Both the Hill AFB and Tyndall AFB sites were contaminated for several years before the bioventing studies, and both were anaerobic. It is possible that nitrogen fixation under these conditions provided the required nutrients. In any case it shows that nutrient addition is not always required.

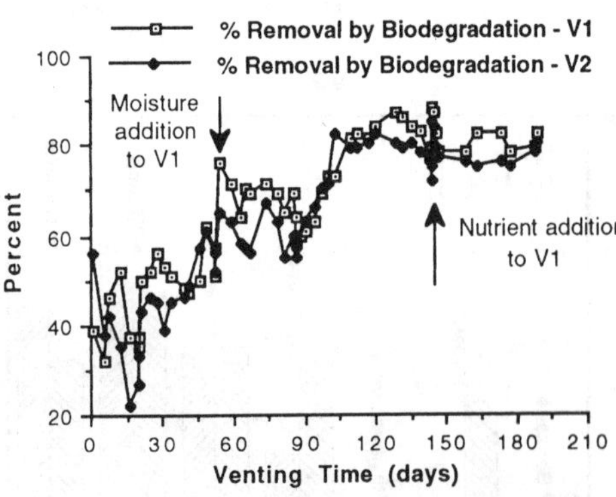

Figure 8. Comparison of the Percent of Total Hydrocarbon Removal Attributed to Biodegradation (Oxygen Basis) in Treatment Plots V1 and V2 During the Field Study at Tyndall AFB

3 CONCLUSIONS

Bioventing is an emerging technology that combines the degradation potential of enhanced biodegradation and the physical removal processes of soil venting. This technology uses in situ forced aeration of contaminated soil to provide oxygen for microorganisms to biodegrade organics. The difference between soil venting and bioventing is that bioventing seeks to maximize biodegradation with minimum volatilization while the objective of soil venting is volatilization. Bioventing reduces the levels of all hydrocarbons while traditional soil venting removes low-molecular-weight hydrocarbons. Addition of moisture and nutrients at two field applications did not increase the effectiveness of bioventing, but soil temperature was found to significantly affect biodegradation rates. Several conceptual bioventing designs for various contaminated site conditions were presented including a conceptual approach to prevent buildup of CO_2 and volatile gases in subsurface structures.

4 REFERENCES

Aggarwal, P.K., J.L. Means, & R.E. Hinchee 1991. Formulation of nutrient solutions for in-situ bioremediation, in R.E. Hinchee and R.F. Olfenbuttel (eds.), In-Situ Bioreclamation, Butterworth-Heinemann, Stoneham, MA, pp 51-66.

Anonymous 1986. "In-situ reclamation of petroleum contaminated sub-soil by subsurface venting and enhanced biodegradation." Research Disclosure, No. 26233, 92-93.

Atlas, R.M. 1986. Microbial degradation of petroleum hydrocarbons: an environmental perspective, *Microbiol. Rev.* 45: 180-209.

Bennedsen, M.B., J.P. Scott, & J.D. Hartley 1987. Use of vapor extraction systems for in situ removal of volatile organic compounds from soil, Proceedings of National Conference on Hazardous Wastes and Hazardous Materials, Washington, D.C., 92-95.

Conner, J.S. 1988. Case study of soil venting, *Poll. Eng.* 7:74-78.

Dupont, R.R., W. Doucette, & R.E. Hinchee 1991. Assessment of in situ bioremediation potential and the application of bioventing at a fuel-contaminated site, in: In Situ and On-Site Bioreclamation, R.E. Hinchee & R.F. Olfenbuttel (eds.), Butterworth-Heineman, Stoneham, MA, pp 262-282.

Ely, D.L. & D.A. Heffner 1988. Process for in-situ biodegradation of hydrocarbon contaminated soil, U.S. Patent No. 4,765,902.

Hinchee, R.E. & M. Arthur 1991. Bench-scale studies of the soil aeration process for bioremediation of petroleum hydrocarbons. *J. Appl. Biochem. Biotech.* 28/29:901-906.

Hinchee, R.E., D.C. Downey, J.K. Slaughter, & M. Westray 1989. Enhanced biorestoration of jet fuels; a full scale test at Eglin Air Force Base, Florida, Air Force Engineering and Services Center Report ESL/TR/88-78.

Huling, S.G., B.E. Bledsoe, & M.V. White 1990. Enhanced biodegradation utilizing hydrogen peroxide as a supplemental source of oxygen: A laboratory and field study. EPA/600-290-006, 48 pp.

Lee, M.D. et al. 1988. Biorestoration of aquifers contaminated with organic compounds, CRC Critical Reviews in Env. Control. Vol. 18, 29-89.

Miller, R.N., R.E. Hinchee, & C. Vogel 1991. A field-scale investigation of petroleum hydrocarbon biodegradation in the vadose zone enhanced by soil venting at Tyndall AFB, Florida, in: In Situ and On-Site Bioreclamation, R.E. Hinchee and R.F. Olfenbuttel (eds.), Vol. 1, Stoneham, MA: Butterworth Publishers.

Miller, R.N. 1990. A field scale investigation of enhanced petroleum hydrocarbon biodegradation in the vadose zone combining solid venting as an oxygen source with moisture and nutrient additions; Ph.D. Dissertation. Utah State University, Logan, Utah.

Oak Ridge National Laboratories 1989. Soil characteristics: data summary, Hill Air Force Base Building 914 fuel spill soil venting project, an unpublished report to the U.S. Air Force.

Ostendorf, D.W. & D.H. Kambell 1989. Vertical profiles and near surface traps for field measurements of volatile pollution in the subsurface environment, Proceedings of NWWA Conference on New Techniques for Quantifying the Physical and Chemical Properties of Heterogeneous Aquifers, Dallas, TX.

Staatsuitgeverij 1986. Proceedings of a workshop, 20-21 March, 1986. Bodembeschermingsreeeks No. 9; Biotechnologische Bodemsanering, pp. 31-33. Rapportnr. 851105002, ISBN 90-12-054133, Ordernr. 250-154-59; Staatsuitgeverij Den Haag: The Netherlands.

Texas Research Institute 1980. Laboratory scale gasoline spill and venting experiment, American Petroleum Institute, Interim Report No. 7743-5:JST.

Texas Research Institute 1984. Forced venting to remove gasoline vapor from a large-scale model aquifer, American Petroleum Institute, Final Report No. 82101-F:TAV.

van Eyk, J. & E. Vreeken 1988. "Venting-mediated removal of petrol from subsurface soil strata as a result of simulated evaporation and enhanced biodegradation." *Med. Fac. Landbouww.* Riiksuniv. Gent, 53(4b), 1873-1884.

van Eyk, J. & C. Vreeken 1989a. "Model of petroleum mineralization response to soil aeration to aid in site-specific, in situ biological remediation." In Groundwater Contamination: Use of Models in Decision-Making, Proceedings of an International Conference on Groundwater Contamination; Jousma et al. (eds.); Kluwer Boston/London, pp 365-371.

van Eyk J. & C. Vreeken 1989b. "Venting-mediated removal of diesel oil from subsurface soil strata as a result of stimulated evaporation and enhanced biodegradation." In Hazardous Waste and Contaminated Site, Envirotech Vienna, Vol. 2, Session 3, ISBN 389432-009-5; Westarp Wiss: Essen; pp 475-485.

Ward, C.H. 1988. A quantitative demonstration of the Raymond process for in-situ biorestoration of contaminated aquifers, Proceedings of NWWA/API Conference on Petroleum Hydrocarbons and Organic Chemicals in Groundwater, 723-746.

Wilson, J.T. & C.H. Ward 1986. Opportunities for bioremediation of aquifers contaminated with petroleum hydrocarbons, *J. Ind. Microbiol.*, 27:109-116.

Subsurface Contamination by Immiscible Fluids, Weyer (ed.) © 1992 Balkema, Rotterdam. ISBN 90 6191 175 3

Transport of ammonium and nitrate ions in hydrocarbon-contaminated soil columns

N. Benazon, P. Lafrance & J.-P. Villeneuve
Institut National de la Recherche Scientifique (INRS-Eau), Québec, Que., Canada

R. R. Simard
Agriculture Canada, Station de Recherches, Québec, Que., Canada

ABSTRACT: The effect of residual subsurface hydrocarbon (kerosene) contamination on the transport of an electron acceptor (nitrate) and a nutrient (ammonium ion) used in in-situ bioremediation was investigated. Soil column experiments were conducted using uncontaminated and contaminated soil and a one-dimensional advection-dispersion transport model was used to predict the breakthrough curves. The presence of kerosene contamination was found to decrease the volumetric water content which in turn increased the pore-water velocity. An earlier breakthrough time for NO_3^- relative to the uncontaminated soil was therefore predicted by the model and corresponded well with the experimental results. However, because the exchange isotherm involving NH_4^+ was not affected by the presence of petroleum, no significant difference in NH_4^+ transport was predicted or observed.

1 INTRODUCTION

Contamination of groundwater and soil by petroleum hydrocarbons is of particular environmental concern due to the widespread use of fuels and solvents and their routine storage in underground reservoirs. It is estimated that hundreds of thousands of subsurface gasoline leaks from storage tanks and associated piping may presently be occurring in the United States (Hoag and Marley, 1986).

Subsurface contamination by petroleum can potentially exist in 4 phases (Wilson and Brown, 1989): (1) mobile free product (phase-separated), (2) residual soil-bound contamination (sorbed-phase), (3) contamination dissolved in groundwater (dissolved-phase), and (4) vapour-phase. The dissolved-phase contamination is of particular concern because soluble petroleum compounds can pose health problems even at very low concentrations and because the mobility of groundwater may spread the contamination over large areas. Sorbed-phase contamination, however, is relatively immobile but is often a continuous source of dissolved groundwater contamination due to fluctuations in the groundwater table and/or groundwater recharge. Therefore, following free-product recovery, the focus of remediation efforts should be on the reduction of the sorbed-phase contamination, thereby simultaneously reducing the dissolved-phase contamination problem.

In-situ bioremediation has been used successfully to clean up numerous sites contaminated by petroleum hydrocarbons (Lee et al., 1988; Staps, 1989). Unlike conventional methods based on pumping and treating, bioremediation addresses the residual soil-bound contamination. However, field and laboratory experience has shown that a deficiency of electron acceptors such as oxygen and/or nitrate as well as nutrients such as phosphorus and nitrogen limits the biodegradation rate of hydrocarbon compounds (Lee et al. 1988). Thus, during in-situ bioremediation these compounds are generally supplied to the subsurface via injection wells or infiltration galleries. However, the success of this approach depends on how well these compounds are distributed in the subsurface following injection. This in turn depends on the intrinsic chemical properties of the added solutes and on the hydrogeological and geochemical characteristics of the formation. The design of efficient bioremediation systems could be greatly improved by the use of computer models that predict solute behaviour in groundwater. Such models could then be used to help optimize the number and location of the injection wells or infiltration galleries.

Over the last 20 years, extensive research has been done on solute transport in soil systems. Mathematical representations of the phenomena of advection, dispersion, ion exchange, and biological degradation have also been developed (Anderson, 1979; Naymik, 1987). Recently, these equations have been used to predict oxygen, nitrate and nutrient transport, as well as the decrease in dissolved-phase contamination during in-situ bioremediation (Widdowson et al.,1988; Borden and Bedient, 1986). However, these models have not taken into consideration the effect of sorbed-phase hydrocarbons on solute transport.

Accordingly, the objective of this study was to determine the effect of residual hydrocarbon (kerosene) contamination on the transport of an exchanging cation (ammonium) and a non-exchanging anion (nitrate) in water-saturated soil column experiments. Ammonium ion (NH_4^+) was selected because it is a popular nitrogen source in petroleum biodegradation. Since nitrate-based respiration has recently received attention as a potential degradation mechanism that can predominate under anaerobic conditions (Widdowson et al. 1988; Major et al., 1988), nitrate transport in contaminated soil was also examined. Kerosene was chosen as the petroleum contaminant due to its relatively low volatility and its use in jet fuel.

2 THEORETICAL

The one-dimensional movement of a non-reactive solute in a homogeneous medium under steady-state water flow conditions is typically described by the classical advection-dispersion equation given by:

$$\frac{\partial C}{\partial t} = D \frac{\partial^2 C}{\partial z^2} - V_o \frac{\partial C}{\partial z} \tag{1}$$

where C is the flux-averaged solute concentration (mg/l), V_o is the pore-water velocity (cm/h), z is the length (cm), t is time (h) and D is the longitudinal dispersion coefficient (cm^2/h). If molecular diffusion is neglected, the longitudinal dispersion coefficient may be approximated by $D=\alpha V_o$ (Selim et al., 1987) where α is the longitudinal dispersivity (cm).

When an exchanging solute is considered, an additional term must be added to account for the interactions between the compound and the soil medium. If S (mg/kg) denotes the amount of solute adsorbed per kilogram of soil, the transport equation can be written in the following form:

$$\frac{\rho_b}{\theta} \frac{\partial S}{\partial t} + \frac{\partial C}{\partial t} = D \frac{\partial^2 C}{\partial z^2} - V_o \frac{\partial C}{\partial z} \tag{2}$$

where ρ_b is the soil bulk density (g/cm^3), Θ is the volumetric water content (cm^3/cm^3), and the term ($\partial S/\partial t$) is the rate of solute adsorption.

When the exchange reaction is assumed to be much faster than the bulk flow of the fluid, the local equilibrium assumption (LEA) is valid (Rubin, 1983). Adsorption may then be described by the commonly used Freundlich equilibrium isotherm given by equation (3):

$$S = K C^N \tag{3}$$

where K and N are equation constants.

If the functional relationship between the equilibrium adsorbed and solution-phase concentration is linear, equation (3) may be reduced to:

$$S = K_d C \tag{4}$$

where K_d is the distribution coefficient (l/kg of soil). Equation (4) may then be substituted into equation (2) to yield:

$$R \frac{\partial C}{\partial t} = D \frac{\partial^2 C}{\partial z^2} - V_o \frac{\partial C}{\partial z} \tag{5}$$

where R is the retardation factor defined as:

$$R = 1 + \frac{\rho}{\theta} K_d \tag{6}$$

A thorough discussion of the appropriate boundary conditions used to solve the above equation for flux-averaged concentrations pertinent to finite systems is given by van Genuchten and Parker (1984). Based on their recommendations, the analytical solution of equation (5) as derived by Lapidus and Amundson (1952) was used to calculate the effluent curves. The program SOLUTE (Beljin, 1985) which contains the above-mentioned solution was used to determine the best-fit dispersion coefficients and to calculate the breakthrough curves resulting from the column experiments.

3 EXPERIMENTAL METHODS

A sandy Orthic Humic Gleysol representative of many Quebec aquifers was sampled from the C horizon of an agricultural area located in Richelieu County, 50 km North-East of Montreal. The particle size analysis, mineralogy, pH, organic matter content, metal oxyhydroxide content, and cation exchange capacity (CEC) were obtained using the methods described by Simard et al. (1989). The soil sample was initially Ca-saturated with 0.2 M $CaCl_2$ and equilibrated several times with 0.0033 M $CaCl_2$ (ionic strength = 0.01 M). The soil was then air dried, mixed carefully by hand, and gently sieved through a 2 mm sieve. Particles greater than 2 mm represented less than 1 g/kg of soil and were discarded.

To determine the kinetics of NH_4^+ adsorption, triplicate 2-g samples were placed in 50 ml polypropylene centrifuge tubes with 20 ml of 0.01 M NH_4NO_3. The samples were then equilibrated on a reciprocating shaker for periods of 0.25 to 64 h. At the end of the equilibration period, the supernatant was recovered by

Table 1. Physical and chemical parameters for the column experiments

		Soil	
		Uncontaminated	Contaminated
ρ_b	g/cm^3	1.6	1.6
α	cm	0.08	0.08
Θ	cm^3/cm^3	0.39	0.35
V_o	cm/h	2.7	3.2
K_d	l/kg	0.67	0.67

centrifugation and kept for NO_3^- and NH_4^+ analysis.

The adsorption isotherms for NH_4^+ on uncontaminated and contaminated soil were determined from transient displacement experiments similar to the method outlined by Jardine et al. (1988) rather than from standard batch experiments. This technique was used to better simulate residual petroleum contamination conditions encountered in the field.

For the adsorption isotherm on uncontaminated soil, air dried Ca-saturated soil was packed uniformly into small vertical plexiglass columns measuring 5.0 cm in length and 2.5 cm in diameter. The soil columns were saturated with at least 7 pore volumes of 0.0033 M $CaCl_2$ using a peristaltic pump. Mixed solutions of NH_4NO_3 and $CaCl_2$ at predefined molar ratios and at constant ionic strength (0.01 M) were then pumped upwards through the columns at a constant flux (2.6 cm^3/ h). The molar ratios of NH_4NO_3 used were 0.2, 0.27, 0.33, 0.47, 0.6, 0.67, 0.73. Continuous flow of each NH_4NO_3 solution through each column was maintained until the effluent concentration equalled that of the influent. The effluent was collected using a time-regulated fraction collector (Gilson, model 203) and subsequently analyzed for NH_4^+ to obtain the corresponding breakthrough curve (BTC). Since complete breakthrough was obtained for each NH_4^+ concentration, the area above each curve is proportional to the amount of NH_4^+ adsorbed in the column. Knowing the area above the curves, initial solution concentrations, flowrate, and weight of soil in the columns, a functional relationship between adsorbed and solution-phase concentration at constant ionic strength was determined.

The adsorption isotherm of NH_4^+ in petroleum-contaminated soil was obtained using the previously described procedure with the following additional steps: (1) after presaturating the columns with 7 pore volumes of 0.0033 M $CaCl_2$, kerosene was added to the top of the columns and allowed to pass freely until no more porewater was displaced; (2) free-phase petroleum was then removed from the columns by pumping at least 7 pore volumes of 0.0033 M $CaCl_2$ solution upwards, thereby leaving only residual soil-bound contamination. The rest of the experiment proceeded as described previously.

The transport of NO_3^- and NH_4^+ in uncontaminated and contaminated soil was characterized in plexiglass columns measuring 21.3 cm in length and 5.1 cm in diameter. The columns were packed with soil and saturated with 0.0033 M $CaCl_2$ as described for the small column transient displacement experiments. The volumetric water contents (Θ), bulk densities (ρ_b), and pore-water velocities (V_o) were measured by conventional direct methods and are listed in Table 1. The soil was contaminated as described previously with the following exception: the solution in the column was lowered gradually over 12 hours during the application of kerosene to simulate a falling water table.

A mixed solution (ionic strength = 0.01 M) containing NH_4NO_3 and $CaCl_2$ at a NO_3^- concentration of 289 mg/l and an NH_4^+ concentration of 84 mg/l and was pumped upwards through the columns until the effluent concentration equalled that of the influent. The effluent was collected as a function of time and then analyzed for NO_3^- and NH_4^+ to determine the BTCs. The NO_3^- BTCs were used to describe the transport behaviour of a non-exchanging solute while the NH_4^+ BTCs were used to characterize the transport of an exchanging solute. Assuming that NO_3^- is a non-adsorbing tracer (Everts et al., 1989), the NO_3^- BTCs were also used to determine the longitudinal dispersivity values of the systems.

The NO_3^- and NH_4^+ solution concentrations for the above experiments were measured using a Dionex 4000 ion liquid chromatography system with a conductivity detector. All samples were analyzed immediately after collection to avoid microbial transformations. Since the duration of the experiments was short (less than 70 h), nutrient losses due to biodegradation were not observed. All experiments were conducted at 296 K.

4 RESULTS AND DISCUSSION

The physical, mineralogical, and chemical properties of the soil are presented in Table 2. As shown, the clay portion of the soil is dominated primarily by vermiculite with lesser quantities of chorite, mica, and interstratified minerals.

Results illustrating the adsorption kinetics for NH_4^+ on uncontaminated soil are plotted in Figure 1. The NH_4^+ concentrations decreased very rapidly during the first two hours and thereafter decreased only slightly with time. Two hours is much less than the residence time in all column experiments. Hence, it can be assumed that the reaction is fast relative to the bulk fluid flowrate and, therefore, local equilibrium as described by Rubin (1983) is valid.

The observed NH_4^+ retention reaction is believed to be the result of NH_4^+ exchange on vermiculite. Although other soil constituents such as organic matter and Al and Fe oxyhydroxides are known to contribute to the soil CEC, their relative content in the sediment studied is low and hence are unlikely to significantly affect NH_4^+ retention.

The adsorption isotherms for NH_4^+ on uncontaminated and contaminated soil are

Table 2. Physical and chemical properties of the soil.

Sand	Silt g/kg	Clay	Mineralogy Silt 50-2 µm	Mineralogy Clay < 2 µm	pH	Organic Matter g/kg	Fe g/kg	Al g/kg	CEC cmol/kg
956	20	24	*FHQInM	*VCMIn	6.7	0.6	1.9	0.5	1.6

* C, chlorite; F, feldspar; H, hornblende; In, interstratified minerals; M, mica; Q, quartz; V, vermiculite

presented in Figure 2. For the uncontaminated soil, the exchange may be approximated by a linear relation up to an equilibrium solution concentration of about 110 mg/l. Since in in-situ bioremediation, concentrations greater than 110 mg/l are rarely observed, a linear isotherm was considered in this transport study for solution concentrations less than this value. The least squares best-fit value of K_d on the uncontaminated soil was 0.67 l/kg of soil. For the contaminated soil, the adsorption isotherm was also approximately linear at equilibrium solution concentrations of 0 to 110 mg/l. The least squares best-fit value of the K_d was 0.71 l/kg of soil.

A Student t-Test confirmed that there was no significant difference between the two K_d values. Thus, ammonia exchange was not affected by the presence of residual kerosene. The value of 0.67 l/kg was then used as the distribution coefficient in both uncontaminated and contaminated soils (Table 1).

The observed BTCs for the non-exchanging NO_3^- for uncontaminated and contaminated soil were well described by equation (5), as illustrated in Figure 3. The longitudinal dispersivity (α) values that provided the best fit of the BTCs (Table 1) were low implying flow homogeneity in the columns. The results show that the longitudinal dispersivity was not altered by the presence of sorbed-phase kerosene contamination.

However, as shown in Table 1, the presence of kerosene reduced the volumetric water content of the soil from 0.39 to 0.35 cm^3/cm^3 which in turn increased the pore-water velocity from 2.7 to 3.2 cm/h. Hence, an earlier break-through time for NO_3^- relative to the uncontaminated soil was obtained. The time needed to achieve one-half of the influent concentration occurred at 7.6 h for the uncontaminated soil and at 6.6 h for the contaminated soil (Figure 3). The ratio (r) corresponding to the change in volumetric water content (r = 0.39/0.35 = 1.11) was similar to the ratio corresponding to the change in breakthrough time (r = 1.15).

Experimental BTCs for NH_4^+ for the uncontaminated and contaminated soil are shown in Figures 4a and 4b, respectively, along with model predictions. The breakthrough of NH_4^+ was well predicted by the model between 0 and 30 h. However, the model did not accurately simulate the tailing observed in the experimental results at t>30 h. The deviations at later times may be due to a lack of complete local equilibrium, possibly as a

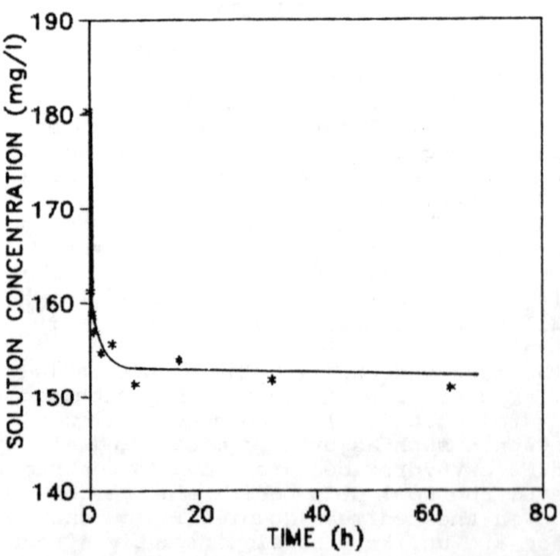

Fig.1 NH_4^+ concentration vs. contact time for uncontaminated soil.

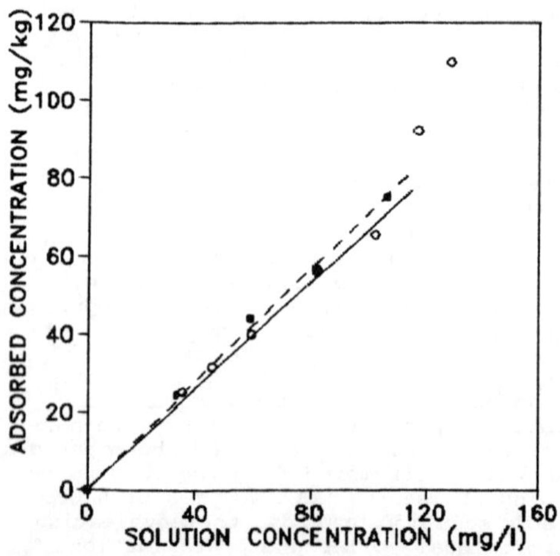

Fig.2 Asorption isotherms for NH_4^+ for uncontaminated (—o—o—) and contaminated (-●-●-) soil. The solid and dashed lines are the least squares best-fit calculated isotherms based on equation (4).

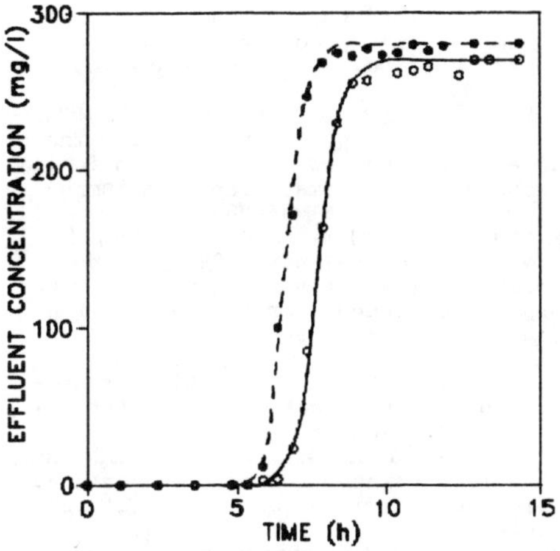

Fig. 3 Breakthrough curves (BTCs) for NO_3^- for uncontaminated (—o—o—) and contaminated (- ●-●-) soil. The solid and dashed lines are model predictions using equation 5.

result of (1) the presence of immobile zones which could act as sinks to ionic diffusion during the displacement process (van Genuchten and Cleary 1979), (2) the presence of a small percentage of slowly exchanging sites as observed at contact times greater than 2 hours (Figure 1), and (3) clay layer collapse due to NH_4^+ exchange on vermiculite which could result in non-instantaneous exchange as the soil sites become saturated (Nommick and Vahtras, 1982). However, since the purpose of this study was to determine the effect of sorbed-phase kerosene contamination on solute transport rather than to accurately model NH_4^+ transport, improvements to the model were not attempted.

The experimental BTCs for NH_4^+ for the uncontaminated and contaminated soil are illustrated in Figure 5. Unlike NO_3^-, the change in breakthrough time of NH_4^+ caused by the increase in pore-water velocity is very small relative to the overall breakthrough time of this strongly adsorbing ion. Therefore, because the adsorption isotherm (K_d value) was unaffected by the presence of kerosene, the BTCs for NH_4^+ for uncontaminated and contaminated soil are almost identical.

5 SUMMARY AND CONCLUSIONS

The effect of sorbed-phase petroleum contamination on the transport of NO_3^- and NH_4^+ was investigated using soil column experiments. Solute transport through water-saturated soils was described using the classical advection-dispersion equation and assuming a linear adsorption relation for NH_4^+. Model and experimental results corresponded well for both NO_3^- and NH_4^+. However, the tailing observed in the experimental breakthrough curves

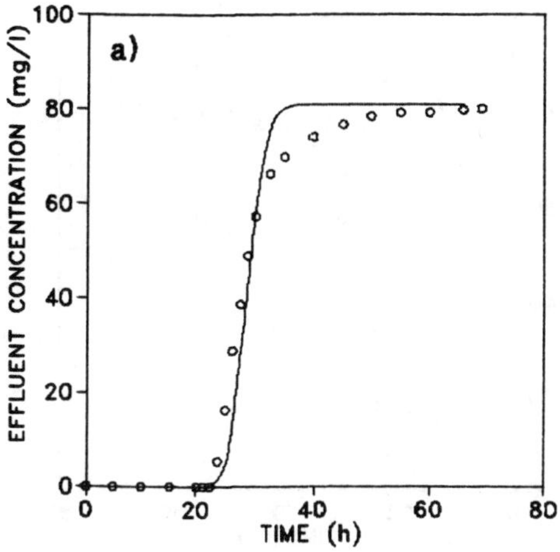

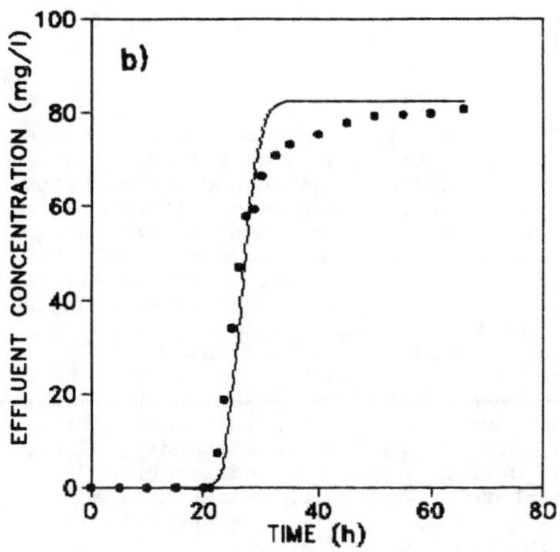

Fig. 4 BTCs for NH_4^+ for uncontaminated (a) and contaminated (b) soil. The solid lines are model predictions using equation 5.

for NH_4^+ at later times was not predicted by the model.

Sorbed-phase kerosene contamination did not alter the longitudinal dispersivity of the soil in the column. However, the presence of kerosene decreased the volumetric water content which in turn increased the pore-water velocity. The increase in pore-water velocity resulted in a faster NO_3^- breakthrough time for the contaminated soil relative to the uncontaminated soil.

Unlike the NO_3^- breakthrough curve, the NH_4^+ breakthrough curve for the contaminated soil was not affected by the presence of kerosene and remained almost identical to the one obtained for the uncontaminated soil. This result was obtained because adsorption, which greatly influences breakthrough time of exchanging solutes, was unaffected by the presence of the hydrocarbon.

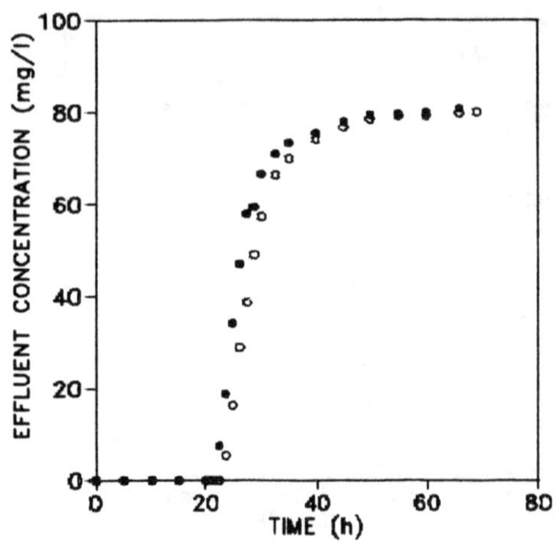

Fig. 5 Experimental BTCs for NH_4^+ for uncontaminated (o) and contaminated (●) soil.

To further verify these findings, field experiments or studies on columns of undisturbed uncontaminated and contaminated soils should be undertaken. If similar results are obtained, then the presence of residual petroleum contamination will have to be taken into account in models simulating the transport of the non-exchanging NO_3^- used in bioremediation. Further studies on other compounds used in bioremediation such as PO_4^{2-} and O_2 should also be conducted to establish their transport behaviour in petroleum contaminated aquifers. A better understanding of the transport of nutrients and electron acceptors in hydrocarbon-contaminated aquifers will help improve the efficiency of in-situ bioremediation.

REFERENCES

Anderson, M.P. 1979. Using models to simulate the movement of contaminants through groundwater flow systems. Crit. Rev. Environ. Control 9(2):97-156.

Beljin, M.S. 1985. A program package of analytical models for solute transport in groundwater "SOLUTE". International Ground Water Modeling Center, Indianapolis, Indiana.

Borden, L.C. and P.B. Bedient 1986. Transport of dissolved hydrocarbons influenced by oxygen limited biodegradation: 1. Theoretical development. Water Resour. Res. 22:1973-1982.

Everts, C.J., R.S. Kanwar, E.C. Alexander Jr., and S.C. Alexander 1989. Comparison of tracer mobilities under laboratory and field conditions. J.Environ. Qual. 18:491-498.

Hoag, G.E. and M.C. Marley 1986. Gasoline residual saturation in unsaturated uniform aquifer materials. J. Environ. Eng. 112(3):586-604.

Jardine, P.M., G.V. Wilson, and R.J. Luxmoore 1988. Modeling the transport of inorganic ions through undisturbed soil columns from two contrasting watersheds. Soil Sci. Soc. Am. J. 52:1252-1259.

Lapidus, L. and N.R. Amundson 1952. Mathematics of adsorption in beds. VI. The effect of longitudinal diffusion in ion exchange and chromatographic columns. J. Phys. Chem. 56:984-988.

Lee, M.D., J.M. Thomas, R.C. Borden, P.B. Bedient, and C.H. Ward 1988. Biorestoration of aquifers contaminated with organic compounds. Crit. Rev. Environ. Control 18(1):29-89.

Major, D.M., C.I. Mayfield, and J.F. Barker 1988. Biotransformation of benzene by denitrification in aquifer sand. Ground Water 26(1):8-14.

Naymick, T.G. 1987. Mathematical modeling of solute transport in the subsurface. Crit. Rev. Environ. Control 17(3):229-250.

Nommick, H. and K. Vahtras 1982. Retention and fixation of ammonium and ammonia in soils. In F.J. Stevenson (ed.), Nitrogen in Agricultural Soils, Madison, Wisconsin.

Rubin, J. 1983. Transport of reacting solutes in porous media: relation betbetween mathematical nature of problem formulation and chemical nature of reactions. Water Resour. Res. 19(5):1231-1252.

Simard, R.R., C.R. de Kimpe, and J. Zizka 1989. The kinetics of nonexchangeable potassium and magnesium release from Quebec soils. Can. J. Soil Sci. 69:663-675.

Staps, J.J.M. 1989. International evaluation of in-situ biorestoration of contaminated soil and groundwater. Final Fellowship report for the Third International Meeting of the NATO/CCMS pilot study on "Demonstration of Remedial Action Technologies for Contaminated Land and Groundwater", Nov.6-9,Mont real, Canada, p. 1-23.

van Genuchten, M. Th. and J.C. Parker 1984. Boundary conditions for displacement experiments through short laboratory soil columns. Soil Soc. Sci. Am. J. 48:703-708.

van Genuchten, M.Th. and R.W. Cleary (1979). Movement of solutes in soil: computer-simulated and laboratory results. In G.H.Bolt (ed.), Soil Chemistry: B. Physico-Chemical Models, New York. p. 349-386.

Widdowson, M.A., F.J. Molz, and L.D. Benefield 1988. A numerical transport model for oxygen- and nitrate-based respiration linked to substrate and nutrient availability in porous media. Water Resour. Res. 24(9):1553-1565.

Wilson, S.B. and R.A. Brown 1989. In situ bioreclamation: a cost effective technology to remediate subsurface organic contamination. Ground Water Monitoring Rev. 9:173-179.

Selim, H.M., R. Schulin and H. Fluhler 1987. Transport and ion exchange of calcium and magnesium in an aggregated soil, Soil Sci. Soc. Am. J., 51(4), pp. 876-884.

8 Case histories

8.1 Field investigations

Subsurface Contamination by Immiscible Fluids, Weyer (ed.) © 1992 Balkema, Rotterdam. ISBN 90 6191 175 3

Characteristics of the migration of immiscible fluids in glacial deposits and dolomites in Niagara Falls, New York

Charles R. Faust & Charles P. Spalding
GeoTrans, Inc., Sterling, Va., USA

Steven J. Wamback
Great Lakes Geological & Environmental Sciences, Angola, N.Y., USA

ABSTRACT: The investigation and startup of remediation at four former chemical disposal sites provide insight into the behavior of immiscible fluids in glacial deposits and fractured dolomites. Approximately 365,000 tons of waste were disposed at the four sites located in or near Niagara Falls, New York. Of this amount, up to 10% could separate as liquids immiscible in water. The chemical composition and the physical properties of the wastes at the four sites varies. Generally, the immiscible fluid observed in the glacial deposits and bedrock is more dense than water (a factor of 1.2 to 1.5). The viscosity of the immiscible fluid ranges from that of water to tar-like substances. Extensive borings and wells have demonstrated that the movement of the free liquid is controlled strongly by the geometry of geological features such as fractures and joints, lithologic boundaries, and grain size. Groundwater flow directions are of secondary importance on immiscible fluid migration at these sites. The immiscible fluids have migrated further laterally (up to 2,000 feet) in the dolomite than in the overlying glacial deposits. During the next several years, aggressive pumping programs will be implemented to control and recover immiscible fluids in the groundwater below these sites. These programs will generate data on immiscible/water recovery ratios, groundwater pumpage, groundwater elevations, and groundwater quality. This new data will be valuable in assessing the feasibility of mitigating large-scale groundwater contamination by pump-and-treat technologies.

1 INTRODUCTION

Inexpensive hydroelectric power has attracted industry to Niagara Falls, New York since the late 1800s. In addition, large chemical manufacturing firms have operated plants in Niagara Falls for most of this century. Throughout that period, the disposal of chemical wastes by shallow land burial has been common practice. As a result, numerous hazardous waste landfills are located in the Niagara Falls area.

Several of the landfills in the Niagara Falls area received wastes relatively insoluble in water. At the four landfills discussed here, S-Area, Love Canal, 102nd Street, and Hyde Park (see Figure 1), chemicals were disposed at rates and in volumes sufficient to remain in the groundwater as separate liquids. These immiscible liquids are more dense than water (referred to as DNAPL for dense nonaqueous phase liquid). Table 1 summarizes the chemicals and wastes disposed at the sites. Significantly more DNAPL forming chemicals were disposed at Hyde Park and S-Area than at Love Canal and 102nd Street.

Groundwater and site remedial investigations conducted at the four sites have provided significant insight into the migration of DNAPL over a long term. Hundreds of wells and miles of trenches have permitted delineation of the extent of NAPL and pathways for migration. The nature of NAPL migration, thus shown, is complex. During the same period that much of this data has been collected, mathematical models have been developed to simulate movement of immiscible liquids in groundwater (many of which are described elsewhere in this volume). The discussion that follows attempts to compare data requirements for a simplified mathematical model of immiscible flow and the reality of what can be observed in the field.

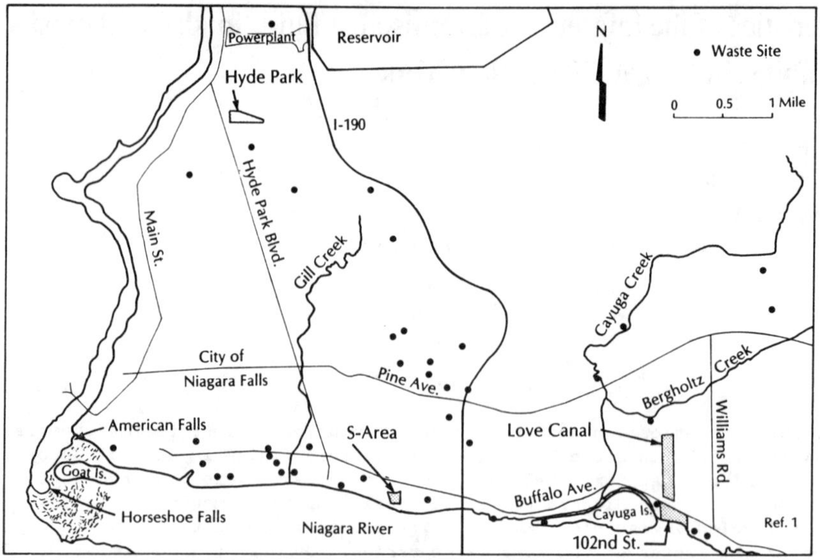

Figure 1. Locations of waste disposal sites in the Niagara Falls area (Koszalka et al., 1985).

Table 1. Estimated quantities and types of buried wastes at Love Canal, Hyde Park, S-Area, and 102nd Street (Interagency Task Force, 1979).

Type of Waste or Category	Estimated Quantity (Tons) Love Canal	Hyde Park	S-Area	102nd St.
Chlorobenzenes	2,000	16,500	19,900	--
Benzylchlorides	2,400	3,400	1,600	--
Benzoylchlorides	800	6,200	3,300	—
Thionyl Chloride	500	—	4,100	—
Trichlorophenol	200	3,300	200	—
Liquid Disulfides, MCT, and Chlorotoluenes	700	2,600	2,200	—
Misc. Chlorinations	1,000	1,600	400	—
Metal Chlorides	400	100	900	—
Misc. Acid Chlorides	400	1,200	400	—
BHC Cake Incl. Lindane	6,900	2,000	—	300
Dodecyl Mercaptans, Chlorides & Misc. Org. Sulphur Compounds	2,400	4,500	600	—
Sulfides/Sulfhydrates	2,000	6,600	4,200	—
C-56 & Derivatives	—	5,600	17,400	—
Thiodan	—	1,000	700	—
HET Acid	—	2,100	500	—
Na Hypophosphite Mud	—	1,000	—	20,000
BTFs & Derivatives	—	8,500	—	—
Dechlorane	—	200	—	—
Calcium Fluoride	—	400	—	—
Mercury Brine Sludge	—	100	—	—
Inorganic Phosphorus	—	100	—	1,300
Organic Phosphorus	—	4,400	200	LT 100
Phenol Tars	—	—	800	—
Brine Sludge & Gypsum	—	—	—	53,200
Tetrachlorobenzene	—	—	—	2,327
BHC, Trichlorophenol, Trichlorobenzene, & Benzene	—	—	—	2,000
Na Chlorite Black Cake	—	—	—	18,673
Graphite	—	—	—	742
Lime Sludge	—	—	—	22,978
Brine Sludge	—	—	—	67,186
Miscellaneous	2,000	7,300	5,700	2,200
Estimated Total	22,000	80,000	63,000	200,000

2 DATA REQUIREMENTS FOR MATHEMATICAL MODEL

To fully describe the migration and fate of NAPL in the subsurface is an overwhelming task. Not only must one assess the migration of immiscible fluid and gas phases, but also one must quantify exchange of NAPL components across each phase and fluid interface. Even if we limit ourselves to the description of immiscible fluid transport, the task is not simple for applications at the field scale.

Faust et al. (1989) present a simplified set of equations that describe the flow of immiscible liquids in the subsurface. These are developed by assuming pressure gradients in the air phase are negligible. This eliminates the need for describing the movement of the air phase. The equations are posed in terms of NAPL pressure, p_n, and water saturation S_w.

The simplified equations require specification of aquifer properties (intrinsic permeability and porosity); fluid properties (densities and viscosities); and saturation dependent fluid/rock properties (capillary pressures between phases and relative permeabilities of each phase). The complete specification of the model requires information on phase pressures and phase saturations throughout the areas and depths of interest.

A discussion of the functional forms of relative permeability and capillary pressure is beyond the scope of this paper. However, it should be noted that at the field scale, each of the functional relationships as well as permeability are likely to show significant spatial variability. In addition, the functional relationships at measurable scales in the field are unlikely to correlate well with laboratory measurements.

3 GEOLOGIC SETTING

For the purpose of this report we may consider the geology of the Niagara Falls area to be subdivided into overburden, the Lockport Dolomite, and the formations below the Lockport Dolomite. The bedrock in the Niagara Falls area consists of generally flat-lying sedimentary rocks that regionally dip gently to the south at approximately 30 feet per mile. The Lockport Dolomite, considered the top of the bedrock in this area, directly underlies a thin layer of unconsolidated deposits (overburden) consisting of glacial till and lacustrine clay, silt and gravelly sand, and fill. Immediately underlying the Lockport in this area is the Rochester shale. Below the Rochester shale are the Irondequoit and Reynales limestones. The general geology of the area is illustrated in the block diagram for the Hyde Park area (Figure 2).

As already noted, the bedrock of the Niagara Falls region is overlain by a thin veneer of Quaternary (Pleistocene) deposits (overburden) ranging from just a few meters in thickness near the Niagara Gorge and Escarpment to greater than ~100 feet thick south of the City of Niagara Falls (Feenstra, 1981). These unconsolidated deposits consist of an unsorted basal till overlain by glacio-lacustrine clays and silts (Calkin and Brett, 1978). Johnston (1964) described the highly fractured and weathered bedrock surface which grades almost imperceptibly into the overlying till and, together with the till, constitutes a single highly transmissive hydrostratigraphic flow regime.

The overburden sediments were deposited as the last Wisconsin ice sheet withdrew from the Niagara Region 13,000 years ago. The oldest and lowermost of these is the basal till which consists of broken subangular pieces of the weathered Lockport

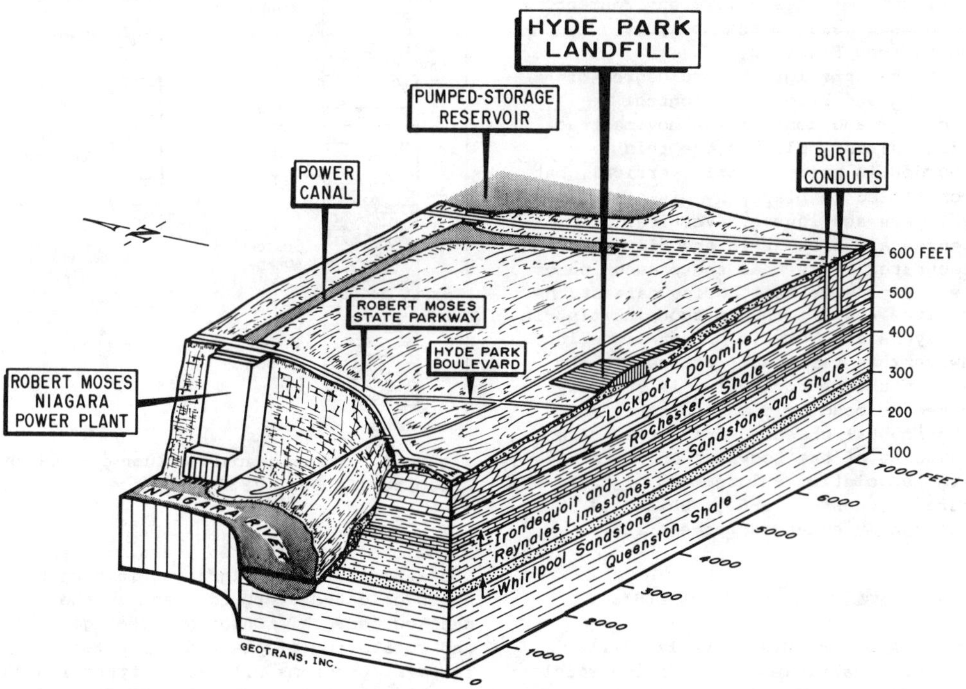

Figure 2. Generalized block diagram showing geologic formations and topographic features in the vicinity of the Hyde Park landfill.

Dolomite bedrock as well as more rounded boulders, pebbles, and gravel abraded and plucked by the glacier from outcrops further north. These larger clasts are suspended in an unsorted mixture of sand, silt, and clay-sized particles characteristic of tills in general.

As the ice sheet melted back from the region, great bodies of water accumulated along the abating edge of the glacier. The steady rainout of fine sediments liberated from the melting ice sheet and from local streams into these glacial lakes gave rise to the extensive clays and silts that comprise much of the overburden in the Niagara Region (Calkin and Brett, 1978). Lake Tonawanda is a post-glacial lake that is thought to have persisted near the present site of Niagara Falls until about 1,000 years ago. Two outlets of this lake were Devils Hole and Bloody Run near the present location of the Hyde Park landfill.

The Lockport formation is the uppermost and most significant water bearing unit in the Niagara Falls area (Johnston, 1964). It is also the only bedrock unit that, to date, shows evidence of NAPL migration. The Lockport formation was deposited in a shallow marine environment. The faunal assemblage indicates a warm environment. Five members make up the Lockport Dolomite (see Figure 3).

Both the stratigraphy and degree of weathering and fracturing control groundwater and contaminant movement at the four sites. Flow in the thin overburden is predominantly vertical, but is controlled in places by dis-continuous clay layers and lenses. Vertical fractures are present in the clay, but are concentrated at or near the present water table. Extensive soil boring data at the four sites suggest that discontinuities in clay layers are more important migration pathways than vertical fractures in the clay. In the bedrock, flow is strongly affected by bedding plane features and sub-horizontal sheeting fractures. The upper zones of the Lockport Dolomite have undergone solution enlargement of these features (Johnston, 1964), which has significantly enhanced permeability.

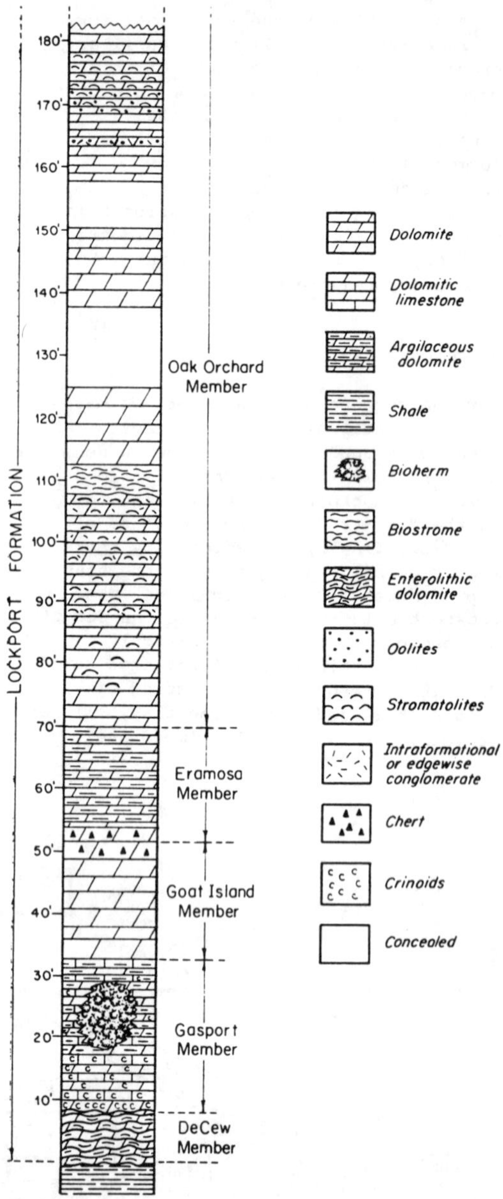

from Zenger, 1965, p39

Figure 3. Stratigraphic column of Lockport Dolomite (Zenger, 1965).

4 SITE INVESTIGATIONS AND ANALYSIS

At the S-Area and Hyde Park landfills, sufficient quantities of water-immiscible fluids have been available to migrate in the subsurface significant distances from source areas. Extensive field investigations over the last ten years have shown the extent of NAPL migration in both the unconsolidated overburden and in the shallow bedrock. Ongoing investigations and monitoring of planned and existing remedial programs will add further insight to the nature of NAPL migration.

At the Love Canal and 102nd Street landfills, the quantities of NAPL are less than at S-Area and Hyde Park. At Love Canal and 102nd Street, most of the NAPL is contained within the overburden (see Pinder et al., 1992, for a discussion of NAPL at Love Canal). The primary focus of this report is NAPL migration below S-Area and Hyde Park.

4.1 Hyde Park setting

According to estimates in Table 1, approximately 80,000 tons of chemical wastes were disposed at the Hyde Park landfill between 1954 and 1975. Much of this waste was disposed in liquid form in the landfill. Because the landfill was excavated to bedrock in some areas, both the overburden materials and the Lockport Dolomite have been contaminated.

The overburden in the vicinity of the site is generally very low in permeability with the exception of a basal sandy till which immediately overlies bedrock. Above the basal till are a silty till and lacustrine silts and clays. The total thickness of the overburden deposits range from a few feet to over 30 feet.

The Lockport Dolomite in the vicinity of the site ranges from 60 feet to 140 feet as shown in Figure 4. It thins to the northwest primarily due to erosion prior to deposition of the overlying glacial materials.

In addition to migration in the natural geologic materials, migration of contaminants in the fill material also is important. The fill material of the Hyde Park landfill is highly nonuniform. It consists of rock fragments, wood, cinders, paper, porcelain, graphite, concrete, styrofoam, rubber, and fine gravel in a matrix of gray to brown sand. Based on available data, the average thickness of the nonuniform fill layer is about 21 feet, with an overall range of about 14 to

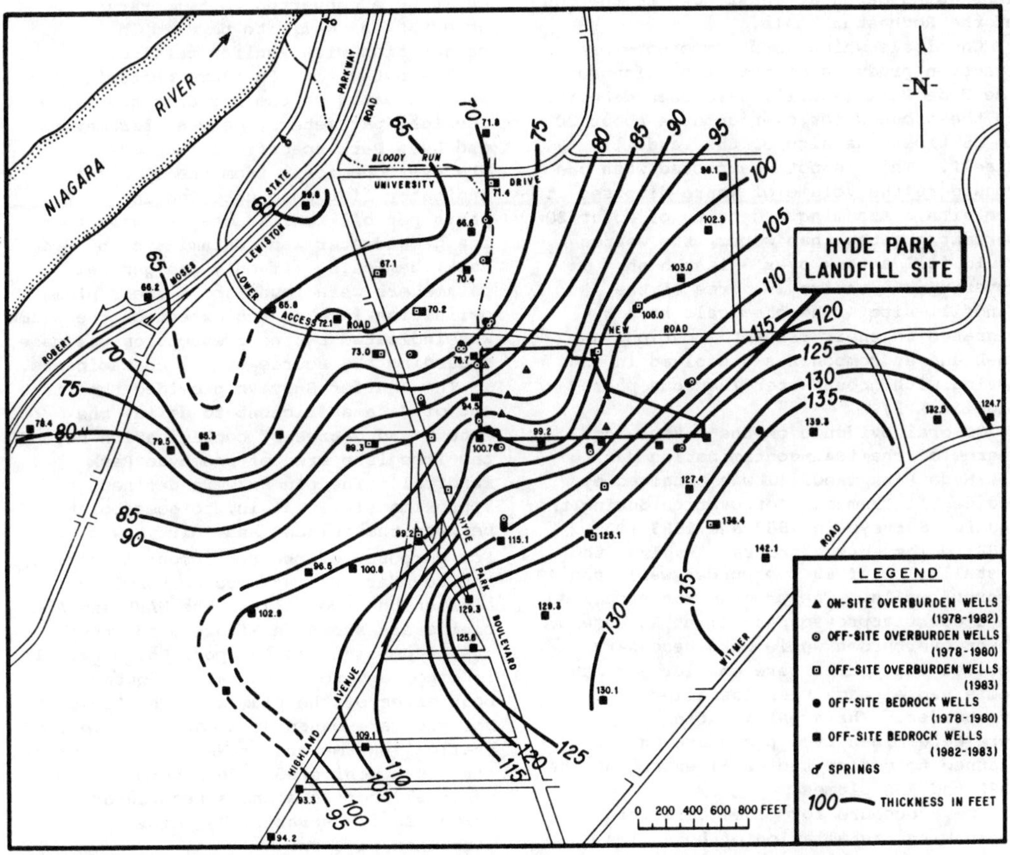

Figure 4. The thickness of the Lockport Dolomite in the vicinity of the Hyde Park landfill.

32 feet. Based on these characterizations, the fill material is thought to be significantly higher in permeability than the surrounding clays and silts.

The groundwater system in the Hyde Park area can be idealized as a sequence of flat-lying, water-bearing zones sandwiched between water-saturated materials of low permeability, all of which are bounded on three sides by drains (the Niagara River Gorge, the forebay canal, and buried conduits). Precipitation infiltrates the low permeability overburden before entering the permeable, fractured upper layer of the Lockport Dolomite, the uppermost water-bearing zone. In the water-bearing zones much of the groundwater moves laterally toward the three boundaries. Between the water-bearing zones, groundwater moves slowly downward to the next lower water-bearing zone. Groundwater elevations measured in wells have shown a consistent trend of decreasing water-level elevation with increasing depth at least down to the top of the Rochester Shale.

Chemicals which were components or reaction products of the waste disposed at the Hyde Park landfill have been detected in the groundwater over an area about 20 to 25 times the size of the landfill itself. This is not unexpected when one considers the volume of waste disposed at the site. Assuming a density of about 20 percent greater than water, the wastes would fill a tank over 3 feet high covering the entire 15 acres of the landfill site. The chemicals have migrated to their present locations as NAPL and as components dissolved in and moving with groundwater (aqueous phase liquid--APL).

General evidence on the extent and degree of chemical contamination due to the Hyde Park landfill was obtained by Occidental Chemical Corporation during the Aquifer Surveys in 1982 and 1983 (OCC, 1983). The survey program involved the installation of 13 overburden wells and 49 bedrock wells. The program was specified in a court approved Settlement Agreement. Fewer overburden wells were necessary because the plumes were smaller and many wells had already been installed in the overburden. The areal locations of wells were determined by a procedure that was planned to define the areal extent of the NAPL and APL plumes.

The procedure for plume definition began with installation of boreholes located at the perimeter of the landfill and continued with installation of additional boreholes outward from the landfill in ten directions. If the survey well showed presence of chemical parameters above specified levels (called survey levels), a new well was installed about 800 feet away along a vector extending outward from the landfill. Along each vector, survey wells were installed at 800 foot intervals until a well showed no chemical parameters above survey levels. A survey well was then installed midway between the outer two wells to define the APL plume extent. The vertical extent of chemicals was determined by sampling at 15-foot intervals during coring operations. Only zones that yielded significant amount of water to pumping were sampled. In a similar manner, the extent of the NAPL plume was refined by installing survey wells along a vector between the outward well showing NAPL and the next well not showing NAPL. Nearly 300 points in the bedrock were sampled by this procedure. After testing and sampling, the wells were grouted. Before grouting, some of these wells were converted to temporary observations wells to be used in conjunction with aquifer testing.

The extent of APL plume contamination was evaluated on the basis of two types of chemical parameters, general parameters, and Hyde Park specific indicators. Wherever chemicals from the specific indicator list were detected above 10 parts per billion, it was presumed that the groundwater was contaminated by Hyde Park chemicals. If only the general parameters were found above their plume definition levels, the extent of the plume was indicated but no presumption was made regarding the source of the contaminants.

The Aquifer Surveys provided the first set of data sufficient to define the extent and degree of contamination beyond the immediate area of the Hyde Park landfill. The plumes were defined approximately; that is, to some point between the "clean" and "dirty" wells. For purposes of the Settlement Agreement, the boundary was assumed to be midway between the two wells. The NAPL and APL plumes are shown in Figure 5 for the overburden and the Lockport Dolomite. It is important to note that the outer boundaries of the plumes in the Lockport Dolomite are based upon maximum levels of chemical parameters or NAPL that occur at any depth monitored. In Figure 6, the vertical profile along a section of boreholes illustrates the complex migration pattern.

4.2 Hyde Park analysis

One concern raised in selecting a remedial approach for NAPL in the bedrock was the rates and amounts of NAPL that would be removed by pumping wells. This concern was addressed using an immiscible flow model and explicit sensitivity analysis. The approach and results of the analysis was described in Faust et al. (1989). Only a brief summary is given here.

The numerical model used in the analysis was based on equations (1) and (2). Cylindrical coordinates were used to simulate flow toward a single pumping well. Only flow in the Lockport Dolomite was considered. Also, the initial NAPL distribution was assumed to be at static equilibrium. Table 2 summarizes the major assumptions used in the analysis. The computational grid for the finite-difference model consisted of 10 layers in the vertical direction (ranging from 2.5 to 10 feet in thickness) and 20 columns in the radial direction (first block 1 foot from well and expanding at factor of 1.76 outward).

The results of each sensitivity analysis were expressed in terms of NAPL recovered during 180 days of pumping and the ratio of NAPL recovered to total fluid pumped. Of 39 simulations, 24 assumed uniform conditions - variations were tested for viscosity, relative permeability, and capillary pressure functions, permeability, degree of anisotropy, and effective porosity. For runs assuming uniform properties,

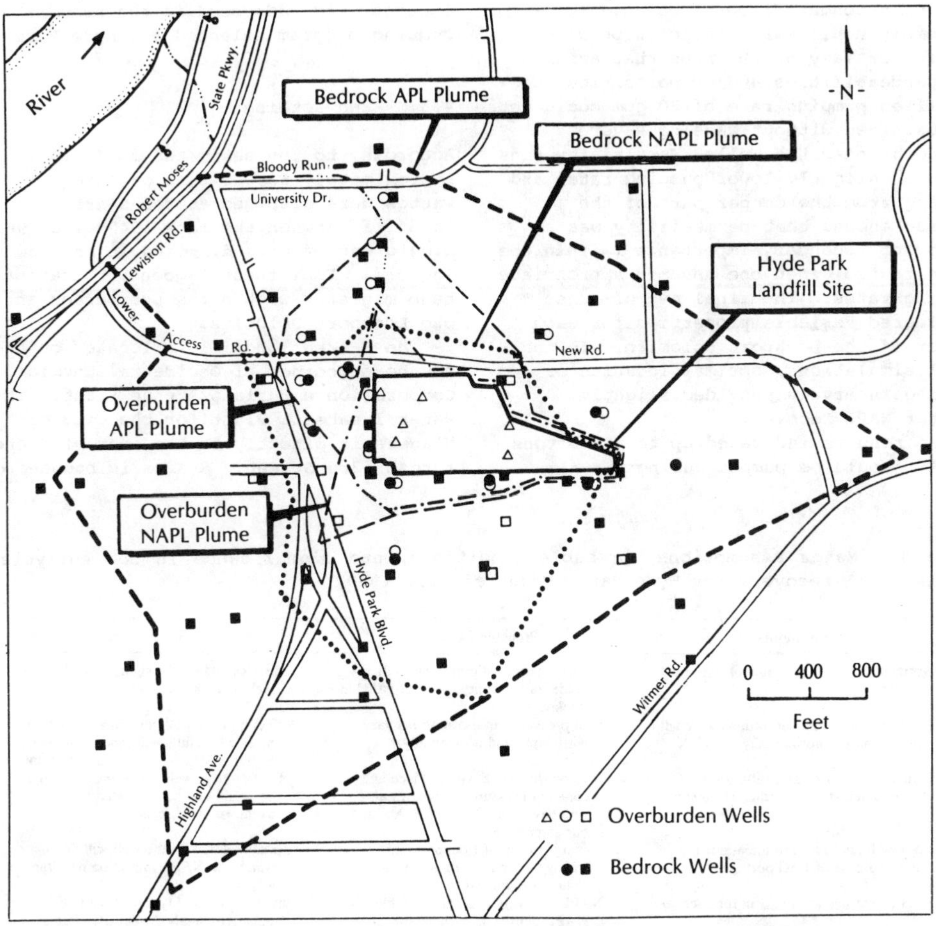

Figure 5. Boundaries of APL and NAPL plume in overburden and bedrock groundwater in the vicinity of the Hyde Park landfill.

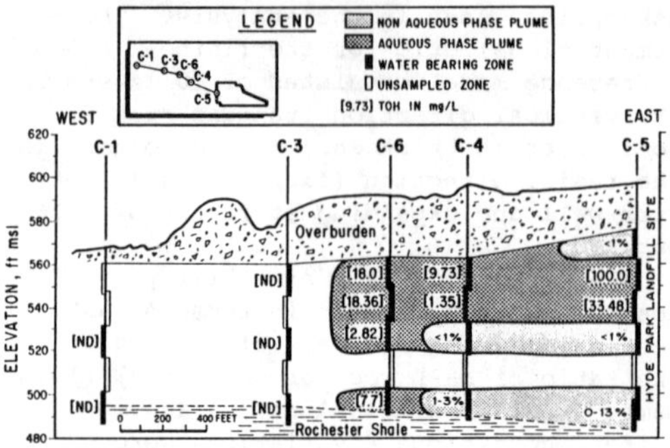

Figure 6. NAPL occurrence in vertical section of Lockport Dolomite west of Hyde Park.

viscosity was the most important factor affecting NAPL recovery. The relative permeability and capillary relationships were of secondary importance unless extremely nonlinear relationships were used. For many of the runs that assumed low permeabilities and/or porosities, the specified pumping rate of 20 gpm could not be sustained without causing severe drawdowns near the wells. Sensitivity runs based on slightly lower pumping rates and pumping from the deeper part of the aquifer showed that permeability was a factor of secondary importance (within the range tested) when one assumed appropriate pumping rates. The final set of runs considered variable properties for each member of the Lockport Dolomite. In the final simulations, optimal location of pumping intervals provided slightly greater NAPL recovery.

The results indicated up to 3,150 tons of NAPL could be pumped during the first six months operation of a single well under optimistic assumptions. However, little NAPL would be produced under pessimistic assumptions. That is quite a large predictive range! The more useful results of the analyses were that it demonstrated the following:

1. Viscosity, a relatively easily measured parameter, should be measured for NAPL to provide better estimates of NAPL recoveries,
2. Wells should be operated at pumping rates that do not significantly dewater the aquifer adjacent to the wells,
3. Wells should be open primarily to permeable zones containing the highest saturations of NAPL, and
4. NAPL recovery rates should be monitored frequently on a well-by-well basis to aid in predicting future NAPL recovery rates and to decide when to delete and/or add new wells. These factors were addressed in the remedial pumping program selected for Hyde Park.

4.3 S-Area setting

According to the estimates in Table 1, approximately 63,000 tons of chemical wastes were disposed at the S-Area landfill between the late 1940s and 1961. Liquid wastes were disposed in shallow lagoons. From these lagoons, contaminants have migrated within the overburden and in the Lockport Dolomite.

The S-Area landfill is located on the southeast corner of Occidental Chemical Corporation's Buffalo Avenue Plant. The Water Treatment Plant for the City of Niagara is directly to the east of S-Area across 53rd Street. S-Area is bounded on

Table 2. Major assumptions, probable conditions, and significance in the analysis of pumping well recovery for Hyde Park (Faust et al., 1989).

Assumption	Probable Conditions	Significance
Horizontal hydrologic units of uniform thickness	general bedrock dip of about 0.01 in Hyde Park vicinity, units of variable thickness	effect should not be significant on near-well performances
Fluid pressures and saturations vary with depth but not horizontally	fluid pressures and saturations vary with depth and horizontally	effect on near-well performance should be small, results will tend to predict higher NAPL recoveries if significant
Constant pressures and saturations maintained at outer radius of 2000 m	seep boundaries at Gorge Place and constant pressure boundaries at Forebay Canal, Niagara River, and buried conduits	effect on near-well performance should be small, can be evaluated by examination of results
No-flow boundaries are maintained at top and base of Lockport	recharge at top of Lockport and leakage to underlying Rochester shale are significant	horizontal flow to the well will dominate water and NAPL migration near the well
Fluid properties are uniform for each run	NAPL viscosity is highly variable, NAPL density is variable but generally greater than water density	viscosity variability tested with sensitivity analysis, density effects should be less significant
Hydrologic properties are uniform in each layer, but variable with depth	hydrologic properties vary with depth and horizontally	tested with sensitivity analysis

the north by Adams Avenue and on the south by the Robert Moses Parkway. On the west, it is bounded by property owned by Niagara Mohawk Power Corporation. Figure 7 shows the location of S-Area.

S-Area is partially situated on a reclaimed portion of the Niagara River. Consequently, the uppermost unconsolidated unit consists of fill material dumped to reclaim land from the Niagara River. The uppermost part of this fill material is highly nonuniform. It generally consists of rock fragments, wood, cinders, and fine gravel in a matrix of gray to brown sand. Near the Robert Moses Parkway, it is comprised primarily of shot rock. Here, shot rock refers to fragments of dolomite rock ranging in size from fine gravel to a few feet in diameter. Based on available data, the average thickness of the nonuniform fill layer is about 12.5 feet with an overall range of about 8.5 to 17.5 feet. Below the nonuniform (heterogeneous) fill, a layer of sands and silts overlie glacially derived clay and till. The average thickness of this sandy layer is about 16 feet. The thickness of the clay/till strata is highly variable. Data from boreholes indicate that it is absent in a small area south of the lagoons. In the northern portion of the site, it has a thickness of up to 18 feet. These glacial units overlie bedrock.

The uppermost bedrock unit is the Lockport Dolomite. As seen elsewhere in the Niagara frontier, the Lockport Dolomite is characterized by a highly fractured upper zone 10 to 15 feet thick. Below this upper zone, the Lockport is less fractured. The total thickness of the Lockport in the vicinity of S-Area is approximately 170 feet. The geologic unit immediately underlying the Lockport Dolomite is the Rochester Shale, which, in the vicinity of S-Area, has an estimated thickness of about 60 feet.

Significant water bearing units in the vicinity of S-Area include the upper fill material and sandy layer of the overburden and the upper zone of the Lockport Dolomite. The upper fill and sandy layer of the overburden can be considered as a single hydraulic zone. The clay/till layer retards groundwater flow between the overburden and bedrock permeable zones. In places, the clay/till may be absent (see Figure 7). Such discontinuities (holes) allow groundwater to flow rapidly down to the Lockport Dolomite. Water bearing zones within the Lockport Dolomite also occur below the upper 10-15 foot zone. However, these lower zones tend to

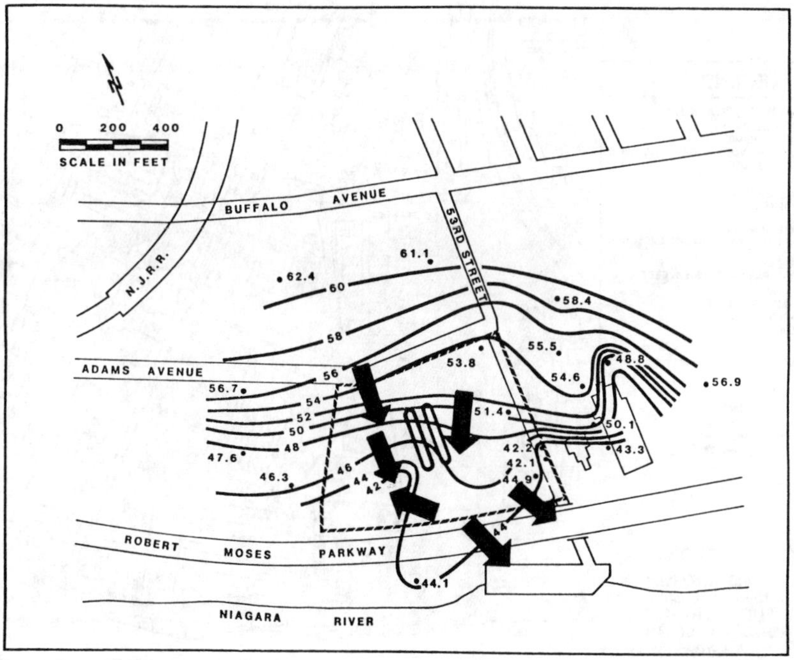

Figure 7. Location of S-Area, elevations of the top of clay, and inferred NAPL migration patterns.

yield much less water to wells than the upper zone. The water bearing zones within the Lockport are weakly interconnected by steep fractures that are approximately vertical. The Rochester Shale, which underlies the Lockport Dolomite, has a very low permeability.

The groundwater flow systems beneath S-Area can be idealized as consisting of two major lateral flow zones and a few minor lateral flow zones, with vertical flow zones separating them. The major flow zones are the overburden fill and sand and the upper Lockport Dolomite. Groundwater elevation measured in these two zones indicate downward flow from the overburden to the Lockport Dolomite. The water levels in the Lockport are generally a few feet lower than the water levels in the overburden. The general flow direction in the overburden is north to south, toward the river. Groundwater flows in nearly the opposite direction in the bedrock (southeast to northwest). There is insufficient water level data to verify the flow direction in the lower Lockport Dolomite. However, based on the general condition in the Niagara Frontier, the lateral flow direction in the permeable zones is probably to the northwest with downward groundwater flow in between permeable zones.

Contaminants from the S-Area landfill have migrated significant distances in the overburden and bedrock. A precise delineation of S-Area contaminants is difficult to obtain because of other sources of contamination in the vicinity. Nonetheless, surveys conducted by Occidental Chemical Corporation from 1986 to 1988 defined the extent of NAPL migration and provided extensive groundwater quality data beyond the NAPL plume. The survey was similar to the Hyde Park survey already discussed, except that survey boreholes were located over a specified area without provision for refining the definition of either the NAPL or APL plumes.

In addition to providing a three-dimensional description of the extent and degree of contamination, the surveys provided a detailed geometry of the Lockport formation (OCC, 1988). Figures 8, 9, and 10 show the survey results. Figures 8 and 9 present maps of the NAPL extent in the uppermost 15 feet of the Lockport Dolomite and in the Goat Island

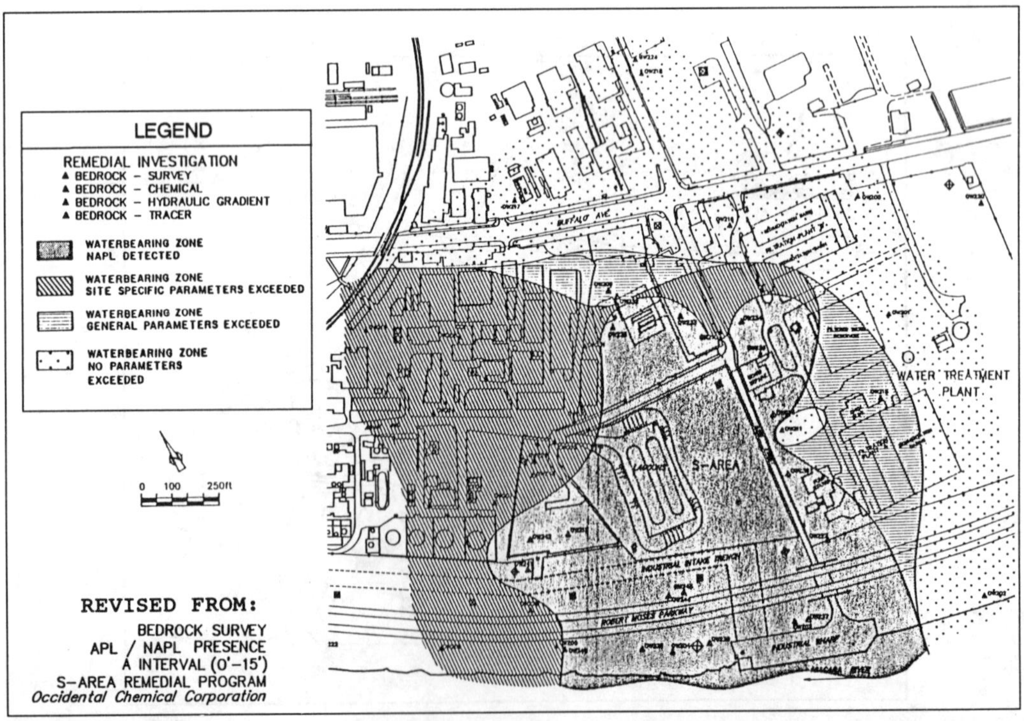

Figure 8. NAPL plume in the uppermost 15 ft of the Lockport Dolomite of S-Area (OCC, 1988).

Formation. In Figure 10, an east-west profile shows the vertical complexity of the NAPL migration.

4.4 S-Area analysis

Prior to the recent surveys at S-Area, the extent of NAPL in the bedrock was not known. It was believed that NAPL would tend to flow toward the river along bedding planes that dipped about 30 feet per mile. The groundwater gradient in the bedrock was northwest, but very low (1.25/1000). Given the potential for NAPL to move downward through a discontinuity in the clay layer under the site, a worst case analysis of potential NAPL migration was performed. The analysis was described by Faust et al. (1989) and a brief summary follows.

The simulation of NAPL migration was performed using a three-dimensional finite difference model based on equations (1) and (2). The model consisted of four layers (vertical), 23 columns (aligned with the dip of the bedrock and the direction of groundwater flow), and 9 rows (parallel to the strike of the bedrock). The conceptual model assumed that NAPL and water leaked through a small discontinuity in the clay layer for about 30 years. The leakage rate of NAPL was assumed to be 110 tons per year. Following 30 years of leakage, the NAPL is assumed controlled by an on-site remedy. The NAPL at S-Area has been observed to have a low viscosity and a high density, which were set at 1.0 and 1.5 times those of water, respectively. Other data and assumptions required for the model are presented in Faust et al. (1989).

The model results showed that the dense NAPL moved downward to the base of the Lockport Dolomite. The results also showed that NAPL tended to move preferentially down dip, but in all directions away from the source area. Figure 11 illustrates the extent of migration through time. As can be seen, the highest saturations are predicted for the lower layers of the model. This is likely due to differences in assumed porosity and relative permeability for the upper and lower layers.

Most of the simulated results were anticipated. Based on simple calculations of the available NAPL volume and the

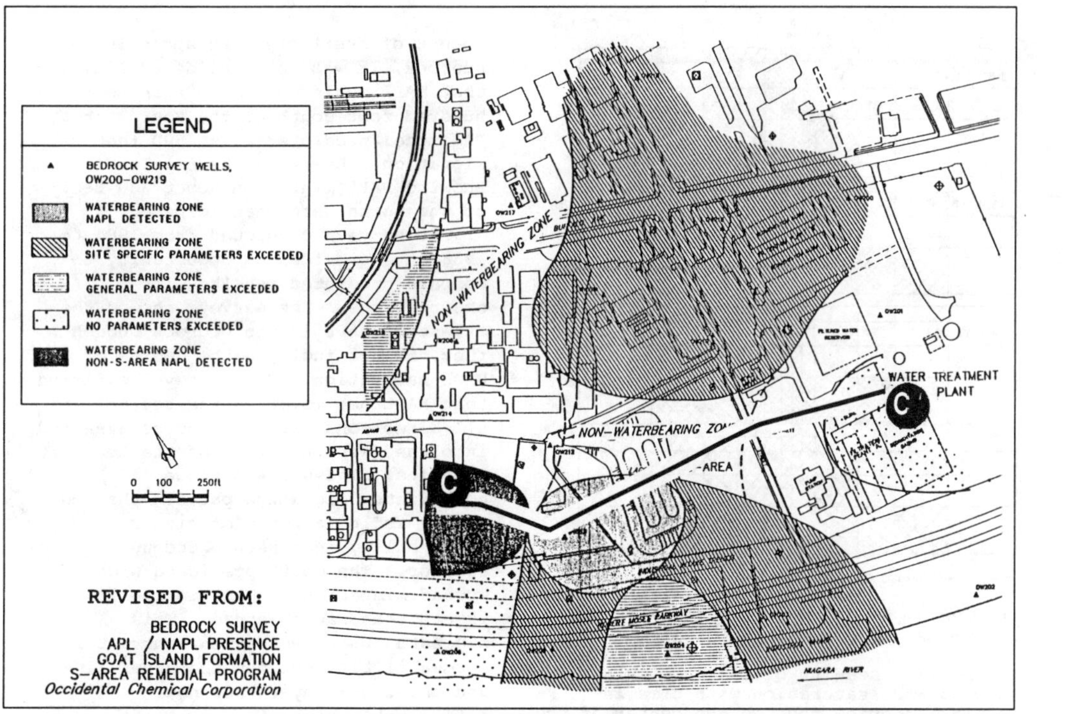

Figure 9. NAPL plume in Goat Island member of the Lockport Dolomite at S-Area (OCC, 1988).

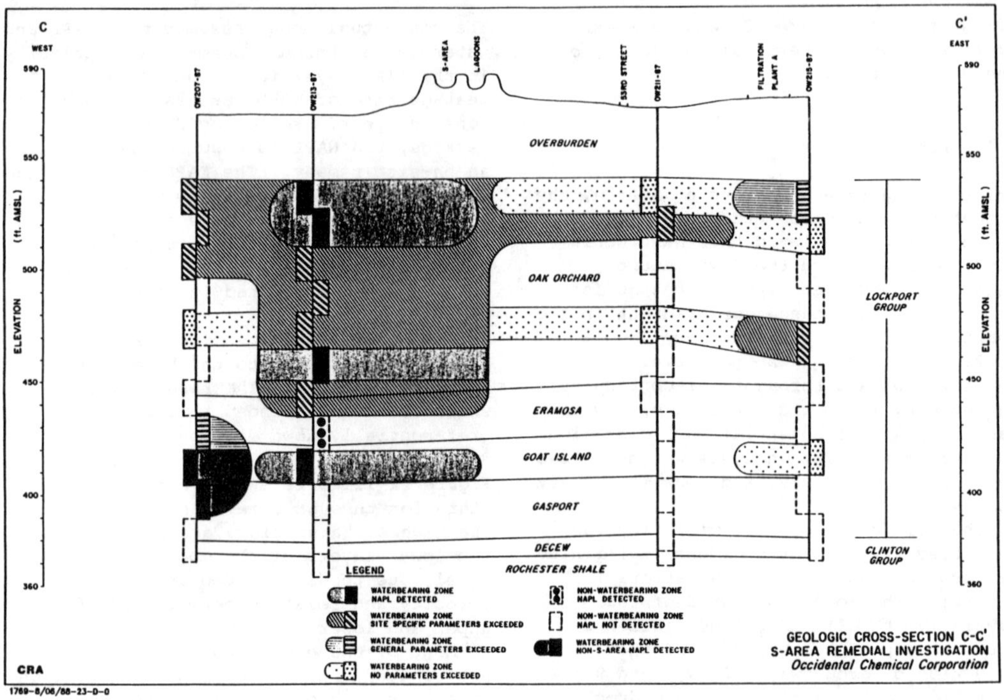

Figure 10. Cross-section through S-Area showing geology and NAPL observations (OCC, 1988).

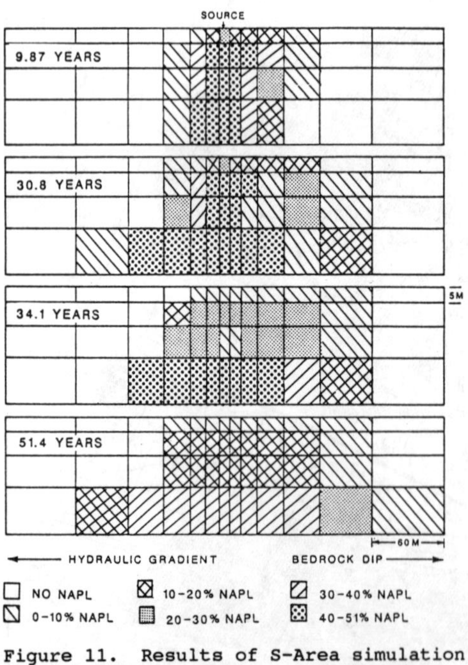

Figure 11. Results of S-Area simulation showing NAPL saturation with time in vertical section through the center of the source area and aligned with bedrock dip (Faust et al., 1989).

amount of available pore space in the bedrock, it was not considered likely that NAPL would move more than several hundred feet south of the source area. The model predicted about 500 feet of migration. However, the results showing NAPL migration about 300 feet to the north were unanticipated.

The surveys conducted from 1986 to 1988 have provided an opportunity to compare predicted results with observations. The surveys showed the extent of NAPL in the bedrock though they did not indicate the degree of NAPL saturation. The surveys confirmed that NAPL has moved in the bedrock in all directions from the source area and into the deepest layers of the Lockport Dolomite. The surveys showed a complicated migration pattern that most likely reflects zones of higher permeability in a heterogeneous aquifer. The model predicted wider spread migration of NAPL in the lower layers than was observed. This suggests that the assumed contrast in permeabilities between the upper layer and the lower layers was not sufficiently high.

5 SUMMARY

Investigations over the past ten years of groundwater contamination at several former waste disposal sites in the Niagara Falls area have generated substantial amounts of information regarding NAPL migration at the field scale. Because most of the studies have been conducted under court ordered agreements, much of this information is available to the public. In the next few years, aggressive groundwater pump-and-treat remedial programs will begin at these sites. These programs will be monitored routinely for characteristics of NAPL production. This new information will provide insight into NAPL migration under pumping conditions. Such information does not now exist (or is not known to exist). It will be useful for assessing the feasibility of NAPL containment and recovery programs.

This paper shows that there exists a wide gap between the type of data required for theoretical models and the types of data collected at the field scale. It is unlikely that this gap will be closed. In the most general case, it is necessary that numerous properties (relative permeability, viscosity, capillary pressures, etc.) be known at many points in the aquifer. If it were feasible to measure these properties, it would still be too expensive. Unfortunately, the methods to measure these properties and NAPL saturation in situ are either not well documented or not available.

The types of information that will be collected in the field will consist of qualitative observations, production decline curves (for NAPL), and indirect measurements such as provided by geophysical techniques. For research development, this suggests that either future models must be simpler or methods must be derived to make use of field scale data. Otherwise predictions of NAPL migration or recovery will be subject to a high level of uncertainty.

REFERENCES

Calkin, P.E. & C.E. Brett, 1978. Ancestral Niagara River drainage stratigraphic and paleontologic setting. Geol. Soc. Am., 89:1140-1154.

Crowley, D.J., 1973. Middle Silurian patch reefs in the Gasport member (Lockport formation), New York, Am. Assoc. of Petr. Geologists, 57:283-300.

Faust, C.R., J.H. Guswa & J.W. Mercer, 1989. Simulation of three-dimensional flow of immiscible fluids within and below the unsaturated zone. Water Resources Research, 25(12):2449-2464.

Feenstra, B.H., 1981. Quaternary geology and industrial minerals of the Niagara-Welland area, Southern Ontario. Ontario Geological Survey Open-file report 5361:260.

Interagency Task Force on Hazardous Wastes, 1979. Draft report on hazardous waste disposal in Erie and Niagara Counties, March.

Johnston, R.H., 1964. Groundwater in the Niagara Falls area, New York, State of New York, Conservation Department. Water Resources Division, GW-53.

Koszalka, E.J., J.E. Paschal Jr., T.S. Miller & P.B. Duran, 1985. Preliminary evaluation of chemical migration to groundwater and the Niagara River from selected waste-disposal sites. USEPA-905/4-85-001, March.

OCC, 1983. Hyde Park - Bloody Run aquifer survey and testing program.

OCC, 1988. Assessment of the extent of APL/NAPL migration from the S-Area in the Lockport Bedrock, S-Area Remedial Program.

Pinder, G.F., R.M. Cohen & J.Feld, 1992. DNAPL migration from the Love Canal landfill, Niagara Falls, New York. This volume.

Zenger, D.H., 1965. Stratigraphy of the Lockport formation (Middle Silurian) in New York State. New York State Museum and Science Service, 404 p.

Subsurface Contamination by Immiscible Fluids, Weyer (ed.) © 1992 Balkema, Rotterdam. ISBN 90 6191 175 3

DNAPL migration from the Love Canal landfill, Niagara Falls, New York

George F. Pinder
University of Vermont, Burlington, Vt., USA

Robert M. Cohen
GeoTrans, Inc., Herndon, Va., USA

Jodi Feld
New York State Department of Law, N.Y., USA

ABSTRACT: Large quantities of Dense Non-Aqueous Phase Liquid (DNAPL) were deposited in drums and as free liquids at the Love Canal landfill in Niagara Falls, New York, by the Hooker Electrochemical Company between about 1942 and 1953. DNAPL and other chemical wastes were buried from the landfill base, typically 8 to 16 ft below ground, to within a few ft of the ground surface. Shallow soils at the site consist of approximately 5 ft of silt and sand underlain by 6 ft of fractured stiff silty clay which grades into a layer of soft silty clay. Remedial investigations indicate that: (1) a water table mound caused outward overland and subsurface flow of contaminated fluids from Love Canal, and (2) pressure gradients, largely due to the thickness of DNAPL in the landfill, resulted in DNAPL migration to adjacent residential properties predominantly through the fractured silty clay layer. Much of the aqueous phase contamination identified in the immediate landfill area may be derived from dissolution of the migrating DNAPL.

1 INTRODUCTION

Love Canal is located in the City of Niagara Falls in northwestern New York State (Figure 1). The canal was promoted by William T. Love who, in 1894, initiated its excavation as part of a plan to produce cheap electricity by connecting the upper and lower parts of the Niagara River. The objective was to use the power so generated to develop a model industrial city. Excavation of the canal had barely begun, with only about 3000 ft dug north of the Niagara River, when Love's dream ended in about 1897 due to economic recession and the development of alternating current which could be transmitted over great distances. As a result, the canal was abandoned uncompleted.

Between about 1942 and 1953, Hooker Electrochemical Company (Hooker, now known as Occidental Chemical Corporation or OCC) disposed of an estimated 22,000 tons of chemical wastes at the Love Canal site. DNAPL was entrained within these wastes. Numerous problems developed at the site including subsidence of the landfill surface and exposure of drums and chemical residue, runoff of contaminated fluids from the landfill to adjoining properties, unpleasant chemical odors which were cited by residents as a cause of illness and discomfort, chemical migration to basements adjacent to the landfill, and chemical migration into and through the local sewer system (USEPA, 1982). Numerous remedial investigations were conducted at the site during the late 1970s and early 1980s and a hydraulic containment remedial system utilizing drains and improved cover was installed beginning in 1978.

Many of the early remedial investigations focused on the migration of dissolved chemicals from Love Canal. Extensive drilling along the leachate collection system drain alignment and elsewhere between 1977 and 1982, however, revealed the presence of DNAPL in soil at numerous locations beyond the landfill. More recent investigations conducted for a long-term monitoring program and ongoing litigation have documented the presence and characteristics of DNAPL in the landfill. The purpose of this paper is to discuss the disposal of DNAPL at Love Canal, observations of DNAPL beyond the landfill, recent measurements of DNAPL elevations, composition, and properties in the landfill, and the mechanisms of DNAPL migration to offsite properties.

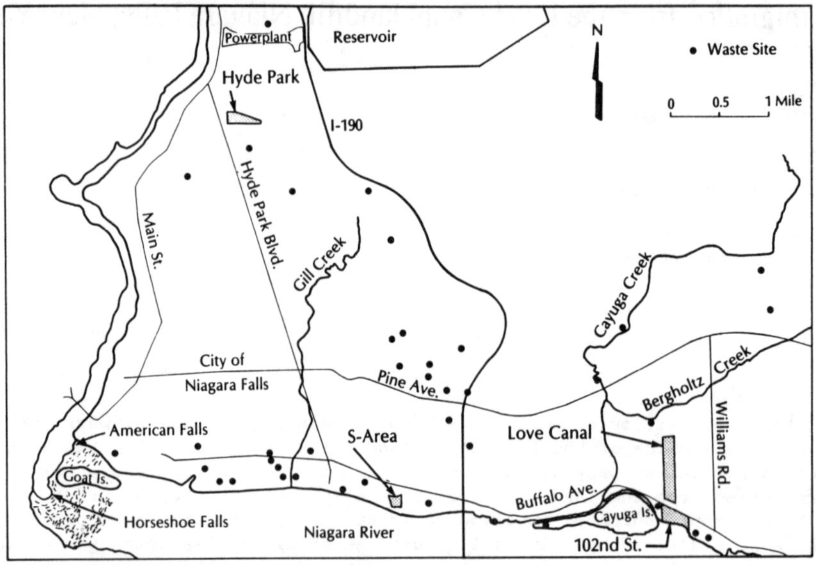

Figure 1. Locations of the Love Canal landfill and other waste disposal sites in Niagara Falls (from Cohen et al., 1987).

2 SITE GEOLOGY

The geology of the Love Canal area is illustrated schematically in Figure 2. Approximately 30 to 40 ft of unconsolidated glacial sediments overlie the Lockport Dolomite at the site. The youngest sediments consist of fill and silts and sands deposited when Lake Tonawanda occupied the area at the time of the Wisconsin glaciation. In the neighbourhood of the canal, these sediments form a layer approximately 5 ft thick and are typically encountered between 569 and 574 ft above MSL. These silts and sands are underlain by older, varved glaciolacustrine silty clay sediments. The upper 6 ft of the silty clay between about 563 and 569 ft are stiff and contain interconnected prismatic fractures and thin silty laminations. The stiffness of the upper silty clay is hypothesized to be the result of dewatering and consolidation above a previously lower water table. Transitionally below the stiff silty clay are 6 to 14 ft of soft silty clay above 2 to 20 ft of glacial till. Beneath the till, the Lockport Dolomite is a fractured aquifer of regional extent.

The Love Canal area is characterized by poor natural drainage due to several factors including: (1) its relatively flat topography, (2) the presence of subsoils of low to moderate permeability, (3) the shallow depth of nearby rivers and streams, (4) the humid climate, and (5) the absence of a well developed natural drainage network. As a result, the site historically experienced a high water table and drainage problems.

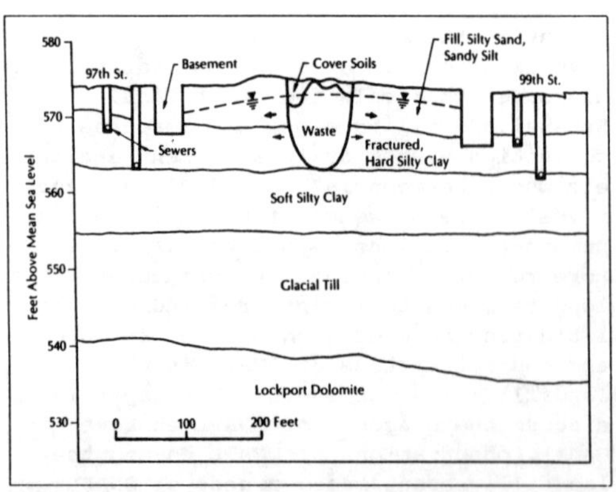

Figure 2. Schematic E-W geologic cross-section through Love Canal (from Cohen et al., 1987).

3 WASTE DISPOSAL HISTORY

Prior to waste disposal activities in the

1940s, the canal lay undisturbed, except for its use as a swimming hole, on the 230-ft wide Love Canal property. Based on historic air photographs and recent drilling, the canal excavation was approximately 3000 ft long, 40 to 100 ft wide, and 7 to 16 ft deep (E.C. Jordan, 1987).

Love Canal was first used for waste disposal in about 1942 when Hooker received permission from the owners of the property, Niagara Power and Development Company, to deposit fill at the site. Disposal of chemical wastes began in the northern end where approximately 450 linear ft of the canal were filled between about 1942 and 1946 (Figure 3). According to OCC (1985), the protocol for burial of wastes in the northern section involved the construction of dams composed of flyash, rubble and clayey soils across the canal. Historic air photographs and a 1941 map indicate the presence of two major dams across the northern sector. The water level in the canal in December 1941, a very dry year, was reportedly 6 ft below grade (Hooker, 1941). Although much of the drummed chemical residues were dumped from trucks into the canal, some drums were emptied of their liquid contents and returned to the Hooker plant. Wastes were typically buried to within a few feet of grade with no compaction and were then covered with two to four feet of flyash and soil.

The disposal operation moved to the southern end of the canal in about 1946, a year before Hooker purchased the Love Canal property. According to OCC (1985), disposal of chemical residues in the southern sector proceeded in excavated and dammed off sections from south to north. Hooker buried chemical wastes in the abandoned canal, which was deepened and widened in some areas, and in separate disposal pits which were excavated beyond the canal perimeter (Figure 3). A Hooker worker who participated in the disposal operation testified that the waste pits were typically 25 to 30 ft deep and up to 50 ft wide and that the pits were filled by rolling drummed liquids from trucks, which frequently caused drums to rupture, and by adding flyash and soil with a bulldozer (Gibson, 1984).

Having concluded that "the Love Canal property [was] rapidly becoming a liability because of housing projects in the near vicinity" (Klaussen, 1952), Hooker sold the site for $1 in April 1953 to the local Board of Education for construction of an elementary school. The property deed stated ". . . that the

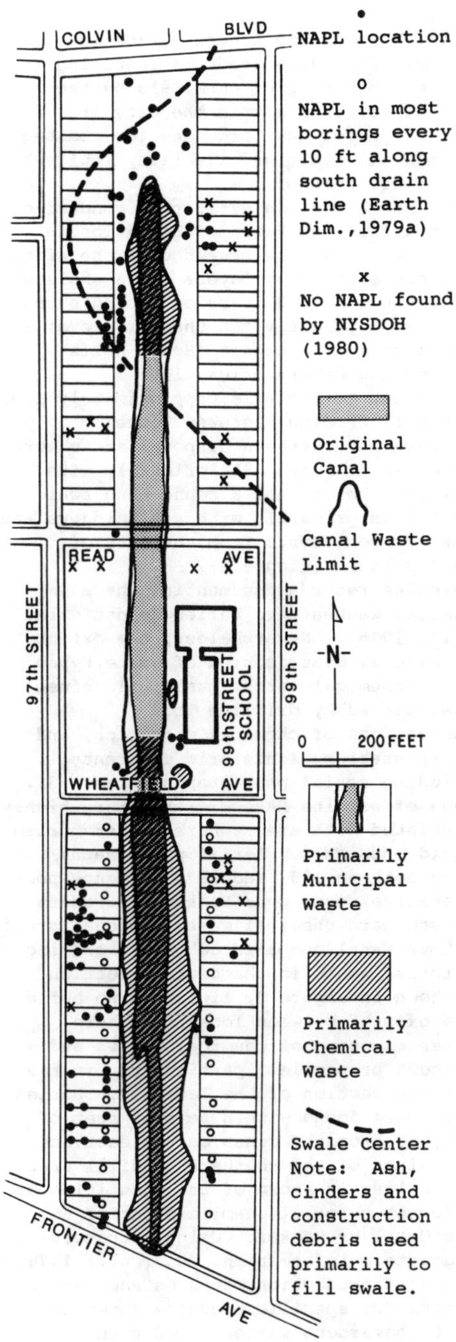

Figure 3. Approximate extent of buried chemical and municipal wastes at Love Canal and relationship to observed NAPL.

premises . . . have been filled, in part or in whole to the present grade level thereof with waste products resulting from the manufacturing of chemicals by the grantor at its plant in the City of Niagara Falls, New York, and the grantee assumes all risks and liability incident to the use thereof."

A geotechnical investigation conducted on behalf of the Board of Education in 1953 documented the presence of the silty sand and silty clay layers and a shallow water table 1 to 3 ft below ground surface that was apparently "at the same level as the water in the canal" (Hough Soils Engineering Laboratories, 1953). Subsequent to the sale, approximately 1000 linear ft of canal between Hooker's southern and northern disposal areas were filled in 1953 and 1954 primarily with municipal refuse and a connecting swale was filled primarily with ash, cinders and construction debris from the City of Niagara Falls (Figure 3).

Precise records documenting where chemical wastes were buried do not exist (Cull, 1978). Nevertheless, the extent and general distribution of waste types (i.e., chemical versus municipal refuse) is evidenced by drilling logs, observations of chemical migration, and interpretation of historic documents including aerial photographs (Figure 3). Areas of offsite NAPL migration are highly correlated with areas where Hooker buried liquid chemicals. Aerial photographs taken between 1953 and 1978 evidence poor vegetative cover over Hooker's northern and southern chemical waste disposal areas at Love Canal compared to that overlying municipal waste in the central section. As shown in Figure 3, little or no buffer zone existed between locations where Hooker disposed of chemical wastes and adjacent properties, particularly in the southern section of the landfill. Buried drums were found within several feet of backyard property lines on the east and west sides of the southern landfill area.

Even less is known of the quantities of different types of chemicals buried at Love Canal by Hooker. Using extremely fragmentary information, OCC (Cull, 1978) estimated production tonnages and residue factors for specific products known to create hazardous wastes. OCC then allocated the calculated wastes to specific landfill sites, sometimes on an arbitrary basis (Cull, 1978). As a result of this analysis, OCC estimated that Hooker buried 21,800 tons of chemical wastes at Love Canal between 1942 and 1952 (Table 1).

Table 1. Estimated types and quantities of chemical wastes disposed of by Hooker at Love Canal based on Cull (1978) and Hooker process sheets.

Waste Type and Contents	Form	Total (tons)
Misc. Acid Chlorides	L,S	400
Thionyl Chloride (Possibly includes chlorinated toluenes, chlorinated styrenes, sulfonated toluene, sulfites, sulfates and sulfur)	L,S	500
Misc. Chlorinations (Wastes from various chlorination processes, including, but not limited to, waxes, oils, napthalenes and aniline)	L,S	1000
Dodecyl Mercaptan (DDM) (Wastes from production of DDM and lauryl chloride)	L,S	2400
Trichlorophenol (Wastes from production of 2,4,5-TCP; possibly including 2,4,5-TCP, TCDD, tetrachlorobenzene, dichlorophenols, and ethers)	L,S	200
Benzoyl Chlorides and Benzotrichlorides (Possibly including benzotrichloride, benzoyl chloride, benzoic acid, benzaldehydes, benzyl chloride, ring-substituded chlorocarbons and tarry polymers)	L,S	800
Metal Chlorides	S	400
Liquid Disulfides (LDS/LDSN/BDS) and Chlorotoluenes (Possibly including chlorotoluenes, toluene, and aromatic disulfides)	L	700
BHCs (Possibly including alpha-BHC, beta-BHC, delta-BHC, Lindane, and heptachlorocylclohexane)	S	6900
Chlorobenzenes (Possibly including monochlorobenzene, dichlorobenzenes, trichlorobenzenes, tetrachlorobenzenes, pentachlorobenzene, hexachlorobenzene and benzene)	L	2000
Benzyl Chloride (Possibly including dibenzyl compounds, benzyl alcohol, chlorinated benzyl chlorides, benzyl chloride and related polymers)	S	2400
Sodium Sulfides and Sulfhydrates	S	2000
Miscellaneous (10% of above)		2000
	TOTALS:	21,800

L refers to Liquid waste under normal conditions.
S refers to Solid waste under normal conditions.
L,S refers to sludge or combination of liquid and solid wastes.

4 FIELD OBSERVATIONS -- 1953 TO 1978

The first monitoring well at Love Canal was installed in 1977. Measured water-level and chemistry data at the site are virtually nonexistent for the period prior to the late 1970s. A historical record comprised of air photographs, complaints, newspaper articles, Hooker memoranda, Health Department inspection notes, Police reports, witness depositions, etc., provides significant information, however, regarding conditions at the Love Canal site during and after the period of disposal. Among other conditions, the historical record indicates that:

1. The disposal of large quantities of liquid chemicals in drums at Love Canal resulted in drum failure due to consolidation and corrosion. This, in turn, caused subsidence of cover materials, creation of 'potholes' and caved areas, exposure of chemicals including NAPL at the surface, emission of chemical vapors, and increased recharge to the landfill.
2. The waste disposal practices created an irregular landfill surface that was elevated above surrounding properties along portions of the site. Combined with the shallow water table, humid climate and poor drainage capacity, this promoted the development of a water table mound at the site and periodic leachate seeps at the landfill perimeter.
3. Chemical wastes, including NAPL in drums, were buried as high as the original ground surface elevation at some locations.
4. Basement drains and sewers surrounding the Love Canal property acted both as hydraulic sinks and large monitoring wells. NAPL and chemical odors were reported at various homes and sewer locations.

The observed conditions at Love Canal demonstrated the need for buffer zones, monitoring, hydraulic control of contaminated fluids, and institutional controls on disposal practices and site security at hazardous waste landfills.

5 RECENT DOCUMENTATION OF NAPL MIGRATION

Many remedial investigations were conducted at Love Canal during the late 1970s and early 1980s, due in part to Emergency Declarations by the New York State Health Commissioner in 1978 and President Carter in 1980. During these studies, more than 1000 soil borings were drilled, a few hundred monitoring wells were installed, and thousands of environmental samples (of groundwater, soil, stream water and sediment, basement sump water and sediment, sewer water and sediment, and air) were analyzed for Love Canal chemicals.

Water level measurements indicate that a groundwater mound caused outward flow from the landfill through overburden to adjacent properties and downward flow toward bedrock prior to construction of a leachate collection system in 1978 and 1979. For example, fluid elevations measured in 10 wells during April and May, 1978 document that "the groundwater table [was] located immediately below or at the land surface" of the southern section of the landfill causing overland flow and the development of leachate ponds in low areas along the landfill perimeter (Conestoga-Rovers and Associates, 1978). Most of the early hydrogeologic studies concluded that outward chemical migration from the landfill had occurred preferentially through the silty sand layer due to the shallow water table mound in the landfill, and contained little explicit discussion of NAPL migration (Calspan, 1977; Conestoga-Rovers and Associates, 1978; F.C. Hart Associates, 1978).

The earliest observations of NAPL migration from the landfill were made in basements and backyards by local residents. During the first major offsite drilling program in which soil borings were augered every 10 ft along the proposed barrier drain line in backyards adjacent to the southern sector of the landfill in late 1978, Earth Dimensions (1979a) discovered that an oily film appeared to displace the water film in the fractures in the stiff silty clay layer at most of the augered sites. Pungent odors were reported at nearly all of the drill sites. The observations of DNAPL in fractures in the stiff silty clay resulted in a decision to lower the leachate collection drain system to beneath this layer.

Numerous other subsurface investigations have documented the extent of NAPL migration from the Love Canal landfill. NAPL observation locations are shown on Figure 3.

The most detailed study of NAPL in soils adjacent to Love Canal was conducted jointly by Earth Dimensions (1980) and the New York State Department of Health (1980). Continuous split-spoon samples were taken at 64 locations in the immediate vicinity of the landfill. Detailed logs were prepared to document the vertical distribution of soil types, chemical odors and NAPL; and samples were analyzed for Love Canal chemicals. Based on this and other studies, it is clearly evident that DNAPL migration occurred from Love Canal to adjacent residential properties and that the migration is characterized by the following attributes:

1. DNAPL migration is highly correlated with the locations where Hooker disposed of chemical wastes and is most extensive at the southern sector of Love Canal.
2. Where present, DNAPL was commonly found in fractures and in thin silty laminations in the stiff silty clay layer.
3. DNAPL was rarely found in deep root channels in the upper soft silty clay. This layer generally prevented DNAPL penetration.
4. DNAPL was occassionally found near the base of the shallow silty sand layer.
5. DNAPL was found in some basement sumps.
6. The upper surface of DNAPL beneath residential properties west of the southern section of Love Canal in 1980 was typically at an elevation between 565 and 567.5 ft and ranged from 564 to 570 ft.

6 EXAMINATION OF DNAPL IN THE LANDFILL

In 1986, 21 borings and 15 piezometers were completed in the waste disposal area (Figure 4) to provide information regarding fluid elevations in the landfill, the nature of waste materials, and the nature and integrity of soils below the wastes (E.C. Jordan Co., 1987). Consistent with knowledge of the disposal practices at Love Canal, the waste materials found along the length of the landfill were diverse and lacked stratigraphic continuity. Elevations of the aqueous phase and DNAPL surfaces measured in the canal wells, the canal bottom, and DNAPL encountered in split-spoon samples taken during drilling are given in Table 2.

As shown by referencing Table 2 to Figure 4, the observed waste types are consistent with historical information regarding the distribution of Hooker chemical waste and City municipal waste. NAPL was encountered in split-spoon samples taken from the Hooker chemical waste disposal areas from below the bottom of Love Canal to elevations ranging between 567 and 574 ft. Ten years after constructing a leachate collection system with many laterals between the main drains and the landfill, DNAPL surface elevations measured in Love Canal wells ranged from approximately 566 to 573 ft.

Pumping experiments were conducted recently in the canal wells by GeoTrans (1988) and J.R. Kolmer and Associates (1990) to examine the distribution, properties, and composition of fluids in the landfill. Typically, 3 to 5 gallons of fluid (equivalent to about 10 to 20 well volumes) were extracted from a subset of the 2-inch diameter wells during the study.

GeoTrans (1988) sequentially lowered the pump intake to examine fluid density stratification and measured the density and viscosity of NAPL samples. Some stratification of DNAPLs was observed in a few of the wells. Results of density and viscosity measurements and a summary of the fluid types pumped from each well are provided in Table 3. NAPL densities ranged from 1.09 to >1.50. NAPL viscosities ranged from similar to water to very thick. At several wells, 3 to 6 gallons of DNAPL were withdrawn without any aqueous phase liquid.

J.R. Kolmer and Associates (1990) sampled the fluid surface using capillary tubes and a skimming vial to determine the distribution of LNAPL in the wells. Although a "thin film of floating material" was found in CW-80, CW-90, and CW-105, insufficient volumes were

Table 2. NAPL elevations in the Love Canal wells based on fluid level measurements (GeoTrans, 1988; and J.R. Kolmer and Associates, 1990) and split-spoon samples (E.C. Jordan Co., 1987; and Earth Dimensions, 1986).

	---- In Well ----				
Well ID	Fluid Surface (Ft>MSL)	DNAPL Surface (Ft>MSL)	Canal Bottom (Ft>MSL)	NAPL in Split Spoon Samples (Ft>MSL)	Main Waste Type
CW-10	571.3	571.3	563.5	562.9-568.3	Chemical
CW-20	572.5	572.5	560.9	562.2-567.0	Chemical
BRM-10	570.5	570.5	562.9	563.4-571.4	Chemical
CW-30	567.3	None	563.9	None	Municipal
CW-40	--	--	560.5	561.9-562.7	Municipal
CW-50	568.7	None	558.5	None	Soil fill
CW-60	567.6	566.4-566.8	557.9	556.9-571.4	Chemical
CW-70	569.1	566.5	558.8	558.8-571.3	Chemical
CW-80	570.0	567.3-569.6	560.4	558.5-572.5	Chemical
CW-90	570.2	569.4	560.4	560.1-569.1	Chemical
CW-105	574.1	573.1	566.7	562.7-574.2	Chemical
CW-108	571.2	--	565.2	562.7-563.2	Soil fill

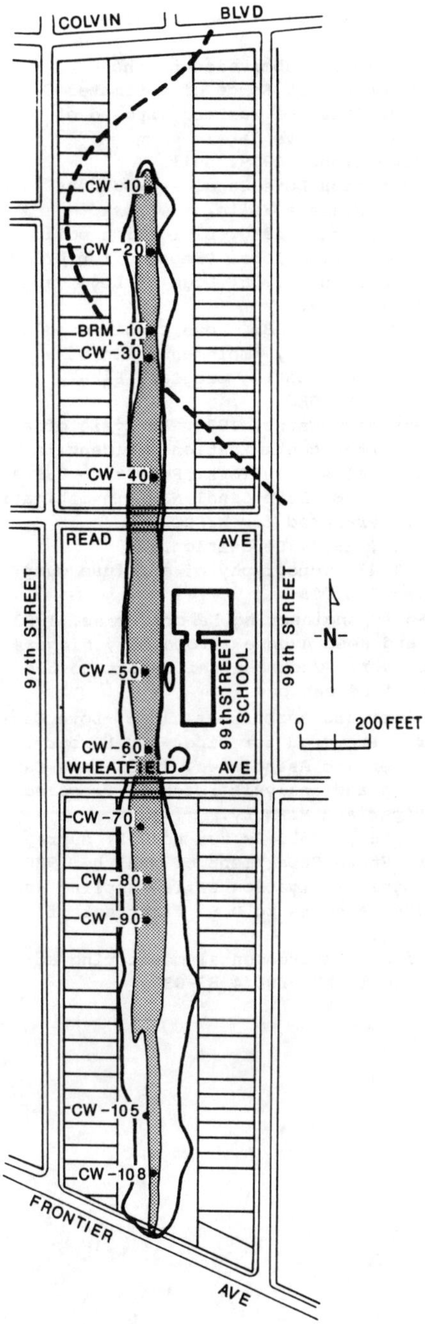

Figure 4. Locations of Love Canal landfill wells.

Table 3. Density and viscosity measurements of NAPL sampled from the Love Canal wells (GeoTrans, 1988; and J.R. Kolmer and Associates, 1990).

Well ID & Sampling Date	NAPL Density	Absolute Viscosity (cps)	Sample Temp. (C)	Fluid Sampled from Landfill Wells
CW-10 11/15/88	1.15-1.18	75-81	60-62	4-5 gals pumped (all DNAPL)
CW-20 11/15/88	1.12	<16.6	60	4-5 gals pumped (all DNAPL)
BRM-10 11/16/88	1.09-1.12	44-50	60-64	3-4 gals pumped (all DNAPL)
CW-60 11/1/88	1.21-1.25	21-36	61-62	6 gals pumped (all DNAPL)
CW-70 11/2/88	>1.50	<23	53-54	Approx. 3 gals pumped (more water than DNAPL).
CW-90 11/17/88	1.18-1.20	<18	61-62	1.5 gals pumped (DNAPL and water)
CW-105 11/16/88	1.09-1.22	37-270	59-62	Approx. 3 gals pumped (mainly DNAPL)
CW-108 10/18/89	1.08	<16	71	4 gals pumped (approx. 30% DNAPL and 70% water)

collected to make any density measurements.

7 INTERPRETATION OF DNAPL MIGRATION

Historical records and recent observations document that large quantities of DNAPL were deposited in the abandoned Love Canal between 1942 and 1953. As a result of pressure gradients created in large part by the substantial thickness of dense immiscible fluid in the landfill, DNAPL migrated into and through the fractured silty clay layer. Fractures and silty laminations in the silty clay layer conducted DNAPL towards 97th and 99th Streets. Drains associated with homes and sewers acted as hydraulic sinks and, to a variable degree, captured NAPL and dissolved chemicals.

Issues pertaining to chemical migration from Love Canal have been the subject of litigation since 1979. An alternative theory has been proposed to explain the presence of DNAPL on the east side of 97th Street adjacent to the southern section of the landfill (J.R. Kolmer and Associates, 1988). According to this theory, DNAPL: (1) seeped through fill used to replace excavated chemical wastes and into the Frontier Avenue storm sewer which was constructed across the southern tip of the landfill in 1968 by New York State contractors; (2) was retained on the sewer water surface; (3) floated on the sewer water surface up the Frontier Avenue and 97th Street storm sewers during rising Niagara River stage events; (4) was deposited along the storm sewer walls as the water level receded; (5) exfiltrated the sewer pipe through joints at an elevation of approximately 566 ft; and (6) migrated eastward, but not westward, from the storm sewer location onto the east 97th Street properties. This theory will probably be a focal issue in a trial that is scheduled to begin in September 1990.

DNAPL that migrated to offsite properties was a moving contaminant source from which chemicals dissolved into groundwater and entered soil gas and basement air by volatilization. Much of the aqueous phase contamination in the immediate landfill area may have derived from dissolution of migrating DNAPL.

Since installation of the tile-drain system at Love Canal, approximately 30,000 gallons of DNAPL have been collected and stored in holding tanks at the onsite leachate treatment plant. Much of the DNAPL appears to have flowed via the many French drain laterals installed directly into the landfill. The recent pumping experiments indicate that it may be feasible to recover large quantities of DNAPL directly from wells in the landfill if incineration or some other destructive technology can be demonstrated to be cost-effective at a site where long-term hydraulic containment will be needed in any event.

REFERENCES

Calspan, 1977. Characterization and abatement of groundwater pollution from the Love Canal landfill, Niagara Falls, New York. Final report to the City of Niagara Falls.

Cohen, R.M., R.R. Rabold, C.R. Faust, J.O. Rumbaugh, and J.R. Bridge, 1987. Investigation and hydraulic containment of chemical migration: four landfills in Niagara Falls. Civil Engineering Practice, Vol.2, No.1.

Conestoga-Rovers and Associates, 1978. Phase I pollution abatement plan -- upper groundwater regime, Love Canal chemical landfill, Niagara Falls, New York.

Cull, J., 1978. Submission to the Interagency Task Force of estimates and quantities of wastes disposed of by Hooker at Love Canal.

Earth Dimensions, 1979. Soil investigation Love Canal -- southern section, Niagara Falls, New York.

Earth Dimensions, 1980. Litigation soil sampling boring logs. Prepared for NYSDOH.

Earth Dimensions, 1986. Logs of Love Canal landfill wells.

E.C. Jordan Co., 1987. Love Canal remedial project task V-C, implementation of a long-term monitoring program. Final report to NYSDEC.

F.C. Hart Associates, 1978. Analysis of a ground water contamination incident in Niagara Falls, New York. Report to USEPA.

GeoTrans, 1988. Love Canal NAPL investigation report. Prepared for NYSDOL.

Gibson, J., 1984. Deposition.

Hooker, 1941. Topography of a refuse dump. Drawing 24510GA.

Hough Soils Engineering Laboratories, 1953. 99th and Reed Avenue (Proposed), Niagara Falls, N.Y., Geotechnical report to the Board of Education.

J.R. Kolmer and Associates, 1988. Love Canal report. Prepared for Piper and Marbury.

J.R. Kolmer and Associates, 1990. Love Canal sampling and analysis report. Prepared for Piper and Marbury.

Klaussen, B., 1952. Letter to R.L. Murray.

New York State Department of Health, 1980. Love Canal litigation soils sampling study.

OCC, 1985. Answers to U.S. first set of interrogatories.

USEPA, 1982. Environmental monitoring at Love Canal. EPA-600/4-82-030.

Subsurface Contamination by Immiscible Fluids, Weyer (ed.) © 1992 Balkema, Rotterdam. ISBN 90 6191 175 3

Coal tar and creosote contamination in Ontario

K.G. Raven
Intera Technologies Ltd, Ottawa, Ont., Canada

P. Beck
Intera Technologies Ltd, Toronto, Ont., Canada

ABSTRACT: Coal tar and creosote are complex mixtures of mutually miscible organic chemicals including polycyclic aromatic hydrocarbons (≈80%), monocyclic aromatic hydrocarbons (≈15%) and phenolics or tar acids (≈5%). These immiscible liquids were produced or used at 60 industrial sites in Ontario. Recent investigations conducted at 25 of these sites indicate subsurface occurrence of coal tar or creosote at all sites. The coal tar or creosote is found within sands and gravels or in fractures in clay and bedrock in accordance with the theory of multi-phase flow for credible ranges for media and fluid properties. Aqueous phase contamination at most of the sites is a minor concern because of the low solubility and mobility of the hydrophobic compounds comprising coal tar and creosote and the lack of groundwater use in the immediate vicinity of the sites. Maximum concentrations of most polycyclic aromatic hydrocarbons in groundwater may be calculated at less than 1 ug/L to a few mg/L considering cosolvent effects and the weight fraction of individual compounds in the immiscible fluid. The principal environmental concerns at coal tar and creosote sites relate to migration and exposure of immiscible liquids. Two case studies, one of a former coal gasification plant and one of a former tar distillation plant are described to illustrate the chemistry, fate and environmental impacts of coal tar and creosote.

1. INTRODUCTION

Coal tar and related creosote are dense immiscible liquid contaminants that were widely produced and used in Ontario for over 100 years from as early as 1850. These immiscible fluids were associated with manufactured gas plants, tar distillation works and wood preservative plants. Recently completed inventory studies have identified 56 of these sites in 40 different communities throughout Ontario. The vast majority of these sites ceased operation over 35 years ago and only recently has the potential environmental significance of these sites been recognized in Ontario and elsewhere in Canada. Coal tar and creosote are complex mixtures of polycyclic aromatic hydrocarbons (PAH) and monocyclic aromatic hydrocarbons (MAH) several of which are documented as hazardous to human health.

2. HISTORICAL REVIEW

2.1 *Plant Operations*

Understanding the historical operation of manufactured gas plants, coal tar distillation works and creosoting wood preservative plants is an important first step in studying coal tar and creosote waste sites. Descriptions of historical operations provide insight to the waste generation, handling and disposal operations and when site plans are available, the likely location where coal tar and creosote wastes may be found.

Manufactured gas plants produced gas for illumination and heating in Ontario from about 1850 to the late 1950s by carbonization of coal and oil. Subsequent availability of electricity and natural gas, caused many of the manufactured gas plants to be uneconomical and close down, some as early as 1890, many in the 1920s and 1930s and almost all by the late 1950s. In Ontario, most of the gas plants started out as coal carbonization plants and those plants that continued into the 1920s and 1930s were either converted to water gas plants or had water gas facilities added to the site. These plants ranged in size from 0.2 to 14 ha and were often adjacent or close to surface waters bodies.

In the coal carbonization process, coal was heated in coke ovens or retorts in the absence of oxygen to drive off volatile components and the resultant gas was collected, condensed to remove tars, purified and stored on-site in large gas holding tanks (gasometers). Figure 1 shows the layout of a typical manufactured gas plant. Coal tars were generated from coal carbonization at rates on the order of 40 L/m^3 of coal and about several hundred to a thousand litres per day for a medium to large size plant. These tars were accumulated in the gas condensers, purifiers and relief gas holders and were often stored in underground tar wells. Prior to the establishment of the coal tar distillation industry in Ontario in the early 1900s, the by-product tar had limited resale value and was likely a nuisance at most gas plant sites.

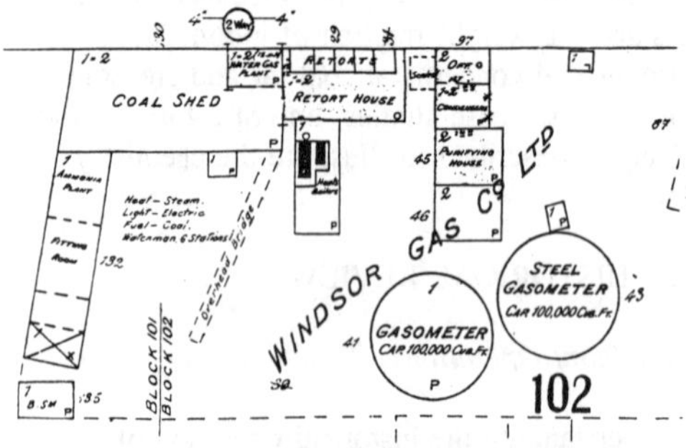

Fig. 1 1917 fire insurance plan for Windsor gas works

Water gas plants produced a gas with superior heating value by reacting steam with incandescent coke and by cracking oil introduced to the gas stream with a carburetter. This gas process produced slightly higher volumes of tar which was chemically dissimilar to coal tar, mainly lacking tar acids (ie. phenolics) and nitrogenous compounds. These tars were less useful as chemical feedstocks to the tar distillation industry and were likely used as on-site fuel and sold as road and roofing tars and to coatings manufacturers.

The tar distillation industry was first operational in Ontario at the turn of the century using coal tars from gas works but was not on a firm footing until by-product coke ovens were established in 1911 in association with the steel industry. Coal tar distillation was conducted in batch or continuous stills and in the 1920s and 1930s produced roofing tars, road tars, creosote, light, middle and heavy oils and pitch. Some of these plants are still in operation today. Average plant size was 1 - 3 ha and principally consisted of many above ground storage tanks.

The creosoting wood preservative industry like the tar distillation industry was not firmly established until about 1911. Typical creosote wood preserving facilities included creosote and tar tanks, steel treatment vessels, boiler houses and retort houses. Following heating and pressurized treatment of wood in the steel treatment vessels, wood was removed and allowed to dry prior to shipment. In addition to creosote drippings and spills, waste sludges and tars were likely disposed on site in pits or wells. Creosoting plants were large facilities averaging 10 - 20 ha with coal tar and creosote wastes likely remnant in the treatment areas and near storage tanks.

Manufactured gas plants, tar distillation works and creosote wood preserving facilities generated, used or stored large volumes of dense immiscible liquids and often for periods in excess of 50 years. These statistics in conjunction with the early operating period, indicate these sites likely contain dense immiscible liquids.

2.2 *Inventory of Sites*

Studies commissioned by Ontario Environment, have identified 41 manufactured gas plants, 11 tar distillation works and 4 creosote wood preservative plants in Ontario (INTERA, 1987, 1988). The inventory process identified the exact location, period of operation, nature of operation and layout of facilities for these sites. In compiling this information several historical sources were useful in identifying and characterizing the sites. Identification sources included Dominion Bureau of Statistics reports, Canadian census and trade index reports and provincial and municipal directories. Site characterization sources included fire insurance plans, historic airphotos and local archival maps, photographs and articles.

Fire insurance plans, such as those generated by Charles E. Goad Co. from 1870 and later by the Underwriters Survey Bureau to 1965 are of greatest use in characterizing sites and therefore identifying areas of potential immiscible fluid sources. These maps (e.g. Figure 1) are available for almost all of the identified sites.

3. COAL TAR AND CREOSOTE WASTES

3.1 *Occurrence*

The results of investigations completed to date show that coal tar or creosote are present at residual saturation, as dispersed ganglia or as pools of free product at all sites studied, and therefore migration and exposure of these immiscible fluids are an environmental concern. At manufactured gas plant sites residual tars have been found to be associated with underground remnants of main and relief gas holders as well as in the vicinity of gas condensing and purifying facilities. At many of these sites it appears that little, if any, dismantling of underground tanks and wells was undertaken during site decommissioning. The available data for tar distillation works show that immiscible fluids as coal tars, creosotes and a variety of light to heavy oils are remnant at these sites in the areas of former storage tanks and loading facilities. Similar results are available for the creosote wood preservative plants that show remnant tars and creosotes in the subsurface near former storage and treatment areas.

At many of the investigated coal tar and creosote waste sites, plumes of immiscible fluid contamination extend vertically and laterally from the assumed source locations indicating historic and possibly present day migration of immiscible fluids. These liquids have been observed within sands and gravels and within fractures in clay and sedimentary rocks. At several sites located adjacent to surface water bodies, investigations indicate migration of immiscible fluids to river or lake sediments. The probability of present day migration of immiscible fluids at undisturbed sites appears unlikely due to the many years that have passed since these plants ceased operation and the surface sources were removed. However, if such sites are disturbed through inadvertent excavation or pumping, migration of immiscible fluids is possible. This potential for migration is difficult to assess because of the difficulty in identifying free dense immiscible fluids in the subsurface through conventional augering and soil sampling. Often the immiscible fluids observed in recovered soil samples may be indicative of residual saturations and not of fully saturated conditions (Villaume, et al. 1983). Consequently it is possible to both underestimate and overestimate the volume and migration potential of dense immiscible fluids at these sites.

3.2 *Chemistry*

Coal tar and creosote, which is derived from distillation of coal tar, are complex mixtures of several groups of mutually miscible organic chemicals including PAH, MAH and phenolics or tar acids. The exact composition of any particular coal tar or creosote is a function of the chemistry of the bituminous coal feedstock and the nature of the carbonization, distillation, condensing and purifying operations. For example, typically coal tar from retort coal gas plants contained up to 80% PAH, 15% MAH and 5% phenolics. Water gas tars have similar chemistry but contain less phenolics and less nitrogenous PAH and are less dense than coal gas tars although still heavier than water. Creosotes are chemically similar to but less dense than coal tars and are derived from the lighter fraction of distilled coal tar oils.

PAH are the principal components of coal tars and creosote and are compounds consisting of two or more fused benzene rings. There are hundreds of PAH in coal tar and creosote and only selected EPA priority PAH are routinely analyzed in environmental samples. Six of the heavier EPA priority PAH, including benzo(a)-pyrene, benzo(a)anthracene, benzo(b)-flouranthene, benzo(k)flouranthene, dibenzo(a,h)anthracene and indeno (1,2,3-cd) pyrene are designated carcinogenic by the International Agency for Research on Cancer. In addition to EPA priority PAH, coal tar and creosote contain nitrogenous PAH or azaarenes such as quinoline, carbazole and acridine. Certain of the azaarenes are considered more carcinogenic than the most active PAH (Lee et al. 1981). The major routes of human PAH absorption are through inhalation and cutaneous exposure.

3.3 *Environmental Fate*

Non-Aqueous Phase Liquids

The fate of coal tars, creosote and other dense non-aqueous phase liquids (DNAPLs) in the subsurface is determined by the complex interplay of gravity forces, capillary forces and viscous shear forces. Above the water table three distinct phases (air, water, coal tar) may exist whereas below the water table a two phase model (water, coal tar) is appropriate. Above the water table coal tar or creosote readily sinks to the water table through fractures and intergranular porosity as the immiscible fluid typically has greater affinity for the solid (e.g. is a wetting fluid) than the air. Once below the water table the fate of coal tar or creosote is controlled primarily by the distribution of capillary pressures, P_c, which are defined as (Berg, 1975):

$$P_c = 2\sigma \cos \theta / r_p$$

where σ is the interfacial tension between the immiscible fluid and water, θ is a wettability term describing the contact angle formed by the immiscible fluid against a solid in the presence of water and r_p is a mean effective radius of a pore. For a water wet medium the contact angle is assumed to be zero and $\cos \theta$ equals one. Coal tar will sink through water-filled pores only when the pressure due to gravity and viscous shear frnom groundwater flow exceed the capillary pressure within the pore. Neglecting groundwater flow, the critical height of an immiscible phase necessary to overcome capillary pressure can be determined for values of immiscible fluid density, interfacial tension, contact angle and pore size. For tar density of 1150 kg/m^3 as measured on Ottawa coal tar and a representative range of interfacial tensions of 0.022 - 0.045 N/m (Villaume et al. 1983), the calculated critical height is shown in Figure 2 for a water wet system as a function of different mean grain sizes for porous media and fracture apertures for fractured media. Figure 2 was determined assuming rhombohedral packing of uniform size grains (Berg, 1975) and a vertical rough-walled fracture with effective opening 2b (Kueper and McWhorter, 1989). Figure 2 shows that heights of immiscible fluids of several metres to tens of metres are necessary to allow migration of coal tar into unfractured fine grained deposits of silt and clay. Consequently, these deposits may act as vertical barriers to tar migration. If the silts or clays are fractured with openings of 10-100 μm, critical tar heights of tens of centimetres to a metre are necessary for immiscible fluid entry into such fractures. Figure 2 also shows that tar and creosote will readily migrate in coarse sands and gravels requiring critical heights of only a few centimetres.

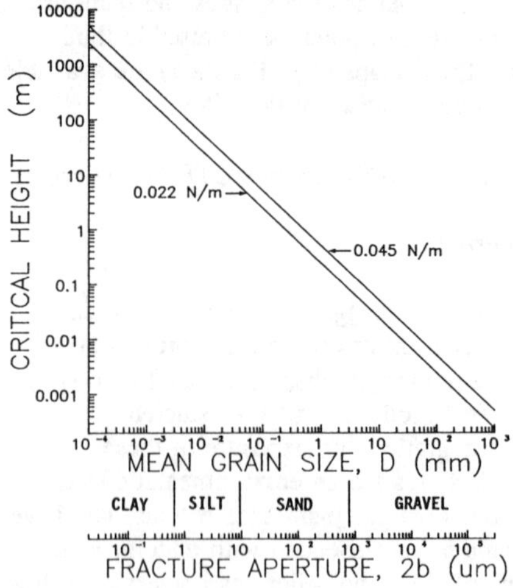

Fig. 2 Critical height of DNAPL necessary to invade pores and fractures

Immiscible fluid migration is also influenced by groundwater flow. As described by Berg (1975), the hydraulic gradient, i, necessary to induce horizontal movement of a horizontal pool of immiscible fluid of length L, is:

$$i = 2\sigma / L r_p$$

This relation shows that larger pools of immiscible fluid are more easily mobilized than smaller pools. This is because of the increased pressures in the immiscible fluid caused by groundwater flow over the fluid. The above relation can be coupled with gravity pressure terms to show the hydraulic gradients necessary to induce movement of immiscible fluids on sloping low permeability surfaces. For hydraulically induced flow of immiscible fluid down a sloping surface, a pressure balance requires that the combined gravity and hydraulic terms exceed the capillary pressure term. For tar properties of 0.022 N/m interfacial tension, and 1150 kg/m^3 density in a coarse sand of mean grain size of 1 mm, the resulting hydraulic gradients as a function of slope angle from the horizontal and pool length are shown in Figure 3. This figure shows that coal tar pools of 3 m or less may readily migrate down slopes as low as 5° - 10° under low hydraulic gradients (i.e. 0.0001 - 0.001).

For up slope hydraulically-induced flow of coal tar to occur it is necessary that the hydraulic

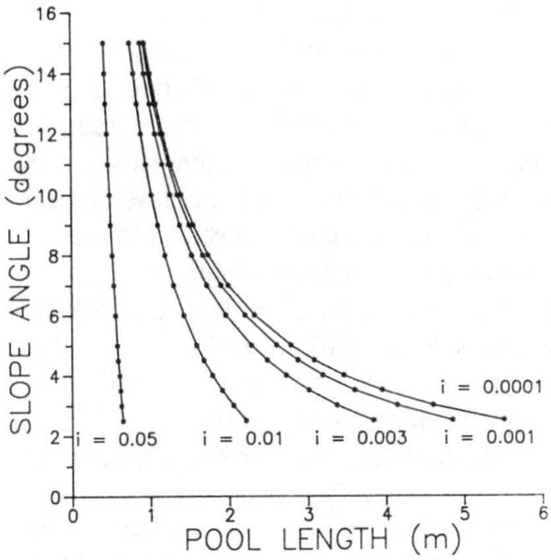

Fig. 3 Hydraulic gradients necessary to induce downslope movement of DNAPL pools

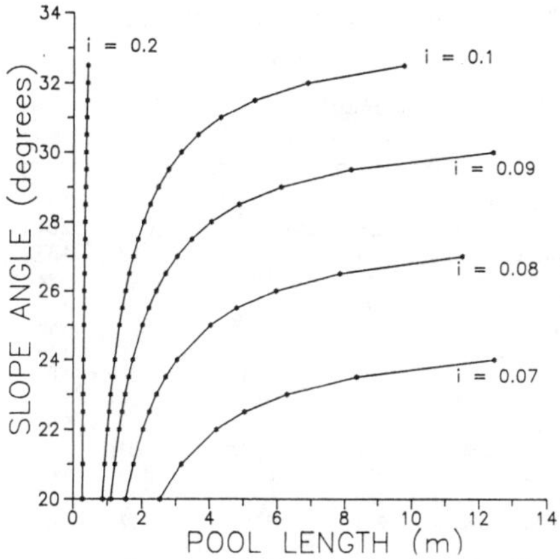

Fig. 4 Hydraulic gradients necessary to induce upslope movement of DNAPL pools

pressure term exceed both the gravity and capillary pressure terms. For the conditions described above, the resulting plot is shown in Figure 4. Figure 4 shows that migration of coal tar up slopes of 20° or less requires hydraulic gradients in excess of 0.07 for tar pool lengths of 3 m or less. Figures similar to 3 and 4 are useful in evaluating potential for horizontal migration of tar under conditions of natural groundwater flow and in assessing the efficacy of pumping systems for recovery of immiscible fluids.

Aqueous Phase Contamination

Immiscible coal tars and creosotes are sparingly soluble in water and, therefore, are long-term sources of aqueous phase contamination. As discussed earlier, coal tars and creosote are composed principally of PAH with smaller quantities of MAH and phenolics. The solubility and mobility of PAH are dependent on the structure of individual compounds. The aqueous solubility and mobility of PAH in groundwater systems decreases as the molecular weight increases. Table 1 summarizes the solubilities and octanol-water partition coefficients for selected PAH. Table 1 lists the solubilities for PAH as solids and as liquids. The solubilities for solids are those solubilities usually quoted in available literature (e.g. MacKay and Shiu (1977)). However, pure PAH are solids at groundwater temperatures and the formation of liquid coal tars consisting of these PAH results in a mutual suppression of the melting points of the individual organic compounds and higher liquid solubilities (Shiu et al. 1988). As shown in Table 1, the liquid solubilities are 3 - 240 times greater than the solid solubilities.

By using an analogy with Raoult's Law, maximum aqueous concentrations of PAH in groundwater can be approximated from the product of the mole or weight fraction of the individual PAH in a coal tar and the liquid solubilities (Shiu, et al. 1988). Because most individual PAH represent between 0.1% and 10% of a particular coal tar or creosote, the maximum aqueous concentration can be expected to be approximately 0.1% to 10% of the respective liquid solubilities. Lower concentrations can occur if high groundwater flow rates exist.

The mobility of hydrophobic organic compounds such as PAH can be approximated using the octanol-water partition coefficient, K_{ow}, and the fraction of organic carbon, f_{oc}, present in soils. The log K_{ow} of selected PAH range from 2.04 to 6.7 (Table 1, EPA (1983)). Carcinogenic PAH have log K_{ow} values greater than 5.0 The fractional organic carbon contents of granular soils in Ontario may range from 0.00017 to 0.00108 (Gillham et al. 1987). Using the correlation developed by Schwarzenbach and Westall (1981), velocity retardation factors, R, of between 14 and 85 can be determined for compounds with log K_{ow} of 5.0 and the quoted range of f_{oc}. Consequently most PAH are relatively immobile as aqueous phase contaminants in granular deposits. Of the EPA priority PAH only naphthalene would be expected to migrate any distance from a PAH source in

Table 1: Solubility and octanol-water partition coefficients for selected PAH

Compound	Molecular Weight	S_S[1] mg/L	S_L[2] mg/L	Log K_{ow}[3]
Naphthalene	128	31.7	111.0	3.37
Quinoline[4]	129	60,000	60,000	2.04
Acenaphthene	154	3.9	19.0	4.33
Fluorene	166	1.9	15.0	4.12
Carbazole[4]	167	--	--	3.29
Anthracene	178	0.073	5.8	4.45
Acridine[4]	179	--	--	3.40
Pyrene	202	0.13	2.2	4.88
Chrysene	228	0.002	0.34	5.61
Benzo(a)anthracene	228	0.014	0.30	5.61
Benzo(b)fluoranthene	252	0.0015	0.038	6.7
Benzo(k)fluoranthene	252	0.0008	0.061	6.6
Benzo(a)pyrene	252	0.0038	0.12	6.04
Dibenzo(a,h)anthracene	278	0.0005	0.12	6.0

1 Solid solubility from MacKay and Shiu (1977) and May et al. (1978).
2 Liquid solubility calculated at 25°C from relation of Shiu et al. (1988).
3 From Verschueren (1983) and EPA (1983)
4 Nitrogenous heterocyclic compound

granular deposits.

Table 1 also shows limited available data for nitrogenous heterocyclic PAH including quinoline, acridine and carbazole, compounds which have been shown to be present in creosote and coal tar and in groundwater at coal gasification, wood preserving and tar distillation facilities (Pereira et al., 1983; Kuhn and Suflita, 1989). Because heterocyclic aromatic hydrocarbons possess lower octanol-water partition coefficients and likely higher solubilities and because they are persistent and toxic, they are potentially more hazardous environmental contaminants than their homocyclic analogues.

Coal tars and creosote also contain small percentages of MAH including benzene, toluene and xylene and phenolics that are more soluble and mobile than PAH. However the potential environmental hazard of these compounds are mitigated by the documented biotransformation of the compounds under both aerobic and anaerobic conditions in aqueous systems.

4. CASE STUDIES

4.1 *Former Manufactured* Gas Plant

This manufactured gas plant operated adjacent to a river from about 1865 to the early 1950s utilizing retort coal gasification and carburetted water gas facilities. At the time of the site investigation the former plant site had been redeveloped to an institutional building with a surrounding paved parking lot. Consequently, some contaminated soil to depth of about 2 m was excavated and removed from the site. Extensive fire insurance plan coverage was available for this site from 1889 that documented the development of the site and the location of several old gas holding tanks and other plant structures. Historical borehole information was also available for this area as a result of redevelopment of the plant site and surrounding properties. This information provided an indication of the subsurface geology and coal tar contamination of soils.

The site plan (Figure 5) shows the location of former gas plant structures, the existing buildings, the exploratory boreholes and monitoring wells and the extent of the non-aqueous and aqueous phase contamination determined from soil and groundwater sampling. Figure 6 is the geologic cross section drawn through the centre of the site parallel to the groundwater flow direction and shows the vertical extent of the non-aqueous plume.

Figures 5 and 6 show an extensive plume of non-aqueous phase contamination that extends from the southern half of the former plant site to the southeast across the street to the adjacent river. Coal tar was also found within the 13.7 m diameter brick gasometer located in the northwest corner of the site. Non-aqueous phase contamination includes visually contaminated soils with residual saturation, dispersed ganglia and pools of coal tar. The distribution of immiscible liquids at this site is controlled by the surface of a dense clay till that slopes at an average angle of 4° from the northwest to the southeast across the site and flattens below the river south of the site. The non-aqueous plume contamination is present at depths of 2 - 8 m within glacial-fluvial coarse sands and gravels and sandy fill that overly the till. Groundwater at the site is unconfined in the sand and gravel and fill and flow occurs from the northwest to the southeast with average hydraulic gradients ranging from 0.04 in the northwest to 0.0006 in the southeast.

The distribution of coal tar shown in Figures 5 and 6 is consistent with the physical principles presented in Figures 2, 3 and 4 in that the coal tar has migrated vertically downward and horizontally through a coarse sand and gravel along a sloping till surface assisted by groundwater flow. Although pools of coal tar exist south of the site adjacent to the river and

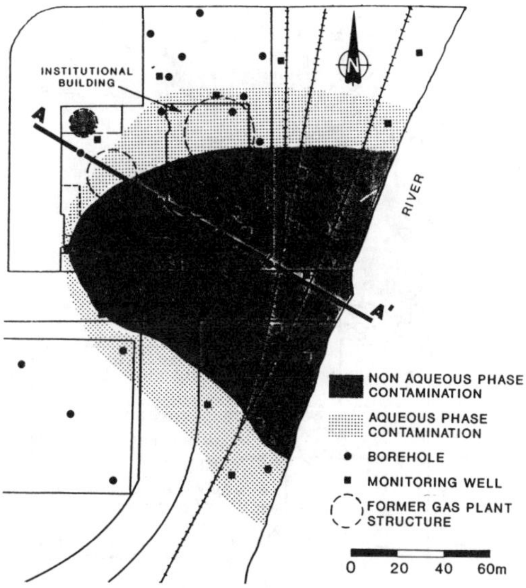

Fig. 5 Plan of former coal gasification plant site

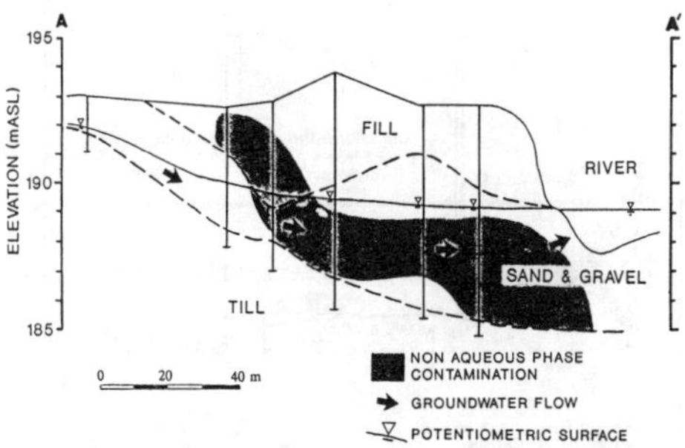

Fig. 6 Geologic cross section of former coal gasification plant site

likely 2 - 3 m below the river bed, upward migration of tar from these pools to the river is unlikely. Preliminary diving inspections and sampling of the river bottom confirm these inferences with no visual evidence of coal tar or PAH present within the shallow river bottom sediments adjacent to and downstream of the site.

Aqueous phase contamination at this site is restricted to the zones of identified separate phase contamination and a 25 m halo surrounding this zone. Within this zone groundwater contains 10s to 100s of ug/L of benzene, toluene, ethylbenzene and xylenes and the lighter PAH - naphthalene, acenaphthlene and fluorene. Low levels ($\leq$ 0.5 ug/L) of some five and six ring PAH were also detected in this zone. Outside this zone within the halo, groundwater contamination is reduced by factors of 10 to 100 for PAH and MAH. Only low concentrations of phenols (<45 ug/L) were detected in groundwater in both zones. Because the zone of non-aqueous phase contamination extends to the river, discharge of PAH and MAH contaminated groundwater to the river will occur at this site for several decades. However because of large flow rate in the river there is presently no measurable degradation of water quality in the river from this site.

4.2 *Former Tar Distillation Plant*

This former tar distillation plant operated from 1922 to 1949 adjacent to a river and used a batch still and several large above ground storage tanks. The facility operated in association with a coal gasification plant located immediately north of the site. The property was used for storage of tar products from 1949 and the facilities were dismantled in 1969. A 1.37 m diameter storm sewer transects the west half of the site and drains to the river. A rectangular high rise apartment building with three levels of underground parking was constructed over most of the former plant structures in 1973. An adjacent apartment building and garage was constructed immediately east of this building in 1980. Because of the parking garages, most of the original soil and coal tar contamination from the plant were excavated and removed from the site. However, in 1986 tar seepages into the lower most parking garage prompted a detailed site investigation to define the extent of contamination and the potential for migration of contamination to the River via the storm sewer line and the natural soils.

Historical airphotos and a 1956 fire insurance plan provided a description of site facilities and drum storage areas at the site. However, the area of these plant structures were excavated and a reconnaissance soil gas survey using a field portable gas chromatograph was completed around the existing building to obtain a preliminary indication of organic contamination. Because of the predominance of clay fill at the site this survey was only partly successful and indicated elevated levels of benzene at one location west of the centre of the apartment building.

A subsequent test pit excavation, drilling and sampling program was completed at the site. The site plan (Figure 7) shows the location of former plant structures, existing buildings, boreholes, monitoring wells and the extent of the non-aqueous and aqueous phase contamination as determined from soil and groundwater sampling. Figure 8 shows a generalized geologic cross section which is oriented parallel to the groundwater flow direction and shows the full vertical and lateral extent of the non-aqueous phase plume.

Figure 7 shows an elongate plume of non-aqueous phase contamination that likely extended from the area of the storage tanks located at the north end of the site to the south toward the river. With excavation and construction of the apartment building, migration of immiscible liquid occurred beside and below the building in two distinct lobes. Most of the non-aqueous phase plume is contained within a sand fill unit located at 55-57 m elevation as shown in Figure 8. This earlier sand fill unit is hydrogeologically equivalent to and continuous with the recent granular fill underlying the existing apartment buildings. A low permeability silty clay underlies these granular deposits and controls the vertical migration of immiscible fluid at the site. A sand and clay fill consisting of ash, gravel and landfill debris overlies the sand fill unit. Groundwater flow at the site is within the sand fill from the southwest to the northeast toward the deeper tile drains of the 1980 building. The average hydraulic gradient is 0.02. Because the clay surface is flat or slopes slightly upward (1-2°) in the direction of groundwater flow, there is potential for migration of tar underneath the building. However, the limited observed migration below the building (Figure 8) suggests that there is insufficient tar present in the sand fill to form large enough pools for migration to occur. Similarly, although there was pooling of a small quantity of tar in the coarse sand bedding material underlying the storm sewer, there is insufficient volume of tar present to overcome residual saturations and cause migration of tar to the River.

Aqueous phase contamination at the site is characterized principally by elevated levels of naphthalene (up to 7500 ug/L) and benzene (up to 8400 ug/L) and other PAH and MAH at lower concentrations. The aqueous phase PAH concentrations (i.e. naphthalene - Figure 7) are greatest in the area of the non-aqueous phase contamination and rapidly decrease in the direction of groundwater flow. Figure 7 indicates movement of naphthalene of about 10 m from the non aqueous phase plume. Assuming this aqueous movement occurred over a 15 year period since the building was constructed, an average velocity for naphthalene of 0.0018 m/day is determined. This velocity compares to an interstitial groundwater velocity of 0.057 m/day determined using Darcy's Law and indicates a retardation factor, R, of 31. For the range of f_{oc} quoted earlier and the log K_{ow} for naphthalene from Table 1, the predicted R value using the Schwarzenbach and Westall correlation would be 2 to 7. This suggests either higher f_{oc} contents in

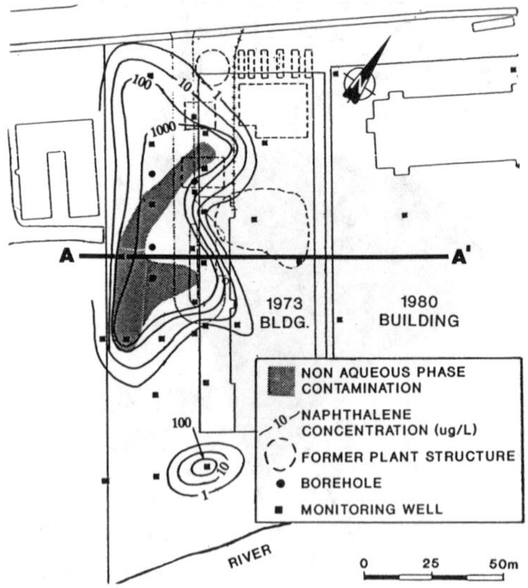

Fig. 7 Plan of former tar distillation plant site

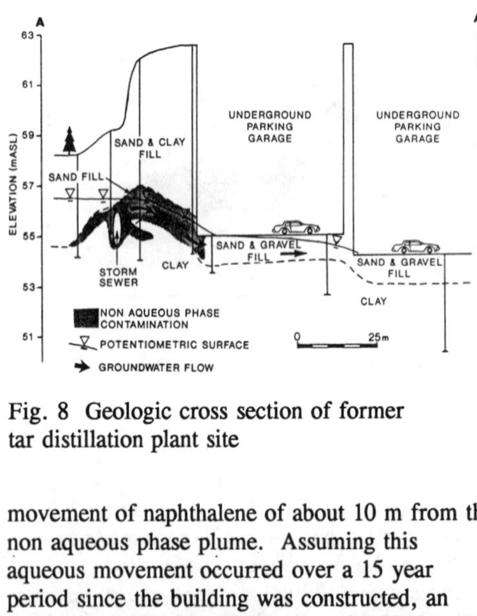

Fig. 8 Geologic cross section of former tar distillation plant site

the sand fill or other processes than sorption may be attenuating naphthalene at the site (ie. volatilization or biotransformation).

A sample of tar collected at this site contained EPA priority PAH at weight fractions of a few percent including: naphthalene 3.1%, fluorene 1.4%, chrysene 0.32%, benzo(a)pyrene of 0.31% and dibenzo(a,h)anthracene 0.035%. Assuming their weight fractions approximate mole fractions, the tar composition data can be used to calculate the maximum PAH concentrations of groundwater in contact with the tar (Shiu et al., 1988). The resulting calculated maximum PAH concentrations and measured maximum PAH concentrations in groundwater at the site are within a factor of two (eg. naphthalene 3600 (calculated) vs 7500 (measured) ug/L, fluorene 214 vs 160 ug/L, chrysene 1.1 vs 0.67 ug/L, benzo(a)pyrene 0.38 vs 0.27 ug/L and dibenzo(a,h) anthracene 0.04 vs <0.1 ug/L). The similarity between the calculated and measured PAH concentrations supports this approach to calculate maximum concentrations of PAH in groundwater if samples of the immiscible fluid can be collected and analyzed. The approach can also be used to determine whether immiscible fluids are present when elevated groundwater concentrations are measured and the weight fractions of components of immiscible fluids can be approximated.

5. SUMMARY

Available information from site investigations conducted at coal tar and creosote sites in Ontario indicate that all identified sites will likely contain dense immiscible fluids consisting of polycyclic and monocyclic aromatic hydrocarbons. These immiscible fluids may be present as residual saturations, dispersed ganglia or as pools resting on low permeability surfaces. Migration of immiscible fluids downward and to a lesser extent horizontally has been observed within sands and gravels and fractured clays and sedimentary rock. The migration of these dense immiscible fluids is in accordance with theoretical predictions based on two phase flow principles.

Localized groundwater contamination of PAH and MAH compounds occurs at the coal tar and creosote sites surveyed in this paper. The migration of dissolved PAH is significantly retarded in agreement with the hydrophobic nature of the compounds. Although groundwater contamination is localized, the presence of immiscible fluids of low solubility means that these sites may continue to be sources of groundwater contamination for decades. The maximum concentration of PAH in groundwater at these sites may be predicted using the measured or approximated weight fractions and liquid solubilities of PAH in coal tar or creosote.

The principal environmental concerns at coal tar and creosote sites relate to migration and exposure of the immiscible liquids. Environmental impacts at these sites have occurred through disturbance and exposure of immiscible fluids by excavation and pumping and by subsurface migration to adjacent surface waters. At many of the older sites, it is unlikely that migration of immiscible fluid is ongoing today due to the many years that have passed since the immiscible fluid sources were removed.

REFERENCES

Berg, R.R. 1975. Capillary pressures in stratigraphic traps. *American Association Petroleum Geologists Bull* 59(6): 939-965.

Environmental Protection Agency (EPA) 1983. *Treatability Manual. Volume 1. Treatability Data.* Office of Research and Development, USEPA, Washington, D.C., 20460. February.

Gillham, R.W., S.F. O'Hannesin, C.J. Ptacek & J.F. Barker 1987. *Evaluation of small scale retardation tests for BTX in groundwater.* PACE Report No. 87-3, September. Petroleum Association for Conservation of the Canadian Environment, Ottawa.

Intera Technologies Ltd. 1988. *Inventory of industrial sites producing or using coal tar and related tars in Ontario, Volumes I and II.* Report prepared for Ontario Ministry of the Environment, Toronto, Ontario.

Intera Technologies Ltd. 1987. *Inventory of coal gasification plant waste sites in Ontario, Volumes I and II.* Report prepared for Ontario Ministry of the Environment, Toronto, Ontario.

Kueper, B.H. & D.B. McWhorter 1989. The behaviour of dense, non-aqueous phase liquids (DNAPLs) in fractured media, *Proc. Focus Conference on Eastern Regional Groundwater Issues:* 423-434, October 17-19, Kitchener, Ontario, National Water Well Association, Dublin Ohio.

Kuhn, E.P. & J.M. Suflita 1989. Microbial degradation of nitrogen, oxygen and sulphur heterocyclic compounds under anaerobic conditions: studies with aquifer samples. *Environmental Toxicology and Chemistry* 8: 1149-1158.

Lee, M.L., M.V. Novotny & K.D. Bartie 1981. *Analytical chemistry of polycyclic aromatic compounds*, Academic Press, New York.

Mackay, D. & W.Y. Shiu 1977. Aqueous solubility of polynuclear aromatic hydrocarbons. *J. Chem. Eng. Data* 22(4): 309-402.

May, W.E., S.P. Wasik & D.H. Freeman 1978. Determination of the solubility behaviour of some polycyclic aromatic hydrocarbons in water. *Analytical Chemistry* 50: 997-1000.

Pereira, W.E. C.E. Rostad, J.R. Garbarino & M.F. Huit 1983. Groundwater contamination by organic bases derived from coal-tar wastes. *Environmental Toxicology and Chemistry* 2: 283-294.

Schwarzenbach, R.P. & J. Westall 1981. Transport of non-polar organic compounds from surface water to groundwater. Laboratory sorption studies, *Environ. Sci. Technol.* 15: 1300-1367.

Shiu, W.Y., A. Maijanen, A.L.Y. Ng & D. Mackay 1988. Preparation of aqueous solutions of sparingly soluble organic substances: II. multicomponent systems - hydrocarbon mixtures and petroleum products. *Environmental Toxicology and Chemistry* 7: 125-137.

Verschueren, K. 1983. *Handbook of environmental data on organic chemicals.* 2nd Ed., Van Nostrand/Reinhold Co., New York.

Villaume, J.F., P.C. Lowe & D.F. Unites 1983. Recovery of coal gasification wastes: an innovative approach. *Proc. Third National Symp. on Aquifer Restoration and Ground Water Monitoring:* 434-445. National Water Well Association, Worthington, Ohio.

Subsurface Contamination by Immiscible Fluids, Weyer (ed.) © 1992 Balkema, Rotterdam. ISBN 90 6191 175 3

Occurrences of coal tar and contaminated groundwater at three sites in Ontario

A.R. Lotimer, D.W. Belanger & R.B. Whiffin
CH2M Hill Engineering Ltd, Waterloo, Ont., Canada

ABSTRACT: This paper discusses the occurrence of coal tar and contaminated groundwater at three sites in Ontario selected to illustrate migration of coal tar related contaminants in various hydrogeologic environments. The case study sites include a fluvial sand aquifer, glacial till, and fractured limestone bedrock. In each case, coal tar was remnant of a former coal gasification plant that had been abandoned for over thirty years.

In the sand aquifer and glacial till, coal tar was found onsite near the original points of disposal. In the fractured limestone, coal tar was found to have migrated downgradient towards a nearby surface water body. Groundwater contamination with polycyclic aromatic hydrocarbons (PAH), light aromatic hydrocarbons (BTEX), and phenolics is essentially limited to the immediate vicinity of coal tar at all three sites. The combined effects of low contaminant solubility, adsorption, volatilization, and natural biodegradation are inferred to account for the limited migration of these contaminants in groundwater at the three sites.

1 INTRODUCTION

It is generally recognized that coal tar originating from former coal gasification plants and coke ovens can impact the environment through migration of immiscible fluid and the release of soluble contaminants to groundwater, surface water, or air (ERT, 1984). Given that the coal gasification process was a primary generator of manufactured gas from the mid 1800's through to the 1950's with at least 150 recorded gas plants in Canada (Carleton University, 1987), and over 1000 in the United States (ERT, 1984), coal tar is potentially a significant and widespread contamination problem in North America.

In Ontario, former coal gasification plant sites are being investigated to evaluate the presence and impacts of remnant coal tar and associated wastes in a province wide program initiated by the Ontario Ministry of the Environment. The authors have been directly involved in detailed hydrogeological investigations at nine of these sites. Based on this experience, three sites have been selected to illustrate observed contaminant migration patterns in different hydrogeologic environments and show that these patterns are generally consistent with basic migration theory for coal tar.

2 COAL TAR AND ASSOCIATED CONTAMINANTS

Several waste types can be encountered at former coal gasification plant sites including tars, sludges, liquors, spent iron oxides, ashes, slag, cinders, off-grade coal, and coke (ERT, 1984). Polycyclic aromatic hydrocarbons (PAH) are the principal chemical constituents forming coal tars although additional chemical constituents include light aromatics (benzene, toluene, xylene, and ethylbenzene or BTEX), phenolics, inorganic nitrogen, inorganic sulphur, and trace metals (ERT, 1984). Of these groups, PAH have received the most attention due to the fact that several of these compounds including benzo(a)-pyrene, benzo(a)-anthracene, benzo (b+k)fluoranthene, dibenzo(a,h)-anthracene, and indeno(1,2,3,-cd)pyrene are reportedly carcinogenic (CCME, 1989). Coal tar is a dense non-aqueous phase liquid (DNAPL). Physical properties include densities in the range of 1.12 to 1.45 mg/L, viscosities as high as 4368 cp (10°C), and surface tension at approximately 43 dynes/cm (24°C)(Intera, 1987a). Light non-aqueous phase components (LNAPL) associated with coal tar or the coal gasification process can also exist at former plant sites.

3 CONTAMINANT MIGRATION

Contaminant migration originating from coal tar can be considered as a four phase problem: an aqueous phase, a DNAPL, a LNAPL, and a vapour phase.

The aqueous phase component consists of dissolved contaminants released to groundwater from both DNAPL and LNAPL. Migration of the dissolved contaminants is controlled by solubility, adsorption, biodegradation, volatilization and matrix diffusion.

Solubility of the PAH, light aromatics (BTEX), and phenolic compounds is dependent upon molecular weight and structure. In general, solubility decreases with increasing molecular weight and structural complexity within each group of compounds. However, variations in solubility (cosolvency effects) can occur when more than one compound is present. Adsorption occurs when compounds adhere to organic material and mineral surfaces in the geologic deposits. The degree of partitioning between the dissolved phase and the adsorbed phase of hydrophobic compounds can be related to the octanol-water partitioning coefficient and the fraction of organic carbon present in the geologic material (Schwarzenbach and Westall, 1981). This relationship shows that the degree of retardation by adsorption increases with increasing octanol-water partitioning coefficient and increasing fraction of organic carbon (Whiffin and Belanger, 1989). Adsorption to mineral surfaces can also be significant, particularly in material with a high clay content and low fraction of organic carbon.

Biodegradation of the lower molecular weight PAH, light aromatics, and phenolics occurs quite readily through the action of enzymes produced by bacteria, fungi, and algae. The biodegradation process involves breakdown of an organic compound to carbon dioxide through one or more intermediate steps or daughter products. Appropriate microbial populations and environmental conditions (i.e. redox, pH, temperature, nutrient availability) are required and deficiency in one or more of these factors will limit the rate of biodegradation. Reported biodegradation half-lives for the light PAH range from 9 to 15 days (NRC, 1983). The rate of volatilization is also dependent upon environmental conditions, and physical and chemical properties of a compound (vapour pressure, Henry's Constant). Compounds with low vapour pressures are more readily volatilized, and reported volatization half-lives for the light PAH are less than 100 hours (NRC, 1983). Matrix diffusion can also be expected to act as an important attenuation mechanism in fractured rock or clay geological environments.

The lower molecular weight compounds such as BTEX, the phenolics, and the lighter PAH (i.e. naphthalene) have higher solubilities and moderate to low octanol-water partitioning coefficients (K_{ow}) (Verschueren, 1983; Coover and Sims, 1987; and Versar, 1979), both of which favour greater mobility of these compounds in groundwater flow systems. On the other hand, this same mobility may also enhance biodegradation by facilitating more contact with suitable microbial populations and environmental conditions (Ehrlich et al, 1982).

The higher molecular weight compounds which include the carcinogenic PAH have very low solubilities and high octanol-water partitioning coefficients. As a result, these parameters are usually considered relatively immobile due to their low solubility and affinity for adsorption. However, other factors such as cosolvency effects and competition for sorption sites can complicate this simplified interpretation.

Coal tar migration as a DNAPL is governed by the interaction between capillary pressures, gravitational forces, and viscous forces (Villaume et al, 1983). Under hydrostatic conditions, coal tar migration will only occur when the coal tar mass is sufficient to overcome capillary pressures (Villaume, 1985). Whiffin and Belanger (1989) use Hobson's Formula to show that under hydrostatic conditions, the height or thickness of coal tar required for flow varies from less than 1 mm in gravel to 10's of mm in sand to over a metre in silt and clay. Clearly, migration of coal tar as a separate phase will be negligible in unfractured clay or silt, and quite significant in coarse sands and gravels or fractured rock. In situations where the coal tar mass is greater than pressures exerted by groundwater flow, it is possible for migration to occur in a direction different from the local groundwater flow. This might occur when coal tar has penetrated to a low permeability stratigraphic unit sloping in a direction opposite to the prevailing groundwater flow.

Lighter immiscible fluids (i.e. light oils) associated with coal tar or coal gasification can migrate as a separate phase on the water table, generally following the water table gradient.

Vapour phase migration is also important for the light aromatic compounds (BTEX) and lighter PAH (i.e. naphthalene) associated with coal tar.

4 APPROACH TO SITE INVESTIGATIONS

The overall approach to investigations undertaken at former coal gasification plant sites focused on identifying the presence of coal tar and assessing contamination of groundwater and surface water. The usual first step was to assemble available historical information which included fire insurance maps and old air photographs. This information is important in showing the layout of the former coal gasification plant in view of the fact that most of these plants were abandoned over 40 years ago and may have changed configuration several times over their history. The historical information was very useful in showing the locations of gas holders and other structures which are likely sources of residual coal tar.

Available geologic, hydrogeologic and engineering information was also evaluated to anticipate subsurface conditions which might have a significant control on migration of DNAPL or other contaminants. This information was used in planning the drilling program to intercept the most likely zones of contamination (i.e. accumulations of coal tar in the vicinity of gas holders).

Drilling and soil sampling through hollow stem augers and rock coring were used to identify the presence of coal tar and associated contamination. Soil/rock samples were described geologically and classified as either visually contaminated with coal tar, trace contaminated based on odour, or non-contaminated. Selected samples from each of these classifications were submitted for analysis of PAH. In addition, a sample of pure coal tar or heavily contaminated soil was also submitted for analysis of PAH.

Monitor wells were installed in selected boreholes to obtain groundwater level data and samples at both on-site and offsite locations. Groundwater samples were collected for analysis of PAH, BTEX, and phenolics in order to assess the presence and migration of these coal tar related contaminants in the aqueous phase.

In addition, water and sediment samples were collected from adjacent storm sewers to evaluate the possible migration of contaminants to the storm sewer network through groundwater seepage. Water and bottom sediment samples were collected at nearby surface water sources to assess the impacts of coal tar related contamination from either groundwater seepage or storm sewer discharge.

5 BEHAVIOUR IN DIFFERENT HYDROGEOLOGICAL ENVIRONMENTS

Results of separate studies at former coal gasification plants in Ontario (Figure 1) overlying a sand aquifer

(Owen Sound), a clay-silt till (St. Thomas), and Palaeozoic limestone bedrock (Kingston) are presented below. As the intent of this paper is primarily hydrogeological in nature, these case histories focus on coal tar migration and groundwater contamination with little or no mention of the sewer components of each study.

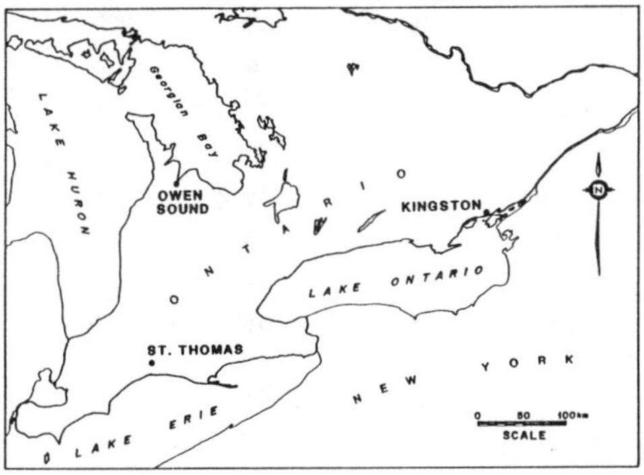

Figure 1: Regional location of study sites

Fluvial sand aquifer

The former coal gasification plant in Owen Sound, Ontario operated initially as a retort coal gas plant and later a coke oven plant from 1888 to 1947 (Intera, 1987b). The site overlies a fine to medium grained sand aquifer of suspected fluvial or beach origin and is located within 50 metres of the Sydenham River (Figure 2). The subsurface deposits were identified by borehole drilling to depths ranging between 5 and 15 metres beneath the property. Coal tar waste was identified near the former gas holders at depths ranging from 2 to 5 metres. No evidence of LNAPL was detected in the boreholes. Monitor wells were installed downgradient near the river, in the vicinity of onsite coal tar and upgradient of the coal tar wastes. Water levels recorded in the wells demonstrate that groundwater flow occurs across the site towards the Sydenham River under lateral hydraulic gradients of approximately 0.005. Average groundwater velocities are on the order of 0.04 m/day based on an average measured hydraulic conductivity of 2.1 m/day and an assumed porosity of 0.3. Calculated advective travel time from the edge of the coal tar wastes to the river is approximately 3 years.

Groundwater analyses are summarized in Tables 1 (PAH), 2 (BTEX), and 3 (phenolics). All three sets of analyses show that groundwater contamination associated with the coal tar waste is only significant in the immediate vicinity of the waste. The PAH are present in the onsite monitor well MW#5 at relatively high concentrations. The highest concentrations are measured for the lower molecular weight PAH such as naphthalene. The higher molecular weight PAH including the reported carcinogenic compounds such as benzo(a)pyrene are present at much lower concentrations, which can be attributed to their lower solubility and greater potential for adsorption. No PAH were detected in an upgradient monitor well (MW#1) and only trace levels of three of the lighter more mobile PAH were detected in a well (MW#4) approximately 38 metres downgradient of the coal tar.

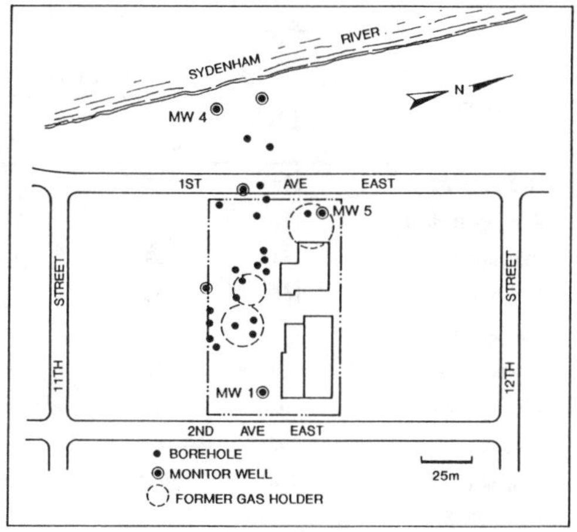

Figure 2: Owen Sound site plan

All BTEX and phenolic components are present at significant levels in monitor well MW#5 onsite, which is consistent with their moderate to high solubilities. Essentially no BTEX components were detected in either the upgradient or downgradient monitor wells. Given the groundwater velocity and travel time calculated above, these results demonstrate that attenuation processes such as biodegradation, adsorption and volatilization of these components are significant. The presence of 5 phenolic compounds at trace levels in downgradient well MW#4 suggests that attenuation mechanisms described above for BTEX also appear to exert an influence on migration of phenolics, but to a lesser degree.

Clay silt till

The St. Thomas coal gasification plant (Figure 3) was reportedly in operation from 1877 to 1935, first as a retort gas facility and later a water gas facility (Intera, 1987b). The site is located over 1 km southeast of Kettle Creek which drains to Lake Erie.

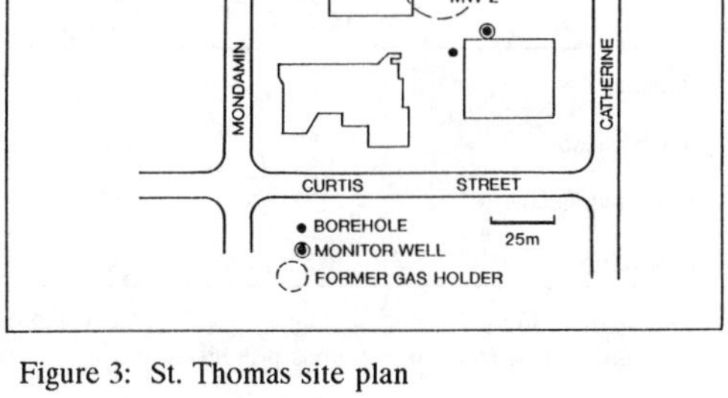

Figure 3: St. Thomas site plan

Table 1: PAH levels in groundwater

Compound	Molecular[1] Weight	Fluvial Sand (Owen Sound)		Clay-Silt Till (St. Thomas)		Limestone (Kingston)	
		Onsite (MW#5)	Downgradient (MW#4)	Onsite (MW#3)	Downgradient (MW #5)	Onsite (MW3A)	Downgradient (MW 6A)
Naphthalene	128	999	<	881	<	210	1100
Acenaphthylene	152	25	1.5	5	<	260	2500
Acenaphthene	154	64	8.7	17	<	16	100
Fluorene	166	32	0.31	12	<	48	370
Phenanthrene	178	28	<	18	<	78	1000
Anthracene	178	10	<	5	<	19	290
Fluoranthene	202	5	<	4	<	4.5	160
Pyrene	202	4.7	<	3	<	5.8	280
Benzo(a) anthracene	228	0.25	<	0.36	<	0.6	110
Chrysene	228	0.28	<	0.30	<	0.5	75
Benzo(b+k) fluoranthene	252	0.21	<	0.14	<	0.2	66
Benzo(a) pyrene	252	0.08	<	<	<	0.2	65
Indeno(1,2,3-cd) pyrene	276	<	<	<	<	<	26
Dibenzo(a,h) anthracene	278	<	<	<	<	<	11

* all analyses in ppb   < below detection limit   [1] from ERT (1984)

Table 2: BTEX levels in groundwater

Compound	Fluvial Sand (Owen Sound)		Clay-Silt Till (St. Thomas)		Limestone (Kingston)	
	Onsite (MW#5)	Downgradient (MW#4)	Onsite (MW#3)	Downgradient (MW #5)	Onsite (MW3A)	Downgradient (MW 6A)
Benzene	263	<	1408	<	65000	26000
Toluene	309	<0.05	375	0.9	24000	22000
m & p-Xylene	536	<0.04	35	<	4000	5700
Ethylbenzene	909	<	34	<	2900	10000
o-Xylene	2715	<	60	<	1700	2100

* all analyses in ppb   < below detection limit

Table 3: Phenolics levels in groundwater

Compound	Fluvial Sand (Owen Sound)		Clay-Silt Till (St. Thomas)		Limestone (Kingston)	
	Onsite (MW#5)	Downgradient (MW#4)	Onsite (MW#3)	Downgradient (MW #5)	Onsite (MW3A)	Downgradient (MW 6A)
Phenol	5	<	26	0.2	211+	102+
o,m,p Cresol	29	0.6	10	0.5	NA	NA
Dimethylphenols	129	0.5	75	0.1	NA	NA
Resorcinol	0.84	<	<	<	NA	NA

* all analyses in ppb   < below detection limit   NA not analyzed
\+ assumes all phenolic compounds present as phenol

The site is underlain by fill (clay-silt, sand, cinders/ash) which ranges in thickness from less than 1 metre to over 6 metres in the vicinity of the former gas holder locations. The native geological deposits consist of discontinuous glaciolacustrine clay-silt and clay-silt till (Port Stanley Till). Vertical fractures were observed in the native clay-silt deposits. A former ravine extending northwest of the site for several hundred metres had been filled with material which reportedly included some waste from the coal gas plant. Prior to the investigation there was some concern that coal tar was present in the buried ravine or that the fill material present in the ravine was acting as a natural pathway for the migration of coal tar and contaminated groundwater.

The drilling results show that residual coal tar contamination is essentially confined to an area in the vicinity of the former gas holders. Depths of the coal tar range from about 2 to 5 m below surface. No LNAPL was encountered in any of the boreholes. Only a very trace level of coal tar contamination (based on odour) was detected in a small interval at one location in the buried ravine. The ravine was found to contain granular fill and reworked clay-silt till to depths ranging from 6 to 12 metres in the areas tested.

Groundwater flow conditions in both the shallow fill and native till/lacustrine deposits are somewhat uncertain due to thick unsaturated sections in the buried ravine and the slow response times of monitor wells. Several of the monitor wells encountered either unsaturated or perched water table conditions. Groundwater elevation data, however, supports the interpretation of groundwater flow to the northwest.

The groundwater results for this site are also summarized in Tables 1 (PAH), 2 (BTEX), and 3 (phenolics) adjacent to the data for Owen Sound.

Elevated levels of all three groups of chemical compounds appear to be limited to groundwater in the immediate vicinity of coal tar encountered onsite (MW#3). PAH levels are similar to those reported for the Owen Sound site, although at somewhat lower concentrations. The lighter PAH compounds are present at much higher levels which is consistent with the Owen Sound data and the solubility/sorption information discussed above. No PAH were detected in wells located on the upgradient edge of the coal tar (MW#2) or downgradient in the buried ravine (MW#5).

All BTEX compounds were detected in well MW#3 adjacent to the coal tar wastes, and essentially absent in the upgradient and downgradient monitor wells.

The phenolics data is somewhat complicated by the presence of several compounds in the trip and lab blanks at levels similar to those reported for the upgradient and downgradient monitor wells. However, the data clearly show that phenolics contamination is significant adjacent to the coal tar (MW#3).

Limestone

The former Kingston coal gasification plant (Figure 4), located adjacent to Lake Ontario and Anglin Bay (Cataraquai River) was reportedly in operation from 1848 to the early 1950's and was one of the oldest and longest operating gas plants in Ontario (Intera, 1987b).

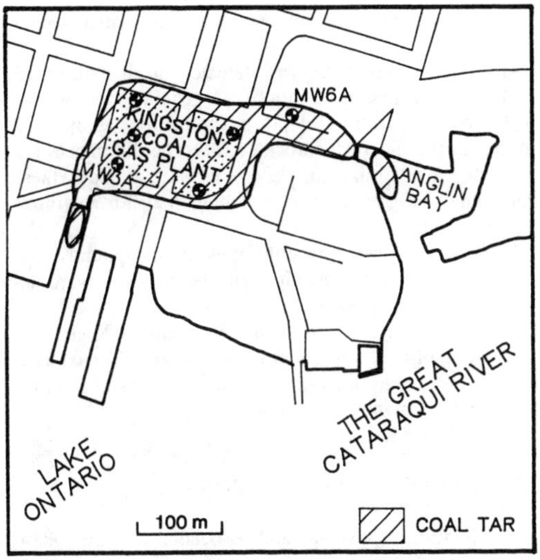

Figure 4: Kingston site plan

The site overlies a thin layer of fill which covers Palaeozoic limestone bedrock of the Gull River Formation. The fill material is variable, consisting of clay, sand, gravel, cinders, ash, coal, asphalt and other materials. The Gull River is typically a grey, finely crystalline limestone with thin shale seams and styolites. The bedrock surface slopes towards the northeast and east, generally in the direction of Lake Ontario and the Cataraquai River. The bedrock is fractured and contains open bedding planes. The bedding planes generally occur along shale seams at spacings of about 0.6 metres and locally dip towards the east. The major joint sets are all vertical with strikes of 220°, 278°, and 313° and spacing of about 1 to 3 metres.

Coal tar was found in five boreholes drilled onsite, one borehole offsite, and in sediments in Anglin Bay and Lake Ontario. Onsite the coal tar was primarily found in the bedrock but also in the surface fill in the area of the gas holding tanks. In the bedrock, the coal tar was often found coating the bedding planes or as globules attached to the surface of the fractures. Free coal tar did not appear to have penetrated the rock matrix although cores had a distinct coal tar odour. From these observations, it is interpreted that the free coal tar has migrated by gravity flow along the bedding planes and fractures while aqueous phase migration has occurred in the bedding planes, fractures, and rock matrix.

All boreholes and monitoring wells encountered free coal tar including the offsite well (MW#6A). The offsite well is located immediately down slope from the main gas holding tank which was found to be a major source of coal tar. An adjacent sanitary sewer excavated into bedrock is thought to collect contaminated groundwater and coal tar and direct it to the area of MW#6A.

The groundwater analyses (Tables 1,2,3) show that PAH levels are high and for MW#6A exceed the solubility limit for compounds such as benzo(a) pyrene and benzo(a)anthracene indicating either the presence of a separate phase in the sample or cosolvency effects. As at other sites, the lower molecular weight PAH compounds are found in higher concentrations than the higher molecular weight PAH.

All BTEX compounds were detected in all monitoring wells at very high concentrations. A general trend was for higher concentrations at depth in the bedrock as opposed to that in the shallow bedrock or fill. The differences are likely due to volatilization. In groundwater at depth, these compounds have little or no contact with air for volatilization to occur.

Total phenols were the only phenolic compounds analyzed at this site. Relatively high phenol levels were detected due to the presence of free coal tar.

PAH were also detected in the sediments of Anglin Bay and in one of the harbours in Lake Ontario. The concentrations were quite high on the order of 100's of ppm for some of the carcinogenic compounds.

6 SUMMARY AND CONCLUSIONS

The migration of coal tar and associated aqueous phase contaminants in groundwater has been a key consideration in the investigation of former coal gasification plants in Ontario.

The major chemical groups of environmental concern are the polycyclic aromatic hydrocarbons (PAH), light aromatic hydrocarbons (BTEX) and phenolics.

Case studies of plants overlying a fluvial sand aquifer in Owen Sound and a clay-silt till in St. Thomas show that coal tar migration has remained within the original site boundaries. A similar study of a former gas plant overlying limestone bedrock in Kingston shows that free phase coal tar (DNAPL) occurs across the site and has migrated a considerable distance downgradient through fractures. Other factors being equal, it would appear that for the three geological environments examined, coal tar migration occurs most readily in fractured limestone, followed in order by fine to medium sand, and clay-silt till.

Groundwater analyses at the Owen Sound and St. Thomas sites indicate that groundwater contaminated with PAH, BTEX and phenolics is essentially limited to the immediate vicinity of coal tar identified onsite. No PAH were detected in downgradient monitor wells at the St. Thomas site and only trace levels of the lower molecular weight compounds were detected in the underlying sand aquifer downgradient of the Owen Sound coal tar wastes. At the Owen Sound site, significant attenuation of all three groups of organic chemicals by processes including biodegradation, adsorption, and volatilization is inferred through general comparison of observed contaminant levels with calculated groundwater velocities. At St. Thomas, the absence of any offsite migration is attributed mainly to the low permeability of the underlying clay-silt till. Elevated levels of PAH and very high levels of BTEX in groundwater downgradient of the Kingston site is correlated at least in part with the presence of free phase coal tar which has also migrated downgradient. The results of all three studies suggest that lateral migration in groundwater of PAH, BTEX, and phenolics originating from residual coal tar is very limited due to solubility and attenuation mechanisms which operate in all hydrogeological settings.

REFERENCES

Canadian Council of Ministers of the Environment (CCME). 1989. Interim Guidelines for PAH Contamination at Abandoned Coal Tar Sites. Queen's Printer for Ontario Pub. No. CCME-TS/WM-TRE004.

Carleton University, Resources Development Research Centre. 1987. National Overview of Abandoned Coal Gasification Works in Canada. Environment Canada.

Coover, M.P. and R.C. Sims 1987. The Effect of Temperature on Polycyclic Aromatic Hydrocarbon Persistence in an Unacclimated Agricultural Soil. *Hazardous Waste and Hazardous Materials* 4, (1): 69-81.

Ehrlich, G.G., D.F. Goerlitz, E.M. Godsy, and M.F. Hutt 1982. Degradation of Phenolic Contaminants in Ground Water by Anaerobic Bacteria: St. Louis Park, Minnesota. *Ground Water* 20 (6): 703-710.

Environmental Research and Technology (ERT). 1984. Handbook on Manufactured Gas Plant Sites. Utility Solid Waste Activities Group. Superfund Committee, Washington, D.C.

Intera Technologies Ltd. 1987a. Lees Avenue Hydrogeologic Study. Volume 1. Ontario Ministry of the Environment.

Intera Technologies Ltd. 1987b. Inventory of Coal Gasification Plant Waste Sites in Ontario. Volume 1. Ontario Ministry of the Environment.

National Research Council Canada (NRC). 1983. Polycyclic Aromatic Hydrocarbons in the Aquatic Environment: Formation, Sources, Fate and Effects on Aquatic Biota. NRCC No. 18981.

Schwartzenbach, R.P. and J. Westall 1981. Transport of Non Polar Organic Compounds from Surface Water to Groundwater. *Environ. Sci. Tech.* 15 (11): 1360-1367.

Versar, Inc. 1979. Water-Related Environmental Fate of 129 Priority Pollutants. Vol. 1 and 2. U.S. EPA. Washington, D.C. EPA 440/4-79-029.

Verschueren, K. 1983. *Handbook of Environmental Data on Organic Chemicals*. 2nd Edition. New York: Van Nostrand/Reinhold Co.

Villaume, J.F., P.C. Lowe and D.F. Unites 1983. Recovery of coal gasification wastes: an innovative approach. *Proc. 3rd Nat. Symp. on Aquifer Restoration and Groundwater Monitoring:* 434-445, Ohio: NWWA.

Villaume, J.F. 1985. Investigations at sites contaminated with dense, non-aqueous phase liquids (NAPLS). *Groundwater Monitoring Review* 5 (2): 60-74.

Whiffin, R.B. and D.W. Belanger 1989. Impacts of Former Coal Gasification Plants in a Number of Hydrogeologic Environments. *Proc. NWWA Focus Conf. on East. Reg. Groundwater Issues*: 137 - 150, Kitchener.

Subsurface Contamination by Immiscible Fluids, Weyer (ed.) © 1992 Balkema, Rotterdam. ISBN 90 6191 175 3

Hydrogeological investigation of coal tar contamination at a former coal gasification facility

E.H.Gerencher & D.E.Konasewich
Envirochem Special Projects, Inc., North Vancouver, B.C., Canada
K.A.Morin
Morwijk Enterprises Ltd, Vancouver, B.C., Canada

ABSTRACT: This paper presents the results of the first phase of an environmental investigation to characterize subsurface contamination and hydrogeologic patterns at a historical coal gas manufacture property in Victoria, B.C.

An archival search was conducted before initiation of field activities to provide documentation of operational practices at the former coal gas facility. This information was used to guide the environmental investigation. Site investigations consisted of surface water and sediment sampling of the harbour, drilling of boreholes and sampling of cuttings, installation of piezometers and groundwater sampling.

The production of manufactured gas from coal produced many waste products including byproduct tars and oils; spent oxide waste containing lime, iron oxides, cyanides and sulfur; and ash. Byproduct tars and oils are immiscible fluids, consequently their groundwater flow patterns are more complex than soluble contaminants. Major stratigraphic zones and groundwater elevations were mapped to enable assessment of likely subsurface contaminant pathways. Results to date have shown that the coal tar is migrating downslope along the top of an impermeable silt-clay layer toward the harbour. Existing subsurface structures will obviously influence both the presence and migration behavior of coal tar residues. The complexity of subsurface migration of immiscible fluids rendered the use of borehole information inadequate for the assessment of contaminant pathways. A geophysical survey has been proposed to provide a more continuous description of subsurface conditions.

1. INTRODUCTION

From 1862 to 1952, coal gas was manufactured by the Victoria Gas Company near the waterfront in Victoria, B.C.. The production of manufactured gas from coal produced many waste products including "coal tar" residues and sludges (which contain polycyclic aromatic hydrocarbons and other hydrocarbon compounds), spent oxide and lime waste, and ash. Although coal tar from the site was sold for various uses, such as creosote and road tars, significant quantities were known to have been stored and/or disposed of on the property. At least some portion of the spent oxide which adsorbed sulfur and nitrogen impurities from the manufactured gas was disposed on-site. Likewise, some portion of ash and coke from the burnt coal was disposed on-site.

The production of manufactured gas was discontinued in the 1950's and the facility was converted to a storage and distribution center for propane gas. A public utility operated the propane distribution facility until 1988, at which time the operation was sold to a private company. Concern regarding company liability resulting from historical contamination prompted the utility to initiate an environmental investigation of the property prior to sale of the facility. Envirochem Services Special Projects Inc. was retained to conduct the environmental investigation. After preliminary results indicated significant contamination at the site resulting from historical industrial practices, the utility elected to sell the operation as planned but retained ownership of the property. Continued ownership would provide the utility a greater degree of flexibility and control in handling remedial procedures. This document presents the results of the first phase of an environmental investigation to characterize hydrogeologic patterns and subsurface contamination by coal gas manufacture wastes at the former Victoria Gas Company property.

2. SITE HISTORY

Following the incorporation of the Victoria Gas Company in 1860, construction of the Victoria Gas Works was completed in 1862. The operation provided coal gas as an energy source for fuel and light in the city of Victoria. An archival search was initiated to provide documentation of operational activities at the property. Documents dating back to 1862 provided operational information for the facility, including the following excerpts:

1909 Reference is made to an increased number of retorts (42), and three gas holders (2 old ones each holding 24,000 cu. ft. and one of more recent date with a capacity of 216,000 cu. ft.) Two small tanks were used to accommodate tar but the bulk of the tar was stored in the tanks of the small gas holder. "The liquor at present runs to waste." Reference is made to a contract for the whole output of tar which "will improve matters as far as tar is concerned". The amount of coal gas sold in FY1907-1908 was 33,865,500 cu. ft.

1919 The carbonizing plant now consisted of 58 retorts, of which 8 were "in reserve". The amount of coal gas sold in FY1918-1919 was 81,910,800 cu. ft.

1928 Use of the horizontal retort facility (now consisting of 60 stop ended horizontal retorts), was discarded in favor of a new vertical retort operation, due to the advent of electricity and a consequent change in gas sales from a volumetric to a calorific basis.

The operation included:

- a condensing plant, consisting of horizontal pipe condenser attached to north side of new retort house, and six annular vertical condensers 4' diameter x 25' high with reversible valves
- one ammoniacal liquor scrubber 30' x 5' and one same sized scrubber for washing the ammonia from the gas
- four elevated boxes filled with native oxide of iron (obtained from Squamish, P.G.E. Rd.) for removal of sulphur, with a hydraulic cover lifting ram, with overhaul moveable carriage, all contained in a two story brick building. Also three circular steel purifiers situated in the yard.

1934 Reference is made to the "diversion of water gas tar from condensers to well".

1939 Coal gas volume: 173,252,300 cu. ft. in FY1938-1939

1946 A new water gas plant was added to prepare water gas from diesel oil.

1949 Plans developed to install 16-30,000 gal. capacity propane storage tanks to "supplement plant capacity and provide backup for the aging manufactured gas plants".

1952 Shutdown of vertical retort operation with total conversion to a storage and distribution center for propane gas. The gas was brought in by barge from the mainland.

3. METHODOLOGY

3.1 Identification of Contaminant Sources

The objectives of this study were to identify the severity and extent of vertical and horizontal contamination by coal gas manufacture wastes and to evaluate groundwater migration patterns at the site. The locations of potential sources of contamination such as horizontal and vertical coal gas retorts, gasometers where coal tar was reportedly stored, and oil storage tanks, were determined by means of CAD generated overlay maps. The overlays were based on original drawings prepared in 1911, 1953 and 1977. Using a common reference point and scale six primary areas were identified where contamination most likely originated (Figure 1):

1) oil tank storage area
2) vertical retort house area
3) gasometer I and coal tar storage area
4) gasometer II and coal tar storage area
5) gas purification area
6) horizontal retort house area

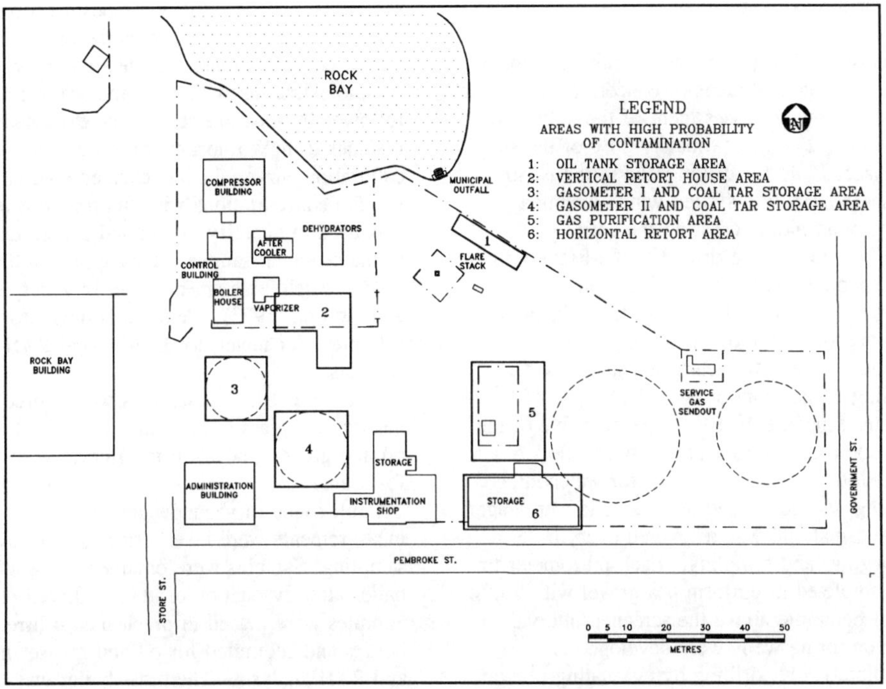

Figure 1. Victoria Gas Plant – Areas with high probability of contamination.

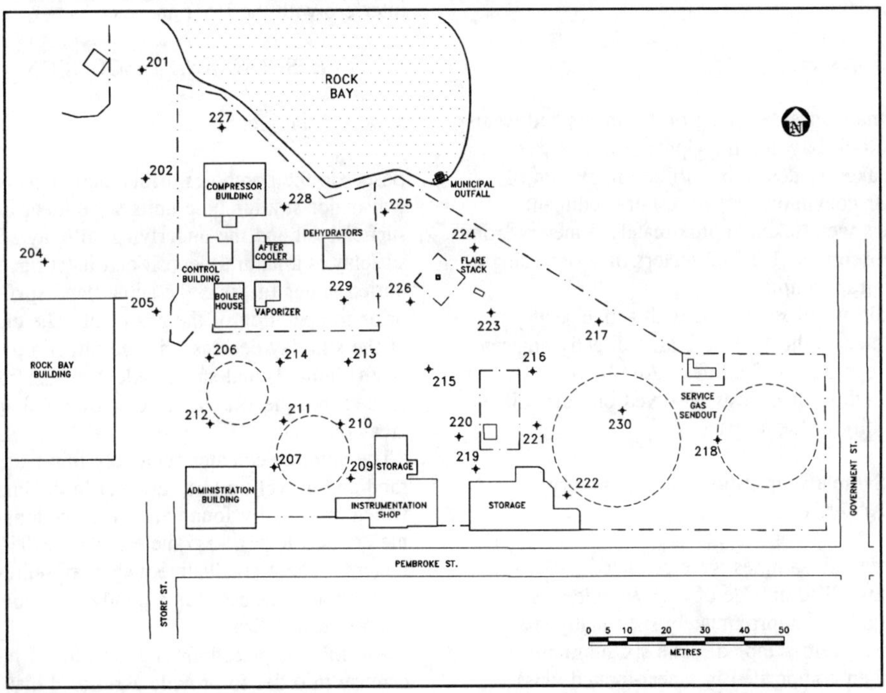

Figure 2. Locations of boreholes.

3.2 Drilling Investigation

A grid of 30 drill holes was developed to focus on the six potential areas of concern as well as to provide an overall assessment of the industrial site (Figure 2). Approximately 20% of the site was inaccessible for drilling due to the existing propane distribution system which included underground piping crossing the entire yard.

The boreholes were drilled to the first confining layer encountered at an average depth of 7.6 meters. Monitor wells were installed in twenty-four boreholes to allow measurement of hydrogeological parameters and groundwater sampling. The monitoring wells consist of 5-cm-diameter schedule 40 PVC pipe with five of the wells completed with stainless steel pipe. A 1.5 meter slotted section was used for water intake.

Drilling was conducted using a solid stem auger. Augered material was recovered every 0.75 m for logging, and sampling. Each piezometer tip was completed in uniform pea gravel with 0.6 to 1 m of bentonite above the screened interval. The monitoring wells were developed immediately after drilling by evacuating several well volumes of water to remove fine grained material from the screened section. Local values of hydraulic conductivity were determined in several wells using single-well-recovery ("slug") tests.

3.3 Foreshore

Reconnaissance sampling of sediments and water from Rock Bay and the Upper Harbour was undertaken to determine off-site migration of coal tar contaminants. Foreshore sediment samples were taken approximately 6 meters from the shoreline in 1.2 to 2 meters of water using a ponar grab sampler.

An old wood stave pipe with active seepage was noted in the foreshore area directly adjacent to the operational facilities. An obvious hydrocarbon sheen was observed on the sediment in vicinity of the seepage.

3.4 Sample Preparation and Laboratory Analysis

Augered soil samples were collected in 0.75 m intervals. The outside of each sample was discarded and approximately six equally spaced locations were sampled. The six subsamples were then composited in a heat-treated glass sample jar. All sediment composites were submitted for an initial screening test for oil and grease concentrations. The test consisted of sample extraction with a hexane-acetone mixture followed by measurement of the extracted residues using a fluorometric procedure. A calibration curve which correlated coal-tar residue concentrations with fluorometric readings was used to quantify the oil and grease residues in the sediment sample. This approach has been used for field assessment of oil in water (Keizer and Gordon, 1973). Selected composites were submitted for metal, total sulfur and PAH analysis.

Foreshore sediment samples were stored in heat-treated glass sample jars and submitted for oil and grease analysis using fluorescence, and PAH analysis.

A full round of groundwater level measurements were taken prior to groundwater sampling. Samples were obtained using a brass bailer after evacuation of at least 3 well volumes. Samples were placed in precleaned 4 litre glass bottles and submitted for oil and grease, metals and PAH analysis. The brass bailer was thoroughly cleaned with hexane and acetone prior to sampling to avoid cross contamination. The solvents were selected because of their cleaning effectiveness and because they would not interfere with the PAH analysis.

4. RESULTS AND DISCUSSION

4.1 Stratigraphy

Based on the boreholes drilled across the site, two major stratigraphic units were identified: the surficial fill and the underlying silt/clay. The silt/clay is a natural deposit and its surface, now buried under fill, was the actual land surface prior to operation of the gas plant. The elevation of the silt/clay decreases in the direction of Rock Bay (Figure 3) and small valleys appear to have eroded into the silt/clay (see arrows and shaded areas on Figure 3).

The silt/clay contains relatively thin beds of sand and gravel up to a few feet in thickness, which are usually found around elevations 0 meters (sea level), -1.5 meters, and -3.05 meters. These conductive beds may represent preferential channels for groundwater movement within the silt/clay.

The fill was placed around the turn of the century in order to provide additional dry surface

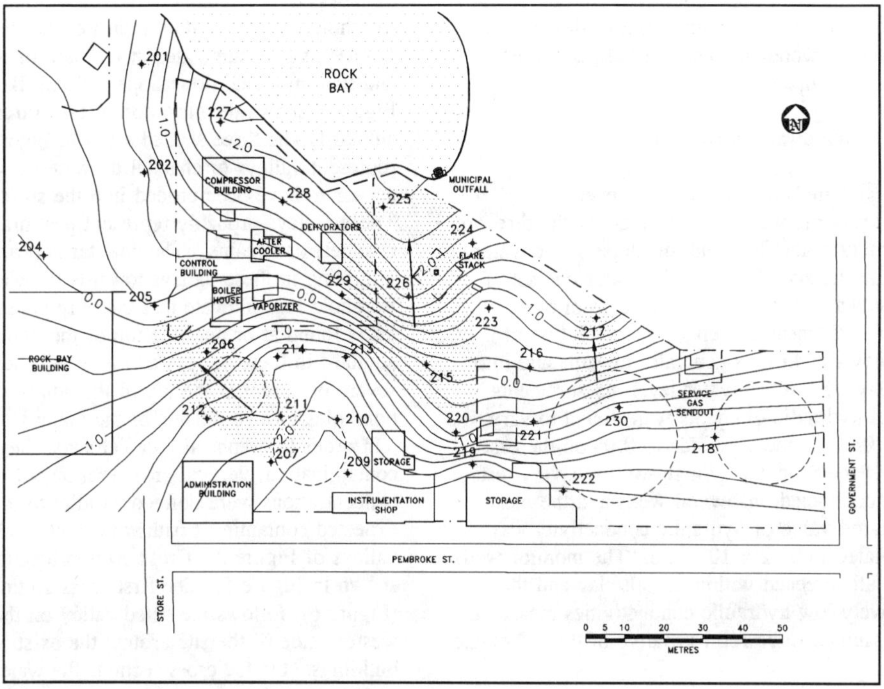

Figure 3. Elevation of upper aquitard surface (metres).

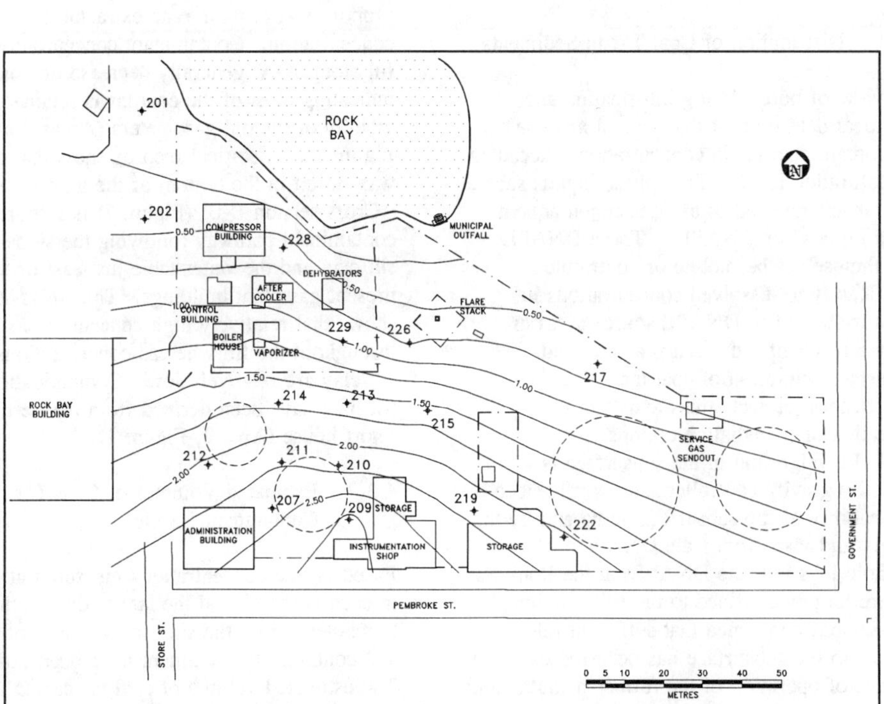

Figure 4. Water-table contours.

for the gas plant operation. The fill consists of random mixtures of natural sandy silt textured sediments, wood fragments, brick, coal dust, coke and clinkers.

4.2 Groundwater Movement

Several rounds of groundwater level measurements were taken to identify the direction of groundwater flow and the slope of the water table at the site. Figure 4 illustrates one set of groundwater contours calculated from these water level measurements. The direction of groundwater flow is generally towards and into the Bay. Based on water level measurements lateral hydraulic gradients were found to vary from 0.02 on the west side to 0.05 on the east side of the property. Single-well-recovery tests were conducted on several wells and the mean measured value of hydraulic conductivity was calculated to be 2×10^{-7} m/s. The monitor wells were all screened within the silt/clay and the relatively low hydraulic conductivities measured are assumed to be representative of this lithologic unit.

4.3 Coal Tar Contaminant Occurrence and Migration

4.3.1 Distribution of Coal Tar in Sediments

A review of borehole log information and analytical data suggest that several areas at the site contain coal tar in concentrations exceeding soil saturation levels. Free phase liquids such as coal tar are referred to as "dense non aqueous phase liquids" or DNAPLs. These DNAPLs may themselves be mobile or contribute significantly to dissolved concentrations in groundwater. The DNAPL source areas are likely a result of industrial practices and subsurface migration of coal tar. The identification of such areas assists in the evaluation of remediation options.

DNAPL migration in the subsurface is primarily gravity controlled and is influenced by permeability of the subsurface strata and by the density and viscosity of the DNAPL. The borehole logs indicate DNAPLs at the Rock Bay site are largely confined to the fill. Sediment analyses has confirmed that sufficient release of coal tar to the subsurface has occurred over the decades of operation of the former manufactured gas plant to have contaminated the entire thickness of the relatively thin unsaturated zone. Borehole log information indicates that the DNAPL has accumulated at the base of the fill and then migrated downslope to Rock Bay. Figure 3 shows the elevation of the silt/clay aquitard. The three shaded areas in Figure 3 represent valleys or channel depressions which appear to have been eroded into the silt/clay. These valleys probably represent past drainage channels or streams. The coal tar appears to migrating in these valleys towards the waterfront. Borehole logs indicate that the pure-phase coal tar is found at or near the top of the uppermost aquitard in these valleys and is usually less severe or absent at the top of the aquitard in boreholes located outside of the valleys.

In order to graphically depict subsurface contamination, vertical cross-sections of coal tar concentrations were defined parallel to the expected contaminant pathways, that is, the filled valleys of Figure 3. Cross section locations are shown in Figure 5. The first cross-section, A-A' (Figure 6), follows the filled valley on the western side of the site around the existing buildings. On the cross-section, the water table and silt/clay decrease in elevation towards Rock Bay (Borehole #202) whereas the surface of the fill rises. The size of a data point (the "X") is proportional to its hexane extractables concentration. Contaminant concentrations in the fill along A-A' generally decrease downslope indicating most of the coal tar is retained in the vicinity of the old gasometers (Areas 3 and 4 in Figure 1). Historical records show that coal tar was stored in the bottom of the gasometers.

Cross-section B-B' (Figure 7) is a short contaminant pathway following the slope of the silt/clay and the water table just east of the present gas-plant buildings. This cross-section shows that relatively high concentrations exist throughout the fill whereas concentrations in the silt/clay are relatively low. Contamination in the fill may have been derived from the vertical retort house (Area 2, Figure 1).

4.3.2 Estimated Volumes of Coal Tar Contaminated Soils

Based on the concentrations measured at intervals in each borehole and the lateral distribution of boreholes across the site, the volumes of coal tar and contaminated sediment have been estimated. The estimated volume of coal tar currently in the subsurface is approximately 330,000 litres, which

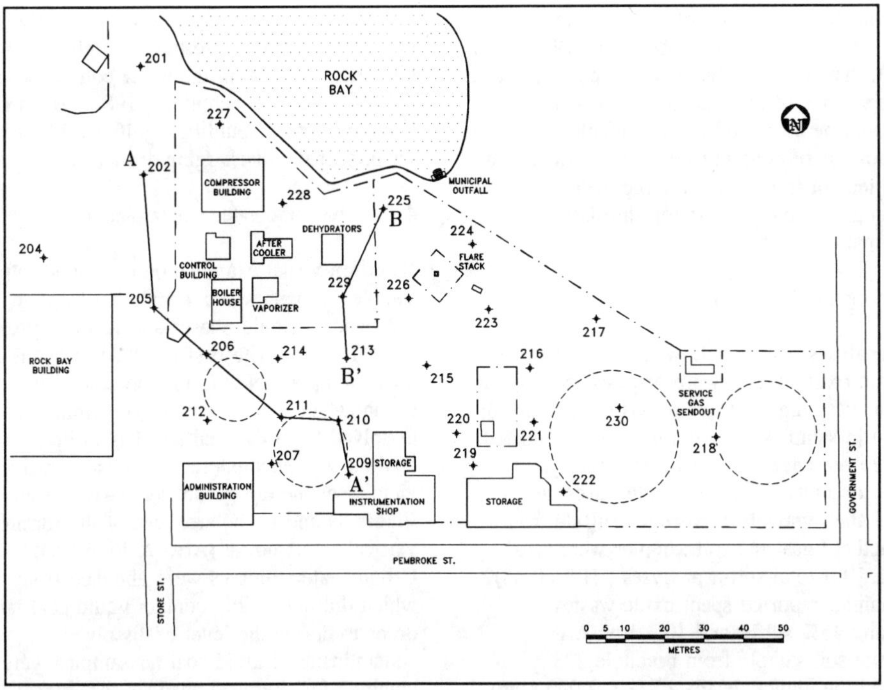

Figure 5. Cross-section locations – contaminant distribution.

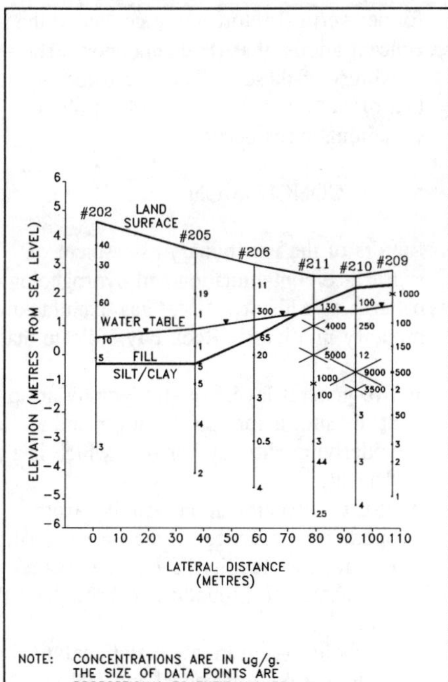

Figure 6. Cross-section A–A'.

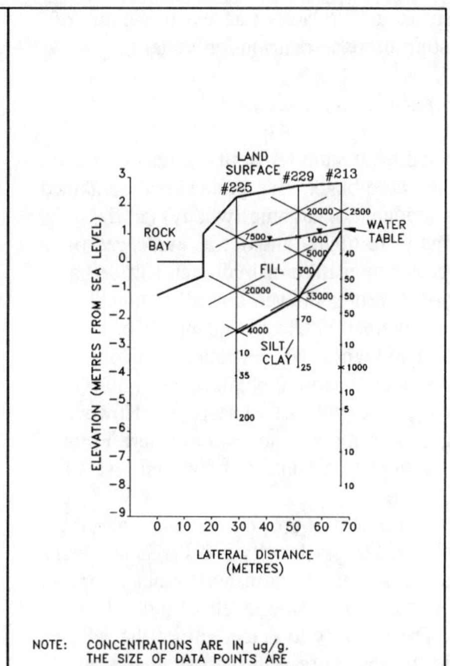

Figure 7. Cross-section B–B'.

is equivalent to about 3/4 of a year's production of coal tar in later years of operation (1930's and 1940's). The original quantity in the subsurface and the volume of coal tar that has reached Rock Bay cannot be estimated from available data.

The volume of contaminated fill is estimated at several tens of thousands of cubic meters, depending on the concentration threshold of contamination.

4.4 Inorganic Residues

Metal analyses of borehole sediments suggested that spent oxide wastes were disposed on-site. In addition, very high concentrations of calcium (up to 54%) in samples from borehole 229 indicated the likely presence of lime wastes from gas purification units. Because oxide boxes were used for the removal of hydrogen sulfide from manufactured gas, the soil samples were also submitted for total sulfur analyses. Hill (1945), for example, reported spent oxide wastes containing 48% sulfur and 10% iron. A composite soil sample from borehole 228 contained the highest level, 4.9% (49,000 ppm) total sulfur confirming the use of this area for disposal of spent oxide wastes. Detailed delineation of sulfur distribution between boreholes 228 and 229 cannot be made at this time because of the lack of access in vicinity of the existing propane distribution system.

4.5 Foreshore Contamination

Bottom sediment samples on the foreshore of Rock Bay were black, fine-grained and contained varying amounts of organic (woody) debris. The sediments were highly anoxic, as evidenced by a strong odour indicative of hydrogen sulfide gas.

Analytical results showed that all sediment samples contained detectable quantities of PAH's. The highest concentration was found in a beach sediment sample obtained from an old wood stave pipe with active seepage. However, a background sample collected elsewhere in the Upper Harbour also contained elevated levels of PAH.

Although oil sheens are, on occasion, observed in vicinity of the gas facility and PAHs are found in foreshore sediments, foreshore water samples did not contain detectable levels of individual PAH's. The inability to detect any of the selected PAHs in the foreshore waters to an analytical detection limit of 0.001 mg/L despite the discharge of contaminated groundwater suggests that dilution at the foreshore is high and biological impacts in the water column would be minimal (Konasewich et al., 1982). However, benthic impacts could be significant in the vicinity of the discharging groundwater.

4.6 Groundwater Contamination

Based on visual examination of groundwater samples in bailers and sample bottles, two independent liquid phases appear to be present in several wells. The first is a black viscous coal tar forming a DNAPL and the second is groundwater which contains contaminants dissolved from the coal tar phase. Free-phase coal tar was encountered in wells located in vicinity of the old gasometers (Areas 3 and 4, Figure 1) and in the area east of the former vertical retort house (Area 2, Figure 1). Groundwater samples were obtained from wells which did not visibly contain liquid coal tar in order to define the level of dissolved contaminants. Liquid coal tar samples were also obtained for chemical analysis. Analytical results showed that all samples contained detectible concentrations of PAH compounds. Groundwater samples obtained from wells in vicinity of the old gasometers and east of the former vertical retort house contained the highest concentrations of PAH compounds. The proximity of these wells to the foreshore suggests that offsite migration of dissolved phase contamination is occurring.

5. CONCLUSION

Results of the preliminary assessment of subsurface contamination and hydrogeologic patterns at a historical coal gas manufacture property in Victoria Rock Bay indicate that:

- two general layers of stratigraphy are present at the site: a surficial fill layer and an underlying silt/clay horizon, which is a natural deposit;
- coal tar contamination resulting from operational practices at the former coal gas manufacture facility have contaminated sediments and groundwater throughout the site;
- borehole log information and analytical data show that both dissolved and DNAPL coal tar contamination are present at the site;

- the site is extensively contaminated with an approximate volume of 330,000 litres of coal tar;
- There is evidence for PAH migration from the site to Rock Bay on the basis of PAH analyses of sediments and groundwater. However, foreshore water samples from Rock Bay did not contain detectable levels of specific PAHs. As a result, biological impacts are probably minimal in the water column, but may be more significant in the benthic zone.

The proposed second phase of this investigation will begin with a surface geophysical survey to provide a more continuous survey of subsurface conditions, specifically the contact of the fill and silt/clay. Based on this information and the study results to date, additional monitoring wells will be installed in order to define groundwater movement through the fill into Rock Bay and to delineate the configuration of the DNAPL plume.

Concern regarding the migration of dissolved coal tar derived contaminants from the manufactured gas plant property to Rock Bay has prompted the requirement for a remediation plan to control subsurface migration of contaminants to Rock Bay. A pump and treat system is the most viable alternative for such remediation. Extraction wells will be installed and modelling and design of a pump and treat system will be undertaken following completion of the proposed second phase of the investigation.

6. REFERENCES

B.C. Ministry of Environment. 1989. British Columbia Standards for Managing Contamination at the Pacific Place Site. Ministry of Environment, Waste Management Program, Victoria, B.C.

Hill, W.H. 1945. Recovery of Ammonia, Cyanogen, Pyridine and Other Nitrogenous Compounds from Industrial Gases. In H.H. Lowry (ed), Chemistry of Coal Utilization. New York: John Wiley and Sons.

Keizer, P.D. and D.C. Gordon. 1973. Detection of trace amounts of oil in sea water by fluorescence spectroscopy. J. Fish. Res. Bd. Can. 30: 1039-1046.

Konasewich, D.E., P.M. Chapman, E. Gerencher, G. Vigers and N. Treloar. 1982. Effects, Pathways, Processes and Transformation of Puget Sound Contaminants of Concern. National Oceanic and Atmospheric Administration, Office of Marine Pollution Assessment, Boulder, Colorado, 357 pp.

Subsurface Contamination by Immiscible Fluids, Weyer (ed.) © 1992 Balkema, Rotterdam. ISBN 90 6191 175 3

The use of existing, pumped boreholes to investigate solvent pollution of a deep sandstone aquifer

David N. Lerner & John W. Lloyd
Hydrogeology Research Group, University of Birmingham, UK
Michael O. Rivett
Hydrogeology Research Group, University of Birmingham, UK (Presently: Waterloo Centre for Groundwater Research, Ont., Canada)

ABSTRACT: The major water supply aquifers of the UK have deep, consolidated lithologies. Existing boreholes, often constructed many decades ago and used for water supply, must often be used for sampling groundwater, with several attendant difficulties. The first regional survey of chlorinated solvent pollution in a UK aquifer was undertaken using 59 existing boreholes, mostly within industrial premises. Separate sampling protocols had to be developed for each. Results were affected by pumping rate differences between boreholes, changing patterns of pumping within each borehole and migration of pollution within the aquifer. It was not possible to discover any vertical variations of concentration, nor to discover how widespread pollution was outside industrial premises. Nevertheless, it was possible to see that solvent pollution was widespread, and to correlate concentrations with landuse and hydrogeological conditions, and to discern trends at some sites.

1 THE PROBLEM

Chlorinated solvents are common pollutants of groundwater in the UK (Rivett et al., 1990a); in fact they are probably the most widespread organic groundwater pollutants.

The two major aquifers of the UK are the Chalk and the Permo-Triassic sandstones which provide 57% and 18% of the nation's groundwater supply. Both aquifers are relatively thick. For example the Permo-Triassic sandstone aquifer under Birmingham is 80-150 m in its unconfined zone, with groundwater between 2 and 50 m below ground (Figure 1). Drilling investigation boreholes is therefore expensive. In addition, the financial and legal climates are such that detailed investigations of groundwater pollution are rare, whether for research, regulatory investigation or compensation

In general British hydrogeologists must make do with existing boreholes. What can they interpret from samples collected from boreholes

- with long open sections
- continuous or variable pumping
- often unknown construction details
- in often non-ideal locations?

These issues are discussed in relation to a recently completed survey of organic contamination of the Birmingham aquifer, which sought to identify the effects of urban development on groundwater (Rivett et al., 1990c).

This paper is confined to the results on chlorinated solvents. These pollutants were so widespread that it was possible to make some interpretation of the regional patterns of pollution, to identify some of the land use and hydrogeological controls, and to investigate trends.

2 BACKGROUND TO BIRMINGHAM STUDY

2.1 Hydrogeology

The Triassic Sandstone sequence may be

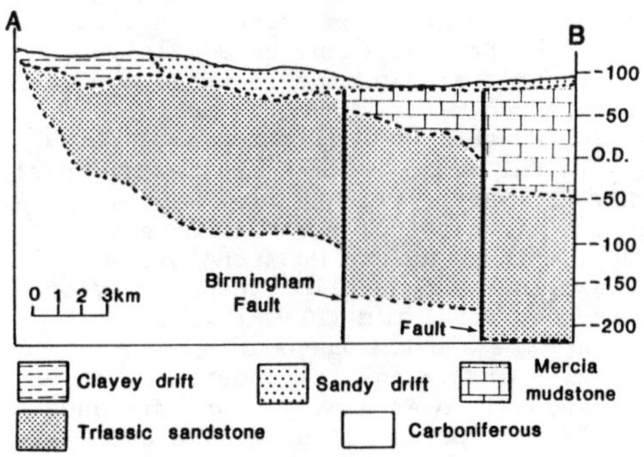

Fig. 1 Cross section of the Birmingham aquifer (section line shown on Figure 2)

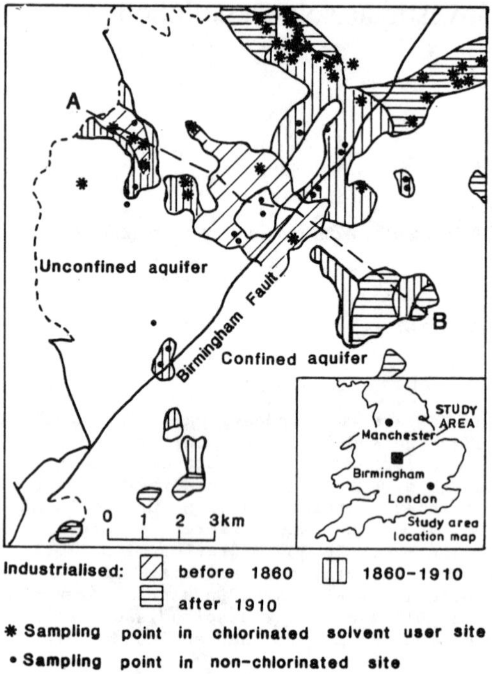

Fig. 2 The Birmingham study area showing the area of unconfined aquifer, industrial zones and sampling points

regarded as a single aquifer, and is unconfined over the greater part of the study area (Figures 1 and 2). The sandstones dip to the south-east and are confined by the Mercia Mudstone Group to the east of the Birmingham Fault. Carboniferous strata underlie the Triassic aquifer and are generally assumed to be the effective aquifer base. The sandstone thickness can reach 200 m but is usually in the range 80-150 m. Mercia Mudstone thicknesses range from 20 m to in excess of 100 m.

The Triassic Sandstones show lithological characteristics ranging from fine grained sandstone with occasional mudstone intercalations to a massive, medium to coarse grained sandstone with interbedded marls. Pumping tests indicated that hydraulic conductivities are in the range 1-8 m/d (10^{-5} - 10^{-6} m/s). (Rivett 1989). Fissure flow may be important, particularly near boreholes. The confining Mercia Mudstone Group usually has a hydraulic conductivity of less than 0.001 m/d (10^{-8} m/s).

The type and thickness of glacial drift deposits overlying the sandstones varies tremendously over a very short distance. They are comprised of glacial clays, sands, gravels, alluvium and peat. The thickest (up to 40 m) and least permeable drift deposits are on the west of the unconfined aquifer (Horton, 1975).

2.2 Land use

The entire outcrop of the unconfined aquifer is now urbanised; about 70% was urbanised by 1900. The industrial zones are shown in Figure 2, and in 1975, comprised about 2200 individual premises on the unconfined aquifer. About 60% of industrial land is occupied by metal-working industries.

2.3 Sampling and analysis

Water samples were collected during 1986-88 from 59 supply boreholes (Figure 2). They ranged from 20 to 300 m deep, with most between 80 and 150 m. All the boreholes were in industrial or commercial premises as the aquifer has not been used for public supply since the 1920's. Separate sampling protocols were developed for each borehole, as discussed below in Section 3.

Samples were analysed for volatile chlorinated solvents by pentane extraction, followed by gas chromatography with electron capture detection (Eklund et al., 1978).

3 SAMPLING PROTOCOLS

The usual objective of groundwater sampling is to obtain a sample representative of the water in the aquifer, that is excluding the water that has been standing in the borehole. A popular protocol is to pump from the borehole and sample when the discharge has reached stable values of pH, conductivity and temperature (Barcelona et al., 1985). Unfortunately low concentrations of non-ionic organics will not influence these easily measured parameters (Keely and Boateng, 1987). An alternative protocol is to flush out a fixed number of borehole volumes.

Difficulties with both protocols are illustrated in Figure 3. During flushing of the first 10 boreholes volumes from borehole B, trichloroethene, (TCE) varies little (304-336 µg/l) but conductivity rises rapidly. TCE concentrations rise over the next 5 hours to 440 µg/l while conductivity only rises slowly. After 3 days (not shown) both parameters were similar to the final values in Figure 3a.

Borehole C (Figure 3b) shows a fivefold rise in TCE during the removal of 7 borehole volumes and concentrations were still rising at the end. Conductivity also rose throughout, but more slowly, and had nearly stabilised by the end. More examples are given by Rivett et al. (1990b).

In most cases, neither flushing a pre-set number of borehole volumes nor pumping until stable conductivity was observed would have resulted in stable solvent concentrations. Borehole effects dominate intitially but, as pumping continues, water is drawn from an increasingly large volume which contains a three dimensional pattern of pollution and of permeability pathways. Only in aquifers with uniformly distributed pollutants will longer pumping automatically lead to stable concentrations.

Each of 59 boreholes was analysed as shown in Figure 3. However, individual sampling protocols were developed for each borehole by chosing sampling times (flushing volumes) for all later visits. These were either when solvent concentrations were stable, or after the maximum flushing allowed by the associated water storage tanks.

Even when a sampling protocol has been established, observed concentrations sometimes vary, not because of changes in groundwater quality, but because of changes in the way the borehole is used. This is discussed in Section 4.2.

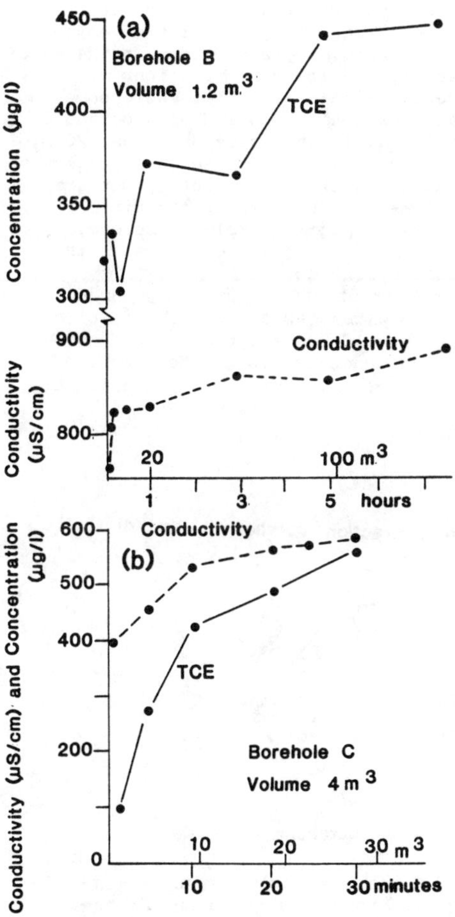

Fig. 3 Development of sampling protocols for two existing, pumped boreholes

4 EXTENT OF, AND CONTROLS ON, SOLVENT POLLUTION

4.1 Regional pattern

The maximum concentrations in each borehole of trichloromethane (TCM), 1,1,1-trichloroethane (TCA), tetrachloromethane (TeCM), trichloroethene (TCE) and tetrachloroethene (TeCE) are summarised in Table 1. 40% of boreholes exceed the UK limit for TCE in drinking water (30 µg/l); fortunately the aquifer is not used for public supply.

Table 1. Summary of chlorinated solvent concentrations, showing percentage of boreholes with maximum in specified range.

Range	TCM	TCA	TeCM	TCE	TeCE
Below detection	47	54	63	22	56
0.02-1.0 µg/l	36	24	35	16	35
1.1-10 µg/l	17	9	2	19	5
10.1-100 µg/l	-	8	-	13	2
101-1000 µg/l	-	5	-	27	2
1000-5500 µg/l	-	-	-	3	-

The relative occurrence of the five solvents is related to their pattern of consumption (Rivett et al., 1990c).

Figure 4 shows the regional distribution of TCE. The data are too sparse and the sampling points are probably too closely linked to the potential pollution sources for it to be possible to contour a regional pattern of concentrations. To discover how far each pollution plume extends, and indeed whether they form a regional block of polluted water would require many new

4.2 Effects of pumping regime

The high frequency with which solvent pollution is observed may be surprising in view of the point pollution sources, the high groundwater volume (15-25% porosity, 80-150 m thickness), and low regional groundwater velocities (5-25 m/year). These factors would encourage small zones of pollution around each source. There are boreholes which are not polluted despite potential sources nearby, but they are in a minority. Most boreholes are polluted because they are pumped, and so sample a large volume of aquifer. This is illustrated in Figure 5 which shows travel

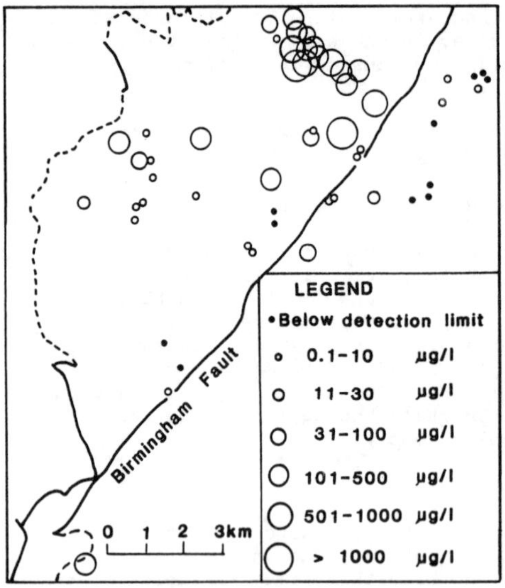

Fig. 4 Maximum concentrations of TCE in the Birmingham Aquifer.

times to high and median abstraction boreholes, calculated using the program RESSQ (Kelly and Tsang, 1983). In the ideal conditions assumed in the model, a solvent spillage on the 5 year isochrone would pollute the borehole from 5 years after spillage until all the solvent had dissolved in passing groundwater. Thus a pumping borehole records an accumulation of spillage within its catchment; an observation borehole will only detect spillages near the upgradient flowline.

Following on from this, if a borehole has a variable pumping regime it may show variations in concentrations which do not necessarily reflect changes in the amount of pollution in the aquifer. The observed variations may just reflect variable movement of pollution plumes in response to variations in pumping. An example is shown in Figure 6 of a borehole which had been unused for 10 years until abstraction started in October 1987 at 230 m^3/d. Solvent concentrations rose markedly throughout 15 months of sampling, although there was no evidence of recent pollution events. Rivett et al. (1990b) give other examples.

4.3 Controls on solvent pollution

There is a close correlation between solvent use on a site and the concentrations observed in any on-site borehole. 54% of solvent using sites had total solvent concentrations over 100 µg/l; the highest concentration in a non-user's site was 40 µg/l.

Surface geology had some control on solvent concentrations. Groundwater which had some protection (by drift cover, or the confining Mercia Mudstone in the east) generally had low concentrations of solvents (Figure 7). There were some notable exceptions including a borehole with 900 µg/l total solvents under 20 m of Mercia Mudstone. In these cases, solvents may have penetrated the cover, but are more likely to have short circuited via the side of poorly completed boreholes, or to have been illegally disposed of in disused boreholes.

A number of other factors, such as depth to water table or depth of solid casing, would be expected to influence observed concentrations. However they were too closely correlated to thickness of protective cover to be separately analysed.

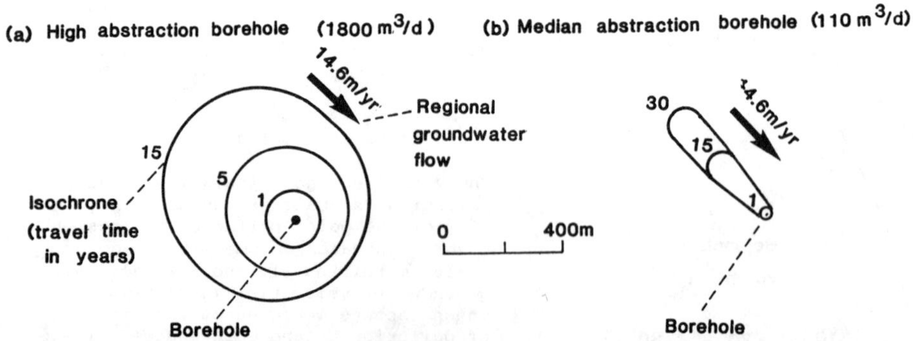

Fig. 5 Isochrones for high and median abstraction boreholes, indicating the areas of aquifer sampled. (Both cases assume kinematic porosity 25% aquifer thickness 90 m, gradient 0.002, conductivity 5 m/d [5.8×10^{-5} m/s]). Kinematic porosity is the ratio of Darcy velocity to average tracer velocity (de Marsily, 1986)

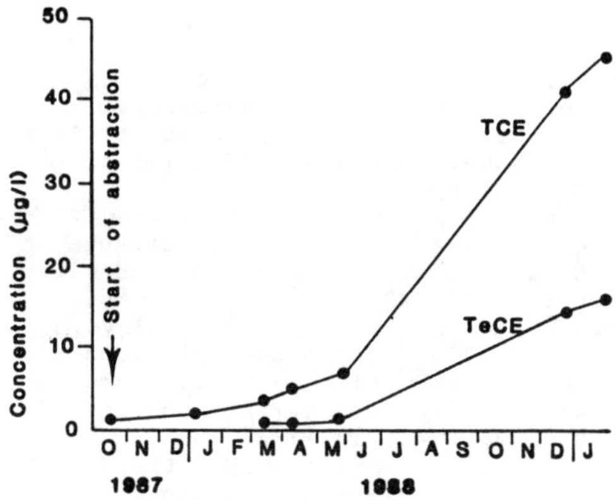

Fig. 6 Apparent trend in solvent concentrations at borehole G

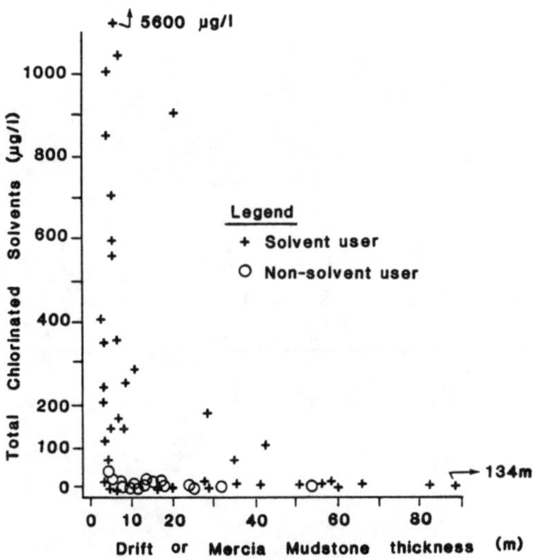

Fig. 7 Relationship between solvent concentration and thickness of protective strata

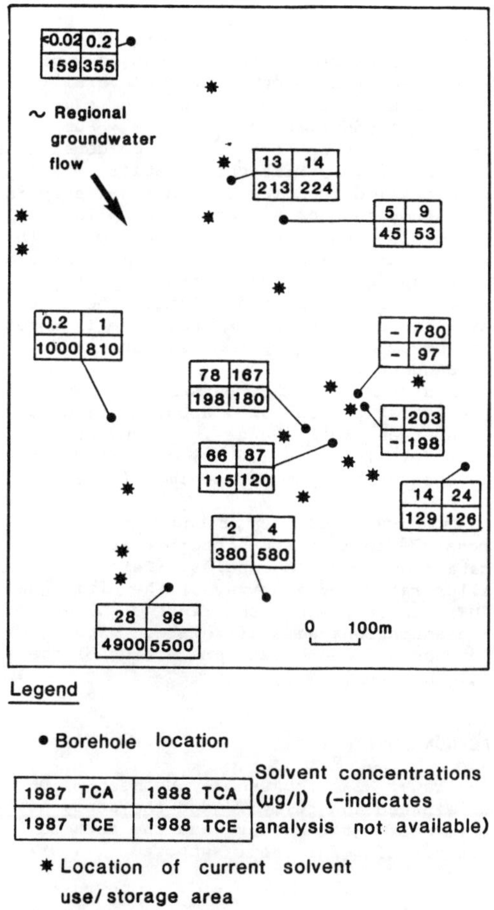

Fig. 8 Relative trends in TCA and TCE concentrations in a small area of the Birmingham aquifer

4.4 Trends

Some boreholes showed trends in solvent concentrations which would not be due to changes in pumping regime because pumping was constant. In general TCA concentrations rose relative to those of TCE. Figure 8 illustrates this relative trend for one small area of the unconfined Birmingham aquifer. Mainly TCA is now used at the locations marked, whereas TCE was used at these and other locations in the past. At the nine boreholes sampled in both 1987 and 1988, TCA concentrations went up by an average of 220%, while TCE concentrations only increased by 20%.

These results correlate with the changing pattern of solvent use in the UK, with TCA consumption rising since the mid-1960's, and TCE consumption falling. Thus, while it may be possible to dismiss TCE pollution of groundwater as a legacy of an older, dirtier age, TCA spillages are probably continuing to the present time.

5 CONCLUSIONS

Investigations into groundwater quality must often use existing water supply boreholes, particularly for regional studies of deeper systems. The variability of construction and pumping regimes, and their often non-ideal locations can lead to difficulties in interpreting data from such boreholes. The advantage of using pumped boreholes is that they sample a large volume of aquifer, indicating some "average" condition of groundwater. Thus they give a more accurate picture of the overall condition of groundwater than the same number of observation boreholes. Pumped boreholes cannot however provide details about spatial variations, whether horizontal or vertical. The first step in any study must be to develop sampling protocols individually for each borehole. Samples are taken throughout a long period of flushing and concentrations plotted against time or pumped volume. The time at which concentrations stabilise is noted and becomes the sampling time on subsequent visits.

Even with a good protocol, observed concentrations can show apparent trends without any real change in groundwater quality. This usually arises from changes in the overall pumping regime of the borehole.

With good sampling protocols, it is possible to make some interpretation of data from pumped boreholes. This is illustrated for a survey of the Birmingham (UK) aquifer, where chlorinated solvent concentrations were correlated with landuse and geological protection of the main aquifer.

ACKNOWLEDGEMENTS

The Water Research Centre, Medmenham, UK, originated and fully funded the study of Birmingham groundwater, and the authors are grateful for their support.

REFERENCES

Barcelona, M.J., Gibb, J.P., Helfrick, J.A. & Garske, E.E. 1985. Practical guide for groundwater sampling. EPA.600/2-85/104. 169 pp.

Eklund, G., Josefsson, B. & Roos, C. 1978. Determination of volatile halogenated hydrocarbons in tap water, sea water and industrial effluents by glass capillary gas chromatography and electron-capture detection. J. High Res. Chrom. Chrom. Commun., 7: 34-40.

Horton, A. 1975. The engineering geology of the Pleistocene deposits of the Birmingham district. Inst. Geol. Sci., Report 75/4.

Keely, J.F. & Boateng, K. 1987. Monitoring well installation, pumping and sampling techniques. Ground Water, 25: 300-313.

Keely, J.F. & Tsang, C.F. 1983. Velocity plots and capture zones of pumping centres for groundwater investigations. Ground Water 21: 701-714.

de Marsily, G. 1986. *Quantitative hydrogeology*. Academic Press.

Rivett, M.O. 1989. Organic contamination of the Birmingham aquifer. PhD thesis, School of Earth Sciences, University of Birmingham, UK. 413 pp.

Rivett, M.O., Lerner, D.N. & Lloyd, J.W. 1990a. Chlorinated solvents in UK aquifers. J. Inst. Water Environ. Manag. 4: 242-250.

Rivett, M.O., Lerner, D.N. & Lloyd, J.W. 1990b. Temporal variations of chlorinated solvents in abstraction wells. Ground Water Monitoring Review, X(4), 127-133.

Rivett, M.O., Lerner, D.N., Lloyd, J.W. & Clark, L. 1990c. Organic contamination of the Birmingham aquifer. J. Hydrol. 113: 307-323.

Subsurface Contamination by Immiscible Fluids, Weyer (ed.) © 1992 Balkema, Rotterdam. ISBN 90 6191 175 3

What happens to diesel oil spilt on a fissured Chalk aquifer?

R.P.Ashley
Hydrogeology Research Group, University of Birmingham, UK (Presently: Cambridge, UK)
D.N.Lerner & J.W.Lloyd
Hydrogeology Research Group, University of Birmingham, UK

ABSTRACT: A study was carried out of the unsaturated zone beneath a ten year old diesel fuel spill site on the major Chalk aquifer of south and east England. The objectives were to determine the present distribution of oil, and to identify changes in oil composition since the spill. A borehole was drilled through the site and cores and pore water extracts were analysed. Specimens were tested for pore size distribution and examined by SEM to characterise the pore structure. The results demonstrated that the main accumulations of oil were within 5-10 mm of the fracture surfaces, with very low concentrations in the body of the chalk beyond these zones. Total hydrocarbon concentrations have been greatly reduced since the spill, probably both by dissolution of the lower molecular weight components, and by bacterial degradation. The conclusions have important implications for the investigation and clean-up of oil spillages in fissured aquifers.

1 INTRODUCTION

This paper describes aspects of a study of a diesel fuel spillage onto the outcrop of a major aquifer (Ashley, 1990). The study was particularly desirable because the spillage had had considerable investigation at the time of occurrence ten years earlier, and hence presented the opportunity to identify changes in oil distribution and chemistry over that period. Furthermore, the Chalk aquifer has been intensively studied in connection with nitrate pollution, and the physical and chemical processes operating in the unsaturated zone are well understood.

2 BACKGROUND

2.1 The Chalk of South-East England

The Chalk is a Cretaceous limestone of eastern and southern England. It provides about one sixth of total British public water supply, although locally it is frequently the sole water source. The outcrop of the Chalk is protected over a large area by a cover of low-permeability Boulder Clay of glacial origin, but at crucial locations the cover may be thin or absent.

The Chalk is characteristically a dual porosity/permeability aquifer. The primary system is formed by microfossilliferous sediment, with extremely small pores and pore throats, the latter typically in the range 0.1-1.0 µm. Bulk porosities may be of the order of 40% or more (Carter & Mallard, 1974). The secondary system is a network of fissures, at spacings ranging between 0.5 m and 10 m, and with openings in the range 0.2 mm to 10 mm (Reeves, 1979). The network may extend down to a smaller scale, possibly to a spacing range between 0.5 mm and 30 mm and an opening range between 1 µm and 20 µm (Reeves, 1979), although direct evidence for such microfissures is sparse. The large-scale fissures are frequently enlarged by solution associated with the dominant direction of groundwater flow, or with seasonal water table fluctuation (Reeves, 1979; David, 1966).

The primary system is thus of low hydraulic conductivity (10^{-5} md^{-1} to 10^{-3} md^{-1}) and high porosity. The small pore size ensures that the pore space is almost entirely filled with water (97% to 98%), even in the unsaturated (vadose) zones. The secondary system by contrast is of higher hydraulic conductivity (1 md^{-1} to an extreme of 1000 md^{-1}) and may occupy only 1% of bulk rock volume. The fissures are free-draining in the unsaturated zone.

The Chalk is chemically of high purity, often consisting of more than 98% calcium carbonate (Hancock, 1975). The organic matter content is typically less than 0.05% by weight, of bitumen and humic matter (Pacey N, Univ. of Birmingham, 1986, pers. comm.).

2.2 Pollution of the Chalk

The Chalk has been intensively studied over the last two decades following recognition of its vulnerability to pollution by nitrate of agricultural origin. This research has led to the development of a model for the behaviour of soluble pollutants in the unsaturated zone (see Foster et al., 1986, for a review).

The essential features of the model are:

1 Input of soluble contaminants to the unsaturated zone via recharge, entering by way of successively large fissures with increasing intensity and duration of rainfall.

2 Dynamic interchange by a diffusion mechanism of contaminants between recharge water in the fissures and water in the microporous matrix.

3 Development of a vertical profile of contaminant concentrations in pore waters, representing fluctuations of recharged concentrations over a number of years.

4 Slow downward migration of the profile in response to the chromatographic effect of diffusive interchange between fissure water and pore water, and slow movement of pore water in the matrix.

The model, although developed for the purpose of understanding processes affecting soluble pollutants in the unsaturated zone, is sufficiently general to be of relevance to the behaviour of many other pollutants, even those of low solubility such as mineral oils.

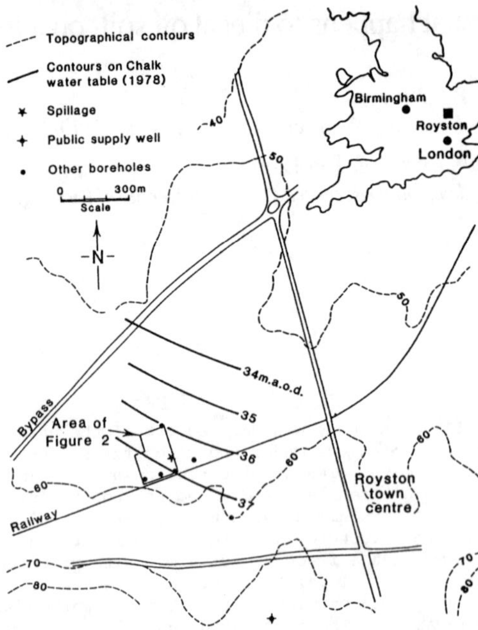

Fig. 1 Location of Royston oil spill

2.3 The Royston diesel oil spillage

In 1976 an incident at an oil depot at Royston, Hertfordshire, resulted in the discharge of 455 m^3 of diesel fuel from a storage tank. Pumping from the ponded oil recovered 323 m^3, but the remaining 132 m^3 infiltrated to the subsurface (Hunter-Blair, 1978). Figure 1 and Figure 2 show the site location and layout respectively.

At the site a thin layer of sandy drift overlies the Middle Chalk (Figure 3). At a depth of 18 m the Melbourn Rock occurs, a hard band of partly recrystallised chalk 3 m to 5 m thick, which marks the base of the Middle Chalk. The underlying Lower Chalk is more marly, particularly for the first few metres, which are referred to as the Plenus Marls. A further hard band, the Totternhoe Stone, occurs at greater depth.

The water table in 1976 lay at about 30 m below ground level, although that was abnormally low and in 1986 the water table was at 24 m to 25 m below ground level. All boreholes at the site were completed as open-hole designs, and thus water level measurements represent the composite piezometry in the strata penetrated. Regional groundwater flow is from south to north, and in 1976 part of the water supply for the town of Royston was obtained from a borehole to the south (Figure 3). Hard bands such as the Melbourn Rock and the Totternhoe Stone are often highly fractured and more transmissive than surrounding strata, and there was concern that downward migrating oil

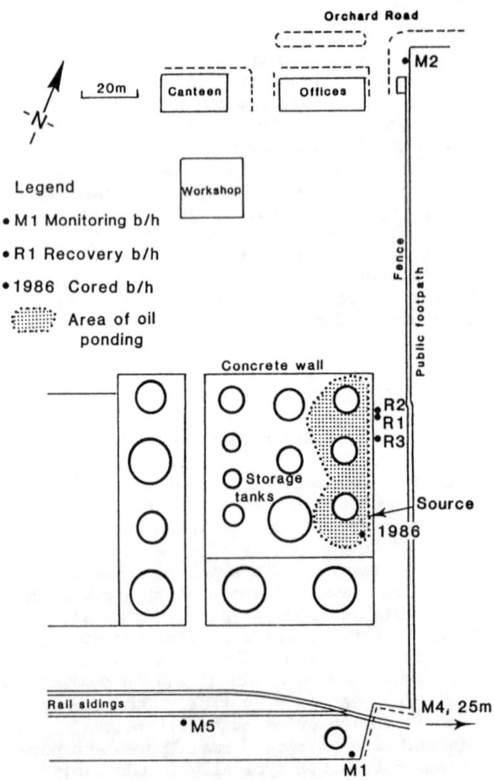

Fig. 2 Royston spill site plan

would be intercepted by the Melbourn Rock and transmitted southwards to the public supply borehole.

Figure 2 shows the boreholes that were constructed to monitor oil migration and permit oil recovery. Recovery operations continued until 1982, when it was concluded that, although the public supply borehole had been closed as a precautionary measure, there was little remaining risk to local water supplies since no migration of oil had been detected beyond the immediate area of the spillage. There remained considerable concern at the small quantity of oil that had actually been recovered (probably little more than 10%), and the research described in this paper was intended to answer some of the outstanding questions, as well as to identify changes in the situation a decade after the original spillage.

3 METHODOLOGY

3.1 Field methods and sampling

The investigation centred around a single, continuously cored borehole to 40 m depth, located immediately adjacent to the source of the spillage (Figures 2 and 3: borehole 1986). It was drilled by the air-flush rotary method, and to prevent the introduction of contaminants all down-hole equipment was steam cleaned before being brought to site, and also flame-cleaned before each "tripping-in" operation. Drilling tool joints were not lubricated during drilling, and the introduction of water to the airstream was not permitted.

The cores, of approximately 110 mm diameter, were retrieved in 1.5 m or 3 m long core barrels lined with "Mylar" plastic sleeving, in which the cores were removed undisturbed. The cores, wrapped in Mylar, were additionally sealed on site in polythene tubing, after division into lengths of approximately 1 m for long term storage. At the end of each day the cores were transferred to storage maintained at a

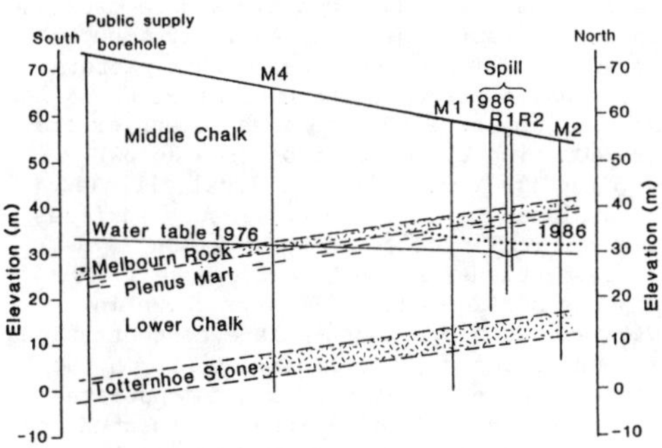

Fig. 3 Schematic vertical section of Royston spill including lithological log

temperature no higher than 4°C. After completion of the borehole, all the cores were transferred again to permanent storage at a similar low temperature. Individual core lengths were then selected and transferred to the laboratory for analysis.

3.2 Extraction and analysis techniques

The core samples were extracted and chemically analysed according to three main processes:

1. Material was scraped from fracture surfaces in weights ranging from 5 g to 15 g, to provide highly specific data representative of particular locations. For all lithologies except for the hard, recrystallised Melbourn Rock it was possible to use a scraping tool blade of the "Surform" type, commonly used for planing wood, to remove from the fissure surface successive layers of 2 mm to 3 mm thickness. The distinction between first scrapings was often blurred by irregularities in the fracture surface, but the deeper scrapings genuinely represent greater distances from the original surface.
2. Chalk samples of 0.5 kg to 1 kg weight were selected from intact pieces of core from which the fractured ends had been removed using a rock vice. Analyses of this material are representative of the chalk matrix distant from visible fissures.
3. Core samples of approximately 3 kg weight were selected and centrifuged to provide analyses of pore water and centrifuged residue. Wherever possible chalk within 2 cm of a fissure surface or the outer core surface was trimmed off, and thus these samples reflect conditions relatively distant from fissures, and unaffected by drilling influences. Samples were partially crushed, and centrifuged in 200 g batches at 4500 rpm for 40 minutes, according to the technique of Barber et al. (1977).

Solid chalk samples were extracted by grinding with an equal quantity of anhydrous sodium sulphate and eluting a small amount of the resulting dry powder (equivalent to 5 g to 10 g of sample) with 25 ml of the appropriate solvent: pentane for gas chromatographic (GC) analysis and 1,1-2-trichlorotrifluoreothane for infrared (IR) analysis. Pore water samples were extracted into solvent by the liquid-liquid method. All extracts were concentrated by slow evaporation under reflux down to a volume of 2 ml to 4 ml. Further concentration of pentane extracts (down to 100 µl to 300 µl) for GC analysis was carried out by slow evaporation of solvent under a fine stream of nitrogen.

A Fourier Transform Infrared Spectrometer was used according to the method of Whittle et al. (1980) and McCrum & Whittle (1982) to estimate the total hydrocarbon concentration of the sample. GC analysis to measure concentrations of individual hydrocarbons was performed using a chromatograph fitted with a 30 m non-polar capillary column and a flame ionisation detector. A limited number of samples, principally pore water extracts, were analysed by gas chromatography/mass spectrometry (GC/MS). For GC analysis it was necessary to add an internal standard of eicosane and anthracene in acetone to the chalk or water sample.

The porosity and pore size distribution of a number of selected chalk samples was investigated partly by mercury injection porosimetry, and partly by scanning electron microscopy (SEM). The SEM images also provided useful data on the mineralogy and petrography of the chalk at different depths.

4 RESULTS

4.1 Characteristics of the Chalk at Royston

The borehole penetrated the lithologies shown in Figure 3, including the hard, crystalline Melbourn Rock from approximately 18 m to 21 m. The topmost 3 m to 6 m were considerably weathered and fragmented. Below the Melbourn Rock the strata were noticeably marly. The water table was at 24.05 m, and had a layer of dark and turbid oil 0.45 m thick above it.

The porosity averages about 41%, except for a section from just above and through the Melbourn Rock, where it averages 27% (Table 1). The average pore size (represented by the D_{50} pore diameter) decreases just above the Melbourn Rock but remains low below it.

Table 1. Summary of porosity and pore size data.

Depth (m)	Nr samples	Porosity (%) Mean	Porosity (%) Std.Dev.	Median pore dia. (µm) Mean	Median pore dia. (µm) Std.Dev.
1.76 -15.11	11	41	3	0.9[1]	0.08[1]
15.95 -21.33	5	27	3	0.5	0.09
21.91 -31.35	7	42	2	0.6	0.09

Note 1 Excluding one anomalously high value

The SEM images reveal that the decreases in porosity and pore diameter just above the Melbourn Rock are caused by increasing recrystallisation of calcite, thus filling in pores. Below the Melbourn Rock there is little sparry calcite but a higher content of clay minerals, which have not reduced the bulk porosity, but have reduced the average pore size.

4.2 Distribution of residual hydrocarbons

Only a small proportion of the samples were analysed by IR for Total Hydrocarbons (TH) as well as by GC. The IR measurements were therefore used to obtain a correlation between TH and the GC Total Peak Area (TPA). The correlation (with a coefficient of about 0.95 from 45 samples) thus enabled the TPA to be used as a measure of TH.

Table 2 contrasts the estimated TH concentration in the first scraping from each fissure, with the TH concentration in the chalk matrix samples distant from fissures, for the same three zones identified in Table 1. The reduction in concentration from the fissures to the matrix is at least an order of magnitude. The reduction is in fact much greater, since the matrix in the first interval includes one anomalously high value of 580 mg/kg. Excluding that value the average is about 25 mg/kg.

Table 2. Comparison of Total Hydrocarbon (TH) content of fissures and matrix.

Depth (m)	Fissure surfaces Nr. samples	Fissure surfaces Mean TH (mg/kg)	Matrix Nr. samples	Matrix Mean TH (mg/kg)
1.76 -15.11	23	680	10	80 (25[1])
15.95 -21.33	11	110	3	0
21.91 -31.35	11	14	4	0

Note 1 Excluding one anomalously high value

Figure 4 compares the hydrocarbon content of individual scrapings from fissure surfaces. For convenience the graphs were prepared from data on n-alkane concentrations, which could be directly obtained from GC analyses. As noted below, errors due to differences in the n-alkane fraction of total hydrocarbons would not account for most of the observed variations. The data clearly show a decrease in concentration with successively "deeper" scrapings. The exceptions to this pattern occur where the absolute concentration is low and is subject either to measurement errors, or reduction by degradation (see below).

Since the bulk of the residual oil has accumulated on or close to fissure surfaces, the vertical profile of hydrocarbon concentrations may best be prepared from fissure surface data. This is shown in Figure 5, again using n-alkane concentrations for convenience. The profile confirms the pattern shown in Table 2, and demonstrates that the residual oil in the unsaturated zone remains mainly above the Melbourn Rock.

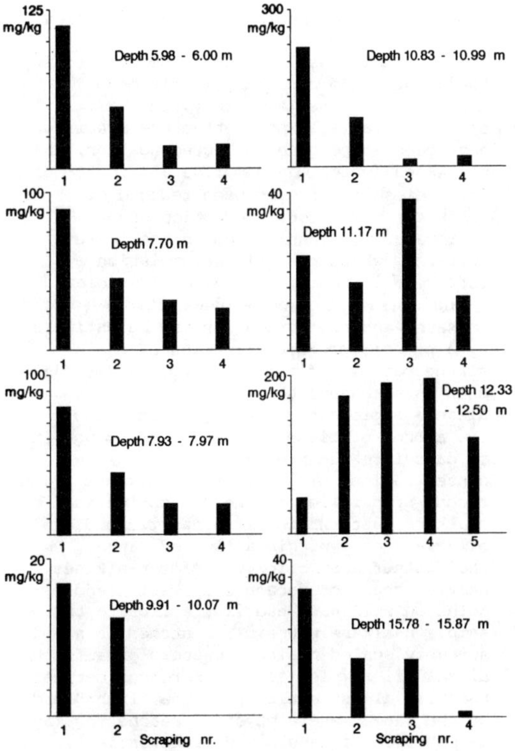

Fig. 4 Concentration of hydrocarbons on fissure surfaces

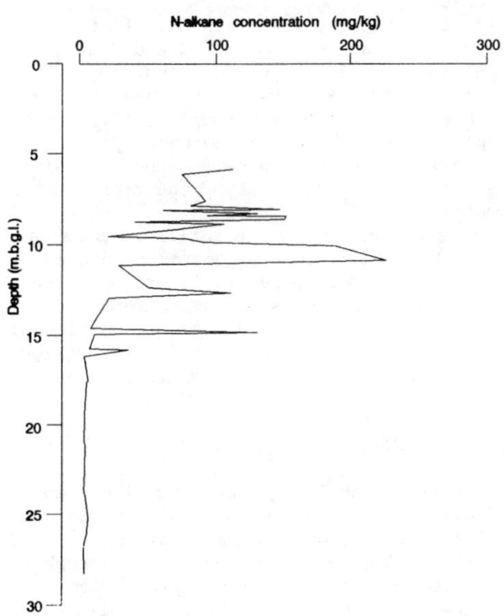

Fig. 5 Vertical distribution of hydrocarbons

4.3 Composition of residual hydrocarbons

The composition of fresh diesel fuel as revealed by GC analysis is dominated by n-alkanes with between roughly 8 and 24 carbon atoms. The other readily identified hydrocarbons are the branched alkanes referred to as isoprenoids. Due to the relative resistance to biodegradation of the isoprenoids, a simple measure of the degree of degradation is calculated as the ratio of concentration of an n-alkane to the concentration of the closest isoprenoid on the GC trace: the n-heptadecane/pristane ($n\text{-}C_{17}/i\text{-}C_{19}$) and n-octadecane/phytane ($n\text{-}C_{18}/i\text{-}C_{20}$) ratios may be used for this purposes (Blumer & Sass, 1972). The ratios in the fresh diesel are approximately known from analyses performed at the time of the spillage in 1976:

$n\text{-}C_{17}/i\text{-}C_{19}$	2.7
$n\text{-}C_{18}/i\text{-}C_{20}$	2.9

Table 3 contrasts the degradation indices for fissure surfaces with those for chalk matrix material distant from fissures. Only a few data points are available for the chalk matrix since in a number of samples no oil hydrocarbons could be detected. Nevertheless, there is evidence of enhanced biodegradation on fissure surfaces compared to the matrix. The biodegradation on the fissure surfaces appears to be approximately uniform however: for those fissures from which successively deeper multiple samples were collected, the index ratios only decreased on average by 12% (8 samples).

Table 3. Comparison of degradation in fissures and matrix.

Depth (m)	Fissure surfaces Nr. samples	Mean Index[1] A	B	Matrix Nr. samples	Mean Index[1] A	B
1.76 -15.11	44	1.43	1.35	5	2.13	1.90
15.95 -21.33	14	0.73	0.66	1	1.61	2.94
21.91 -31.35	9	0.07	0.07	4	–[2]	–[2]

Note 1 Index A = $n\text{-}C_{17}/i\text{-}C_{19}$
Index B = $n\text{-}C_{18}/i\text{-}C_{20}$
Note 2 Hydrocarbons were not detected in any of the four samples analysed

The data in Table 3 also suggest that degradation on the fissure surfaces has increased with depth, and in fact little of the characteristic n-alkane structure of diesel appears where the concentration of hydrocarbons was lowest.

The pore water extracts similarly showed no recognisable diesel oil characteristics, other than a large number of compounds that eluted from the GC column at the same temperature as diesel oil hydrocarbons. On examination by GC/MS many were found to be organic acids, esters, ketones and phenols, and related compounds derived from aromatic and alkyl aromatic bases. Many could be correlated with GC peaks in partly degraded chalk extract samples, and it is concluded that they are diesel oil biodegradation products.

5 DISCUSSION

5.1 Migration of oil

The research demonstrates that the oil which entered the ground at the time of the original spillage migrated along the major fissures down to the water table, where an unknown, but considerable, volume remains. Within the unsaturated zone the oil appears to have accumulated on the fissure surfaces, mainly within the thin skin of chalk sampled, 5 mm to 10 mm thick. It is to be expected that some of the pore structure near the surface will be saturated only partly with water, and capillary action would draw the oil into air-filled pores. The maximum accumulation of oil occurs above 15 m where the pore throat size is greatest. Although the smaller accumulations of oil coincide with high degradation indices, the degree of degradation does not appear to be sufficient to account for the present low concentrations.

An explanation must be sought however, for the small amounts of oil that are found in the chalk matrix at a distance from fissures. This oil appears to be relatively fresh and undegraded in spite of its low concentration. The entry pressure required for oil to displace water from pores of 1 µm diameter or less is in excess of that exerted by a 25 m column of oil, which is not possible at this depth. An alternative mechanism to combine solution in water, transport by advection or diffusion followed by sorption, and produce the characteristic n-alkane pattern, is difficult to envisage.

The possibility remains that the oil has migrated from major fissures into the microfissures postulated by Reeves (1979). Such microfissures would necessarily be limited in extent or highly irregular, or else they would be readily noted during core handling. A water-filled fissure of 10 µm width requires a column of oil only 1.25 m high for oil entry, and a system of such fissures at 10 mm intervals could hold about 400 mg/kg of oil. Bearing in mind the approximations involved, this estimate within range of those measured (80 mg/kg average, or 25 mg/kg excluding extreme values).

5.2 Physical weathering and biodegradation

The original diesel oil that accumulated in the unsaturated zone has clearly been degraded. On the (arbitrary, but conservative) assumption that the isoprenoids have been subject to no biodegradation, the n-alkanes (which may comprise 25% to 50% of a fresh diesel oil) have been reduced by at least 40%. The total reduction of hydrocarbon concentrations on the fissure surfaces, where the main accumulation occurred, is thus probably considerable.

The overall losses by evaporation, solution in water and biodegradation may be estimated by a process of reconstruction of the n-alkane pattern of the oil sample. The first step is to assume that the n-alkane/isoprenoid ratio is a true measure of the amount of biodegradation, and to use it to calculate the predegradation concentrations of individual n-alkanes. The second step relies on the assumption that the n-alkane patterns in different fresh diesel oils vary only within a limited range. On the further assumption that for n-alkanes heavier than octadecane evaporation and solution will have had little impact, the sample n-alkane pattern is matched to a suitably scaled n-alkane pattern of a fresh diesel oil sample. The differences between them lie almost entirely at the "light" end of the range, where physical weathering may be expected to have acted preferentially.

Figures 6 and 7 give examples of this reconstruction process from a fracture surface sample and a chalk matrix sample respectively. Although selected to demonstrate the method, they also show the characteristically fresh appearance of matrix oil samples, and the weathered and degraded appearance of fissure oil samples. By these estimates and assuming that the same pattern is shown by other hydrocarbons, the fissure samples overall have lost between 30% and 50% of their weight by physical weathering, with a smaller percentage lost by the matrix samples.

It is not clear why the matrix samples have been subjected to less biodegradation than the fissure samples, but it is presumed that the fissure surface is an environment where the supply of oxygen, and possibly nutrients, are better supplied from occasional recharge than the microfissures of limited opening.

6 CONCLUSIONS

Oil infiltrating into an aquifer with a dual hydraulic system such as the Chalk migrates preferentially along the major fissure routes. Oil entry to the bulk of the rock is restricted by the high excess pressures required to displace the water already present.

Oil accumulates mainly on the fissure surfaces within a "skin" 5 mm to 10 mm thick, where the pore structure is only partly

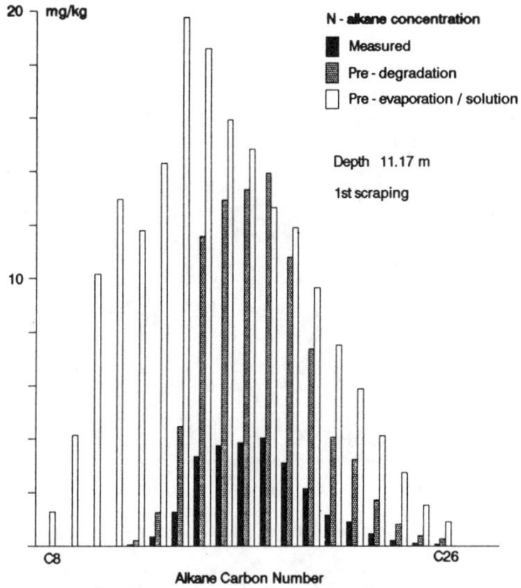

Fig. 6 Composition of hydrocarbons on fissure surfaces

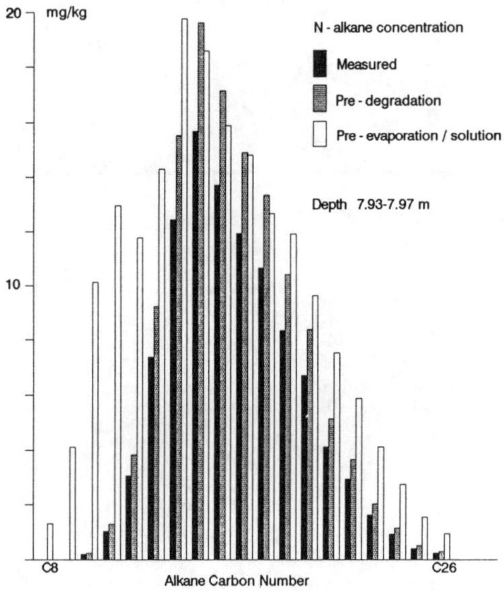

Fig. 7 Composition of hydrocarbons in chalk blocks

water-filled. Pore size determines the pressure required to displace water from the pores and thus controls the degree of accumulation.

Oil may migrate in small quantities along microfissures with widths of the order of 10 μm and spacing of the order of 10 mm.

Physical weathering (solution and evaporation) may remove 30% to 50% of the hydrocarbons in the fissures. Biodegradation may remove considerably more. Biodegradation in the microfissures appears to be more limited.

The distribution of oil in the unsaturated zone found during this study has clear implications for investigations of spillages. Specific sampling techniques need to be devised that identify the environments of accumulation, since bulk sampling provides a misleading picture.

The study indicates that even 10 years after an oil spillage, the pollutant has not migrated far away from major fissures, and is thus still accessible to remedial techniques, such as flushing and enhanced degradation.

ACKNOWLEDGEMENTS

This research was supported by funds from the Natural Environmental Research Council (studentship) and the Anglian Water Authority (cored borehole).

REFERENCES

Ashley, R.P. 1990. Mechanisms controlling oil behaviour in the unsaturated Chalk. Unpub. PhD thesis, School of Earth Sciences, Univ. of Birmingham, UK.

Barber, C., P.J. Maris and K. Knox. 1977. Groundwater sampling: the extraction of interstitial water from cores of rock and sediments by high speed centrifuge. Technical Report TR54. Water Research Centre.

Blumer, M. and J. Sass 1972. Oil pollution: persistence and degradation of spilled fuel oil. Science 176: 1120-1122.

Carter, P.G. and D.J. Mallard 1974. A study of the strength, compressibility and density trends within the Chalk of south-east England. Q.J. Eng. Geol. 7: 47-56.

David, S.N. 1966. Initiation of groundwater flow in jointed limestones. Nat. Speleolog. Soc. Bull. 28: 111-118.

Foster, S.S.D., L.R. Bridge, A.K. Geake, A.R. Lawrence and J.M. Parker 1986. The groundwater nitrate problem. Hydrogeology Report 86/2, British Geological Survey.

Hancock, J.M. 1975. The petrology of the Chalk. Proc. Geol. Ass. 86: 499-535.

Hunter-Blair, A. 1978. Oil pollution of a Chalk aquifer - a case history. In Proceedings of the International Symposium on Groundwater Pollution by Oil Hydrocarbons, Prague.

McCrum, W.A. and P.J. Whittle 1982. Infrared determination of petroleum oil Part II. Extraction from water and removal of interferences. Analyst 107: 1081-1085.

Reeves, M.J. 1979. Recharge and pollution of the English Chalk: some possible mechanisms. Eng. Geol. 14: 231-240.

Whittle, P.J., W.A. McCrum and M.W. Horne 1980. Infrared determination of petroleum oil: new approaches to the calculation. Analyst 105: 679-684.

Subsurface Contamination by Immiscible Fluids, Weyer (ed.) © *1992 Balkema, Rotterdam. ISBN 90 6191 175 3*

Field investigation of chlorinated solvent pollution of groundwater in Coventry, UK

E.Gosk
Hydrogeology Division, Geological Survey of Denmark, Copenhagen, Denmark

P.K.Bishop, M.W.Burston & D.N.Lerner
Hydrogeology Research Group, School of Earth Sciences, University of Birmingham, UK

ABSTRACT: A regional survey of groundwater quality in the Coventry area of the UK indicated that urban groundwaters were widely contaminated with chlorinated solvents, trichloroethene being most ubiquitous at a maximum concentration of 1100 μg/l. The area is underlain by a multi-layered geological sequence and required investigation into the vertical distribution of solvents. A detailed investigation was performed at an industrial site incorporating several field methodologies. An initial soil gas survey indicated gaseous solvent concentrations as high as 2000 μg/l. Specially designed investigation boreholes were constructed and a series of groundwater sampling methods applied. These included the use of portable and dedicated sampling systems. Similar vertical trends in solvent ratios were obtained indicating a degree of compatibility between the methods used. However, variations in maximum trichloroethene concentrations suggested some negative bias using the dedicated system. Vertical trends indicate that 1,1,1-trichloroethane pollution of groundwater abstracted in Coventry is likely to increase in the future. Spatial trends suggest that regional correlation of pollution between boreholes is precluded.

1 INTRODUCTION

Until recently, there has been a lack of data on the occurrence of organic contaminants in UK groundwaters. In response to documented contamination of aquifers underlying urbanised areas elsewhere in the world (Cavallero *et al.*, 1985; Fusillo *et al.*, 1985) and the lack of UK data, studies of the Birmingham (Rivett *et al.*, 1990) and Coventry aquifers have recently been conducted to assess the impact of long established urban environments upon the underlying groundwaters.

The study area described in this paper is centred on the city of Coventry, and is defined by geological and catchment divide boundaries. The total study area comprises approximately 200 km² of which around 32% is urban. Coventry is a major city within the English Midlands of the UK (Figure 1) and has a long industrial heritage. Chlorinated solvents have been used in industry since their development in the 1930s, mainly trichloroethene (TCE) and 1,1,1-trichloroethane (TCA) for metal degreasing. The use of these solvents has led to groundwater pollution. This paper describes both regional and vertical variations in solvent pollution, with particular emphasis on the methodologies employed in a specific site investigation.

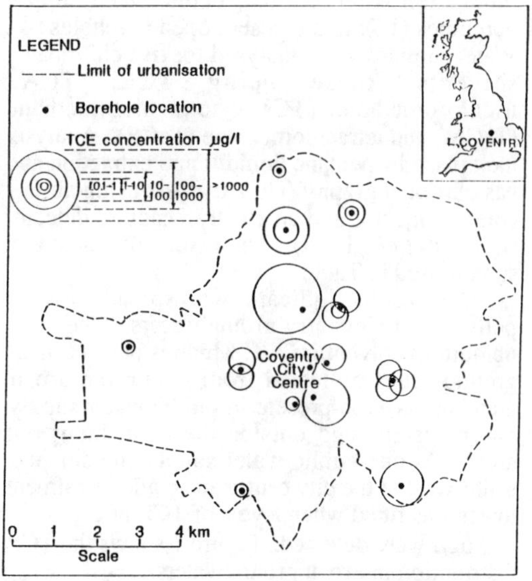

Fig. 1 Regional distribution of TCE in urban groundwaters

2 GEOLOGY AND HYDROGEOLOGY

Coventry is underlain by predominantly Carboniferous strata, the most important unit being the Coventry Sandstone Formation which consists of interbedded sandstones and mudstones of around 520 m total maximum thickness. Intercalation between units varying in thickness typically between 0 and 10 m provides a multi-aquifer system. The units were deposited in a fluvial environment (Besly, 1988) and are lenticular in shape, resulting in them being discontinuous and difficult to correlate over relatively short distances. The sandstones are fissured and overall formation transmissivity estimates (based on packer and pumping test analyses) are generally in the range $1x10^{-3}$ to $5x10^{-3}$ m^2/s.

Existing production boreholes typically penetrate deep into the multi-layered aquifer; for example, 187 m at the industrial site described in this paper.

3 REGIONAL SOLVENT POLLUTION

An initial search of borehole records indicated over one hundred within the study area, from just a few metres to over 200 m in depth. However, it was found that many of the boreholes had been lost due to site redevelopment and ultimately samples were collected and analysed from thirty-nine. Boreholes fell into four main categories: industrial production boreholes (13); rural production boreholes (6); public water supply boreholes (16); and disused, open boreholes (4). Water samples were analysed for five chlorinated solvents, these being TCE, TCA, trichloromethane (TCM), tetrachloromethane (TeCM) and tetrachloroethene (TeCE). Analysis took place by pentane liquid/liquid extraction and gas chromatography (GC) analysis using an on-column injector and an electron capture detector (Eklund *et al*., 1978). The results obtained are summarised in Table 1.

The results indicate widespread solvent pollution of Coventry groundwaters. The most ubiquitous solvent is TCE which is present in all groundwaters collected from within the urban area, and is even present in public water supply boreholes situated outside the main industrial areas. At one public water supply site situated south-west of the city centre an aeration treatment tower was fitted when levels of TCE of over 170 μg/l were detected. Figure 1 shows the TCE distribution in urban groundwaters.

1,1,1-trichloroethane was detected in all but one of the boreholes within the urban area. Concentrations were regionally much lower than

Table 1. Summary of results from regional sampling.

Description	TCE	TCA	TCM	TeCM	TeCE
% boreholes with >1 μg/l solvent	77	62	49	36	39
World Health Organisation (WHO, 1984) guideline value	30	-	30	3	10
% boreholes >WHO guideline value	34	-	0	16	11
Maximum concentration (μg/l)	1100	270	21	80	53
Mean concentration (μg/l)	67	10	2	5	4
Median concentration (μg/l)	10	1	1	0	1

TCE (Table 1). The probable reason for the relatively low TCA levels is that in the UK, TCA use only became widespread (generally replacing TCE) in the mid 1970's, so giving a longer interval for TCE pollution to occur.

Trichloromethane, TeCM and TeCE all occur less frequently than TCE and TCA and at generally lower concentrations. Trichloromethane has only ever been used in the UK on a small scale as an industrial solvent and so its occurrence in groundwater may be related to leakage of chlorinated mains water. Tetrachloroethene is rarely used in the UK as a metal degreaser but is a major dry cleaning fluid. None of the sites sampled were occupied by dry cleaners and so the occurrence of TeCE at a few sites may be due to impurities in other solvents. Tetrachloromethane, like TCM, has never been used as a major industrial solvent, although it was popular with smaller industries for cleaning components by wiping with rags and was also used in fire extinguishers. Since the toxicity of TeCM became known in the 1950's, its use was limited to the pharmaceutical industry and for the synthesis of fluorocarbons, neither of which are carried out in Coventry. The highest TeCM level found was at an old fire station where extinguishers containing TeCM were tested. High concentrations were found at some industrial sites where there is no present day use of TeCM and so contamination must be from past use. High levels of TeCM on these sites suggest that once solvents enter the subsurface they remain in the groundwater system causing pollution. If this is the case then even with the recent decline in the use of TCE, it is likely to remain a major pollutant of Coventry's groundwaters for many years.

4 SITE INVESTIGATION

The regional survey indicated significant spatial variations in chlorinated solvent pollution of groundwaters, which appear to be related, to an extent, to past and present land use. To compliment this survey and determine vertical variations in the system, a detailed investigation of an industrial site was performed. This investigation also presented the potential to apply various specially designed soil gas and groundwater sampling methods.

4.1 *Site description*

The industrial site described is situated within urban Coventry, approximately 1.5 km from the city centre. The site has been occupied by industry for about a century and chlorinated solvents have been used for approximately fifty years for liquid and vapour degreasing of metal (Department of Trade and Industry, 1990). The main solvent use switched from TCE to TCA about a decade ago, in response to national solvent use trends (Herbert *et al.*, 1986). The solvents were, until recently, stored in two above ground storage tanks (Figure 2). Barrels were filled manually at this location and transported to points of usage around the site. Prior to this study, the only data on solvent pollution pertaining to the site was of mixed groundwater pumped from the production borehole (Figure 2); TCE and TCA concentrations were in the order of 240 and 7 μg/l, respectively.

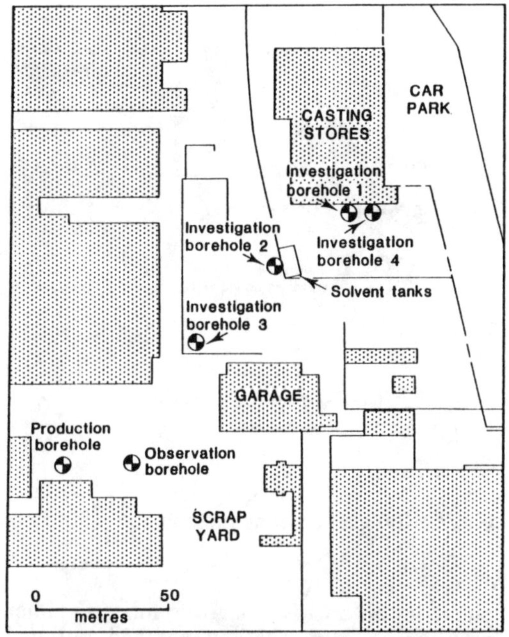

Fig. 2 Area of industrial site studied in detail

4.2 *Soil gas survey*

Surveying soil gas for volatile organic compounds is a rapid and convenient method of detecting subsurface contamination (Devitt *et al.*, 1987). In this study soil gas sampling was used to identify the most polluted area of the industrial site where investigation boreholes were drilled. A surveying method was developed by the authors and has been described in detail by Bishop *et al.* (1990). Briefly, a 1.1m stainless steel tube was driven into the ground and soil gas was drawn into a glass sampling bulb via teflon (PTFE) tubing (Figure 3). The contents of the sealed bulb were then transported to a laboratory for analysis using a GC fitted with an electron capture detector, 15 m non-polar 0.53 mm internal diameter capillary column and a direct on-column injector. The method was found to be very sensitive (detection limits <0.01 μg/l) and gave results comparable with those of an established on-site method (Marrin and Thompson, 1987).

A reconnaissance survey of accessible points around the entire site discovered pollution to be mainly by TCA (maximum concentration ~2000 μg/l) and TCE (maximum concentration ~50 μg/l). The highest solvent concentrations were detected in the area of the site immediately surrounding the solvent storage tanks (Figure 2). A detailed follow-up survey of soil gas in this area enabled contouring of solvent concentrations; the TCA distribution is shown in Figure 4. The origin of the soil gas pollution

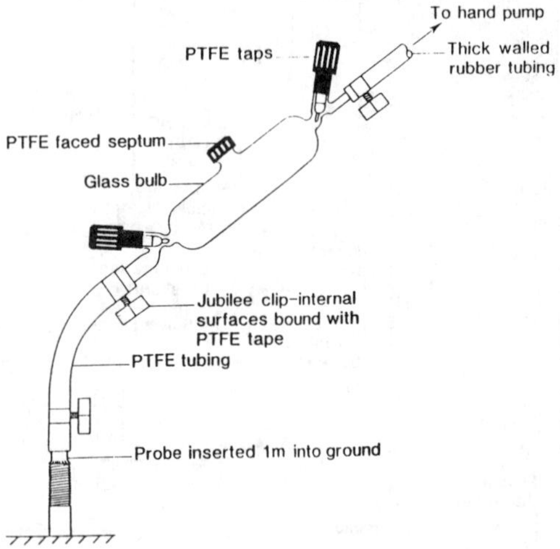

Fig. 3 Soil gas sampling method

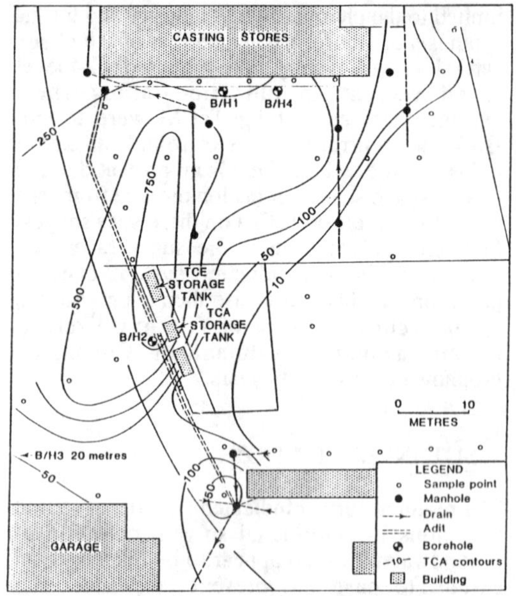

Fig. 4 1,1,1-trichloroethane concentrations in solvent tanks area

plume in this area is thought to relate to either partitioning from shallow perched polluted groundwater and/or solvent vapour movement in the unsaturated zone, possibly conducted along the lines of drains and pipes (Bishop *et al.*, 1990). However, the main result of the soil gas survey was to define a substantially polluted area of the site where, in the absence of other information, the drilling of investigation boreholes was most appropriate.

4.3 *Borehole construction*

Four investigation boreholes were drilled on the site in the solvent tanks area (Figure 2) and two methods of completion were used. Construction details of the three deep boreholes are shown in Figure 5; borehole 4 was only drilled to 7 metres below surface (mbs) and is not discussed in this paper. The site observation borehole (Figure 2) did not offer the potential for detailed vertical profile sampling.

Boreholes 1 and 3 were completed as open, lined boreholes with alternating slotted and plain thermoplastic casing sections placed adjacent to the major sandstone and mudstone units, respectively. Although the documented slumping nature of the mudstone units was anticipated to create the necessary hydraulic seals between individual aquifer units outside the plain casing, the downwards movement of mudstones to slotted casing sections had to be prevented. This was achieved by incorporating double rubber collars, or grommets, in the borehole annulus above screened casing sections (Figure 5).

Borehole 2 was completed with a dedicated sampling system, four gas-driven water samplers being placed adjacent to major saturated sandstone units and two gas samplers being placed adjacent to unsaturated units (Figure 5). The construction of the dedicated sampling system, described in detail by Bishop *et al.* (in press), was novel in that all functional units were housed within protective plain water well casing (Figure 6). This facilitated the backfilling operation outside the casing, which was achieved using sand, gravel, and bentonite to form the hydraulic seals (Figure 5). A key element of the four water sampling units was that they incorporated vibrating wire piezometers (Figure 6), enabling the measurement of the vertical hydraulic gradient in the borehole.

4.4 *Groundwater sampling methods and results*

Groundwater sampling from the three investigation boreholes described was performed in a variety of ways. There were two main objectives to the sampling:

1. To test the applicability of the sampling methods and particularly to compare results from the two open boreholes with those from the single dedicated borehole.

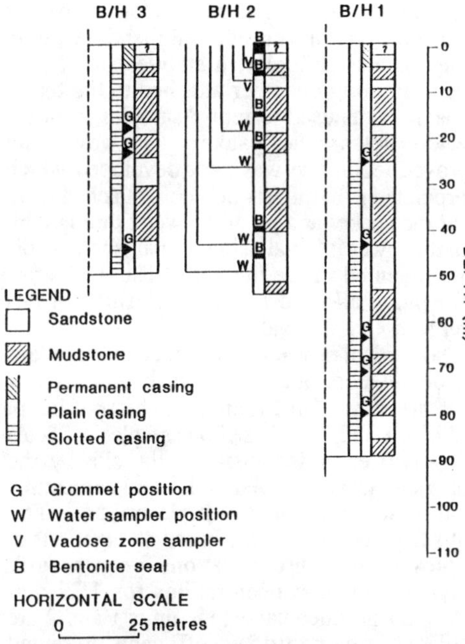

Fig. 5 Construction details of investigation boreholes

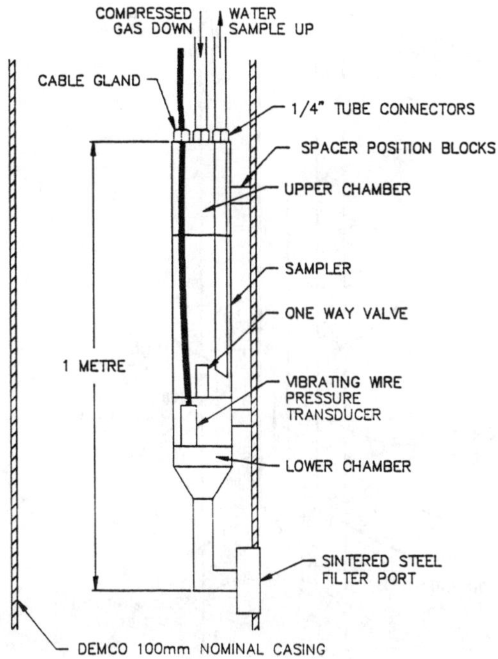

Fig. 6 Cross-section of gas-driven water sampler

2. To determine vertical solvent pollution profiles and the spatial pollution distribution over the relatively short distances between the boreholes.

Sampling from the dedicated system installed in borehole 2 was relatively simple and could be completed within an hour on each visit to the site. Compressed nitrogen was used to return groundwater samples to the surface, where they were collected at a purpose-designed manifold. Gas samples from the two upper unsaturated sandstone units were collected in glass bulbs and analysed in the same manner as for soil gas (Bishop *et al.*, 1990). A readout unit plugged into the same surface manifold was used to read groundwater heads from the four vibrating wire piezometers. The heads of the three individual sandstone units responded in varying degrees to the pumping cycles from the nearby production borehole (Figure 2), and recorded a downwards hydraulic gradient in the range 0.3 to 0.4.

Boreholes 1 and 3 were sampled using portable systems based on submersible pumps, gas-driven units and packers. A packer was used to divide the borehole into upper and lower sampling intervals.

Two techniques were used to collect samples from the upper zone above the packer. The first technique was to sample a narrow interval above the packer, with high rate pumping from the zone above eliminating vertical mixing in the sampling interval (Figure 7). Since a high downward hydraulic gradient existed at the site, it was important that the high rate pumping was sufficient to reverse the downward flow in the sampling zone. The second technique used above the packer was known as separation pumping. The principle of this method is based on the creation of a water divide in the borehole by diverting flow to two pumps, one near the water surface and one near the bottom of the sampling zone (Figure 8). The water entering the borehole around the water divide is assumed to originate from an equivalent level in the aquifer. By adjusting the rates of the two pumps, the position of the water divide could be adjusted to the required sampling level. In borehole 1, sufficient space existed to use a heat pulse flowmeter to locate the water divide and a sampling pump to return water directly from this level (Figure 8a). However, borehole 3 was 25 mm narrower in diameter and the vertical distribution of pollutants was calculated using simple flow/mass balance equations (Figure 8b). The upper and lowermost sampling interval concentrations were given directly from

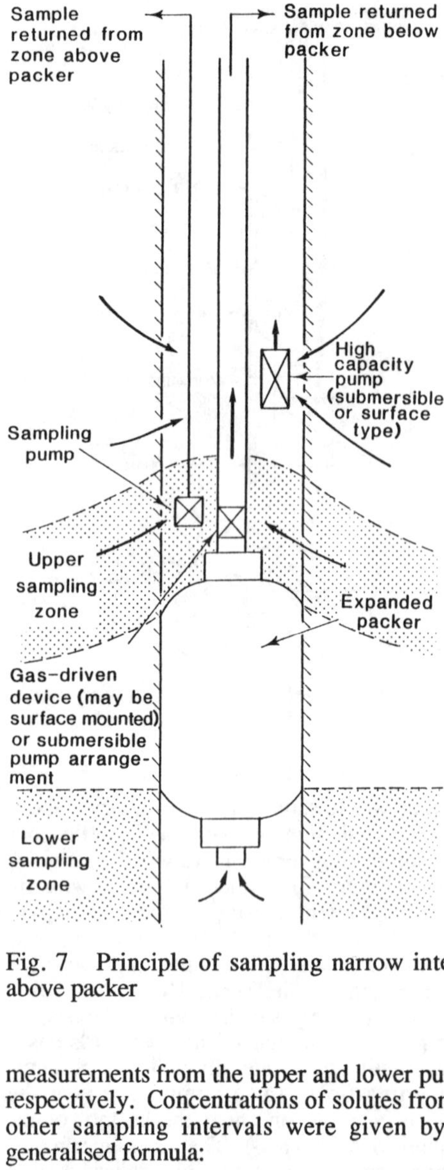

Fig. 7 Principle of sampling narrow interval above packer

measurements from the upper and lower pumps, respectively. Concentrations of solutes from the other sampling intervals were given by the generalised formula:

$$c_i=\{(Q_iC_i+Q'_{i-1}C'_{i-1})-(Q_{i-1}C_{i-1}+Q'_iC'_i)\}/2q_i$$

where Q and C represent flow rate and solute concentration from the upper pump, Q′ and C′ represent flow and concentration from the lower pump, q and c represent groundwater inflow and concentration from the sampling interval, and i represents the interval sampled. A similar approach has been adopted by Tellam and Eyre (1990) to define borehole inflow rates and conductivities.

Samples were returned from below the packer using both gas-driven and submersible pump arrangements. Initially, a stainless steel gas-driven sampling chamber attached to the top of the packer was used to return samples from the interval below the packer. However, an improved technique was later developed which incorporated a 50 mm suction hose attached to the top of the packer and pumping water through this from the lower interval using a small submersible pump mounted inside the hose. The 0.5 l/s rate of the pump allowed improved flushing of this lower sampling interval.

The vertical profiles of solvent concentrations in groundwaters obtained from boreholes 1, 2 and 3 are shown in Figure 9. The profiles in boreholes 1 and 3 are based on samples collected during two extended visits to the site by the Geological Survey of Denmark to implement the portable sampling methods described. The results from borehole 2 are based on eight sets of samples collected over a one year period. Equivalent water concentrations for TCE and TCA in unsaturated sandstone units 1 and 2 are calculated using partition coefficients presented by Eastwood *et al.* (in press).

Both TCE and TCA are detectable through the entire depth of each borehole. The most noticeable similarity between the individual profiles is that TCA concentrations exceed TCE concentrations at depths of up to 25 mbs. Below this TCE concentrations exceed those of TCA. The most probable explanation for this phenomenon relates to the history of solvent use on the site. Trichloroethene was first stored in this area approximately 20 years ago and has therefore had longer than TCA (introduced about ten years ago) to penetrate vertically into the system. It is also possible, however, that TCE penetration is enhanced by its relatively high density and low viscosity in relation to TCA (Schwille, 1988). The vertical penetration of both solvents is likely to be enhanced by the high vertical hydraulic gradient demonstrated.

The different sampling methods used yield comparable trends in the TCE:TCA ratio with depth. However, differences exist between the maximum concentrations attained by TCE in boreholes 1 and 3, and borehole 2. The reason for this is likely to be associated with the sampling technique, since the nylon tubing used in the dedicated system was subsequently shown to have significant sorption effects when relatively long lengths were used (Barcelona *et al.*, 1985; Bishop *et al.*, in press). It is therefore likely that TCE concentrations do attain the same maximum values between aquifer units 3 and 5 in all three boreholes.

Finally, with respect to spatial variations in

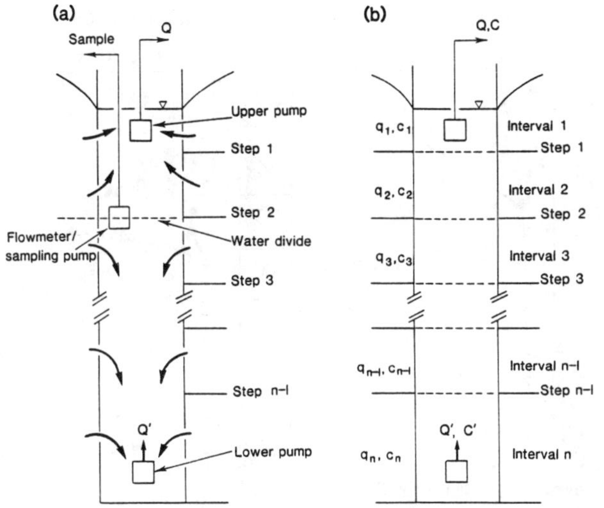

Flow balance	Mass balance
$Q_1 = q_1$	$Q_1C_1 = q_1c_1$
$Q_2 = Q_1 + q_2$ $Q'_2 = Q'_1 - q_2$	$Q_2C_2 = Q_1C_1 + q_2c_2$ $Q'_2C'_2 = Q'_1C'_1 - q_2c_2$
$Q_3 = Q_2 + q_3$ $Q'_3 = Q'_2 - q_3$	$Q_3C_3 = Q_2C_2 + q_3c_3$ $Q'_3C'_3 = Q'_2C'_2 - q_3c_3$
$Q_{n-1} = Q_{n-2} + q_{n-1}$ $Q'_{n-1} = Q'_{n-2} - q_{n-1}$	$Q_{n-1}C_{n-1} = Q_{n-2}C_{n-2} + q_{n-1}c_{n-1}$ $Q'_{n-1}C'_{n-1} = Q'_{n-2}C'_{n-2} - q_{n-1}c_{n-1}$
$Q'_{n-1} = q_n$	$Q'_{n-1}C'_{n-1} = q_nc_n$

Fig. 8 Principle of separation pumping

solvent concentrations between the three investigation boreholes, borehole 3 showed high concentrations of both TCE and TCA throughout its entire depth. Anomalously high chloride and TeCE concentrations in this borehole suggest that the aquifer in this location is more severely polluted than in the environs of boreholes 1 and 2. This highlights the significant spatial variations that may occur in solvent pollution over distances of some tens of metres on a single industrial site.

5. SUMMARY AND CONCLUSIONS

A regional survey of groundwater from existing production boreholes in the city of Coventry, UK, showed them to be polluted by chlorinated solvents. Pollution by the solvents TCE and TCA is most severe, TCE concentrations reaching as high as 1100 μg/l. The pollution is related to the use of chlorinated solvents by industry in the city and is likely to persist for many years.

Investigation of solvent pollution at an industrial site was performed using several specially developed field methods. A successful method was developed for sampling soil gas where samples were collected in glass bulbs in the field and transported to a laboratory for analysis. The soil gas survey delineated a localised area of the site, locating the bulk solvent storage tanks, which was severely polluted by both TCA and TCE. Three deep investigation boreholes were drilled in this area. Boreholes 1 and 3 were constructed as open boreholes, lined with alternating sections of plain and slotted casing. Borehole 2 was equipped with a novel system of dedicated water and gas samplers, and vibrating wire piezometers.

A variety of methodologies were employed to sample groundwaters from the boreholes. Borehole 2 was sampled by returning groundwater using compressed nitrogen, and unsaturated zone gas was sucked into glass bulbs using a similar method to that used for soil gas sampling. Boreholes 1 and 3 were sampled using portable systems based on submersible pumps, gas-driven units and packers. Packers were used to divide the boreholes into upper and lower sampling intervals. Two techniques were used to collect samples from above the packer. The first consisted of sampling a narrow interval directly above the packer, eliminating mixing in this zone by pumping. The second technique involved separation pumping, whereby a combination of pumps were used to separate flow upwards and downwards in the borehole. Level-determined samples were then either returned directly from the located water divide by a sampling pump, or a mass balance calculation was used to calculate the pollutant distribution. Samples were returned from the interval below the packer by either gas-driven or submersible pump arrangements.

The vertical profiles obtained from these boreholes showed a general decrease in TCA:TCE ratio with depth. This is interpreted as

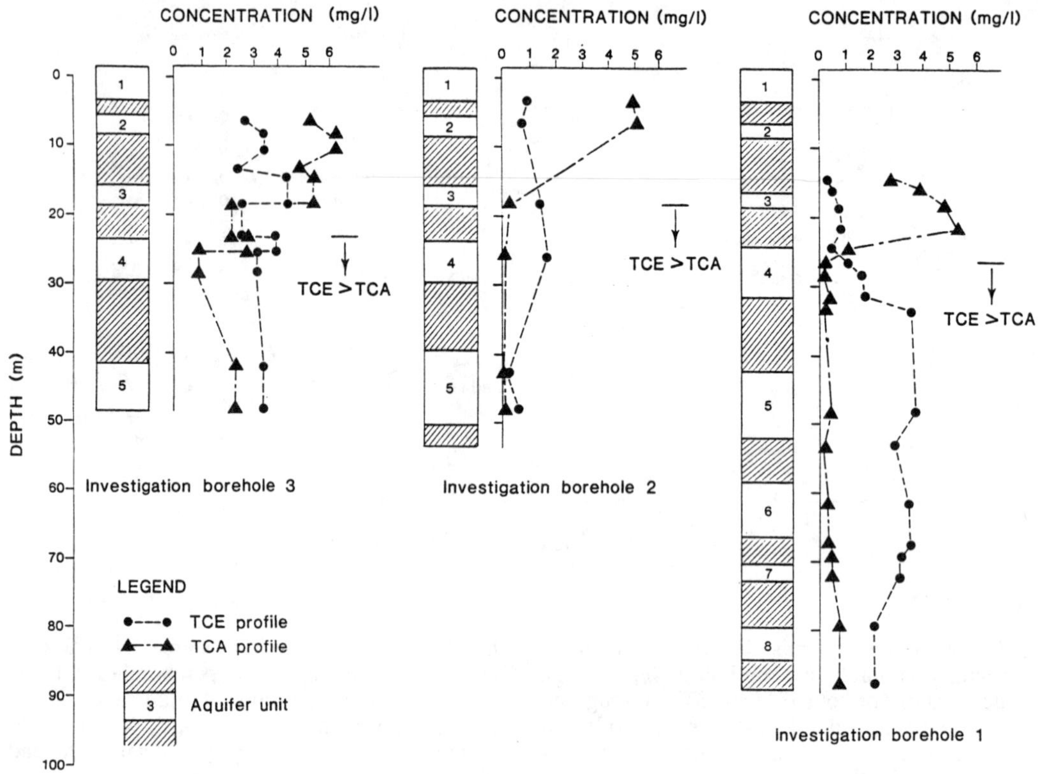

Fig. 9 Vertical chlorinated solvent profiles.

TCE pollution having had longer to penetrate deep into the multi-layered aquifer. These similar trends are thought to indicate a significant degree of compatability between the sampling methods used. However, relatively low TCE concentrations in samples returned from depth in borehole 2 are thought to be related to sorption onto the nylon tubing used in the dedicated system.

The vertical distribution of pollutants discovered on the industrial site, coupled with significant spatial variations over relatively short distances have obvious implications for the interpretation of the regional pollution study. The vertical profiles, in which TCA dominates over TCE in shallow aquifer units, suggest that rising levels of TCA will be observed in abstracted groundwater in the future as TCA pollution penetrates deeper into the multi-layered system. Rising levels of TCA in groundwater abstracted in Birmingham have been noted by Rivett *et al.* (1990). The significant variations in spatial pollution over very short distances suggest that regional correlation of solvent distributions between boreholes on different sites is precluded, particularly in light of the relatively complex multi-layered geology. As in Birmingham, the pollution of individual boreholes is likely to be most readily related to the use of the site on which they are located, the geology they penetrate, and their construction details (Rivett *et al.*, 1990).

ACKNOWLEDGEMENTS

This study forms part of a tri-national project entitled Coventry Groundwater Investigation and involving the University of Birmingham, the Geological Survey of Denmark and the Bureau de Recherches Géologiques et Minières, France. The project is funded by the Commission of the European Communities under contract no. EV4V-0101-C(BA). Thanks are due to Rasmus Jacobsen and Eric Clausen for assistance in the field, and the occupiers of the industrial site described for their co-operation.

REFERENCES

Barcelona, M.J., J.A. Helfrich, and E.E. Garske. 1985. Sampling tubing effects on groundwater samples. *Analytical Chemistry* 57: 460-464.

Besly, B. 1988. Palaeogeographic implications of late Westphalian to early Permian red-beds, Central England. In: B. Besly and G. Kelling (Editors), *Sedimentation in a syn-orogenic basement complex*: 200-221. Blackie.

Bishop, P.K., M.W. Burston, D.N. Lerner, P.R. Eastwood. 1990. Soil gas surveying of chlorinated solvents in relation to groundwater pollution studies. *Quarterly Journal of Engineering Geology, London* 23: 255-265.

Bishop, P.K., M.W. Burston, T. Chen, and D.N. Lerner. 1991. A low cost dedicated multi-level groundwater sampling system. *Quarterly Journal of Engineering Geology, London*, 24: 311-321 (in press at submission).

Cavallaro, A., C. Corradi, G. De Felice, and P. Grassi. 1985. Underground water pollution in Milan and the Province by industrial chlorinated-organic compounds. In: J.F. de L.G. Solbé (Editor), *Effects of land use on fresh waters* : 68-84. Chichester: Ellis Horwood.

Department of Trade and Industry. 1990. *Chlorinated solvent cleaning*. London: Her Majesty's Stationery Office.Devitt, D.A., R.B. Evans, W.A. Jury, T.R. Starks, B. Eklund, A. Gnolson and J.J. Van Ee. 1987. *Soil gas sensing for detection and mapping of volatile organics*. Ohio: National Water Well Association.

Devitt, D.A., R.B. Evans, W.A. Jury and T.R. Starks. 1987. *Soil gas sensing for detection and mapping of volatile organics*. Dublin: National Water Well Association.

Eastwood, P.R., D.N. Lerner, P.K. Bishop, and M.W. Burston. 1991. Identifying land contaminated by chlorinated hydrocarbon solvents. *Journal of the Institution of Water and Environmental Management*, 5: 163-171 (in press at submission).

Eklund, G., B. Josefsson, and C. Roos. 1978. Determination of volatile halogenated hydrocarbons in water by glass capillary gas chromatography and electron capture detection. *Journal of High Resolution Chromatography and Chromatography Communications* 7: 34-40.

Fusillo, T.V., J.J. Hochreiter, and D.G. Lord. 1985. Distribution of volatile organic compounds in a New Jersey coastal plain aquifer system. *Ground Water* 23: 354-360.

Herbert, P., P. Charbonier, L. Rivolta, M. Servais, F. Van Mensch and P. Campbell. 1986. The occurrence of chlorinated solvents in the environment. *Chemistry and Industry* 24: 861-869.

Marrin, D.L. and G.M. Thompson 1987. Gaseous behaviour of TCE overlying a contaminated aquifer. *Ground Water* 25: 21-27.

Rivett, M.O., D.N. Lerner, J.W. Lloyd and L. Clark 1990. Organic contamination of the Birmingham aquifer, UK. *Journal of Hydrology* 113: 307-323.

Schwille, F. 1988. *Dense chlorinated solvents in porous and fractured media*. Chelsea, Michigan: Lewis.

Tellam, J.H. and S.F. Eyre. 1990. Estimating rates and electrical conductivities of borehole inflows using a logging/two-level pumping technique. In: F.G. Bell, M.G. Culshaw, J.C. Cripps and J.R. Coffey (Editors), *Field Testing in Engineering Geology*: 335-343. Geological Society of London Special Publication No. 6.

World Health Organisation. 1984. *Guidelines for drinking water quality: Vol. 1, Recommendations*: 62-64. Geneva: WHO.

Subsurface Contamination by Immiscible Fluids, Weyer (ed.) © 1992 Balkema, Rotterdam. ISBN 90 6191 175 3

Monitoring systems at industrial sites: The Uiterburen case

A.G.Veltkamp
Nederlandse Aardolie Maatschappij b.v., Environmental Department, Assen, Netherlands

ABSTRACT: The "Nederlandse Aardolie Maatschappij" (NAM) exploits some 800 onshore oil and gas producing sites. During the last few years the company has formulated its policy on soil management for these sites. Part of this policy is to investigate soil quality to identify the risks of contaminated soil and groundwater and to take clean-up measures in order to control the risks. Monitoring systems play an important role at the various stages in this process. In this paper, the need and development of monitoring systems are discussed. Special emphasis will be paid to hydrogeological aspects. On the basis of a case history (natural gas plant Uiterburen) the approach, details of the system and practical experiences are presented. Monitoring systems are shown to be useful and profitable at industrial sites.

1 INTRODUCTION

The exploration and production of oil and gas is a major onshore/offshore industry in the Netherlands. The "Nederlandse Aardolie Maatschappij" (NAM) exploits some 800 onshore oil and gas production sites and 17 offshore gas platforms.
Production and maintenance activities may cause soil contamination at these sites.

During the last few years NAM has formulated its policy of soil management for the onshore sites. Soil management requires a broad scale of information. Monitoring systems provide valuable information on the quality of the soil, including descriptive data, analyses and evaluation. These data are used for implementing decisions concerning soil protection. For many reasons, regular monitoring of soil quality on and around industrial sites is necessary. These include: soil protection, management, clean-up activities and/or cost reduction. Monitoring requires routine systems which should be operated over a long period.

The following elements of monitoring systems are important:
- lay-out;
- collection of data;
- data storage and processing;
- organization and management.

For an optimal monitoring system these elements should be well coordinated. In this paper attention will be paid to the lay-out of monitoring systems and collection of data. The lay-out of these systems depends on the purpose of monitoring and the nature of the collected data. The nature of these data is connected with the type of contamination and the migration possibilities (geophysical and hydrogeological aspects). These aspects differ between industrial sites. Therefore a monitoring system is site specific.

Industrial site monitoring in the broadest sense includes monitoring of the state of the surrounding environment taking into account the geophysical, hydrogeological, ecological and economic aspects.

This paper deals with the need for monitoring systems and the development of those systems. Special emphasis will be paid to hydrogeological aspects. On the basis of a case situation (natural gas plant Uiterburen), the approach, details of the system and practical experiences will be presented.
Monitoring activities, which are not dealt with in this paper include the monitoring of the quality of the influent and effluent of treatment installations, the monitoring of infiltration discharges and monitoring of soil moisture content.

2 NEED FOR MONITORING

During the production process on industrial sites contaminants may enter the environment by several means (i.e. leakage, waste and calamities). These emissions can cause soil and groundwater contamination resulting in unacceptable levels which form a risk for human health and/or cause adverse effects on the environment, capital goods and industrial infrastructure. This situation calls for actions to be taken in order to restore the quality of soil and groundwater to an acceptable level.

The sequence of production process, emissions, contamination and (unacceptable) risks requires a constant check of soil and groundwater quality to see if any measures should be taken. Monitoring systems provide the most important information on and around industrial sites. The lay-out of these local, site specific systems strongly depends on the purpose of monitoring (see Table 1).

All kinds of monitoring systems can be developed for the different purposes. The implementation is usually done in phases.

Limited care of the environment on and around industrial sites may result in great economic damage as the costs of remedial actions increase when the distribution of contamination in the soil spreads (horizontally as well as vertically).

Table 1. Purposes of monitoring

	MONITORING SYSTEMS		
	production - emissions - contamination - risks - remedial actions →		
MONITORING PURPOSES	warning	setting priorities	restoration
	protection	planning	adjusting clean-up processes
	prevention	strategy	aftercare
	control	determine methods of removal	evaluation
	cost reduction		

The next figure shows the costs of remedial actions for the natural gas plant at Uiterburen, as a function of time since contamination first occurred.

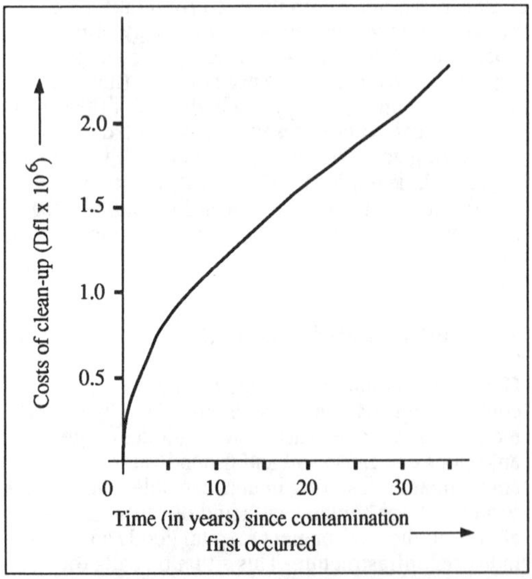

Figure 1. Costs of clean-up as a function of time

The contamination at the Uiterburen site dates from 15 to 20 years ago and the costs of clean-up amount to Dfl. 1,500,000.00. The costs of a preventive groundwater monitoring system for such a gas plant amount to only Dfl. 10,000.00.

If a preventive monitoring system had been installed at the start of the gas production, one would have identified the contamination at an earlier stage. Therefore adequate measures could have been taken at a lower price than the Dfl. 1,500,000.00 mentioned above. Monitoring systems (preventive or warning) would have been profitable in this case. Generally speaking, monitoring systems may prove to be a good investment for most industrial sites.

3 DEVELOPMENT OF MONITORING SYSTEMS AT INDUSTRIAL SITES

As mentioned in the previous section, the development of monitoring systems at industrial sites depends on the purpose of monitoring.

The design of the systems relates to the nature of the parameters involved. In this respect two elements are important: the type of contaminant, and the geophysical interaction between contaminant and soil (migration possibilities).

As part of NAM's Environmental Policy, soil investigations are carried out at most of the NAM industrial sites. These investigations are carried out in different phases. Table 2 gives a general view of the different types of investigation.

Table 2. Types of soil and groundwater investigations

INVESTIGATION PURPOSE	
First investigation	Determine the presence of soil contamination at the industrial site
Supplementary investigation	Determine the risk as a function of soil and groundwater quality, migration possibilities, land and water use
Remedial investigation	Design of remedial action
Implementation of remedial actions	Restoration of soil and ground-water quality to an acceptable level
After-care	Control of desirable soil and water quality

The change-over from one investigation to the next, is based on several decision making criteria, i.e. quality levels, migration possibilities, land/water use and apparent risks.

In the process of these investigations, the number and type of monitoring systems increase. With each phase in the investigation, knowledge of the contamination level improves. This makes it possible to, successively, arrange the lay-out of monitoring systems in an optimal way.

Because the type of contaminants and the geophysical aspects differ, the lay-out of monitoring systems is generally site specifical designed. On the basis of a case-study at the gas plant Uiterburen the various steps in the development of a monitoring system will be demonstrated.

4 MONITORING HISTORY GAS PLANT UITERBUREN

The gas production site Uiterburen is situated in the province of Groningen in the northern part of the Netherlands (see Figure 2).

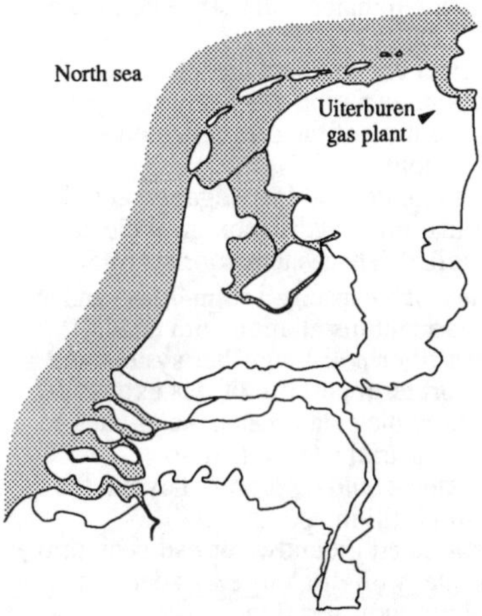

Figure 2. Location of gas plant Uiterburen

Geologically this part of the Netherlands is situated in the holocene, sea clay area of northern Netherlands. At the plant natural gas is produced and treated (± 15 million m^3/day). The gas plant includes a well field (10 wells) and a treatment installation. A general view of the gas producing site Uiterburen is shown in Figure 3.

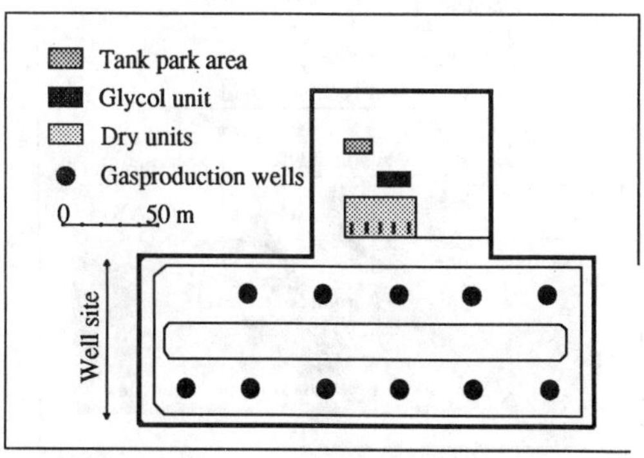

Figure 3. General view of gas plant Uiterburen

The natural gas is treated in 5 low temperature units. The condensate, separated from the gas is a mixture of hydrocarbons (C8 to C20), partly consisting of volatile aromatic hydrocarbons (e.g. benzene), which are easily soluble in water (max. 1780 mg/l).

The first soil investigation at this plant, carried out in 1986, revealed that a considerable part of the soil and the groundwater was contaminated with these volatile aromatic hydrocarbons.

The hydrogeological situation (Fig. 4) consists of a sandy aquifer of 5 to 8 meters thickness, covered by a holocene layer of about 1 meter of clay. As a result of construction activities, this clayey layer is not present everywhere. In order to prepare the site for building, the surface level was raised approximately 1 meter by adding a layer of sandy material. The base of the aquifer system is formed by a thick pot clay layer. This pot clay has a very low permeability, i.e. $5{,}78.10^{-10}$ m/sec.

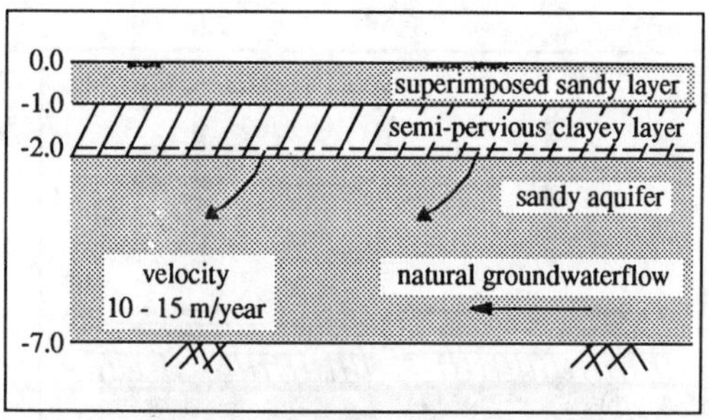

Figure 4. Hydrogeological cross-section of the subsoil

The gas plant Uiterburen is an infiltration area. Part of the precipitation penetrates the holocene clayey layer into the underlying sandy aquifer.

Investigations at the plant aswell as outside of it, showed that a major part of the plant area was a source of contamination. Especially in and around the tank park area, high concentrations of aromatic hydrocarbons were found in the upper clay layer. Vertically, the contamination plume had migrated downward into the aquifer. Because aromatic hydrocarbons are easily soluble in water, the contamination had spread horizontally through the aquifer by natural groundwater flow. Figure 5 shows the horizontal extent of benzene contamination in 1986, and Figure 6 shows a cross-section through the contamination plume.

Both figures indicate that the contamination is found beyond the industrial site, showing concentration levels above the current official Dutch quality levels ("ABC levels"= Volatile aromatic hydrocarbon; VAH). This may cause risks for the use of groundwater.

In a first soil investigation the following monitoring elements were used: drilling operations

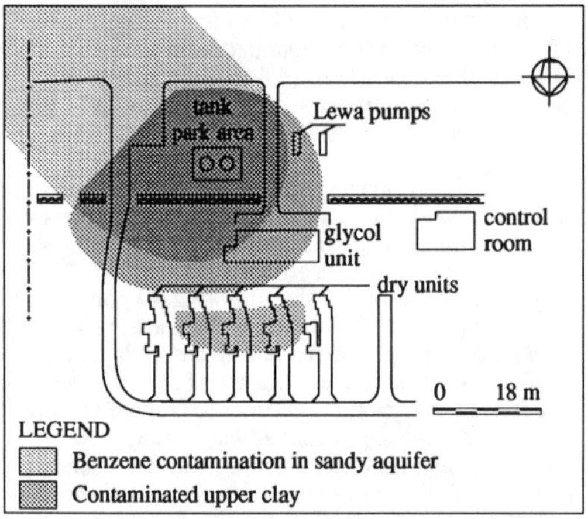

Figure 5. Horizontal extent of benzene contamination in aquifer and upper clay layer

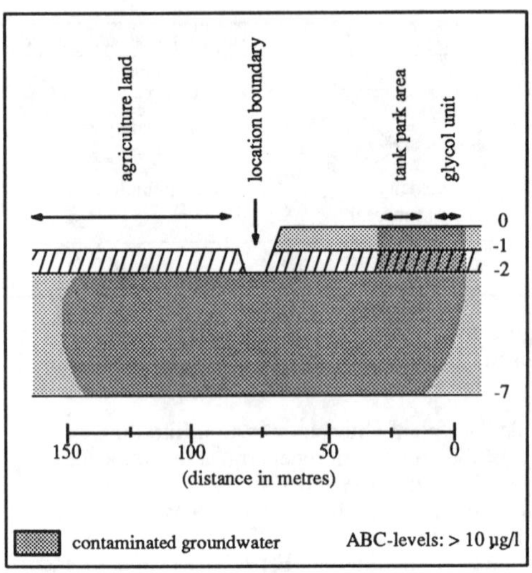

Figure 6. Cross section through contamination plume

to 3 meters below surface level, description of soil material, soil sampling and analyses, groundwater sampling and analyses , and determination of groundwater level and flow pattern.

Due to the fact that contamination had accumulated beyond the gas plant at unacceptable levels, a supplementary investigation was carried out in 1987. The purpose of this investigation was to determine the exact extent of the soil and groundwater contamination and the risks involved. Based on the results of the fist investigation, a choise of groundwater sampling locations was made, and the number of sampling locations (both soil and groundwater) was increased.

The supplementary investigations indicated that the state of contamination was undesirable and that remedial actions should be taken to clean up the contaminated soil and groundwater in order to reduce the risks and control contamination.

The strategy chosen for the remedial action was presented later by Veltkamp et al. 1988 and Satijn et al. 1988.

The selection of remedial actions was based on contamination level, subsoil quality, hydrological situation, available time and costs. Special attention was paid to investigate the development of specific clean-up techniques which are adapted to the particular circumstances at NAM-sites.
Briefly, the following remedial actions were taken:
1. Removal of contaminated soil at the site as far as possible;
2. Hydrogeological containment and clean-up of the contaminated groundwater in the sandy aquifer;
3. In situ biorestoration of the contaminated soil under the installations.

Based on hydrogeological investigation and the state of contamination, a hydrogeological clean-up system was installed. The system consists of 5 extraction wells (total capacity 150 m^3/day) and a groundwater treatment installation with an air stripper/compost filter/polishing filter system and a biorotor. The short term purpose of this extraction system was to draw the cotaminated water back to underneath the industrial site, so that no contaminated water would be present beyond the site. This system is still in operation.
After the contaminated groundwater had been drawn underneath the site, well discharge was decreased in order to control groundwater flow.

Measures for stimulation of the in situ biorestoration were taken. These measures include an infiltration system and a soil venting system for aerating the unsaturated zone (see Figure 7).

The infiltration system consists of a series of horizontal drains. These drains were placed just above the contaminated clay layer. Treated

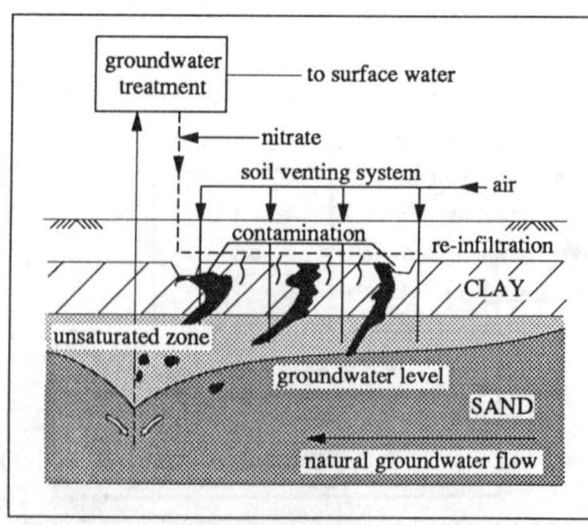

Figure 7. Scheme of biorestoration

groundwater from the biorotor is used for infiltration and nitrate is periodically added as a nutrient. Infiltration leads to better moisture conditions in the soil. The soil venting system was installed in order to obtain better aerobic conditions for in situ biorestoration.

In order to obtain adequate registration of the progress of the various processes, a monitoring system was installed.
On the basis of the monitoring data, decisions could be made to improve the remedial actions. Partly, a selection of existing monitoring wells from earlier investigations was used but additional monitoring elements were also installed as:

- monitoring wells;
- soil moisture observation wells;
- soil vapour sampling apparatus.

An outline of the entire monitoring system is presented in Figure 8.

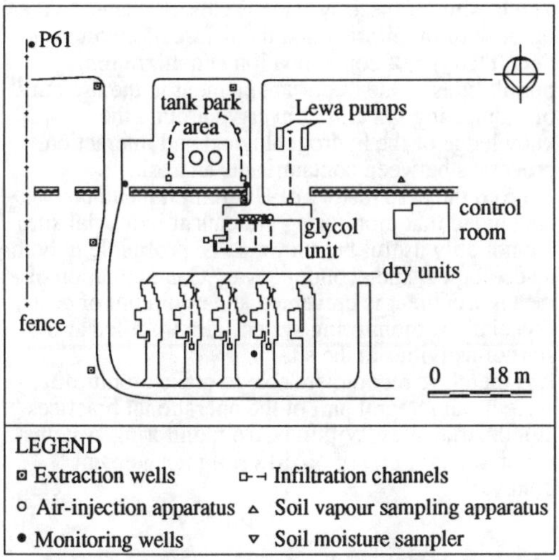

Figure 8. Outline of the monitoring system at Uiterburen natural gas plant

5 SOME PRACTICAL EXPERIENCES

The monitoring of a contamination plume implies that groundwater samples need to be taken on a regular basis, while physical parameters, such as groundwater levels and yields, must also be monitored and recorded.

This section deals with the practical findings at the Uiterburen gas production site.

The groundwater levels of the piezometers on and outside the Uiterburen plant were and still are regulary recorded. On the one hand this serves to define the groundwater flow pattern for the extraction regime employed, in order to discover whether the contamination plume is within the sphere of influence of wells. On the other hand, it serves to determine the thickness of the unsaturated zone. This is important for in situ biorestoration. The measured data indicated that the selected well configuration and the employed extraction regime meet both aspects.

An important element of clean-up operations is the determination of changes in contamination levels in both the monitoring wells and the extraction wells. This is important to ascertain whether and at what rate contaminants flow in the direction of the wells. On the basis of these findings, the extraction regime can be adjusted or optimized.
If the contamination has decreased to an acceptable level this may lead to a decision to reduce the groundwater extraction regime. Figure 9 gives the aromatic hydrocarbon concentrations in two monitoring wells.

At one point in time the contamination level of the groundwater at Uiterburen was below the acceptable level. Furthermore the contaminated groundwater had been drawn underneath the gas production site. On the basis of these findings the extraction regime was reduced from 150 m^3/day to approximately 30 m^3/day.
An interesting phenomenon was recorded during the monitoring programme, namely that the concentration levels of VAH in the extraction well water were many times higher than the concentration levels in some monitoring wells. These monitoring wells were situated within the sphere of influence of the soil venting system. Special experiments were carried out at these wells. One experiment dealt with prolonged pumping and frequent sample-taking. The samples were analysed for VAH and oxygen level. Figure 10 gives a survey of the results.

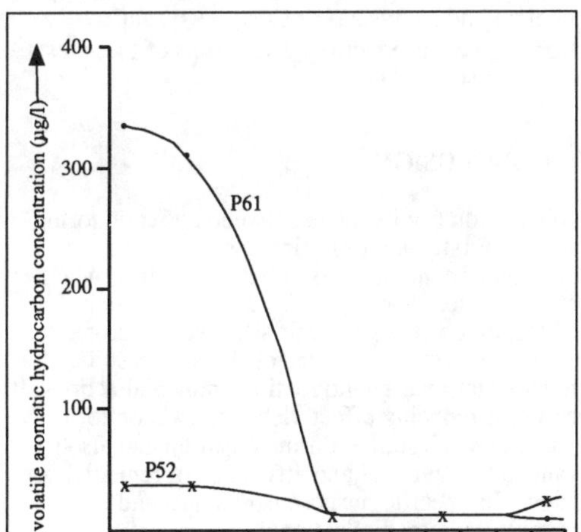

Figure 9. Volatile aromatic hydrocarbon concentrations (VAH) in two monitoring wells

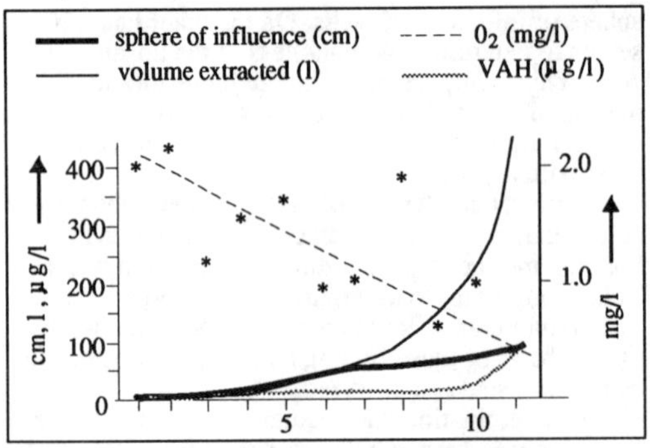

Figure 10. Pumping experiment at monitoring well 60, Uiterburen, d.d. 14-06-1989. The water extracted was analysed for VAH and oxygen content. The volume extracted and sphere of influence of the extracted water were also recorded.

From the changes of the levels of VAH and the oxygen level, one becomes aware of the fact that the water sampled from the monitoring wells is to a high degree aerobic, which is favourable for the destruction of VAH. The more water is extracted from the monitoring wells, the less aerobic it becomes and the more VAH it contains. The results of this experiment show that it is of the utmost importance to collect representative groundwater samples i.e. representative of the actual quality of groundwater in natural conditions.

In order to observe the in situ biorestoration it is essential that the biological activity is monitored. This was achieved in a number of ways, i.e. by counting the number of micro-organisms, soil air sampling and analysis (O_2, CO_2, VAH) and recording heat production. The results of these tests will be presented later.

6 CONCLUSION

A distinction can be made between two monitoring systems. Firstly, a monitoring system for the detection of contaminants at industrial sites (whether before or after clean-up). This system of monitoring is directed towards preventions as well as protection. Its purpose is to quickly detect changes in soil and groundwater quality and to direct remedial actions. It has a cost-reducing effect. Secondly, a monitoring system that is installed during clean-up and also to evaluate the outcome and efficiency of remedial actions. Its specific purpose is to adjust and to optimize the remedial actions.

The results and performance of the monitoring system used in the clean-up at the natural gas plant Uiterburen showed the following:

The recording of groundwater levels and the monitoring of contamination levels in the groundwater were very important for the evaluation and decision making concerning the clean-up of contaminated groundwater. On the basis of these data, decisions could be taken for the adjustment and optimization of the extraction regime. The location of the monitoring wells is determined by the hydrogeological conditions and the soil structure. It is essential that groundwater collected represents the groundwater quality at that particular location. Disturbing influences should be avoided.

Another element which came to light in the case study of Uiterburen was that soil vapour quality measurements, temperature measurements and microbiological investigations proved to be excellent techniques by which to measure the increased biological activity due to the degradation of contaminants (VAH). In situ biorestoration proved to be feasible in a practical way.

In summary, monitoring systems at industrial sites are essential for many reasons, i.e. protection, prevention, control, planning, evaluation and adjustment of clean-up systems.
The lay-out of these systems is closely related to the purpose of monitoring and the nature of monitoring data (the type of contamination and migration possibilities). One important element in the lay-out of monitoring systems converges around the knowledge of the hydrogeological and interaction processes between contaminants and soil.

From the case history of Uiterburen it can be concluded that monitoring systems at industrial sites are not only useful but may also be profitable in both an ecological and economic way. Contamination of the environment is prevented at a minimum of costs, especially if monitoring systems are installed at the start of activities at the site.
It is therefore recommendable to make monitoring systems an integral part of the operational practices at industrial sites. Within NAM monitoring systems are an integral part of NAM's soil management policy.

REFERENCES:

Veltkamp, A.G. & Jacobs, R.P.W.M. 1988. Soil management, an industry-specific approach. In Wolf, K., Van den Brink, W.J. & Colin, F.J. (ed.), Conference paper Contaminated Soil, Hamburg.

Satijn, H.M.C. & De Boks, P.A. 1988. Biorestoration, a technique for remedial action on industrial sites. In Wolf, K., Van den Brink, W.J. & Colin, F.J. (ed.), Conference paper Contaminated Soil, Hamburg.

Subsurface Contamination by Immiscible Fluids, Weyer (ed.) © 1992 Balkema, Rotterdam. ISBN 90 6191 175 3

Observations on the phase separation of penta/tetra chlorophenol following a spill to a high velocity aquifer

J.W.Atwater
University of British Columbia, Vancouver, B.C., Canada

ABSTRACT: A spill of a dissociated solution of sodium penta and tetra chlorophenate occurred into a highly permeable aquifer, located in an alluvial fan. A hydrogeological investigation was carried out to locate the contaminant plume culminating in a recovery operation. The spill broke into at least two plumes as a result of the geology. One plume moved in the order of 2.6 metres per day, the other at about 1.3 metres per day. Partitioning of one phase of the contaminant likely occurred and would provide an explanation for a shift in the ratio of the spilled constituents. While partitioning would provide the sink for the chlorophenates the differences in per cent dissociation resulted in a difference in the availability for partitioning and subsequent uptake of the two chlorophenols.

1 INTRODUCTION

In most situations where an organic contaminant enters a groundwater there are certain expectations as to the behaviour within the aquifer. Those expectations are based on a knowledge of the specific gravity, solubility and organic partitioning characteristics of the contaminant as well as a knowledge of the organic content and hydrogeologic characteristics of the aquifer. When some 18000 litres of a 0.75 percent solution of sodium penta/tetrachlorophenate (Na-PCP/TTCP) leaked into a high rate fluvial aquifer there were certain expectations as to how the spilt material would behave based on at that time, limited chemical and aquifer information. As it turned out what was observed was only in part what was expected. Presented in this paper is a brief description of the pesticide spill including the geology of the site, observations as to the chemistry and behaviour of the chlorophenate solution and some conjecture as to what possibly occurred.

2 BACKGROUND

On October 27 and 28, 1978 approximately 18000 litres of a dilute solution of a Na-PCP/TTCP leaked from a newly constructed dip tank. The dip tank was located at a sawmill;l adjacent to and some 120 metres from the Okanagan River in Penticton, British Columbia. The location of the site is shown on Figure 1. The solution, a registered pesticide, was a formulation used to prevent sap staining of freshly sawn lumber destined for overseas shipment. Stacks of lumber were lowered into the tank full of the solution and the moved out into the storage yard. The concentrated formulation contained 16.38 % sodium tetrachlorophenate (NaTTCP), 7.62 % sodium pentachlorophenate (NaPCP) and 2 % sodium tetraborate as a buffer. For dipping, the concentrate was diluted and at the time of the leak was about 3 % with a specific gravity of about 1.007. The solution that leaked through the tank walls contained 5000 mg/l of NaTTCP and 2500 mg/l of NaPCP. A test pit dug 2.6 metres from the tank revealed the water table to be some 5.2 metres below the ground surface and that the intervening material to be coarse

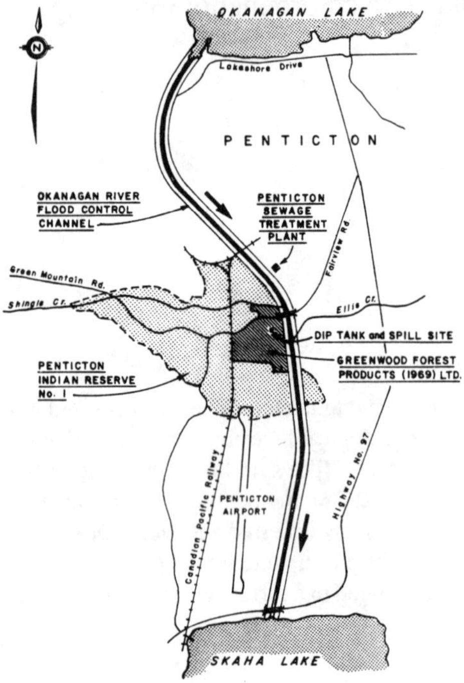

Fig. 1 Location and site plan

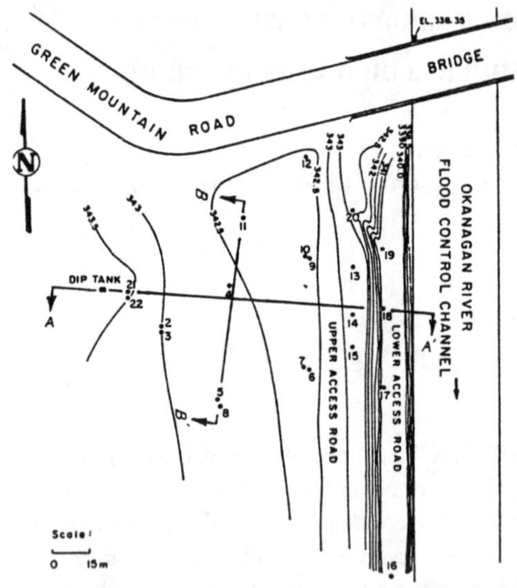

Fig. 2 Piezometer locations and site topography

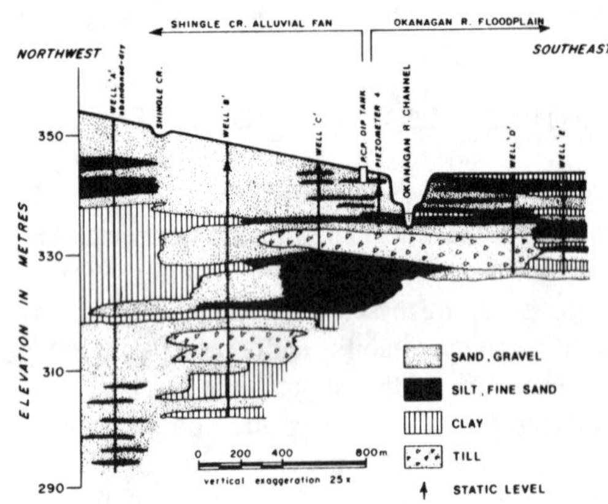

Fig. 3 Cross section of regional geology

sands and gravels. No evidence of the leaked solution was found. A hydrogeological investigation was carried out to locate and determine the concentration of the contaminate in the groundwater. The investigation was carried out in two phases over a period of 48 days and culminated with a decision to install recovery wells in order to prevent the chlorophenates entering the Okanagan River.

It was found that the chlorophenate flowed down through some 5.2 metres of unsaturated material to a shallow 2.1 thick aquifer isolated from underlying sands and gravel by a silt clay layer. The underlying sands and gravel were in turn overlying an aquifer that was confined at about 12 metres depth and was locally developed for domestic water supplies. Presented in Figure 2 is a plan of the site showing piezometer locations and site topography. Figure 3 is a cross section constructed from available domestic well logs showing the near region geology, only Well "C" was on the site. As can be seen from the section the spill was into an alluvial fan. Presented in Figure 4 are two cross sections constructed from the piezometer well logs. With the exception of two wells the piezometers were screened across the water table. As shown the aquifer is primarily made up of sands with the upper most portion in gravel. In general the material was coarser and better sorted moving towards Shingle Creek which at this time was some 100 metres north of the spill. The discharge point for the aquifer was the Okanagan River. Figures 5 and 6 are plots of the groundwater elevations prior to and after the start

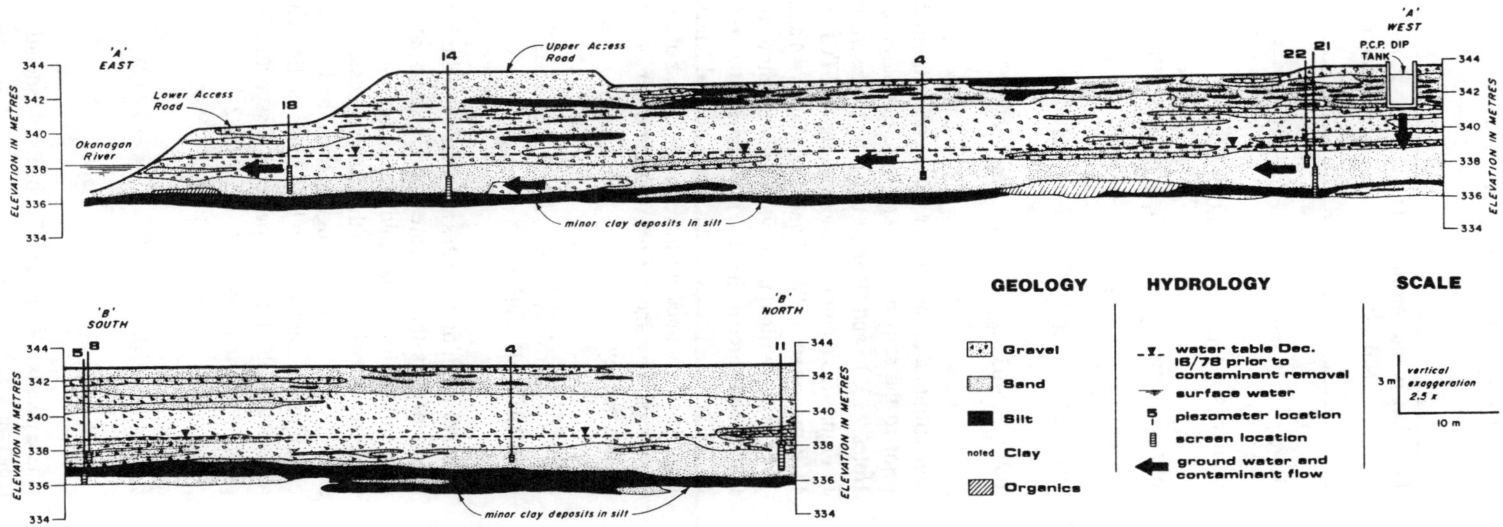

Fig. 4 Cross sections of site surficial geology

up of the recovery wells respectively. As can be seen from Figure 5 the flow was almost due east from the dip tank before swinging south-east adjacent to the river. It is felt that there may have been a slight northerly trend in the area of the dip tank in response to the cleaner aquifer material however there were not enough wells put in to provide adequate definition. From Figure 6 it can be seen that the gradients upstream of Production well # 2 changed but not appreciably with the pumping rate of ~1.35 l/sec. Slug tests carried out on holes 2 to 12 suggested hydraulic conductivities in excess of 100 to 1000 metres per day. The gradient (dh/dl) between the dip tank and the river was 0.0075 and coupled with the suggested hydraulic conductivities indicated a very fast groundwater flow. Velocities were never measured directly however calculations two ways provide an indication of the range of velocities. Pumping tests carried out on Production Wells 1 and 2 gave hydraulic conductivities(K) for the aquifer of 111 and 130 metres per day respectively. Patterson and Leibscher (1987) found a porosity (n) of 38 % for aquifer material adjacent to the Okanagan River. Given K, dh/dl and n the average linear groundwater velocity (v) was calculated to be 2.19 to 2.57 metres / day using the equation

$$V = V_s / n = K * dh / dl \quad (1)$$

where V_s is specific yield. No attempt was made to estimate an effective porosity. A parallel calculation using the arrival time of the Na-PCP/TTCP front at the Okanagan River of 47 days and a travel distance of 122 metres gave a value of 2.59 metres per day which is in reasonable agreement with the above high end value. The arrival time at the river is bracketed to within a day. The chlorophenate moved in the aquifer as a series of pulses. Presented in Table 1 are peak concentrations of NaPCP measured at various points along with calculated travel rates and where applicable the ratio of the pumping rate to rate of the leak ie. the chlorophenate was assumed to entered the aquifer over a 2 day period at a rate of 9000 litres per day.. The pumps would have drawn in extraneous water diluting the plume concentration however to what extent is unknown. Using the ratio one can arrive at an order of magnitude estimate of the peak

Table 1. Measured peak concentrations and travel rates.

Location	Concentration NaPCP	Travel Rate	Pump Ratio
River	18 mg/l	2.6 m / day	—
Well # 13	50 mg/l	> 2.0 m / day	3.5
PW # 1	20 mg/l	1.4 m / day	9.5
PW # 2	25 mg/l	1.3 m / day	13.
PW # 2	63 mg/l	1.15 m / day	13.

concentration in the aquifer. Peak concentrations prior to the start of recovery were measured in Holes # 13, 15 and 18 at 51 to 61 mg/l. Presented in Figure 7 is a plot of concentrations of TTCP to PCP measured at PW # 1 and 2 and Wells # 4 and 13. The curve fitted to the values gives a slope of 0.98 as compared to the ratio at the time of the spill of 2.1 to 1. The peak and front arrival times suggest two plumes one travelling in excess of 2.0 metres per day and the other in the order of 1.3 metres per day.

3 CONTAMINANT AND AQUIFER CHARACTERISTICS

The sodium chlorophenates are weak acid salts. At elevated pH's the salts are entirely dissociated with solubilities in the hundreds of thousands of mg/l. As the pH decreases the chlorophenate ions start to pick up hydrogen ions and form chlorophenol molecules which have a limited aqueous solubility. The per cent dissociated is given by the relationship (NRCC 1982)

$$\% \text{ diss.} = \frac{100 \times 10^{-pKa}}{10^{-pH} + 10^{-pKa}} \quad (2)$$

where pKa is the negative log of the dissociation constant.

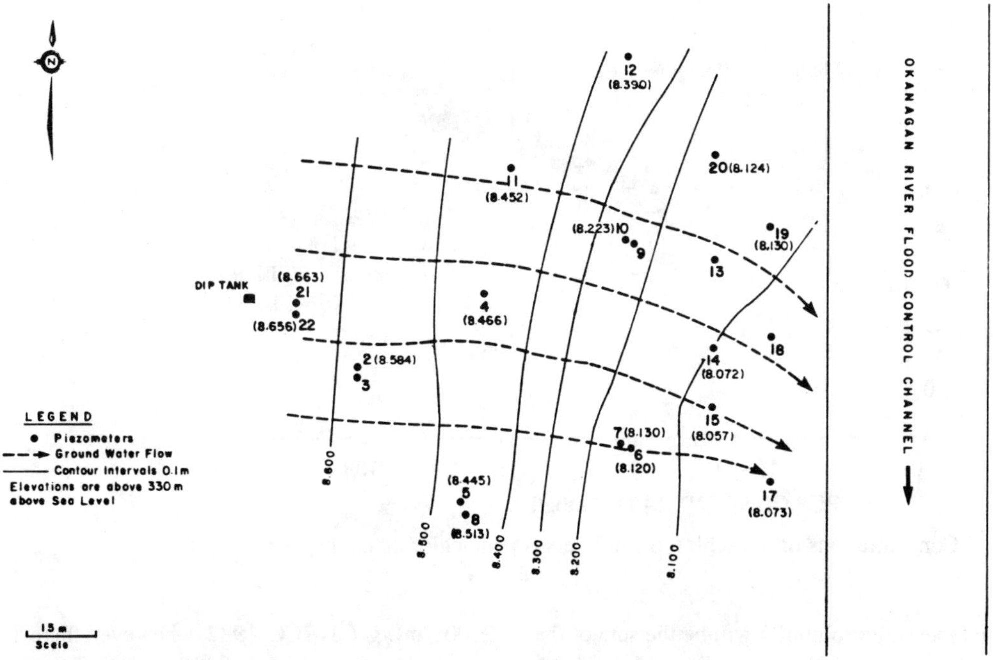

Fig. 5 Groundwater contours prior to recovery

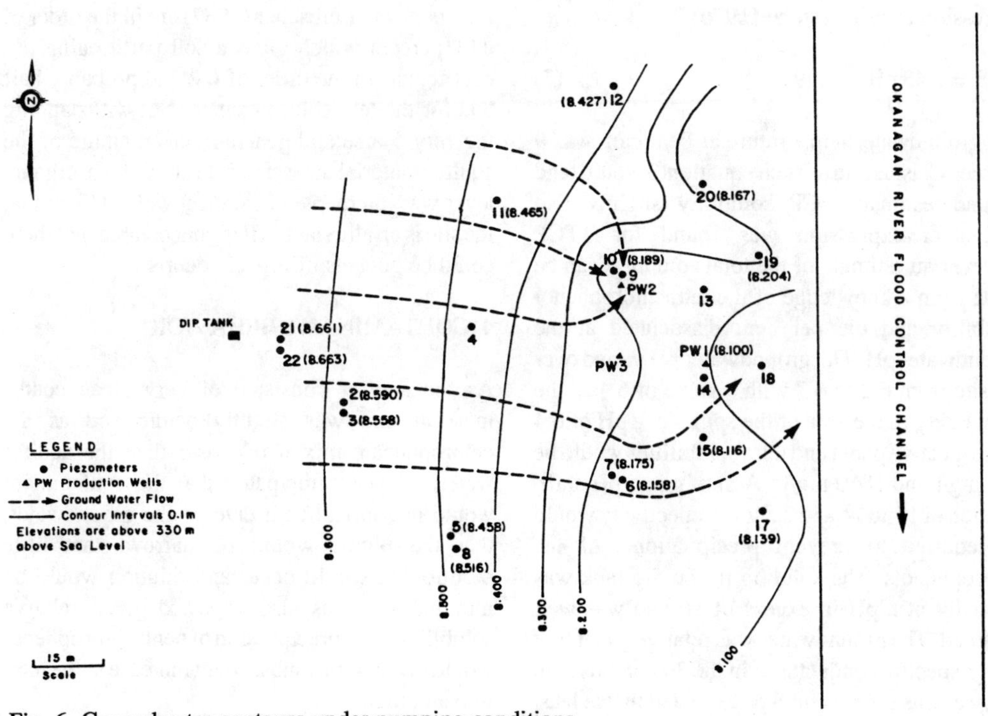

Fig. 6 Groundwater contours under pumping conditions

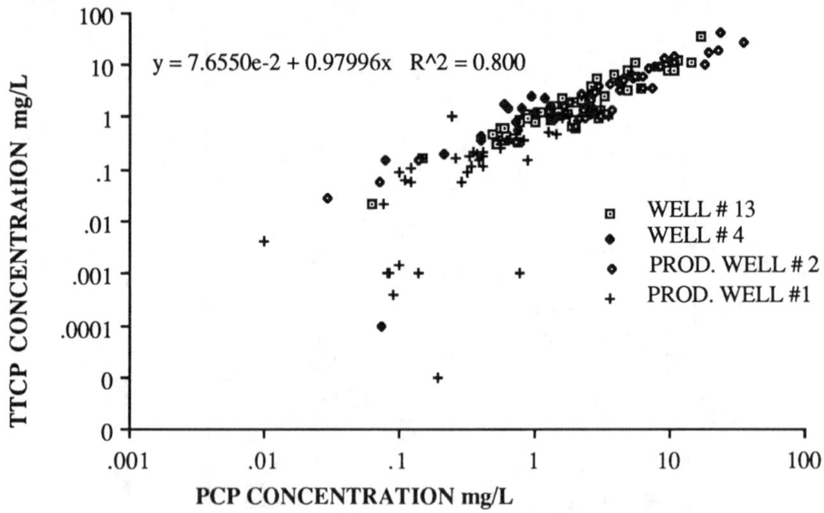

Fig. 7 Concentrations of tetrachlorophenol versus pentachlorophenol

Total measured solubility will be the sum of the chlorophenate and chlorophenol and for PCP at 5 degrees Celsius is given by the the following expression after Firestone (1976)

$$\log S = 1.43\,pH - 6.89 \qquad (3)$$

The groundwater temperature at Penticton was 9 degrees Celsius and as such equation 3 would tend to underestimate PCP solubility slightly. No equivalent expression was found for TTCP however an estimate of the total solubility can be made from a knowledge of the tetrachlorophenol solubility and the per cent dissociated at the groundwater pH. The groundwater pH varied over the site from 6.2 to 6.7 with a value of 6.4 as the most likely in the area of the spill. At a pH of 6.4 the respective penta and tetra solubilities would be 183 mg/l and 1810 mg/l. Assuming no pH shift, dilution of 13 to 14 and 2.5 to 3 respectively would be required to prevent precipitation of the chlorophenols. The solution in the dip tank was normally at a pH in excess of 10 and was well buffered. The groundwater was relatively salt free with a specific conductance in the 300 umhos / cm range and as such it could be expected that at least at the time of the initial mix the pH would be above background. The chlorophenols will partition to aquifer organics with an estimated K_{om} for PCP of 2200 mL/g (NRCC 1982) however there is negligible partitioning of the chlorophenates. Estimates of the aquifer organic content from Patterson and Leibscher (1987) are in the order of 0.01 percent which gives a soil partitioning, K_S coefficient in the order of 0.2 and probably half that for the tetrachlorophenate. Not withstanding the very coarse and generally clean nature of the aquifer material at least one discontinuous organic layer was intercepted in several wells. Given that the site is an alluvial fan it is conceivable that there could be pockets of organic debris.

4 CONTAMINANT BEHAVIOR

As the aquifer consisted of very clean coarse material and was slightly acidic and as the chlorophenate mix was denser then the aquifer water; it was anticipated that the contaminant would not spread as it entered the groundwater, that the plume would be narrow, that little partitioning would occur and dilution would be minimal. It was also expected given relative solubilities that precipitation of pentachlorophenol would occur with a subsequent increase in the tetra to penta ratio..

What is believed to have happen is that the contaminant flowed vertically down through the unsaturated zone in a narrow stream with most but not all of the contaminant going into the lower sandy portion of the aquifer as a result of the solution having a specific gravity of about 1.007. A quantity of the contaminant remained in the upper gravel portion of the aquifer which was more permeable and consequently the water moved at a higher rate. The contaminant seen at the river and in Well #13 moved at the higher rate. It is felt that the geology accounted for most of the differences in the observed contaminant travel rates with the late arriving peak seen in Production Well # 2 moving through sand. Some partitioning would have occurred but because of two forms of the contaminant a clearly defined retardation would not have occurred. The most likely explanation for the shift in the tetra / penta ratio is because of partitioning. Given that the adsorption coefficient for PCP is higher than TTCP one might expect that more PCP would partition to the aquifer solids. However in the given situation because of differences in the dissociation constants (4.2×10^{-6} for tetra as compared to 1.2×10^{-5} for penta (Jones 1981)) and the relative concentrations at the time of the spill, about four times as much TTCP would be available for partitioning. A comparison of amounts of recovered material to that spilled showed that about 60 per cent of the penta was recovered as compared to 30 per cent for the tetra.

5 SUMMARY

A spill occurred of a dissociated solution of a registered pesticide into a high rate aquifer located in an alluvial fan. The spill broke into at least two plumes as a result of the geology. Partitioning likely occurred and would provide an explanation for a shift in the ratio of the spilled constituents. While partitioning would provide the sink for the chlorophenates the differences in per cent dissociation accounted for a difference in the availability of the two chlorophenols for partitioning and conjectured uptake.

6 ACKNOWLEDGEMENTS

The groundwater investigation was carried out under the direction of Mr. Hugh Leibscher of Environment Canada and the author who at that time was employed by the Environmental Protection Service of Environment Canada The success of the project was due to the efforts of a great number of people in Environment Canada, the B.C. Ministry of the Environment, the City of Penticton and Greenwood Forest Products.

7 REFERENCES

Firestone, D. 1977. Chemistry and Analysis of pentachlorophenol and its contaminants. *FDA By-lines* 2: 57 - 89.

Jones, P.A. 1981. *Chlorophenols and their impurities in the Canadian environment*. Ottawa . Minister of Supply and Services Canada.

NRCC. 1982. *Chlorinated phenols: criteria for environmental quality*. Ottawa. National Research Council Canada.

Patterson, R.J., & H.M.Liebscher. 1987. Labratory simulation of pentachlorol / phenate behaviour in an alluvial aquifer. *Water Poll. Res. J. Canada.* 22 : 147 - 155.

Subsurface Contamination by Immiscible Fluids, Weyer (ed.) © 1992 Balkema, Rotterdam. ISBN 90 6191 175 3

Practical aspects of hydrogeology studies for hydrocarbon migration in British Columbia

R.Allan Dakin, Andrew T.Holmes & David J.Tiplady
Piteau Associates Engineering Ltd, North Vancouver, B.C., Canada

ABSTRACT: Over the last three years, there have been numerous leaks and accidental spills of a variety of hydrocarbon compounds into the ground in southern British Columbia. In the course of searching for and delineating the flow paths of these fugitive hydrocarbons, the writers have reached conclusions with practical implications that may interest other investigators.

In the following, a brief description is presented for four case histories, and one or two unique aspects of either the investigative approach or interpretation of results are described.

The examples involve leaks of gasoline and spills of xylene and ethylene dichloride. Descriptions are given of buried underground services acting as gasoline traps, a drawdown cone around a nearby well causing an apparent bifurcation of a floating hydrocarbon plume and fluid density causing distortions in interpretation of piezometric head.

At one site, measurement of thickness of hydrocarbon saturated sand in drill core was compared with thickness of floating hydrocarbon fluid in a completed monitoring well. After applying appropriate corrections to the hydrocarbon thickness measured in the well, this value was generally in close agreement with that measured in the saturated sand sample.

BACKGROUND

When investigating the fate of hydrocarbon fluids that have leached into the subsurface, there are many factors to consider. These could include: groundwater flow direction; groundwater velocity; degree of heterogeneity of sediments; octanol water partitioning coefficient of fluid; density of fluid; organic content in sediments; water table levels; etc. However, there could also be other factors that may not always be very obvious, such as the effects of pumping local water wells and flow to buried pipes, and presence of other leaked organics in the sediments. In the following, we have briefly described four case histories of leaks and spills that have occurred in southern and southwestern areas of British Columbia, and have highlighted some of the practical aspects of the hydrogeological investigations.

Due to the sensitive nature of many of these case histories, the exact locations of the sites have not been given.

Case History No. 1: Gasoline Migration Into Buried Utility Conduits

While investigating leaks from underground storage tanks at gasoline stations in two areas in each of the cities of Squamish, Richmond and Victoria, and one in Kelowna, it was found that in each case the gasoline had seeped into manholes and conduits used to house buried telephone cables in the nearby streets. These conduits are usually 100mm diameter PVC pipes, and are arranged in duct banks, with manholes spaced every 150 to 250 metres to provide access (see Fig. 1). The 50mm diameter telephone cables are normally inserted into these ducts after construction, and can at any time be removed for repair or replacement without

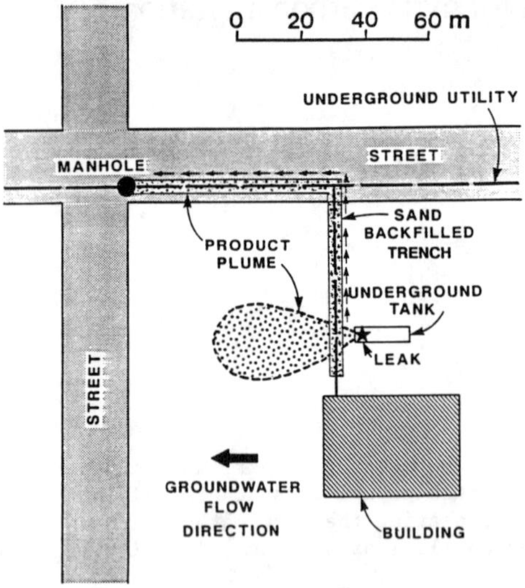

FIGURE 1. PLAN OF A TYPICAL GASOLINE PLUME IN AN AREA WITH HIGH HIGH WATER TABLE AND PERMEABLE SERVICE TRENCHES.

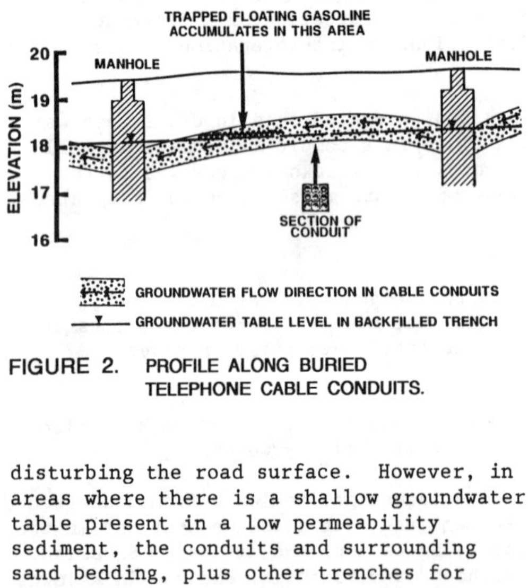

FIGURE 2. PROFILE ALONG BURIED TELEPHONE CABLE CONDUITS.

disturbing the road surface. However, in areas where there is a shallow groundwater table present in a low permeability sediment, the conduits and surrounding sand bedding, plus other trenches for sewer and water services, provide a preferred path for the seeping gasoline (see illustrations in Figs. 1 and 2).

Depending on the particular elevation and geometry of any particular site, groundwater could flow into and along the pipes or backfill material at any time of the year. Gasoline floating on top of the water table can also permeate the unsealed joints in the conduit pipes and migrate in the annulus along side the telephone cable. However, whenever the conduit pipe dips below the water table, the undissolved floating gasoline products (float) will concentrate at the prevailing water table. In such circumstances, it is only a matter of time before the gasoline weakens the sheathing material wrapped around the telephone cables, allowing water to seep in and to render the cable useless for telephone transmission. As the water table rises and falls, the float could migrate up and down the cable, thereby damaging a longer length of cable. In relatively flat areas, the resultant flow path could be quite different from the regional groundwater flow direction. Cost of repairs resulting from this type of damage could range up to many hundreds of thousands of dollars.

This example illustrates the need to research the complete history of any site being investigated, and especially to locate all buried pipes or backfilled trenches that could distort the natural flow direction.

Case History No. 2: Gasoline Contamination of Municipal Groundwater Source

A gasoline tank had been leaking for many years before a gasoline taste was detected in one of the city's water wells, located in south central British Columbia. This 56 L/s production well taps a 20m thick section of an areally extensive aquifer with hydraulic conductivity of 10^{-4} to 10^{-3} m/s.

Eight years after the leak was first discovered, the well is still contaminated and cannot be used. A series of drilling programs has been conducted to delineate the source of gasoline and extent of the plume. The probable extent of the gasoline plume is illustrated in Fig. 3, and from this it is apparent that: i) the probable source is one or both of the two underground storage tanks (UST) in or close to the plume; ii) the plume has split, with one tongue (Plume No. 1) being drawn towards the municipal well, and the other tongue (Plume No. 2) migrating with the natural groundwater flow. As this well is located in a relatively flat area, is completed in a very permeable aquifer, and is only used intermittently, plotting of water table elevations and assessment of plume migration when the pump was not being operated could have led to erroneous conclusions on direction of plume migration. However, when the influence of the well's drawdown cone is superimposed

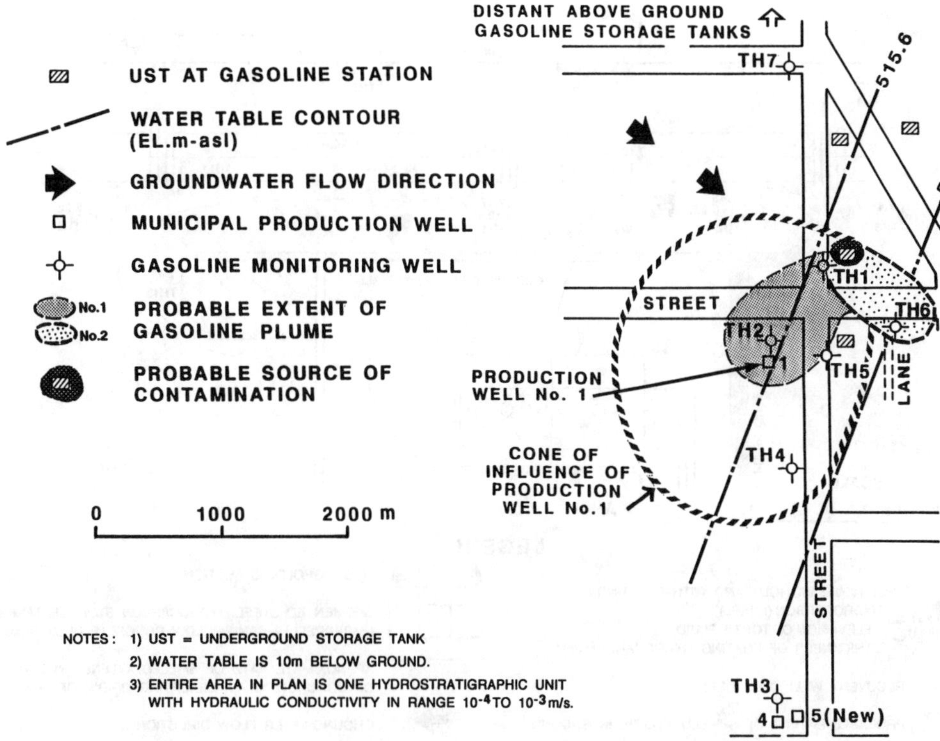

FIGURE 3. PLAN OF GASOLINE PLUMES IN VICINITY OF MUNICIPAL WELLS.

onto the prevailing groundwater flow direction, the split plume can be fully explained.

Case History No. 3: Xylene Spill

In 1989, approximately 4,500 litres of an xylene product was spilled onto the concrete floor of an elevated platform; about 80% of the xylene flowed over the western edge of the platform, and about 20% went over the northern edge. Once over the edge, the xylene flowed down on top of a sand fill and seeped into the ground. Much of it ponded on top of a shallow water table in the sand (see Fig. 4).

The sand is medium to fine grained, and was originally placed as a 2m thick fill on top of low permeability floodplain silts (see Fig. 5). Depending upon the time of year, between 0.5 and 1m of the sand is saturated. The water table was mounded near the spill site, but regionally was almost flat, with a gradient of about 0.1m per km towards the southwest.

Xylene is a colourless, flammable liquid with a specific gravity of about 0.86, and solubility in water at 10°C is about 150 mg/L.

The spill area was in and around an important railway spur line and the property owner required that rail cars be able to pass along the line at all times, following about one hour's notice. With these constraints in mind, a relatively portable method of drilling and sampling of test holes was required.

The system selected was developed at the University of Waterloo, and involves the use of a hand held Ponjar (mechanical) hammer to drive a special head that attaches to a diamond drill rod. A 66.4mm diameter aluminum tubing (electrical conduit) is inserted inside the drill rod, along with a special piston sampler. At the xylene spill site, the drill rods were driven into the ground to depths ranging from 2 to 3m. Between 1 and 1.5m lengths of core inside the aluminum tubing were withdrawn from the drill rods, after each drive.

This new piston sampler allowed high quality core samples to be collected from

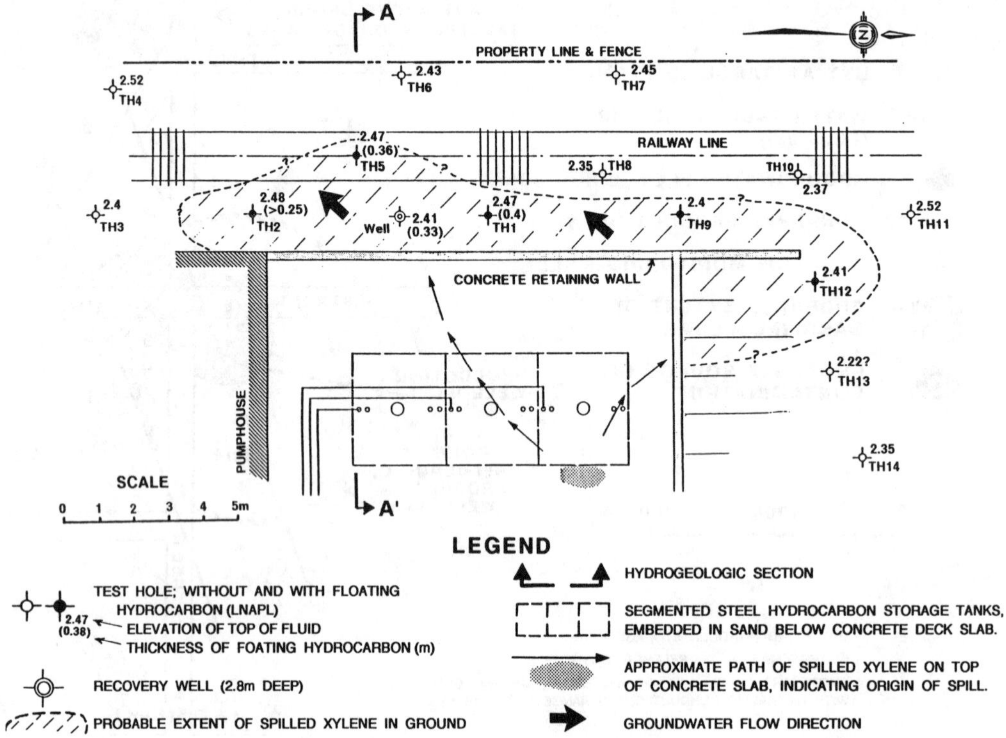

FIGURE 4. PLAN OF XYLENE SPILL SITE

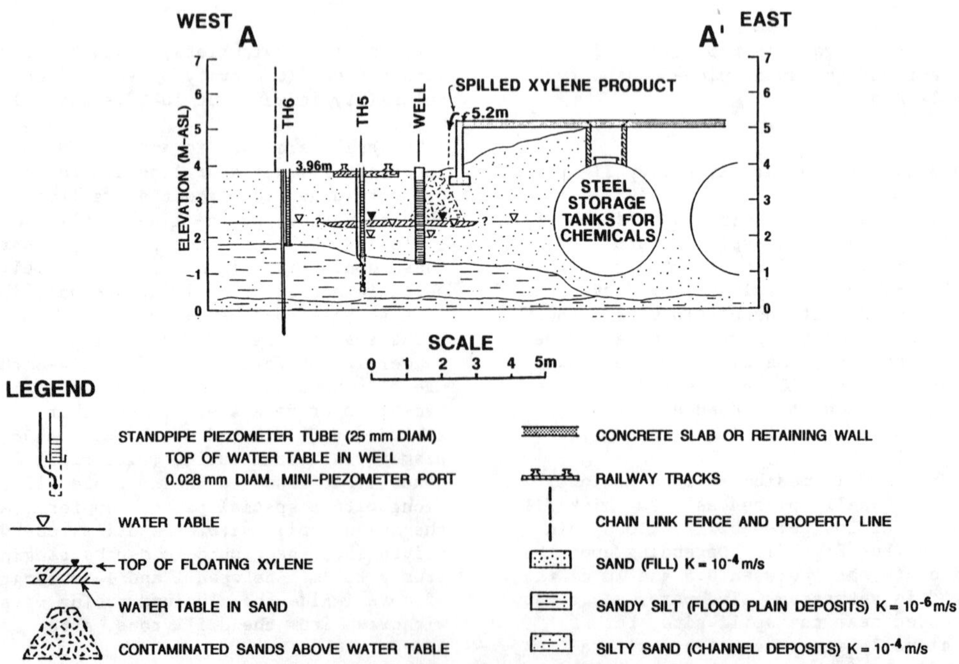

FIGURE 5. SECTION THROUGH XYLENE SPILL SITE

the sand and underlying silty sand. In addition to its portability, this drilling method had the advantage of causing minimal mechanical disturbance of the sand and saturating fluids, thereby enabling a very precise identification of sample depth; it was also possible to determine the position of the top of both xylene and water saturated sand within a few millimeters of accuracy. This was done by drilling holes into the side of the tubing and withdrawing a small amount of fluid.

When each hole was completed, a slotted PVC pipe was inserted in the hole. A few days after drilling was completed, the thickness of xylene floating on top of the water table in the monitoring tubes was measured, using the tape and paste method. The apparent thickness of the product was between 0.33 and 0.4m at the centre of the spill area. However, as is usually the case, drainage of hydrocarbon fluid from the top of the capillary fringe on top of the water table in the sand into the monitoring tube depressed the water surface, until a density equilibrium was established (see Fig. 6); this resulted in measurement of an apparent hydrocarbon thickness which was greater than the true value.

Calculation of true product thickness (T) under even hydrostatic conditions in a porous medium is a complex process (Abdul et al, 1989; Lenhard and Parker, 1989; and Testa and Paczkowski, 1989), and is usually estimated by dividing the apparent thickness measured in a monitoring well (h_w+H_f) by a constant value (C), with values often ranging between 4 and 6, depending on the density of the hydrocarbon (Fig. 6).

Based on visual inspection of drill core, the thickness of the oil capillary fringe (h_c) in the fine to medium coarse sand at the site was about 60mm, and the thickness of free xylene (h_f) in some of the drill cores was about 30mm; hence, using the simple formula, the apparent thickness of product observed in the piezometers:

$$T = h_w + h_f =$$
$$4\,(h_c + h_f) = 4\,(60 + 30)/1000 =$$
$$(0.09 \times 4) = 0.36\text{m}.$$

This agrees fairly well with field data for thickness of hydrocarbon measured in the tubes, which ranged from 0.33 to 0.40m.

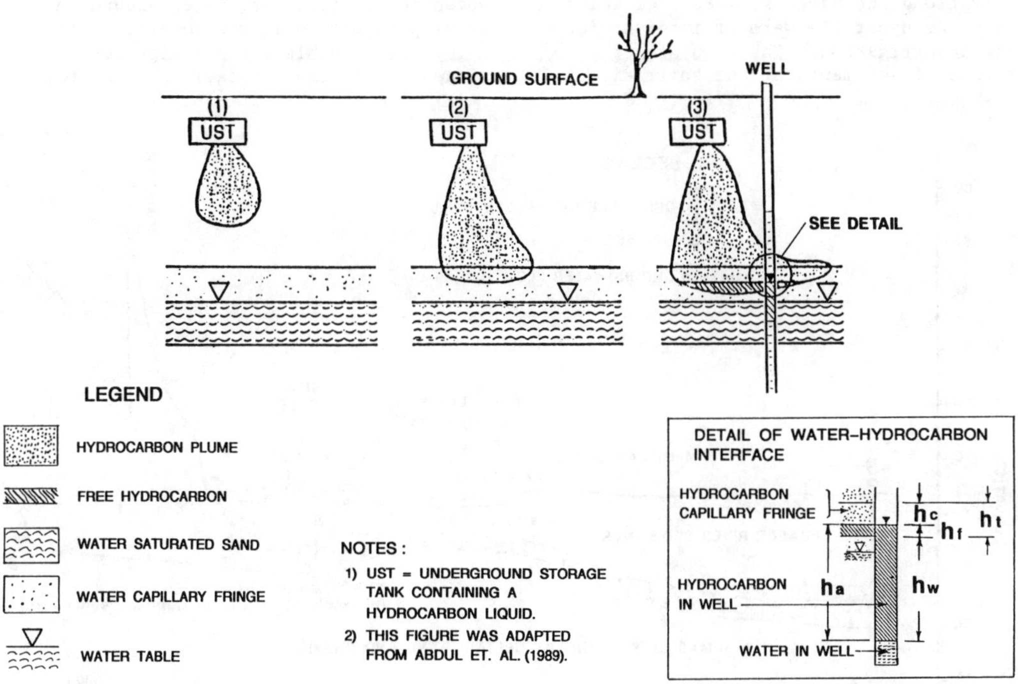

FIGURE 6. THREE STAGES OF HYDROCARBON MIGRATION ONTO A SANDY AQUIFER

Case History No. 4: Delineation of an Ethylene Dichloride Contaminant Plume in a Gravel Aquifer

On February 15, 1986, derailment of a CN Rail freight train near Fort Langley, B.C. resulted in a 247,500 litre spill of ethylene dichloride (EDC). Within a few hours, the spilled chemical had seeped into the ground. Piteau Associates were retained to supervise an investigation to determine the direction and rate of movement of the EDC in the ground.

Ethylene dichloride (or 1,2-dichloroethane) is a heavy (density 1.267 @ 10°C), colourless and flammable liquid (flash point at 13°C), with a moderately low viscosity (1 mPa.s) (Montgomery and Welkom, 1990; Environment Canada, 1984). As it is moderately soluble in water (8,490 mg/L), it was not surprising to find that there were two EDC plumes in the gravel aquifer underlying the spill; one was a dissolved plume travelling towards the Fraser River about 350m away at a maximum rate of about 2.5m/day, and the other was a liquid non-aqueous plume that migrated slowly along the bottom of the aquifer.

The spill area is underlain by more than 100m of unconsolidated sediments (Armstrong and Hickock, 1986), of which only the upper 15m were of interest for the investigation. The sequence of surficial sediments at the spill site is shown in Fig. 7. From ground surface, the sequence consists of a 0.5 to 1.0m thickness of a silty fine sand covering from 4.5 to 7m of wood waste fill, which overlies 3.0 to 5.5m of silty sand. The silty sand is underlain by 3 to 9m of sand and gravel which, in turn, is underlain by clayey silt. None of the monitoring holes drilled for the study penetrated more than about 1.2m into the basal unit.

A Becker hammer rig was used to drill the twenty-one test holes. Selection of materials for piezometer construction posed a problem, as PVC softens when in contact with high concentrations of EDC. The only readily available alternative to PVC was black steel pipe, stainless steel or polyethylene tubing. Some of the earlier holes were completed with 31mm diameter steel piezometers, and five holes located upgradient or at some distance from the spill site were constructed of 37mm diameter flush coupled, threaded PVC. All of the remaining piezometers were constructed using a bundle type piezometer design, as illustrated in Fig. 8. A centre steel piezometer constructed from 25mm diameter steel pipe with a slotted bottom was used for the deep piezometer in each hole. Polyethylene tubes, with a 9mm I.D. and a 150mm long stainless steel woven screen tip, were taped around the steel pipe before it was lowered into the hole. The stainless steel tips were placed at different locations along the

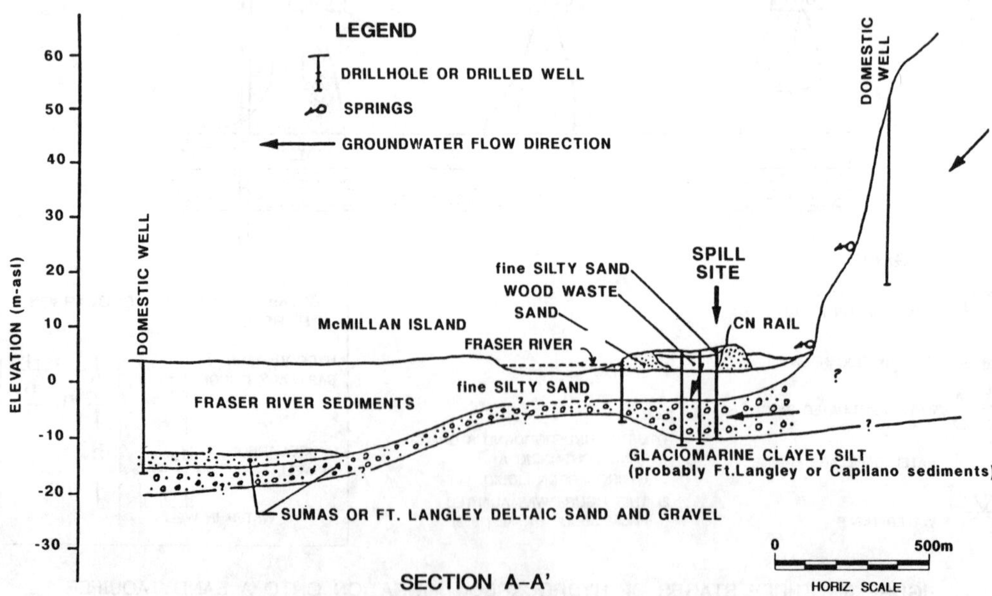

FIGURE 7. REGIONAL SECTION THROUGH ETHYLENE DICHLORIDE SPILL SITE.

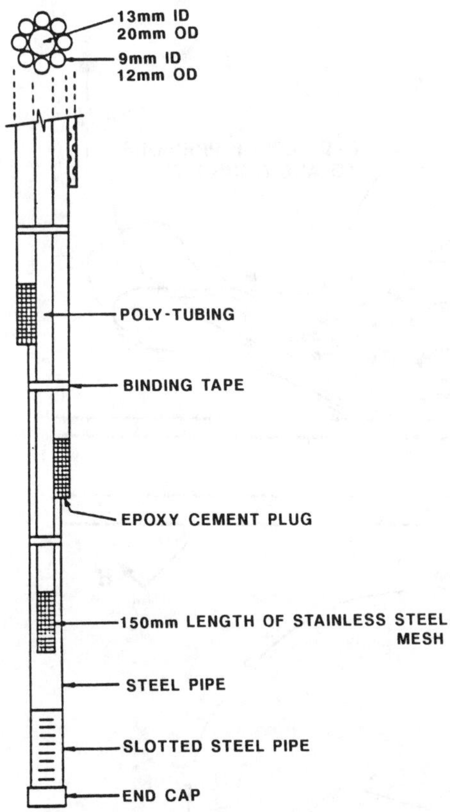

FIGURE 8. BUNDLE TYPE MULTI PORT PIEZOMETERS

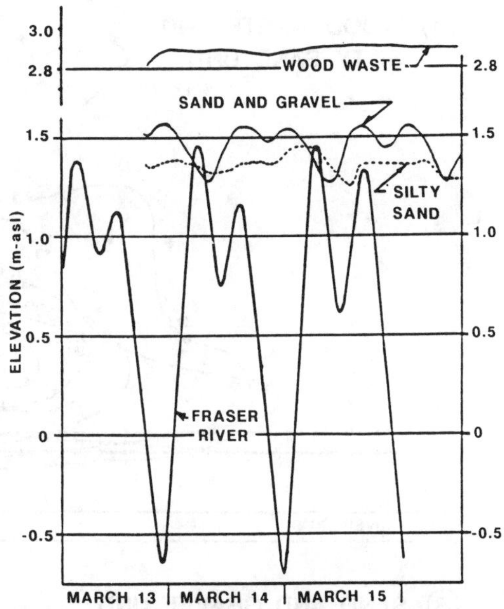

FIGURE 9. HYDROGRAPH SHOWING RELATIONSHIP BETWEEN FRASER RIVER LEVEL AND PIEZOMETRIC LEVELS UNDER SPILL SITE

pipe so that a profile could be carried out. While pulling the drill rods out of the hole, small amounts of bentonite slurry were pumped into the hole in the interval between each piezometer tip, to provide a seal against vertical migration of groundwater in the hole annulus. Uniform sand was placed in the intervals where the piezometer tips were located, except in the sand and gravel aquifer.

Water levels measured in the overlying wood waste were higher than that in the underlying aquifer, indicating a perched water bearing zone (Figs. 7 and 9). Water levels in the deep piezometers were observed to fluctuate in response to tidal activity in the Fraser River. The relationship between the water levels measured in the sand and gravel aquifer, the silty sand unit, the wood waste unit and the Fraser River for a three day period in March 1986 is shown in Fig. 9. As illustrated in this figure, there is a varying but persistent downward hydraulic gradient between the wood waste and the sand and gravel aquifer.

Groundwater flow in the aquifer is generally in the northwest direction, although for brief periods during flooding of the Fraser River, groundwater flow may be to the southwest (Dakin and Holmes, 1988). The average velocity of groundwater flow in the sand and gravel aquifer was estimated to range between 0.7 and 1.3m/day.

A series of plans and sections are presented in Figs. 10 and 11, showing contours of EDC concentration at various times after the spill occurred. The distribution of the EDC in the wood waste and upper silty sand unit, approximately four weeks after the spill, is shown in Fig. 10A. It is evident from this figure that initial migration of the EDC was in the east-west direction, along the orientation of an old slough.

Within the area of high EDC concentration, some samples of free EDC were obtained, indicating that pockets of dense non-aqueous phase EDC (DNAPL) were accumulating at the base of the wood waste in some areas. However, as illustrated in Fig. 11, these pockets existed for only relatively short periods. Eleven weeks after the spill, concentrations greater

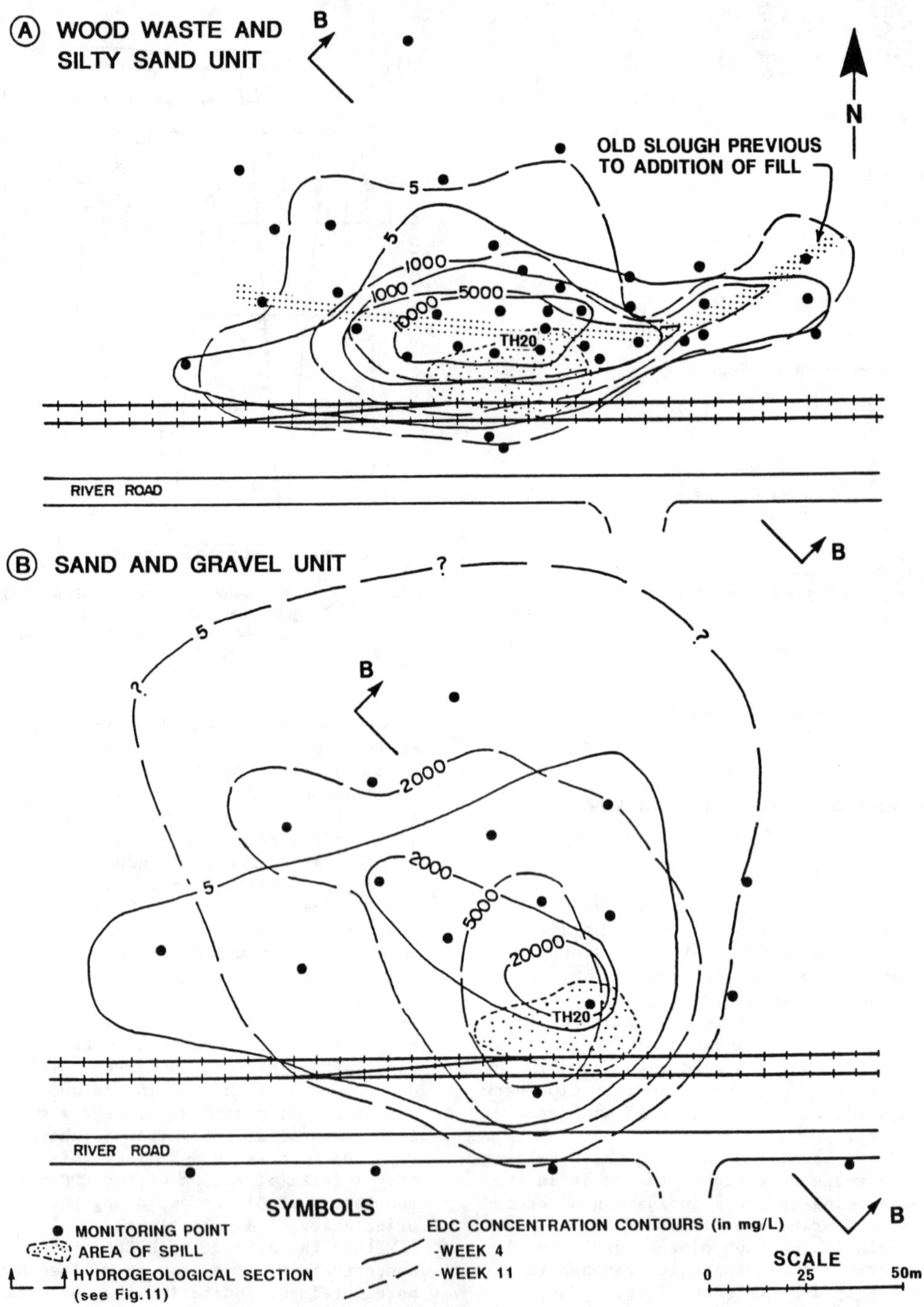

FIGURE 10. PLAN OF ETHYLENE DICHLORIDE PLUMES

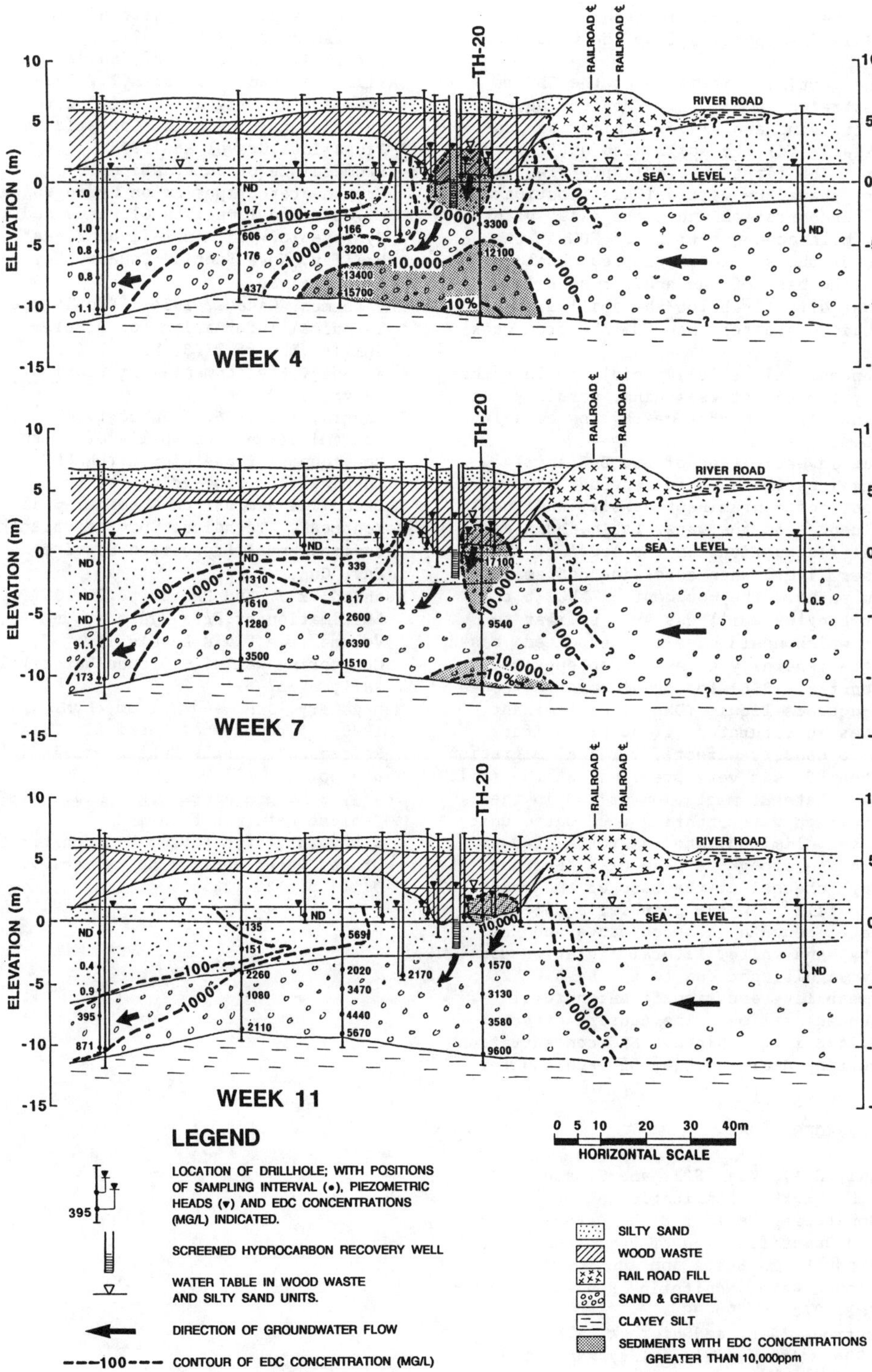

FIGURE 11. CONCENTRATION OF ETHYLENE DICHLORIDE THROUGH SPILL SITE.

than the solubility limit for EDC were rapidly disappearing from the wood waste unit.

The vertical migration of the EDC is illustrated in Fig. 11. Monitoring data for the sand and gravel aquifer was not available for the spill area until week four. The first available profile (week four) shows the vertical distribution of EDC concentration under the site, with very high concentrations of EDC at the base of the aquifer. Samples obtained from the base of the aquifer at a location almost directly below the spill area contained greater than 70% EDC from week four through to week seven.

Concentrations fell rapidly to less than 1% by the end of week nine, possibly due to installation of a nearby low capacity recovery well.

The investigation of the EDC spill emphasizes the need to consider the effects of contaminant density and local influences on the water table, such as tidal fluctuations in the level of the Fraser River, when assessing contaminant flow paths. The movement of EDC at the Fort Langley spill was very similar to that which would have been expected, based on the generally accepted concepts (Feenstra, 1986) for development of dense non-aqueous liquid (DNAPL) contaminant plumes in saturated, granular aquifers. Due to density effects, vertical migration of the EDC was very prevalent at the spill site. Lateral migration of EDC in the unconfined water bearing wood waste unit was very limited, and was restricted to migration along the base of a "perched" slough channel at the base of the wood waste fill. In the granular aquifer unit located beneath the wood waste and the silty sand units, lateral migration was very significant due to the high permeability and significant natural groundwater flow. The density effect resulted in the highest EDC concentrations occurring near the base of each aquifer.

REFERENCES

Abdul, S.A., Kia, S.F. and Gibson, T.L., 1989. "Limitation of Monitoring Wells for the Detection and Quantification of Petroleum Products in Soils and Aquifers". Ground Water Monitoring Review. Vol. 9/2, p. 90-99.

Armstrong, J.E. and Hickock, S.R., 1986. "Surficial geology maps of New Westminster and Vancouver Map Sheets 148A and 1496A" (1:50,000 scale incorporates all or parts of NTS map sheets 92G/2, 926G/3, 92/G6, and 92/G7). Geol. Surv. of Canada.

Dakin, R.A., and Holmes, A.T., 1988. "Monitoring, Migration and Control of an Ethylene Dichloride Contaminant Plume in a Gravel Aquifer". Proceedings of International Groundwater Symposium on Hydrogeology of Cold and Temperate Climates and Hydrogeology of Mineralized Zones". Halifax, Nova Scotia, Canada, May 1-5. p. 41-50.

Environment Canada, 1984. "Environment Technical Information for Problem Spills (Enviro TIPS) Manual for Xylene", Environmental Protection Service, 108 pp.

Feenstra, S., 1986. "Subsurface Contamination from Spills of Dense Non Aqueous Phase Liquid (DNAPL) Chemicals". Proceedings of the Technical Seminar on Chemical Spills, Montreal. Technical Services Branch, Environmental Protection Service, Env. Can. 22 pp.

Lenhard, R.J. and Parker, J.C., 1989. "Estimation of Free Hydrocarbon Volume from Fluid Levels in Monitoring Wells". Ground Water Vol. 28/1, p. 57-67.

Montgomery, J.H. and Welkom, L.M., 1990. "Groundwater Chemicals Disk Reference". Lewis Publishers Inc. 640 pp.

Testa, S.M. and Paczkowski, M.J., 1989. "Volume Determination and Recoverability of Free Hydrocarbon". Ground Water Monitoring Review, Vol. 9/1. p. 120-128.

8.2 Investigation and remediation

Subsurface Contamination by Immiscible Fluids, Weyer (ed.) © 1992 Balkema, Rotterdam. ISBN 90 6191 175 3

Reflections on polychlorinated biphenyl and polychlorinated benzene movement at Regina, Saskatchewan

Moir D. Haug
Department of Civil Engineering, University of Saskatchewan, Saskatoon, Sask., Canada

James Atwater
Department of Civil Engineering, University of British Columbia, Vancouver, B.C., Canada

ABSTRACT: A large spill of Polychlorinated Biphenolys (PCB's) occurred at Federal Pioneer Limited's Regina Plant in 1976. The City of Regina is underlain by a relatively shallow aquifer which supplies a significant proportion of its drinking water. A remedial measures plan was developed to contain this spill within the boundaries of the site, arrest any vertical migration of PCB's and clean up any contaimination which had spread to the adjacent property. Following the implementation of the plan. An ongoing program of monitoring and investigation was undertaken to ascertain the hydraulic isolation of the site, whether or not there was any subsequent movement of PCB, and thirdly to determine the efficiency of the underlying Condie Silts as a barrier to downward movement of the PCB. The conclusion of this work is that while there has been movement of PCB's in the overlying clays, there is no evidence of movement in the silt barrier, and that the Condie Silt will provide an effective barrier to the movement of PCB's under porous flow conditions.

1 INTRODUCTION

The development of remedial measures to contain and control the spread of hazardous waste is one of the most challenging tasks in waste management. Many of these wastes are chemically active, present a threat to uncontaminated groundwater and must be contained for considerable lengths of time. Presented in this paper is a summary of the background to development, installation, monitoring and analysis of remedial measures designed to control a spill of a transformer fluid Inerteen 70-30. Some aspects of this investigation and construction of the remedial works have been previously reported by Roberts et al. 1982, Schwartz et al. 1982, Atwater 1984, Haug et al. 1988, and NNRC 1980. Throughout this paper the fluid is referred to as PCB's, however implicit with that usage is the understanding that there was up to 30% combined tri & tetrachlorobenzene in the mix. Presented secondly and in detail are the results from deep hole drilling to ascertain whether there has been any subsequent movement of the transformer fluid with the construction of the remedial works and thirdly, the assessment of the underlying Condie Silts as a barrier to downward movement of the transformer fluid.

2 BACKGROUND

In 1976, Canada's first major known spill of PCB's occurred at Federal Pioneer Limited's Regina Plant (Figure 1). This plant used a dielectric consisting of tri- and tetra- chlorobenzenes (TCB's) in the manufacture of transformers. The fluid was stored in an outside tank and pumped to the plant through an underground line. The exact date on which this line ruptured is not known, and the total quantity of PCB's released is also not known. However, a review carried out by the National Research Council of Canada (NRCC 1980) estimated that up to 21,000 litres of oil was involved. The specific gravity of this oil was 1.5 with a water solubility of 4 - 5 x 10^{-2} mg/l at 20^{0} C (NRCC 1980). At the time of the spill, little was known about the health hazards of PCB's or their threat to the environment. Even less was known on how to effectively contain such a spill. The manufacturer responded to the spill by pumping off free oil, and covering the spill area with a thin asphalt seal.

In 1977 the Canadian Government prohibited the use of PCB's for most applications. This increased awareness of the potential problems associated with PCB's, and in 1978 public concern was expressed over the security of the City of Regina's drinking water, as approximately one-third of the city's drinking water underlies the site. These concerns eventually led to the formation of an ad hoc panel of scientists under the umbrella of the National Research Council of Canada. This panel was requested to evaluate information on the nature and consequences of the spill, and prepare a public report. The report (NRCC 1980) identified the major areas of contamination, the principle mechanism for oil movement, and recommended that all of the contaminated soil under the plant be eventually removed and deposited in a hazardous waste facility. The NRC study team recognized that

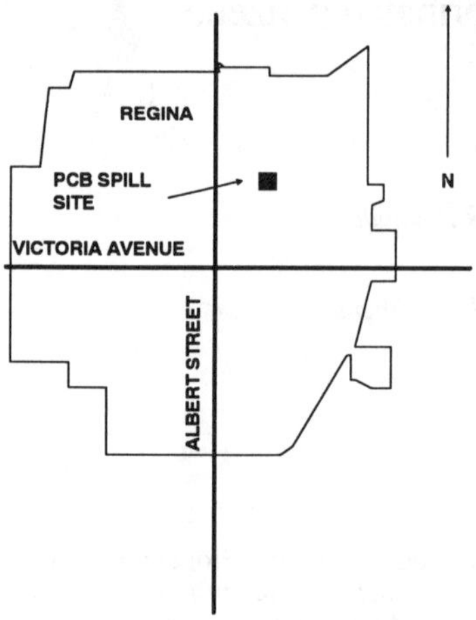

Figure 1 : Location of Spill Site

while downward movement of water could occur, attenuation would in all likelihood prevent contamination of the aquifer unless the silt layer was breached by some subsequent activity. It was this unknown as to what the future held that prompted their recommendation for removal.

The major areas and the depths of contamination are shown in Figure 2. The manufacturing plant is situated on a lot approximately 105 m wide and 190 m long. Old concrete pads at the rear of the lot mark the location of a previous plant which burned in 1974. The natural slope of the site changes near the rear of the plant, with water draining to the north and south. The location of heaviest concentrations of PCB's were found at the north-east corner of the plant near the location of the buried underground line. PCB's migrated out from this location to the west along an underground power line trench, south along the east limit of the building and east to an old transformer oil sump. Other areas of contamination were found near the railroad tracks, where the oil was unloaded.

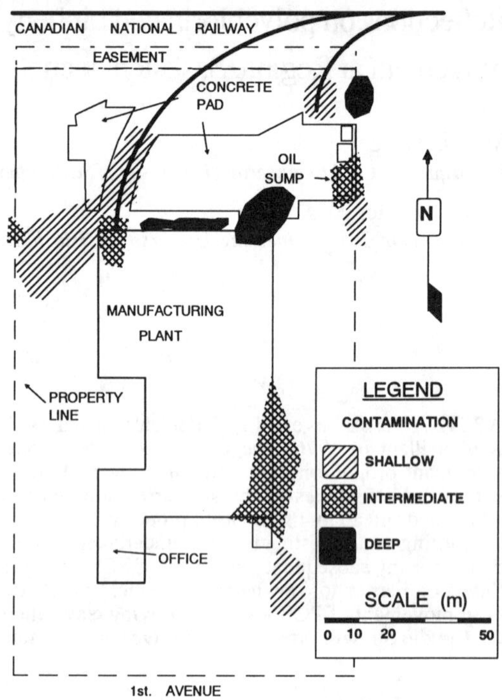

Figure 2 : Location of Major Areas of Contamination (after Haug, et al., 1988)

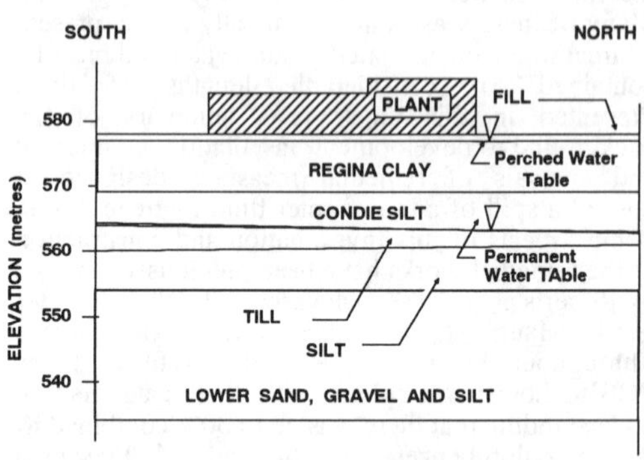

Figure 3 : Stratigraphic Section Across Site Showing Location of Water Table and Primary Material Units.

3 SITE CHARACTERISTICS

Christiansen (1979) carried out an investigation to determine the geology and stratigraphy underlying the site. This information is presented in Figure 3. It shows a north-south cross section through the site. The surface of the site is covered with approximately 1 m of fill. Underlying this fill is a weathered clay referred to as Regina Clay. This highly plastic clay was formed in a large glacial lake during the recession of the last ice age. Its depth varied from 7.3 to 8 m under the site. Prolonged periods of drying have resulted in severe desiccation of this clay, significantly increasing its permeability. The clay is underlain by 3 to 7 m of silt designated as the Condie Silts. Christiansen (1979) found some indications of weathering in the upper portion of this silt deposit. The silt is underlain by 1 to 5 m of soft till, which is underlain in turn by 5 to 6 m of hard till. Underlying the till is a deposit consisting of from 16 to 22 m of interglacial silts and sands. The Regina Aquifer is in this zone, approximately 35 m below the surface. However, the presence of the Regina aquifer was not confirmed under the site. According to Christiansen (1979), an unknown collapse stratigraphically controlled aquifer varying in depth from 37 to 59 m is present under the site. The investigation by the National Research Council and the Ground Engineering team confirmed Christiansen's observations were applicable at this site. In addition, gypsum and calcite crystallization subsequent dissolution, resulted in the formation of a zone of increased horizontal permeability in the clay, Atwater (1984).

At least two different water tables underly the site. The first is a perched water table located in the fill. The second "main" is located near the base of the silt.

4 REMEDIAL MEASURES

The objective of the remedial measures plan was to develop a strategy to contain the spill within the boundaries of the site and to prevent further downward migration of PCB's.

The fundamental principle on which this strategy was based, involved severely limiting and potentially reversing the hydraulic gradient in the fill and clay beneath the site. Since the supply of PCB's had been cut off, the primary downward driving force would be the downward movement of moisture through the clay. This is not to suggest that the mechanism of PCB movement is only in the dissolved aqueous phase. Rather, it is assumed, somewhat simplistically, that the free PCB in the clay had come to static equilibrium and the only situation under which "heads" were likely to change would be due to the addition of water. This downward movement is driven by water in the fill replenished by precipitation through the year. The goal of the remedial measures plan was to seal the surface of the site preventing the recharging of the perched water table and forcing a reversal of the hydraulic gradient. To aid in this reversal, the perimeter of the site was ringed with a cutoff wall extending down through the clay to the silt. Shallow sumps were installed to dewater the fill. Figure 4 shows an east west cross section through the zone of deepest contamination. This figure conceptually illustrates how the cutoff wall, surface seal and dewatering combined to enhance the containment strategy.

The primary consideration in selecting the depth of the cutoff wall was the prevention of lateral water flow in the upper few metres. At the time of the installation of the wall, this flow was expected to be concentrated in the fill and clay. However, the presence of weathering in the upper portion of the silt raised questions about the potential lateral movement of water and PCB's in this zone. Thus, in an attempt to ensure the effectiveness of the lateral containment system, the cutoff wall was extended approximately 1.5 m into the silt.

The remedial measures plan also involved excavation of contaminated soil on the fringe and outside the boundaries of the site. Soil containing more than 50 ppm PCB's was excavated in this process. The contaminated soil was transported to an interim storage facility established on the east side of the plant. It was intended that this facility would hold the contaminated soil until a permanent facility was constructed for such purposes in the Province of Saskatchewan or until it could be safely transported out of the Province.

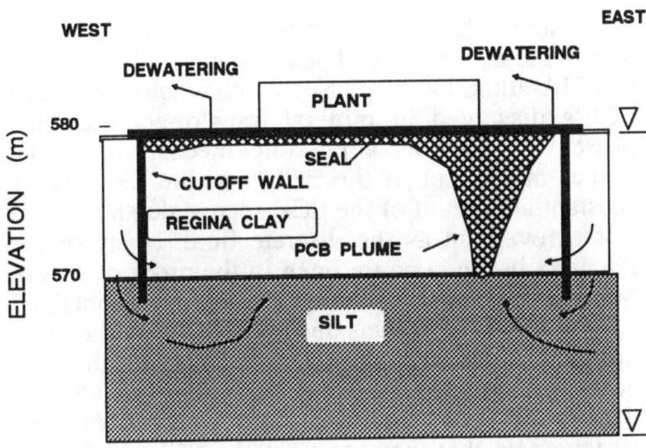

Figure 4 : Section Across Site Through Center of PCB Plume

The effectiveness of the measures to limit the infiltration of water and dewater the site was monitored with a network of piezometers spread over the site. The volume of water collected in the sumps was recorded and its PCB concentration checked periodically. An in depth presentation of the piezometer monitoring was presented by Haug et al.,1988.

The total volume of water collected in the sumps varied from 125,000 L in 1983 to less than 30,000 L in 1987 (Haug et al.,1988), largely depending on precipitation and the degree of tightness of the surface seal. While a considerable effort was directed towards improving the surface seal and reducing infiltration, it became increasingly evident that lowering the sump yield to "zero" would be virtually impossible. The maximum tightness achieved by the surface seal for this heavy industrial site was approximately 98 %, based on sump yield as a functional precipitation.

The PCB concentration in the water collected from the sumps generally ranged from 1 to 500 ppm. The concentrations usually increased as the total volume of water collected in the sumps decreased. While some high concentrations (approximately 5000 ppm) were found in oil collected in the sumps, the PCB concentrations in the sump water were generally low.

5 PCB MOVEMENT

PCB movement on the site was by four mechanisms;

as a discrete fluid, as an emulsion with water, bound to soil and as a dissolved phase in water. There was in all likelihood a 5th phase, which would have been PCB's dissolved in mineral transformer oil. This phase was seen as a possible mechanism in the deeper movement of the PCB's within the clay. A substantial amount of the PCB seen at depth resulted from movement as the discrete fluid. Many of the fractures in the clay are open in the order of 1 mm. Many, if not most, in the area of the leak contained PCB's, however in the same area fractures that were not visible but were apparent on handling the core contained PCB's. At some fracture size PCB's would not enter, however there were no obvious fracture sets that were so tight the PCB's were not present. There was low level PCB contamination of much of the site, particularly the rear portion, as a result of product handling and the earlier fire. However, the heavy contamination was believed to have resulted from the single event of unknown, but less than a 2 year, duration. The pipe that leaked was located in the bottom of the dirty gravel fill which underlies the building footprint which, in turn, interconnected with the overlying clays and gravel filled cable trenches. The cable trenches, in turn, interconnected with porous zones in the clay, resulting from either fractures or calcite and gypsum solubility.

To answer these questions two programs were initiated; one was to be ongoing deep hole drilling in the area of the deep contamination and the second program was to determine the characteristics of the silt. This second investigation focused on establishing the hydraulic conductivity and the retardation factors for the Condie Silt in an attempt to provide data which could be used in prediction of rates of PCB movement.

5.1 Deep Hole Drilling

As a consequence of the fracturing in the clay and the apparent vertical variation in PCB concentrations, it was felt that several sets of samples would be required over a period of time in order to determine whether or not there was movement. Three sets of three holes were drilled in the summers of 1982, 1983 and 1984. Presented in Figure 5 is the location of the holes relative to the building. Specific locations and hole identifications including earlier holes in the area are shown on Figure 5. The hole locations were set out in a triangular pattern approximately 3 metres (10 feet) apart. The intent was each year to offset the new set in an equal amount which baring pipe and wires was more or less accomplished. The holes were drilled using a 150 mm (6 inch) hollow stem auger. Samples were collected using Shelby tubes driven ahead of the augers. All equipment used for drilling was steamed clean or scrubbed, varsol and hexane rinsed and were dried before drilling between holes and upon completion. The Shelby's were extruded on site, the cores logged, split and samples for

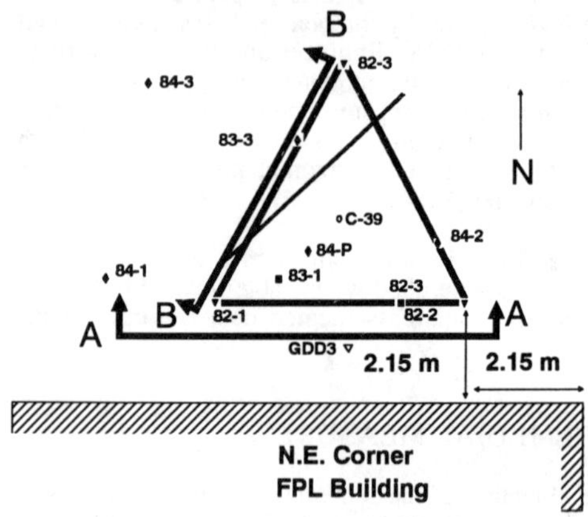

Figure 5 : Location of Deep Hole Monitoring

analyses taken from the exterior of the core. Recovery was usually good. All equipment coming in contact with the soil taken for analysis was cleaned, acetone and hexane rinsed, baked for 24 hours and wrapped in likewise cleaned aluminum foil prior to being brought on site. Analysis following solute extraction was carried out by the Saskatchewan Research Council. Detection limits moved from 1 mg/Kg for 1982 and 1983 to 0.1 mg/Kg for the 1984 sampling. All holes were backfilled using granular bentonite with care taken to avoid cavities. The detailed geology in the area of the deep contamination and spill site is shown on Figures 6 and 7 with the results of the PCB and PCB analysis shown on Figure 8 and 9.

The Regina Clay at this location, rather than being a single stratigraphic unit, is made up of an upper clay zone and low silt clay zone of varying composition. The upper clay consists in turn of two zones, one being a band containing large numbers of calcite and gypsum crystals as well as cavities where the crystals have resolubilized. This is referred to as blocky clay on Figures 6 and 7. The remaining clay normally, but not exclusively, is primarily the blocky clay except for the fractures, is unaltered. The upper meter of the clay did not show presence of fractures and it is unlikely that in the absence of the blocky texture that downward movement from the gravel fill would have occurred. This point is best illustrated by Holes 82-2, 83-2 and 84-2 as shown in Figures 6 and 8. For Hole 84-2, where the blocky clay is in contact with fill, high PCB values as seen in the upper portion in contrast to the other two holes. Once the blocky zone is reached, higher values are seen which are believed to be the result of horizontal movement. Below one meter fracturing was generally continuous throughout the clay stopping at the self contact with only 2 fractures, noted in the top 1/2 metre of the silt. The exception in the clay was in hole 84-2 in the 4-6 m depth range where there were no fractures, and very little measured PCB's. Except where the blocky texture exists there was no visual evidence of PCB penetration from the fractured surfaces. There was no evidence of free water on the fractured surfaces, however it is likely that the fluid that was observed, would have contained some water. During the 1983 drilling the base of the clays were found to be saturated with transformer fluid. As a result of that finding, a purge well was drilled in 1984, fitted with a vacuum pump and over the next few years approximately 5 litres of oil were removed.

The purge well consited of a 100 mm diameter 2.1 m long collection chamber passing through the base of the fractured clay and upper most zone of the silt. At the base of this collection chamber was located a 0.6 m long sealed steel casing reservoir situated 0.75 m into the silt. During the first two years of its operation, this well collected over 13 L of water and 2 L of PCB contaminated oil. The PCB concentration in this oil was typically in the range of 25,000 ppm.

The deep hole PCB concentration data as presented in Figures 7 and 8, do not, because of spatial variation, readily lend themselves to an assessment as to whether or not there had been PCB movement subsequent to completion of the mitigation works.

In general, the vertical variability seen in any given hole can be related to a difference in geology over the depth in question. The differences may be fracture density or fracture alteration, porosity of the clay (blocky texture) or facies change silt, clay, sand, etc. Again, in a general sense, the geology is reasonably homogenous in the horizontal plane. In viewing the data in a horizontal section, one could expect that the highest PCB concentrations should be seen near the point of the leak with decreasing concentrations as you move away. Assuming no PCB movement in the last three years, the time of drilling becomes irrelevant and it should only be the location of the hole that determines the concentration. That model would hold for horizontal spread due to water movement in the fill and also for PCB's moving down vertical fractures to the silt-clay contact and then spreading. It does not hold if there has been significant movement along dipping fractures in which case the distribution would be like a cone with the possibility of higher PCB concentration further away horizontally from the point of the leak.

The data shown in Figure 10 (the 2.74 and 3.66 m depth section) suggests that there has been continued PCB spread within the upper clay zones. In reflection, this could be expected, since the clay in this section is porous and connects directly with the fill. The dewatering has tended to pull water out from the point of the leak in this local, enhancing the PCB spread.

Figure 11 showing the 4.57 and 5.49 m level as an example in the clay and Figure 12 for the silt, would suggest that there has been no recent movement of PCB's or so little movement that the spatial and temporal relationship has been maintained. The 7.32 m section data, Figure 13, is inconclusive. The data at the 8.23, 8.54 and 8.84 m sections, Figure 14 and 15, suggest that there has been a spread of PCB's since the monitoring began. There are a number of oxidized fractures, 45-60° in this section since they are limited in their vertical range, it is unlikely that the cone effect eluded to earlier is causing an anomally. These fractures may, however, have allowed for continued spreading under residual conditions after the start of the remedial works. The apparent stability above and below this section makes it difficult to visualize what else could have happened in this section, however these fractures may connect directly to the fill at some distance.

Trichlorobenzene analyses were carried out during the 1983 and 1984 programs on the lower samples, 8 meters and down. While the data is limited, it appears that the ratio of trichlorobenzene (TCP) to PCB increases with depth with the ratio at about the product level near the clay-silt contact where the silt-clay was seen to be saturated. The ratio shift seen in the clay from less than 5 % may be due to preferential retention of the PCB on the clay surface due to proportioning differences or possibly due to differences on the surface tension of the two products. It may also be possible that there has been subsequent volatilization of the PCB. Below the silt-clay-clay contact, the ratio of PCB to TCP approaches 1:1 and in some cases is closer to 1:2. While the actual concentrations are low, the increased ratio does suggest that water played some role in moving the PCB and TCB into the silts since the increased ratios are explainable only by differences in solubility and subsequent partitioning. The penetration of PCB and TCB for at least 2 meters is believed to be fracture controlled because of the time period and to have occured prior to the start of dewatering since there was no evidence of water during the deep drilling. The movement in the upper silts may be aided by the presence of continuous pebble zone (Figure 6 and 7) which generally coincides with the bottom of oxidation.

6 CHARACTERIZATION OF THE SILT

Cores from the Condie Silt were collected from the property immediately to the west of the plant in April of 1984. The samples were split with half going to the University of Saskatchewan for determination of coefficients of permeability and with the remainder sent to the Oregon Graduate Centre for determination of the organic content of the silt and the effective partioning coefficients and retardation factors for three characteristic tetra, penta and hexachloropenyls. The 4,506 chlorine isomers make up the bulk of the PCB in Aroclor 1254 which is the PCB mix in Inerteen 70-30.

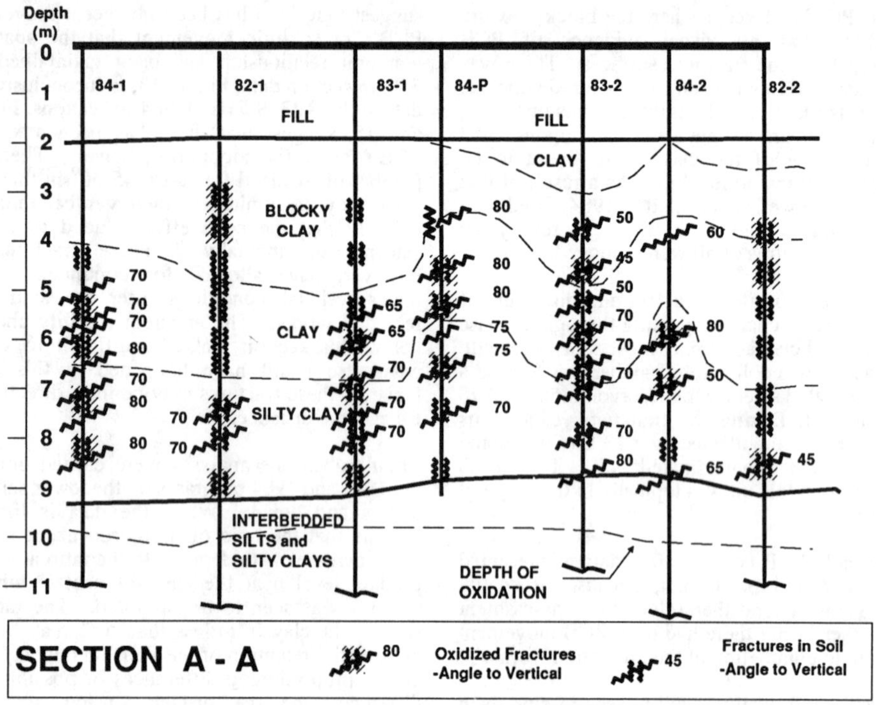

Figure 6: Geological Cross Section A-A, Deep-Hole Monitoring Program

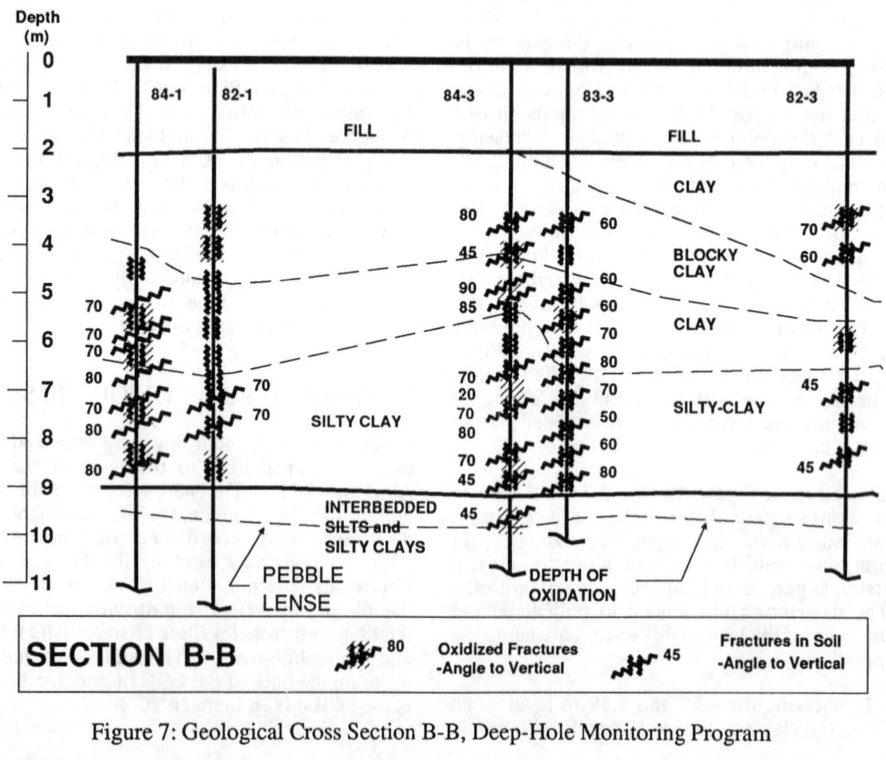

Figure 7: Geological Cross Section B-B, Deep-Hole Monitoring Program

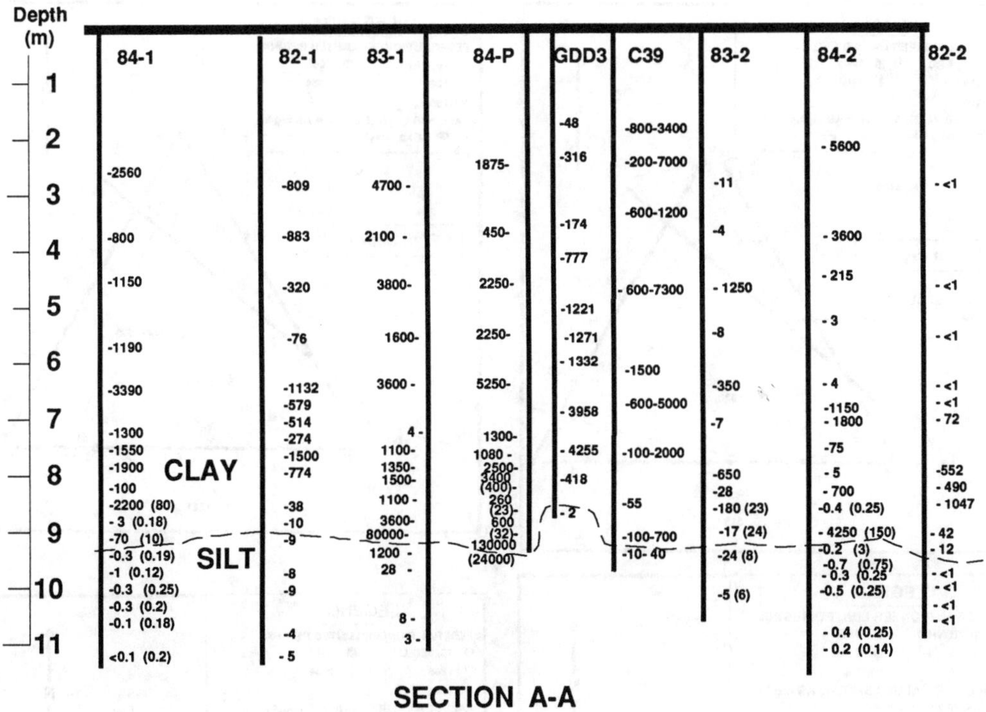

Figure 8: PCB and TCB ( ) Concentrations in mg/kg along Cross-Section A-A

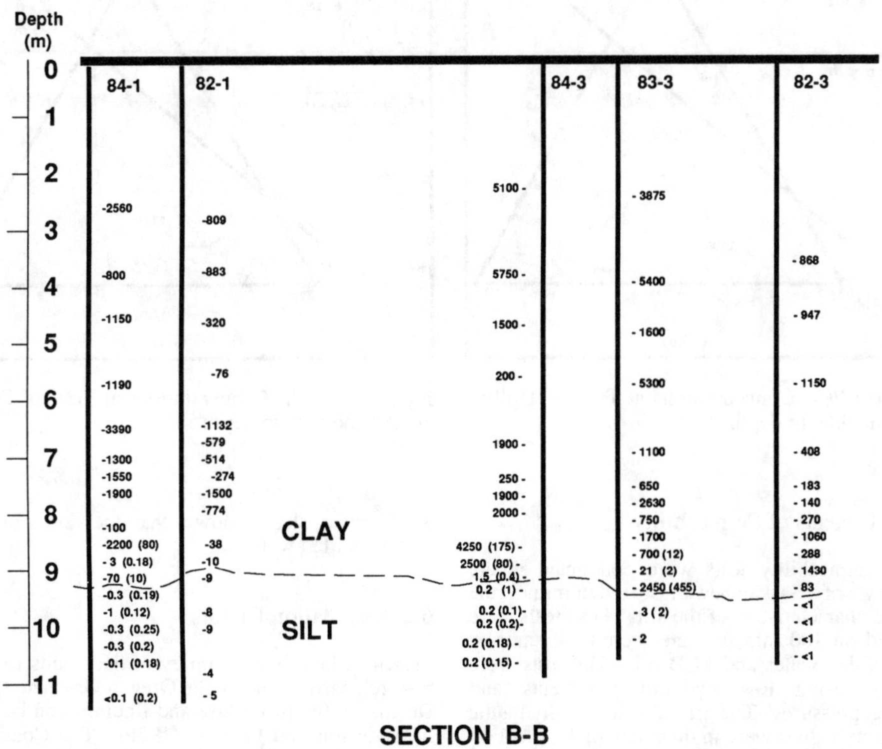

Figure 9: PCB and TCB ( ) Concentrations in mg/kg along Cross-Section B-B

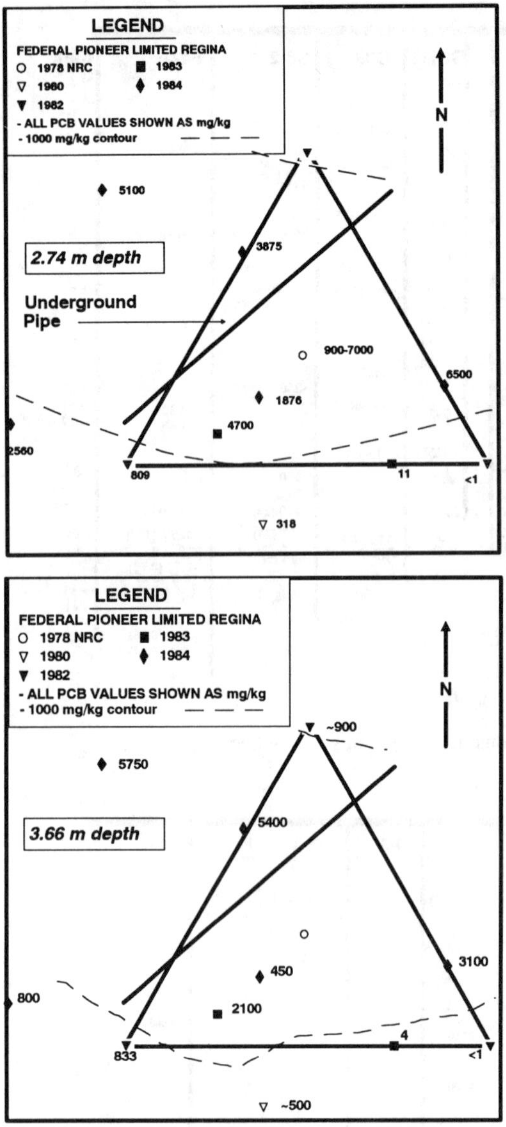

Figure 10 : PCB Concentrations at Time of Drilling at 2.74 and 3.66 m depth

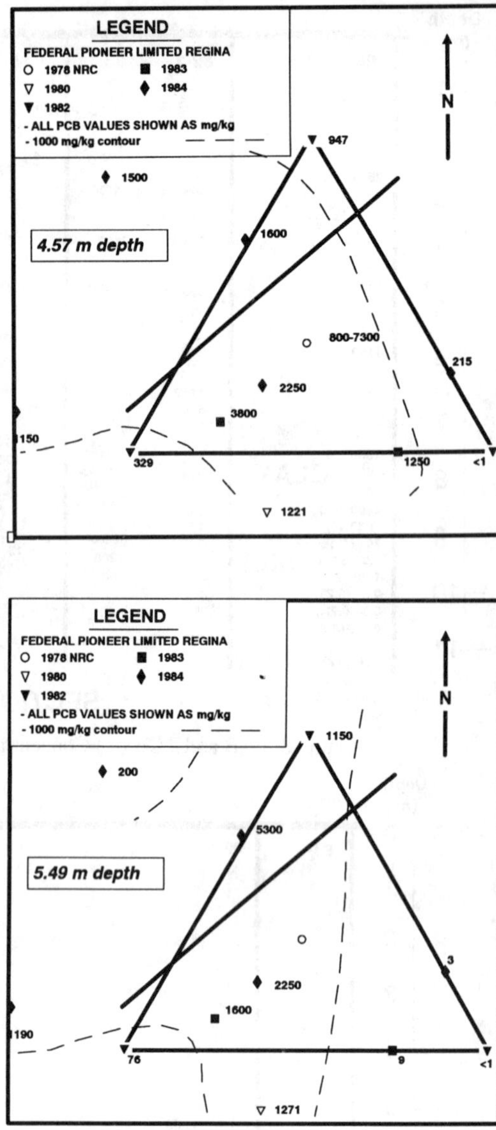

Figure 11 : PCB Concentrations at Time of Drilling at 4.57 and 5.49 m depth

6.1 Coefficients of Permeability

Triaxial permeability tests were conducted at the University of Saskatchewan to determine the hydraulic characteristics of the silt. These tests were conducted on 100 mm diameter cored silt samples using distilled water and PCB oil. The tests were conducted using low hydraulic gradients and confining pressures. The transformer oil hydraulic conductivity values were in the range of 1.11 to 1.45 x 10^{-8} cm/s, slightly lower than the values obtained using distilled water.

6.2 Retardation Factors

Presented here is a summary of the results from the research carried out by the Oregon Graduate Centre. Details on the procedure and findings can be found in Anderson and Pankow (1986). The Condie Silt

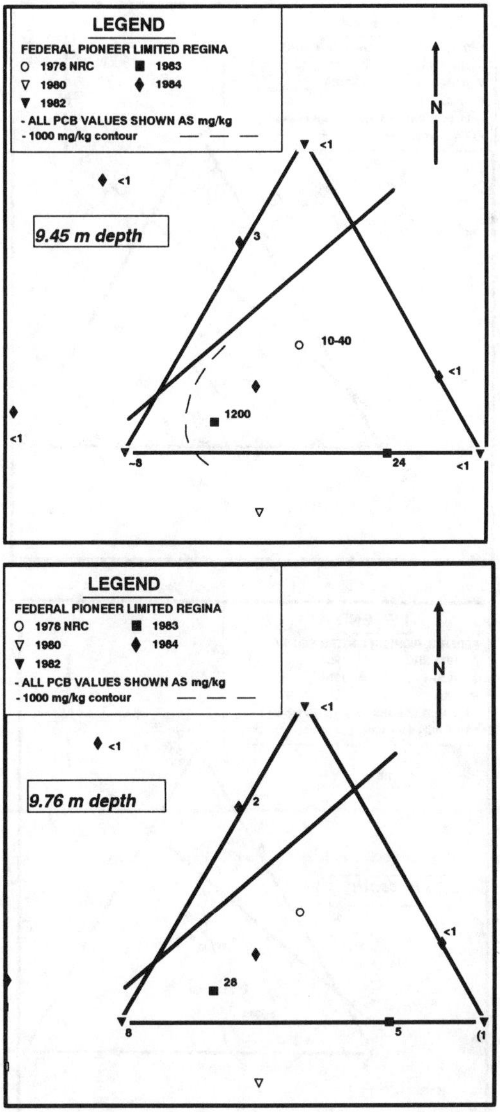

Figure 12 : PCB Concentrations in the Silt, 9.45 and 9.76 m

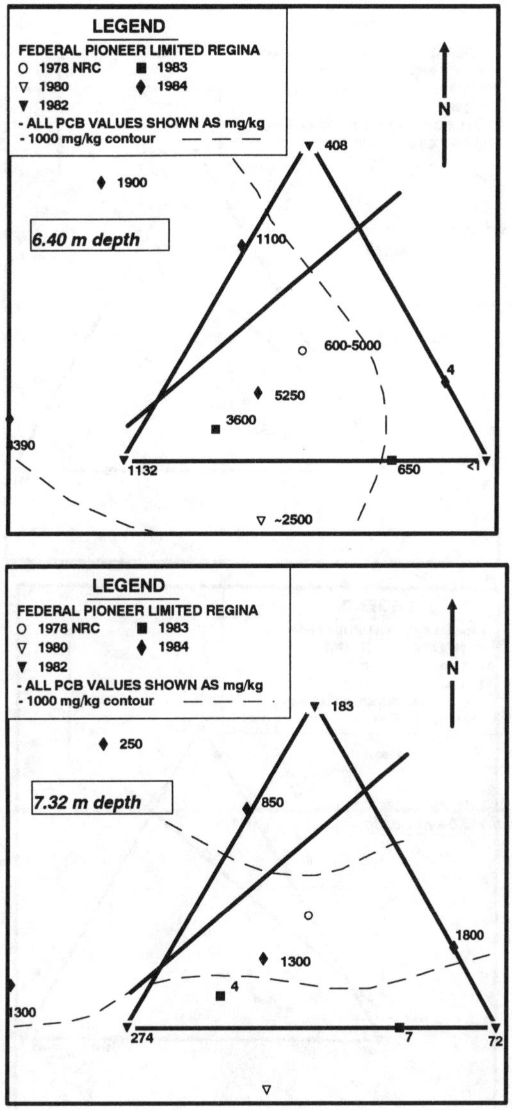

Figure 13 : PCB Concentrations at Time of Drilling at 6.40 and 7.32 m depth

organic fraction was found to be 0.0036 ±0.00088 with no statistical difference over the 4.2 metre depth of core. The partioning coefficient (Kp) for the three PCB measured were; 5000 mL/g, 9400 mL/g and 26000 mL/g for 2,2'55' - tetrahclorobiphenol, 2,2', 45,5' - penta chlorbiphenol and 2,2'4,4',5,5' hexachlorobiphenal respectively, which resulted in corresponding retardation factors (R) of $2.7x10^4$, $5.0x10^5$ and $1.4x10^5$. Calculated R values for 1,2,4 trichlorobenzene and 1,2,4,5 tetrachlorobenzene were 120 and 600 using the measured foc and literature Koc values. For the PCB's, the measured Kp were in excess of what would be calculated using foc and Koc alone. It was felt by Anderson and Pankow (1988) that the high clay content of the silt was effectively increasing the PCB partioning. If that is the case, one might expect higher retardation factors for the chlorbenzenes.

As was stated earlier, measurable but very low concentrations of PCB's were found in the Condie

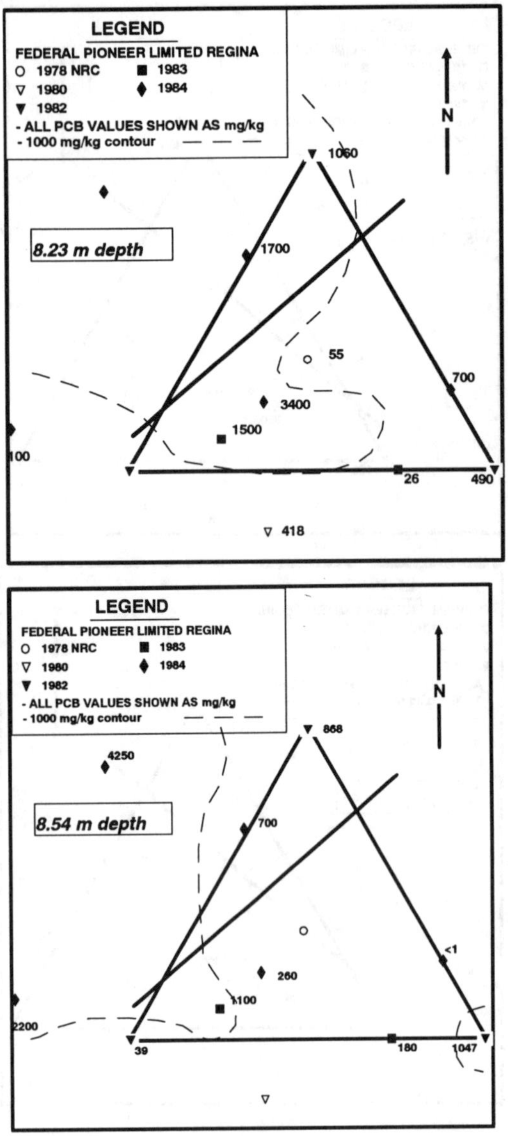

Figure 14: PCB Concentrations at Time of Drilling at 8.23 and 8.54 m depth

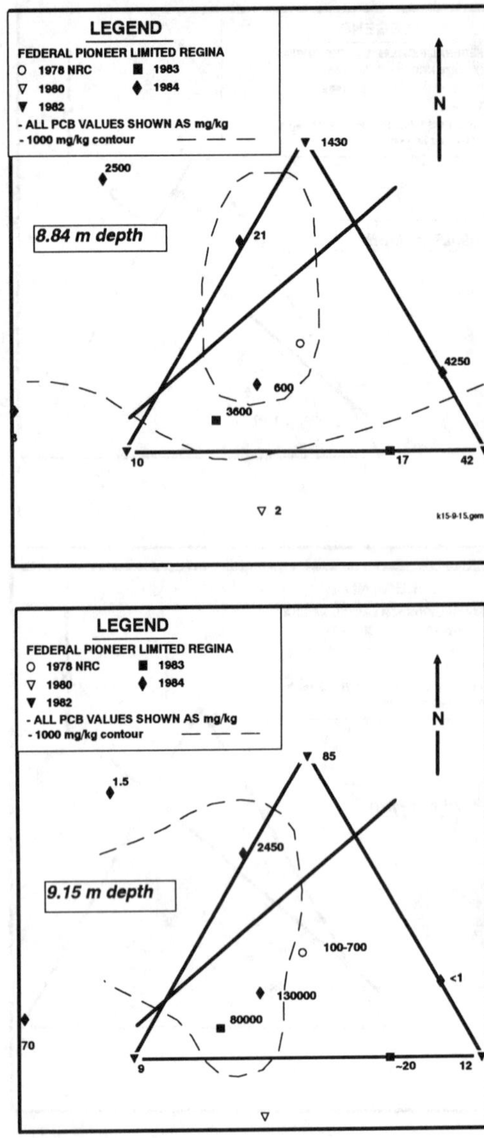

Figure 15 : PCB Concentrations at Time of Drilling at 8.84 and 9.15 m depth

Silt at least 2 metres and probably deeper as several holes terminated with measurably concentrations. The work done by Anderson et al (1986) and by Haug (1985) would suggest the porous movement in whatever formthrough the silt, could not account for the depth of penetration even if one was to consider that the PCB's resulted from events earlier than the spill date.

The only explanation that appears viable is fracture flow with subsequent diffusion into the mass from the fractures. However there were only 2 fractures seen in the Condie Silt, both within 0.5 metres of the clay silt contact, and no other evidence of discontinued sites. At the termination of the monitoring there was no substantial explanation for the observed depth of penetration.

7 SUMMARY AND CONCLUSIONS

The investigations into this spill found that the PCB saturated oil moved rapidly downward through fractures in the clay. Only two years after the spill occurred, PCB's had migrated downward 10 m.

The deep hole drilling and monitoring program found that while there has been movement of PCB's in the overlying clays, there is no "real" evidence of continued movement in the silt barrier. The movement in the clays appears to be primarily horizontally outward at the base of the clay from the vicinity of the source of the spill.

The permeability and retardation factor data appears to suggest that the Condie Silt could provide an effective barrier to further movement of the PCB. It is estimated that under a unit head, penetration of the silt should, by convective porous flow, be in the order of 1-2 mm per year. Much faster rates of penetration have occurred. However, this rapid penetration was most likely aided by fractures in the silt, fractures which are are believed to be limited in extent and restricted to the top of the silt.

The magnitude of this spill and its potential impact on the City of Regina's drinking water are serious. Thus, it is essential that the containment system implemented at this site be adequately maintained until such time as the contaminated soil can be satisfactorily dealt with. It is also important that the monitoring program continue to ensure the effectiveness of the system and to warn against any sudden increase in the downward movement of PCB's through the silt.

REFERENCES

Anderson, M.R. and Pankow , J.F. 1986. Case study of a Chemical Spill: Polychorinated biphenyls (PBC's), 3. PCB Sorption and Retardation in Soils underlying the site. Water Resources Research, Vol22 ,No7, pp.1051-1057.

Atwater, J. 1984. A case study of a PCB spill revisited. Water Resources Research, Vol.20 No.2,pp.317-319

Chritiansen, E.A. 1979. Regional Geology of the PCB Spill Site Regina, Saskatchewan, Saskatchewan Environment Report No. 0031-002

Haug, M.D.1985. Hydraulic conductivity of PCB's in Condie Silt.Department of Civil Engineering,University of Saskatchewan Report p.6

Haug, M.D., Atwater J., Knight, R.B., Kozicki, Paul and Lissey, A. 1988. "Implementation of Remedial Measures to Contain a Sill of PCB's. Conference Proceedings, 2nd International Conference on Case Histories in Geotechnical Engineering, St. Louis, Missouri, June 15.

NRCC 1980. A Case Study of a Spill of Industrial Chemicals-Polychorinated Biphenyls and Chloronated Benzenes. National Research Council of Canada Publication No. 17586

Roberts, J.R., J.A. Cherry and F.W. Schwartz 1982, A Case Study of a Chemical Spill: Polychlorinated Biphenyls (PCB's) 1, History, Distribution and Surface Translocation, Water Resources Research, Vol. 18, pp. 525-534.

Schwartz, F.W., J.A. Cherry and J.R. Roberts 1982, A Case Study of a Chemical Spill: Polychlorinated Biphenyls (PCB's) 2, Hydrogeological Conditions and Contaminant Migration, Water Resources Research, Vol. 18, pp. 535-545.

Subsurface Contamination by Immiscible Fluids, Weyer (ed.) © 1992 Balkema, Rotterdam. ISBN 90 6191 175 3

Field studies of nonaqueous phase liquids at two sites in California's Santa Clara Valley

Guy C. Patrick & Anthony S. Burgess
Golder Associates Inc., Redmond, Wash., USA

ABSTRACT: Data developed from hydrogeologic field programs are evaluated for two industrial sites contaminated with nonaqueous phase liquids (NAPL) in California's Santa Clara Valley. At the first site, contamination by light NAPL and miscible solvents resulted from leaks from former underground storage tanks at a paint manufacturing facility. At the second site, dense NAPL contamination resulted from releases of chlorinated solvents at a group of semiconductor manufacturing facilities. Hydrogeologic conditions at both sites are characterized by laterally continuous and discontinuous alluvial sand and silt aquifer zones, separated by more laterally extensive estuarine and marine silt and clay aquitards. The variable lithology, changing piezometric conditions, the presence of higher permeability paleochannels, chemical and biochemical transformation processes, diffusion into clays, and dissolution has resulted in irregular chemical distribution. The selection of appropriate remedial measures requires consideration of the consequences of long-term dissolution of NAPL, chemical diffusion from clays into mobile groundwater, and the re-mobilization of NAPL effected by changes in groundwater elevations.

1. INTRODUCTION

The Santa Clara Valley of California, or Silicon Valley as it is more commonly known, emerged in the early 1960s as a centre of high technology, primarily related to the development and manufacture of semiconductors and other electronics and computer-related equipment. These industries historically have required the use of large quantities of a variety of organic solvents in manufacturing processes. As a legacy of the industrial chemical storage and handling methods in common practice through the 1960s and 1970s, the subsurface environment has been contaminated with a range of organic chemicals including both light (e.g., xylenes, petroleum naphthas, and other non-chlorinated hydrocarbons) and dense (trichloroethene, 1,1,1-trichloroethane, and other chlorinated solvents) nonaqueous phase liquids (NAPL). Such industrial sites contaminated with organic chemicals account for over 25 Superfund and 90 non-Superfund sites currently under investigation and remediation in the valley (CRWQCB 1989).

The NAPL chemicals of concern typically have low solubilities, although solubility limits are generally one or two orders of magnitude greater than drinking water guidelines, when released to the subsurface as a separate phase, they are likely to become immobilized by capillary forces in soils and remain in the subsurface for years to decades unless some form of remedial action is undertaken. The chlorinated solvents are of particular concern because, unlike the less dense petroleum hydrocarbon-based compounds, they are denser than water and, if released in sufficient volume can migrate deep within a groundwater system where they can remain as a long-term contamination source.

In this paper, two case studies are presented that illustrate the complexity of the hydrostratigraphic environment in the Santa Clara Valley, and data are presented that suggest probable transport and fate processes of NAPL and dissolved chemicals through this environment. The sites, which are located about 10 kilometers from one another, share similar histories in that investigations were initiated at both sites in the early 1980s, and investigations and remedial actions have proceeded more or less continuously from that time to date. While subsurface conditions are similar at both sites, they differ in that chemicals accidentally released at one site were primarily light NAPL, while chemicals released at the other site were primarily dense NAPL.

2. REGIONAL HYDROGEOLOGIC SETTING

The Santa Clara Valley extends from approximately 100 km south of San Francisco Bay in San Benito County, California north towards and under San Francisco Bay. It is flanked by the Santa Cruz Mountains to the south and west, and the Diablo Range to the north and east (Figure 1). The valley is underlain by Pleistocene and Holocene sediments that accumulated over the last two to three million years within a structural depression between the two mountain ranges. The bedrock, which outcrops within the mountains, descends to depths of up to 460 m below the valley floor. The sedimentary sequence above the bedrock has been determined by alluvial and marine depositional processes in response to rising and falling sea levels during repeated glacial episodes in northern latitudes during the last Ice Age (Helley et al. 1979). During periods of low sea-level stands, streams descending from the mountains carried coarser-grained sediments and deposited these materials in alluvial fans along the flanks and forebay regions of the valley, and in stream channels that descended onto the flat

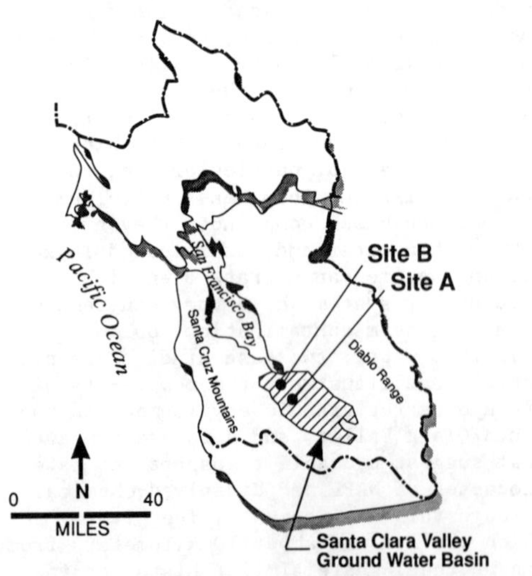

Figure 1. Santa Clara groundwater basin sites

baylands, where they downcut through existing sediments. During periods of glacial melting, sea levels rose resulting in deposition of clays and silts in marine and estuarine environments over the valley floor, and over the deeply incised stream channel sediments.

This depositional history has resulted in an extremely complex stratigraphy of interbedded coarse-and fine-grained sediments exhibiting a large degree of variability in both thickness and lateral continuity.

Groundwater in sufficient supply for domestic and industrial use occurs within the alluvial sediments overlying bedrock. Within the area where most of the electronics industries have located, groundwater occurrence can be divided into a deep and highly productive aquifer zone in which the majority of municipal production wells are completed, and an upper zone which historically has served the needs of individual farms and small industry (Iwamura 1980). The two zones, referred to here as the Upper Zone and Confined Zone, are separated by a confining aquitard layer that is laterally extensive, and ranges in thickness from about 6 to over 30 m (Wilson and Iwamura 1989).

3. SITE A - PAINT MANUFACTURING FACILITY

3.1 Site Description and History

Site A is a paint and coatings manufacturing facility located on relatively flat baylands between the City of San Jose and San Francisco Bay (Figure 1). A history of chemical use and initial subsurface characterization studies at the site have been described previously (Oliveira and Sitar 1985), and are briefly summarized here.

The facility was constructed in the mid-1950s on lands that had been used previously for agricultural purposes. By 1971, a total of thirteen 15000-litre underground solvent tanks and a 1900-litre underground gasoline/diesel fuel tank were in place.

Chemicals stored in the tanks included gasoline, petroleum naphthas, natural fish oils, ketones, xylenes, and toluene. By virtue of their chemical properties, these liquids stored in the tanks can be classified as either light NAPL (petroleum hydrocarbon based compounds) or as miscible product (ketones).

Tank leaks were first discovered at the site in March 1983. Use of all tanks was discontinued at that time, and all tanks were emptied. Subsurface investigations were initiated shortly thereafter, and site characterization and remedial action programs have continued from that time to date.

In addition to chemical releases at the paint manufacturing facility, contamination by chlorinated and non-chlorinated chemicals has resulted from a number of other sites in the area.

3.2 Summary of investigations

Investigative efforts have concentrated on defining the lateral and vertical extent of chemicals. In a phased approach, a total of 45 soil boreholes (excluding those in which wells have been installed) have been drilled on and adjacent the property, and 46 groundwater monitoring and nine groundwater extraction wells and four soil gas monitoring wells have been installed to investigate contamination related to the paint manufacturing facility. These programs have proceeded concurrently with those initiated by other facilities on adjacent properties.

All drilling programs have been performed using either hollow-stem auger or mud-rotary methods. During initial phases of investigation, 200 mm diameter hollow-stem augers were employed, and generally undisturbed soil samples were retrieved for inspection and chemical analysis using a modified California "split-spoon" sampler.

In later phases of investigation, soil samples were retrieved using continuous coring techniques, which generally yielded 100 percent sample recovery. A 1.5 m long by 100 mm diameter CME core barrel was used with hollow-stem augering methods, while a 910 mm long by 75 mm diameter Pitcher sampler enclosing a steel Shelby tube was used with mud-rotary methods.

Depending on location and expected chemical concentrations in the subsurface, 50 mm or 100 mm diameter groundwater monitoring and 100 mm or 150 mm diameter extraction wells have been completed using either PVC or stainless steel casing and screen. Stainless steel has been used in areas where MIBK (methyl isobutyl ketone, or 2-methyl pentanone) concentrations, which range up to 0.3 percent in groundwater at the site, may be sufficient to alter or damage the PVC.

In areas where the potential exists for cross-contamination between upper and lower aquifer zones, those wells completed in deeper zones have not been drilled or installed until a larger diameter steel conductor casing is grouted in place to seal off the upper zones.

Representative groundwater samples have been retrieved using several methods including submersible bladder-type pumps fitted with Teflon tubing, Teflon or stainless-steel bailers and, for extraction wells, downhole submersible electric pumps. NAPL product samples have been recovered using clear acrylic bailers.

In addition to groundwater wells, four 50 mm diameter PVC soil gas wells have been installed in soils above the water table to determine concentrations of volatile gases, and to evaluate the effectiveness of soil gas extraction as a potential remedial technology at the site.

Subsurface hydrologic testing has consisted of aquifer pumping tests and rising head slug tests. Four constant discharge pump tests have been performed on one A-Zone and three B-Zone wells, with pumping periods ranging from three to seven days. Rising head slug tests have been performed in 21 monitoring wells.

Chemical analyses have consisted of on-site and off-site analyses of soils, groundwater and soil gases. On-site analyses have used semi-quantitative methods for screening purposes to direct the field programs and to aid in defining the magnitude of contamination. Methods used have included head space analysis of soils using gas chromatography, head space analysis of soils and groundwater using a calibrated OVA, and soil gas analysis using gas chromatography.

With the on-site methods, where volatile chemical concentrations are quantified in the head space of sample vials, concentrations must be converted to an estimated concentration in soils. These estimates assume that chemical partitioning equilibrium has been achieved between the soil solids, water and head space. Because the partition coefficients are not determined explicitly for each soil sample and are likely to vary among samples, reported concentrations are not directly comparable with more quantitative methods that involve an extraction step, such as the purge and trap method, which yield total chemical concentrations in the soil sample.

All groundwater samples, at least ten percent of all soil samples, and chemical product (NAPL) samples have been submitted to an off-site laboratory for analysis. Because of the variety of organic compounds detected at the site, GC/MS analyses are typically conducted in the open scan mode to identify and semi-quantify concentrations of all volatile chemicals detected as peaks on the chromatograms.

3.3 Initial Remedial Actions

Following the initial phases of investigation in 1984 and 1985, a tank excavation and replacement program was implemented at the site. All 14 underground tanks were removed, and four double-walled underground tanks complete with leak detection systems were installed as replacements. During backfilling, three 150 mm diameter stainless-steel extraction wells were installed in the backfill to recover groundwater from the A-Zone.

Concurrent with the tank excavations, a 12 m long by 5.5 m deep groundwater interceptor trench was excavated immediately north of the former tank locations. Similar to the tank excavations, the trench was backfilled with a permeable pea gravel overlain with compacted clay, and a 150 mm diameter stainless-steel extraction well was installed in the trench backfill for A-Zone groundwater recovery.

The four A-zone extraction wells, together with five deeper B-zone extraction wells have served as a groundwater recovery system to contain and remediate contaminated groundwater at the site. Groundwater pumped from the wells is conveyed to an on-site aerobic fixed-film bioreactor, which is capable of degrading all organic chemicals of concern to non-detectable levels. Treated water is conveyed to the paint manufacturing facility where it is used as non-contact cooling water prior to discharge under permit to the sanitary sewer.

3.4 Data summary and evaluation

Lithologic information has been generated from over 220 boreholes and wells completed at and in the immediate vicinity of the site. Over 500 soil and 300 groundwater samples have been retrieved and analyzed for organic chemicals. A stratigraphic cross section of subsurface conditions that extends in an east-west direction across the site is presented in Figure 2.

3.4.1 Stratigraphy

Stratigraphic conditions underlying the site are typical of near-surface conditions underlying the baylands of the Santa Clara Valley. As illustrated in Figure 2, the upper approximately 10 feet of soils consist of silts and clays, which are referred to regionally as a "clay cap" (Wilson and Iwamura 1989). The cap materials are weathered, and desiccation cracks, rootlets and root holes are commonly observed in core samples.

Individual or groups of sand units may be hydraulically interconnected and typically will behave as laterally continuous aquifers separated by aquitards. At this site, four such zones have been

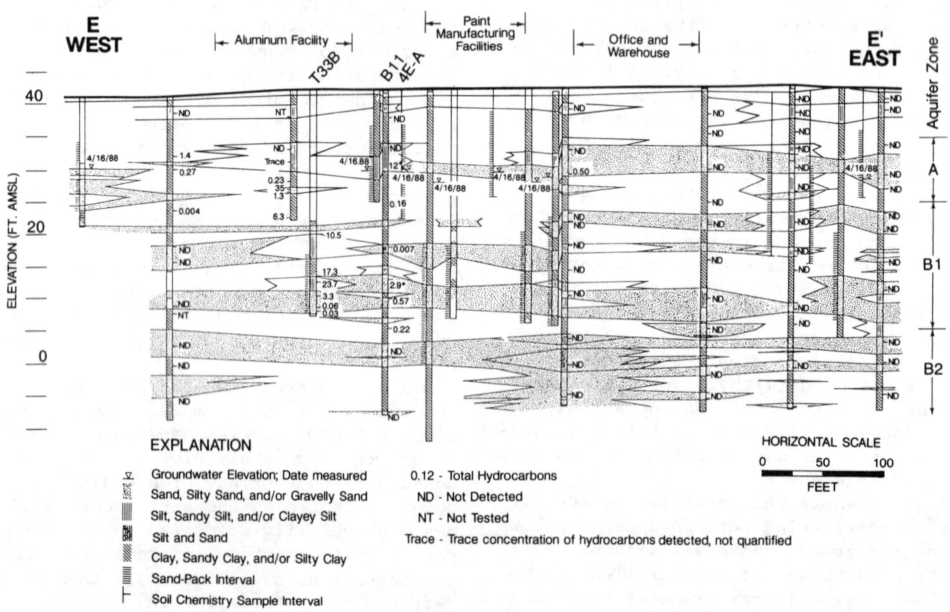

Figure 2. Stratigraphic cross section At Site A

identified. Respective zone designations and approximate depths below ground surface are summarized as follows:

A-Zone	3 to 6 m
B1-Zone	6 to 12 m
B2-Zone	12 to 18 m
C-Zone	Greater than 18 m

Based on a review of regional hydrogeologic conditions, the C-Zone designation at this site likely represents deeper aquifer zones within the Upper Zone described by Iwamura (1980).

Aquitards separating the zones also appear to be laterally continuous, although the presence of features such as paleochannels may breach the aquitard layer, providing a "window" between aquifer zones. While such windows may be present in deep aquitards, no such features have been observed at the site along the aquitard separating the A and B1 Zones. The aquitards, at least in the upper nine metres of sediments at the site, typically contain rootlets and other organic materials, and frequently individual blades of partially decomposed grasses can be observed in core samples.

Based on aquifer pumping and slug tests performed at the site, hydraulic conductivities of the A Zone and B Zone are variable, and both are estimated to range from about 10^{-4} to 10^{-6} m/s. Analyses of B-Zone aquifer test data also revealed that the aquitard separating the A and B1 Zones behaves as a leaky system. Estimates of aquitard hydraulic conductivity from pumping tests are about 10^{-8} m/s, which contrast with measured laboratory values of 10^{-9} m/s. Such differences, which are commonly observed at other sites in the Santa Clara Valley, suggest that secondary porosity features such as desiccation cracks or root holes within the aquitard materials are contributing to vertical groundwater flow.

3.4.2 Hydrologic Flow Regime

Groundwater flow in each of the aquifer zones is generally directed to the north, towards San Francisco Bay, and vertical hydraulic gradients are generally downward from the A- to B- to C-Zone aquifers. However, local stratigraphic conditions within each aquifer zone effect significant variations in both flow direction and the magnitude of the hydraulic gradient.

Water levels fluctuate seasonally in response to rainy winter months and long dry summers. During site characterization activities from 1985 through 1989, water levels have declined at the site approximately nine feet in each aquifer zone, due to regional drought conditions.

Historically, water levels in the region have declined and then recovered at other times. A historic minimum occurred in the 1960s as a result of overpumping (California Department of Health Services et al. 1984), and water levels also declined several feet during the mid-1970s during a drought period (Sitar et al. 1987).

Data from September 1989 illustrated in Figure 3, depicts localized flow directions from the site both to the east and to the north, and a convergence of flow further to the north. Hydraulic gradients flatten east and northeast of the site, suggesting the presence of a more permeable south-north trending paleochannel at that location. Subsequent drilling east of the facility revealed the presence of relatively permeable gravels in the A-Zone along the east property boundary.

Fewer data are available for the B-Zone aquifers. Interpretation has been complicated by the variations in monitoring well screen length and sand-pack intervals.

3.4.3 Vertical and lateral extent of chemicals

A variety of chemicals have been identified in soils at the site to depths up to 13.4 m below ground surface. The dominant chemical groups identified are petroleum hydrocarbons and the ketone MIBK. The hydrocarbons include aromatic and several

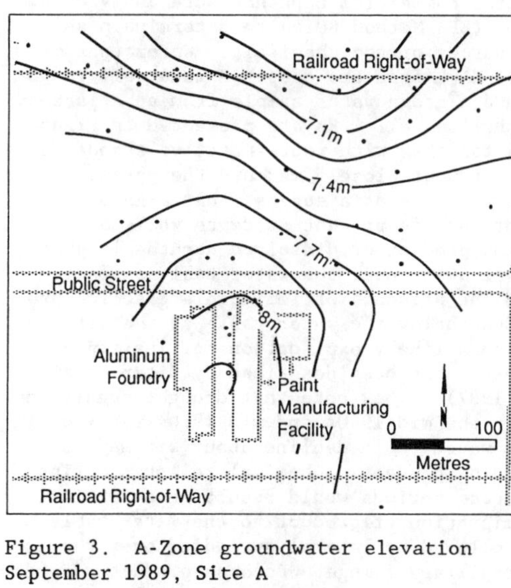

Figure 3. A-Zone groundwater elevation September 1989, Site A

branched and cyclic aliphatic hydrocarbons that are primarily constituents of petroleum naphthas and fuels.

The vertical distribution of total hydrocarbons, as determined from soil analyses, is typified on the cross section in Figure 2, which traverses the north property boundary of the paint manufacturing and adjacent facilities. Where hydrocarbons were detected, concentrations appear relatively high near the April 1988 water table at an elevation of 9 m (30 ft) above mean sea level (AMSL), but are greatest approximately 4.5 to 6 m (15 to 20 ft) below the water table at elevations of 3 to 4.5 m (10 to 15 ft) AMSL. Chemicals were not detected in soils in the deeper B2-Zone, below an elevation of 1.5 m (5 ft) AMSL (approximately 10.7 m below ground surface).

On-site GC head-space analysis of soils indicated that the hydrocarbon compositions were substantially different between the two depths, suggesting that the origins of the chemicals were different. At the deeper location, a B-Zone monitoring well (T-33B) was installed, with a completion interval as shown in Figure 2. During well development, floating free product was recovered from the well. Given the vertical profile of hydrocarbons detected in the soils and the known stratigraphy at T-33B, product recovery from the well indicated that NAPL was present about 4.5 to 6 m (15 to 20 ft) below the water table, and that the NAPL was likely trapped stratigraphically in a sand or silt underlying clay.

Groundwater and NAPL product samples, including gasoline, diesel fuel, and several commercial naphthas were analyzed by GC (EPA Method 8015) to determine possible sources of the chemicals. Chromatograms of NAPL from the deeper B-zone well T-33B and a groundwater sample from an adjacent shallow well 4-EA are presented in Figure 4 together with representative standards that most closely matched the chromatograms. The data suggest that a gasoline product is present at depth while a naphtha product or dissolved naphtha is present at or near the water table surface.

The presence of immiscible gasoline product below the water table at the site, and a likely explanation for its emplacement, has been described by Sitar et al. (1987). They note that drought conditions in the mid-1960s and mid-1970s resulted in a water table decline about six metres below present. Chemical releases during these periods would result in a vertical migration of product to the water table followed by lateral spreading over the capillary fringe. Because product was

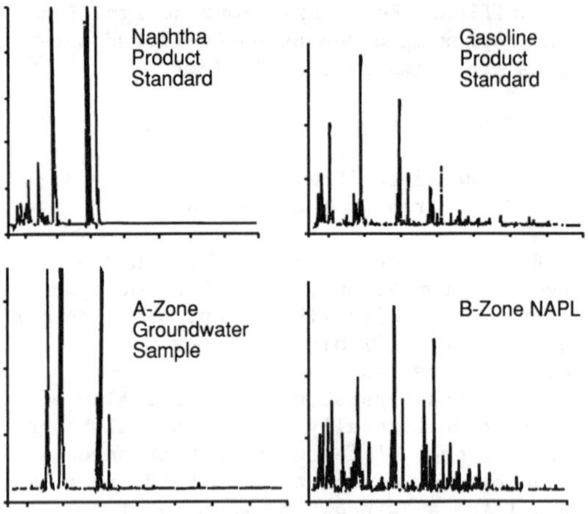

Figure 4. Gas Chromatograms

found below laterally continuous clays, it is probable that the secondary porosity features, such as the rootlets and root holes observed during drilling, can transmit the NAPL.

The lateral extent of hydrocarbon chemicals in the subsurface is depicted in Figure 5, which presents areas where greatest total hydrocarbon concentrations have been detected in soils. The data suggest that hydrocarbons have "pancaked" from source areas at both the paint and aluminum facilities. The predominant migration to the east and north is coincident with the potentiometric surface in the A zone (Figure 3), suggesting that the slope of the water table has controlled NAPL migration at the site.

Maximum concentrations of the more prevalent chemical constituents in soil and groundwater can be used to infer the presence of NAPL in soils, but rarely are sufficient to confirm the presence of NAPL. For example, maximum total hydrocarbon concentrations in soils from T-33B, where NAPL was detected in the well following well installation, were 23.7 mg/kg (Figure 2). The soil sample was retrieved from a depth that likely corresponds to the depth or very close to the depth where NAPL hydrocarbons are located. If NAPL were present in the soil sample analyzed, percent concentrations (ie. several thousand mg/kg) would be expected. Because the vertical sampling intervals, as shown in Figure 2, were of the order of 0.3 to 0.6 m, it appears that the vertical scale of the NAPL distribution is much smaller, perhaps of the order of about 0.3 m or less, and very likely on the order of centimetres.

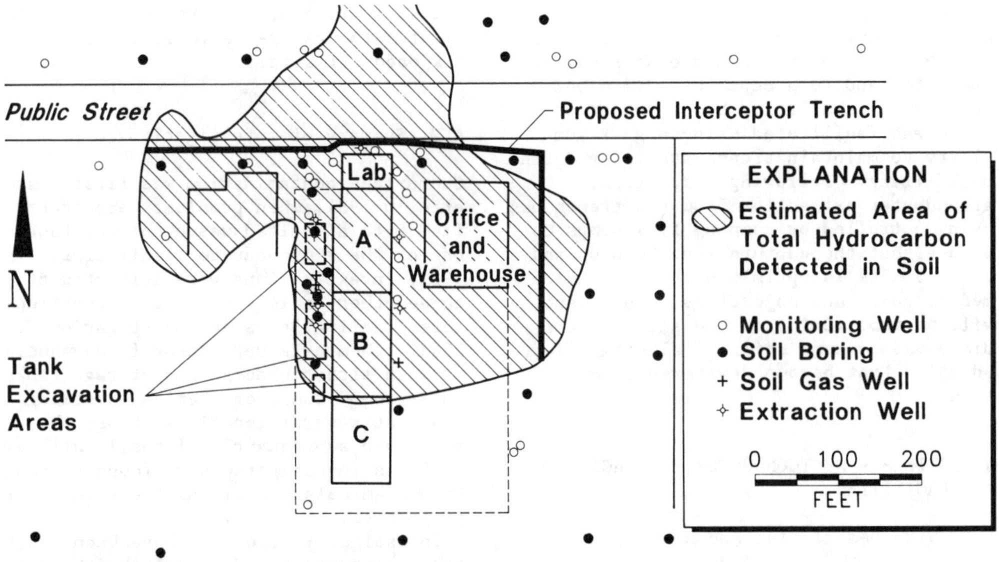

Figure 5. Estimated area of total hydrocarbons in soil at Site A

3.5 Future remedial activities

A review of existing remedial activities implemented at the site indicated the need for further action. Soil excavation and tank removal in 1985 eliminated "hot spots" in soils in the immediate vicinity of the former tanks, but clearly did not remove entire chemical sources in soils. The shallow trench installed immediately north of the former tanks had been rendered ineffective because of decrease in groundwater elevation. The existing groundwater recovery wells had limited capture, and, together with the shallow trench were inadequate to contain all chemicals on site.

An effective remedial system was desired that would address the full lateral and verical extent of chemicals underlying the site, including the presence of chemical product in soils at depths down to six metres below the water table, high chemical concentrations in soils above the water table, and chemicals in both permeable sands and less permeable clays. Chemicals that had migrated off site to the north were to be addressed separately by off site remedial measures.

Because chemicals were present in clays at depth, it was postulated that chemical diffusion and desorption into mobile groundwater were occurring and could continue for years to decades. Neither conventional nor innovative technologies were identified that could effectively remove the constituents in place from the clays over the short term. Thus a containment system was required that could operate for several years to decades, that would provide remediation over the long term, and that would be compatible with potentially promising technologies that may become available in future.

An additional objective of the on-site remedial system was to ensure that chlorinated solvents from adjacent sites would not be drawn on to the site. Trichloroethene was of particular concern because, if extracted in groundwater, it could adversely affect the operation of the bioreactor of the onsite treatment facility.

Following an alternatives evaluation, it was concluded that a site containment option that included product recovery and groundwater removal and treatment was appropriate for the site. Pilot studies were performed to assess the effectiveness of multiple recovery wells versus interceptor trenches for site containment. In addition, a soil venting pilot study (also referred to as in situ vacuum extraction) was performed to evaluate this technology for remediating chemicals from vadose zone soils.

The pilot test to evaluate multiple recovery wells indicated that optimum well spacings on the order of 1.5 m or less were required to ensure groundwater capture. It was, therefore, concluded that a groundwater interceptor trench was more appropriate for the site.

The trench, which was completed in summer 1990, extends approximately 200 m in length along the north and east sides of the site, and to a depth of 10.7 m below surface. Installation below the water table was facilitated using a guar gum slurry to maintain trench stability during backfilling. Following completion, groundwater extraction from the trench has been controlled by pumping from sumps to ensure that the capture zone both on and off the site is optimized to meet the remedial goals and objectives. Soil venting will be implemented on the site to remediate vadose zone soils, including affected soils that become dewatered by the trench.

4. SITE B - SEMICONDUCTOR MANUFACTURING FACILITIES

4.1 Site Description and History

Site B is a group of semiconductor and electronics manufacturing facilities located on relatively flat baylands approximately 10 kilometers west-northwest of the paint manufacturing facility. The site occupies approximately 120 hectares, and includes facilities owned or operated by about 20 companies and includes three Superfund sites (U.S. EPA 1989).

The semiconductor industry began locating in the area in the late 1950s. Certain of the industries re-located in the early 1980s, and currently the site is a mix of operational and vacant facilities.

Organic chemicals used in greatest quantities historically have included trichloroethene (TCE), 1,1,1-Trichloroethane (TCA), and 1,1,2-Trichloro-1,2,2-Trifluoroethane (Freon-113). These have been favoured in the electronics industry as solvents and degreasing agents for cleaning metal and other surfaces because of their high volatility and density, low viscosity and boiling point, and chemical stability. Because of health concerns (TCE is a suspect carcinogen; Schaumberg 1989 and references therein), in the late 1970s TCE was largely replaced by TCA, which has similar chemical properties but is considered less of a health risk.

Other chemicals used in relatively large quantities by some of the facilities have included xylenes, dichlorobenzenes, and phenols.

Chemical storage and handling practices of raw and used solvents have included transfer to and from bottles, drums and above-ground or underground storage tanks. Waste solvents and solvent mixtures were often collected in drums or tanks prior to off-site disposal. Currently, the majority of solvents are recycled rather than disposed off site.

4.2 Summary of investigations

Subsurface contamination was first discovered in 1981, when problems were identified with a vaulted waste solvent tank at one of the semiconductor facilities. In 1982, investigations were initiated at three other facilities, and an additional three facilities began investigations in 1983. Chemicals were identified associated with either underground storage tanks, underground sumps or chemical handling areas at several locations across the site, and subsequently virtually all facilities investigated were found to contain chemicals underlying their respective properties.

Investigative programs have been undertaken separately and jointly by the various companies involved. Programs have included borehole drilling, monitoring and extraction well installation, aquifer testing, laboratory grain size and permeability testing, and soil and groundwater chemical analyses. Soil gas surveys have been performed at the site in at least two facility locations, and a soil vacuum extraction pilot study was performed at one of these facilities (Hochmuth et al. 1988).

To date, over 500 groundwater monitoring wells have been installed across the site, and an additional 460 or more exploratory boreholes have been drilled. Other site investigative programs have included a survey of abandoned agricultural wells to determine their potential to act as conduits between shallow and deep aquifers, a surface soil sampling program, and a surface water and storm sewer sampling program. At some locations, monitoring wells have been completed at depths of more than 150 m to assess contamination of the deep aquifer zones possibly resulting from abandoned "conduit" wells.

Investigative methods are similar to those employed at the paint manufacturing facility, and have included borehole drilling and well installation programs using mud-rotary and hollow-stem auger methods, geophysical logging, and detailed visual logging and sampling using continuous coring techniques or modified California split-spoon samplers. Over 4,600 representative groundwater samples have been retrieved and analyzed for volatile organic chemicals. Other groundwater chemical testing has included analyses for dissolved metals, general water quality parameters, and coliform bacteria.

Soil analyses have been performed on over 4,100 samples using either EPA Methods 8240 or 8270, or using less quantitative field screening methods using head space analysis. All sampling, data collection and data review have been conducted in accordance with quality assurance and quality control programs established and approved by both the companies and regulatory agencies involved.

4.3 Hydrostratigraphy

Hydrostratigraphic conditions underlying the site are similar to those encountered at the paint manufacturing facility (Site A), and reflect the general complexity of the hydrostratigraphic environment in the Santa Clara Valley. In recognition of highly variable stratigraphic and lithologic conditions, detailed information was compiled at certain facilities at the site through a deep exploratory borehole program, where boreholes were placed generally every 23 m along facility boundaries and drilled to the B/C Aquitard at depths of about 50 m. For these detailed studies, soil cuttings were retrieved every 0.3 m along each borehole, and logged separately in the field by two hydrogeologists and undisturbed soil samples were retrieved from each borehole generally every 1.5 to 3 m for chemical analysis.

Stratigraphic conditions developed from the detailed borehole program that illustrate aquifer and aquitard zones are presented on the cross section in Figure 6. The sands tend to thicken and thin, and frequently pinch out when observed in section. Respective aquifer zone depths are as follows:

A-Zone	6 m to 14 m
B1-Zone	15 m to 23 m
B2-Zone	23 m to 34 m
B3-Zone	37 m to 49 m
C-Zone	61 m to 73 m
DEEP Zone	greater than 73 m

The A Zone is either semiconfined or unconfined depending on location and depth to the water table, which varies in depth across the site from about 4.6 m to 7.6 m. The B Zone and deeper aquifers are all confined.

Aquitards separating the aquifer zones are designated accordingly from the surface down as the A/B, B1/B2, B2/B3, and B/C and DEEP Aquitards. The B/C Aquitard is at least 7.6 m thick and continuous across the site, and correlates with the major regional confining aquitard that separates the Upper and Confined Aquifers described by Iwamura (1980). The upper aquitards generally vary from a metre to several metres in thickness and appear to be laterally continuous, although they may be breached by paleochannel features providing "windows" through an aquitard. However, similar to Site A, no windows have been identified through the A/B Aquitard in any boreholes drilled at the site.

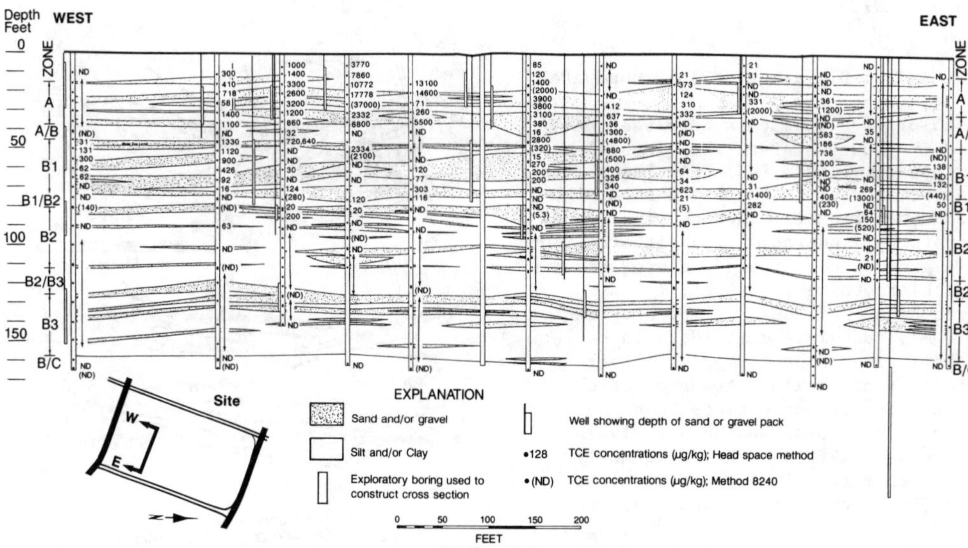

Figure 6. Stratigraphic cross section at Site B

Hydraulic properties of the aquifer and aquitard zones have been determined from aquifer pumping and slug tests, and from laboratory permeability tests. Estimates for the aquifer zones range from about 10^{-3} to 10^{-5} m/s for the A and B1 zones, and 10^{-4} to 10^{-7} m/s for the deeper B2 and B3 zones. Because of the lateral variability in thickness and consistency of the sands and the presence of paleochannel deposits, it is reasonable to infer that aquifer zone properties can vary up to an order of magnitude or more over lateral distances of a few tens of metres.

Laboratory estimates of aquitard hydraulic conductivities range from about 10^{-8} to 10^{-10} m/s, while field results suggest maximum values up to 10^{-7} m/s. As observed at Site A, higher permeability results from field tests suggest that secondary porosity features within the aquitards may contribute to vertical groundwater flow.

Present day lateral hydraulic gradients within the A and B zones are generally to the north towards San Francisco Bay. Within the C and Deep Aquifers, groundwater flow direction is controlled by regional pumping, and is generally directed toward active high production wells to the northeast of the site.

Vertical hydraulic gradients are either upward or downward among the aquifer zones, depending on location, regional pumping and seasonal recharge conditions. It is likely that, similar to Site A, the water table historically has been lower than at present as a result of regional overpumping and drought conditions in the 1960s and 1970s.

4.4 Chemical Distribution in Groundwater

Over 25 organic chemicals have been detected in groundwater at the site. The chlorinated solvents TCE, TCA, Freon-113, are detected most frequently and in highest concentrations. Other chemicals that appear to be associated with either TCE or TCA in groundwater, include cis or trans 1,2-dichloroethene (1,2-DCE), 1,1-dichloroethene (1,1-DCE), 1,1-dichloroethane (1,1-DCA), and vinyl chloride. These latter chemicals can be formed by either chemical or biochemical transformation processes in the subsurface (Vogel and McCarty 1987a, Vogel and McCarty 1987b, Wilson et al. 1986). Other frequently detected chemicals include tetrachloroethene (PCE), chloroform (which may be either a subsurface chemical transformation product or contaminant introduced during drilling), and 1,2-dichlorobenzene (1,2-DCB).

Chemicals detected most frequently are concentrated in the A and B1 Zones where the bulk of chemical mass resides, with fewer detections in the underlying B2, B3, C, and Deep aquifer zones. Chemicals detected in the C and Deep Aquifer zones are postulated to have entered these zones through abandoned agricultural wells at the site, which were completed across the B, C, and Deep Zones. The suspect wells have been identified and sealed.

Of all chemicals detected, TCE is most prevalent and has served as an indicator chemical to identify the extent of subsurface contamination at the site and areas requiring remediation. The chemical distributions in the A and B1 zones, as estimated from comprehensive groundwater sampling rounds in 1986 and 1987, are presented in Figure 7. The plumes emanate from multiple source areas across the site, and extend for at least 1000 m downgradient to the north, towards San Francisco Bay. Both plumes are estimated to have impacted in total about 6 x 10^9 litres of groundwater. Conservatively low mass estimates of dissolved TCE mass in the plumes are about 14,350 kg, which is equivalent to approximately 9800 litres of original NAPL, or 47 fifty-five US gallon drums (Mackay and Cherry, 1989). Based upon the continuing existence of high-concentration areas, the amount of TCE originally released to the subsurface is likely to be higher than estimated above.

The bulk of the TCA observed at the site appears to be concentrated in the vadose and A zones, and has not impacted the

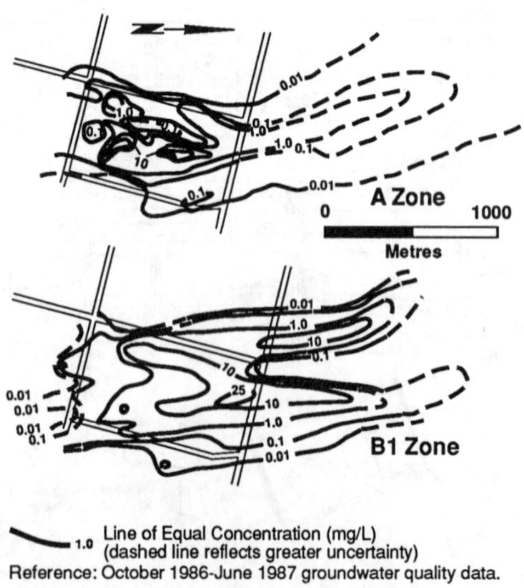

Figure 7. TCE distribution at Site B

deeper B1 zone to the same extent as TCE. Because TCA largely replaced TCE in the late 1970s, its distribution may be a consequence of factors such as smaller volumes released, higher water table conditions at the time of release, and a reduced permeability of soils to TCA because of the presence of TCE.

Because the solubility of TCE at typical groundwater temperatures is about 1100 mg/L, concentrations approaching this value can sometimes be used to infer the presence of a NAPL source in soils. However, these elevated concentrations are rarely observed at the site, and typical maximum concentrations in the vicinity of probable chemical source areas range from less than ten to a few hundreds of mg/L TCE. Of over 500 monitoring wells installed at the site, maximum TCE concentrations of 1000 mg/L have been reported at only one monitoring well, which is completed within one of the potential TCE source areas.

As illustrated in Figure 6, chemical concentrations in subsurface soils are highly variable both laterally and vertically, and may vary vertically at the scale of a few centimeters in soils and laterally at the scale of a few meters. However, monitoring wells are typically screened vertically over a meter or more of aquifer materials, and generally intersect single or multiple sand units within an aquifer zone. Groundwater samples withdrawn from such wells will be representative of depth-averaged dissolved concentrations. Such averaging will tend to bias maximum dissolved concentration data toward relatively lower values.

Maximum TCE concentrations observed at the site, which are typically much lower than the aqueous solubility of TCE, also likely reflect the nature of chemicals originally released at the site. Chemicals may have been released as NAPL product mixtures or as constituents dissolved in waste water. Where NAPL mixtures were released, the aqueous solubility of individual constituents may be either enhanced or reduced, depending on the composition of the mixture. For example, co-solvents such as isopropanol may increase aqueous solubilites of other constituents (Zenon Environmental 1986), whereas hydrophobic organics that are similar in structure, such as TCA in a mixture with TCE, may decrease each component's solubility approximately in accordance with Raoult's Law (Banerjee 1984).

4.5 Chemical Distribution in Soils

Chemical detection frequencies within aquifer zone soils are similar to those identified from groundwater data. Soil chemistry data also provide information on chemicals detected in aquitard zones separating the aquifers. Both TCE and TCA are present in the aquitards separating impacted aquifer zones, with highest concentrations in the A/B aquitard that separates the A and B1 aquifer zones.

With the possible exception of one vadose zone soil sample containing 6300 mg/kg of TCA, none of the soil chemical concentration data from over 4,100 samples tested provide conclusive evidence that NAPL was or is present in the subsurface at the site. For example, the maximum TCE concentration detected at the site of 190 mg/kg, is at least an order of magnitude below expected concentrations, which could range up to several thousand mg/kg if NAPL were present (Schwille 1984, 1987).

Because soil chemistry data are retrieved from discreet depths, by examining concentration-depth profiles the data can provide useful information on chemical distributions in the vicinity of potential source locations. Two examples are illustrated schematically for TCE in Figure 8. Although no single soil sample concentration is sufficiently high to indicate the presence of NAPL, the compiled data suggest that TCE has concentrated in both locations at the base of the A-Zone aquifer, and has penetrated the A/B aquitard into the B1 Zone. At one of the potential source locations, the data suggest that the TCE has continued vertically down into at least the B2 zone at a depth of over 27 m (90 ft), whereas at the other location the TCE may be confined to the upper half of the B1 zone, at a depth of about 24 m (80 ft).

4.6 Remedial Activities

Remedial activities at the site have included source removal and source control actions. Source removals have included removal and replacement of underground tanks and sumps, and soil excavation. Control actions have included groundwater extraction and treatment to hydraulically contain sources and remediate groundwater, and soil-bentonite slurry walls to physically contain sources.

Groundwater extraction and treatment programs have been initiated at most facilities at the site. These wells are operating to control chemcial sources, and a regional groundwater remediation program is currently under design.

The time to effect cleanup is estimated to be tens to hundreds of years. Such long estimated cleanup times have prompted some facilities at the site to implement cost-effective source control measures that are compatible with potentially applicable remediation technologies that may become available in future. To address containment of potential chemical sources in soils, four slurry walls have been installed at the site. Three are completed around three separate facilities, and extend into the A/B aquitard at a depth of approximately 12 m. The fourth slurry wall, which has been described by Burgess et al. (1988), encompasses a single facility, and is completed at a depth of 30.5 m. This latter wall is not keyed into an aquitard but rather, is designed to contain chemicals emanating from all on-site sources by creating inward and upward hydraulic gradients within the contained area. The wall effectively reduces the volume of groundwater that must be pumped and treated, while pumping within the contained area maintains the desired hydraulic gradients. Recent reviews of the compatibility of clays and dissolved or NAPL chemical constituents (e.g., Middleton et al., in press), suggest that the walls may prove to be an effective long-term method of maintaining site containment.

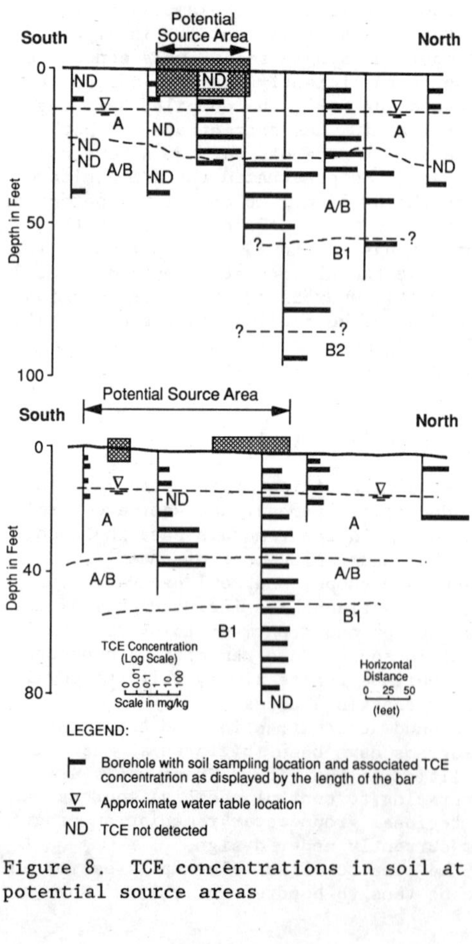

Figure 8. TCE concentrations in soil at potential source areas

5. SUMMARY AND CONCLUSIONS

Near-surface hydrostratigraphic conditions in the Santa Clara Valley provide a complex medium through which both light and dense NAPLs have infiltrated and dispersed. Subsurface chemical distributions within these sediments are typically irregular, and vary both vertically and laterally at scales that are not necessarily compatible with the scales of typical groundwater and soil sampling programs, which often are not sufficiently detailed to detect the presence of NAPL by direct interception. Changing water table conditions resulting from periodic droughts have been instrumental in dispersing chemicals through these sediments.

It is also evident that, in addition to vertical migration through "windows" in aquitards, NAPL penetrated the near-surface aquitard materials through features such as fractures, root holes, or dessication cracks. Consequently, these shallow aquitards cannot be considered as barriers to vertical chemical transport.

For remediation, pump and treat systems alone may take centuries to achieve clean up. Containment options such as slurry walls, which are being implemented at some sites, can potentially control NAPL dissolution and dispersion. Treatment costs will be minimized because of the reduced volume of groundwater that must be treated. Also, the schemes are likely to be compatible with future cleanup technologies. As an additional advantage, facilities where these options have been implemented can remain fully operational, and can offer a fully implemented site containment system as a physical and financial asset.

REFERENCES

Banerjee S. 1984. Solubility of organic mixtures in water. Envir. Sci. & Tech. 19:47-52.

Burgess, A.S., G.S.Laird, M.S.Leonard 1988. Proc. 2nd Internat. Conf. on Case Hist. in Geotech. Eng. June 1-5, St. Louis, MO.

California Department of Health Services, California Regional Water Quality Control Board, Santa Clara County Public

Health Department, Santa Clara Valley Water District & U.S.EPA 1984. Groundwater and drinking water in the Santa Clara Valley: a White Paper, October.
CRWQCB 1989. South Bay Site Management System Milestone Reports July-September 1989, California Regional Water Quality Control Board, San Francisco Bay Region.
Helley, E.J., L.R.Lajoie, W.E.Spangle & M.L.Blair 1979. Flatland deposits of the San Francisco Bay Region, California-their geology and engineering properties, and their importance to comprehensive planning. U.S.Geol. Survey Prof. Paper 943.
Hochmuth D.P., K.S.Udell, E.G.Lappala & J.L.Jasperse 1988. Soil venting of volatile organic chemicals from low permeability soils at a site in Santa Clara County, California. Amer. Inst. Chem. Eng. 1988 Summer Nat. Meeting, August 21-24, Denver, CO.
Iwamura, T.I. 1980. Saltwater intrusion in the Santa Clara County baylands area, California. Santa Clara Valley Water District. September.
Mackay, D,M. & J.A.Cherry 1989. Groundwater contamination: pump-and-treat remediation. Env. Sci. & Tech. 23:630-636.
Middleton, T.A., J.A.Cherry & R.M.Quigley (in submission). A review of neutral organic fluids on the permeability of clays. Submitted to Can. Geotech. Jour.
Oliviera, D.P. & N. Sitar 1985. Ground water contamination from underground solvent storage tanks, Santa Clara, California. Proc. Fifth Nat. Symp. & Exp. on Aquifer Restoration & GW Mon., Columbus, Ohio, May 21–24.
Schwille,F. 1987. Dense chlorinated solvents in porous and fractured media – model experiments. Translated by J.A. Pankow. Lewis Publishers:Chelsea, MI.
Schwille, F. 1984. Migration of organic fluids immiscible with water in the unsaturated zone. In B.yaron, G. Dagan, J. Goldsmith (eds.), Pollutants in porous media. Springer–Verlag: Berlin.
Sitar, N., J.R.Hunt & K.S.Udell 1987. Movement of nonaqueous liquids in groundwater. Proc. Geotech. Prac. for Waste Disp. '87/GT Div. ASCE/Ann Arbor, MI. June 15–17.
U.S.EPA 1989. EPA announces cleanup plan for MEW Study Area. U.S.EPA Region IX, San Francisco. July.
Vogel, T.M. & McCarty P.L. 1987a. Rate of abiotic transformation of 1,1–dichloroethene from 1,1,1–trichloroethane in groundwater. Jour. Contam. Hydrol. Vol. 1.
Vogel, T.M. & McCarty P.L. 1987b. Abiotic and biotic transformations of 1,1,1–trichloroethane under methanogenic conditions. Envir. Sci. & Tech. 21:1208–1213.
Wilson, L.D. & T.I.Iwamura 1989. Standards for the constructionn and destruction of wells and other deep excavations in Santa Clara County. Santa Clara Valley Water District, July.
Wilson, B.H., G.B. Smith & J.F.Rees 1986. Biotransformations of selected alkylbenzenes and halogenated aliphatic hydrocarbons in methanogenic aquifer material: a microcosm study. Envir. Sci. & Tech. 20:997–1001.
Zenon Environmental 1986. A study of enhanced dissolution for the in situ remediation of dense non–aqueous phase liquid (DNAPL) chemicals in the subsurface. Report to Environment Canada, Ontario Regional Office, Toronto. November 3.

Subsurface Contamination by Immiscible Fluids, Weyer (ed.) © 1992 Balkema, Rotterdam. ISBN 90 6191 175 3

Case history: A kerosene spill at Strasbourg-Entzheim International Airport, France

L.Zilliox & O.Razakarisoa
Institut de Mécanique des Fluides, Université Louis Pasteur, URA CNRS 854, Strasbourg, France

G.Rinck
BRGM, Service Géologique Régional Alsace, Strasbourg, France

ABSTRACT : The discovery in 1971 of a serious leakage of aviation fuel led to extensive works in order to understand the nature of the pollution and to implement remedial actions. From results obtained by coredrill investigation and different studies, the oil spreading was limited by natural factors and there was no apparent serious effect of the oil on the water-table due to geological and hydrogeological characteristics of the local area. It must be recognized that geological and hydrogeological conditions of Strasbourg Entzheim case are particular but not unique. The methods used to deal with the spill, could provide a useful background for the treatment of another inland oil spill of this nature.

1 HISTORY OF THE SPILL

At the beginning of the year 1971, the drinking water supply well of the Entzheim International Airport meteorological station was found to have a strong odour of oil. The station is located a few hundred meters from the fuel storage depot. Strasbourg Entzheim International Airport is situated to the north of Entzheim village which is 15 km southwest away from Strasbourg (Figures 1 and 2).

In June 1971, kerosene (aviation turbo fuel TR4) was identified in this well. Consequently, it was concluded that the spill or successive spills had resulted from overflow of kerosene from the buried tanks in one of the fuel storage depots, probably during loading. The precise volume of spilled turbo fuel is not known. The specific gravity of the kerosene is 0.76. A through examination of the well showed that a 60 cm thick kerosene layer was lying on the groundwater surface.

The actions taken to deal with the Entzheim spill were initiated and coordinated by the Regional Water Resources Management Service (SRAE Alsace). Various organizations were commissioned to participate and give guidance on actions to be taken. Major contributors have been the Service Geologique d' Alsace et de Lorraine (SGAL) which is the regional branch of the Bureau de Recherches Géologiques et Minières (BRGM), the laboratoire d'Hydrologie de la Faculté de Pharmacie and the Institut de Mécanique des Fluides of Strasbourg University.

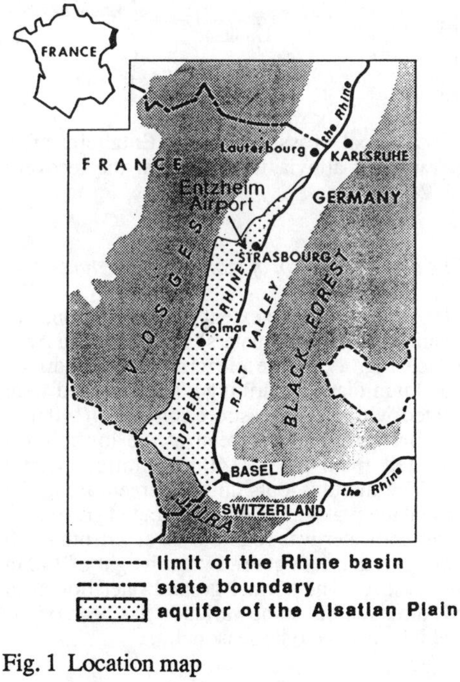

Fig. 1 Location map

2 THE EXTENT AND THE MECHANISM OF THE POLLUTION

2.1 *Hydrogeology of the site*

The region affected by the spill of turbo fuel lies within the Upper Rhine rift valley water-bearing formation and consists of Rhine alluvial deposits.

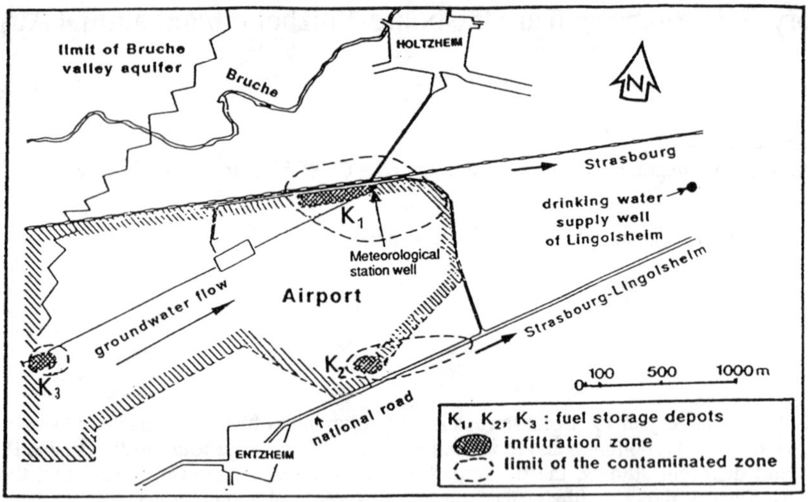

Fig. 2 Strasbourg-Entzheim case. Location map

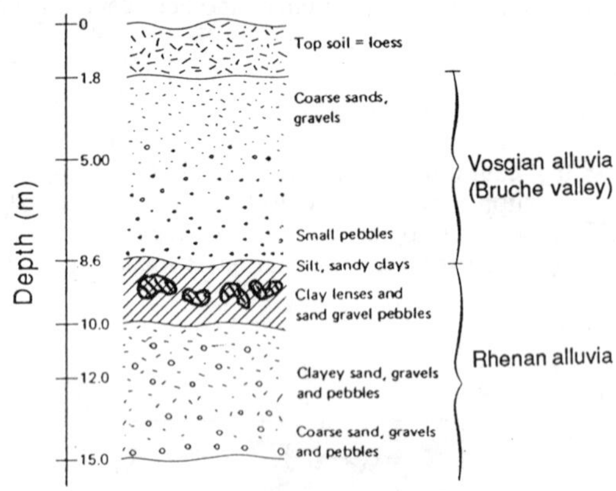

Fig. 3 Typical cross-section in Entzheim area. Stratigraphy of alluvia formations at the recovery well 272-1-51

This aquifer of the Alsatian plain is approximatively 160 km long and 25 km wide with a thickness on average of 100 m but reaching 200 to 250 m in some sectors. The annual replenishment of the aquifer is typically 1.3 billion m^3, with annual extraction amounting to 0.45 billion m^3 (domestic, agricultural and industrial use), about 30% of the annual replenishment. The water-table is 10 to 12 m below the surface between Colmar and Basel, and about 3 to 4 m below the surface near Strasbourg.

A typical description of the subsurface strata at the Entzheim site is given by a cross-section at the recovery well 272-1-51 shown in Fig.3. The water-bearing formation at this point is 30 to 40 m thick and corresponds to sands, gravels and pebbles of Rhenan origin. This lies beneath 8 to 10 m of Vosgian alluvial deposits of finer grain size (Bruche valley), with 1 to 2 m of top-soil (loess) at the surface. The water-table at Entzheim lies 10m below the surface and fluctuates during the year by ± 0.5 m.

The hydrogeological characteristics of the aquifer, as measured at the recovery well are on average as follows :

Hydraulic conductivity	K : about 2x 10^{-3} m/s
(Permeability	k : about 200 darcy)
Transmissivity	T : about 10^{-1} m^2/s
Storage coefficient	S : about 0.01

Above the aquifer, the Bruche alluvial deposits are 10 to 25 times less permeable (K: 10^{-4} m/s; k: 10 darcy) than the Rhine deposits, and they may exhibit large variations both in the horizontal and vertical directions. Some clay lenses are encountered between these two geological formations, at a depth of between 8 and 10 m (Fig. 3).

2.2 *The implementation of monitoring methods*

In order to determine the extent of the pollution, five observation wells (piezometers) were drilled in April 1972 (see Fig. 4). These piezometers were drilled with 600 mm diameter drilling equipment and were completed with a full plastic casing (134/140mm) for the first 9m, and a screen at the lower level constructed from a 134/140mm slotted plastic casing. The annulus outside the screen along the perforated interval was filled with graded gravel.

The groundwater was found to be at a depth of 10 m, corresponding to a groundwater height of 140 m above sea level. The normal groundwater flow is in a NNE direction but some variation has been observed, which appears to be seasonal.

A kerosene layer (60 cm thick) was found only in the piezometer 272-1-37. The extracted cuttings from this piezometer all had a strong odour of hydrocarbon from 6.5 m deep. Free kerosene was identified at a depth of 8.5 m.

In each piezometer, groundwater samples were taken for full chemical analysis before and after pumping for 30 min. with a submersible pump immersed one meter below the static water-table. Before pumping, water samples did not show any iridescence or visible trace of hydrocarbons, apart from well 272-1-38 where a very slight hydrocarbon odour was detected. Water samples taken after pumping were analyzed for hydrocarbons by the carbon tetrachloride-infrared technique (CCl_4-IR, without florisil). Results ranged between 0.3 and 0.1 mg/l corresponding to the detection limit of the method. (From some routine water quality analyzes, deterioration of groundwater quality was observed downstream of the depot. It was suggested that this probably originated from secondary phenomena due to hydrocarbon oxidation).

2.3 *Delimitation of the contaminated area*

A coredrill investigation completed in June/July 1973 was carried out to improve the delineation of the spilled kerosene in the unsaturated area between the ground surface and the water-table.

This programme involved drilling 41 augerholes (150 mm diameter, total length 401 m) and the taking of core samples every 1.5 m from depth of 2 m down to the water-table (10 m). Kerosene was detected both by smell and chloroform extraction followed by UV-fluorescence examination. It was observed that the odour quickly diminished, completely disappearing within one hour of sampling. The location of the test-borings are given in Fig. 4.

From this investigation, it was tentatively concluded that the kerosene body did not rest uniformly on the water-table, and that it was generally located at a depth between 5 to 8.5 m above both the zone of groundwater level fluctuations and the capillary fringe. Diagramatic representation of these results (Fig. 4 and 5) shows that the polluted area stretched over some 1km in length and some 400 m in width in the direction of groundwater flow. The contaminated surface was therefore estimated to be about 400000 m^2 in area based on those samples showing a marked (distinct) presence of hydrocarbons. This area would decrease to 160000 m^2 if the area considered as polluted was confined to those samples showing significant presence of hydrocarbons.

The corresponding estimation of the total amount of kerosene spilt was in the range between 3000 m^3 and 6000 m^3, assuming an equivalent layer of 0.5 to 1 m thick with kerosene (porosity: 15%; air content: 50%; water content: 25%; kerosene content: 25%). That means 20 to 40 liters of kerosene per m^2.

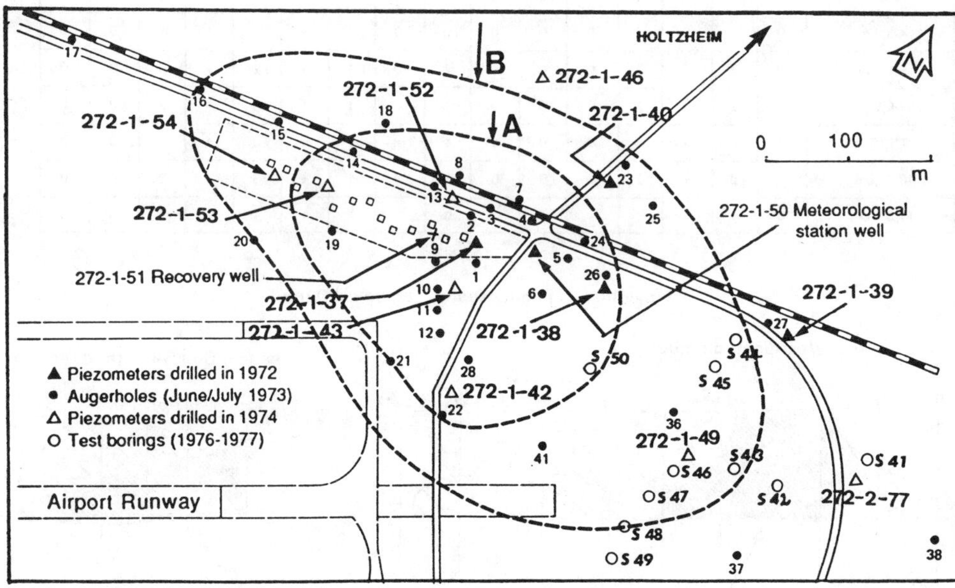

Fig. 4 Location of piezometers and test-holes (after Concawe (1981))
A: Contaminated zone of 160000 m^2 for significant extraction (see Fig.5).
B: Contaminated zone of 400000 m^2 for distinct extraction (see Fig.5)

The conclusions from the programme of 1973 were later confirmed after drilling eight piezometers in 1974 (see Fig.4 for location and Table 1 for results), and after some additional test borings in 1976/1977 to the east of the depot. A search for hydrocarbons did not detect any evidence of pollution, apart from test hole number 50 located in the pollution zone (Table 2). This appears to be consistent with the earlier interpretations of the spill zone. However, it must be remembered that the spill most probably took place over a long period of time (several years) and should be regarded as a series of consecutive spills occuring under different conditions, i.e. different water contents of the unsaturated zone (Concawe (1981)).

2.4 *Interpretation of the pollution mechanism*

From local hydrogeological considerations and chemical analysis of groundwater samples taken from piezometers of the site, and which did not reveal dissolved aromatic traces from kerosene, these studies suggested that a discontinuity was likely to exist in water content due to the two-layered structure of the site and the presence of some clay lenses between these units (Vosgian and Rhenan units with very different permeabilities - see Fig.6): This particular structure may favour the formation of hanging capillary zones over which the infiltrated kerosene was spread out. Thus, the impregnation body of the kerosene did not directly reach the capillary fringe of the groundwater, or the contact between the kerosene and the water-table was very partially established because the bottom part of the Vosgian formation was highly water-saturated and therefore constitutes a natural barrier to the downward migration of kerosene. Figure 6 shows a probable configuration of the impregnation body on account of the geological cross-section and water content. This mechanism should explain the practical lack of dissolved substances in wells drilled

Table 1. Analysis of coredrills from piezometers drilled in 1974.
I : Detection of kerosene by odour
II : Extraction by $CHCl_3$ and UV fluorescence (scale 0: no oil detected; 1: trace; 2: positive; 3: yellow colour of the extract)

	No. of Piezometers															
	272-1-42		272-1-43		272-1-46		272-1-49		272-1-52		272-1-53		272-1-54		272-2-177	
Depth of Drill	I	II	I	II	I	II	I	II	I	II	I	II	I	II	I	II
2-4 m	0	1	0	0	0	0	0	0	0	0	+	1	0	1	0	0
4-5,5 m	0	1	0	2	0	0	0	0	0	0	+	1	+	1	0	0
5,5-7 m	0	1	0	3	0	0	0	1	0	1	+	2	+	3	0	0
7-8,5 m	+	2	+	3	0	0	0	1	+	1	+	3	+	3	0	0
8,5-10 m	+	2	+	3	0	0	0	1	+	2	+	3	+	3	0	0
10-11,5 m	+	2	+	2	0	0	0	1	0	1	+	3	+	2	0	0
11,5-13 m	0	1					0	0			+	1	+	2	0	0

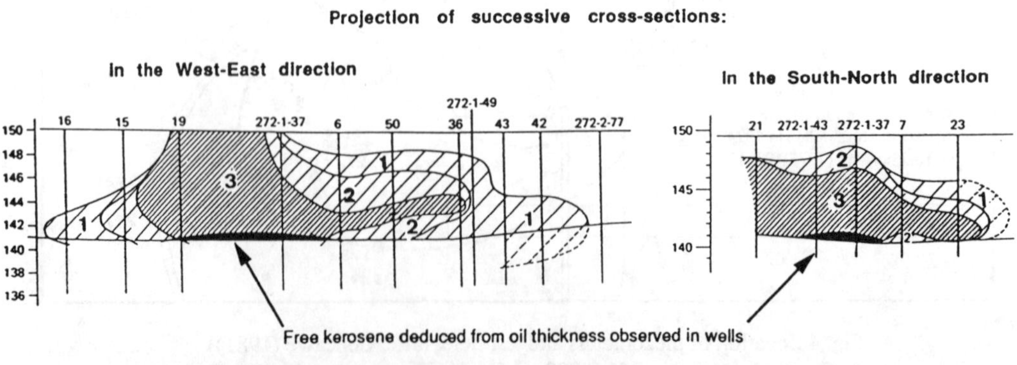

Fig. 5 Cross-sections of contaminated zones. 1: Traces of hydrocarbons, 2: Distinct extraction, 3: Significant extraction. (After Concawe (1981)

Table 2. Analysis of coredrills (1976-1977)
I : Detection of odour
II : Extract by CCl_4 followed by UV spectrometry

Depth of Drill	42		43		44		→ 49		50		51	
	I	II	I	II	I	II	I	II	I	II	I	II
2-4 m	0	0	0	0	0	0	0	0	0	0	0	0
4-5,5 m	0	0	0	0	0	0	0	0	+	0	0	0
5,5-7 m	0	0	0	0	0	0	0	0	+	+	0	0
7-8,5 m	0	0	0	0	0	0	0	0	+	+	0	0
8,5-10 m	0	0	0	0	0	0	0	0	+	+	0	0
10-11,5 m	0	0	0	0	0	0	0	0	+	+	0	0

around the impregnation body on the one hand, and the small amount of recovered oil during the pumping in the recovery well on the other (cf. § 3).

The existence of this natural barrier (suspended capillary water) preventing the transfer of soluble hydrocarbons in groundwater, incites us to be very careful for the execution of new wells on the site: in fact, drilling of this "protective screen" yields the eventual inflow of the oil on the water-table (Schiegg (1977); Zilliox et al (1978); Zilliox & Muntzer (1985)). This situation was occured at Strasbourg-Entzheim case.

From the hydrodynamical point of view, it also must be taken into account the estimation errors of the oil layer present on the water-table, which is often deduced from oil thicknesses observed in wells: Because of the incidence of capillary pressures on interfaces, the thickness of oil spread out over the water-table is not the same as the thickness of oil measured in the wells. This was postulated by Van Dam (1967) and Zilliox & Muntzer (1975). Figure 7 shows that at equilibrium, in the case where the continuity of the oil layer thickness is ensured, the difference between the thickness of oil in a borehole and the one spread out over the water-table may be important (H_2-H_1).

3 REMEDIAL AND RECOVERY MEASURES

Each week from 1972, kerosene was pumped from the meteorological station well. It was observed that the lower the groundwater level, the greater was the inflow and consequently the greater the recovery of kerosene. A total volume of some 10 m^3 was recovered after 18 months (August 1973). During that period, the level in the well fluctuated between 60 to 80 cm and it was not possible to correlate the water level recorded in the well with the groundwater level in the adjacent formation. During the summer of 1973, pumping tests were performed in the well using a two-pump system in order to improve the recovery rate and to allow optimisation of the design of the permanent recovery installation. The well (Fig. 8) is 10.55 m deep and is equipped with seven concrete sleeves (diameter 1.2 m) and four concrete sleeves (diameter 0.75 m), in the upper and lower parts respectively

Diametrically opposed holes (diameter 12 cm) are set in each sleeve and the bottom of the well is opened. The presence of gas in the well precluded a detailed examination, although a fireman entered the well and observed kerosene seepage at the joint between the second and third sleeves (lower part, 8.8 m deep). After pump out, both kerosene and water reentered the well over a period of time (Fig.9). The total fluid column stabilised at 60 cm after 5 h, with a kerosene thickness of 13 cm. The level then gradually increased with time,

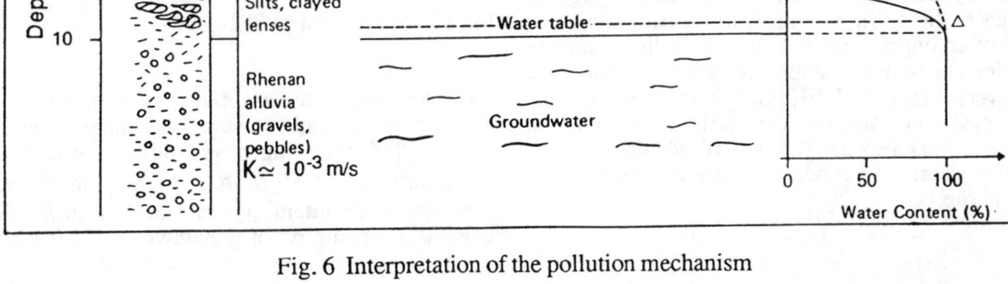

Fig. 6 Interpretation of the pollution mechanism

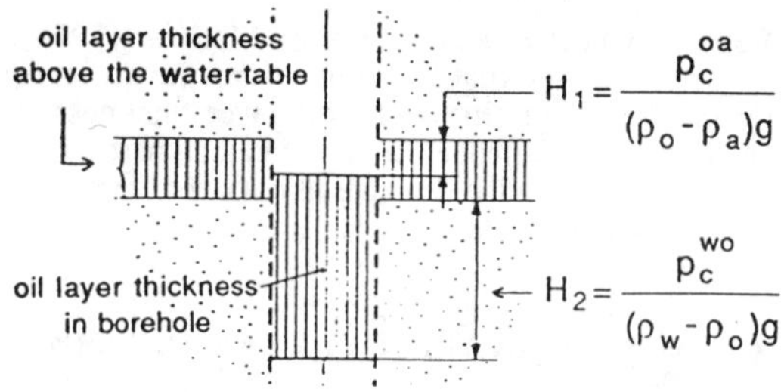

Fig. 7 Oil thickness in a borehole and adjacent formation. P_c^{oa} and P_c^{wo} are capillary pressure differences between oil and air, and between water and oil respectively. ρ_a, ρ_o, ρ_w are density of air, oil and water; g is acceleration due to gravity.

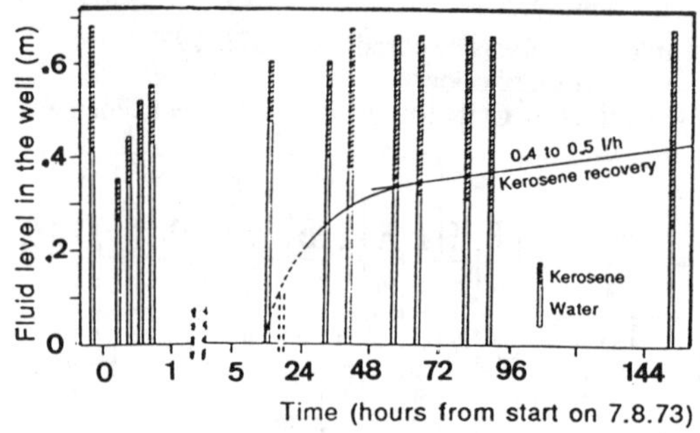

Fig. 9 Recovery test in the meteorological station well (272-1-50)

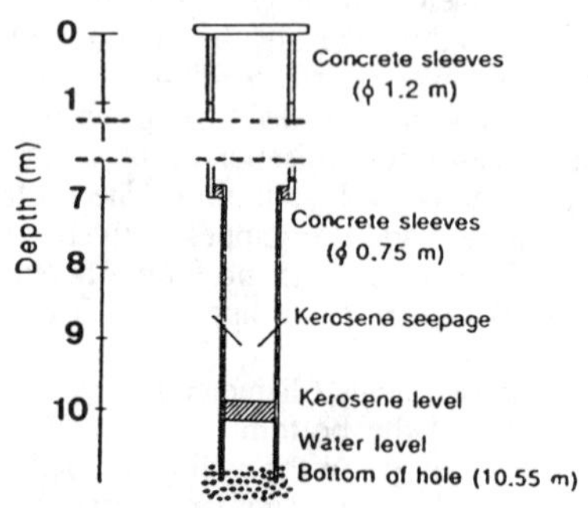

Fig. 8 The meteorological station well (272-1-50)

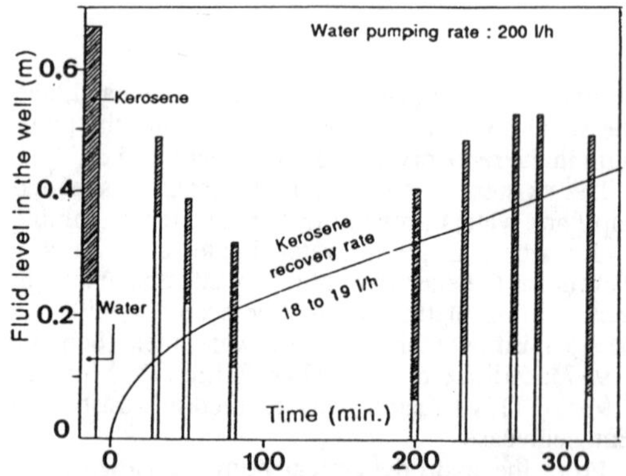

Fig. 10 Pumping test in the meteorological station well (272-1-50)

corresponding to an inflow rate of 0.4 to 0.5 l/h. When pumping groundwater from the well at a rate of 200 l/h, a 30 cm drawdown was created and kerosene flowed into the well at a rate of 15 to 18 l/h (Fig.10). However, after continuous pumping for one week, the kerosene inflow rate had fallen to 6 - 7 l/h at which it remained stable. In two weeks, 1.5 m^3 of kerosene was extracted.

3.1 *Technique adopted*

Following these tests, several options for permanent recovery installations were studied, such as digging a trench, the drilling of a row of wells, or the insertion of an impervious wall. It was finally decided to implement hydrodynamic protection by drilling a recovery well (272-1-51, Fig.11) and to reinject the extracted water back into the aquifer upstream of the depot. It was also decided to drill additional piezometers in order to conduct a more complete survey of the site.

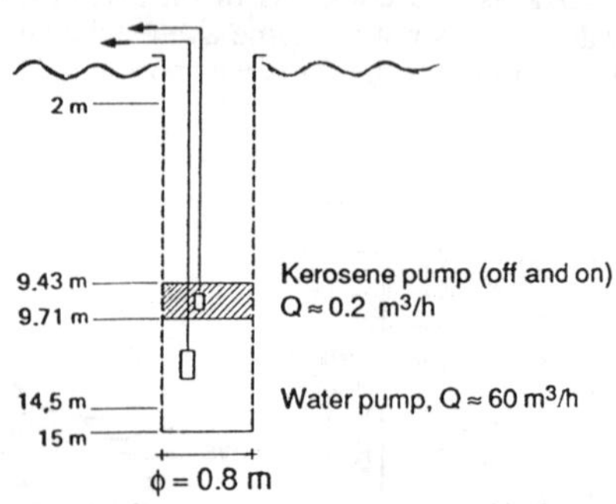

Fig.11 Recovery well 252-1-51

The recovery well was drilled in the depot in 1974 and put into operation with a two-pump system in March 1975. Pumping of groundwater was continuous at a rate of 60 m^3/h, and pumping of kerosene was intermittent at a rate of 0.20 m^3/h. The continu-ous pumping of groundwater was intended

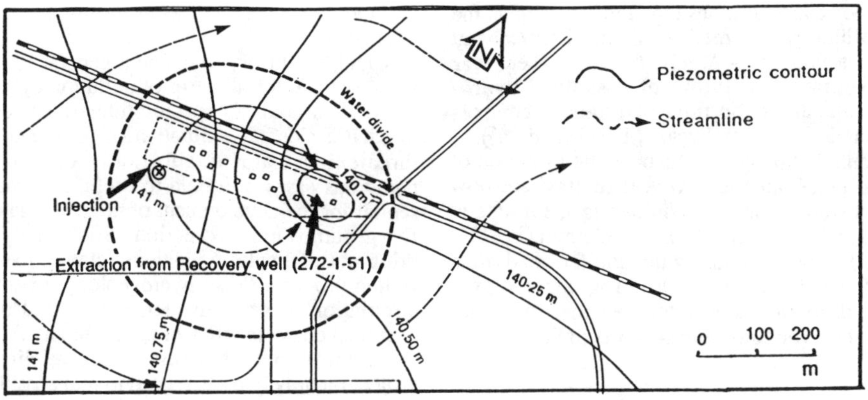

Fig. 12 Hydrodynamic protection of the site

to create a stable cone of depression of the water-table, thus draining the kerosene to the recovery well, and preventing it from flowing downstream of the polluted site. The extracted water was reinjected 300 m upstream into two trenches, with the objective of protecting the site hydrodynamically (see Fig.12).

During the three years 1975 to 1977, water was pumped continuously, but kerosene pumping was performed only during periods of low groundwater level when the thickness of the kerosene layer in the well reached 2 m.

3.2 *Results*

The water and kerosene levels, and volumes extracted up to the end of 1977 are shown in figure 13. The average recovery yield (ratio or pumped kerosene to pumped water) over the three-year period was 0.6×10^{-4}. Volumes of extracted kerosene were 58 m^3 in 1975, 40 m^3 in 1976, and only 4 m^3 in 1977. It was observed that kerosene inflow towards the well was very slow during that year probably as a result of the higher groundwater level (1977 was a wet year), and to decline in the water pumping rate (57 m^3/h) over the first six months due to plugging of the screen. The pump was cleaned in August 1977 and then put into operation at a rate of 90 m^3/h.

These data clearly indicate that the amount of oil recovered is small relative to the estimated spill volume. It must be concluded that significant amounts of oil remain in a stable situation. Since no odour or taste problems have been identified in the normal drinking water wells in the area downstream of Entzheim. It may also be concluded that the lack of pollution of the groundwater has largely resulted from the position of the oil in the formation although the pumping programme may also have contributed (Concawe (1981)).

3.3 *Current situation*

Kerosene pumping from the recovery well was stopped in April 1980, but the controls of drinking water supply wells are ensured until May 1985. Actually, the problem is that to know what are the risks of groundwater contamination after stopping the recovery measures of the zone of influence of the drinking water supply well. In order to isolate the well from the contaminated domain, the water pumping rate must be reduced. If an increase of the production is required, it is possible to change exploitation sites because the aquifer of the Alsatian plain is vast.

If the water pumping rate is maintained at 4000 m^3/day, the zone of influence does not include the contaminated area and the pumping does not attract the dissolved hydrocarbons which are transported downstream and diluted without threatening the drinking water supply wells. The observed total

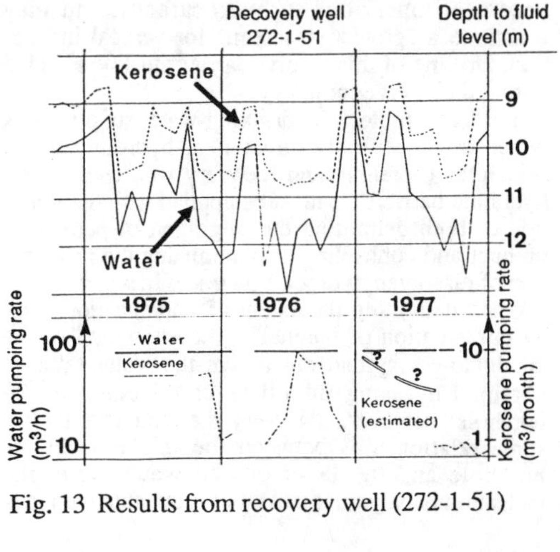

Fig. 13 Results from recovery well (272-1-51)

hydrocarbon concentrations remained lower than the detection limit of the method: 0.1mg/l (IR/solvent extraction techniques - AFNOR T 90203). There was no unfavourable evolution of dissolved hydrocarbon contents at the time of quaterly controls carried out during several years (BRGM (1984)).

New controls taking into account the evolution of analytical procedures and AFNOR method are now envisaged with the idea of comparing the results to the new AFNOR standard (Morel & Courtot (1981)).

It should be noticed that in the aquifer of Alsatian Plain, after pumping from drinking water supply wells, the distributions of water are carried out without treatment. The Entzheim sector is in that case.

4 CONCLUSIONS

The quantities of oil involved were difficult to estimate because of the nature of the spill. Assessment of available geological and hydrogeological data supplemented by test borehole studies led to a general understanding of the probable extent of the spill and its relationship to groundwater. This understanding was an essential step in defining remedial methods to recover and control the amount of oil in the ground.

From observations of Strasbourg-Entzheim case pollution, arguments and conclusions are as follows:

At the prevention or intervention level, it is essential to make a good survey of the site about the nature and the structure of the aquifer (local geology) as well as the existence and circulation of groundwater reserves. Namely, it is important to take into account the role played by the structure of the vadose zone on the formation of the impregnation body. In case of sedimentary soils made of layered structure with very different permeabilities, the discontinuity of water contents or the water saturation degree of these layers will determine various behaviours of hydrocarbon infiltration. Then, a thin material layer with very low permeability and high water content becomes impervious to hydrocarbons, and may constitute a "protective screen" for vertical infiltration. Drilling of this natural barrier must be avoided during the intervention.

In order to protect the drinking water supply wells which are potentially threatened, hydrodynamical principles (borehole and recovery well techniques, hydraulic barriers...) must be applied with the intention of both delimiting the migration of petroleum product and controlling or neutralizing the propagation of dissolved hydrocarbon traces in water.

When delimiting the extent of impregnation body by implantation of boreholes, the estimation of the oil thickness spread out above the water table is obtained by using the oil layer thickness in the borehole: however, it is very difficult to establish some relationships between the thickness in the borehole and the layer on the water-table; the indication is only qualitative but a quantitative interpretation may yield erroneous results.

Whatever the reservoir type, it must be noticed that it would be illusory to define a general intervention strategy valid for all cases of pollution by hydrocarbons. On account of specific fullness and evolution of each accidental infiltration of petroleum product threatening water quality, only the analysis of numerous cases can yield to useful information essential for the development of necessary knowledge for groundwater protection. Such analysis will bring out particular conditions relating to sites and spilled products (local hydrogeology, existence and circulation of groundwater, nature and physical-chemical characteristics of the pollutant...), and will show the variety existing in the application of protection principles as related to real possibilities of intervention and restoration of the contaminated area.

REFERENCES

BRGM 1984. Etude documentaire sur la dépollution des nappes polluées par les hydrocarbures. *Rapport du BRGM* , 84 SGN 198 - eau, Orléans, F.

Concawe 1981. A review of the investigation of a kerosene spill at Strasbourg-Entzheim airport, France. *Report n°8/81*, den Haag.

Morel, C. & P. Courtot 1981. Dosage d'hydrocarbures dans l'eau et le sédiment marin par infrarouge et chromatographie sur colonne capillaire. *Revue de l'Institut Français du Pétrole*, 36(5): 629-666.

Schiegg, H.O. 1977. Methode zur Abschätzung der Ausbreitung von Erdölderivaten in mit Wasser und Luft erfüllten Böden. *Thesis, Technischen Hochschule Zürich*, Switzerland.

Van Dam, J. 1967. *The migration of hydrocarbons in a water bearing stratum.* London : Ed. P. Hepple.

Zilliox, L. & P. Muntzer 1975. Effects of hydrodynamics processes on the development of ground water pollution study on physical models in a saturated porous medium. *Progress in Water Technology,* 7 (3/4), 561-568, GB: Pergamon Press.

Zilliox, L., Muntzer, P. & C. Gatellier 1978. Pollution des eaux souterraines par les produits petroliers. *Bull.etin du BRGM* (2è série), III (4): 317-326, F.

Zilliox, L. & P. Muntzer 1985. Pollution des eaux souterraines par les hydrocarbures. *Journées Techniques AGHTM*, 115-136, Roma: Ed. Federgasacqua.

Subsurface Contamination by Immiscible Fluids, Weyer (ed.) © 1992 Balkema, Rotterdam. ISBN 90 6191 175 3

The application of (bio)venting for in situ subsoil restoration following a hydrocarbon spill at a retail gasoline station: Design of the system

J.van Eyk & C.Vreeken
Delft Geotechnics, Netherlands

ABSTRACT: A system was designed to demonstrate the bioventing and venting approach as a methodology to enchance volatilisation and mineralisation of mineral oil products in the vadose zone. The system was designed to allow the inocculation of microbes and the intermittent introduction into the subsoil of chemical fertilizers, if and when required.

1 INTRODUCTION

The technique of venting is being used in many instances to remove volatile petroleum products from the vadose zone. Particulary petrol (saturated vapour pressure of 0,38 ato at 20°C) can pose problems when it has been spilled to the ground in built-up area's. However, in addition to the possibility to enchance the evaporation of volatile organics, venting can also be used to enchance convective air transport in soil to stimulate the oxygen-dependent petroleum biodegradation in the vadose zone.

The possibilities for bioventing were prompted by lysimeter studies on microbially-mediated gasoil removal from sandbodies (Tibbetts, 1982). In the absence of bioventing, an alomost 100% removal was achieved in a period of 8 years. A feasibility study was subsequently carried out to asses the venting technique for enhancing volatilisation and/or biodegradation (Van Eyk and Vreeken, 1988 and 1989a), as well as a large-scale field experiment (Van Eyk and Vreeken, 1988 and 1898b), to validate the feasibility study. The next step is to further develop this methodology in a real spill situation.

After careful consideration and screening, a contaminated location of a gasstation was selected as a test case to further develop and demonstrate the feasibility of the venting approach in practice. The demonstration project was financially supported by the Department for the Environment (Grant no. MJZ 20D 8037).

2 PURPOSE OF THE PROJECT

As a result of significant differences with respect to physical properties, like volatility, viscosity and adsorption, it has up to now been impossible to accomplish complete removal of a complex product like petrol or diesel oil from subsurface soil strata by applying one remedial technique only. Soil venting is in principle an appropriate technique to achieve complete residual product removal since, it anchances volatilisation of the light petroleum fractions and stimulates venting-assisted biodegradation of the heavier petroleum products.

When large quantities of petroleum products have been spilled into the ground, a skimming pump for the removal of liquid products will make it possible to remove the bulk quantity in a relatively short period of time, soil venting can subsequently be used as a medium to longterm remedial strategy, pending contaminant levels (guidelines outlined by Dutch law on soil remediation indicate, that mineral oil levels for soil with an organic matter content of up to 2% should be reduced to 10 mg kg^{-1} dry soil and for groundwater to 50 $\mu g\ l^{-1}$. In addition, individual values for benzene, toluene, and the xylenes, should be equal to or smaller than 0,2 $\mu g\ l^{-1}$).

The purpose of this project is to demonstrate that it is feasible to comply with these guidelines, using this particular in situ methodology.

3 SITE SELECTION AND CHARACTERISATION

The site to be selected for remedial action using venting-assisted evaporation and biodegradation had to meet some specific criteria.

The hydraulic permeability of the soil had to be at least equal to or smaller than 10^{-5} ms^{-1}, because smaller values would result in exceedingly long cleanup times.

The groundwater table had to be at least two metres below groundsurface, because the economics of an in situ technique, using venting-assisted evaporation and biodegradation, when compared to excavation becomes progressively favourable only, with increasing depth of the groundwater table.

The site eventually selected for this project, is shown in figure 1.

3.1 Degree and extent of contamination

3.1.1 Soil

Analysis results for borings B1 to B4 are shown in table 1 (B5 to B10 were only slightly contaminated.

It appeared that a considerable spillage had occured in the area of B1 and that concentration levels were decreasing towards B3 and B4.

Although neither the cashier in the shop, nor personnel in the carshop had complained about faul smells, some extra handborings have been carried out in the garage. These borings are indicated by the numbers 11, 12 and 13.

The analysis results are shown in table 2

The results showed that only boringnumber 13, which is close to B3 (c.f. table 1), has significantly elevated levels.

Table 1 Results of borings B1 to B4 with respect to mineral oil and BTX in mg kg^{-1}

Component	B1 0.0-3.0	B2 0.0-3.0	B3 0.0-3.0	B4 1.0-3.0
Benzene	< .5	130	39	5.4
Toluene	< .5	860	250	130
Ethylbenzene	< .5	280	96	60
Xylenes	< .5	2100	740	430
CH-total	< 50	6900	2300	1300
Mineral oil	< 100	$15 \cdot 10^3$	1300	320

Table 2 Analysis results of handborings carried out under the garage floor in mg kg^{-1}

Component	Boring number and depth below ground surface		
	No. 11 1.0-1.5	No. 12 1.5-2.0	No. 13 1.0-1.3
Benzene	0.15	3.7	28
Toluene	0.56	5.6	270
Ethyl-benzene	0.09	1.1	81
Xylenes	0.75	4.9	460
Mineral oil	< 20	< 20	850

3.1.2 Groundwater

Analysis carried out on groundwater samples obtained from approximately 15 monitoring wells, confirm more or less the contaminant situation as was found via soil sample analysis. The greatly elevated levels in monitoring well no 11 confirmed the elevated levels found in borings B2 and B4 and indicates the presence of mobile oil in the capillary fringe.

Because the sampling from some monitoring wells seemed to suggest that the aquifer had been contaminated up to approximately 10m below groundsurface, samples were also obtained with the groundwater sampling probe. The latter device is more reliable, because groundwater samples are obtained directly. After having reached the required depth of penetration, removal of the outer casing, when in position, exposes the filter, which reduces the chances for contamination from soil-strata located outside the sampling area.

However, analysis of probe no GW2, did indicate elevated levels for aromatics, particularly for benzene. It was decided therefore to install two pumping wells, one with the filter at 2,80m below GS and the other with the filter at 9,80m below GS.

3.2 Soil structure and composition

The phreatic aquifer zone up to 12m below groundsurface is composed of medium-fine to medium-coarse sand, followed by fine-silty sands and clay deposits (Twente Formation).

A pulse boring carried out up to 20m below groundsurface showed the presence of a 1m thick loamy layer at 6m below groundsurface. This layer did not show up, however, in other borings carried out further than 6m below groundsurface. From observation wells it appeared that the downward vertical hydraulic gradient is very small. Because of these observations, the phreatic aquifer zone has been considered as one and the same deposit. The permeability results obtained in the laboratory of a handboring obtained in the vicinity of the fuel supply pumps, and carried out to evaluate the possibilities for in situ remediation by venting, are shown in table 3.

Table 3 Permeability results obtained in the laboratory from a handboring 1,2 to 1,5m below groundsurface

Porosity	Permeability
45.3 %	$2.14 \cdot 10^{-4}$ m/sec.
42.7 %	$1.92 \cdot 10^{-4}$ m/sec.
40.8 %	$1.52 \cdot 10^{-4}$ m/sec.

The different porosities reflect the different degrees of compaction which have been applied in the laboratory. In addition, a number of size distribution curves were obtained from Begemann borings B1 to B10 (figure 1) which were subsequently also used to calculate hydraulic permeabilities.
The results of the size distribution curves are shown in table 4.

Table 4 Results of size distribution curves

Boring number	Metres below ground surface	d_{10} µm	d_{50} µm	d_{60} µm
1	0.00-1.80	36	160	180
2	1.00-1.80	80	175	200
3	1.00-1.80	75	165	185
4	0.00-1.80	53	150	175
5	0.00-1.80	75	160	185
6	0.00-1.55	63	160	185
7	0.00-1.80	75	140	160
8	1.00-1.80	80	165	185
9	1.00-1.80	75	160	180
10	0.00-1.80	36	160	190

To calculate the hydraulic permeability, the equations of Seelheim and Beyer were used:

$$K_w = 0.00357\ (d^{50})^2 \qquad \text{(Seelheim, 1880)}$$

$$K_w = c\,(d_{10})^2 \qquad \text{(Beyer, 1964)}$$

in which:

K_w = water saturated permeability
c = a constant, which depends on the uniformity and compaction of the soil sample.
d_{10}, d_{50} = percentage of soil mass with particles smaller than d_{10} or d_{50}.

The hydraulic permeabilities calculated on the basis of the data of table 4, according to Seelheim (S) and Beyer (B), are shown in table 5. The value of c in Beyer's equation is based on the natural compaction (second graph in figure 2).

Table 5 Calculated hydraulic permeabilities based on size distributions.

Boring depth below ground surface	d_{10} µm	d_{50} µm	K_w (S) m/s	K_w (B) m/s
0.00-1.80	36	160	$8 \cdot 10^{-5}$	$1.2 \cdot 10^{-5}$
aver.1-10	65	156	$8.7 \cdot 10^{-5}$	$4.2 \cdot 10^{-5}$
1.00-1.80	80	175	$1.1 \cdot 10^{-4}$	$6.4 \cdot 10^{-5}$

Comparison of tables 3 and 5 shows that the experimental values obtained in the laboratory are higher than the calculated permeabilities based on grain size distributions (table 5).
To calculate airtransport in the vadose zone, a saturated water permeability of $1{,}0 \cdot 10^{-4}$ m/s was selected as a middle value between the calculated and measured values.

4 MODEL CALCULATIONS FOR SYSTEM DESIGN

4.1 Groundwater withdrawal

To prevent contaminated plume movement and to clean up contaminated groundwater, pumping rates had to be calculated.
Contaminated groundwater had migrated 27m away from the planned location of the pumping well (figure 3). In other words, 27m was critical with respect to the radi-

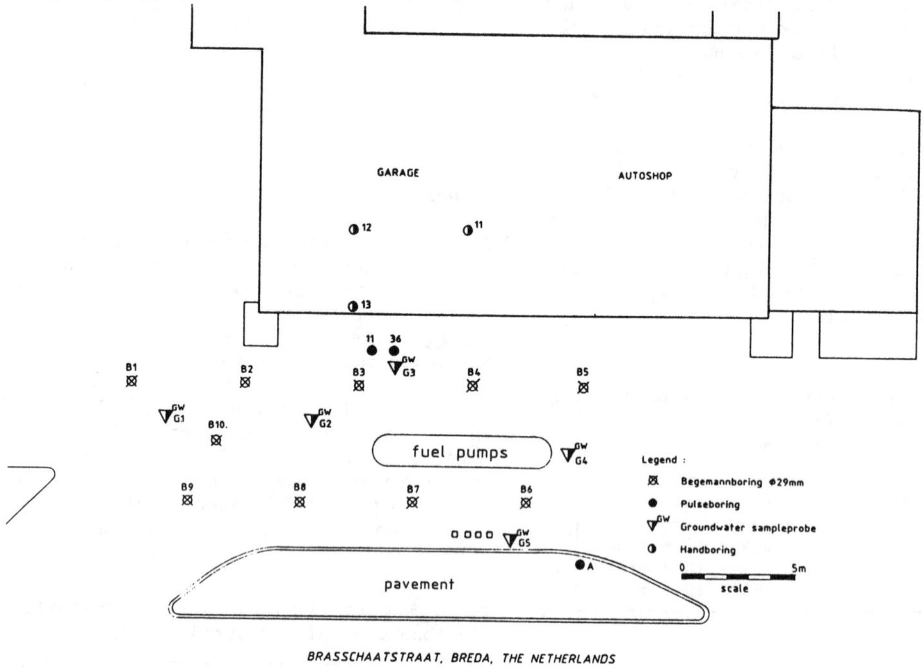

Figure 1 Plan view of retail gasoline station showing the locations for soil and groundwater sampling.

Table 6 Results of BTEX analysis for groundwatersamples

Component μg/l	Monitoring well no 11[a)]	no 36[b)]	Groundwater sample probe number and depth below GS GW2 9.0	GW3 9.5	GW4 7.5	GW5 9.5
Benzene	$40 \cdot 10^3$	280	3.2	< 0.2	< 0.2	< 0.2
Toluene	$85 \cdot 10^3$	1400	1.7	< 0.5	< 0.5	< 0.5
Ethylbenzene	$8.4 \cdot 10^3$	150	< 0.5	< 0.5	< 0.5	< 0.5
Xylenes	$58 \cdot 10^3$	980	0.8	< 0.5	< 0.5	< 0.5
CH-total	$200 \cdot 10^3$	3300	< 20	< 20	< 20	< 20

a) filter at 1.60 - 4.10 - GS
b) filter at 9.10 - 11.10 - GS

us of influence of the pumping well and the withdrawal area therefore had to contain at least this point. To calculate the radius of influence, groundwater transport was modelled with the computer model called SLEAM (Strack, 1989). The model was verified with measured data on groundwater and surface water levels. This model was used to calculate the effect of the radius of influence on static water levels under the following assumptions: the phreatic aquifer is confined to the impermeable base at 11m below groundsurface; the nett rainwater infiltration rate amounts up to 0,15m per year and a k_d value of 250 m^2 per day.

These parameters were used to calculate the effect of pumping on static water levels. The diameter of the well, backfilling inclusive was 0,2m. Several scenario's were calculated using pumping well rates of up to 5 m^3 per hour. The calculations indicated that a pumping rate of 5

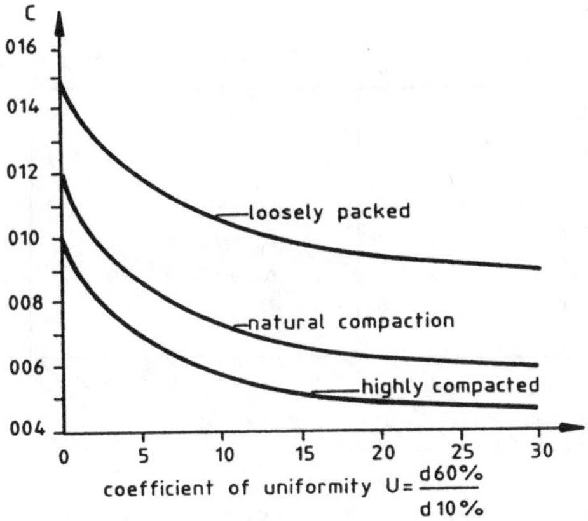

Figure 2 Value of c in Beyer's equation as a function of uniformity and grain compaction.

m^3 was sufficient to control plume movement.

4.2 Vapour transport calculations

4.2.1 Introduction

To obtain insight into the parameters required for te design and (lenght of) operation of the ventilation system, the Hydrology Contaminant Transport Model was used (Vreeken and Sman, 1989). With this HCT-model, Vreeken and Sman showed that transport of air in the vadose zone can be described by combining the basic equation describing single-phase flow in a porous medium and Darcy's law in three dimensions. If the water content of the soil remains constant, the resulting flow equation is the same as that which describes water flow in a completely saturated system. Transport of gaseous contaminants, finally, can be described using the advection-dispersion equation (Vreeken and Sman, 1989).

4.2.2 Model characterisation

Size distribution curves showed that the soil was composed of medium fine sand. Stroosnijder's data (Stroosnijder, 1976) were used to quantify the variation in water content as a function of depth to the water table. With respect to medium fine sand, we made the following assumptions:

- porosity 0,35
- air-filled porosity at 2 m above groundwater table 0,27
- air-filled porosity at 1,0 m above groundwater table 0,18
- air-filled porosity at 0,5 m above groundwater table 0,05

It has been assumed that the soil up to

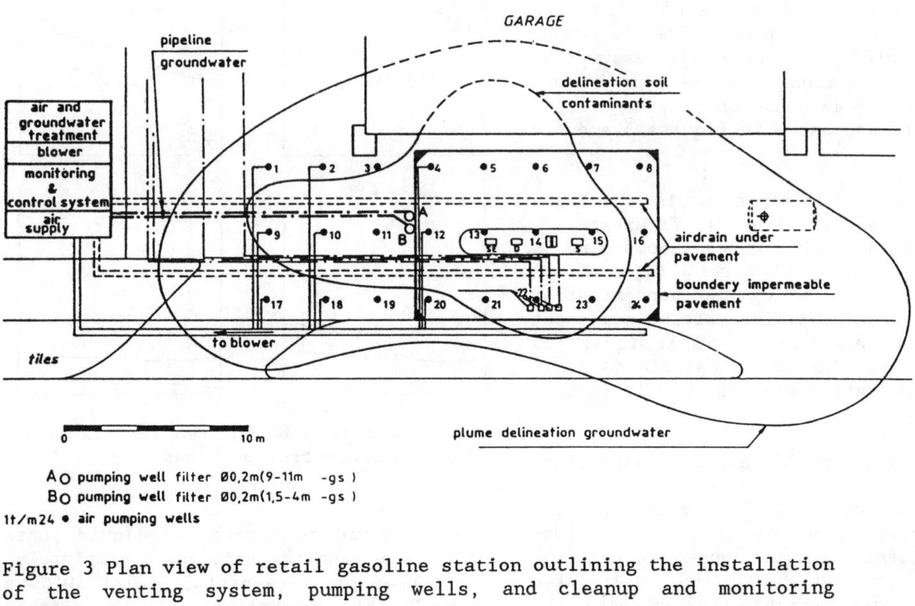

Figure 3 Plan view of retail gasoline station outlining the installation of the venting system, pumping wells, and cleanup and monitoring systems.

0,50 m above the groundwater table, is completely impermeable to air. Beyond o,5 m, a value of 0,2 has been assumed.

To calculate the air permeability of the soil, a modification of the Carman-Kozeny equation was used (Vreeken and Sman, 1989).

On the basis of the hydraulic permeabilities presented before (tabel 5), a value of $8,7*10^{-4}$ m/s was calculated for soil air-permeability.

In addition to air transport as a result of the application of a pressure gradient, gaseous transport will also be enchanced as a result of diffusion.

Toluene was used as a key component in these calculations. The diffusion coëfficient in soil can be calculated according to

$$D = D_0 \epsilon_a^{1,5}$$

in which D_0 represents the diffusion coefficiënt in air (Marshall and Holmes, 1979) and ϵ_a the air-filled porosity.

For toluene, the diffusion constant was calculated to be $D = 0,7*10^{-6}$ m²/s.

The basic scheme used for the model calculations is represented in the following diagram, for which it was assumed that air wells are 3 m apart in a square grid.

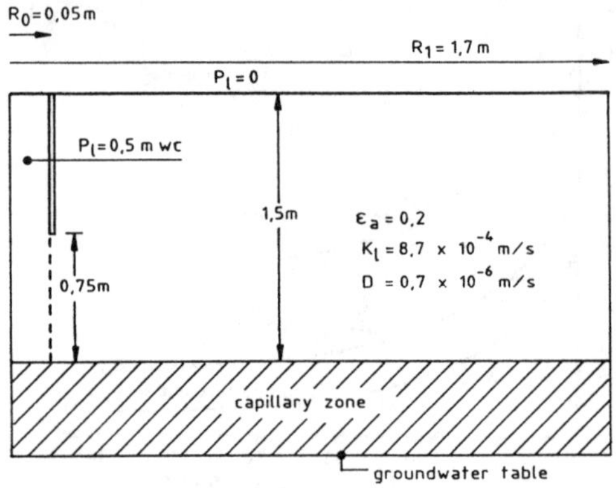

Figure 4 Computational scheme to calculate air transport in soil.

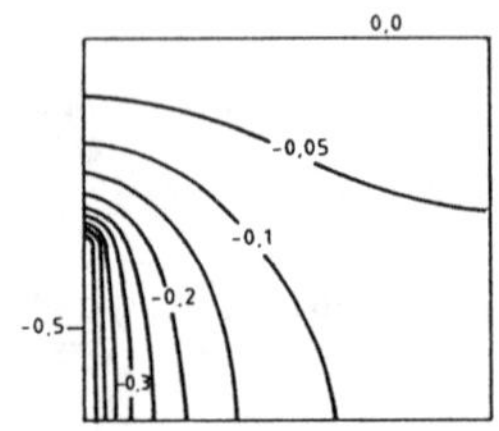

Figure 5 Calculated isobars in soil (in metres water).

4.2.3 Results of model calculations for air transport

Model calculations were carried to calculate the purging of the air-filled soil pores as a result of convective air transport and diffusion under the assumption that the contaminant to be purged was present in the gasphase only.

The results are shown in figures 5 and 6 a to d.

The direction of air-flow is at right angles to the given lines in figure 5.

The degree to which a gaseous component is purged from the soil pores, is shown in figure 6 a to d using iso concentration lines of 1% and 10%. The gaseous component in the soil area above the 1% (10%) iso concentration line has a calculated residual concentration smaller than 1% (10%).

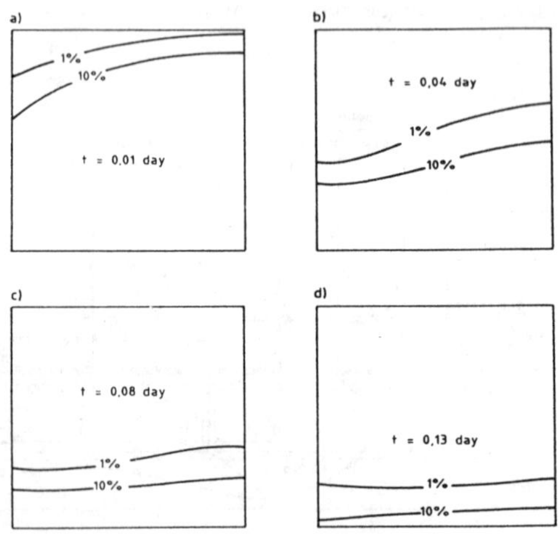

Figure 6 a to d Calculated results showing iso concentration lines.

4.2.4 Evaluation of rate of evaporation

Calculations with the above mentioned HTC-Model showed, that the total pore volume is completely purged once every four hours. On the bases of this calculated result it was possible to estimate the time required to remove the liquid phase of toluene from the soil pores completely. The vapour pressure of toluene at 10°C is 14 mm Hg. The concentration in the gasphase (air) can be calculated, using

$$C_g = \frac{P_A.M}{R.T}$$

in which:

C_g = concentration in gasphase (kgm^{-3})
P_A = vapour pressure (Pa)
M = Molecular weight
R = gas constant (8,21 m^3.Pa/mol.K)
T = absolute temperature.

When applied to toluene, it can be shown that

$$C_g = \frac{1842}{8,21} * \frac{0,092}{283} = 0,073kgm^{-3}$$

As the air-filled pore space per m^3 soil was estimated to be 0,2 m^3, the amount of toluene which evaporates per m^3 can be shown to be: 0,2*0,073 = 0,0146 kg or 0,0876 kg per m^3 soil per day.

At a local estimated residual toluene level in soil of 11 kgm^{-3} (boring B2), complete evaporation would require approximately 125 days.

These calculations predict, that pure toluene to a very large degree can be removed from soil, within one year.

As the vapour pressure of the xylenes is approximately four times lower than the vapour pressure of toluene. the evaporation is also approximately four times slower.

4.2.5 Evaluation of maximum attainable biodegradation rates

Although the obligatory participation of atmosheric oxygen may not be axiomatic (Vogel, 1986, Zeyer, 1986), petroleum biodegradation can only take place in the presence of atmosperic oxygen (van Eyk, and Vreeken, 1989b).

In addition, as hydrocarbons are devoid of oxygen, aerobic mineralisation of petroleum products requires more oxygen than the primary products of photosynthesis like sugars.

Calculations carried out for the design of the venting system (compare 4.2.2. and figure 4), show that a ventingratio of six can be realized under the circumstances as outlined in 4.2.2. One m^3 of air contains approximately 0,286 kg oxygen per m^3. With a nett pore volume of 20%, the oxygen supply amounts to 6*0,286*0,20 or 0,34 kg of oxygen per m^3 per day. Assuming complete mineralisation of hydrocarbons [$(CH_2)_n$] to carbondioxide and water, 0,34 kg oxygen should sustain complete oxydation of 100 gram of hydrocarbons per m^3 per day (63 mg kg^{-1} day^{-1}). Because of the temperature regime in the subsoil as well as other limiting factors, such rates are unlikely to be realized.

The biodegradation rate of approximately 10 mg kg^{-1} day which was found in a large scale field experiment, was significantly lower than this figure (van Eyk and Vreeken, 1989b).

5 DESIGN OF THE SYSTEM

5.1 Installation of the (bio)venting system

Pumping air wells were installed according to the results of the calculated design specifications. Air wells were connected to a pipeline which in turn was connected to a blower. Monitoring equipment was installed to control and measure gasflux and gaspressures.

As the groundsurface has an impermeable pavement (see below), access of air was obtained via two horizontal drains, which were installed below the pavement (see figure3).

5.2 Installation of an impermeable paving

An impermeable paving is an integral part of this venting system for several reasons. First of all, the venting system had to be protected from any form of adverse effect or damage.

In addition, because the petrolstation has on-going business during the period of remedial action, migration of fresh spilled products around the fuel supply pumps into the area to be claened up had to be excluded.

Finally, it should be noted for the paticular lay-out and design of our (bio)venting system, that it will not be removed from the location. Because of that, it can be used as a remedial technique for both past and possible future petroleum spills.

5.3 Installation of air and groundwater cleanup system

Cleanup of groundwater, stripped-air and air-purged from soil, is achieved by using an oil/water separator, a stripping-tower, and a biogas or compost filter.

6 APPLICATION OF FERTILIZERS AND MICROFLORA

The two horizontal air wells (drains) which were installed below the pavement to supply air via the coarse sandy top layer to the vertically installed air wells, were incorporated into the venting system to enable additional fertilizers to be flushed into the soil, if required.

Fertilizer application into the soil was described before (van Eyk and Vreeken, 1988)

7 MINOTORING AND CONTROL

Progress of the remediation process will be monitored via vented-air sampling, groundwater samples and soil samples.

8 RESULTS

Data on the progress of soil cleanup by (bio)venting, will be the subject of a separate paper.

REFERENCES

Beyer, W., 1964: Zur Bestimmung der Wass erdurchlässigkeit von Kiesen und Sanden aus der Kornverteilungskurve. Berlin, Wasswirtsch.-Wassertechn. H6, 1964.

van Eyk, J. and C. Vreeken, 1988: Venting mediated removal of petrol from subsur face soilstrata as a result of stimu lated evaporation and enchanced biodegradation. Med. Fac. Landbouww. Rijksuniv. Gent, 53(4b) p. 1873-1884.

van Eyk, J. and . Vreeken, 1989a: Model petroleum mineralisation response to soil aeration to aid in site-specific, in situ biological remediation. Groundwater Contaminants: Use of models in Decision - Making, pp 365-371, Eds. Jousma et al.. Proceedings of an Inter national Conference on Groundwater Contamination. Boston/London, 1989.

van Eyk, J. and C. Vreeken, 1989b: Venting-mediated removal of dieseloil from subsurface soilstrata as a result of stimulated evaporation and enhanced biodegradation. Envirotech Vienna, vol. 2 session 3. Hazardous waste and contaminated sites p. 475-485. Essen: Westarp Wiss, 1989 ISBN 3-89432-009-5.

Marshall, T.J. and J.W. Holmes, 1979: Soil Physics, p. 273. Cambridge University Press, Cambridge, 1979.

Strack, O.T.L.: Groundwater Mechanics 1989, Presence Hall Inglewood Cliffs, New Yersey 07632.

Stroosnijder, L. 1976 Infiltratie en herverdeling van water in grond. English Summary PUDOC Centrum voor landbouwpu blicaties en landbouwdocumentatie Wageningen, The Netherlands, 1976 ISBN 90-220-0596-8.

Tibbets, P.J.C. 1982: The analysis of oil in sand from four lysimeters in Katwijk, The Netherlands. A COOW-CONCAWE Draft report no 2870/227/1/925. TUD The Netherlands.

Vreeken, C. and H.T. Sman, 1989: The use of a Hydrology contaminant transport model for the prediction of the effect of air-stripping on the in situ clea ning of contaminated soil. In: Groundwater Contamination: Use of Models in Decision - Making. pp 329-327. Eds. Jousma et al. Proceedings of an Inter national Conference on Groundwater Contamination. Kluwer Boston/London, 1989.

Vogel, T.H.M. and D. Grbic - Galic, 1986: Incorporation of oxygen from water into toluene and benzene during anaerobic feremntative transformation. Applied Environ. Microbiol. 52, pp 200-202.

Zeyer, J., E.P. Kuhn and R.P. Schwarzenbach, 1986: Rapid microbial mineralization of toluene and 1,3-dimethylbenzene in the absence of molecular oxygen. Appl. Environ. Microbiol. 52; pp. 944-947.

Seelheim, I., 1880: Zeitschrift für Analytische Chemie 19.

Subsurface Contamination by Immiscible Fluids, Weyer (ed.) © 1992 Balkema, Rotterdam. ISBN 90 6191 175 3

Subsurface contamination and remediation at Alberta sour gas plants: An overview

P.E. Hardisty, T.L. Dabrowski & L.S. Lyness
Piteau Engineering Ltd, Calgary, Alb., Canada

R. Scroggins
Environment Canada, Ottawa, Ont., Canada

P. Weeks
Husky Oil Ltd, Calgary, Alb., Canada

J.F. Wells
Integrated Environments Ltd, Calgary, Alb., Canada

ABSTRACT: A study of subsurface contamination monitoring and remediation data from 55 Alberta sour gas plants was undertaken through sponsorship of the Canadian Petroleum Association and Environment Canada. Study objectives were to determine the most commonly occurring subsurface contamination situations, classify them by contaminant type, source, and geologic host, and evaluate the level of remediation technology being applied in the province. Of the sour gas plants reviewed which had monitoring systems in place, only one did not exhibit some form of subsurface contamination. In the majority of cases contamination was local and restricted to plant-site areas. Dominant sources of contamination were process water ponds, process area, on-site landfills, and sulphur block areas. The most common groundwater contaminants were main ions such as sulphates and chlorides, and dissolved organics. Contamination by free phase hydrocarbons, notably condensate, was present at several facilities. The majority of contaminated saturated horizons were of moderate hydraulic conductivity (10^{-5} to 10^{-8} m/s), typically consisting of sandy till deposits common in Alberta. Currently, subsurface remediation systems are installed at five gas plants. Pump and treat schemes are the common remediation methods employed. A case history of free phase condensate contamination recovery is presented.

1. INTRODUCTION

The province of Alberta is fortunate to have an abundance of natural gas, much of which is associated with varying concentrations of hydrogen sulphide gas (H_2S). Sour gas plants remove hydrogen sulphide from the natural gas stream through a variety of processes, producing elemental sulphur and clean gas. There are presently more than one hundred and fifty sour gas plants operating in Alberta. Plants range in capacity from as little as 11,000 m^3/day to more than 17,000,000 m^3/day of raw gas (Oilweek, 1987). The oldest facilities have been in operation since the early 1950's, and several new complexes are now in the design stage. This fact is reflected in the wide range of process types and plant designs present in Alberta.

Due to the nature of the processes involved in sour gas processing and the wastes and by-products produced, sour gas plants pose a potential threat to local groundwater quality. Soil horizons and the unsaturated zone may also be affected. Possible sources of contamination at plant sites include free phase and dissolved hydrocarbon products (such as condensate), process water and chemicals (such as amines, glycols, and degradation products), produced waters (brines, saline and brackish water), solid wastes and sludges, seepage waters, surface runoff (from sulphur blocks, process and loading areas), and active or abandoned landfills on site.

Recognizing the need to protect groundwater from deleterious effects resulting from plant activities, the Canadian Petroleum Association, in conjunction with Environment Canada, sponsored this study into subsurface contamination and remediation at Alberta sour gas plants (Piteau Engineering Ltd. 1990). The main objective of the study, upon which this paper is based, was the compilation of knowledge of sour gas plant related subsurface contamination situations and remediation systems in Alberta, and the determination of the most common types of contaminant situations present at these facilities. In addition, a brief review of legislation pertaining to groundwater monitoring and protection in Alberta is presented. Subsequent phases of this work are to include design and implementation of one or more remediation technology demonstration projects at Alberta sour gas plants.

2. AVAILABLE DATA

This study was based on data provided by Alberta Environment, and consisted of documents submitted to the Standards and Approvals Division by plant operators pursuant to the Alberta Clean Water Act (1971). In all, information was obtained for 54 sour gas plants.

The quality and completeness of information contained in the reports were quite variable. In some cases basic hydrogeological data such as groundwater flow directions, geological logs and piezometer locations, were not available. Of the 54 plants considered, sufficient data for adequate appraisal of subsurface contamination situations were available for 32. Four of the remaining plants had no monitoring systems installed. Data contained in reports from five of the plants were insufficient for any meaningful appraisal of groundwater contamination. Information from the remaining 13 plants allowed partial contamination assessment.

The variability in data reporting reflects the differences in approach among plant operators in Alberta with respect to existing groundwater monitoring requirements in the province.

3. METHODOLOGY

3.1 Assessment of Contamination

The available data for each gas plant were reviewed and where possible groundwater contamination situations identified. Presence of contamination was determined using standard techniques, such as comparison of up-gradient and down-gradient water quality and presence in groundwater of direct indicators such as gas condensate, petroleum fuels, and plant process chemicals. No assessment of the seriousness of contamination was undertaken, as in the majority of cases, insufficient data were available for this task.

3.2 Classification of Findings

If sufficient data were available, the source, type and geologic host of contamination were determined for each contamination situation. From these three pieces of information Contamination Situation Classifications, or CSC's were assigned, according to the categories shown in Table 1.

Table 1. Contamination situation classification system.

Contaminant source categories
1. Process area
2. Sulphur block
3. Surface runoff
4. Process water ponds
5. Product loading facility
6. Landfill
7. Injection well
8. Other
9. Undetermined
Contaminant type categories
A. Free hydrocarbons
B. Dissolved organics
C. Main ions
D. Sulphur products
E. Metals
F. Other
Contaminated zone categories
I. Unsaturated zone
Saturated zone:
II. High hydraulic conductivity (K) $K > 10E-5$ m/s
III. Moderate K; $10E-05 > K > 10E-08$ m/s
IV. Low K; $K < 10E-08$ m/s

For the purposes of this study, a contamination situation is defined as: "an occurrence of subsurface contamination at a given sour gas plant, distinct from other occurrences at the same plant in that it has a different source, or the contamination is found in a different hydrogeological unit".

It is possible for several CSC's to exist at a given plant. For example, a facility may have a small sulphate plume in the uppermost groundwater bearing zone (sandy clay till layer of low hydraulic conductivity) extending from the sulphur block area, and a larger plume consisting of high concentrations of chlorides and dissolved organics (high TOC and sulfolane identified in trace organics scan) originating from the evaporation pond, in another part of the same uppermost aquifer, but also found in a deeper bedrock aquifer of moderate hydraulic conductivity.

Consulting Table 1, three CSC's could be developed for this scenario:

2 D IV: sulphur products from sulphur block in low K zone;
4 BC IV: ions and dissolved organics from evaporation pond in a low K zone;
4 BC III: ions and dissolved organics from evaporation pond in a moderate K zone.

Once so classified, contamination situations at various gas plants were compared and grouped with the assistance of a computerized database system developed for the study. In this way, common trends in subsurface contamination at Alberta sour gas plants were identified.

There exists in the data, however, a slight correlation between the completeness of the monitoring data available for a given facility and the number of contamination situations identified there. Small monitoring networks which did not cover all areas of possible contamination may have failed to detect all groundwater contamination present. In light of this, the data presented as a result of this review should not be taken solely as an indication of the relative care with which plants have controlled and disposed of plant wastes and by-products, but also of the relative degree to which they have attempted to assess the subsurface contamination present at their sites.

4. GROUNDWATER CONTAMINATION

Of the 45 plants for which information allowed assessment of the presence of groundwater contamination, only one showed no signs of impact of plant activities on groundwater quality. Table 2 shows the number of plants at which one or more CSC's were determined. Two or more contamination situations were identified at 36 of the 45 plants.

Table 2. Number of contamination situations determined at gas plants.

No. CSC's	# of Plants	Cumulative #
4 or more	10	10
3	12	22
2	14	36
1	8	44
none	1	45

4.1 Sources of Contamination

There were sufficient data available at 42 plants to determine the source of groundwater contamination. Table 3 shows the number of plants with at least one contamination situation originating from each of the source categories.

Thirty-three of the 42 plants (78.6%) had at least one contamination situation originating from the process water/evaporation pond. Other common sources of groundwater contamination were process areas and on-site landfills. The sulphur block area was also identified as a frequent source of contamination.

Table 3. Number of plants with at least one contamination situation originating from the given source.

Source	# plants	percentage of plants
Proc. water ponds	33	78.6
Process area	23	54.8
Landfill	20	47.6
Sulphur block	16	38.1

4.2 Types of Contamination

There were sufficient data available at 44 plants to determine the types of contaminants in groundwater. Table 4 shows the number of plants with at least one contamination situation involving each of the six major contaminant groups.

Main ions and dissolved organics were the most commonly identified groundwater contaminants. Contamination by sulphur products, notably sulphate and in some cases acid seepage waters, was identified at 17 of the 45 plants. Free phase condensate was identified at five plants.

Table 4. Number of plants where the indicated contaminant was identified at least once.

Type	# plants	percentage of plants
Main ions	42	93.3
Dissolved organics	41	91.1
Sulphur products	17	37.8
Free hydrocarbon	5	11.1
Metals	3	6.7

Contaminants in groundwater often occurred in combinations. For instance a particular plume emanating from an evaporation pond may have contained main ions such as chloride and sulphate, dissolved organics, and metals. All such combinations were recorded during the data review phase. Table 5 shows a breakdown of the occurrence of the more common combinations. Among gas plants in the study set, half had at least one situation where groundwater was contaminated by dissolved organics and main ions.

Table 5. Number of plants where the indicated contaminant combinations were identified at least once.

Combination	# Plants
Main ions + dissolved organics	22
Main ions + dissolved organics + other	19
Sulphur products only	15
Main ions only	13
Free hydrocarbon + dissolved organics + main ions	3

Dissolved organics combined with main ions was the most frequently occurring contaminant grouping. It must be noted that findings are partially affected by the analytical schedules which the various operators chose to run of their groundwater samples. If certain parameters were not analyzed for, they could not be detected, and would not appear in the CSC's. If more comprehensive analyses were to be done, additional contaminant types may have been identified.

4.3 Zone of Contamination

Data with which the location of groundwater contamination could be determined were scarcest in the reports provided by Alberta Environment for this study. All of the 13 plants for which only partial CSC's could be generated lacked sufficient data to classify the zone of contamination. In most instances, borehole logs or piezometer construction details were not supplied by operators, making it impossible to determinewhich groundwater-bearing zone was being sampled. As a result, the nature and hydraulic properties of the geologic host of groundwater contamination could be determined at only 32 of the 54 plants in the study group.

Zones were classified according to their average hydraulic conductivity, to provide some indication of the ability of contamination to migrate away from source. Unsaturated zone (soil) contamination was provided a separate classification (Type I). Table 6 shows the number of plants at which contamination occurs in each of the three saturated zone classifications. The majority of groundwater contamination situations at Alberta sour gas plants seem to occur in zones of moderate hydraulic conductivity (10E-5 K 10E-8 m/s), represented by such materials as intertill clayey sand and silt layers and fractured bedrock common in Alberta.

Table 6. Number of plants with at least one contamination situation occurring in the indicated zone.

Zone	Hydraulic Conductivity Range (m/s)	# Plants
II	10E-05	7
III	10E-05 - 10E-08	20
IV	10E-08	6

Although directly related to the surficial geology of Alberta, this breakdown does serve to confirm that most groundwater bearing zones which are impacted by sour gas plant operations are not extremely hydraulically conductive. Twenty-six of 32 plants had contamination in groundwater bearing zones whose hydraulic conductivity was less than 10E-5m/s. This helps to put the situation at Alberta sour gas plants into perspective. In the majority of cases, the estimated advective transport rates of groundwater contamination were not very high, however consideration must also be given to the influences of fracture permeability, contaminant-matrix interactions and the accuracy of hydraulic conductivity data provided in reports. Detailed analysis of contaminant transport rates at the various gas plants was beyond the scope of this study, and should be determined as part of risk analysis exercises recommended for each gas plant.

4.4 Contamination Situations

Of 83 contamination situations identified at 32 plants, the most common CSC was 4 BC III (main ions and dissolved organics in moderate K zones, originating from process water ponds). The other most common CSC's are shown in Table 7.

Table 7. Most common contamination situations at Alberta sour gas plants.

CSC Source	Type	Zone	# Occurrences
4	BC	III	6
2	D	III	6
1	BC	III	5
4	C	III	3
4	BCF	III	3
1	BC	II	3

Situations involving dissolved organics and main ions, alone or associated with other aqueous phase contaminants such as metals and nitrogenous compounds, originating from process water ponds and located in moderate K groundwater bearing zones occur at 15 of the 32 plants. This type of situation is clearly the most common at sour gas processing facilities in the province.

Five situations were identified involving free phase hydrocarbon contamination in moderate or high hydraulic conductivity aquifers, from a variety of sources. Although this number represents only about one-tenth of the surveyed plants, the relative concern attached to this type of contamination makes it of particular interest. Free phase hydrocarbon contamination is difficult and expensive to remediate, and very low levels of hydrocarbon render water unfit for human or animal consumption.

5. GROUNDWATER REMEDIATION

5.1 Data Available

Information was obtained on subsurface remediation programs at five sour gas plants in Alberta. Of these, two were primarily soil remediation operations (plant decommissionings), and three were groundwater remediation schemes currently installed at operating facilities.

5.2 Case History

The groundwater remediation program at Plant P-192 was initiated by plant personnel in 1986, in response to discovery of a free-phase condensate plume downgradient of the plant site. During that year four 114 mm OD diameter wells were installed to recovery condensate and contaminanted groundwater originating from the process area and evaporation ponds. These attempts at using single conventional pumping wells to extract LNAPL met with very limited success.

Subsequently, a more detailed hydrogeological study of the plant site and surroundings was completed. This study refined understanding of groundwater flow patterns at the site and the distribution and properties of the various hydrogeologic units at the site, and accurately delineated the extent of the plume. The condensate plume was found to extend about 1.5 km downgradient of the plant, attaining apparent thicknesses of up to 1.5 m. The new information indicated the need for an improved contaminant recovery system.

In 1989, five 130 mm test wells were placed downgradient of the plant, near the downgradient edge of the condensate plume. These wells were tested to provide information on the hydrogeologic properties of the contaminanted aquifer, which consists of interbedded fractured shale, siltstone and sandstone bedrock. These wells will be fitted with dual pump systems designed expressly for the recovery of light non-aqueous phase liquids (LNAPL) from the groundwater surface. An intetrated groundwater recovery, treatment and disposal program is now being designed for implementation sometime in 1991.

An area of muskeg approximately 750 m downgradient of the plant site was also found to have condensate contamination. An interception trench is planned for 1990/91 to contain the plume in this area.

The site characterization and remediation programs conducted at Plant P-192 were found to be some of the most thorough and advanced of all plants involved in the study.

5.3 Discussion

Subsurface remediation efforts at Alberta sour gas plants to date have included a variety of techniques. Where geologic conditions were suitable, large diameter passive collection systems have been used to recover free-phase hydrocarbon. Attempts to recover free phase condensate in more permeable aquifers using single well pumping schemes have, not unexpectedly, been relatively unsuccessful. Application of dual-pump scavenger systems to recover free phase condensate and groundwater separately is being considered at several facilities. Traditional pumping well methods are being applied for the recovery of groundwater with dissolved contaminants, and the control of plume migration. Deep well injection is the favoured method for disposing of contaminated groundwater.

The number of plants at which remediation operations are known to be presently underway is relatively small, considering the number of operational plants in the province. In many cases, remedial action has been deemed unnecessary due to geologic conditions and relative isolation of plants, far from any nearby groundwater users, water courses or population centres. In some instances, remediation has been deemed impractical due to limitations of available technology.

However, rapid advances in the understanding of processes governing contaminant migration in heterogeneous geologic media, and new developments in remedial technologies should improve our ability to remediate difficult sites. One consideration to date has undoubtedly been the relatively high cost of subsurface remediation and the fact that at present no regulations or guidelines are available for these operations in Alberta.

At present, facility operators, in cooperation with Alberta Environment, are taking it upon themselves to clean-up subsurface contamination which they consider to be unacceptable. As could be expected, different operators view similar contamination situations with different levels of concern, and react accordingly.

The available data seem to indicate that some groundwater remediation operations have proceeded on an ad-hoc basis, with insufficient preliminary site characterization and contaminant delineation resulting in a piecemeal approach to system design. This often leads to higher overall costs, lower ultimate clean-up levels, and longer clean-up times.

The need for clean-up must also be determined. Risk analysis techniques provide a way of assessing the likelihood of contamination impacting the public or the environment. Results of the risk analysis can then be provided to regulators to ensure a case-by-case assessment of the need for remediation at sour gas plants.

Gas processing facilities in the province are found in such diverse locations, and represent such a wide range of hydrogeological and climatic conditions, that universal application of a single set of rigid criteria is not recommended. A case-by-case consideration of facilities would help to ensure the economical application of the limited resources available for site clean-up. Companies could then direct available funds toward the most serious problems.

6. LEGISLATION

At present in Alberta, issues relating to groundwater contamination are dealt with under the Clean Water Act. This piece of provincial legislation deals with all aspects of water contamination and protection, and therefore includes groundwater. The legislation and regulations are supplemented with guidelines for specific industries. Guidelines have been published for Alberta brine storage reservoirs, waste water management standards for gas processing facilities, waste water effluent guidelines for Alberta petroleum refineries, industrial landfill guidelines, and other topics.

Under the present regulations, sour gas plants in Alberta require a licence to operate from Alberta Environment, which requires operators to install groundwater monitoring systems around all sources of potential contamination. The minimum requirements include one upgradient and two downgradient piezometers. The operator is encouraged to submit a report providing details of the proposed monitoring system, which Alberta Environment can assess, and approve or modify.

In practice, this system suffers from several deficiencies. Operators are not provided with specific guidelines on how a monitoring system should be set up, what it should consist of, and how it should be operated. In some cases, operators have neglected to monitor for contamination from sulphur blocks and process areas. The results of this study show that the process area and sulphur block are two of the most common sources of groundwater contamination at sour gas plants. In addition, no specific reporting requirements on the groundwater monitoring system are specified. This has resulted in Alberta Environment receiving reports of widely varying quality and completeness. Many operators retain qualified hydrogeological consultants to install the monitoring system and provide a detailed report on the systems and the groundwater quality at the plant site, even though such thoroughness is not explicitly required under the terms of the licence. Others, however, choose to satisfy only the minimum requirements, and the tasks of monitoring system installation and reporting are sometimes not performed by qualified personnel.

Alberta Environment is now formulating a new set of groundwater monitoring guidelines for the province as a whole, to be applied to all industries and facilities which have the potential to contaminate groundwater. These guidelines are to include detailed information on groundwater assessment and monitoring requirements, including numbers and locations of piezometers, and recommended sampling schedules. In addition, the need for an explicit statistically-based definition of groundwater contamination will be assessed. Alberta Environment will hitherto require submission of detailed reports, including full presentation and interpretation of results, with which the situation at a given gas plant can be evaluated, and the need for remedial action assessed.

At present, the need for groundwater remediation is assessed on the basic of monitoring results submitted by operators. Information from each plant is reviewed by Alberta Environment, and if necessary, remedial measures are suggested to the operator. With this system, special conditions present at each plant can be taken into account when deciding on remediation approaches, rather than applying a rigid set of criteria to all plants, regardless of location, proximity to other users of groundwater, and the results of risk analysis. In practice, however, the success of this system is dependent on the quality of monitoring reports submitted by operators. Historically, there may have been a tendency for operators who fulfilled only the minimum requirements, and had perhaps not investigated all areas of potential contaminations, may have appeared to have fewer cases of contamination. However, under the new regulations, operators renewing licences will come under much more strict scrutiny.

The new guidelines, which may be available as early as mid 1990, should substantially improve the overall quality of groundwater monitoring at sour gas plants, and help to identify those facilities where remediation is required.

7. CONCLUSIONS

The most common sources of groundwater contamination at Alberta sour gas plants were process water ponds and on-site landfills. Dissolved organics and main ions were the most common contaminants, although free phase condensate LNAPL was present at several plants, and is seen as being of prime concern. Most contamination occurs in geologic horizons of relatively low hydraulic conductivity. The number of Alberta sour gas plants at which remediation programs are underway is relatively small, and the general level of technology being employed is low. This study supports the need for the new set of proposed guidelines being formulated for subsurface monitoring at industrial facilities in Alberta. It is recommended that industry and government consider developing a set of guidelines for the design and implementation of remediation programs.

REFERENCES

Government of the Province of Alberta, 1971. The Clean Water Act. Edmonton.

Oilweek, 1987. Gas Processing Plan Capacities. Jan. 26, 1987 Issue.

Piteau Engineering Ltd., 1990. Effectiveness of Subsurvace Treatment Technologies at Alberta Sour Gas Plants, Phase I: Assessment of Subsurface Contamination and Remediation at Alberta Sour Gas Plants. Report to Canadian Petroleum Association and Environment Canada.

Subsurface Contamination by Immiscible Fluids, Weyer (ed.) © 1992 Balkema, Rotterdam. ISBN 90 6191 175 3

Light non-aqueous phase liquid hydrocarbon occurrence and remediation strategy, Los Angeles Coastal Plain, California

Stephen M.Testa
Applied Environmental Services, Laguna Hills, Calif. & University of Southern California, Los Angeles, Calif., USA

ABSTRACT: California has an enriched history of oil exploration and production dating back to 1876. The Los Angeles coastal plain was a prime location for the construction of numerous refineries, tank farms, and major pipeline corridors in the early 1920's, reflecting the close geographic proximity to both major production zones and the harbor. The coastal plain is currently occupied by a large number of oil refineries, tank farms, pipelines and underground storage tanks. Pools of light non-aqueous phase liquid hydrocarbon (LNAPL) product are widespread in the subsurface, occurring as isolated perched layers of limited extent and as pools overlying the water table. Some of these LNAPL pools encompass tens to hundreds of acres in extent, and a minimum estimated volume of 1.5 million barrels. Remediation of LNAPL was initiated in the late 1970's at some sites, and accelerated by regulatory order in the mid-1980's. The development of short- and long-term remediation strategies will require a perspective and understanding on a regional scale. The regional hydrogeologic setting serves as an important tool in the designation of aquifer(s) for beneficial use, determination of level of remediation warranted, implementation of remediation strategies, prioritization of limited manpower and funds, and overall future management.

INTRODUCTION

California has an enriched history of oil and gas exploration and exploitation dating back to 1876 (Rintoul, 1990), the first year of commercial production. California comprises 56 counties of which 30 are known to produce oil and gas. Eighteen of the 30 counties produce oil while the remaining 12 counties produce gas. Generally, south from 37° North Latitude, hydrocarbon occurrence is predominantly oil. Specifically, most oil production in Southern California occurs within four basins: the Los Angeles Basin, Ventura Basin, San Joaquin Basin, and Santa Maria Basin. The Los Angeles Basin (encompassing the Los Angeles coastal plain), which has experienced the most urban growth and development, is known to have over 10 giant oil fields.

The coastal plain is about 50 miles in a northwest-southeast direction, and about 15 to 20 miles in width from the Pacific Ocean to the base of the Puente Hills and Santa Ana Mountains. Encompassing about 775 square miles, the coastal plain is characterized by low relief and slopes gently seaward. The low relief is interrupted by the Newport-Inglewood structural zone (NISZ). The NISZ is characterized by a northwesterly trending line of gentle topographic prominence extending about 40 miles in length (Barrows, 1974). At ground surface, the NISZ is marked by the common alignment of low hills and coastal mesas which are broken by six gaps.

The Los Angeles coastal plain prior to the turn of the century was composed of large Spanish land grants used primarily for agricultural purposes (Grenier and Gillingham, 1987). With the discovery of oil, numerous refineries were constructed in the Los Angeles coastal plain during the early 1920's in close proximity to both production areas as well as taking advantage of nearby shipping facilities of the Los Angeles Harbor. The majority of these refineries continued to expand their operations and areal extent through to the late 1940's. Location of major oil fields throughout the Los Angeles coastal plain are shown in Figure 1. Location of major refineries, tank farms, and pipeline corridors within the Los Angeles coastal plain is shown in Figure 2. Not shown in Figure 2 are the numerous underground storage tanks in which documented releases have been reported. Although many of these facilities were initially isolated and located in moderately to heavily industrialized areas, with the encroachment of development many are now in close proximity to densely populated light commercial and residential zones. An increased awareness of environmental issues associated with oil fields and their subsequent redevelopment is presented by Dezfulian (1988), Garcia and Henry (1989), Lovegreen (1989), and Testa (1991a).

Presented in this paper is discussion of the historical background, regulatory history, regional and site specific geologic and hydrogeologic setting, and LNAPL occurrence. Also discussed are the numerous factors that play an important role in the approach to both short- and long-term remediation, and a brief synopsis of current and planned strategies.

REGULATORY HISTORY

Several refineries, tank farms, and other petroleum-handling facilities are included on the United States Environmental Protection Agency (EPA) National Priorities List or are being regulated under the Resource, Conservation and Recovery Act (RCRA). Such facilities present several potential subsurface environmental concerns reflecting approximately 80 years of operation. These concerns include:

o Soils containing elevated hydrocarbon concentrations such that the soil may be considered a hazardous waste;

o LNAPL hydrocarbon pools which have leaked from reservoirs, tank farms, and pipelines, and serve as a source of soil contamination and a continued source for groundwater contamination;

o Dissolved hydrocarbons in groundwater which may adversely affect water-bearing zones considered of beneficial use or drinking water supplies; and

o Accumulation of vapors which could pose a fire or explosion hazard.

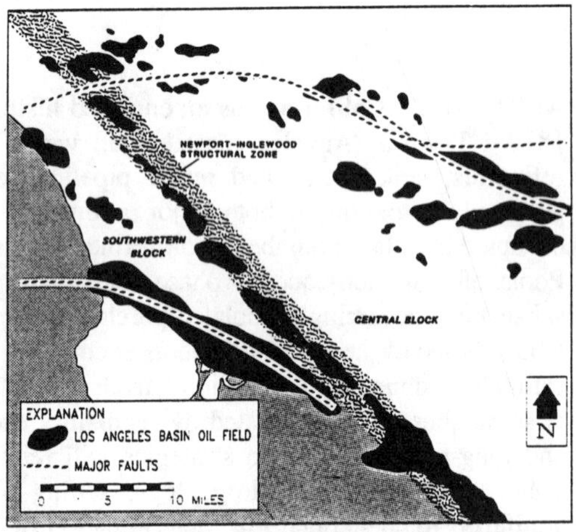

Figure 1. Location of major oil fields in relation to the Newport-Inglewood Structural Zone and other regional faults (after Testa and Winegardner, 1990)

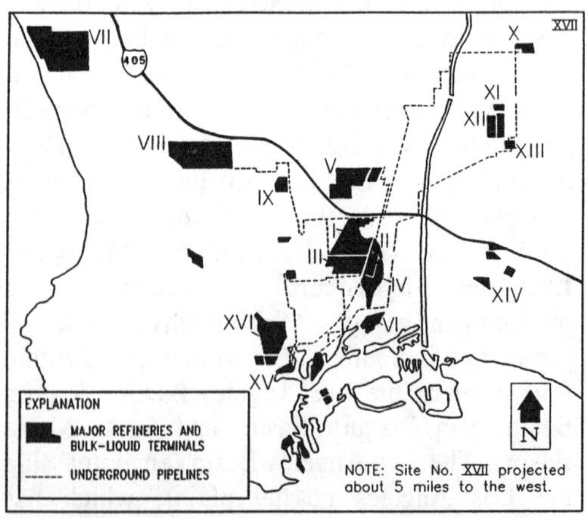

Figure 2. Location of major refineries, tank farms, and pipeline corridors (after Testa and Winegardner, 1990)

In early 1985, oil droplets were evident on beach sands near high-priced beachfront real estate just west of a refinery. The product evidently leaked from the nearby refinery situated near the waterfront, migrated downward to the shallow water table, and then moved laterally with the regional groundwater flow direction toward the ocean. Explosive concentrations of vapors were also detected at the bottom of a construction pit being excavated near the refinery and several nearby homes. A massive pollution problem was then recognized by the regulatory community. At this time, an estimated although exaggerated 6,000,000 barrels of product were thought to exist beneath the majority of the 1,000 acre 80-year-old refinery.

The highly visible presence and potential hazard to public health, safety and welfare prompted a minimum of 17 oil refineries and tank farms to be designated as health hazards by the California Department of Health Services (DOHS). This designation reflected the potential and actual subsurface occurrence of leaked hydrocarbon product derived from such facilities during 60 years of operations that has migrated through the subsurface, resulting in the presence of LNAPL pools overlying the water table.

APPLICABLE REGULATIONS

Although several of these refineries were listed as hazardous waste sites with assessment and remediation being required under RCRA, the majority of such facilities fell under the Los Angeles Region of the California Regional Water Quality Control Board's Order No. 85-17 adopted in February, 1985. This order was the first to address large-scale regional subsurface environmental impacts by the petroleum refinery industry. This order required, in part, assessment of the subsurface presence of hydrocarbons and other associated groundwater pollutants which may affect subsurface soils and groundwater beneath such facilities. Specifically, the following items were included:

o Characterization of subsurface geologic and hydrogeologic conditions;

o Delineation of LNAPL pools including chemical characterization, and areal extent and volume;

o Implementation of LNAPL recovery;

o Overall aquifer restoration and rehabilitation of dissolved hydrocarbons and associated contaminants; and

o Eventual soil remediation of residual hydrocarbons.

At the time this order was issued, some refineries had already commenced LNAPL hydrocarbon recovery programs, although limited in scope. In these cases, the large volumes of LNAPL present and its initial ease in recovery proved economically beneficial.

HYDROGEOLOGY

Hydrogeologic Setting

The Los Angeles coastal plain is divided into four groundwater basins (CDWR, 1961). The Santa Monica Basin and West Coast Basin are located southwest of the NISZ; the Central Basin and Hollywood Basin are situated northeast of the NISZ. This paper focuses upon the West Coast Basin and the western portion of the Central Basin.

The coastal plain is punctuated by a northwesterly-trending series of low hills along the NISZ. The NISZ is the major structural feature in the area and consists of an echelon of faults, anticlines, and domes, and is marked by a series of low hills. The NISZ separates the West Coast Basin from the Central Basin.

Initial studies concerning groundwater beneath the coastal plain were reported by Mendenhall (1905a, 1905b, 1905c, and 1909). Since these earlier studies the most comprehensive information on aquifers beneath the coastal plain is presented by Poland et al. (1948), Richter (1950), Piper et al. (1953), Poland et al. (1956), Poland et al. (1959), Zielbauer et al. (1962), and the Los Angeles County Flood Control District (LACFCD, 1978). The Silverado aquifer, which occurs within the San Pedro Formation, is of very good quality and is the primary source of all groundwater known to be pumped in the West Coast and Central basins.

The predominant areas occupied by major petroleum-handling facilities are shown in Figure 2. Aquifers beneath the area encompassing site nos. I through IV situated in the West Coast Basin, in descending order, are the Gaspur aquifer, Gage aquifer ("200-foot sand"), Lynwood aquifer ("400-foot sand") and Silverado aquifer as depicted in Figure 3. Beds of relatively low permeability soils which act as aquiclude or confining layers separate the aquifers in some but not all places. Pertinent to the occurrence of LNAPL is the Gage aquifer (Testa et al. 1989). In this portion of the coastal plain, the hydrogeologic regime is influenced by numerous factors which all play a role in the approach to remediation (Testa, 1989). These factors include regulated pumpage from the Silverado aquifer, and continued artificial recharge by injection associated with the Dominguez Gap Barrier Project into the Gaspur, Gage and Lynwood aquifers. The NISZ also has a significant multifaceted impact on groundwater occurrence, quality, and usage. The impact of the NISZ and hydrogeologic effects of pumpage and reinjection is discussed by Testa et al. (1988) and Testa (1991b).

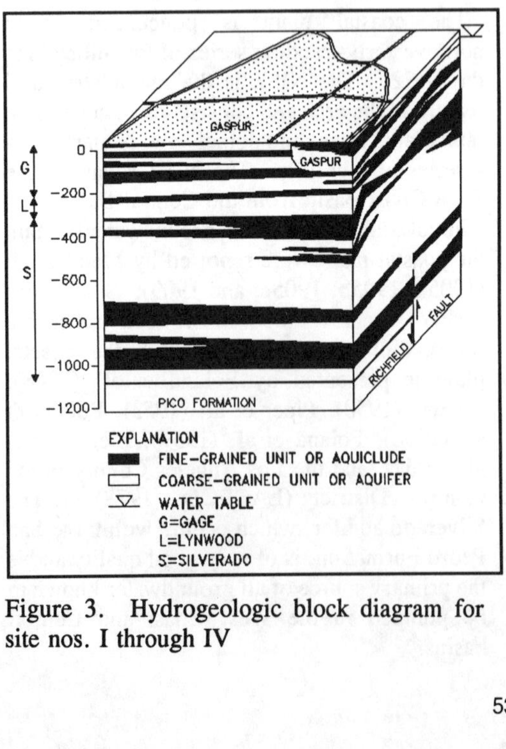

Figure 3. Hydrogeologic block diagram for site nos. I through IV

The area encompassed by site no. VII is situated near the western margin of the West Coast Basin. This portion of the coastal plain is underlain by dune deposits of Holocene age, and marine and continental deposits of Pleistocene age. In descending order, the hydrostratigraphic units are the Old Dune Sand aquifer, the Manhattan Beach aquiclude, the Gage aquifer, the El Segundo aquiclude, and the Silverado aquifer (Figure 4). Pertinent to the occurrence of LNAPL hydrocarbon is the Old Dune Sand aquifer. This portion of the West Coast Basin is also influenced by artificial recharge.

Site nos. X through XIII are situated in the Central Basin northeast of the NISZ. This area is underlain by the shallow perched and semi-perched aquifers, and eight distinct regional aquifers (CDWR, 1961). In descending order, these aquifers are the Semi-perched, Gaspur, Exposition-Artesia, Gardena, Gage, Hollydale, Jefferson, Lynwood, and Silverado (Figure 5). The Exposition-Artesia and deeper aquifers are the primary water-bearing zones tapped for municipal and industrial uses. The shallow aquifers, primarily the Semi-perched but also to a large extent the Gaspur, are polluted with brines and industrial wastes rendering them unsuitable for domestic use throughout the region. Pertinent to the occurrence of LNAPL in the Central Basin are shallow perched zones of limited lateral extent and the Semi-perched aquifer. The deeper aquifers are generally of

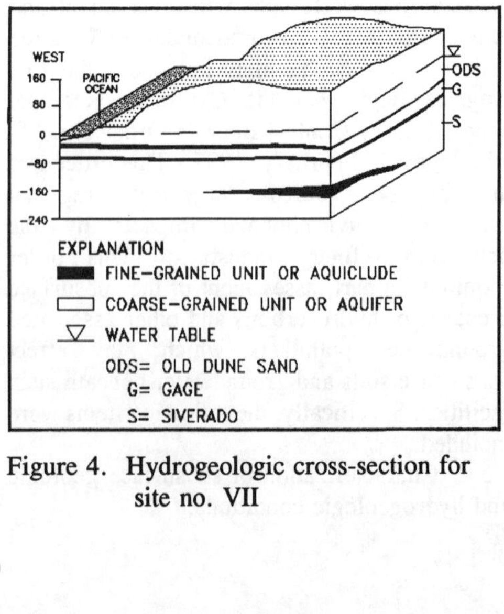

Figure 4. Hydrogeologic cross-section for site no. VII

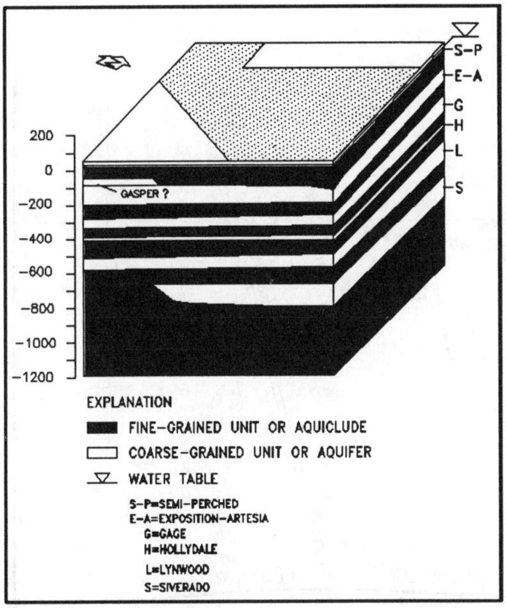

Figure 5. Hydrogeologic block diagram for site nos. X through XIII

good water quality although some saltwater intrusion and localized occurrences of contaminated groundwater has been reported for the Exposition-Artesia aquifer.

FREE HYDROCARBON OCCURRENCE

The spatial distribution and areal extent of major LNAPL hydrocarbon pools beneath the larger refineries and tank farms situated on the Los Angeles coastal plain is shown in Figure 6. The combined areal extent of these LNAPL pools is on the order of 1,500 acres. The estimated minimum volume is on the order of approximately 1.5 million barrels; an estimated maximum volume is on the order of 7.5 million barrels. These rough estimates exclude pipeline, underground storage tanks, and other LNAPL releases. The discrepancy in LNAPL hydrocarbon volume reflects varying approaches taken to estimate these volumes (Testa and Winegardner, 1990).

One of the more complex LNAPL-impacted areas, although not the largest in total volume (see site no. VII), is in the vicinity of site nos. I through IV. Several major refineries, terminals, and major pipeline corridors are located in this area. Groundwater beneath this area occurs within the Gage aquifer under both perched and water table conditions at depths of 40 to 50 and 30 to 60 feet below ground surface, respectively. Likewise, LNAPL also occurs under perched and water table conditions. Some facilities are underlain by several pools which in turn are an accumulation of two or more coalescent pools (Testa et al. 1989). For purposes of this discussion, a pool is defined as an aerially continuous accumulation of LNAPL. Two or more pools which have distinct differences in their respective physical and chemical properties are referred to as coalesced pools. Individual accumulations of relatively uniform product are referred to as subpools since it is inferred that they have coalesced to form aerially continuous occurrences or pools. The occurrence of several pools and subpools at a particular site reflect releases from multiple sources at various times. Many of these occurrences have resulted from very slow releases over long periods of time.

Source identification of dissolved hydrocarbons and refined LNAPL is difficult due to several factors. These factors include numerous microbiological, chemical and physical processes, the complex historical industrial development of the area, numerous property ownership transfers, and the close proximity of crude and petroleum-handling facilities which range from clusterings of refineries, bulk storage tank farms, underground pipelines in industrialized areas, to numerous underground storage tanks associated with gasoline service stations in less industrialized areas. Several methodologies have been used to identify not only the crude and/or refined product type, but also the brand, grade, and in some instances, the source crude (Testa and Winegardner, 1990). These methods have included routine determination of API gravity, development of distillation curves, and analysis for trace metals (notably, lead and sulfur). More sophisticated methods have included gas chromatography, statistical comparisons of the distribution of paraffinic or n-alkane compounds between certain C-ranges,

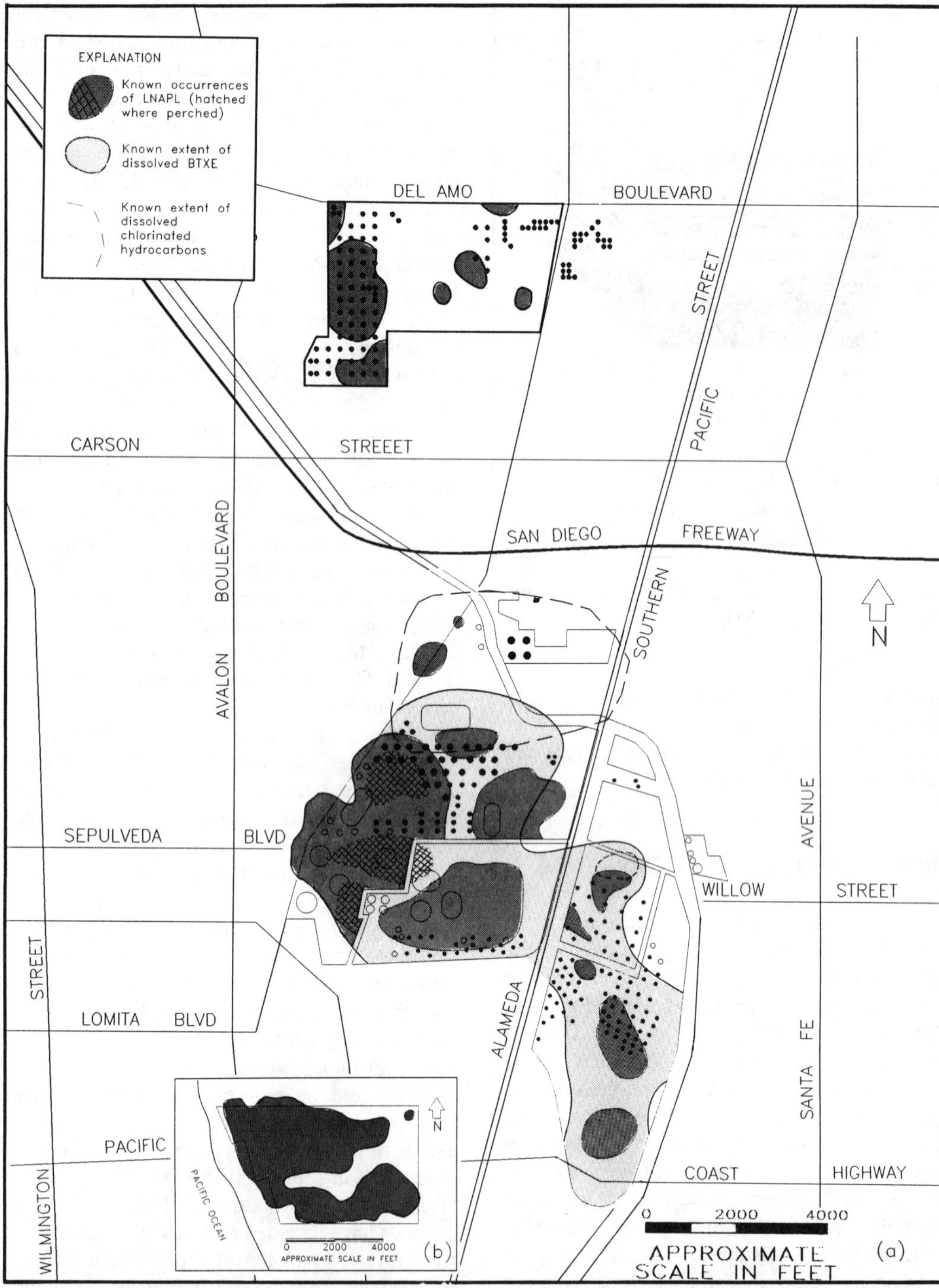

Figure 6. Spatial distribution of LNAPL beneath portions of the West Coast Basin

and determination of certain isotopic ratios of carbon and hydrogen for lighter gasoline-range fractions, and ^{15}N and ^{34}S for heavier petroleum fractions.

Site no. VII as shown in Figure 2 is occupied by an active refinery underlain by the largest volume of LNAPL hydrocarbon beneath a single facility. Groundwater occurs under unconfined water table conditions. The LNAPL hydrocarbon pool is elongated toward the northwest in an east-west and north-south orientation encompassing about 690 acres in areal extent (Figure 6). Apparent LNAPL thickness for this areea ranges up to approximately 12 feet. The LNAPL pool is comprised of a variety of product types including light oil, diesel fuel, gasoline, jet fuel, and reformat.

REMEDIATION STRATEGY

Most major petroleum-handling facilities have implemented aquifer restoration programs with some being further ahead in their respective programs relative to others. Approaches to LNAPL hydrocarbon recovery have included both single- and two-pump type systems. However, the complexities of the regional setting and the numerous factors involved in developing a long-term remediation strategy has veered away from a site specific approach to setting regional objectives and prioritizing these objectives; the primary objective being the protection of the Silverado aquifer and other beneficial use aquifers.

Regulatory status is currently focused on the following issues:

o Complete vertical delineation of LNAPL and its respective dissolved constituents;

o Enhance efficiency and effectiveness of existing LNAPL recovery programs;

o Develop analytical and numerical groundwater models to 1) predict the fate and transport of LNAPL and its dissolved constituents; 2) provide more reliable volume determinations; and 3) design for an optimal groundwater clean-up strategy;

o Identify the extent and source(s) of other organic compounds;

o Develop a source elimination program to detect leakage from above-ground tanks and underground piping in the early stages; and

o Review current available technologies and develop a soil clean-up strategy consistent with the depth and quantity of contaminants present.

In response to these issues outlined above, horizontal delineation of LNAPL is near completion with off-site delineation required at a few sites. The horizontal and vertical delineation of dissolved hydrocarbon constituents has yet to be addressed for the most part. Despite the large volumes of LNAPL released over time, the lateral extent of dissolved hydrocarbon constituents, notably benzene, toluene, ethylbenzene and total xylenes have been limited to distances of 150 to 200 feet hydraulically downgradient. The vertical extent of these constituents has not been fully ascertained.

Certainly if protection of the Silverado aquifer is the primary concern, and the water table aquifer (i.e., Gage, Old Dune Sand, or Semi-perched aquifers) is essentially a write-off, then the intervening aquifer(s) may serve the purpose of a guardian aquifer. Delineation programs to address the vertical extent of dissolved hydrocarbon constituents in groundwater is currently being implemented on a regional basis. In regards to enhanced efficiency and effectiveness of the existing LNAPL recovery programs, the primary factor limiting most recovery programs is the ability of the facilities to handle the generated co-produced water. This reflects antiquated wastewater treatment capabilities in conjunction with existing systems at their maximum capacity.

Two of the larger LNAPL hydrocarbon occurrences, site nos. I and VII, reinject co-produced groundwater into generally the same stratigraphic zone from which it is withdrawn; site no. I reinjects without treatment into the Gage aquifer whereas site no. II reinjects into the Old Dune Sand aquifer. This water is, however, characterized as a waste because of elevated benzene under EPA's recent toxicity characteristic rule. However, a two-year extension has been given to allow oil companies time to comply with this rule so that aquifer clean-up efforts involving the return of benzene-affected groundwater via reinjection wells will not be readily interrupted. Although this relieves certain facilities from immediately complying with this ruling, many of these LNAPL hydrocarbon recovery programs will continue for tens of years, thus, despite the reasonableness of the extension, future regulatory requirements are uncertain.

Only rudimentary modeling has been performed to date to predict the fate and transport of dissolved hydrocarbons in groundwater over time. In regards to volume determinations, there is still much uncertainty concerning actual volume present.

Other organic compounds, notably chlorinated hydrocarbons, are not derived from petroleum-handling facilities and, thus, reflect old landfill cells and chemical-related industry-type sources. Source elimination programs are currently being developed on a site-specific basis.

Clean-up strategies for hydrocarbon-affected soil will most likely be the last issue addressed.

CONCLUSION

Within the NISZ, structures such as folds and faults are critical with respect to the effectiveness of the zone to act as a barrier to the inland movement of salt water. A nearly continuous set of faults are aligned along the general crest of the NISZ, notably within the central reach from the Dominguez Gap to the Santa Ana Gap. The position, character and continuity of these faults are fundamental to the discussion of groundwater occurrence, regime, quality, and usage. In addition, delineation and definition of aquifer inter-relationships with a high degree of confidence is essential. The multi-faceted impact of the NISZ is another aspect of the level of understanding required prior to addressing certain regional groundwater issues. A primary issue is which aquifers are potentially capable of being of beneficial use versus those that have undergone historic degradation. Those faults which do act as barriers with respect to groundwater flow, may, in fact, be one of several factors used in assigning a part of one aquifer to beneficial use status as opposed to another. A second issue, based on beneficial use status, is the level of aquifer rehabilitation and restoration to be required in association with the numerous LNAPL recovery and aquifer remediation programs being conducted within the Los Angeles coastal plain.

Overall understanding of the regional hydrogeologic conditions, including proper delineation of the relationship between the various aquifers, is essential to implementing both short- and long-term regional aquifer restoration and rehabilitation programs (Testa, 1989). The Silverado aquifer is recognized presently as the primary source for municipal supplies in the area. The NISZ, where many of the oil fields beneath the coastal plain are situated, has a significant multi-faceted impact on groundwater occurrence, quality, usage, and remediation strategy (Testa et al. 1988). Present groundwater management programs include fresh water injection as part of the seawater intrusion barrier project, and regulated pumpage which is limited by court order. Future land use and site-specific concerns such as wastewater handling all play a role in the approach to remediation, thus, remediation strategies need to be based on a case-by-case basis as well as the regional situation.

ACKNOWLEDGEMENTS

Appreciation is expressed to my wife, Lydia, for typing of this paper, and Tony Carmouche' for preparation of the technical illustrations.

REFERENCES

Barrows, A.G., 1974, A review of the geology and earthquake history of the Newport-Inglewood structural zone, southern California: California Division of Mines and Geology Special Report, No. 114.

California Department of Water Resources (CDWR), 1961, Planned utilization of the ground water basins of the coastal plain of Los Angeles County, Appendix A: Ground Water Geology.

Dezfulian, H., 1988, Site of an oil-producing property: Proceedings of the Second International Conference on Case

Histories in Geotechnical Engineering, Vol. 1:43-49.
Garcia, D.H. and E.C. Henry, 1989, Environmental considerations for real estate development of oil well drilling in California: In Environmental Concerns in the Petroleum Industry (edited by Testa, S.M.), Pacific Section of the American Association of Petroleum Geologists Symposium Volume: 117-127.
Grenier and Gillingham, R.C., 1987, California legacy: the Watson family: The Watson Land Company.
Los Angeles County Flood Control District (LACFCD), 1978, Coastal plain well location map: shallow aquifer.
Lovegreen, J.R., 1989, Environmental concerns in oil-field areas during property transfers: In Environmental Concerns in the Petroleum Industry (edited by Testa, S.M.), Pacific Section of the American Association of Petroleum Geologists Symposium Volume: 129-158.
Mendenhall, W.C., 1905a, Development of underground waters in the eastern coastal plain region of southern California: United States Geological Survey Water Supply Paper, No. 137.
Mendenhall, W.C., 1905b, Development of underground waters in the central coastal plain region of southern California: United States Geological Survey Water Supply Paper, No. 138.
Mendenhall, W.C., 1905c, Development of underground waters in the western coastal plain region of southern California: United States Geological Survey Water Supply Paper, No. 139.
Mendenhall, W.C., 1909, A phase of ground water problems in the West: Economic Geology, Vol. IV, No. 1:35-45.
Piper, A.M., A.A. Garrett, and others, 1953, Native and contaminated groundwaters in the Long Beach-Santa Ana area, California: United States Geological Survey Water Supply Paper, No. 1136.
Poland, J.F. and A. Sinnott, 1959, Hydrology of the Long Beach-Santa Ana area, California: United States Geological Survey Water Supply Paper, No. 1471.
Poland, J.F., A.A. Garrett, and A. Sinnott, 1948, Geology, hydrology and chemical character of ground waters in the Torrance - Santa Monica area, Los Angeles County, California: United States Geologic Survey Open File Report.
Poland, J.F., A.M. Piper, and others, 1956, Ground water geology of the coastal zone, Long Beach - Santa Ana area, California: United States Geologic Survey Water Supply Paper, No. 1109.
Richter, R.C., 1950, Geology of the West Basin, Los Angeles County, California: California State Department of Public Works, Division of Water Resources, unpublished preliminary report.
Rintoul, W., 1990, Drilling through time. California Department of Oil and Gas, Sacramento, California.
Testa, S.M., 1989, Regional hydrogeologic setting and its role in developing aquifer remediation strategies: In Proceedings of the Geological Society of America 1989 Annual Meeting Abstracts with Programs, Vol. 21, No. 6: A96.
Testa, S.M., D.M. Baker, and P.L. Avery, 1989, Field studies on occurrence, recoverability and mitigation strategy for free phase liquid hydrocarbon: In Proceedings of the American Association of Petroleum Geologists Pacific Section Symposium on Environmental Concerns in the Petroleum Industry (edited by Testa, S.M.): 57-81.
Testa, S.M. and D.L. Winegardner, 1990, Restoration of petroleum- contaminated aquifers: Lewis Publishers, Chelsea, Michigan.
Testa, S.M., E.C. Henry, and D. Hayes, 1988, Impact of the Newport- Inglewood structural zone on hydrogeologic mitigation efforts - Los Angeles Basin, California: Proceedings of the National Water Well Association of Ground Water Scientists and Engineers FOCUS Conference on Southwestern Groundwater Issues: 181-203.
Testa, S.M., 1991a, Environmental aspects relating to redevelopment of oil field properties within the California regulatory framework; In proceedings of the Hazardous Materials Control Research Institute National Research and Development Conference on

the Control of Hazardous Materials, Anaheim, California: 4-10.
Testa, S.M. 1991b, Elevation changes associated with groundwater withdrawal and reinjection in the Wilmington area, Los Angeles coastal plain: In proceedings of the International Conference on Subsidence, Houston Texas, in press.
Zielbauer, E.J., Kues, H.A., Burnham, W.L., and Keene, A.G, 1962, Dominguez gap barrier project, geologic investigation: Los Angeles County Flood District.

Subsurface Contamination by Immiscible Fluids, Weyer (ed.) © 1992 Balkema, Rotterdam. ISBN 90 6191 175 3

Investigation and remediation of an industrial site with a hundred year history at Düsseldorf, Germany

Karl Hoffmann & Andreas B. Eberstein
Büro Dr. K. Hoffmann, Essen, Germany

ABSTRACT: A 100 year old industrial site with an area of 45 ha had to be cleaned and made available for redevelopment. Onehundredandtwentyfive polluted sites were found and 180,000 t of contaminated ground had to be excavated and cleaned on site. For that purpose a variety of soil and groundwater cleaning systems have been installed at the site.

1 INTRODUCTION

An area which has been used by a pipe manufacturer for over 100 years has to have subsurface pollution by all the chemical compounds having been part of the production process. Today many of those substances are known to be harmful to man and the environment.

The area investigated is a former industrial site of approximately 45 ha in Düsseldorf, West Germany. The area is located about 2 km from downtown Düsseldorf. The factory owned by the Mannesmann company was closed in December 1987 and sold.
As former proprietor Mannesmann had to pay all the costs for remediation measures. Another aspect of the sale was a time-restriction, about 70 % of the area had to be cleaned by June 30th, 1989, which left two years for demolition of the buildings, excavation of foundations, and remedial measures.

With the local regulatory authorities a work scheme was set up.

1) cleaning of the interior of the factory-buildings
- dismounting of machinery
- cleaning of pipe- and drainage-system
- removal of contaminated concrete

2) clearance from the local authorities for demolition

3) demolition of buildings above ground
- tearing down of buildings
- sorting of debris

4) excavation of floor and foundations down to 3.5 m below surface

5) removal of contaminated soil found during excavation down to groundwater level

6) planning of further remediation measures

7) refill of excavations with recycled concrete

8) final clearance from the local authorities for redevelopment

The cleaning, demolition, and excavation was carried out under the close supervision by geologists, with up to 5 geologists permanantly at the site. The idea was to put greater emphasis on the surveillance of excavation than on extensive investigations beforehand.

No legislation on remediation of former industrial sites and also no offically fixed permissible concentration-values for pollutants exist in Germany. Therefore a contract under public law was signed by Mannesmann and the city of Düsseldorf in order to provide a guideline for the remediation measures. In this contract permissible concentrations for the following pollutants were stated for soil:

-hydrocarbons (mineral oil)
-chlorinated hydrocarbons
-heavy metals

During the process of cleaning additional agreements had to be made due to the discovery of further pollutants. That led to a list of permissible concentrations (table 1):

Table 1: Permissible concentrations to be left in the ground

pollutants		saturated soil [mg/kg]	unsaturated soil [mg/kg]		[mg/l] *
hydrocarbons (oil)		500,0	1000,0		
chlorinated hydrocarbons		0,1	0,1		
benzene, toluene, xylene (BTX)		n. f.	2,5		
PAH		n. f.	10-20		
PCB (28, 52, 101, 138, 153, 180)		n. f.	5,0		
heavy metals under unsealed surfaces	As	n. f.	50,0	and	0,050
	Cd	n. f.	20,0	and	0,005
	Pb	n. f.	600,0	and	0,050
	Ni	n. f.	500,0	and	0,050
	Cr	n. f.	800,0	and	0,050
	Cu	n. f.	500,0	and	0,100
	Hg	n. f.	10,0	and	0,001
heavy metals under sealed surfaces	As	n. f.	50,0	and	0,100
	Cd	n. f.	20,0	and	0,050
	Pb	n. f.	600,0	and	0,500
	Ni	n. f.	500,0	and	0,500
	Cr	n. f.	800,0	and	1,000
	Cu	n. f.	500,0	and	1,000
	Hg	n. f.	10,0	and	0,005

n. f. : not fixed
* : dissolved in water

For groundwater the goal was to reach equivalent concentrations upgradient and downgradient of the site in question.

One important part of the contract was cleaning of all contaminations on site to avoid any exporting of contaminated soil off the premises.

2 INVESTIGATIONS

2.1 Anamnesis

Even though the greatest emphasis for the discovery of contaminated areas was put on the supervision of all the demolition activities considerable time had already been spent on investigations carried out beforehand.

First a study was done by Mannesmann based on their knowledge of the former use of potential contaminats at different locations. Mannesmann also questioned former employees on indications for possible contaminations.

This study pointed out 34 areas with potential subfurface pollution.

2.2 Investigation on site

Based on the results of the first study our office was asked in 1986 to conduct field investigations, to supervise the demolition and excarvation, and to design and supervise the remediation.

The first four investigation phases carried out before demolition concentrated on groundwater and the 34 areas of potential contaminations. Table 2 shows the number of piezometers, sampling-holes, and samples taken. Of the first 34 sites only 20 showed contaminations which had to be cleaned.

Table 2: Investigations in phase 1 to 4

phase	number of piezometers	number of sampling-holes	number of soil- and watersamples
1	5	32	65
2	0	53	192
3	5	58	146
4	12	52	174

2.3 Investigations and observations during demolition

Demolition of the buildings started in summer of 1987. In the process of excavating the foundations of former buildings 105 additional contaminated sites have been found due to the close supervision.

2.4 Investigation results

Three major contaminants have been found, chlorinated hydrocarbons, hydrocarbons and polynuclear aromatic hydrocarbons. The chlorinated hydrocarbons mainly consist of trichlorethene. The hydrocarbons are a combination of mineral oils and converted mineral oil products. In the production process (hot tubes submerged in oil) and due to the ageing of the oil products, the hydrocarbons underwent an oxidizing process and transformed to polar compounds. The composition-pattern of the 16 EPA-PAH showed differences, depending on the area, i.e. former type of application.

Additionaly contaminations of BTX (benzene, toluene, and xylene), PCB (polychlorinated biphenyls), and heavy metals have been found.

Table 3: Contaminated areas

# of areas	volume [m3]	contamination	maximum concentration [mg/kg]	average concentration [mg/kg]
98	33,000	mineral oil	120,000	10,000
10	6,000	PAH	3,000	100
6	12,000	chlor. hc	30,000	5
10	7,000	heavy metals	/	/
1	4,000	PCB	350	20
125	62,000			

At ten of the oil-contaminated areas contaminations had reached the groundwater table (about 6 m below surface). At some sites polluted soil was found down to 15 m below surface, 9 m below groundwater table, with up to 30 cm of oil in free phase on the water in boreholes. at one site trichlorethene reached into the saturated zone resulting in a 350 m long plume of dissolved trichlorethene downgradient with concentrations of up to 120 mg/l.

Table 3 shows the 125 contaminated sites found so far, the volume of single sites ranging from 5 to 15,000 m³. It also gives the concentrations of different contaminants.

3 SELECTION OF REMEDIATION MEASURES

In the spring of 1988 the results of the investigations and the knowledge of the first period of demolition and excavation made it possible to set up a remedial strategy tailored to the site.

The criteria for the selection of the remedial measures were:

- enviromental efficency
- technical feasibility
- reliability
- time of application
- secondary effects
- influence on third parties
- costs.

These topics guided the evaluation of the measures to be taken to deal with the contaminated areas themselves as well as the selection of the most suitable cleaning systems.

4 SELECTED REMEDIATION MEASURES

It was decided to excavate all polluted ground to 1 m above groundwater. For that purpose an area of about 10,000 m² was prepared to store up to 40,000 m³ of contaminated material on site.

To keep track of the excavated soil a list of 8 contaminant-groups was set up, consisting of single-pollutants-groups and multiple-pollutants-groups. The excavated soil of the 125 contaminated areas was registered and followed through the whole cleaning process. That way it was possible to keep track of the complete flow of masses, from excavation to reuse after cleaning.

The treatment of soil-pollutions reaching below groundwater was based on the downstream effects of the contamination. Therefore, seven areas had to be excavated up to a depth of 15 m below surface.

In 4 cases the groundwater table had to be lowered by pumping water from installed wells. Three sites were excavated by installing sheeting pile casings and then lowering the groundwater table. With both methods the raised water was cleaned and discharged into the sewer system.

The next step was the selection of suitable cleaning systems. For the on-site treatment of excavated polluted soil three systems were selected:

- high-pressure soil washing
- high-temperature treatment (incineration)
- on-site vapor extraction in pits.

For the in-situ cleaning of contaminated areas three systems were used:

- in-situ stripping
- hydraulic cleaning
- vapor extraction

To clean contaminated groundwater the following methods were selected:

- underground low pressure vapor extraction
- pumping and stripping of water above ground
- pumping and skimming of light pollutants off the groundwater.

Table 4 shows the selected remediation measures in respect to the contaminants and the location of the contaminated soil or water.

4.1 High-pressure washing

The contaminated soil is treated solely by water in a mobile machine consisting of 50

Table 4: Remediation measures in respect to location

		mineral oil	chlorinated hc	PAH	heavy metals
soil above groundwater	on site	excavation washing/incineration	excavation vapor extraction in pits	excavation washing/incineration	excavation deposition under sealed surface
	in situ	/	vapor extraction	/	sealing of surface
soil below groundwater	on site	excavation washing/incineration	/	/	/
	in situ	hydraulic measures	/	hydraulic measures	/
groundwater		hydraulic measures	underground low-pressure vapor extrac.	/	/

containers. The soil is sucked through 4 rings with highpressure jets. The contaminants are removed from the grains by water at a pressure of 350 bar. In the jetrings the material is accelerated to over 900 km/h and is then 'shot' against metal plats also breaking off contaminants. Afterwards the soil is washed and run through screens. In that process the washing-water with the pollutants and grains smaller than 30 um is separated. The pollutants with particles smaller than about 3 um are floatated and skimmed off the water. The remaining particles are sedimented.

Ninety percent (90 %) of the water is returned to the washing process, 10 % go through a water treatment plant, which is part of the system, and into the sewer system. The discharged water is replaced by fresh water. The amount of water which has to be replaced depends on the amount of so soluble substances in the soil. For an average input of 45 t/h 15 t of fresh water is needed. The costs for cleaning are 180 - 240 DM/t (105 - 140 $US/t).

So far 52,000 t of contaminated soil with a total contents of 99 t of mineral oil and 515 kg of PAH have been washed. The outcome of cleaned soil was 46,000 t with average concentrations of 240 mg/kg oil and 4.9 mg/kg PAH (table 5).

Table 5: Concentration of pollutants in the washing process

	Soil, dry [t]	Mineral oil [mg/kg]	PAH (EPA) [mg/kg]
Soil, contam.	51.240	1.936	10,1
Soil, cleaned	46.108	243	4,9
SFK	3.016	12.316	63,2
FLT	116	20.860	26,0
Hum.	20	6.677	84,0

SFK: sedimented particles
FLT: floatated material
Hum: humus material

The efficency of the washing process was about 85 % for hydrocarbons and about 55 % for PAH. The machine produced 116 t of floatated concentrates with 21 g/kg oil and 26 mg/kg PAH as well as 3000 t of sedimented material with average concentrations of 12 g/kg hydrocarbons and 63 mg/kg PAH. About 90 % of the sedimented material had to be deposited at a dump site. Ten percent (10 %) could be reused on the site. The flotation material was burned at a special incineration plant off site.

In the calculation of masses it was found that 49 % of the hydrocarbons were "lost" within the machine (table 6). It seems that

Table 6: Flow of masses in the washing process

	Soil, dry [t]	Mineral oil sum [kg]	PAH (EPA) sum [kg]
Soil, contam.	51.240	99.182	515
Soil, cleaned	46.108	11.192	225
SFK	3.016	37.146	190
FLT	116	2.420	3
Hum.	20	132	2
Sum	49.260	50.890	420
Losses	1.980	48.292	95
in %	3,9	48,7	18,4

SFK: sedimented particles
FLT: floatated material
Hum: humus material

a large amount of hydrocarbons disintegrated due to a biological activity. The water and the sediment stay in the system for days under ideal conditions. Tests are conducted to verify this assumption.

The limiting factor of the high-pressure washing machine is high concentrations of PAH. It has not been possible to clean off tarlike PAH's.

4.2 High-temperature treatment

The mobile incineration system works with a heatresistant conveyor belt. The contaminated soil is heated to 800 C and subsequently the pollutants are decomposed into harmless compounds. The evolving gas is heated to over 1200 C for over 3 seconds to destroy dioxine. Water is added to the cleaned soil up to the original water contents.

The system was selected to cope with pollutants that could not be treated sucessfully in the high-pressure washing machine. In the first selection stage it was planned to burn the contaminated floatation and sedimentation materials from the washing machine on site as well. But the offical permission by the local authorities was delayed such, that most of the materials to be washed had already been treated and the remains of the washing process had been discharged.

The incinerator is not been erected at the site yet so no results can be presented. But testruns have shown that the maximum capacity is 15 t/h and mineral oil is reduced to less than 10 mg/kg and PAH to less than 1 mg/kg. The costs are about 200 DM/t (120 $US/t), calculated for a six months period and the treatment of 24,000 t of contaminated soil.

4.3 On-site vapor extraction

The soil contaminated with volatile substances like trichlorethene was excavated and arranged into 2 m high pits. Suction pressure is put onto perforated tubes imbedded in the pit. Thus the volatile hydrocarbons are extracted and then absorbed onto carbon filter systems.

About 20,000 t have already been treated this way. Sixty percent (60 %) were cleaned to concentrations of less than 100 ug/kg trichlorethene, the rest reached concentrations of less than 1,000 ug/kg. This method works faster than in-situ treatment and there are no limitations on the redevelopment. Its costs run to about 60 DM/t (35 $US/t).

4.4 Hydraulic cleaning

In some areas the mineral oil contamination reached the saturated zone. If no discharge of hydrocarbons appeared downgradient of the polluted area hydraulic cleaning was applied. The groundwater-level was lowered intermittently to force movable oil to the well where it is skimmed off.

4.5 Vapor extraction

Two areas were cleaned by in-situ vapor extraction, a well known, cost efficent method. The drawback is the long treatment period. One site of approximately 1,500 m³ was cleaned to trichlorethen concentrations below 100 ug/kg, except 40 m³ with 600 ug/kg trichlorethen. One system runs for about 3,500 DM per month (2,000 $US).

4.6 Underground low-pressure vapor extraction

The vapor extraction of chlorinated hydrocarbons out of the groundwater takes place in a well. A low pressure is introduced to the airtight well. Air is drawn through a tube from the outside into the water where it strips off the volatile substances. They are extracted from the well by fans. The trichlorethene is then filtered off. The system worked in a well with good results, reducing the concentration of trichlorethene from 24,000 ug/l to 1,000 ug/l 100 m downgradient within 3 weeks. Later the system was installed 200 m away and failed. The mechanism extracting solvents from the groundwater up to 100 m away can't be explained yet.

The method is cost efficent because no water has to be pumped to the surface and therefore there is no problem of discharging it. The costs are 2,800 DM per month (1,650 $US).

4.7 Stripping

In one case the groundwater was cleaned of trichlorethene by pumping the contaminated water through a strip column above ground. The main problem is the discharge of the water. Reinfiltration was deemed to be hazardous because of the high iron concentration, so it had to be discharged into the public sewer system at costs of about 100 to 480 DM per day (60 - 280 $US/d) just for the discharge. The cleaning-unit costs 14,000 DM per month (8,200 $US).

5 TOTAL COST

The costs for the remedial measures only were estimated in February 1988 to be 1 million DM ($ 600,000), in June 1988 to 20 million DM ($ 12 million), in April 1990 to 35 million DM ($ 21 million) and have now (January 1991) reached 42 million DM (26 million $). They were spent on the following topics (in million DM):

- 17: excavation, construction, transportation
- 13: high pressure cleaning (110,000 t)
- 4: high temperature incinerator (15,000 t)
- 3: cleaning of contaminations by chlorinated hydrocarbons
- 5: consulting,
  - 2.8: chemical analysis
  - 0.7: drilling for soilsamples and 2" piezometers
  - 1.2: consultants on site
  - 0.3: DM expertises

The costs will run to about 100 DM/m² (5.8 $US/sq. ft.) for the remediation measures. That covers the complete cost for the treatment of 125 contaminated areas which have been found and dealt with so far.

The clean-up of the site will be finished in June 1991 having lasted for 4 years. The first new building on the site was already constructed at the end of 1988 in a cleaned area and by January 1991 about 90 % of the area are under new construction.

Subsurface Contamination by Immiscible Fluids, Weyer (ed.) © 1992 Balkema, Rotterdam. ISBN 90 6191 175 3

Sydney Tar Ponds clean up: Hydrogeologic assessment – Coke Oven Complex

F.E. Baechler
Nolan, Davis & Assoc. (N.S.) Ltd, Sydney, N.S., Canada

D.S. MacFarlane
Jaques Whitford & Assoc., Halifax, N.S., Canada

ABSTRACT: Ninety years of steel production at Sydney, Nova Scotia resulted in contamination of groundwater and surface water by coal tar compounds, principally polycyclic aromatic hydrocarbons. Three distinct sources of contamination were delineated over the Coke Ovens site, including: 1) the Domtar plant, characterized by free coal tar DNAPLs, 2) the Benzol plant with LNAPLs and 3) raw coal stockpiles with acid drainage. The zone of maximum organic contamination in the shallow bedrock appears to be transported off site under the Steel Plant, not into populated areas. Due to the presence of DNAPLs within the fractured, porous bedrock, total decontamination of the site does not appear to be feasible. However, selected areas of the site are suitable for partial cleanup.

1. INTRODUCTION

In 1981 a Canada/Nova Scotia Subsidiary Agreement authorized a program to modernize the Sydney Steel Corporation's facilities at Sydney, Nova Scotia (Figure 1). Under the auspices of that program, an investigation was undertaken to assess the extent of coal tar contamination created by the discharge of liquid wastes since the early 1900s, into an estuarine embayment of Sydney Harbour, referred to locally as the Tar Ponds. As a result of elevated polycyclic aromatic hydrocarbons (PAH) levels in lobsters in Sydney Harbour, commercial lobster fishing was banned in the South Arm of the harbour in 1982.

Approximately 600,000 m^3 of coal tar contaminated sediment was found within the Tar Ponds, (Acres 1985). It was also determined that ongoing discharge originated from Coke Oven Brook, within which the Coking facility was located. A preliminary investigation over this site indicated groundwater contamination by coal tar and other related chemicals.

A 10-year, 50 million dollar cleanup of the Tar Ponds was initiated on November 7, 1986. The project is jointly funded by the Nova Scotia Department of the Environment and Environment Canada with project management by Acres International Ltd. A more detailed hydrogeological investigation of the 60 hectare Coke Oven Complex was also undertaken. This paper summarizes the investigation carried out between June 1988 and March 1989.

2. FIELD PROGRAM

A total of 285 metres of drilling was carried out with the installation of 24 piezometers at 16 sites. This was followed by an extensive program of hydraulic testing, and groundwater sampling which keyed in on 16 PAH and 23 heterocyclic nitrogenous compounds (HNC). The investigation focused on the bedrock aquifer to a depth of 20 metres. A domestic water well survey was deemed essential to the assessment of the health impact of PAH releases. Of the 286 properties contacted, a total of 10 operating wells were identified, all of which draw water from what will be identified as the Lower Morien bedrock aquifer, upgradient from the site.

3. CONCEPTUAL FLOW MODEL

The study area is characterized by a humid continental climate. The 30 year normal precipitation is 1400 mm with an air temperature of 5.7°C. There is a low annual moisture deficit of 0.58 cm and a

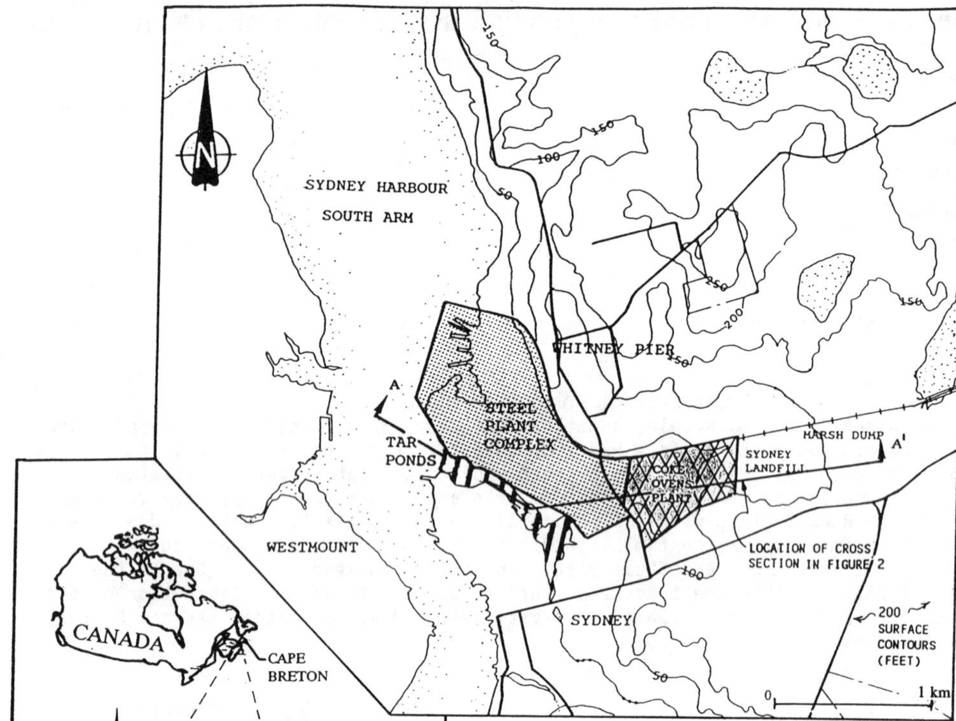

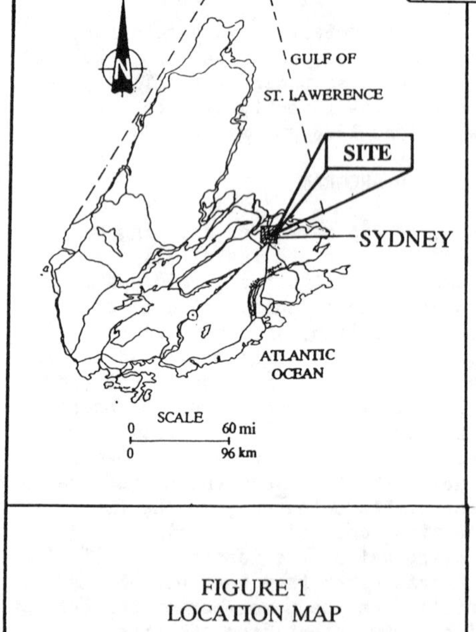

FIGURE 1
LOCATION MAP

large annual surplus of 29.4 centimeters. The Coke Oven facility is located within the 8.4 km² Coke Oven Brook watershed which discharges directly into the Tar Ponds. Specifically it is positioned over the valley floor near the downstream end of the watershed (Figure 1). The brook flows through the centre of the site (Figure 3) at a longitudinal gradient of less than 1 percent. Three natural and three man made tributaries enter the brook as it transits the site. The most notable is Cagney Brook which enters the site from the south, and crosses the site for 550 m by means of an underground culvert

Acres (1985) calculated a theoreterical mean annual flow for the watershed of 0.34 m^3/S, with a mean annual and 100 year flood focus of 6 and 13 m^3/S respectively. Theoretical low mean daily flows were calculated to be in the order of 0.007m^3/S. At this discharge the site transit time was 1 to 2 hours.

Four hydrostratigraphic units (HU) were identified on-site at a macro scale (Figure 2):

A) Fill HU: Fill thickness varied from 0 to greater than 3.7 m, comprising reworked till, coal tar, coke breeze, slag, bricks, concrete and coal. At a macro scale, this unit is best described as a heterogeneous, anisotropic, porous, unconsolidated, non-fractured media. Hydraulic conductivity (K) ranged around 10^{-4} cm/sec.

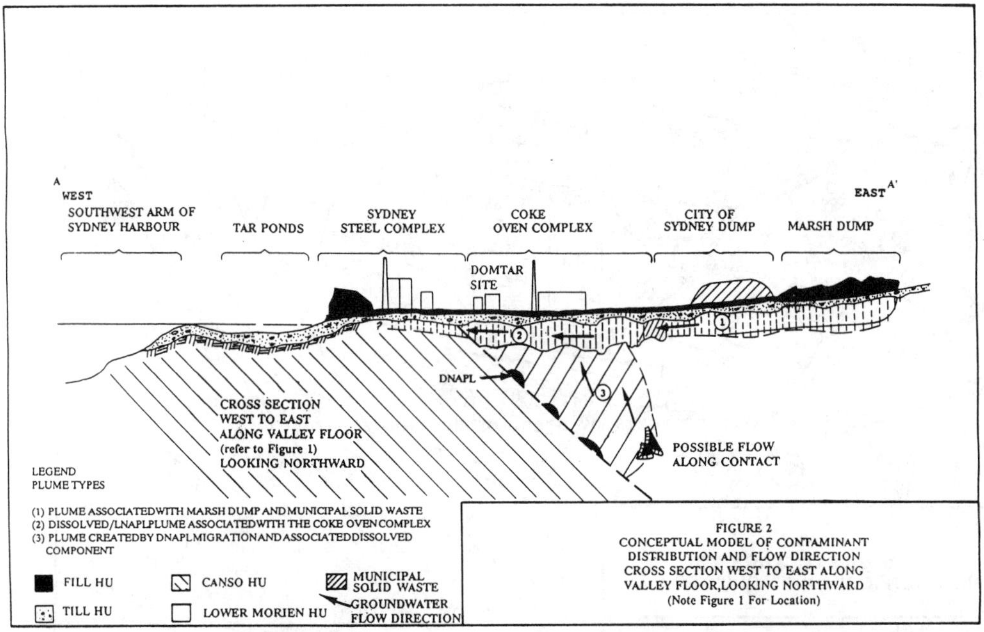

FIGURE 2
CONCEPTUAL MODEL OF CONTAMINANT DISTRIBUTION AND FLOW DIRECTION CROSS SECTION WEST TO EAST ALONG VALLEY FLOOR, LOOKING NORTHWARD
(Note Figure 1 For Location)

B) Till HU: The site is underlain by 2 to 11 m of a stony, sandy silt basal till. It is a homogeneous, isotropic, porous, semi-consolidated, non-fractured media. Matrix porosity varies from 0.29 to 0.34. K values are in the order of 10^{-7} cm/sec.

C) Canso Bedrock HU: This HU subcrops only at the northwestern corner of the site (PN 88-1 and 2), where it comprises a soft mudstone. The unit was also encountered at 18.6 m below the subcrop of the Morien Sandstone at PN 88-5 (Figure 3). It is a heterogeneous, anisotropic, semi-porous, consolidated, fractured media. At two sites, K values derived from slug testing for this study were in the order of 10^{-6} to 10^{-7} cm/sec.

D) Lower Morien Bedrock HU: The study area is predominantly underlain by a grey medium-grained sandstone. The bedrock surface slopes westward under the site at 2.5 percent. Two major depressions were noted including a 4 meter deep zone under under the Benzol Plant area and a more shallow undulation under the Domtar Site. Exposures indicate a well-defined orthogonal joint system. Borehole cores within 0 to 5 m of the subcrop surface exhibited near horizontal fractures; sub-vertical fracturing became dominant with depth. In macro terms, it may be described as a homogeneous, isotropic, porous, consolidated, fractured media. Fracture K values exhibited a geometric mean (n=16) of 4.7×10^{-4} cm/sec; ranging from 1.7×10^{-2} to 7.0×10^{-6} cm/sec. It is suspected that K_h is greater than K_v near the subcrop surface; and that this reverses at depth. Joint mapping suggests the development of anisotropy with larger K's parallel to bedding strike. Matrix porosities range from 5 to 15 percent; fracture porosities are unknown.

Three separate groundwater flow regimes are identified within the Coke Oven Site: 1) a perched, seasonally variable, water table system within the Fill HU in direct hydraulic connection with extensive underground piping infrastructure and Coke Oven Brook;(Figure 3) 2) a local flow system within the underlying highly permeable fractured shallow bedrock; and 3) a regional deep groundwater flow regime which discharges into the Coke Oven Brook Valley. The latter is bounded by a zero discharge boundary at the base provided by the Canso HU, and variable head recharge boundaries to the north, south and east, coincident with topographic divides.

The piezometer surface developed from piezometers installed immediately below the Subcrop Surface in the local flow system is presented on Figure 3 for November conditions. It generally reflects the topography, and is located

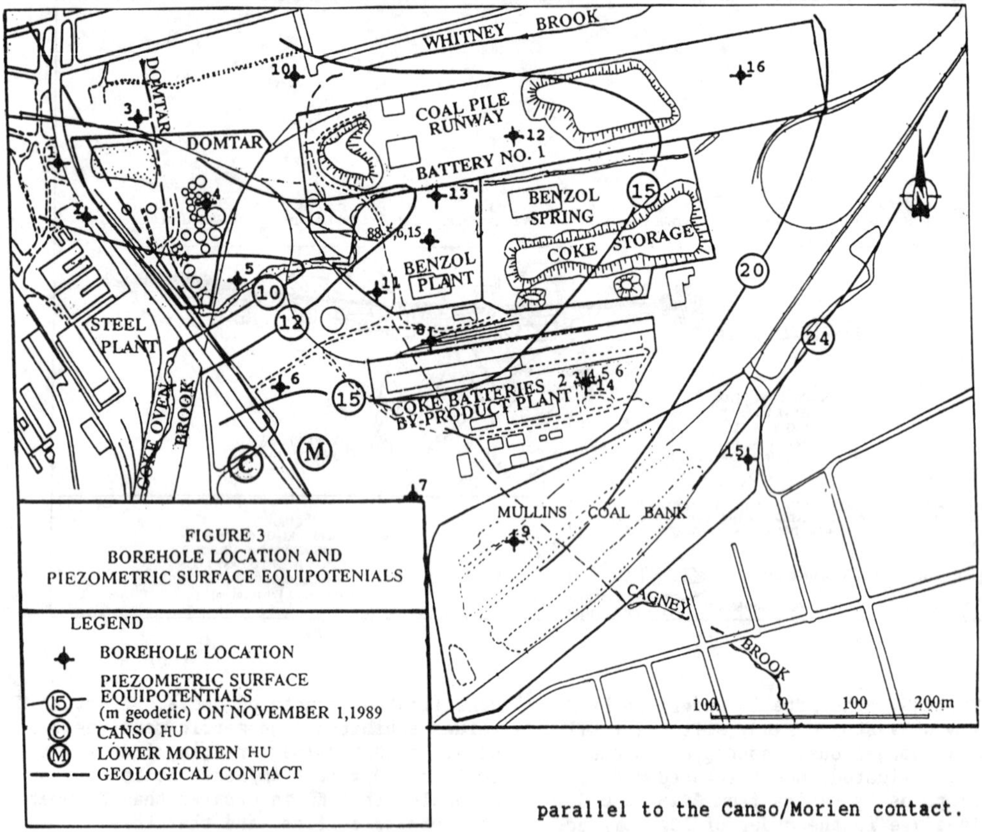

FIGURE 3
BOREHOLE LOCATION AND
PIEZOMETRIC SURFACE EQUIPOTENIALS

within the overburden, exhibiting non-flowing artesion conditions.

The dominant horizontal flow direction is generally west-northwest down the valley at an average gradient of 0.015. Locally the flow appears to converge on Coke Oven Brook and the bedrock depression prior to moving downvalley and off-site to the west. The average linear velocities could range upward to 1 to 9 metres/day depending upon fracture porosities.

The vertical component of flow is characterized by downward gradients of 2-14% between the overburden and shallow bedrock. A small 1-2% downward gradient is observed in the bedrock except at one Site 88-8 which consistently exhibited an upward gradient of 2-3%.

The extensively weathered and fractured bedrock surface appears to act as a drain for local groundwater flow and is considered to be the critical contaminant release pathway from this site.

The physical and chemical evidence to date does not provide any strong evidence of a north-south component of lateral flow parallel to the Canso/Morien contact.

4. CONTAMINANT SOURCE AREAS

The type and location of facilities within the 60 hectare site varied considerably both spatially and temporally over the life of the site. Five source areas covering 30 hectares exhibited a characteristic "fingerprint of contamination" (Figure 3).

A) Mullins Coal Bank: This 10 hectare site was utilized to temporarily store up to 500,000 tonnes of coal from the 1930s to 1970s. The floor of the bank now comprises a thin (less than 0.3 m) layer of compact coal over the Till HU. Surface flow dominates runoff, which is directed into the underdrain reach of Cagney Brook transecting the site. The shallow bedrock exhibited the impact of acidic drainage with a brackish (ion sum 2712 mg/l), Ca-SO_4 type water, a relatively low pH (5.6) and elevated iron (69 mg/l) and manganese (34.6 mg/l). No indicator organics were found.

B) Coal Pile Runway: Since 1920, this 10 hectare area stored up to 150,000 tonnes of raw coal. Between 1900 and 1920, Battery 1, consisting of 400 retort coke ovens, was located in this area. A water table is developed within the 2 to 4 m thick Fill HU. Water levels were observed to rise with rainfall events to ground surface within localized depressions; no surface drains are present. Exfiltration could be directed south toward Coke Oven Brook or west into the culverted reach of Whitney Brook. Vertical infiltration into the bedrock was documented with gradients up to 0.20. Total PAH concentrations in the shallow bedrock ranged from 12.6 to 30.4 µg/l at the middle and downgradient end of the site. Inorganic contamination from acidic drainage was also identified in this zone. It was most pronounced at the former, being characterized by a brackish (TDS 1204 mg/L), corrosive, Ca-SO_4 type water, with a lower than normal (by 2 units) pH of 6.1 and elevated manganese (13 mg/L) and silica (21 mg/L).

C) Coke Batteries 2 to 6/By-Products Plant: Batteries 2 to 6 operated from 1920 to 1988. The 114 oven of Batteries 5 and 6 had an annual capacity of 300,000 tonnes. The shallow water table within the Fill HU is dominated by an extensive network of subsurface piping over the 4.6 hectares. It is heavily stained with organics. However, the bedrock exhibits only minor organic contamination. Analysis of coal tar (Table 1) produced by the facility indicated a total BTEX concentration of 3,000 ppm, due principally to xylenes. The total PAH (16 compounds) concentration was 2,210 ppm; over 50 percent of which was due to naphthalene, phenanthrene and fluoranthene. Total HNC (23 compounds) concentrations were 122 ppm, predominantly carbazole. Large volumes of tar were discharged directly into Coke Oven Brook through the Cagney Brook underdrain. The drain apparently is not deep enough to encounter bedrock. Vertical upward groundwater gradients of 2 to 3 percent should prohibit the migration of dissolved contaminants into the Morien HU. However, it is thought that its porous, brick-lined base would allow DNAPL's to migrate to depth.

D) Benzol Plant: This plant, constructed in the early 1930s, used benzene to make light oil at 1.7×10^5 Lpd. The 2.8 hectare area is underlain by 1 to 3 m of Fill over 9 to 10 m of Till. The water table developed within the fill was located at <1 m below grade; saturated overland flow was noticeable downgradient of the "Benzol spring".

The water table is partially intercepted on the north by an open ditch, which discharges into the Site Drain and eventually Coke Oven Brook. A significant feature at this site was LNAPLs discharging from the Benzol spring into Coke Oven Brook. A sample collected of the LNAPL (Table 1) indicated Benzene (28 ppm), total PAH 790 ppm (greater than 70 percent was naphthalene and Fluorene) and HNC (1.7 ppm). Piezometers noted a decline in concentrations with depth so that PAH and HNC concentrations are approximately 3 orders-of-magnitude lower in the bedrock (PAH - 2.7 µg/l) than in the fill (PAH - 890 µg/l. The exception to this trend was the total BTEX in the shallow bedrock at PN 88-11 of 108,705 µg/l, primarily due to benzene.

E) Domtar Site: A tar recovery plant operated between 1903 and 1963 over 3.3 hectares at the northwest corner of the site. It was constructed to store and sell coal tar produced by the ovens at 80,000 to 90,000 Lpd. A perched water table in the Fill HU governs shallow groundwater flow.

The 0.4 hectare lagoon, allowed storage of coal tar up to 3 m above existing grade. This site was found to be the most heavily contaminated area over the Coke Oven Complex. It originates with the above-grade lagoon and tank farm areas. Prior to 1950, a major road under-pass was present under the railway, northwest of the lagoon which, in all probability, intercepted bedrock. DNAPLs could migrate into this sump and enter the bedrock along the Canso/Morien contact. Free DNAPL of up to 1 m in thickness was noted in four piezometers. A sample (Table 1) indicated total BTEX (65 ppm), PAH (1.24×10^6 ppm) and HNC (48,500 ppm). Within BTEX, xylenes dominated, as in the raw coal tar, but at significantly reduced concentrations. The two PAH compounds with the highest concentration were identical with those in the tar. The top three HNCs were similar to that found in the tar. Both the PAH and HNC's were enriched in the DNAPL up to three orders-of-magnitude.

The physical properties of the DNAPL are quite different from the coal tar; being considerably less viscous (5.8 versus 1006.1 cSt) and exhibiting a lower specific gravity (1.075 versus 1.225). Piezometers near both sources indicated decreasing contamination with depth. However, at PN 88-5 contamination

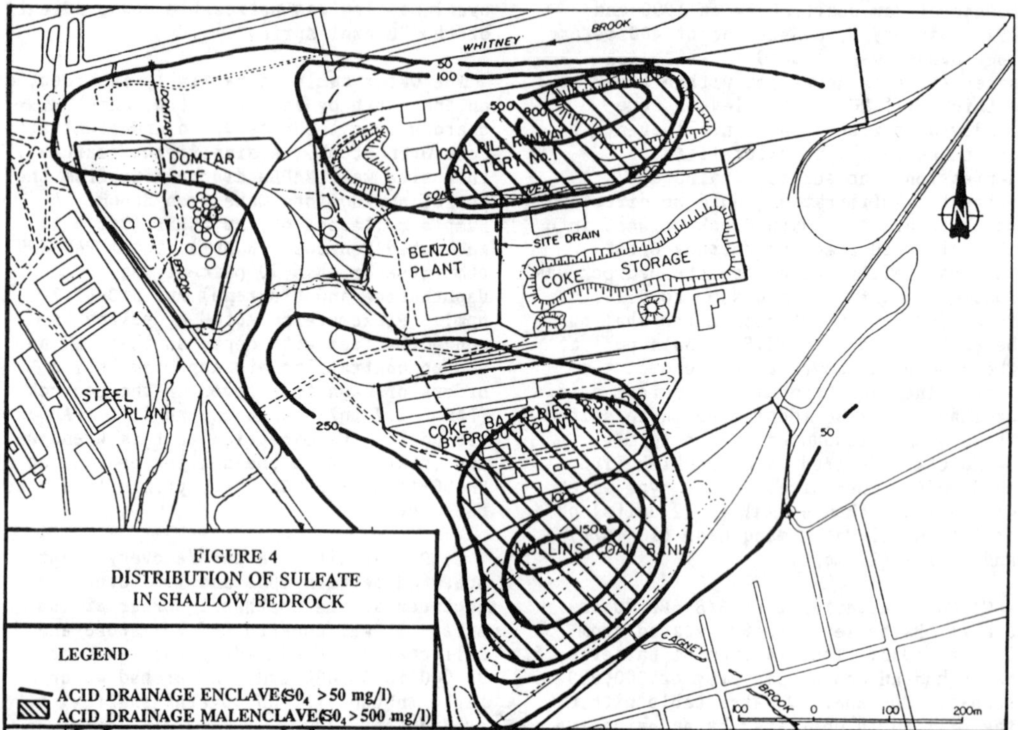

FIGURE 4
DISTRIBUTION OF SULFATE IN SHALLOW BEDROCK

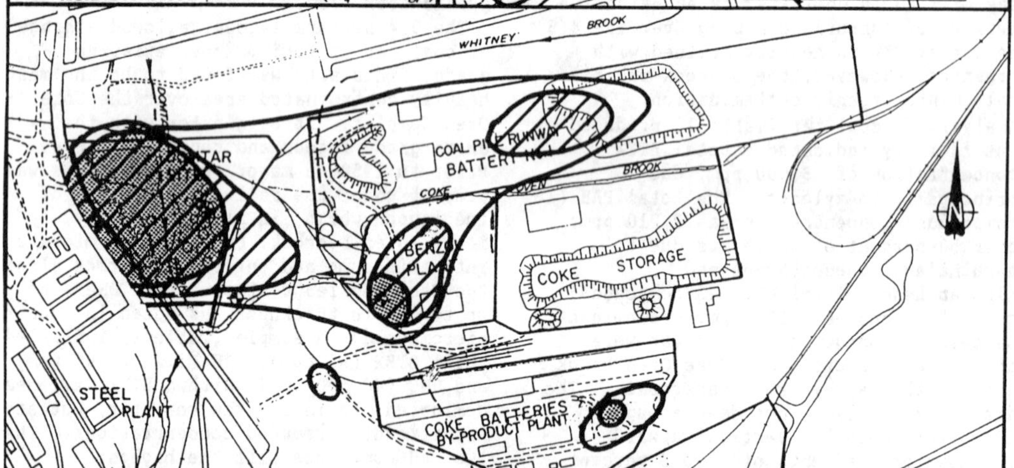

FIGURE 5
DISTRIBUTION OF BENZENE IN SHALLOW BEDROCK GROUNDWATER

increased with depth. The lowest piezometer was set at the Morien-Canso HU contact to detect DNAPL migration. Although no DNAPL was present, BTEX increased in depth from the fill (3.8) to the contact (1116 µg/l); xylenes dominated each site. Total PAH at the shallow bedrock (2.58 µg/l) increased to 1080.95 µg/l at the contact where naphthalene and acenaphthlene dominated. HNCs showed a

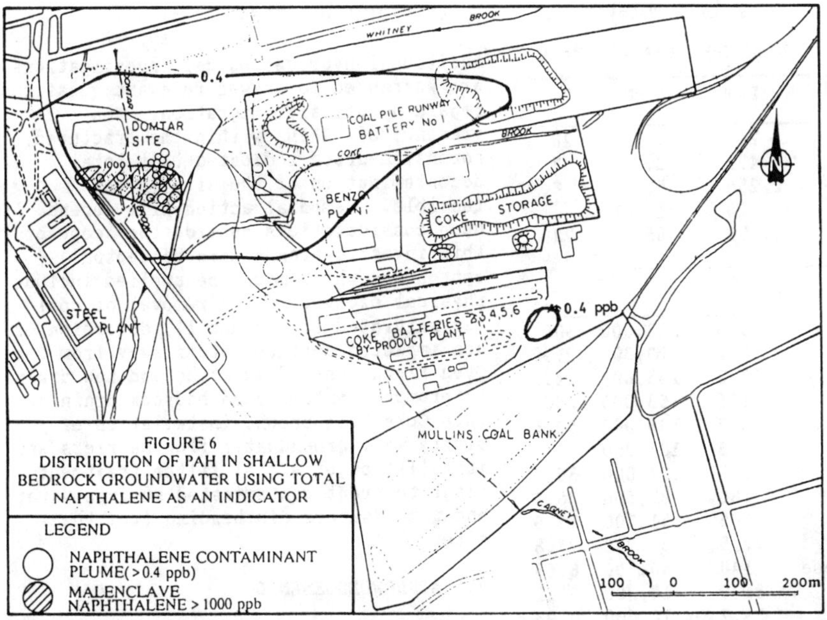

FIGURE 6
DISTRIBUTION OF PAH IN SHALLOW BEDROCK GROUNDWATER USING TOTAL NAPTHALENE AS AN INDICATOR

similiar trend with increases from 3.76 to 199.05 µg/l; carbazole dominated.

5. CONTAMINANT ENCLAVES

The overall extent of contamination is plotted (Figures 4 to 6) for the shallow bedrock since contaminants could move more quickly off site in this zone without discharging into Coke Oven Brook, and could impact on domestic withdrawals. It was recognized that contaminant transport in fractured bedrock is difficult to detect when utilizing the relatively sparse monitoring network employed at this site.

The concentration of SO_4 delineated acid drainage originating from both the Mullins Bank and the Coal Pile Runway (Figure 4). The enclavé was taken as greater than 50 mg/l background; the malenclave at greater than the 500 mg/l potable standard.

Benzene was utilized as an indicator for monocyclic aromatic hydrocarbons. The enclave (Figure 5) delineated at 1.0 µg/l, encompassed primarily the downgradient half of the Coal Pile Runway/Battery No. 1, as well as the Benzol and Domtar Plants. The major areas surrounded the Benzol Plant peaking at 96,000 to 107,700 ug/l and the Domtar site at 6,000 to 24,100 µg/l. The shape of the malenclave in the Domtar Plant area does not agree with the piezometric surface (Figure 3) since it is moving due west onto the Steel Plant, not south toward Coke Oven Brook.

Napthalene was utilized as an indicator for PAHs. The enclave (Figure 6) delineated at 0.4 µg/l, encompassed a similar area to the Benzene enclave. Given the absence of drinking water standards, the malenclave was taken at 1000 µg/l. If the World Health Organization limit for total PAH is 0.2 ug/l, the malenclave expands to encompass the entire enclave.

Total HNC concentrations were monitored only at six sites in the shallow bedrock. The malenclave defined arbitrarily at 1,000 ppb delineated a very narrow zone under the Domtar site.

In summation, the shallow bedrock exhibits significant organic contamination over 18 hectares principally under the processing (Benzol) and storage (Domtar) areas not the coal storage or actual Coking Ovens. However, the old production area, Battery 1, still shows evidence of contamination some 70 years after operation ceased.

To the extent of the investigation, the organic plume appears to be staying within the site boundaries and migrating to the west beneath the Steel Plant. The effect of the Canso HU on limiting westerly contaminant flow, with possible diversion to the north parallel to the contact, is unknown. The limit of organic contamination at depth has not been established.

Table 1. Pertinent Organic Chemistry

COMPOUNDS	COAL TAR	DNAPL	LNAPL
VOLATILE COMPOUNDS	ppm	ppm	ppm
Benzene	247	0.7	28
Toluene	493	2.2	*
Xylenes	2,250	50	*
Ethylbenzene	15	13	
Total BTEX	3,000	65	28
PAH COMPOUNDS			
Naphthalene	573	213,000	309
Acenaphthylene	25.2	29,300	2.98
Acenaphthene	13.8	145,000	14.2
Fluorene	125	153,000	260
Phenanthrene	422	196,000	52.6
Anthracene	64.5	106,000	--
Fluoranthene	255	124,000	56.9
Pyrene	186	87,500	35.8
Bz(a)anthracene	105	51,700	13.9
Chrysene	77.5	37,200	27.8
Bz(b,k)fluoranthene	148	50,100	8.65
Bz(e)pyrene	31.9	10,200	3.11
Bz(a)pyrene	56.7	18,300	0.93
Indenopyrene	32.1	10,100	1.78
Bz(ghi)perylene	29.7	7,950	1.15
Dibenz(a,h)anthracene	63.3	1,750	0.06
Total PAH	2,210	1,240,000	789
HNC COMPOUNDS			
2,4-lutidine	1.5	--	--
3,5-lutidine	0.1	--	--
benzonitrile	0.114	--	--
3,4-lutidine	0.258	--	--
quinoline	8.06	178	--
isoquinoline	5.06	843	--
2-Me-quinoline	2.74	1,400	0.04
7-Me-quinoline	2.14	564	0.06
4-Me-quinoline	2.2	143	0.06
2,6-DiMe-quinoline	1.42	2,390	--
2,4-DiMe-quinoline	1.60	2,070	0.10
DiMe-quinolines	2.26	2,240	0.16
cyanonaphthalene	0.122		0.09
7,8-benzoquinoline	4.48	3,150	--
benzoquinolines	6.48	1,890	0.19
acridine	3.68	1,480	0.17
carbazole	47.8	22,900	0.52
9-anthr.carbonitrile	1.54	622	0.17
anthr.carbonitriles	12.1	2,840	--
benzocarbazoles	3.16	886	--
benzocridines	11.7	3,160	0.08
Me-benzocarbazoles	0.948	849	--
Me-benzocridines	2.38	923	0.07
Total HNC	122	48,500	1.71

-- not detected, less than 0.04 ppm
* present in minor amounts, mostly alkanes (C_{14} and up); similar composition to fuel oil.

6. REMEDIAL MEASURES

Additional hydrogeological investigations are warranted to answer relevant questions regarding off-site migration. The presence of DNAPLs within the fractured, porous, bedrock indicated that total decontamination of the site is not feasible. Remedial action will depend upon considerations outside the scope of this investigation. Some conceptual approaches which might be applied include: physical excavation and removal of coal tar saturated fill and till, diversion of non-impacted surface waters away from Coke Oven Brook, use of existing and new drains to promote contaminated bedrock drainage into Coke Oven Brook, installation of slurry wall groundwater flow barriers and installation of a coal filter treatment plant to treat contaminated surface water and groundwaters discharging from the site.

7. ACKNOWLEDGEMENTS

Funding for this investigation was provided through the auspices of Environment Canada and the Nova Scotia Department of Environment. Acres International Limited are responsible for project management. Drs. J. Cherry and J. Barker of the University of Waterloo, as well as D. Belanger of Interra Technologies Ltd. (now with CH2M Hill Engineering Ltd.) were scientific advisors to the study team and critically reviewed the manuscript. B. Bennett of Acres also provided assistance and review.

8. LIST OF REFERENCES

Acres International Ltd., 1985, Sydney Tar Ponds Study - Final Project Report to the Nova Scotia Department of Environment, 3 vols.

Subsurface Contamination by Immiscible Fluids, Weyer (ed.) © 1992 Balkema, Rotterdam. ISBN 90 6191 175 3

Remediation of subsurface gasoline losses at two retail outlets in Southern Ontario

G. Westerby
Gartner Lee Limited, Markham, Ont., Canada

ABSTRACT: Two cleanup case histories are presented for sites with gasoline losses from retail service station outlets. Both stations are located in urban locations with similar patterns of adjacent land use. The hydrogeological settings and the magnitude of the resulting off-site environmental impacts were, however, very different. Hence, there were significant differences in the approach to the investigation and remediation of the problems. The importance of considering potential alternatives to the traditional "dig it up and truck it away" approach to remediation is stressed. The case histories review not only the environmental and technical issues to be addressed, but also present some of the other aspects of the project which are generally very important, including regulatory requirements, legal issues, and liaison with local residents and the public. The discussion of results has been updated to the summer of 1991 and hence includes data that were not available at the time the paper was originally presented.

1 CASE HISTORY NO. 1

1.1 Background and site setting

This case history reviews the investigation and remediation program following a substantial subsurface loss of gasoline from a retail service station. It is located at the intersection of two major roadways in a high density residential area. It is immediately adjacent to two large high rise residential condominium and apartment buildings. The edge of the underground parking garage for the condominium is located within about 15 m of one edge of the service station property and a children's play ground abuts the other edge of the property. In preliminary discussions with the condominium owner it was found that there were plans to extend the underground garage near the site to construct a maintenance room. Hence, concerns about potential product migration in this area were a high priority. Land use across the intersection includes single family residential, high rise apartments, and some commercial development. A site plan for the service station is provided in Figure 1.

To get a quick first assessment of the physical site setting and an indication of potential impacts of the product loss, all available geological and hydrogeological reports and maps, recent and historical air photos, and subsurface utility plans were obtained and reviewed. These available data were quickly supplemented by an on-site investigation using test pitting and drilling as described later in this paper.

It was found that the site had been developed as a service station in the mid 1950s and prior to that an orchard existed at this location. Considerable filling and subsequent drainage modifications had occurred since the 1950s. The area is now fully serviced and there is no known ground water use in the area. All nearby underground services were located and were found to be above the water table.

In general the site stratigraphy, for the upper 10 m from the surface, includes asphalt or concrete over most of the surface, varying thicknesses of gravelly sand fill, a sandy silt with clay layers, a fractured sandy silt till with layers of fine sand, underlain by stratified silt and sand. The water table was typically found at a depth of about 5 to 6 m below grade within the till. Water level monitoring indicated ground water flow to be generally diagonal across the site from southwest to northeast.

1.2 History and investigation of product loss

The service station owner first suspected a loss of product in February 1989, when a review of the product inventory records

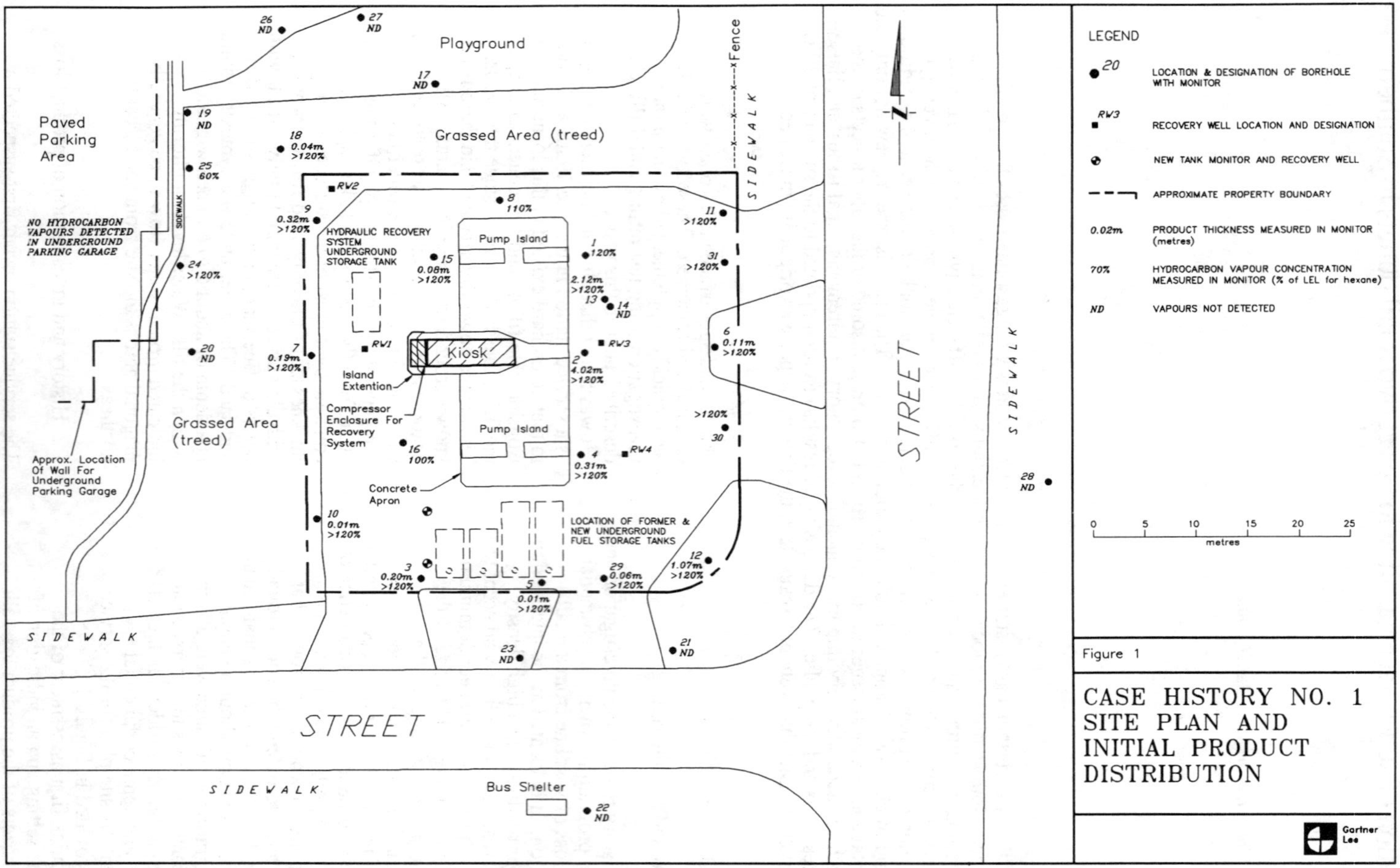

Figure 1

CASE HISTORY NO. 1
SITE PLAN AND
INITIAL PRODUCT
DISTRIBUTION

suggested a possible loss of approximately 10,000 L. Subsequent product recovery suggests that this estimate may have been low. The owner advised the Fuels Safety Branch of the Ministry of Consumer and Commercial Relations (MCCR) and the Ministry of the Environment (MOE). The MCCR issued an order to pressure test the tanks and lines, and to remove all contaminated soils from the site. No pressure loss was reported in the tests on the four tanks but the product delivery lines were found to be corroded and leaking at three different locations close to one of the pump islands.

The confirmation of the product loss was quickly followed by an intensive subsurface investigation using both backhoe test pits and drilled boreholes. Continuous, or nearly continuous, soil samples were taken from all boreholes and logged on-site by visual examination. All soil samples were checked immediately for the presence of volatile organic vapours using a portable photoionization meter.

Selected samples were submitted for laboratory analysis. Fully screened monitoring standpipes were installed in all holes to allow monitoring of fluid levels and soil vapours and to facilitate hydraulic testing and ground water sampling. All monitoring wells were surveyed to establish geodetic elevations. Monitoring well 14 is a piezometer installed through casing set below the water table. Recovery wells, consisting of perforated or slotted 0.3 m diameter corrugated steel pipe, were installed in selected backhoe test pits. The recovery wells ranged in depth from 6.4 to 7.6 m. A total of 31 monitoring wells and 4 recovery wells were installed (Figure 1).

Drilling commenced around the pump islands and stepped out as required to confirm the extent of migration of free and dissolved product and product vapours. This involved meetings and negotiations with private property owners for access to their property and with the Region and the Municipality for authorization to drill and install monitoring wells on their respective road allowances. All accessible subsurface utilities, including sewer catch basins and manholes and utility vaults were checked for the presence of hydrocarbon vapours and none were detected.

The site investigation showed that free liquid product was present on the water table over a large portion of the station area. It ranged in thickness (as measured in the monitoring wells) from barely measurable (less than 1 mm) to in excess of 4 m. The greatest thicknesses were measured close to the area where the line leaks were identified. A fringe area with product vapours in the unsaturated zone was identified, with levels ranging from less than five percent to greater than 100 percent of the lower explosive limit (LEL). The vapours typically extended only a few metres beyond the area where free product was found. Figure 1 shows product thickness and combustible vapour levels measured in the monitoring wells at an early stage of the investigation. The migration of the product appeared to have been limited by the relatively low hydraulic conductivity of the soils. It also appeared as though the pattern of product migration was being influenced by stratigraphic controls (more permeable sand lenses within the till) as well as by the water table configuration.

Monitoring showed that no free product or vapours had reached beneath the road to the south of the station. The monitoring data along the road to the east suggested that there may be a small quantity of product beneath the sidewalk and possibly part of the road. Full delineation of the extent of product migration in this area was limited by the presence of numerous subsurface utilities which precluded the selection of safe drilling locations. Drilling and monitoring well installation across the road showed no evidence of either free product or vapours.

The concerns were an area where free product was identified on the adjacent private property to the northwest, and the measurement of combustible vapours in the subsurface near an underground parking garage to the west. A monitoring program for combustible vapours was implemented in the nearby underground parking garage, both with and without the ventilation system operating. No vapour levels above the instrument detection limit have been identified to date in the garage. Most vapour monitoring referred to in this paper was done with portable combustible gas sniffers calibrated to hexane. The minimum level of combustible vapours that can reliably be detected with this instrument is about 5 percent of the LEL for hexane.

1.3 Site remediation

Following confirmation of the product loss, the MCCR directed the station owner to remove the tanks and contaminated soils from the site. There had not been any widespread off-site product migration but, because of the depth and extent of product distribution in the subsurface, this would have involved a major excavation. It would have covered almost the entire site area and would have had to extend onto both the adjacent private property and the road allowances. The excavation might have involved up to 30,000 t of soil and the landfill tipping fees for soil disposal could have been well in excess of 2 million dollars, even assuming that some segregation of clean soils was possible. The excavation and trucking costs, backfilling, station reconstruction, engineering design and site supervision would have added significant additional costs.

In reviewing other options for site remediation, the factors that had to be considered included:

- the on-site soils had effectively limited the extent of product migration so that there were no significant off-site impacts;
- the costs of excavation and station reconstruction would be very expensive, time consuming and disruptive;
- the unnecessary use of valuable landfill capacity for disposal of large volumes of slightly contaminated soil is not viewed favorably by regulatory agencies or environmentalists; and
- the owner wished to get the station back into service as quickly as possible, while ensuring that the necessary remedial measures were implemented to restore the site to a safe condition.

A range of options were considered, including various combinations of limited excavation and off-site soil disposal, together with a number of methods for containment and recovery of the lost product. Because of the relatively impermeable soils, the substantial accumulation of free product, and the increasingly stringent regulatory controls on air emissions, vapour extraction was not initially considered to be a viable option. Testing of the recovery and monitoring wells indicated that it would be feasible to create a sufficient drawdown to contain and recover a substantial proportion of the lost product using a simple hydraulic recovery operation.

A proposal for a conceptual remedial action plan was submitted to the MCCR and the MOE. The remediation plan was based on a hydraulic recovery operation and ongoing monitoring combined with a limited excavation of the most highly contaminated soils. Because of the low yield of the till, separate water table depression pumps were not considered to be necessary and a total of four recovery wells with single pneumatically operated pumps were proposed. The pump intakes were to be set just below the fluid surface so that they would both skim off the floating product as well as pumping sufficient ground water to create an adequate drawdown cone. Because of the low volumes of fluid that would be pumped and the complexities and timing delays associated with the design and approval of an on-site ground water treatment system, it was proposed that the recovered fluid mix would be pumped into an on-site buried storage tank for removal and treatment by a licensed waste hauler.

The proposal was approved by both regulatory agencies and the MCCR issued a deviation from their order to remove all contaminated soils. Throughout this period of site investigation and remedial plan development and negotiation, a manual recovery program was conducted to remove as much product as possible from the recovery and observation wells. During this period, the volume of recovered product was estimated to be about 3,500 L.

During the design, approval and installation of the hydraulic recovery system, other aspects of the station reconstruction were also undertaken. The old tanks and lines were removed and selected areas of soil were excavated. All tanks and lines were replaced with state of the art equipment for compliance with the Ontario tank upgrading requirements that came into effect at the beginning of 1991.

The recovery wells were deepened and granular filled trenches were extended radially outward from the wells to improve their hydraulic recovery. Concrete manholes were installed around the recovery wells so that they could be completed flush with grade. The manholes were all interconnected with buried conduits through which the control lines, air lines, and discharge lines to and from the pumps were installed. A 23,000 L fiberglass tank was installed as a holding tank and a tank level sensor and system shut down control switch were installed. An extension was constructed on the end of the existing kiosk to house the air compressor, air dryers and filters, and the pump control panel.

The pumping system, including all associated lines and controls, was installed and put into operation about six months after the initial product loss was detected and the following month the station reconstruction was completed and it reopened for business. The hydraulic recovery system has been in operation since that time.

1.4 Results

As with any new system, some initial startup problems were encountered with the equipment but these were addressed by the various equipment suppliers. Over the first few months of operation, some manipulation of the pumping rates and adjustment of pump intake levels was required to balance the system and obtain and maintain an appropriate drawdown cone. The system was set up so that it maintains a drawdown on the order of about 3 m at each of the recovery wells and the combined effect of this pumping is to create a drawdown cone across the central portion of the site. This creates an inward flow from all directions around the site perimeter. Approximately 1,121,000 L of water and product have collected since recovery operations began, and the estimated liquid product recovered is about 15,100 L. The proportion of product to water in the recovered liquids has declined significantly since the recovery operations began and at present it is estimated that less than 0.5 percent of the recovered liquid is product. The content of dissolved hydrocarbons in the recovered water has not been monitored.

The measured product thicknesses have declined greatly across the site and there is no longer any measurable free product in any off–site monitoring wells. A few on–site wells (2, 6, 9 and 13) still show occasional occurrences of free product but it is typically no more than a few millimetres thick. Figure 3A is a plot of product thickness versus time as measured in Monitor 13, located near the thickest part of the product plume. This graph is typical of the response in most monitors and shows that the hydraulic recovery operation has been quite successful.

A number of the on–site monitoring wells still show vapour levels close to LEL. While the system was not designed to reduce vapours in the vadose zone, there has been a significant reduction in off–site vapours on the adjacent private properties. Only one off–site monitor where there had previously been free product present still has low but detectable vapour levels. Several monitors in the vicinity of the underground parking garage previously had petroleum hydrocarbon vapour levels close to the lower explosive limit. Vapours in these monitors are now nondetectable.

1.5 Summary

A detailed subsurface investigation was conducted at the site of a gasoline loss from a service station to delineate the extent of product migration and to develop an understanding of the factors controlling its distribution. Based on the knowledge of the site developed from the investigation, a remedial action plan based on hydraulic recovery was proposed as an alternative to the regulatory agency order to remove all contaminated soils. The plan was approved and implemented and the recovery system is now operating successfully. The station has reopened and the ongoing recovery operation does not interfere with normal station operations. Throughout the project, liaison was maintained between the station owner, the consultant, adjacent property owners, regional and municipal officials, and the regulatory agencies. It was found that this open communication was helpful in assuring that the site investigation and remediation progressed smoothly.

The automated hydraulic recovery system has been in operation for over two years and is now reaching the point of diminishing returns. Only thin, isolated occurrences of free product are now left on–site and the off–site vapour concentrations have been greatly reduced. It is felt that the costs of the ongoing hydraulic recovery and off–site fluid disposal operation are no longer justifiable and that it may now be appropriate to terminate the hydraulic recovery. The installation of a vapour extraction system (VES) as a final stage of the cleanup or as a contingency to be implemented if ongoing monitoring shows it to be necessary, is now being evaluated. A request for the appropriate modifications will be made to the regulatory agencies in the near future.

A long term monitoring program will be continued to ensure that the recovery operation is optimized and that there continue to be no significant environmental impacts or health and safety concerns. The long term monitoring results will be used to assess when the remedial work should be modified or terminated.

2 CASE HISTORY NO. 2

2.1 Background and site setting

This case history reviews the investigation and remediation of another gasoline leak at a retail service station. As in the first case history, this site is also in a residential area. The subsurface setting and resultant product distribution, however, are very different at the second site, resulting in a longer and more complex site assessment and remediation program.

The retail station is located near a local topographic high in an area of glacial outwash deposits. The site is underlain by up to about 14 m of fine to coarse sand and gravel, above a silty clay. The sand and gravel deposits are moderately to highly permeable whereas the

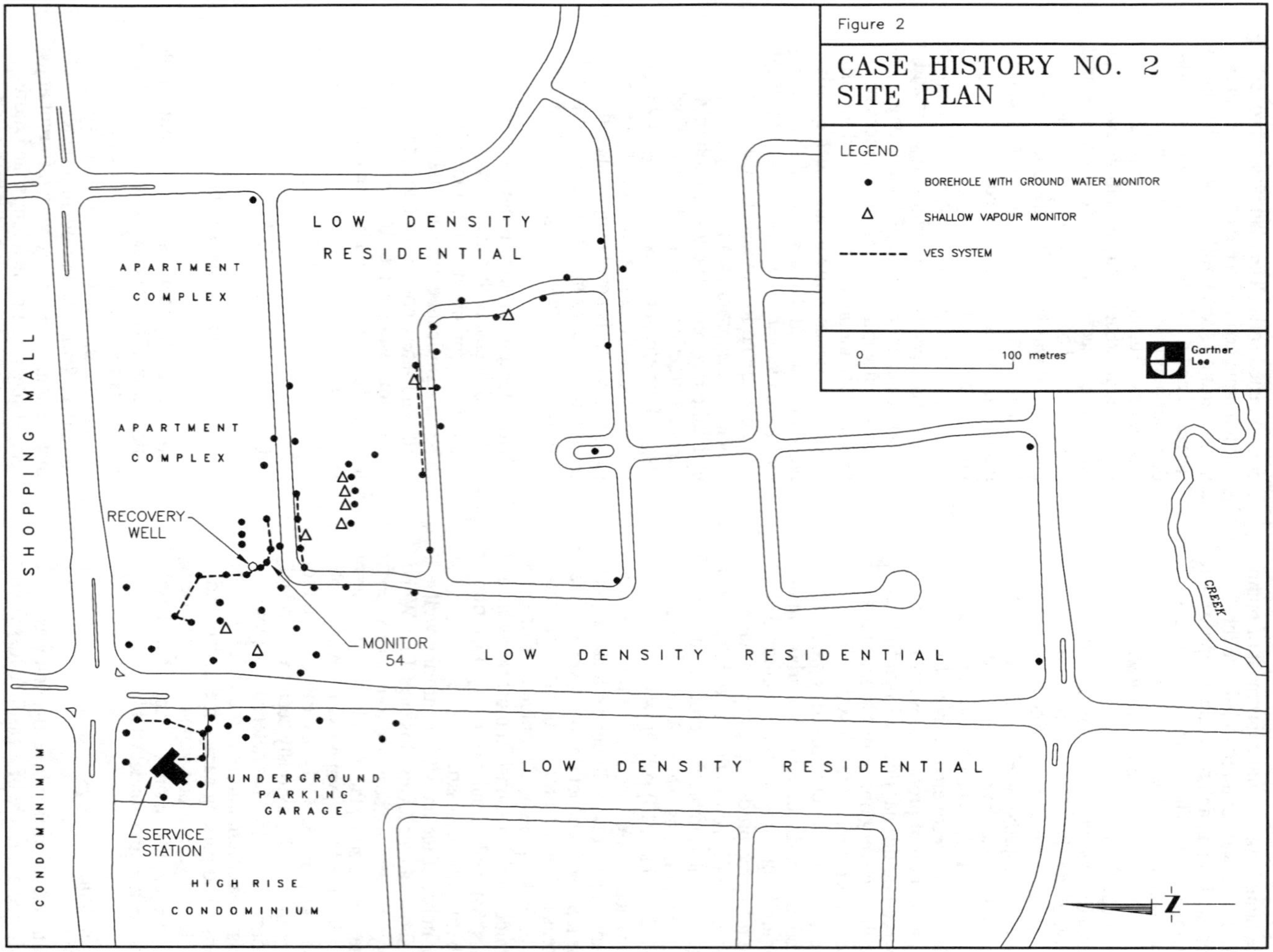
Figure 2
CASE HISTORY NO. 2
SITE PLAN
LEGEND
BOREHOLE WITH GROUND WATER MONITOR
SHALLOW VAPOUR MONITOR
VES SYSTEM
0
100 metres
Gartner Lee
LOW DENSITY RESIDENTIAL
APARTMENT COMPLEX
APARTMENT COMPLEX
SHOPPING MALL
RECOVERY WELL
MONITOR 54
LOW DENSITY RESIDENTIAL
LOW DENSITY RESIDENTIAL
UNDERGROUND PARKING GARAGE
SERVICE STATION
HIGH RISE CONDOMINIUM
CONDOMINIUM
CREEK
N

silty clay layer below has a much lower hydraulic conductivity. Bedrock in the area is shale and, based on auger refusal in several holes, it is inferred to be present a few metres below the silty clay. Bedrock was not confirmed by coring. The water table in the vicinity of the service station was at a depth of about 13 m.

The detailed site investigation program identified a narrow, nearly linear channel with coarse granular deposits overlying the silty clay. This channel was close to the depth of the water table and played a major role in controlling the product migration. Its identification and delineation was very important to the success of the remedial program.

This service station is also located at a major road intersection, immediately adjacent to a high rise condominium with a very large underground parking garage. One segment of the garage was immediately downhill from the service station and within about 30 m of the tank that was found to be leaking. Land use on other corners of the intersection includes another large condominium, a shopping mall (with another gas station), and a high rise apartment building. Large residential subdivisions are located a short distance topographically downhill from the service station. The area is fully serviced and there is no known ground water use in the area. The nearest surface water is a small creek about a kilometre downhill from the station. A site plan for the study area is shown in Figure 2.

2.2 History and investigation of product loss

The station operator reported a suspected product loss from one of the underground storage tanks to the station owner in late September, 1988. The tank was exposed and removed the following day, and a small hole was found beneath the area of the fill pipe. The tank was located in dry sandy soil and no evidence of soil contamination was reported by the contractor at the time of tank removal. The station owner advised the MCCR and the MOE of the suspected product loss.

It was found that the station operator had not been maintaining adequately detailed product inventory records so the volume of the product loss was not accurately established. Based on the available data, however, the volume of the product loss was initially estimated at roughly 60,000 to 70,000 L. Subsequent product recovery operations suggest that this initial estimate was low.

A drilling program was commenced to investigate the product loss the day after the tank was removed. Because of the coarse granular nature of the soils and the relatively deep water table a considerable time was required to drill, sample, and install a monitor in each borehole. To expedite the initial investigation, two drill rigs were mobilized to the site and operated seven days per week.

The early drilling was performed around the site perimeter and in the first eight boreholes drilled, product was found at only one location near the lowest (southeast) corner of the property. At this time, the MOE was advised of the probability that product had migrated off–site. The drilling then stepped out, focusing initially on the area topographically downslope from the station, including the area adjacent to the underground parking garage. Liaison was established with the condominium owner, both to get access for drilling and for regular monitoring of the underground garage.

This early drilling downslope from the station failed to detect any widespread off–site product migration. The drilling then fanned out into a broader pattern that extended across the road onto an apartment building property, and into the residential subdivision several hundred metres beyond. This again involved liaison and negotiation with both individual home owners, the apartment building owners, occupants, lawyers and the property manager, and the regional and municipal authorities. This was necessary to advise them of the situation and to obtain the necessary approvals for drilling and monitor installation. In some cases, it was also necessary to make arrangements for ongoing access for monitoring and remedial operations.

The extended drilling pattern gradually began to reveal the presence of the buried channel mentioned previously. Initially, however, it was difficult to define because it was over 10 m deep and often appeared to be only a few metres to a few tens of metres wide. By late October, 1988 over 50 boreholes had been drilled, each with a ground water and soil vapour monitor installed. While the full extent of the product migration had not yet been defined, the drilling and monitoring results suggested that the buried channel was the primary migration pathway.

The leading edge of the liquid product plume appeared to be on the order of 150 m downgradient from the service station property and it was moving predominantly below the back yard of the apartment building property. There was evidence of some product migration beneath two residential properties along the regional road across from the service station. The liquid product plume had not reached the main area of the subdivision, south of and downgradient from the apartment building. Approximately a month after the investigation began, the free product plume was well enough defined to commence a remedial action plan.

During October, the station owner removed the remaining tanks and piping. No other leaks were found. They were replaced with upgraded systems and permission was received from MCCR to resume operations at the station while the off–site investigation and the remediation work continued. Because of the complex and varying nature of the migration pathways for free product, product vapours, and dissolved product, the subsurface investigations continued in conjunction with the implementation of the site remediation work. This was necessary to better define the full extent of the various phases of the product distribution and to address both potential health and safety concerns of area residents and concerns about potential environmental impacts.

Additional site investigations and monitoring continued through 1989 and into 1990. Drilling was performed in the yards of a number of residential properties to allow more detailed monitoring of the effectiveness of the various remedial measures that have been implemented. A total of 91 boreholes were drilled and monitors were installed in most of these. The borehole locations are shown on Figure 2. The drilling extended more than half a kilometre from the service station to confirm that there was no dissolved hydrocarbon plume reaching the creek south of the service station. Most monitors extend to, and are screened across, the water table to allow monitoring for fluid levels, free liquid product, ground water sampling for dissolved product analysis, and monitoring of vapour concentrations above the water table. A number of shallower monitors have also been installed to depths of 3 to 6 m in the vicinity of, and on, the private properties along the buried channel to confirm that any hydrocarbon vapours present were staying at depth, close to the water table.

The site investigation work produced a good delineation of the small buried channel and confirmed that most of the product migration had been along or very close to this channel. The vapour migration extended further along and somewhat laterally outward from the channel. The plume of dissolved hydrocarbons moved preferentially along the channel but also moved laterally outward from the channel axis.

The maximum extent of free liquid product migration was found to be about 150 m from the service station, just short of reaching a large residential subdivision. No evidence of free product was found beneath the subdivision. Hydrocarbon vapours were found in the unsaturated zone above the water table considerably further downgradient beyond the maximum extent of free liquid product. Vapour concentrations close to or above, the lower explosive limit were found at the water table as far as about 300 m from the service station. Trace vapour concentrations were found as far as 350 m from the station. The shallow monitors mentioned previously were installed to confirm that there was no significant upward migration of vapours toward the basements of the houses. The highest vapour concentrations were found to be restricted to an interval of less then about a metre above the water table.

The water table depth was greatest at the service station where it was found to be approximately 12 to 13 m below grade. Near the upgradient side of the subdivision, the depth to the water table was about 10 m below grade and near the leading edge of the vapour plume it was about 8 m below grade. None of the basements examined extended more than 3 m below grade and no measurable vapours have been detected in any of the residential basements being monitored.

Dissolved hydrocarbons in the 1 to 10 mg/L range have been detected up to about 450 m downgradient of the service station but dissolved hydrocarbon concentrations decrease rapidly beyond the limit of the free product. No detectable (less than 3 µg/L) dissolved volatile organic hydrocarbons have been found in ground water samples taken from three monitors close to the creek, at distances of about 550 to 650 m from the station.

A long term monitoring program for free liquid product, product vapours, and dissolved product was implemented to assess the effectiveness of the site remediation work and to watch for any unexpected changes.

2.3 Site remediation

Because of the evolving nature of our understanding of the extent of the various phases of product migration, the site remediation plan was developed in stages. As areas of concern were identified and delineated, remedial measures were designed and implemented to address them. Whereas excavation was one of the options considered in the first case history, because of the depth and extent of product migration under roads and buildings, excavation was clearly not feasible at this site.

The first stage in the remediation plan was to install a 200 mm diameter recovery well in the back yard of the apartment building property across the street and about 110 m downgradient from the service station. The well was drilled in October 1988, when the leading edge of the plume of liquid product was identified. A meeting was held with the MOE to discuss the proposed hydraulic recovery plan. Approval to start pumping and dispose of the recovered liquids off–site was received from the MOE at the end of October. A pumping test was conducted to determine the well's capacity and a pumping system was installed. Both a water table depression pump and a skimmer pump for removing floating product were used to enhance the rate of product recovery.

A secure fenced compound was constructed around the well and a bermed and vented steel tank was installed above ground for storing recovered product prior to off–site disposal. Large tanker trucks were stationed beside the well to collect the ground water from the water table depression pump. This water contained dissolved hydrocarbons and was trucked to a nearby refinery facility for treatment and disposal.

The use of an air stripper, carbon adsorption, and a range of newer innovative technologies was considered to treat the ground water for removal of volatile hydrocarbons as an alternative to off–site trucking. Because of increasingly stringent MOE restrictions on benzene emissions to the atmosphere and the complexities of the approvals process necessary to discharge the treated effluent to the sanitary sewer, it was decided that off–site trucking was the preferred alternative.

The ground water pumping rate and the rate of product recovery were limited by the stratigraphic restrictions on the available drawdown in the thin, narrow, channel deposits. A second recovery well was considered for installation closer to the subdivision but following a further review of monitoring data in May 1989, it was concluded that a second recovery well was not required.

By the end of May 1989, the recovery of free liquid product at the recovery well had declined to negligible levels so the use of the two pump system was discontinued. A single pump was used to keep the water table depressed and the pump intake was set and the pumping rate manipulated so that any free product drawn into the well would be removed. By this time, free product could no longer be measured in the full tankers of pumped fluid.

By October 1989, free product could only be detected in one monitor and an assessment of site monitoring data suggested that this was situated in a small pool of residual product contained in a stratigraphic trap. This monitor is only about 20 m from the recovery well and appears to be hydraulically isolated from it. Since September 1989, this monitor was pumped manually on a regular basis and the product thickness is showing a slow but significant decline.

A large number of monitoring wells were used to assess the effectiveness of the hydraulic recovery operation. Monitor 54 was installed in an area of the plume where the thickest free product was encountered just downgradient of the recovery well, during the assessment of the need for a second recovery well. Figure 3B is a plot of product thickness versus time in this monitor and provides graphic evidence of the effectiveness of the recovery operation. The total free liquid product recovered from the start of pumping to July 1990 was slightly in excess of 63,000 L.

In light of the fact that there are no ground water users or surface water bodies in the vicinity, and the rapid decline in dissolved hydrocarbon concentrations in ground water downgradient from the area of the original free liquid plume, a request was made to the MOE to terminate pumping operations at the recovery well and monitor ground water quality. This request was approved and the pumping operation was terminated in July 1990.

The second component in the site remediation was to control the subsurface hydrocarbon vapours. Although the monitoring data indicated that the vapours were concentrated just above the water table, well below the basements or subsurface utilities, the residents were concerned about a possible upward migration of vapours that could potentially create a risk.

The use of vapour extraction systems (VESs) for control of subsurface vapour migration was discussed conceptually with the MOE in late October 1988. The use of a VES was considered to be more applicable at this site than at the site discussed in the first case history because of the more permeable soils at this site. While the MOE was in general agreement with the concept, they expressed concerns about the potential benzene loading to the atmosphere. Although the MOE did not have the final document published, they were

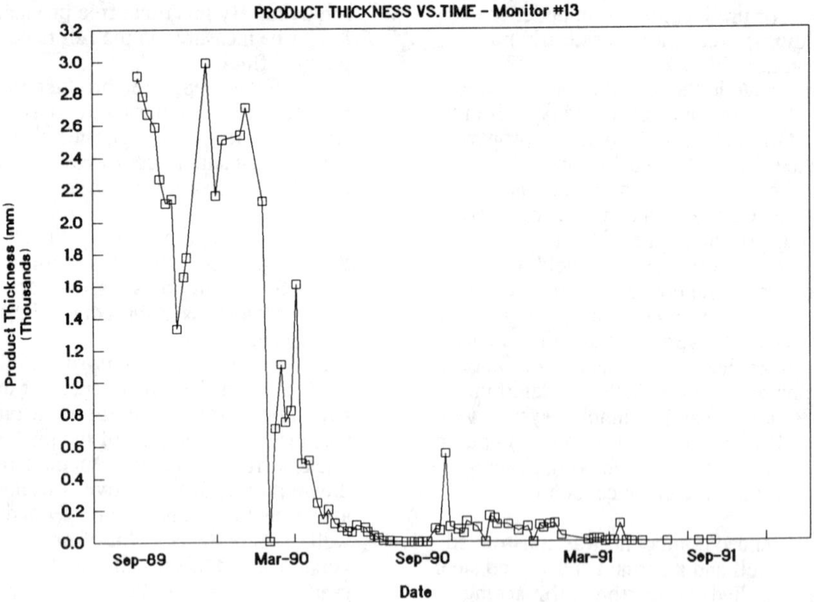

FIGURE 3A - Case History No.1

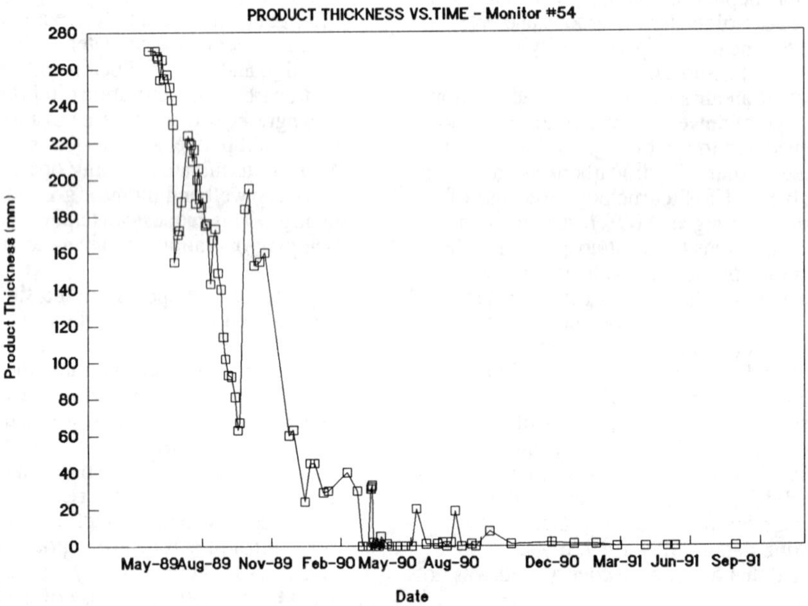

FIGURE 3B - Case History No.2

in the process of tightening restrictions on atmospheric benzene emissions. Hence treatment of the collected vapours using methods such as carbon adsorption or catalytic thermal destruction were considered.

In February and March, 1989, two vapour extraction systems were installed, utilizing a number of the existing monitors that had been installed as part of the site investigation program. One of the systems hooked up 12 monitors on the apartment building property with the highest vapour readings (Figure 2). They were connected by laterals to a regenerative vacuum pump and exhaust stack in the fenced compound where the recovery well was installed. The second system hooked up a number of monitors on the service station property and the pump and stack were installed adjacent to the service station building. All VES components were installed below grade except for the pumps and stacks.

Applications for Certificates of Approval (C of A) were submitted to the MOE to operate the systems but they were not approved because uncertainties about the air emission criteria. This was discussed with MOE in March 1989 but the status of the MOE directive on annual benzene loading rates was not resolved and no further action was taken at that time. By the end of April 1989, however, the monitoring data indicated that vapour concentrations in some of the monitors close to the subdivision were increasing. In mid–May 1989, MOE gave interim approval to operate and monitor the system with no off–gas treatment and it was activated right away. A routine sampling program of the exhaust stack gases was implemented, both to assess the effectiveness of vapour recovery and to calculate atmospheric discharges. The monitoring showed that the hydrocarbons were very rapidly dispersed and there were no complaints about odours from area residents during the operation of the VES.

While the activation of this VES produced some decline in vapour concentrations in monitors near the subdivision, there was still a concern about vapour levels both in the subdivision area and also near the underground parking garage at the condominium immediately adjacent to the service station. Another meeting was held with the MOE in July 1989 and the MOE acknowledged the need to install and activate additional VESs as required. Because of the urgency of the situation, the MOE did not initially require treatment of the exhaust gases. However, a vapour monitoring and sampling program was implemented to assess the need for treatment in the longer term.

The VES at the service station was activated in late July 1989 and another VES was installed and activated in the subdivision area in early August 1989. In October 1989, a fourth VES was installed and activated at a more central location in the subdivision where vapours had been detected. These last two systems were installed on municipal road allowances, utilizing previously installed monitoring wells.

During this time period, the installation of combustible vapour alarm systems in the underground parking garages and in the basements of the houses was considered and details of an integrated alarm system were developed. The alarm system installation was deferred since no vapours were ever found in the shallow monitors in these areas. Detailed laboratory analyses of air samples from selected basements showed only trace amounts of volatile organic carbon which were considered to be representative of background levels.

During September and October 1989, the liaison with homeowners, the apartment and condominium owners, and regional and municipal officials was increased to advise interested parties of the status of the site investigation and remediation. Letters were delivered to property owners and requests were made for access to selected private properties for additional drilling and monitoring. A public meeting was held for area residents early in November 1989 and liaison was established with the local ratepayers group. This public consultation process helped to ease some of the tensions that were developed earlier in the program,.

During December 1989 and January 1990, approval to drill additional boreholes and install monitors on a number of key residential properties was received. A variety of small portable drills were used for this work to minimize damage to the yards. The data from these additional monitors confirmed that the VESs were effectively reducing the subsurface vapour concentrations. The monitoring results were provided to the individual homeowners. Plans were made for another public meeting and a regular newsletter was started to advise the residents of the project status. All of these activities assisted in further improving the relationship with area residents. Somewhat surprisingly, there has been very little media or political attention focused on this project.

Within about two months of activation of the fourth VES, vapour concentrations in monitors throughout the study area showed substantial decreases. By January 1990, vapour concentrations were below the detection limit in most monitors and the highest concentration measured was about 10 percent LEL. All vapour levels remained low to nondetectable from January through March 1990. In late March, a request was made to the MOE to shut down all VESs and monitor vapour concentrations under static conditions. The MOE did not object and all VESs were shut down in the second week of April 1990. An intensive monitoring program was implemented and subsurface vapour concentrations throughout the study area remained low to nondetectable. Monitoring continued at a slightly reduced frequency for over a year.

By July 1991, over a year after the recovery well and VES operations were shut down, the monitoring data continued to show favourable results and a request was made to terminate the remedial and monitoring activities at this site. The initial reaction from the MOE indicated that they were in agreement but final approval for termination has not yet been received.

2.4 Results

The results of our site investigation and monitoring defined a significant movement of product off-site from the location of a leaky tank at a service station. The migration patterns were, to a large degree, controlled by the complex stratigraphy and hydrogeology of the area. Free product was found up to 150 m from the source, product vapours were detected as far as 350 m from the source, and dissolved hydrocarbons in the ground water appear to have migrated up to about 500 m from the source. The investigation of the product distribution in the subsurface was complicated, not only by the complex physical setting, but also by the fact that the various phases of product had migrated under both public roads and numerous privately owned residential properties.

Remedial operations, including both hydraulic recovery and vapour extraction, were designed and implemented and the monitoring results indicate that the cleanup has been very effective. It is estimated that the hydraulic recovery operation has removed over 63,000 L of liquid product and the vapour extraction systems have removed the equivalent of about another 4,000 L. The product recovered in the dissolved phase has not been calculated but would add somewhat to total recovery. The total volume of water recovered was on the order of 18 million litres. These product recovery volumes suggest that the initial estimates of product loss were on the low side.

2.5 Summary

A large scale subsurface investigation involving over 90 boreholes drilled over a period of more than a year was conducted to define the nature and extent of product migration resulting from a leak in a gasoline tank at a service station. The leak resulted in a significant off-site movement of product in liquid, vapour, and dissolved phases. Because of the deep water table and complex hydrogeology, full delineation of the extent of off-site migration was very time consuming.

A remedial action plan based on a combination of hydraulic recovery and vapour extraction was negotiated with and approved by the regulatory agencies. Liaison was established with the regional and municipal authorities and the adjacent property owners. The plan was implemented in stages and now appears to have reached a successful conclusion. In total, it is estimated that over 67,000 L of product was recovered.

3 CONCLUSIONS

A review of these two site cleanup case histories illustrates how the magnitude and complexity of a remediation project can be affected to a large degree both by the local hydrogeological setting and the local land use. The need to properly understand both the hydrogeological setting and all relevant anthropogenic features in both assessing the problem and developing appropriate remedial measures is stressed.

Another very important factor in the development and implementation of remedial measures that is all too often overlooked is the need for early and open liaison, both with the regulatory agencies, nearby property owners, members of the public, and in many cases, the media. Developing these channels of communication early in a project, especially one involving significant off-site impacts, is essential to the smooth running and successful completion of both the investigation and cleanup stages of a project.

Finally, it is emphasized that even though regulatory agencies may still commonly issue orders to "remove all contaminated soil" from a site, full consideration should be given to a range of other viable remediation options. There are numerous tried and proven technologies as well as a range of newer emerging technologies that offer solutions which may not only be more environmentally friendly, but may also be less disruptive and more cost-effective.

9 Legislative and regulatory issues

Subsurface Contamination by Immiscible Fluids, Weyer (ed.) © 1992 Balkema, Rotterdam. ISBN 90 6191 175 3

Legal framework for remediation and prevention of ground water pollution in the Federal Republic of Germany

Christian Vorreyer
Federal Ministry for the Environment, Nature Conservation and Nuclear Safety, Bonn, Germany

ABSTRACT: In the Federal Republic of Germany ground water contamination is considered a disturbance of public safety and order and is therefore prosecutable. Regulatory authorities are obliged to enforce clean-up of ground water contamination. Concurrently, the solicitor general starts proceedings against the polluter. Generally, the polluter has to carry the costs for clean-up. If the polluter cannot be identified or has limited means, then the owner of the land could be held responsible. If the source of the contamination cannot be determined, costs for remedial action are carried by the municipalities. The Water Act states that the former condition of ground water quality must be restored. As this is often impossible, a conflict between reality and law arises. In a compromise, regulatory authorities decide on the remediation goal and the methods to be used. Intending to promote prevention rather than clean-up, German water legislation, in 1956, started to establish precautionary regulations and specifications for private and industrial storage, handling and use of substances hazardous to water. Important points of this legislation are outlined.

1 INTRODUCTION

The legal regulations in the Federal Republic of Germany relative to the prevention and remediation of ground water contamination may be divided into the following sectors:

- Regulations governing the intervention of the public authorities in case of damage
- Regulations in Criminal Law
- Regulations governing liability for damage
- Regulations governing the safety of installations using substances hazardous to water (anticipatory action).

2 REGULATIONS FOR INTERVENTION IN CASES OF DAMAGE

The Federal Republic of Germany is a state that is federally structured. The authorities act within the framework of the Federal Water Act (Wasserhaushaltsgesetz) and the Water Acts of the individual Federal states, known as Länder, in conjunction with the police and administrative law.

The Federal Water Act prescribes that installation in which substances hazardous to water are manufactured, treated, used, stored, transported or filled, must be constructed and run in such a way that no ground water is polluted. In addition, the discharge of substances into ground water may only be permitted if ground water is not detrimentally affected. Ground water, therefore, is subject to particular protection. Thus the pollution of ground water represents an offense against public safety and order and the responsible authority is obliged to take action.

The police and administrative law of each individual state form the basis of the way in which the authorities remediate damage. The regulations vary somewhat from state to state, but the most important regulations are similar throughout the whole country. As a rule, the Lower Water Authority, i.e. the district (Kreis), is responsible for designing the measures to be taken. Water Management Authorities (Wasserwirtschaftsämter) and the relevant state agencies concerned are brought in as consultants.

The necessary measures to remediate damage are to be set in motion by whoever caused the ground water pollution, the so-called "Handlungsstörer", "disrupter by virtue of an action".

The polluter must also bear the costs entailed in this action. The authorities are at liberty to turn to the so-called "Zustandsstörer", or "disrupter by virtue of possession", that is whoever owns the land or the object from which the pollution of the ground water emanates, even if he has nothing to do with the contamination.

In principle, the authorities have the right to

decide freely in such matters. As a rule, though, they turn to the polluter. However, their decision has to take into account the fact that the danger is to be remedied as soon and as effectively as possible.

This may mean the involvement of the land owner, because he may be in a better position financially and because he is the first person able to be identified or contacted when a speedy decision is required, for example, when acute danger is posed to a drinking water supply.

In the Federal Republic of Germany, the "polluter pays principle" is applicable. The land owner, however, is obliged to pursue his demands on the polluter by procedures under civil law. This often leads to serious problems, particularly when the polluter cannot be identified.

The polluter is also responsible for costs entailed for measures taken to determine the extent of ground water pollution. This does not apply to investigations conducted to establish if indeed any ground water pollution is present.

If the measures required to remedy damage are not taken, or are not taken in good time by the polluter, the authorities may take the measures themselves or commission a third party to take them on their behalf. The polluter is obliged to pay the costs thus entailed. If the cause or origin of a case of ground water pollution remains unknown, then the authorities must bear the costs themselves. It is easily understandable that in many cases the speed at which the investigations and rehabilitations measures are carried out is determined by the budgetary constraints to which the authorities are subject.

One question that plays a decisive role in the remediation of ground water pollution is: how clean is clean? The Federal Water Act prohibits any detrimental change in ground water quality, and thus demands the ground water be returned to its original state. Everyone knows, however, that this is not practical. It is not technically possible to eliminate pollution of ground water completely, and even in those cases where this would be possible, it would entail considerable technical and financial costs. When it comes to remedial action for ground water pollution, therefore, the authorities are faced with a mostly irreconcilable conflict between high legal requirements and the reality of the situation.

There are no binding limit values for contaminant concentrations of hazardous substances in ground water in the Federal Republic of Germany, since this would violate the legal requirement of absolute protection for ground water. Thus it is up to the authorities to set rehabilitation objectives in each individual case, taking the following criteria into account:

- the hydrogeological conditions
- the present use of ground water, e.g. the provision of drinking water
- the risk potential of the hazardous substance, i.e its toxicity, degradability and mobility in ground water.

Cost-effectiveness must, of course, also be considered, taking into account the remedial technologies available.

Given these criteria, and in view of the differing expert knowledge and execution capabilities of the water authorities involved, the way in which problems are handled in practice vary greatly.

In many cases, limit values for drinking water are given as remedial objectives. In the case of accidents involving mineral oil, often all required is that the liquid phase be removed from ground water.

There are also cases where more or less arbitrary limit values define the rehabilitation objective or where such an objective is set with a view to one particular water use. This is, of course, extremely unsatisfactory for ground water protection. Several working groups are presently engaged in trying to find satisfactory solutions without allowing ground water protection to fall by the wayside.

3 ENVIRONMENTAL CRIMINAL LAW

Article 324 of the Criminal Code of the Federal Republic of Germany states that "Anyone who pollutes or otherwise detrimentally affects the properties of water shall be liable to a term of imprisonment of up to five years or to a fine".

Thus, whenever ground water pollution occurs, it is up to the police and the public prosecutor's office to determine whether a punishable act has been committed.

The Criminal Code does not prescribe that the administrative authorities are obliged to notify the authorities responsible for criminal prosecution when an environmental offense has been committed. It was feared that such an obligation would adversely affect the trusting cooperation between administration and plant operators required to ensure improved pollution control.

However, almost all the states have now instructed their authorities to inform the public prosecutor's office if a ground water pollution has taken place. This strategy and the increased environmental awareness on the part of the police are reflected in the statistics on police inquiries (fig. 1). The number of enquiries in the environment sector rose from 2,321 in 1973 to 21,116 in 1988. In 1989 more than 25,000 cases were expected. It is worthy of note that in 1988 more than half of all the inquiries were connected with water pollution (fig. 2) (Geppert 1990). These statistics show that legislation on environmental offense is playing a

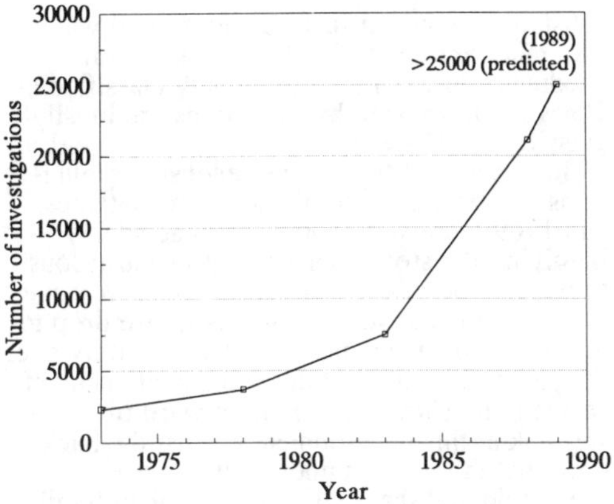

Figure 1. Environmental investigations by police (source of data - Geppert 1990).

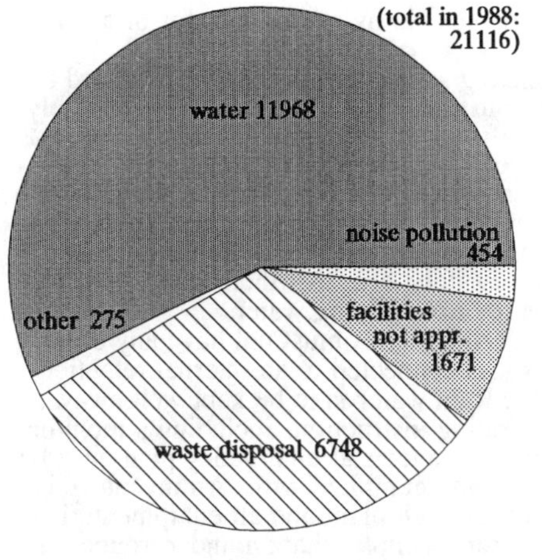

Figure 2. Environmental investigations in 1988 (source of data - Geppert 1990).

substantial role in the field of environmental protection.

In 1986, the Federal Republic of Germany recorded 3025 court decisions on environmental offenses. Of these, some 51 % ended with a conviction, 44 % of the cases were dropped and only 5 % ended with an acquittal. In well over 90 % of the convictions, a fine was imposed.

Discussions are presently taking place in parliament on tightening up legislation on environmental offenses. This includes giving more severe penalties as well as extending the range of activities which may be classified as an offense; soil pollution, for example, is to be introduced as an offense.

4 LIABILITY

The Federal Water Act also contains extensive regulations concerning liability for damage caused by water pollution.

For example, anyone who introduces substances into ground water which change its physical, chemical or biological quality, is obliged to pay any damages incurred by a third party, e.g. a water works. This is also true if substances leaking from an installation reach the ground water, irrespective of whether or not the polluter acted in a intentional or negligent way.

5 PREVENTATIVE REGULATIONS

The importance attached to ground water and the difficulties surrounding the elimination of ground water contamination led, at an early stage, to stringent regulations being introduced for installations using substances hazardous to water.

The 1957 Water Act already contained the provision that substances could only be stored or dumped in such a way that there need be no cause for concern that they might be a source of ground water pollution."Cause for concern" here means that it must, according to human knowledge and experience, be improbable that damage will occur.

In 1976, special regulations were introduced for installation in which substances presenting a hazard to water are stored, filled or transported. They were extended in 1986 to include installations which manufacture, treat and use such substances.

The most important regulation is that all these facilities must be designed, fitted, erected, maintained and operated in such a manner that there need be no cause for concern that they might pollute water. For facilities loading and unloading these substances, the best possible protection of the water against contamination is to be guaranteed. All these facilities must be designed, fitted, erected, maintained and operated in such a manner that they at least comply with the generally acknowledged rules of technology.

In order to meet the stringent demands set by the Federal Water Act, the following basic requirements must be put into practice:

- In principle, installations must be erected and run in such a way that it is impossible for substances presenting a hazard to water to leak. Storage tanks and pipes for substances damaging to water must be leakproof. It must be possible to monitor mobile piping along its entire length.

- Any leaks of substances presenting a hazard to water must be able to be detected reliably and without delay. They must be visible or easily detectable by a leak detector or a leak probe.

- Leaking substances presenting a hazard to water must be retained and disposed of safely. Generally, the installations must be fitted in a leak-proof and stable collecting zone.

- In case of damage, substances which may be contaminated by leaking substances presenting a hazard to water must be retained and disposed of safely, e.g. that there must be catch basins for fire-fighting water.

- Collecting zones must not have any drain unless it is connected to an internal retention system. Rain water is to be kept well away.

- Operating instructions including a monitoring, maintenance and emergency plan must be designed and complied with. An inventory is to be kept for each plant and all equipment. This means, for example, that an underground storage tank, both industrial and private, must be double walled, and that there must be both a leak detector and a system to avoid any overfilling.

Official controls using suitability tests and type approvals have been introduced. All facilities for storage, filling, loading and unloading substances hazardous to water or parts of such facilities as well as the technical safety equipment may only be used if their suitability has been established by the relevant authority. Where they are manufactured in series, they may be officially authorized with respect to type.

The Federal Water Act includes obligations for the operator too. The operator is obliged to entrust certain tasks to specialist firms if he does not have the expertise required to carry them out. He shall be obliged to monitor the facilities continually for leaks and the effective functioning of the safety equipment. Moreover, he shall allow officially approved experts to check, in certain cases, whether installations are in proper condition.

The competent authority may require the operator to take measures for monitoring water and soil, where this is required for the early detection of contamination.

The safety requirements are geared towards the risk potential of the installation, the volume of substances kept there and the properties of these substances.

The Federal Republic of Germany has a "Catalog of Substances Hazardous to Water", which divides substances into four classes depending on the degree of hazard they pose to water:

Class 0: in general, not hazardous to water
Class 1: slightly hazardous to water
Class 2: hazardous to water
Class 3: extremely hazardous to water

This classification is based on the toxicity of substances to mammals, fish and bacteria, and on their biodegradability. The mobility of a substance in the subsurface and ground water is also taken into consideration. Mineral oil products, for example, are mostly classified as Class 2, chlorinated hydrocarbons are mostly classified as Class 3.

The success of these comprehensive regulations can be seen from the accident statistics:

In 1969 there were some 1,300 accidents involving the storage of substances hazardous to water. The year 1976 - from when we have had periodic statistics - saw this figure drop to around 1,000 accidents annually. It is only over the past few years that the number of accidents seems to be showing a slight upward trend. If we look at this development against the backdrop that the use of mineral oil increased thirty-fold and that there are more than 5 million facilities for the storage of substances hazardous to water in the Federal Republic, we can see that the stringent safety provisions are having their effect.

Worthy of note is the assessment of the causes of damage to water resources, which show that in more than 50 % of cases, human error was at the root of the problem.

The best technology is useless in the face of human error. Under this aspect, even 1,000 accidents a year are to many. For ground water, a vital resource, 1,000 accidents are 1,000 too many.

REFERENCES

Geppert, P.P. 1990. Umweltstrafrecht Mitteilunsblatt des Bundesverband des Deutscher Geologen (BDV), Nr. 30 (1/90), p.16.

Subsurface Contamination by Immiscible Fluids, Weyer (ed.) © 1992 Balkema, Rotterdam. ISBN 90 6191 175 3

Alberta's Help End Landfill Pollution (HELP) Project: Prioritization of sites

W.J.Ceroici & H.A.Kerr
Alberta Environment, Edmonton, Alb., Canada

A.M.Hingston
HELP Project, Edmonton, Alb., Canada

ABSTRACT: The HELP Project was initiated by Alberta Environment in September 1986 to inventory, assess and (if necessary) take corrective action at both active and abandoned industrial landfills in Alberta. Information on sites was obtained from a number of sources including media response, government records, questionnaire returns from industry and local governments, direct phone contact with industry and a limited number of on-site visits. Under the Inventory Phase, 682 sites were identified; 213 of the sites are active, 427 are inactive and 42 have been abandoned. A comprehensive ranking system was developed to assign a numerical score to all sites based on impact potential to human health and the environment.

PROJECT HISTORY AND OVERVIEW

In September of 1986, the HELP Project was initiated by Alberta Environment in response to public concerns about past waste disposal practices used by industry. The focus of the project was on industrial landfills. Municipal landfills in Alberta were inventoried and studied in two previous investigations (McLaren Plansearch Lavalin, 1982 and Associated Engineering Services Ltd., 1985).

The HELP Project is structured in three phases; Phase I (completed on June 30, 1987) consisted of;

- conducting a media campaign,
- reviewing existing government landfill records,
- contacting industrial associations,
- classifying industrial sites and developing questionnaires for industry to report on their specific landfill sites,
- development and implementation of a data capture system,
- sending questionnaires to industry and municipal authorities,
- contacting industry and municipalities for onsite assessments,
- development of a ranking system to establish priorities,
- data compilation and interpretation of results.

The Phase I inventory identified 620 sites - 209 active, 374 inactive, and 37 abandoned sites.

The 620 (increased to 682 in 1988) sites were moved into Phase II/III in July 1987 for assessment of any potential hazards to human health or the environment. Phase II/III involved;

- development of site action plans,
- onsite inspection of high priority landfill sites,
- follow-up with noncomplying companies,
- reviewing site assessment results,
- development of a data tracking and management control system,
- researching archives for historical data,
- conducting preliminary investigations of abandoned facilities,
- assessing data from abandoned facilities.

Phase III activities include expansion of the inventory and assessing new sites. An important part of Phase III is to conduct detailed studies and initiate remedial action at orphan sites.

LEGISLATIVE BASIS

The HELP Project is authorized under an overall program founded within the Department of Environment Act. This Act gives the Minister of Environment far reaching powers to control the quality of the Environment. An important power is promulgation of regulations for the purpose of controlling substances detrimental to the environment.

Enactments that have provisions utilized in the pursuit of the HELP Project include the Clean Water, Hazardous Chemicals, Land Surface Conservation and Reclamation and Public Health Acts.

The Clean Water Act is the main instrument for controlling water contamination in Alberta. It empowers the Minister to make regulations prescribing water contamination and maximum permissible concentrations of contaminants in water. Section 17 of the Act contains a prohibition with respect to the deposit of a water contaminant. Subsection B of Section 17 provides that it is an offence to contaminate water. The Minister can direct any work to be done by or on behalf of the Crown to repair or remedy the condition or mitigate any damage to or destruction of life or property. Cost and expenses of the work are recoverable by the Crown.

Permits, pursuant to the Clean Water Act are required for the construction, alteration or expansion of most facilities which produce waste water. Licenses, renewable every three to five years, are required for operation of an industrial facility.

The Hazardous Chemicals Act is important legislation under which the HELP Project operates. The Act:

- allows the government to order a person responsible for a hazardous substance to furnish information in his possession on the substance.
- empowers the Crown to draft regulations for the handling, storage and disposal of hazardous materials.
- allows for the issue of Chemical Control Orders. An order can relate to a landfill if the storage and disposal of hazardous material is involved. An order allows the Crown to take whatever action is required to ensure compliance and to charge the cost of the action to the person to whom the Chemical Control Order is directed.
- facilitates the right of entry, upon reasonable notice to the owner, by an employee of Alberta Environment to any land or premises for the purpose of investigating, inspecting or carrying out tests. A court order allowing entry can be obtained if entry is refused.

The Hazardous Waste Regulation under the Hazardous Chemicals Act controls the transport, storage and disposal of hazardous wastes. Alberta Environment requires that approval be obtained for the construction and operation of any facility for the storage and disposal of hazardous wastes. The regulation also prohibits the disposal of certain substances in landfills.

The Land Surface Conservation and Reclamation Act deals with the conservation and reclamation of land in Alberta. Aspects of the Act relevant to the HELP Project include:

- the ability to designate future land use.
- provision of financial assistance for restoring or reclaiming derelict land (orphan sites in the HELP Project could perhaps be considered derelict land).
- allowing for entry onto land for the purpose of exercising any powers or duties under the Act.
- provision for the issuance of reclamation orders.

The Act is divided into a number of parts, the most relevant being Parts 2 and 3. Part 2 requires approvals for regulated surface operations (conservation) while Part 3 involves reclamation orders and certificates (restoration). Inquiries can be held with respect to operations including landfill sites abandoned prior to August 15, 1978. Costs or a proportion of costs of reclaiming a site are to be borne by the government and by the operator. There is no requirement that the present owner must bear any portion of the costs. The government is required to bear the cost of performing any work where an operator can not be located or is no longer doing business in Alberta. Negotiations can be entered into with the owner to allow the government to perform the work and if necessary to allow the government to purchase all or part of the land.

The Public Health Act gives the province the right to make regulations respecting the location and operation of waste facilities. Probably the most relevant regulation to the HELP Program is the Waste Management Regulation which prohibits a person from depositing hazardous waste into a landfill. Noncompliance with the regulation can result in a fine.

PROGRAM STRUCTURE

In 1987, a contract team of two individuals, was retained by Alberta Environment to implement the HELP Project. The HELP Team worked closely with a Technical Review Committee (TRC) consisting of key individuals from within Alberta Environment. The role of the TRC is to;

- provide technical expertise,
- act as a communication link to Alberta Environment,
- evaluate and implement recommendations of the HELP Team,
- deal with industries which have an operating license.

DATA GATHERING

To identify and inventory industrial landfills a number of approaches were used.

Media Program

In initiating the Program in October 1986, the Minister of the Environment launched a communications strategy to raise public awareness. Posters were distributed and advertisements were placed in newspapers. These coupled with good coverage on radio and television helped to get the public involved. Respondents were encouraged to contact the Project team, via a toll-free number to pass on confidential information regarding sites. While the number of landfills identified by this route was small (approximately 10), receptiveness to public input was an important component of the project strategy.

Government Agencies

Knowledge obtained from other provincial agencies and local government jurisdictions proved to be a useful cross-reference to industrial sources of information.

Within the provincial government, field staff of the Energy Resources Conservation Board (ERCB) and Department of Forestry, Lands and Wildlife were good sources of information on oil and gas industry landfills in remote areas.

Considerable information was also received from questionnaires sent to 361 local jurisdictions ranging in size from summer villages to cities and from the 27 health units in the province. Staff responses and/or file records were used to identify active or closed industrial landfills or major industries using municipally owned facilities. This served as a check against the responses received from the industrial survey.

Industry Survey

The great majority of the landfills reported to the HELP Project were reported by industry in response to a mailed questionnaire.

Initially, industrial activity in Alberta was divided into 13 classifications or industry types (Table 1). Using sources such as the Alberta Manufacturer's Index and trade and membership directories, a total of 1320 companies were identified. However, as not all companies have

Table 1. Industrial classifications

1. Primary agricultural processing
2. Food processors
3. Metal processors
4. Mining & primary mineral processing
5. Wood processors
6. Hydrocarbon processors
7. Petroleum upstream operations
8. Chemical processors
9. Utilities
10. Manufacturing
11. Service & chemical handlers
12. Transportation
13. Special industry group

landfills, the industrial companies were assigned into one of three categories (Table 2).

Table 2. Industry priority ranking

Priority	# of Companies	# of Quest.	%
High[1]	273	273	100
Medium[2]	306	187	61
Low[3]	741	78	11

[1] Landfill which could have a serious impact on health or the environment because of volume and waste type.
[2] Landfill not likely to have an impact on health or the environment.
[3] Not likely to have a landfill or insignificant volume or no threat to health or the environment.

Questionnaires were then developed by the Project Team in liaison with industrial associations and government agencies. The questionnaires were designed to solicit information readily available for a variety of industries in a consistent manner that would be suited to a database. Questionnaires were mailed to the 538 industries in a special folder as a package which included:

1. Covering letter stating deadline for return.
2. Background Information Sheet.
3. Government Records Data Sheet - relevant information from the existing government records (eg. Clean Water License).
4. Questionnaire Guide - a six page point by point guide to assist respondents in completing answers in a consistent manner.
5. Questionnaire - consisting of four pages containing 15 questions designed to collect information on basic company operation, waste management practices/landfill operation and a description of the area within a radius of 1.6 km.

Questionnaires were issued to all the high priority industrial sites. For the medium priority industries, 61% were issued questionnaires including all the larger producers. A sampling (11%) of questionnaires was issued to the low priority group to determine if landfills existed and whether the ranking was comprehensive.

Of the total 538 questionnaires mailed to industry, 453 (84%) were returned promptly, with only 85 nonrespondents. Hydrocarbon processors (including refineries, gas plants, etc.) were most cooperative with a 95% return and this sector alone identified 257 landfills. Certain industrial groups (eg. food processors and transport companies) were ranked low to medium priority as they were less likely to have landfills but returned questionnaires were still typically in the order of 80%.

Response from local government was comparable with 324 out of 388 (83%) questionnaires returned promptly. These sources identified an additional 58 sites while the public contributed a further 10 through the media program.

Archival Search

An archival search in Phase II/III turned up 42 abandoned sites. A field visit was conducted at each of the sites and grab samples of soil and/or water collected. An important part of the process involved interviewing local residents to obtain information about the facilities. Of the 42 sites examined, 13 were determined to be of enough concern to warrant investigation. Six of the sites are former wood preserving operations, while seven are former refineries or oil/ chemical disposal sites. These sites were termed "orphan" since no party could be identified or held responsible for decommissioning. Preliminary soil and groundwater investigations were initiated by the Department at all orphan sites in 1987/88.

A number of interesting points emerge from a preliminary review of the data.

1. As might be expected, the most important sector for landfills is the hydrocarbon industrial group.
2. Over two thirds of the wood preserving, mining and hydrocarbon processing industries have landfills.
3. Industry types with the most landfills on site were those which produced large

amounts of landfillable waste (eg. utilities, upstream petroleum, etc.).
4. Closed sites outnumbered active sites two to one.

DATA TRACKING AND MANAGEMENT CONTROL SYSTEM

An AT personal computer running a dBase III$^+$ software package was used to track the questionnaires issued, store the collected information and collate the data.

Two separate databases were maintained:

1. Company Database
   - contact person
   - plant type
   - number of landfills
   - tracking - issue and return of questionnaires
2. Individual Landfill Database
   - operational procedures
   - waste types
   - receptors
   - suitability

In addition, a paper copy of the company's questionnaire return was retained on file.

To ensure the accuracy of data submitted by industry, twenty were selected at random and visited by Project staff. In all cases the company disclosure accurately reflected the site conditions.

During the inventory the Company Database was used to check on overdue returns and generate prompt letters, followed, if necessary, by telephone contact to ensure return of completed questionnaires.

Upon completion of the survey, a data tracking system was utilized to give an overall summary of the industrial landfills in Alberta. As a continuation of the Management Control System, this also uses dBase III$^+$ software on an IBM AT personal computer. As a result, the inventory can be updated (eg. improvement in operation of landfill or shutdown of facility) as necessary.

The information in the inventory is broken into four types (or fields):

1. Identifier
   - sites listed alphabetically by company
   - file number (industrial type/subtype/group)
   - number of landfills at the facility
2. Management
   - license status, T (true) if licensed
   - status, O (open), C (closed), A (abandoned)
   - status - a numerical coding of the current status, eg. whether remedial action is current, completed or required.
3. Locational
   - location by section, township, range and meridian
   - area - nearest population centre
4. Comment
   - history of operation
   - comments including applicable licence numbers
   - waste type description
   - site design and operation

As noted, the inventory is organized by company alphabetically. A total of 682 sites are listed and the inventory has been public information since October 1988.

HAZARDOUS RANKING SYSTEM

A "hazard potential" system was developed to be applied to all landfill sites identified in the inventory. The method adopted by the HELP Project was modelled on that developed by JRB Associates (1984). This method was used by the USEPA for initial determination of disposal sites hazard potential and was rated by the New Jersey Department of Environmental Protection as "a good reflection of the hazard potential of those (30 in number) sites".

The ranking system adopted was based on four factors as follows:

1. Waste types - an evaluation of the impact potential of the wastes in the landfill.
2. Receptors - human and environmental factors which could be impacted in the surrounding area.
3. Site Suitability - characteristics of the site (eg. soils, hydrogeology, etc.).
4. Design and Operation - measures taken by the operator to lessen the pollution potential.

The factors in each category were evaluated from information supplied in the industrial survey questionnaire. Each factor was given a point score. A "weighting" or multiplier was then applied to that score to give a total for that factor. The factor scores adopted for the HELP Project are as follows:

	HELP Project	JRB Associates
Receptors	40	25
Waste Types	35	25
Site Suitability	25	25
Design/Operation	(30)*	25

* will be deducted from the cumulative scores

Note the scores are similar to those used by JRB Associates but the numbers have been modified to reflect certain influences unique to the HELP Project.

CONCLUSIONS

The HELP Project has provided Alberta Environment with valuable insight into the number and type of industrial landfills/plants in the province. In addition, the project has provided useful information for improving the design and operation of industrial facilities.

ACKNOWLEDGEMENTS

The work described in this paper between October 1986 and December 1988 was largely undertaken by a four person HELP Project Team consisting of Gavin Hamilton (Project Manager), Neil Shelly (Project Engineer), Glenn Berry (Project Geologist) and Cathy Kuchik (Word and Data Processing). This paper is printed with the permission of Alberta Environment.

REFERENCES

Associated Engineering Services Ltd., 1985, Field Investigation of Waste Disposal Sites.Report prepared for Alberta Environment.

JRB Associates, 1984, Rating the Hazard Potential of Waste Disposal Facilities. Report prepared for U.S. E.P.A.

McLaren Plansearch Lavalin, 1982, Identification and Verification of Active and Inactive Land Disposal Sites in Alberta. Report prepared for Alberta Environment.

Subsurface Contamination by Immiscible Fluids, Weyer (ed.) © 1992 Balkema, Rotterdam. ISBN 90 6191 175 3

Author index